T0188825

Lecture Notes in Computer Science   14469

Founding Editors

Gerhard Goos
Juris Hartmanis

Editorial Board Members

The series Lecture Notes in Computer Science (LNCS), including its subseries Lecture Notes in Artificial Intelligence (LNAI) and Lecture Notes in Bioinformatics (LNBI), has established itself as a medium for the publication of new developments in computer science and information technology research, teaching, and education.

LNCS enjoys close cooperation with the computer science R & D community, the series counts many renowned academics among its volume editors and paper authors, and collaborates with prestigious societies. Its mission is to serve this international community by providing an invaluable service, mainly focused on the publication of conference and workshop proceedings and postproceedings. LNCS commenced publication in 1973.

Verónica Vasconcelos · Inês Domingues ·
Simão Paredes
Editors

Progress in Pattern Recognition, Image Analysis, Computer Vision, and Applications

26th Iberoamerican Congress, CIARP 2023
Coimbra, Portugal, November 27–30, 2023
Proceedings, Part I

 Springer

Editors
Verónica Vasconcelos
Polytechnic Institute of Coimbra, Coimbra
Institute of Engineering
Coimbra, Portugal

Inês Domingues 🆔
Polytechnic Institute of Coimbra, Coimbra
Institute of Engineering
Coimbra, Portugal

Simão Paredes 🆔
Polytechnic Institute of Coimbra, Coimbra
Institute of Engineering
Coimbra, Portugal

ISSN 0302-9743   ISSN 1611-3349 (electronic)
Lecture Notes in Computer Science
ISBN 978-3-031-49017-0   ISBN 978-3-031-49018-7 (eBook)
https://doi.org/10.1007/978-3-031-49018-7

This Springer imprint is published by the registered company Springer Nature Switzerland AG
The registered company address is: Gewerbestrasse 11, 6330 Cham, Switzerland

Paper in this product is recyclable.

Preface

The 26th Iberoamerican Congress on Pattern Recognition (CIARP) was the 2023 edition of the annual international conference CIARP, which aims at fostering international collaboration and knowledge exchange in the fields of pattern recognition, artificial intelligence, and related areas with contributions covering a broad spectrum of theory and applications. We are pleased to acknowledge the endorsement of CIARP 2023 by IAPR, the International Association for Pattern Recognition.

Over the years, CIARP has evolved into a pivotal research event, playing a vital role within the Iberoamerican pattern recognition community. As in previous editions, CIARP 2023 brings together researchers and experts from around the world to showcase ongoing research in areas such as Biometrics, Character recognition, Classification clustering ensembles and multi-classifiers, Data mining and big data, Feature extraction, discretization and selection, Fuzzy logic and fuzzy image processing, Gesture recognition, Hybrid methods, Image description and registration, Image enhancement, restoration and segmentation, Image understanding, Image fusion, Information theory, Intelligent systems, Machine vision, Neural network architectures, Object recognition, Pattern recognition applications, Sensors and sensor fusion, Soft computing techniques, Statistical methods, Syntactical methods, Deep learning, Transfer learning, and Natural language processing.

Moreover, CIARP 2023 serves as a platform for the global scientific community to share their research experiences, disseminate novel insights, and foster collaborations among research groups specialising in artificial intelligence, pattern recognition, and related fields.

CIARP has always prided itself on its international character, and this edition received contributions from 21 countries. Among the Iberoamerican contributors were Portugal, Brazil, Spain, Argentina, Chile, Cuba, Ecuador, Mexico, and Uruguay. Other notable submissions came from France, Germany, Ireland, Belgium, India, South Korea, the Netherlands, Czech Republic, Italy, Taiwan, Tunisia, and the USA.

Through a meticulous review process, involving 59 dedicated reviewers who invested significant time and effort, 61 papers were selected for inclusion in these proceedings, reflecting an acceptance rate of 60.4%. All accepted papers achieved scientific quality scores exceeding the overall mean rating. The selection of reviewers was guided by their expertise, ensuring representation from diverse countries and institutions worldwide. We extend our heartfelt gratitude to all members of the Program Committee for their invaluable contributions, which undoubtedly enhanced the quality of the selected papers.

The conference, held in Coimbra Institute of Engineering, Portugal, from November 27 to 30, 2023, featured four days of engaging sessions, tutorials, and keynotes. The keynotes were delivered by João Paulo Papa, Petia Radeva, and João Manuel R. S. Tavares. CIARP 2023 also awarded the Aurora Pons-Porrata Medal, honouring a female researcher for her significant contributions in the field of pattern recognition and related areas. The authors of the Best Paper and the Best Student Paper and the recipient of the

Aurora Pons-Porrata Medal were invited to submit a paper for publication in the Pattern Recognition Letters journal.

CIARP 2023 was jointly organized by the Coimbra Institute of Engineering (ISEC) and the Polytechnic University of Coimbra (IPC). We express our sincere gratitude for their invaluable contributions to the success of CIARP 2023. We would also like to express our gratitude to i2A for their generous sponsorship. Furthermore, we wish to acknowledge the dedication of all members of the Organizing and Local Committees for their dedication in orchestrating an outstanding conference and proceedings.

We extend our special thanks to the LNCS team at Springer for their invaluable support and guidance throughout the preparation of this volume.

Finally, our deepest gratitude goes out to all authors who submitted their work to CIARP 2023, including those whose papers could not be accommodated. We trust that these proceedings will serve as a valuable reference for the global pattern recognition research community.

November 2023

Inês Domingues
Verónica Vasconcelos
Simão Paredes

Organization

Conference Chairs

Inês Domingues   Polytechnic Institute of Coimbra, Coimbra
Institute of Engineering, Portugal
Verónica Vasconcelos   Polytechnic Institute of Coimbra, Coimbra
Institute of Engineering, Portugal

Program Chair

Simão Paredes   Polytechnic Institute of Coimbra, Coimbra
Institute of Engineering, Portugal

Local Committee

Cristiana Areias   Polytechnic Institute of Coimbra, Coimbra
Institute of Engineering, Portugal
Cristina Caridade   Polytechnic Institute of Coimbra, Coimbra
Institute of Engineering, Portugal
Fernando Lopes   Polytechnic Institute of Coimbra, Coimbra
Institute of Engineering, Portugal
Frederico Santos   Polytechnic Institute of Coimbra, Coimbra
Institute of Engineering, Portugal
Luís Santos   Polytechnic Institute of Coimbra, Coimbra
Institute of Engineering, Portugal
Nuno Lavado   Polytechnic Institute of Coimbra, Coimbra
Institute of Engineering, Portugal
Nuno Martins   Polytechnic Institute of Coimbra, Coimbra
Institute of Engineering, Portugal
Teresa Rocha   Polytechnic Institute of Coimbra, Coimbra
Institute of Engineering, Portugal

Technical Support

António Godinho   Polytechnic Institute of Coimbra, Coimbra
Institute of Engineering, Portugal

Aurora Pons-Porrata Award Committee

Elisabetta Fersini   Università degli Studi di Milano-Bicocca, Italy
Gabriella Pasi   Università degli Studi di Milano-Bicocca, Italy
Maria Matilde García Lorenzo   Universidad Central "Marta Abreu", Cuba

Program Committee

Paulo Ambrósio   Universidade Estadual de Santa Cruz, Brazil
Cristiana Areias   Polytechnic Institute of Coimbra, Coimbra
                               Institute of Engineering, Portugal
Amadeo José Argüelles   Instituto Politécnico Nacional, Mexico
Joel Arrais   University of Coimbra, Portugal
Felipe de Castro Belém   Unicamp, Brazil
Bárbara Caroline Benato   Unicamp, Brazil
Rafael Berlanga   Universitat Jaume I, Spain
Mara Franklin Bonates   Universidade Federal do Ceará, Brazil
Susana Brás   IEETA, UA, Portugal
Alceu de Souza Britto   Pontifícia Universidade Católica do Paraná, Brazil
Maria Elena Buemi   Universidad de Buenos Aires, Argentina
Pedro Henrique Bugatti   Federal University of São Carlos, Brazil
Pablo Cancela   Udelar, Uruguay
Jaime dos Santos Cardoso   FEUP, Portugal
Cristina Caridade   Polytechnic Institute of Coimbra, Coimbra
                               Institute of Engineering, Portugal
Jesús Ariel Carrasco-Ochoa   INAOE, Mexico
Pedro Couto   University of Trás-os-Montes e Alto Douro,
                               Portugal
António Cunha   Universidade de Trás-os-Montes e Alto Douro,
                               Portugal
Matthew Davies   SiriusXM/Pandora, USA
Inês Domingues   Polytechnic Institute of Coimbra, Coimbra
                               Institute of Engineering, Portugal
Jacques Facon   UFES, Brazil
Alicia Fernández   Universidad de la República, Uruguay
Gustavo Fernandez Dominguez   AIT Austrian Institute of Technology, Austria
Vítor Manuel Filipe   University of Trás-os-Montes e Alto Douro,
                               Portugal
Luis Gomez   Universidad de Las Palmas de Gran Canaria,
                               Spain

Pilar Gómez-Gil	National Institute of Astrophysics, Optics and Electronics, Mexico
Lio Fidalgo Gonçalves	University of Trás-os-Montes e Alto Douro, Portugal
Teresa Gonçalves	University of Évora, Portugal
Sónia Gouveia	University of Aveiro, Portugal
Michal Haindl	Institute of Information Theory and Automation, Czech Republic
Xiaoyi Jiang	University of Münster, Germany
Martin Kampel	TU Wien, Austria
Sang-Woon Kim	Myongji University, South Korea
Vitaly Kober	CICESE, Mexico
Nuno Lavado	Polytechnic Institute of Coimbra, Coimbra Institute of Engineering, Portugal
Fabricio Lopes	Universidade Tecnológica Federal do Paraná, Brazil
Fernando Lopes	Polytechnic Institute of Coimbra, Coimbra Institute of Engineering, Portugal
Alexei Machado	Pontifical Catholic University of Minas Gerais, Brazil
Luís Marques	Polytechnic Institute of Coimbra, Coimbra Institute of Engineering, Portugal
Nuno Martins	Polytechnic Institute of Coimbra, Coimbra Institute of Engineering, Portugal
Alessandro Bof Oliveira	UNIPAMPA, Brazil
Hélder Oliveira	INESC TEC/University of Porto, Portugal
João Paulo Papa	São Paulo State University, Brazil
Simão Paredes	Polytechnic Institute of Coimbra, Coimbra Institute of Engineering, Portugal
Armando J. Pinho	University of Aveiro, Portugal
Pedro Real	Universidad de Sevilla, Spain
Bernardete Ribeiro	University of Coimbra, Portugal
Teresa Rocha	Polytechnic Institute of Coimbra, Coimbra Institute of Engineering, Portugal
Mateus Roder	São Paulo State University, Brazil
Priscila Saito	Federal University of São Carlos, Brazil
Frederico Santos	Polytechnic Institute of Coimbra, Coimbra Institute of Engineering, Portugal
Jefersson Alex dos Santos	University of Sheffield, UK
Rafael Santos	INPE, Brazil
Ana Sequeira	INESC TEC, Portugal
Catarina Silva	University of Coimbra, Portugal
José Serra Silva	CINAMIL, Portugal

Samuel Silva   University of Aveiro, Portugal
Luis Enrique Sucar   INAOE, Mexico
Alberto Taboada-Crispi   UCLV, Cuba
César Teixeira   University of Coimbra, Portugal
Luis Filipe Teixeira   Faculdade de Engenharia da Universidade do
                               Porto, Portugal
Murilo Varges da Silva   IFSP, Brazil
Verónica Vasconcelos   Polytechnic Institute of Coimbra, Coimbra
                               Institute of Engineering, Portugal

Contents – Part I

Contents – Part II

Deblur Capsule Networks

Daniel Felipe S. Santos⬛, Rafael G. Pires⬛, and João P. Papa$^{(\boxtimes)}$⬛

Department of Computing, São Paulo State University, Bauru, Brazil
`joao.papa@unesp.br`

Abstract. Blur is often caused by physical limitations of the image acquisition sensor or by unsuitable environmental conditions. Blind image deblurring recovers the underlying sharp image from its blurry counterpart without further knowledge regarding the blur kernel or the sharp image itself. Traditional deconvolution filters are highly dependent on specific kernels or prior knowledge to guide the deblurring process. This work proposes an end-to-end deep learning approach to address blind image deconvolution in three stages: (i) it first predicts the blur type, (ii) then it deconvolves the blurry image by the identified and reconstructed blur kernel, and (iii) it deep regularizes the output image. Our proposed approach, called Deblur Capsule Networks, explores the capsule structure in the context of image deblurring. Such a versatile structure showed promising results for synthetic uniform camera motion and multi-domain blind deblur of general-purpose and remote sensing image datasets compared to some state-of-the-art techniques.

Keywords: Deblurring · Remote Sensing · Capsules

1 Introduction

Blur is an undesirable artifact in images, usually caused by physical limitations of the acquisition sensor or by unsuitable environmental conditions. It affects the visual quality of the captured images, conferring bad aspects to them. Quantitatively, it can impose difficulties on computational tasks such as edge detection, segmentation, and image classification. Therefore, deblurring is an important asset in digital image processing and computer vision [6].

The challenge of image deconvolution is to estimate the underlying sharp image I given its blurry counterpart $\tilde{I}$ and an unknown corresponding blur kernel k, aka Point Spread Function (PSF) [2]. Such a process can be summarized by the following relation:

The authors are grateful to the São Paulo Research Foundation (FAPESP) grants #2013/07375-0, #2014/12236-1, #2017/25908-6 (Microsoft Research), and #2019/07665-4, and to the Brazilian National Council for Research and Development (CNPq) grant #308529/2021-9. This study was also financed in part by the Coordenação de Aperfeiçoamento de Pessoal de Nível Superior - Brasil (CAPES) - Finance Code 001.

V. Vasconcelos et al. (Eds.): CIARP 2023, LNCS 14469, pp. 1–15, 2024.
https://doi.org/10.1007/978-3-031-49018-7_1

$$I = k^{-1} * \tilde{I}, \tag{1}$$

where $*$ stands for a two-dimensional convolution operator.

We assume k to be known explicitly before deconvolution in classical image restoration problems. In such a scenario, a set of techniques can restore $\tilde{I}$, including inverse filtering, Wiener filtering [1], recursive Kalman filtering [27], least-squares (LS) filtering, and constrained iterative deconvolution methods [3]. Unfortunately, in many practical situations like remote sensing image processing (e.g., [7,11,17,20,25]), the blur kernel is often unknown, and also little information is available about the sharp image. Therefore, the sharp image I must be directly retrieved from $\tilde{I}$ using partial or no information about the blurring process and the sharp image. Such an estimation problem, assuming a certain degradation model (Eq. 1), is called *blind deconvolution*, where the goal is to learn the blur characteristics without any prior knowledge.

In this work, we propose an end-to-end blind image deblurring approach composed of three steps: (i) blur type classification, (ii) PSF reconstruction, and (iii) deep regularized deblurring. Our proposal, called Deblur Capsule Networks (DbCN), is evaluated over uniform camera motion and cross-domain blur categories. The main contributions of this paper are twofold:

- To propose a fast and efficient blind deblurring approach and
- To explore the potential of capsule structures [16] in the context of image deblurring.

The remainder of this paper organizes as follows. Section 2 presents the blind deblurring techniques related to our proposal, while Sect. 3 introduces the proposed approach. Section 4 describes the experimental design, and Sect. 5 states conclusions and future works.

2   Related Works

Recent advances have shown that deep convolutional neural networks are a powerful tool to image deblurring (e.g., [14,15,22,26,30]), being a natural alternative for filter-based approaches. Nah *et al.* [14] introduced a deep learning-based blind deconvolutional model that avoids problems related to kernel estimation. Their work presents a multi-scale loss function that mimics conventional coarse-to-fine approaches and is trained in a multi-scale end-to-end manner.

Tao *et al.* [22] suggested an approach based on a coarse-to-fine deep learning strategy that restores a blurry image gradually using a pyramidal network architecture arrangement. The proposed blind deblurring method, named Scale-recurrent Network (SRN-DeblurNet), has a simpler network structure and a reduced number of parameters.

The seminal work of Ren *et al.* [15], SelfDeblur, introduces the Deep Image Prior [23] concept to the blind image deconvolution field. It adopts an asymmetric autoencoder and a fully connected neural network to capture the deep priors of the latent clean image and the blur kernel. The authors suggest a joint optimization algorithm to solve the unconstrained blind deconvolution problem.

Wen *et al.* [26] proposed a minimal local intensity-based prior technique, named patch-wise minimal pixels (PMP). Since the PMP of clear images is much more sparse than the blurred ones, such a technique can effectively discriminate between clear and blurred images. With a novel algorithm designed to exploit the PMP sparsity at the deblurring task efficiently, the proposed approach avoids a non-rigorous approximation solution while proposing a computationally efficient technique.

Zamir *et al.* [30] introduced a multi-stage deep learning approach that progressively learns restoration functions for the degraded images. Such a model splits the recovery process into more manageable steps. Each processing stage introduces a per-pixel adaptive processing strategy that leverages in-situ supervised attention to reweight the local network features. The model learns the contextualized features using multiple encoder-decoder architectures and combines them with a high-resolution branch that retains local information. The proposed approach, named MPRNet, delivers substantial performance gains regarding multiple sources of image degradation, including blur.

Zhang *et al.* [32] proposed a new remote sensing image scene classification architecture (CNN-CapsNet). The CNN-CapsNet is composed of a pretrained deep CNN, such as VGG-16 [21], which is used as an initial feature map extractor. Then, the initial feature maps are fed into a CapsNet [16] image classifier. Such approach obtained competitive classification performance.

Recently, Safder *et al.* [18] developed a capsule-based deep network model (BA_EnCaps) to improve wildfire assessment. The BA_EnCaps in-depth capsule network is used to substitute typical contextual thresholding and to accomplish high-quality wildfire segmentation of burned areas in multispectral remote sensing images.

3   Deblur Capsule Networks

Deblur Capsule Networks is an end-to-end three-stage blind image deblurring convolutional neural network composed of blur type classification, PSF reconstruction, and deep regularized deconvolution modules, depicted in Fig. 1. Given an input blurry image $\tilde{I} = k * I$, DbCN first encapsulates the blur kernel k in a vectorized encoded representation c using a routing-by-agreement learning mechanism proposed by Sabour *et al.* [16]. Further, DbCN recovers k while decoding the encapsulated PSF into an occupancy map. In the final stage, the network recovers $I_r = k^{-1} * \tilde{I}$ using an auxiliary noise deep regularization procedure.

3.1   Blur Type Classification

The blur type classification module comprises an encoding convolutional neural network that mimics the first three convolutional blocks of the well-known VGG16 backbone [21] and a Capsule Network (CapsNet) structure [16]. Figure 2

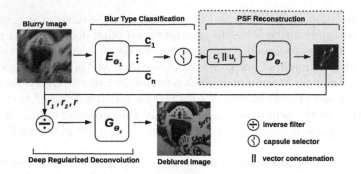

Fig. 1. Overview of the *Deblur Capsule Networks* modules, where component E_{Θ_1} represents the capsule network encoder with set of parameters Θ_1, $c = \{c_1, \cdots, c_n\}$ the set of capsules, u_i a particular PSF occupancy map coordinate, D_{Θ_2} the PSF implicit decoder neural network with set of parameters Θ_2, $r = \{r_1, r_2, r_3\}$ the inverse filter regularization parameters, and G_{Θ_3} represents the deep regularization convolutional neural network with set of parameters Θ_3. The shadowed area runs for each image's coordinate u_i considering the most activated capsule c_j.

illustrates the proposed architecture for blur classification, where the VGG16 feature maps are forwarded to the CapsNet module, which is responsible for generating the PSF capsules using the same routing-by-agreement optimization procedure adopted by Sabour *et al.* [16].

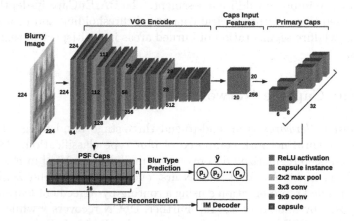

Fig. 2. Overview of the blur type classification module: an input *blurry image* is first encoded by a four-stage VGG16 architecture [21]. The produced features are forwarded to a CapsNet [16] module, which is responsible for generating n PSF capsules. Each capsule encodes one different blur kernel instance, being activated in response to the input *blurry image* that belong to the same domain. The p−th highest value from $\tilde{y}$ (predicted one-hot outputs) indicates the current active capsule.

PSF Encoder Optimization. The PSF capsules encode multiple blur kernel instances in a vectorized representation. The routing-by-agreement optimization procedure [16] connects the input blurry image $\tilde{I}$ to the corresponding $i-$th output capsule vector c_i with the highest activation response. The training step aims to minimize the following margin loss cost function:

$$\mathcal{L}_{margin} = \frac{\sum_{i=1}^{n} \tilde{y}_i F(\epsilon^+, ||c_i||_2^2) \lambda (1 - \tilde{y}_i) F(\epsilon^-, ||c_i||_2^2)^2}{n}, \tag{2}$$

where

$$F(\epsilon, c_i) = max(0, \epsilon - c_i), \tag{3}$$

considering that n is the number of PSF capsules, ϵ^+ and ϵ^- are the upper and lower output limits, respectively, $|| \cdot ||_2$ corresponds to the l^2-norm, λ is a penalization term, and $max(a, b)$ returns the maximum value between a and b. The correct activated capsules are indicated during the network training by $\tilde{y}_i = 1$; otherwise by $\tilde{y}_i = 0$.

3.2   Point Spread Function Reconstruction

Following the work of Chen and Zhang [4], the PSF recostruction module consists of a multi-layer perceptron network (*IM Decoder*), shown in Fig. 3, used to decode capsules into $2D$ blur kernel shape representations.

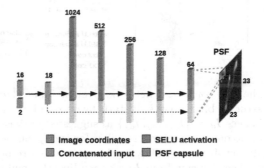

Fig. 3. Overview of the PSF reconstruction module *IM Decoder*: given a PSF capsule instance concatenated with an image's coordinate position u_i, the *IM Decoder* produces the corresponding *PSF* occupancy map. As suggested by Klambauer *et al.* [9], SELU activation functions enable a high-level abstraction representation of the *IM Decoder* network.

During the capsule decodification procedure, the *IM Decoder* scans a grid map $M = \{u_1, u_2, \ldots, u_{h \times h}\}$ of the PSF occupancy image and outputs a probability map indicating whether each position $u_i = (x, y)$ is a member of the encoded blur kernel or not. Notice that we consider a PSF of support size $h \times h$.

PSF Decoder Optimization. The *IM Decoder* optimization encourages each PSF capsule to encode the corresponding blur kernel instantiation parameters, as suggested by Sabour *et al.* [16]. The *IM Decoder* training step aims to minimize the following cost function:

$$\mathcal{L}_{dec} = \frac{\sum_{i=1}^{n} \tilde{y}_i \; MSE(k_i, D_{\Theta_2}(c_i \parallel u_i))}{n}, \qquad (4)$$

where $MSE(\cdot, \cdot)$ refers to the mean squared error function, k_i is the reference $2D$ shape blur kernel[1], symbol $\parallel$ stands for a vector concatenation, and $D_{\Theta_2}(\cdot)$ represents the underlying decoded $2D$ shape blur kernel, where Θ_2 stands for the *IM Decoder* network parameters.

3.3   Image Deep Regularized Deconvolution

Figure 4 illustrates our *deep regularizer network*. The leftmost operation, denoted by the symbol $\div$, is borrowed from the work of Schuler *et al.* [19].

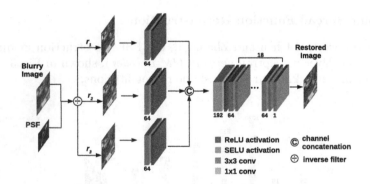

Fig. 4. Overview of the *deep regularizer* network: given an input *blurry image* $\tilde{I}$ and its corresponding reconstructed blur kernel $\tilde{k}$, it starts by deconvolving $\tilde{I}$ into a set of three partially deblurred images $\boldsymbol{I_d} = \{I_{d_1}, I_{d_2}, I_{d_3}\}$ as suggested by Vasu *et al.* [24]. However, we used a direct deconvolutional inverse filter for such a purpose. Each partially deblurred image is deconvolved according to the set $\boldsymbol{r}$ of corresponding regularization hyperparameters. The elements from $\boldsymbol{I_d}$ are then linearly combined in the feature domain space and forwarded to a deep convolutional neural network to produce the *restored image* I_r. Although the *deep regularizer network* resambles a DnCNN [31], its architecture does not make use of residual connections or batch normalization layers, since the SELU activations are sufficient to stabilize the network.

The *deep regularizer* module starts with a quotient layer, which is capable of deconvolving $\tilde{I}$ using the reconstructed blur kernel $\tilde{k}$ according to the standard inverse filtering operation:

---

[1] The reference blur kernel can be easily substituted by an approximated version of the underlying ground-truth PSF.

$$I_d = \mathcal{F}^{-1} \left(\frac{\overline{\mathcal{F}(\tilde{k})}\tilde{I}}{|\mathcal{F}(\tilde{k})|^2 + r} \right), \tag{5}$$

where $\mathcal{F}(\cdot)$ represents the Fourier domain transformer operator and $\mathcal{F}^{-1}(\cdot)$ its inverse. To improve the network's noise removal capacity, we adopt the same approach described in Vasu et al. [24], which demonstrates that combining partially deblurred images can be more advantageous. The remaining parts of the *deep regularizer network* comprise standard convolutional layers used to capture the fine details of high-frequency regions from the blurry image, leading the model to produce the restored image version.

Deep Regularizer Optimization. The *deep regularizer* optimization procedure consists of minimizing the following mean squared error cost function:

$$\mathcal{L}_{rec} = \sum_{j=1}^{m} \frac{(\||I_j - G_{\Theta_3}(\tilde{I}_j)\||_2^2)}{m}, \tag{6}$$

where m indicates the number of input blurry images and $G_{\Theta_3}(\cdot)$ represents the underlying restored image I_r, in which Θ_3 stands for the network parameters. Besides, rather than using constant hyperparameter values, we let the *deep regularizer network* to find the set of hyperparameters $r = \{r_1, r_2, r_3\}$ and Θ_3 that lead to the best I_r outcome. Therefore, our *deep regularizer* works as a complementary regularization technique for the standard inverse deconvolutional filter.

4 Experiments and Analysis

The proposed DbCN was trained using a three-stage optimization routine, and the experiments were subdivided into (i) Synthetic Camera Motion Blur and (ii) Synthetic Multi-Domain Blur categories. The following section gives a better description concerning the DbCN training, test, and evaluation methodology.

4.1 DbCN Optimization Procedure

The DbCN optimization procedure uses Adam [8] technique and the Common Objects in Context (COCO 2015) training dataset[2]. In the first optimization stage[3], the *blur type classifier* and the *IM Decoder* modules are trained jointly using the following loss function:

$$\mathcal{L}_{caps} = \mathcal{L}_{margin} + \alpha \mathcal{L}_{dec}, \tag{7}$$

where $\alpha = 0.1$, $\epsilon^+ = 0.9$, $\epsilon^- = 0.1$, and $\lambda = 0.5$ (Eqs. 2 and 4).

---

[2] http://images.cocodataset.org/zips/test2015.zip.
[3] The first three VGG16 stages, pre-trained over the ImageNet dataset, are kept fixed during the four-stage training process over COCO 2015 dataset.

In the second stage, only the *deep regularizer* parameters Θ_3 and the regularization parameters r are jointly optimized (Eq. 6). In a such training stage, the remaining DbCN parameters (Θ_1 and Θ_2) are kept unchanged. In the third optimization stage, besides adjusting Θ_3 and r parameters once more, we also fine-tuned the *IM Decoder* network parameters Θ_2 by minimizing Eq. 6. Here, the *blur type classifier* parameters Θ_1 are kept fixed.

Each optimization stage uses a mini-batch of size 64 randomly extracted from artificially augmented grayscale image patches[4]. The patches comprise 224×224 pixels, considering stage one, and 40×40 pixels regarding stages two and three. The first training stage comprises a learning rate of 10^{-4}, while the second uses learning rates of 10^{-5} and 10^{-8} for the *deep regularizer network* and for learning r parameters, respectively. Last but not least, learning rates of 10^{-6}, 10^{-6} and 10^{-9} are used for the third-stage training of the *deep regularizer*, *IM Decoder*, and r parameter learning step, respectively. Our experiments take a maximum of 10,000 iteration steps for each training stage convergence. The experiments run on an Intel Core i7-9700K CPU computer using a NVIDIA GeForce GTX 1080 Ti GPGPU.

4.2   Synthetic Camera Motion Blur

We used the Levin *et al.* [10], and the Xia *et al.* [28] AID datasets for blind deconvolution assessments concerning eight synthetic camera motion blur kernels as shown by Fig. 5.

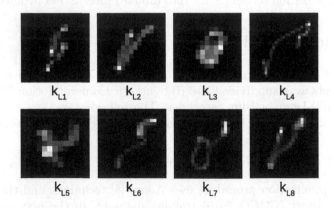

Fig. 5. Overview of the Levin *et al.* [10] synthetic uniform blur kernels, whose support sizes vary from 10 to 25 pixels.

From the Levin *et al.* [10] dataset, we used all of the 32 test pairs of blurred images and their corresponding ground truths. From AID dataset, we generated 960 test pairs based on the following steps:

---

[4] During the DbCN training, the patches are augmented according to random horizontal and vertical flips combined with rotation transforms.

1. To extract central patches of size 224 × 224 from each AID image;
2. To select the four most sharpen clean patches from each AID category (30 categories), based on the Mittal *et al.* [13] average local deviation field measurement criteria;
3. To convolve each group of 120 selected clean patches with one of the eight Levin *et al.* [10] blur kernels.

Even though the blur kernels were synthesized under camera uniform motion assumptions and using the same capture device, each blur kernel has its unique shape and its corresponding support size. Therefore, we can analyze the DbCN behavior under a multi-shape and multi-size blind deblurring task. For we are dealing with a prediction task in the *blur type classifier*, this experiment comprises eight different classes (blur kernels). As mentioned earlier, we worked with patches cropped randomly from the input images. For each training patch, we assigned a particular class randomly.

Quantitative and Qualitative Comparisons. Table 1 presents the average PSNR and SSIM (Structural Similarity Index) values calculated over the Levin *et al.* [10] and the AID datasets. We compared DbCN against state-of-the-art blind deconvolution methods including Ren *et al.* [15], Wen *et al.* [26], Nah *et al.* [14], Tao *et al.* [22], and Zamir *et al.* [30]. Besides, it was necessary to retrain SRN [22], DMCNN [14], and MPRNet [30] using COCO 2015 training dataset with blurry images synthesized over k_L. Notice that each technique was retrained following its standard configuration protocols until convergence.

Table 1. Overall average quantitative results regarding Levin *et al.* [10] and AID test datasets. Time is represented in seconds, PSNR in decibels (dB), and FLOPS in Giga ($\times 10^9$).

	Levin		AID		Time	FLOPS
	PSNR	SSIM	PSNR	SSIM		
Blured	20.84	0.573	17.19	0.281	—	—
PMP	32.55	0.933	22.00	0.678	87.09	—
SelfDeblur	33.77	0.935	22.44	0.691	13.60	—
DMCNN	31.84	0.939	23.85	0.778	0.07	257
SRN	36.39	0.972	30.01	0.946	0.03	137
MPRNet	35.07	0.965	28.94	0.932	0.11	595
DbCN (Ours)	**45.32**	**0.990**	**36.72**	**0.955**	**0.01**	**53**

According to Table 1, DbCN performance was considerably more reasonable than the baselines. In terms of average PSNR, regarding Levin dataset, DbCN achieved improvements of 24.48 dB and 8.93 dB over the blurred (input image) and second-best technique, respectively. Concerning running time, DbCN also demonstrates its superiority, being at least three times faster than the second-best technique from Table 1. In terms of average PSNR, regarding AID dataset, Table 1 shows that DbCN also achieved improvements of 19.53 dB and 6.71 dB over the blurred (input image) and second-best technique, respectively.

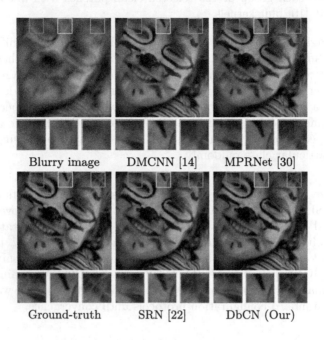

Blurry image   DMCNN [14]   MPRNet [30]

Ground-truth   SRN [22]   DbCN (Our)

Fig. 6. Restored images from Levin *et al.* [10] dataset considering the blur kernel k_{L8}.

Besides the quantitative results, Fig. 6 shows that DbCN produces visual results closer to the ground-truth images in comparison to the top three techniques from Table 1. It can successfully recover high-frequency details from the blurry images, such as hair strands, eyebrows, and eyelids.

4.3   Synthetic Multi-domain Blur

The experiments of this section are based on the work of Yan and Shao [29], which consists of using synthesized uniform blur kernels from Gaussian, Defocus, and Linear Motion domains, shown in Fig. 7.

In our experiments, Gaussian blur kernels k_G are generated with support sizes of $3\sigma \times 3\sigma$, for $\sigma = \{1, 2, 3, 4, 5\}$. Defocus blur kernels k_D are built using support sizes from $R = \{5 \times 5, 7 \times 7, 11 \times 11, 15 \times 15, 19 \times 19, 23 \times 23\}$. Linear Motion blur kernels k_M are generated according to the set of angle-size tuples $A = \{(0°, 3 \times 3), (45°, 11 \times 11), (90°, 29 \times 29), (315°, 19 \times 19), (180°, 49 \times 49), (225°, 5 \times 5), (270°, 13 \times 13), (135°, 39 \times 39)\}$. Here, we used Martin *et al.* [12] Berkeley[5]

---

[5] https://bit.ly/3gZWt2E.

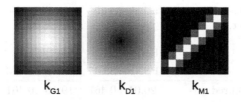

k_{G1} k_{D1} k_{M1}

Fig. 7. Examples of three uniform blur kernels generated according to Yan and Shao [29]. Gaussian blur kernel k_{G1} has support size of $3\sigma \times 3\sigma$ considering $\sigma = 5$. Defocus blur kernel k_{D1} has a support size 23×23, and the linear Motion blur kernel k_{M1} has a support size 11×11 and $45°$ slope.

and the Cheng *et al.* [5] NWPU-RESISC45 datasets. From Berkeley dataset, we generated 750 test pairs of blurred images and their corresponding ground truths based on the following steps:

1. To extract central patches of size 224×224 from the 250 most sharpen Berkeley's clean images, based on the Mittal *et al.* [13] average local deviation field measurement criteria;
2. To convolve each group of 250 clean patches with randomly selected uniform blur kernels from each corresponding domain (k_G, k_D, and k_M).

From NWPU-RESISC45 dataset, we generated 945 test pairs of blurred images and their corresponding ground truths based on the following steps:

1. To extract central patches of size 224×224 from each NWPU-RESISC45 image;
2. To select the seven most sharpen clean patches from each NWPU-RESISC45 category (45 categories), based on the Mittal *et al.* [13] average local deviation field measurement criteria;
3. To blur each group of selected 315 clean patches following same Berkeley dataset test procedure.

The experiments addressed in this section allow us to analyse the DbCN behaviour under a multi-domain and multi-size blind deblurring task, which is more complex than the previous experiments. Once again, for each training patch, we assigned a particular class randomly during the optimization procedure.

Quantitative and Qualitative Comparisons. Table 2 presents the PSNR and SSIM average results computed over the Berkeley and NWPU-RESISC45 datasets. We compared DbCN against Wen *et al.* [26], Nah *et al.* [14], Tao *et al.* [22], and Zamir *et al.* [30]. According to Table 2, DbCN performance was considerably superior in terms of average PSNR, for it achieved improvements of 10.36 dB over the blurred images and 2.22 dB over the second-best technique, regarding Berkeley dataset.

Table 2. Overall average quantitative results regarding Berkeley and NWPU-RESISC45 test dataset.

	Berkeley		NWPU	
	PSNR	SSIM	PSNR	SSIM
Blured	20.01	0.467	19.16	0.401
PMP [26]	22.06	0.600	21.26	0.556
SelfDeblur [15]	23.39	0.670	22.23	0.633
DMCNN [14]	26.94	0.809	25.42	0.778
SRN [22]	28.15	0.869	26.67	0.850
MPRNet [30]	26.70	0.795	25.28	0.756
DbCN (Ours)	**30.37**	**0.883**	**28.72**	**0.866**

With respect to NWPU-RESISC45 dataset, Table 2 also shows that DbCN achieved a superior performance in terms of average PSNR, for it achieved improvements of 9.56 dB over the blurred images and 2.05 dB over the second-best technique.

Concerning the visual quality of the reconstructed images, Fig. 8 shows that DbCN successfully recovered the image's high-frequency details regarding intricate regions such as small leaves, trees' branches, and thin boat candlesticks.

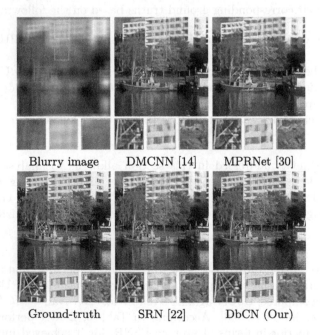

Blurry image   DMCNN [14]   MPRNet [30]

Ground-truth   SRN [22]   DbCN (Our)

Fig. 8. Restored images from Berkeley dataset considering blur kernel k_D of size 15×15.

4.4   Ablation Study

In this section, we conduct an ablation study to better comprehend the contribution of the proposed DbCN third training stage to the final deblurring results
regarding AID and NWPU-RESISC45 test datasets.

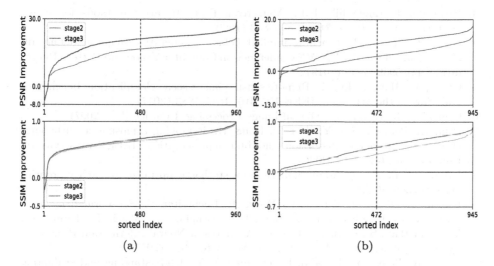

Fig. 9. Average PSNR (upper graph) and SSIM (lower graph) improvements, regarding
(a) AID and (b) NWPU-RESISC45 test datasets.

As shown by Fig. 9, a third training stage increased even more the restoration results. In that, as shown by Fig. 9(a), we obtained gains of 4.57 dB, and
0.027 concerning AID test dataset average PSNR and SSIM results, respectively.
As shown by Fig 9(b), relative to NWPU-RESISC45 test dataset, we obtained
improvements of 3.90 dB, and 0.103 in terms of PSNR and SSIM average results,
respectively.

5   Conclusions and Future Works

This work proposes a novel deep learning-based blind deconvolution technique
composed of three stages: (a) blur type identification, (b) PSF reconstruction, (c)
deep image regularization. According to the experiments, DbCN can effectively
handle the multi-shape and multi-size imposed problems, overcoming state-of-
the-art blind deblurring techniques evaluated over Levin and AID datasets.
DbCN also showed highly accurate results regarding the multi-domain category
concerning Berkeley and NWPU-RESISC45 datasets. DbCN can almost fully
recover the high-frequency regions considering Defocus and Linear Motion blur
models. Besides, the proposed approach is three times faster than the others. We
would also like to investigate the DbCN capacity to address non-uniform camera
motion blur problems in future works.

References

1. Andrews, H.C., Hunt, B.R.: Digital Image Restoration. Advanced monographs, Prentice-Hall, Prentice-Hall Signal Processing Series (1977)
2. Banham, M.R., Katsaggelos, A.K.: Digital image restoration. IEEE Signal Process. Mag. **14**(2), 24–41 (1997)
3. Biemond, J., Lagendijk, R.L., Mersereau, R.M.: Iterative methods for image deblurring. Proc. IEEE **78**(5), 856–883 (1990)
4. Chen, Z., Zhang, H.: Learning implicit fields for generative shape modeling. In: Proceedings of the IEEE/CVF Conference on Computer Vision and Pattern Recognition, pp. 5939–5948 (2019)
5. Cheng, G., Han, J., Lu, X.: Remote sensing image scene classification: benchmark and state of the art. Proc. IEEE **105**(10), 1865–1883 (2017)
6. Gonzalez, R., Woods, R.: Digital Image Processing. Prentice-Hall (2007)
7. Hua, X., Pan, C., Shi, Y., Liu, J., Hong, H.: Removing atmospheric turbulence effects via geometric distortion and blur representation. IEEE Trans. Geosci. Remote Sens. **60**, 1–13 (2020)
8. Kingma, D.P., Ba, J.: Adam: a method for stochastic optimization. arXiv preprint arXiv:1412.6980 (2014)
9. Klambauer, G., Unterthiner, T., Mayr, A., Hochreiter, S.: Self-normalizing neural networks. In: Guyon, I., Luxburg, U.V., Bengio, S., Wallach, H., Fergus, R., Vishwanathan, S., Garnett, R. (eds.) Advances in Neural Information Processing Systems, vol. 30, pp. 972–981. Curran Associates, Inc. (2017)
10. Levin, A., Weiss, Y., Durand, F., Freeman, W.T.: Understanding and evaluating blind deconvolution algorithms. In: 2009 IEEE Conference on Computer Vision and Pattern Recognition, pp. 1964–1971. IEEE Computer Society (2009)
11. Liu, H., Gu, Y.: Deep joint estimation network for satellite video super-resolution with multiple degradations. IEEE Trans. Geosci. Remote Sens. **60**, 1–15 (2022)
12. Martin, D., Fowlkes, C., Tal, D., Malik, J.: A database of human segmented natural images and its application to evaluating segmentation algorithms and measuring ecological statistics. In: Proceedings of the 8th International Conference on Computer Vision, vol. 2, pp. 416–423 (2001)
13. Mittal, A., Soundararajan, R., Bovik, A.C.: Making a "completely blind" image quality analyzer. IEEE Signal Process. Lett. **20**(3), 209–212 (2012)
14. Nah, S., Hyun Kim, T., Mu Lee, K.: Deep multi-scale convolutional neural network for dynamic scene deblurring. In: Proceedings of the IEEE Conference on Computer Vision and Pattern Recognition, pp. 3883–3891 (2017)
15. Ren, D., Zhang, K., Wang, Q., Hu, Q., Zuo, W.: Neural blind deconvolution using deep priors. In: Proceedings of the IEEE/CVF Conference on Computer Vision and Pattern Recognition, pp. 3341–3350 (2020)
16. Sabour, S., Frosst, N., Hinton, G.E.: Dynamic routing between capsules. In: Guyon, I., et al. (eds.) Advances in Neural Information Processing Systems, vol. 30, pp. 3859–3869. Curran Associates, Inc. (2017)
17. Sacramento, I., Trindade, E., Roisenberg, M., Bordignon, F., Rodrigues, B.B.: Acoustic impedance deblurring with a deep convolution neural network. IEEE Geosci. Remote Sens. Lett. **16**(2), 315–319 (2018)
18. Safder, Q., et al.: Ba_EnCaps: dense capsule architecture for thermal scrutiny. IEEE Trans. Geosci. Remote Sens. **60**, 1–11 (2022)
19. Schuler, C.J., Hirsch, M., Harmeling, S., Schölkopf, B.: Learning to deblur. IEEE Trans. Pattern Anal. Mach. Intell. **38**(7), 1439–1451 (2015)

20. Shen, H., Du, L., Zhang, L., Gong, W.: A blind restoration method for remote sensing images. IEEE Geosci. Remote Sens. Lett. **9**(6), 1137–1141 (2012)
21. Simonyan, K., Zisserman, A.: Very deep convolutional networks for large-scale image recognition. arXiv preprint arXiv:1409.1556 (2014)
22. Tao, X., Gao, H., Shen, X., Wang, J., Jia, J.: Scale-recurrent network for deep image deblurring. In: Proceedings of the IEEE Conference on Computer Vision and Pattern Recognition, pp. 8174–8182 (2018)
23. Ulyanov, D., Vedaldi, A., Lempitsky, V.: Deep image prior. In: Proceedings of the IEEE Conference on Computer Vision and Pattern Recognition, pp. 9446–9454 (2018)
24. Vasu, S., Maligireddy, V.R., Rajagopalan, A.: Non-blind deblurring: handling kernel uncertainty with CNNs. In: Proceedings of the IEEE Conference on Computer Vision and Pattern Recognition, pp. 3272–3281 (2018)
25. Wang, X., Zhong, Y., Zhang, L., Xu, Y.: Blind hyperspectral unmixing considering the adjacency effect. IEEE Trans. Geosci. Remote Sens. **57**(9), 6633–6649 (2019)
26. Wen, F., Ying, R., Liu, Y., Liu, P., Truong, T.K.: A simple local minimal intensity prior and an improved algorithm for blind image deblurring. IEEE Trans. Circuits Syst. Video Technol. **31**(8), 2923–2937 (2021)
27. Woods, J., Ingle, V.: Kalman filtering in two dimensions: further results. IEEE Trans. Acoust. Speech Signal Process. **29**(2), 188–197 (1981)
28. Xia, G.S., et al.: AID: a benchmark data set for performance evaluation of aerial scene classification. IEEE Trans. Geosci. Remote Sens. **55**(7), 3965–3981 (2017)
29. Yan, R., Shao, L.: Blind image blur estimation via deep learning. IEEE Trans. Image Process. **25**(4), 1910–1921 (2016)
30. Zamir, S.W., et al.: Multi-stage progressive image restoration. In: Proceedings of the IEEE/CVF Conference on Computer Vision and Pattern Recognition, pp. 14821–14831 (2021)
31. Zhang, K., Zuo, W., Chen, Y., Meng, D., Zhang, L.: Beyond a gaussian denoiser: residual learning of deep CNN for image denoising. IEEE Trans. Image Process. **26**(7), 3142–3155 (2017)
32. Zhang, W., Tang, P., Zhao, L.: Remote sensing image scene classification using CNN-CapsNet. Remote Sens. **11**(5), 494 (2019)

Graph Embedding of Almost Constant Large Graphs

Francesc Serratosa[✉][iD]

Universitat Rovira I Virgili, Catalonia, Spain
francesc.serratosa@urv.cat
https://deim.urv.cat/francesc.serratosa/

Abstract. In some machine learning applications, graphs tend to be composed of a large number of tiny almost constant sub-structures. The current embedding methods are not prepared for this type of graphs and thus, their representational power tends to be very low. Our aim is to define a new graph embedding, called GraphFingerprint, that considers this specific type of graphs. The three-dimensional characterisation of a chemical metal-oxide nanocompound easily fits in these types of graphs, which nodes are atoms and edges are their bonds. Our graph embedding method has been used to predict the toxicity of these nanocompounds, achieving a high accuracy compared to other embedding methods.

Keywords: graph embedding · graph regression · graph classification · metal-oxide nanocompound · chemical 3D-structure · GraphFingerprint · NanoFingerprint

1 Introduction

Chemical metal-oxide nanocompounds are chemical compounds composed of only tens of thousands of atoms and they only have two different types of atoms, oxygen and a metal. Moreover, their internal sub-structures are almost constant. Currently, there are several models to predict their properties, such as the toxicity, which are based on global properties but they do not use the information of their three-dimensional structure. Supported by the European projects NanoInformatix[1] and Sbd4nano[2] we modelled, for the first time, the toxicity of metal-oxide nanocompounds based on their three-dimensional structure. To do so, we represented these structures by attributed graphs.

An attributed graph, in general, is a data structure depicting a collection of entities represented as nodes, and their pairwise relationships represented as

Supported by the European project SbD4Nano (H2020-NMBP-TO-IND-2019-862195), the Spanish project NexPandemics (PID2022-138327OB-I00) and AGAUR research group 2021SGR-00111: "ASCLEPIUS: Smart Technology for Smart Healthcare".

[1] https://www.nanoinformatix.eu.
[2] https://www.sbd4nano.eu.

V. Vasconcelos et al. (Eds.): CIARP 2023, LNCS 14469, pp. 16–30, 2024.
https://doi.org/10.1007/978-3-031-49018-7_2

edges. Attributed graphs have been used for pattern recognition and machine learning for several decades to represent objects [2,5].

Note there is a growing interest in having graph-based techniques applied to machine learning. They are used to represent chemical compounds and then predict their toxicity [6]. This can be attributed to their effectiveness in characterising instances of data with complex structures and rich attributes and also, the goodness of distances between graphs, such as the Graph edit distance [21,22], to capture the dissimilarity between graphs. Nevertheless, this classical framework has the main drawback that computing a graph distance has usually a high computational cost (although the existence of sub-optimal algorithms). This is an important issue since the computational time for learning the model and also for using it could be inadmissible. This is because some chemical compounds could be composed of thousands of atoms and chemical databases (for learning and testing) could have millions of compounds.

Besides, attributed graphs that represent the three-dimensional structure of a metal-oxide nanocompound are composed of almost constant sub-structures and only two types of attributes appear in the nodes. Thus, graphs have not enough internal variability and the graph edit distance becomes useless for being the pair representation and distance not discriminating enough.

To overcome this problem, some researches have applied Graph Embedding, Graph Convolutional Networks or Graph Autoencoders to classify or to predict some properties of chemical compounds [4,15]. These methods learn the local structures and semantics of the attributed graphs and convert them into a latent space, represented as a matrix or vector of Real numbers. Then, given this space, classical machine learning methods can be applied. Note it is key for these methods that not only the involved graphs have to be different (large graph distance) but also having a rich internal variability (composed of several different local structures and attributes). Unfortunately, as commented below, the internal variability of our attributed graphs is very low and it is not possible to use graph alignment methods [3] or the well-know graph edit distance [19].

The aim of this paper is to present a new graph embedding specifically designed for applications of graph regression or graph classification in which attributed graphs are composed of almost constant sub-structures and attributes are almost the same. We have called *GraphFingerprint* to this graph embedding and it has been used for metal-oxide nanocompound toxicity prediction. In the next section, we recapitulate the old graph embedding and graph regression methods, and also we comment the new graph convolutional and graph autoencoder methods. In Sect. 3, we explain in detail our embedding method and in Sect. 4, we explain the application of GraphFingerprint in nanocompound toxicity prediction. We show that our method achieves better accuracy compared to current graph regression and classification methods. Section 5 and Sect. 6 concludes this paper and presents future work, respectively.

2   Graphs and Graph Embedding

Graphs are effective mathematical tools for characterising instances of data with complex structures and rich attributes. For this reason they have been used for automatically classifying objects in pattern recognition or deducing global properties through regression methods for almost fifty years [2,5].

The classical K-Nearest Neighbours algorithm could be used for graph classification or graph regression purposes. In this case, a distance between graphs have to be used, such as the graph edit distance [21,22]. Nevertheless, one downside of these methods is the calculation of graph distances [17,18,20], which would increase the computational time in correspondence to the number of data elements.

With the aim of avoiding the calculation of graph distances, it is usual to move from the graph domain into a Euclidean domain. This conversion is achieved through graph embedding techniques. Some of these techniques were presented some years ago, in which the embedding process did not use trainable parameters [9]. More recently, Graph Convolutional Networks [10,25] and Graph Autoencoders [4,13,24] have emerged as tools to embed graphs. The main difference of these last methods and the old embedding methods is the ability to select the intrinsic features of the graphs through learned parameters.

3   GraphFingerprint: An Embedding for Almost Constant Graphs

GraphFingerprint is an attributed graph embedding defined as a vector of numbers, which has been designed to encode attributed graphs composed of almost constant small substructures. Moreover, nodes are labelled in only two classes and edges are unattributed. This type of graphs appear in some specific and completely different scenarios, such as toxicity prediction of chemical nanocompounds (graphs represent the three-dimensional structure of the compound) or social networks analysis (for instance, nodes are people labelled by their voting tendencies).

More formally. We define a graph $G = (V, E)$ as a set of nodes V and a set of edges E. G_i is the ith node in V and $G_{i,j}$ is the edge in E between the ith node and the jth node in V. Moreover, $\gamma_i = \{N_k, N_p\}$ is the label or attribute of node G_i. N_t is a cardinal attribute, being, $1 \leq t \leq T$. Note given a specific graph, it can only have two types of attributes, N_k and N_p and it is not possible to have three different types of attributes, such as, N_k, N_p and N_n, being $k \neq p \neq n$. For this reason, since now, we classify the nodes of a specific graph in two types: O and M, where class O are the nodes that have attribute N_k and class M are the nodes that have attribute N_p, being $k < p$.

In the following subsection, we explain the input parameters of the algorithm that generates a GraphFingerprint. Then, we move to explain the local substructures, which are the basics of the GraphFingerprint. And we define the GraphFingerprint. Finally, we show some examples of GraphFingerprints generated from nanocompounds.

3.1   Algorithm Input Parameters

Basically, the algorithm has three parameters:

- Shell thickness. A positive real number that defines an external radius of the compound. The atoms (nodes in the graph) in the defined volume are considered to be inside the shell and thus, the atoms that influence on the generation of the GraphFingerprint. It is measured in angstroms (Å). Thus, only the most external nodes of the graph are used to generate the GraphFingerprint. This is because the external atoms are the ones that influence in the toxicity level. In the case that the whole graph is required to be converted into the GraphFingerprint, then a large value could be imposed.
- Maximum node degree (MAX). A natural number that defines the maximum number of edges per node that are considered to generate the GraphFingerprint. This parameter is needed to have a fixed representation of the GraphFingerprint, as described in the next subsection. A larger number makes a larger GraphFingerprint and a larger chance of having more null values in it.
- 3D structure. A file in XYZ format that contains the 3D structural information of the NanoFingerprint. Note that it could be considered to introduce an "emptied" 3D structure, this is, an XYZ file that only contains the nodes in the shell. In this case, the XYZ is smaller and the generation of the GraphFingerprint is faster

3.2   Local Substructures

A *local structure* is a small set of connected nodes in the graph. In this section, we present some *local structures* that are used to define, in an organised manner, the GraphFingerprint.

$O(x)$: It represents a node type O that has x edges.

$M(x)$: It represents any node of type M that have x edges.

$O(x, y)$: A local structure composed of a central node of type O connected to x nodes of type O and y nodes of type of M. Note these x O and y M could be connected to other nodes.

$M(x, y)$: Similarly, a local structure composed of a central node of type M connected to x nodes of type O and y nodes of type M.

$O(x, y) - O(x', y')$: A structure that is composed of two of the previous ones. It is composed of an $O(x, y)$ and an $O(x', y')$ whose central O are connected by an edge.

$M(x, y) - M(x', y')$: In a similar way, it is composed of an $M(x, y)$ and an $M(x', y')$ whose central M are connected by an edge.

$O(x, y) - M(x', y')$: Finally, a similar structure but that connects an $O(x, y)$ and an $M(x', y')$ whose central nodes are connected.

3.3 GraphFingerprint Definition

A GraphFingerprint is a vector of Natural numbers that counts appearances of *local structures*. It is split up in four main sections: Global information, Node information, Edge information and Structural information. Note the number of nodes and edges involved in each local structure increases in each section, that is, in each section, larger local structures are considered.

- **Section 1: Global information**
  The first section only specifies the two attributes of the nodes and counts the appearances of the two types of nodes, MAX. Moreover, it imposes a maximum order of nodes. Nodes that have larger number are not considered. This value is a parameter in the construction of GraphFingerprint and it is useful to impose the length of the GraphFingerprint to be constant in a specific database or application.
  1: MAX (Maximum order of nodes per type of node).
  2: N_k (the attribute of nodes of type O).
  3: N_p (the attribute of nodes of type M).
  4: Number of nodes type O in the graph.
  5: Number of nodes type M in the graph.
- **Section 2: Node information**
  This section is composed of $2 * MAX$ values, where MAX is the maximum number of edges per node.
  1: Number of $O(1)$
  ...

  MAX: Number of $O(MAX)$
  $MAX + 1$: Number of $M(1)$
  ...

  $2MAX$: Number of $M(MAX)$
- **Section 3: Edge information**
  This section includes the information of the local structures $O(x, y)$ and $M(x, y)$. It is composed of $2(MAX + 1)^2$ values.
  1: Number of $O(0, 0)$
  ...

  $MAX + 1$: Number of $O(0, MAX)$
  $MAX + 2$: Number of $O(1, 0)$
  ...

  $2(MAX + 1)$: Number of $O(1, MAX)$
  ...

  $(MAX + 1)^2$: Number of $O(MAX, MAX)$
  $(MAX + 1)^2 + 1$: Number of $M(0, 0)$
  ...

  $(MAX + 1)^2 + MAX + 1$: Number of $M(0, MAX)$
  $(MAX + 1)^2 + MAX + 2$: Number of $M(1, 2)$
  ...

  $(MAX + 1)^2 + 2(MAX + 1)$: Number of $M(1, MAX)$
  ...

  $(MAX + 1)^2 + (MAX + 1)^2$: Number of $M(MAX, MAX)$

– **Section 4: Structural information**
This section includes the information of the local structures $O(x, y) - O(x', y')$, $M(x, y) - M(x', y')$ and $O(x, y) - M(x', y')$. It is composed of $3(MAX + 1)^4$ values.
1. Number of O(0,0)-O(0,0)
...

$(MAX + 1)^2$: Number of $O(0, MAX) - O(0, MAX)$
$(MAX + 1)^2 + 1$: Number of $O(1, 0) - O(1, 0)$
...

$(MAX + 1)^4$: Number of $O(MAX, MAX) - O(MAX, MAX)$
$(MAX + 1)^4 + 1$: Number of $M(0, 0) - M(0, 0)$
...

$(MAX + 1)^4 + (MAX + 1)^4$: Number of $M(MAX, MAX) - M(MAX, MAX)$
$(MAX + 1)^4 + (MAX + 1)^4 + 1$: Number of $O(0, 0) - M(0, 0)$
$(MAX + 1)^4 + (MAX + 1)^4 + (MAX + 1)^4$: Number of $O(MAX, MAX) - O(MAX, MAX)$

The length of the GraphFingerprint depends on the maximum accepted node order MAX, $length(GraphFingerprint_{MAX}) = 5 + 2MAX + 2(MAX + 1)^2 + 3(MAX + 1)^4$. Thus, the theoretical worst computational cost of generating GraphFingerprints is in fourth power of the maximum order of nodes (MAX).

3.4 GraphFingerprint Examples

In this section, some examples of GraphFingerprints are presented. Figure 1 shows the 3D structure of the nanoparticle TiO_2 with size 0.6 nm. The image has been generated by the MATLAB function molviewer. The rest of nanoparticles 3D representations on this work have been represented using this function. The oxygen atoms are represented as red spheres and the Titanium atoms in green. As it can be seen, there are two local structures marked. In green, there is the local structure M(2,1). The central atom (node with id 0) is a titanium, circled in green as well. It has two bonds with two oxygen atoms (bonds formed by the node ids 0-4 and 0-1) and one bond with a titanium atom (bond formed by the node ids 0-5). In yellow, there is the local structure O(0,3). Its central atom (node with id 6) is an oxygen, circled in yellow. It has three bonds with three titanium atoms (bonds formed by the node ids 6-8, 6-5 and 6-3) and any bond with oxygen atoms.

Figure 2 presents the extracted GraphFingerprint (centre) from a TiO_2 of size 0.6 nm (left), and its 3D nanoparticle structure and XYZ format (right). Note we have ommited the null elements of the GraphFingerprint and we have indicated the number of cell in the vector. There are four two-fold coordinated O sites (marked as O(0,2)) and two three-fold coordinated O sites (marked as O(0,3)). Since there is no coordination for Ti, the local environment is not captured by the combination of pairs of O and Ti sites.

Note that the maximum number of bonds selected (second element of the GraphFingerprint), MAX=10, is the same as the maximum number of bonds

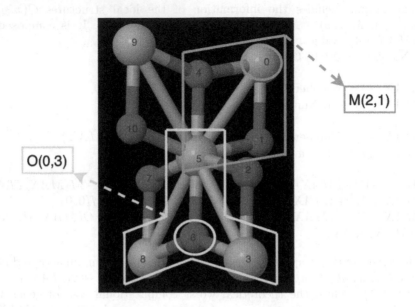

Fig. 1. Nanoparticle TiO_2 with size 0.6 nm. Two local sub-structures have been highlighted.

3D Structure	**GraphFingerprint**	**XYZ format**		
	Shell: 100 MaxBonds: 10 Size: 5.988513 Atomic: 22 O: 6 M: 5 8-> O[2]: 4 9-> O[3]: 2 19-> M[3]: 4 26-> M[10]: 1 29-> O[0,2]: 4 30-> O[0,3]: 2 171-> M[2,1]: 4 218-> M[6,4]: 1 17764-> M[2,1]_M[6,4]: 4 29817-> O[0,2]_M[2,1]: 4 29864-> O[0,2]_M[6,4]: 4 29938-> O[0,3]_M[2,1]: 4 29985-> O[0,3]_M[6,4]: 2	11 O2 Ti Ti -1.8650 O -1.8650 O 0.0000 Ti 0.0000 O 0.0000 Ti 0.0000 O 0.0000 O 0.0000 Ti 0.0000 Ti 1.8650 O 1.8650	-0.0000 0.0000 -1.8650 -1.8650 0.0000 0.0000 0.0000 1.8650 1.8650 0.0000 0.0000	-2.3425 -0.4685 0.4684 2.3425 -1.8740 0.0000 1.8740 0.4684 2.3425 -2.3424 -0.4685

Fig. 2. TiO_2 of size 0.6 nm: Compound, GraphFingerprint and XYZ file.

of the nanoparticle, corresponding to atom 5 (titanium atom connected to 6 oxygen atoms and 4 titanium atoms), and Local Structure M(6,4). Regarding the size, for small structures, the shell includes the whole nanoparticle whatever the thickness is, such as the previous one. As the particle starts to grow, the shell thickness parameter starts to make an effect on the considered atoms. For example, Fig. 3 shows the 3D structure of the nanoparticle TiO_2 with size 1.4 nm. On the left, the whole nanocompound was considered whereas on the right, a shell thickness of 4 Å was imposed. From 123 total atoms, the shell is composed of 116 atoms (thus, the core has 7 atoms).

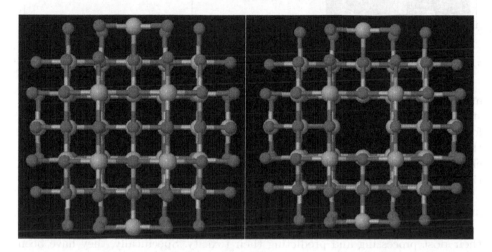

Fig. 3. TiO_2 with size 1.4 (left) and the shell that has thickness of 4 Å of the same compound (right).

Figure 4 presents the GraphFingerprint from the nanocompund TiO_2 with size 0.4 nm, and its 3D nanoparticle structure in XYZ format. In this case, regarding to coordination, there are 6 singly coordinated O sites (marked as O(0,1)) and 1 six-fold coordinated Ti site (marked as M(6,0)). Moreover, the local environment is also captured by the combination of pairs of O and Ti sites: there are 6 singly coordinated O sites (O(0,1)) connected to a six-fold Ti site. Note that the maximum number of bonds selected (second element of the NanoFingerprint), MAX=6, is also the maximum number of bonds of the nanoparticle, corresponding to node 0 (a titanium atom connected to 6 oxygen atoms), and Local Structure M(6,0). In the case, the shell of this nanoparticle describes the whole nanoparticle.

4   Experimental Section

Modelling size-realistic nanomaterials to analyse some their properties, such as toxicity, solubility, or electronic structure, is a current challenge in computational

3D Structure	GraphFingerprint	XYZ format
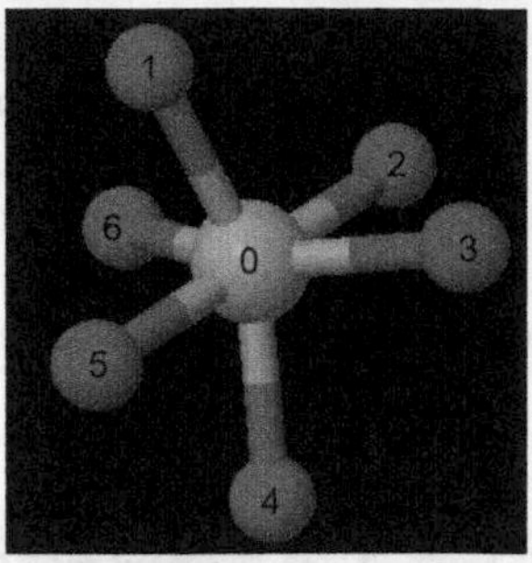	Shell: 100 MaxBonds: 6 Size: 3.845876 Atomic: 22 O: 6 M: 1 7-> O[1]: 6 18-> M[6]: 1 20-> O[0,1]: 6 110-> M[6,0]: 1 5011-> O[0,1]_M[6,0]: 6	7 O2Ti O -1.8650 -0.0000 -0.4685 O 0.0000 -1.8650 0.4684 O 0.0000 0.0000 -1.8740 Ti 0.0000 0.0000 0.0000 O 0.0000 0.0000 1.8740 O 0.0000 1.8650 0.4684 O 1.8650 0.0000 -0.4685

Fig. 4. TiO_2 of size 0.4 nm: Compound, GraphFingerprint and XYZ file.

and theoretical chemistry. The representation of all-atom three-dimensional structure of the nanocompound would be ideal, as it could account explicitly for structural effects. However, the use of the whole structure is tedious due to the huge data management and the structural complexity that accompanies the chemical nanocompound. Developing appropriate tools that allow the quantitative analysis of the three-dimensional structure, as well as the selection of regions of interest such as the most external atoms, is a crucial step to enable efficient analysis and processing of model nanostructures.

GraphFingerprints have been applied to embed chemical nanocompounds for their post processing and predicting their toxicity. Specifically, they have been applied to metal-oxide nanocompounds, which are chemical compounds that have the following three features: 1) They are composed of only some thousands of atoms. 2) They have only two types of atoms, the ones that are oxygen and the other ones that are metals. 3) Their local sub-structures are almost constant. Thus, we performed nanocompound toxicity prediction through graph embedding and regression.

The following sections explain the process of embedding a nanocompound into a GraphFigerprint and the regression applied to GraphFingerprint to deduce their toxicity. Moreover, we also depict a web site from which users can generate GraphFingerprints, compute nanocompound toxicity predictions or other graph-related operations.

4.1 From Metal-Oxide Nanocompound to GraphFingerprint

Metal-oxide nanocompounds are chemical compounds composed of two types of atoms, oxygen and any type of metal, for instance, Al, Cu, Fe, Ti or Zn. The most common metal-oxide nanoparticles are Al_2O_3, CuO, Fe_2O_3, TiO_2, SiO_2 and ZnO. The analysis and prediction of their reactivity and toxicity is crucial for better understanding these compounds and their use in the industry [1]. Given that experimental evaluation of the safety of chemicals is expensive and time-consuming, computational methods have been found to be efficient alternatives to predict their toxicity level.

The three-dimensional structure of chemical nanocopounds can be represented as attributed graphs [16]. Attributed graphs have been applied in machine learning for chemical compound toxicity prediction for a long time [7,8]. In our case, atoms in the nanocompound are represented by nodes and chemical bounds by edges. Due to the specific simplicity of metal-oxide nanocompounds, nodes can be attached with the type of atom (O, Al, Cu, Fe, Ti or Zn) and edges do not have attributes since all bonds tend to be of the same non-covalent type. Specifically, and considering the graph definition in Sect. 3, we define $\gamma_i = \{O\}$ for the oxygen atoms and $\gamma_i = \{Al, Cu, Fe, Ti, Zn\}$ for the metal ones. Then, class O atoms are the ones that $\gamma_i = \{O\}$ and class M atoms are the ones that $\gamma_i = \{Al, Cu, Fe, Ti, Zn\}$. Note the three-dimensional position of the atoms is not represented in the graph.

4.2 Toxicity Prediction Based on Global Features

Some regression models have been reported for the prediction of toxicity of metal-oxide nanocompounds [12]. Two of them have been selected as a ground truth of our method. The first one is based on a linear regression and in the second one is based on a logistic regression.

The first one [14], presented in 2015, is a linear regression model that predicts the lactate dehydrogenase release (LDH)[3] based on the following five parameters: nanocompound size, nanocompound size in water, nanocompound size in PBS, Concentration and Zeta potential[4]. The model was trained with 33 nanocompounds. 24 of them were TiO_2 and 9 of them were ZnO. TiO_2 sizes are 30, 45 and 125 nm, and ZnO size are 50, 60 and 70 nm. The first row of Table 1 shows the results presented in [14], which we recomputed and checked. In that paper, regression quality metrics were the mean square error and the standard deviation[5]. A *one leave out* training method was used to train the linear regression model.

The second one [23], presented in 2021, is a logistic linear regression model for toxicity assessment of metal-oxide nanoparticles using the following nine physicochemical features: Core nanocompound size, nanocompound size in water, surface charge, surface area, Ec, reaction time, percentage of dose, energy and number of oxygen. The method returns two labels: Toxic and Non-toxic. The model was trained with 484 nanocompounds. 18 of them were Al_2O_3 (size 39.7 nm), 18 of them were CuO (size 46.3 nm), 18 of them were Fe_2O_3 (size 42.5 nm), 195 of them were TiO_2 (sizes 10 nm, 20 nm, 21 nm, 25 nm, 30 nm, 33.7 nm, 35 nm, 40 nm and 125 nm) and 234 of them were ZnO (sizes 7.5 nm, 32 nm, 35.6 nm,

---

[3] LDH is a stable cytoplasmic enzyme that is found in all cells. LDH is rapidly released into the cell culture supernatant when the plasma membrane is damaged, a key feature of cells undergoing apoptosis, necrosis, and other forms of cellular damage.

[4] These parameters and the generated LDH were measured in vitro.

[5] The data extracted from [14] is detailed in the Table 1 in the supplementary material.

45.3 nm, 60 nm, 86 nm, 100 nm and 115 nm). The first row of Table 2 shows the results presented in [23], which we recomputed and checked. In that paper, quality metrics were Accuracy, Precision and Recall[6]. A *cross-validation partition* with 10 folders training method was used to train the logistic linear regression model.

4.3   Toxicity Prediction Based on GraphFingerprints

We proceeded to analyse the GraphFingerprint representation given the GraphFingerprints generated by the three-dimensional structures of the nanocompounds used in [14] and also in [23]. GraphFingerprints depend only on the type of nanocompound and its size. For this reason, given the 33 different samples used in [14], only six GraphFingerprints were generated. Moreover, given the 484 different samples used in [23], only twenty GraphFingerprints were generated. Figure 5 shows the process of generating a GraphFingerprint given a nanocompound. Circles represent processes and rectangles represent data.

When these GraphFingerprints were generated, we proceeded to learn a linear regression in the case of [14] and a logistic regression in the case of [23]. We finally performed the same type of experiments as depicted in Sect. 4.2. The second row of Table 1 and Table 2 show the results obtained given this process.

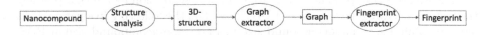

Fig. 5. The process of generating a GraphFingerprint given a chemical nanocompound.

As it was expected, only the structural information of the nanocompound is not enough to predict its toxicity. This is because several nanocompunds with the same size but different parametres appear in the database and obtain different toxicity levels. For instance, in the case of [14], there are eight TiO_2 with size

Table 1. Mean Square error and its standard deviation achieved in: the data and method in [14], regression on the GraphFingerprint generated by the 3D-structure of data in [14] and regressions on the data in [14] concatenated to (represented by symbol '+') the embedding vector generated by [11].

	MSE	STD
Original experiment in [14]	0.12	0.08
Regression on GraphFingerprint	0.76	0.16
Regression on [14]+GraphFingerprint	0.08	0.08
Regression on [14]+GCN	0.11	0.13

---

[6] The data extracted from [23] is detailed in the Table 2 in the supplementary material.

30nm that have LDH from 0.7 to 1.2. (Table 1 in the supplementary material shows the original data). Another example is ZnO with size 32nm that appear in [23] (Table 2 in the supplementary material shows the original data). There are 21 examples of it but 6 of them are toxic and the other 15 none.

Table 2. Classification metrics achieved in: the data and method in [23], regression on the GraphFingerprint generated by the 3D-structure of data in [23] and regressions on the data in [23] concatenated to (represented by symbol '+') the embedding vector generated by [11].

	Accuracy	Precision	Recall
Original experiment in [23]	0.81	0.81	0.98
GraphFingerprint	0.77	0.77	0.78
[23]+GraphFingerprint	0.93	0.92	0.98
[23]+GCN	0.56	0.80	0.88

4.4 Toxicity Prediction Based on Global Features and GraphFingerprints

In this last experiment we concatenated the original samples in [14] or in [23] with their respective embed vectors generated by GraphFingerprint and also the embedding method Graph Convolutional Network [11]. That is, each sample in [14] is composed of a vector of five elements. Then, we concatenated these five elements to the corresponding embedded vector. Similarly was done with the data in [23], but in this case, vectors in [23] have ten elements. Finally, we learned the weights of the regression (linear in [14] and logistic in [23]) and we performed the same type of experiments as depicted in Sect. 4.2.

Rows from the third one to the last one of Table 1 and Table 2 show the results obtained given this process. Considering Table 1, we realise that GraphFingerprints concatenated to the original data obtained the lowest MSE and standard deviation. Moreover, note that some of the methods obtain higher or equal MSE, which means that it is not worth using the embedding the three-dimensional structure in this cases. Considering Table 2, we realise the original data concatenated to GaphFingerprint is the combination that obtains better accuracy, precision and recall.

We conclude, from these three experiments that:

- Structural information alone is not enough for the toxicity prediction (Balanced accuracy, Precision and Recall are lower in the second row than in the first one in both tables.
- Nevertheless, the structural information represented as a GraphFingerprint helps to increase the quality of the toxicity prediction combined to the global features (The third row returns better validation parameters than the first one in both tables).

– Finally, GCN is not able to capture the structural information of the three-dimensional structure of the nanocompound.

5   Conclusions

We have presented a new embedding method, called GraphFingerprint, specifically designed to embed graphs that are composed of a large number of almost similar structures and attributes. This embedding has been applied to predict the toxicity of chemical metal-oxide nanocompounds in a regression scenario and also a classification scenario. Our proposal has been tested with different databases presented in other papers and it has achieved the highest accuracy when combined with global features of the nanocompound. We have seen that the other embedding methods are not able to properly represent the attributed graphs that represent the three-dimensional structure of the nanocompounds. Note we have also realised that the three-dimensional structure, without global features, has not information enough to properly classify the nanocompound or deduce its regression.

Thus, we conclude that, on the one hand, our embedding method applied to classification and regression of nanocompounds achieves the highest accuracy compared to other embedding methods. On the other hand, the three-dimensional structure of the compounds seems not to have information enough to properly deduce the toxicity and, for this reason, the best accuracy appears when the global information is concatenated to the GraphFingerprint.

6   Future Work

We are currently using GraphFingerprints to analyse social networks. Social networks are easily represented as graphs and some of them have our required features. For instance, nodes have only two classes (the two most voted parties in some countries) and usually the local structures (people friendship) are almost constant. We realised the current graph embedding methods do not extract the main information of these graphs, as it happened with the three-dimensional structure of nanocompounds.

References

1. Çetin, Y.A., Martorell, B., Serratosa, F., Aguilera-Porta, N., Calatayud, M.: Analyzing the TiO2 surface reactivity based on oxygen vacancies computed by DFT and DFTB methods. J. Phys.: Condens. Matter **34**(31), 314004 (2022)
2. Conte, D., Foggia, P., Sansone, C., Vento, M.: Thirty years of graph matching in pattern recognition. Int. J. Pattern Recognit Artif Intell. **18**(3), 265–298 (2004), https://doi.org/10.1142/S0218001404003228
3. Cortés, X., Serratosa, F.: An interactive method for the image alignment problem based on partially supervised correspondence. Expert Syst. Appl. **42**(1), 179–192 (2015)

4. Fadlallah, S., Julià, C., Serratosa, F.: Graph regression based on graph autoencoders. In: Krzyzak, A., Suen, C.Y., Torsello, A., Nobile, N. (eds.) Structural, Syntactic, and Statistical Pattern Recognition, pp. 142–151. Springer, Cham (2022)
5. Foggia, P., Percannella, G., Vento, M.: Graph matching and learning in pattern recognition in the last 10 years. Int. J. Pattern Recognit. Artif. Intell. **28**(1), 1450001 (2014). https://doi.org/10.1142/S0218001414500013
6. Garcia-Hernandez, C., Fernández, A., Serratosa, F.: Ligand-based virtual screening using graph edit distance as molecular similarity measure. J. Chem. Inf. Model. **59**(4), 1410–1421 (2019)
7. Garcia-Hernandez, C., Fernández, A., Serratosa, F.: Learning the edit costs of graph edit distance applied to ligand-based virtual screening. Curr. Top. Med. Chem. **20**(18), 1582–1592 (2020)
8. Garcia-Hernandez, C., Fernández, A., Serratosa, F.: Ligand-based virtual screening using graph edit distance as molecular similarity measure. J. Chem. Inf. Model. **59**(4), 1410–1421 (2019), https://doi.org/10.1021/acs.jcim.8b00820
9. Gibert, J., Valveny, E., Bunke, H.: Graph embedding in vector spaces by node attribute statistics. Pattern Recogn. **45**(9), 3072–3083 (2012)
10. Kipf, T.N.: Deep Learning with Graph-Structured Representations. Ph.D. thesis, University of Amsterdam (2020)
11. Kipf, T.N., Welling, M.: Semi-supervised classification with graph convolutional networks. In: 5th International Conference on Learning Representations, ICLR 2017, Toulon, France, April 24–26, 2017, Conference Track Proceedings. OpenReview.net (2017), https://openreview.net/forum?id=SJU4ayYgl
12. Lamon, L., et al .: Computational models for the assessment of manufactured nanomaterials: Development of model reporting standards and mapping of the model landscape. Comput. Toxicol. **9**, 143–151 (2019). https://doi.org/10.1016/j.comtox.2018.12.002, https://www.sciencedirect.com/science/article/pii/S2468111318300847
13. Lin, M., Wen, K., Zhu, X., Zhao, H., Sun, X.: Graph autoencoder with preserving node attribute similarity. Entropy **25**(4), 567 (2023). https://doi.org/10.3390/e25040567
14. Papa, E., Doucet, J., Doucet-Panaye, A.: Linear and non-linear modelling of the cytotoxicity of TiO2 and ZnO nanoparticles by empirical descriptors. SAR QSAR Environ. Res. **26**(7–9), 647–665 (2015). https://doi.org/10.1080/1062936X.2015.1080186
15. Reiser, P., et al.: Graph neural networks for materials science and chemistry. Commun. Mater. **3**, 93 (2022). https://doi.org/10.1038/s43246-022-00315-6
16. Rica, E., Álvarez, S., Serratosa, F.: Ligand-based virtual screening based on the graph edit distance. Int. J. Mol. Sci. **22**(23), 12751 (2021)
17. Serratosa, F.: Fast computation of bipartite graph matching. Pattern Recogn. Lett. **45**, 244–250 (2014)
18. Serratosa, F.: Speeding up fast bipartite graph matching through a new cost matrix. Int. J. Pattern Recogn. Artificial Intell. **29**, 1550010 (2014). https://doi.org/10.1142/S021800141550010X
19. Serratosa, F.: Graph edit distance: restrictions to be a metric. Pattern Recogn. **90**, 250–256 (2019)
20. Serratosa, F.: A general model to define the substitution, insertion and deletion graph edit costs based on an embedded space. Pattern Recogn. Lett. **138**, 115–122 (2020), https://doi.org/10.1016/j.patrec.2020.07.010
21. Serratosa, F.: Redefining the graph edit distance. SN Comput. Sci. **2**(6), 1–7 (2021). https://doi.org/10.1007/s42979-021-00792-5

22. Serratosa, F., Cortés, X.: Graph edit distance: moving from global to local structure to solve the graph-matching problem. Pattern Recogn. Lett. **65**, 204–210 (2015)

23. Subramanian, N.A., Palaniappan, A.: NanoTox: development of a parsimonious in silico model for toxicity assessment of metal-oxide nanoparticles using physicochemical features. ACS Omega **6**(17), 11729–11739 (2021). https://doi.org/10.1021/acsomega.1c01076

24. Wang, J., Liang, J., Yao, K., Liang, J., Wang, D.: Graph convolutional autoencoders with co-learning of graph structure and node attributes. Pattern Recogn. **121**, 108215 (2022). https://doi.org/10.1016/j.patcog.2021.108215

25. Wu, Z., Pan, S., Chen, F., Long, G., Zhang, C., Yu, P.S.: A comprehensive survey on graph neural networks. IEEE Trans. Neural Netw. Learn. Syst. **32**(1), 4–24 (2021). https://doi.org/10.1109/TNNLS.2020.2978386

Feature Importance for Clustering

Gonzalo Nápoles[⊠], Niels Griffioen, Samaneh Khoshrou, and Çiçek Güven

Tilburg University, Tilburg 5037, AB, The Netherlands
g.r.napoles@uvt.nl

Abstract. The literature on cluster analysis methods evaluating the contribution of features to the emergence of the cluster structure for a given clustering partition is sparse. Despite advances in explainable supervised methods, explaining the outcomes of unsupervised algorithms is a less explored area. This paper proposes two post-hoc algorithms to determine feature importance for prototype-based clustering methods. The first approach assumes that the variation in the distance among cluster prototypes after marginalizing a feature can be used as a proxy for feature importance. The second approach, inspired by cooperative game theory, determines the contribution of each feature to the cluster structure by analyzing all possible feature coalitions. Multiple experiments using real-world datasets confirm the effectiveness of the proposed methods for both hard and fuzzy clustering settings.

Keywords: Cluster Analysis · Explainability · Feature Importance

1 Introduction

Artificial Intelligence (AI) is becoming increasingly prevalent in decision-making processes across a variety of domains, sectors and industries. Despite their astounding success, more accurate algorithms often lack transparency due to their complexity. In other words, more accurate algorithms may be associated with higher complexity but lack the ability to explain their reasoning mechanism, while models with a lower complexity may have a higher interpretability but lose out in their predictive accuracy [7]. Certain industries (e.g., finance, insurance, healthcare, and autonomous vehicles) may benefit or require transparency of their models to their human users. This requirement has supported the emergence of the Explainable Artificial Intelligence (XAI) field.

The XAI field can be further subdivided into intrinsically interpretable AI models and post-hoc methods, which are used to explain inherently opaque AI models. These post-hoc methods are used after a model is built and help obtain an explanation for the model's decisions. Post-hoc methods can be further subdivided into model-specific methods, such as aggregated weights for fuzzy cognitive maps and model-agnostic methods. As an example of model-agnostic methods, LIME (Local Interpretable Model-Agnostic Explanations) [17] alters input data and evaluates how the output changes to explain the model. Changes in the

© Springer Nature Switzerland AG 2024
V. Vasconcelos et al. (Eds.): CIARP 2023, LNCS 14469, pp. 31–45, 2024.
https://doi.org/10.1007/978-3-031-49018-7_3

feature values of individual data points help determine which features drive the predicted value for that input. Shapley Additive Explanations (SHAP) values [11] is a method derived from the mathematical foundation of Shapley values, originating from game theory [19]. It measures how individual features contribute to a model's output. SHAP is used for the explainability of the whole model, while LIME focuses on individual data points.

Within unsupervised machine learning, clustering focuses on dividing a collection of instances into a predefined number of clusters, such that instances in the same cluster are more similar than instances in other clusters. Clustering methods are linked to several data science problems such as document classification, image segmentation, object recognition, social network analysis, business analytics, data reduction and big data mining, among others [14]. XAI methods can be applied to a clustering task by transforming it into a supervised task, for example, by using the (cluster) labels of the data. In [3] such an approach is adopted, after which cluster prototypes are generated in the form of multidimensional bounding boxes, and rule-based explanations are obtained. However, the interpretation of these rules remains a challenge for high-dimensional data. Recently, some unsupervised learning models, including k-means clustering [9], were made explainable by rewriting them as neural networks [13] and attributing the cluster assignment of the network to input features.

The literature on interpretable centroid-based clustering is divided into two groups: 1) Indirect explanations focus on building a surrogate model to approximate the clustering allocation. In [6,15] decision trees and neural networks have been used as a proxy to simulate cluster allocation. The critical challenge of this approach is finding a trade-off between the accuracy of the surrogate (faithfulness to the original model) and explainability. 2) Direct approaches focus on explaining the original clustering method. In [9], the outcome of k-means clustering was modeled using neural networks, which allows computing feature importance through a reverse propagation procedure.

This paper proposes two post-hoc methods to quantify the contribution of each feature to the cluster structure obtained with a prototype-based clustering algorithm. The first method uses the distance among cluster centers along each dimension as a proxy for feature importance. The intuition of this method is that more dissimilar dimensions should be responsible for the difference among the clusters. The second method modifies the well-known SHAP method to the clustering setting. It computes how the membership values of data points to each cluster change after marginalizing each problem feature in all possible permutations of features. The numerical simulations using real-world datasets show that both methods perform comparably for fuzzy c-means, while the SHAP modification seems more appropriate for k-means.

This paper is organized as follows. Section 2 describes the fundamentals of hard (k-means) and soft (fuzzy c-means) clustering as our contribution will primarily target these algorithms. Section 3 presents our post-hoc methods for computing feature importance when clustering numerical data. Section 4 describes the experimental methodology and performs a comparative analysis between both methods using real-world datasets. Section 5 provides concluding remarks and some future research avenues.

2   Cluster Analysis

Gathering d-dimensional data points into several groups based on a similarity or distance function is deemed valuable for customer segmentation, image compression, and anomaly detection tasks. Clustering algorithms ensure that data points allocated to a cluster are similar to other points within the same cluster and comparatively dissimilar to data points in a different cluster. Therefore, a desired cluster structure involves a low maximum cluster diameter (intra-cluster distance) and a high minimum distance between clusters (inter-cluster distance). Such optimal behavior is rooted in the assumption that data points within a cluster are distributed in the form of a hypersphere.

Typically, clustering consists of several steps, including selecting or extracting features, measuring the distance between pairs of data points using a predefined distance measure such as the Euclidean distance, grouping the points into clusters, and assessing the output [8]. Clustering can further be divided into *hard* and *soft* clustering. Next, we will briefly describe two state-of-the-art prototype-based algorithms that belong to these categories.

Hard clustering groups data points in such a way that each point is assigned exclusively to one cluster and is not a member of other clusters. K-means is a well-recognized iterative hard clustering method, introduced by MacQueen in 1967 [12] as an improvement on an earlier paper by Steinhaus in 1956 [20]. The algorithm starts by selecting k points from a population of n data points as the initial prototypes. In each iteration, it computes the distance between each data point and these prototypes and assigns each point to the nearest prototype. It then updates the prototypes by aggregating all data points assigned to each prototype. This process is repeated until convergence, aiming to minimize the within-cluster sum of squares as shown below:

$$min \rightarrow \sum_{j=1}^{k} \sum_{x_i \in C_j} ||x_i - z_j||^2 \tag{1}$$

where C_j is the set of data points assigned to the j-th cluster, x_i denotes the i-th data point in that cluster, while z_j is the cluster center or prototype. Minimizing the within-cluster sum of squares ensures that the data points within each cluster are tightly grouped around their respective prototypes, leading to well-separated and compact clusters. Hence, solving this minimization problem helps identify natural groupings or patterns in the data.

Soft clustering, also known as overlapping clustering, is a clustering approach where data points do not fully belong to one cluster, instead they have membership degrees to each cluster. Fuzzy c-means is a popular method developed by J.C. Dunn in 1973 [5] and improved by J.C. Bezdek in 1981 [2]. It relies on the soft clustering principle to "assign" data points to clusters using membership degrees indicating partial membership [4]. In other words, the fuzzy c-means algorithm finds membership values so that the fuzzy clusters are well-separated and compact using the following equation,

$$min \rightarrow \sum_{i=1}^{n} \sum_{j=1}^{c} (\mu_{ij})^m ||x_i - z_j||^2 \qquad (2)$$

where m is the fuzzification coefficient, c is the number of clusters and u_{ij} is the membership degree of point x_i to the cluster with prototype z_j. The reader can note that, while fuzzy c-means also attempts to minimize the within-cluster sum of squares, it does not operate under the assumption that a data point belongs to a single cluster. In this soft clustering algorithm, the centroid of the j-th cluster is computed by using the data points and the fuzzification coefficient as $z_j = \sum_{i=1}^{n}(\mu_{ij}^m \cdot x_i)/\sum_{i=1}^{n}\mu_{ij}^m$. Fuzzy c-means and k-means share the same working principles and they might report comparable results when the data being processed involves well-separated clusters.

3   Proposed Methods

Clustering data points as an unsupervised data mining task has seen wide use. However, there are currently few available methods to determine which features lead to the emergence of the cluster structure given an existing clustering partition. In this section, we propose two methods to determine feature importance under the assumptions that (i) each cluster can be associated with a prototype and (ii) that each data point is associated with membership degrees denoting the extent to which the point belongs to each cluster.

Let $X \subseteq \mathbb{R}^m$ be a collection of data points to be gathered into k clusters using k-means or fuzzy c-means. The clustering method generates a partition $C = \{C_1, \ldots, C_j, \ldots, C_k\}$ such that $C_j \subset X$ is the j-th cluster with prototype z_j. Therefore, it holds that $\bigcup_j C_j = X$, and $\bigcap_j C_j = \varnothing$ regardless of whether the clustering algorithm is hard or soft. In the soft setting, each point will be allocated to the cluster reporting the largest membership degree. Finally, in the hard setting, the membership degree of a point x_i to the j-th cluster can be computed as $\mu_{ij} = ||x_i - z_j||_2 / \sum_l ||x_i - z_l||_2$.

3.1   Prototype-Based Feature Importance

The first method proposed in this paper —termed Prototype-based feature importance (PBFI)— uses the distance between the prototypes along each axis as a proxy for feature importance. This method assumes that dimensions (features) reporting the largest differences between prototypes are more important in distinguishing between discovered clusters than other dimensions. Similarly, if prototypes do not differ significantly along a given dimension, we could infer that the corresponding feature is less important. These assumptions supporting the method are particularly reasonable for clusters having similar densities and spherical-like organization without outliers.

Figure 1 depicts the intuition of this method for a space containing three well-separated clusters described by two standardized numerical features. In this example, the second and third clusters are roughly located at the same height along the y-axis. The histogram of the first feature indicates the presence of three clusters, but the second histogram seems to suggest that there are two clusters. In this figure, we also show the cluster centers (or prototypes), which are located at $z_1 = (-0.01, 1.37)$, $z_2 = (-1.17, -0.7)$ and $z_3 = (1.19, -0.67)$, and the absolute distance between clusters for each dimension. The tuple $(1.16, 2.07)$ indicates that the absolute difference between the first and second clusters along the first dimension is 1.16 and 2.07 along the second dimension. By averaging and normalizing all tuples in the edges, we obtain scores that can be used as a proxy for feature importance. The scores for this example are 0.53 and 0.47 for the first and second features, respectively. Therefore, we can conclude that the second feature is less important than the first one because it does not allow distinguishing between the second and thirds clusters.

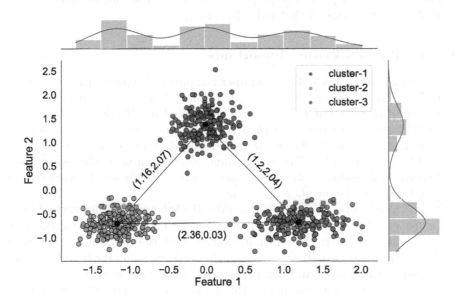

Fig. 1. Histograms, cluster prototypes and absolute distances between prototypes for each problem dimension in a three-cluster space.

Similarly to k-means and fuzzy c-means clustering, this post-hoc explanation method is biased towards high-variance features. To mitigate the variance effect on the results, it is advised to standardize the data before applying the clustering algorithm to have zero mean and unit variance.

The scoring function depicted in Eq. 3 formalizes our first method to determine the importance of the i-th problem feature,

$$\Phi(i) = \phi(i)/\sum_{l=1}^{|F|} \phi(l) \tag{3}$$

such that,

$$\phi(i) = \sum_{j=1}^{k}\sum_{l=1}^{k} |z_j(i) - z_l(i)| \tag{4}$$

where F denotes the set of problem features and $|F|$ is its cardinality, k stands for the number of clusters, while $z_j(i)$ represents the i-th feature value for the j-th prototype. Notice that this feature importance measure reports values in the $[0, 1]$ interval. Moreover, it holds that $\sum_i \Phi(i) = 1$.

In summary, the importance $\Phi(i)$ attached to the i-th feature is a proportion, given as the score $\phi(i)$ for the i-th feature divided by the sum of scores $\sum_l \phi(l)$ of all features in the dataset. The score $\phi(i)$ for a given feature is calculated by summing up for each prototype pairs, the absolute difference between feature values of the i-th feature of that pair of prototypes.

3.2   SHAP-Based Feature Importance

The second method proposed in this paper computes the contribution of a problem feature f_i based on the changes in the membership degrees of data points to the discovered clusters when that feature is marginalized. To do that, we modify the SHAP feature importance method [11] to an unsupervised setting since it allows performing this calculation by considering the effect of a given feature in all possible coalitions. Therefore, it is assumed that suppressing the most important features will lead to more perceptible variations in the cluster structure as quantified through the membership values.

Figure 2 shows an example that illustrates the information this method seeks to capture in a three-cluster problem described by three standardized features. The histograms in the main diagonal indicate that the first feature is unable to induce any cluster structure since its values behave similarly for all clusters, as opposed to the second and third features. The cluster results confirm that using the least important feature and the second one does not allow obtaining a clear cluster structure. However, we obtain the desired clustering as soon as we include the third feature in the calculations, regardless of whether we combine it with the first or the second feature. Figure 3 portrays the clustering results after extracting two principal components (with equations $c_1 = 0.38 \cdot f_2 + 0.93 \cdot f_3$ and $c_2 = -0.01 \cdot f_1 + 0.93 \cdot f_2 - 0.38 \cdot f_3$) and then applying k-means. The clustering results and the equations of the two principal components confirm that the first feature is largely irrelevant for this toy example.

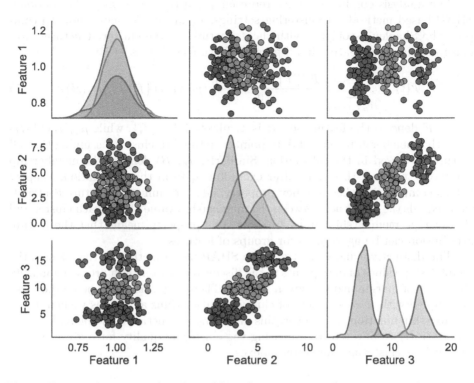

Fig. 2. Density functions of each problem feature across clusters concerning a toy example and pairwise clustering results using k-means.

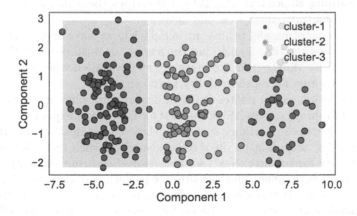

Fig. 3. Clustering results after extracting two principal components from the toy example introduce in Fig. 2 and applying k-means.

The analysis conducted above represents well the intuition of the proposed SHAP-based method for clustering settings. Equation (5) shows how to compute the Shapley associated with the f_i feature for the cluster structure, after marginalizing that feature in all possible feature coalitions,

$$\Psi(i) = \sum_{B \subseteq F \backslash \{f_i\}} \frac{|B|!(|F| - |B| - 1)!}{|F|!} \left(\mu_{xj}(B \cup \{f_i\}) - \mu_{xj}(B) \right) \tag{5}$$

where F denotes the feature set, B is a subset of $F \backslash \{f_i\}$, while $\mu_{xj}(B \cup \{f_i\})$ gives the membership value of data point x to the j-th cluster, assuming that all features are used in the calculation. Similarly, $\mu_{xj}(B)$ denotes the membership value of x to the j-th cluster after excluding f_i from this calculation. Shapley values comply with the efficiency property, i.e., the sum of all feature contributions equals the difference between the membership value of x to each cluster and the average membership value. In short, this property means that the feature attribution can be aggregated for groups of features.

The distinctive characteristic of the SHAP-based method is that it uses the membership values of data points to the discovered clusters as a proxy to quantify the effect of marginalizing a given feature. Therefore, this post-hoc method can be coupled with a wide variety of clustering algorithms as long as we can define membership functions. Moreover, this membership function could be replaced by other strategies devoted to evaluating the clustering quality for a given feature set, such as clustering validation indexes.

4   Experimental Simulations

In this section, we compare the proposed feature importance methods using 12 publicly available datasets containing numerical features from the UC Irvine Machine Learning Repository [1]. Table 1 describes these datasets in terms of the number of features, instances, and classes.

As for the clustering algorithms, for simplicity, we have used both k-means and fuzzy c-means algorithms such that the number of clusters equals the number of decision classes. Moreover, a fixed random seed was used when generating the initial cluster centers for reproducibility purposes.

Another relevant aspect relates to the evaluation procedure itself, which relies on the "pixel-flipping" curves [18] often used for comparing explanation methods that generate feature importance scores. In these curves, features are ranked in descending order according to their relevance scores such that the most relevant features are ranked first. Next, we progressively corrupt the features in the same order they were ranked and measure the performance deterioration. Hence, every time a feature is corrupted, the clustering algorithms are executed using the same number of clusters and the performance scores are calculated. In this paper, performance can be understood as the extent to which the membership of data points to each cluster changes after corrupting a feature, using the original membership values as the reference point.

Table 1. Descriptive statistics (number of features, instances and decision classes) characterizing the datasets used for simulations purposes.

Dataset	Number of features	Number of instances	Number of classes
Appendicitis	7	106	2
Echocardiogram	11	132	2
Ecoli	7	336	8
Glass	9	214	6
Hayes Roth	4	160	3
Heart statlog	13	270	2
Liver disorders	6	345	2
New thyroid	5	215	2
Pima	8	768	2
Vehicle	18	846	4
Wine quality red	11	1599	6
Yeast	8	1484	10

Figures 4 and 5 show the performance measure for each clustering algorithm and feature importance method. The performance values are obtained after corrupting the sorted features in each dataset progressively. The y-axis denotes the performance, and the x-axis gives the feature rank. As such, the first point gives the performance for k-means and fuzzy c-means when using all features. The second point in the curve denotes the performance after replacing the values of the most important feature with its mean. Similarly, the third point denotes the performance after altering the first-ranked and second-ranked features. This procedure is repeated until all features have been corrupted. The areas marked in grey indicate the difference in performance between PBFI and SHAP when using the k-means clustering algorithm, and the area marked in pink indicates the difference when using the fuzzy c-means algorithm.

The results allow us to draw several conclusions. Firstly, the performance deteriorates as more features are corrupted, which is the desired behavior in this experiment, thus confirming the correctness of both algorithms. Secondly, the curves associated with fuzzy c-means often fall below the ones associated with k-means. Thirdly, for most datasets, we can observe that the drop in performance after removing the most important features is larger than those after removing the least important features. This behavior holds for echocardiogram, ecoli, glass, hayes-roth, heart-statlog, liver-disorder, new-thyroid, pima, wine-quality-red and yeast when using the fuzzy c-means algorithm. It is also interesting how the performance falls sharply after removing the top-k features (see the appendicitis dataset, for example). Finally, the proposed feature importance methods reasonably agree in most cases for the fuzzy clustering settings, with PBFI being much simpler and faster than SHAP. In the hard clustering settings, SHAP reports deeper curves for ecoli, glass and yeast, which is a desirable behavior, thus making the SHAP variant the safer choice.

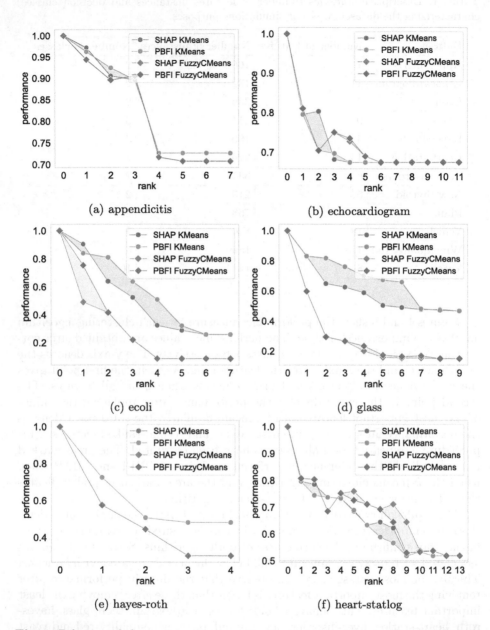

Fig. 4. Behavior of the performance measure for each algorithm as the features are altered in the same order as determined by each feature importance method. Altering a feature means that its values are replaced by the mean. Likewise, the performance quantifies how much the membership values change after one or several features have been modified (part one).

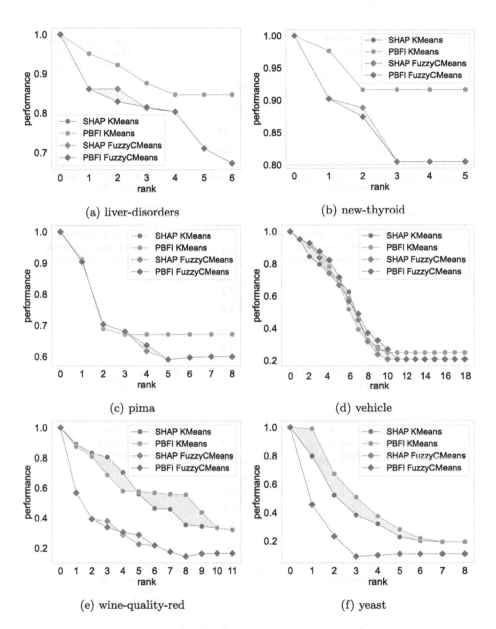

Fig. 5. Behavior of the performance measure for each algorithm as the features are altered in the same order as determined by each feature importance method. Altering a feature means that its values are replaced by the mean. Likewise, the performance quantifies how much the membership values change after one or several features have been modified (part two).

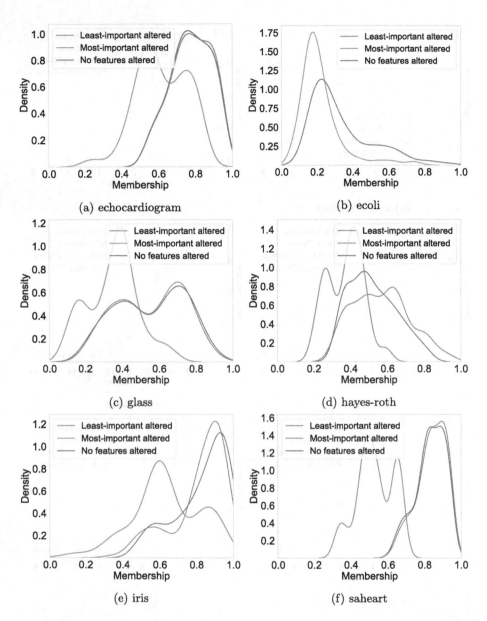

Fig. 6. Distribution of the membership values for the fuzzy c-means algorithm when no features are altered, when the most-important feature is altered, and when the least-important feature is altered. The most-important and least-important features are selected by the SHAP-based feature importance method. Altering a feature means that its values are replaced by the mean.

It is worth discussing why the curves for fuzzy c-means often fall below the ones obtained for k-means. While this might be related to the way of computing the membership values of data points to their clusters, the reason might be in the data itself. If the data is described by overlapping clusters, then hard clustering would not be as effective as soft clustering. However, this conjecture must be validated through exhaustive experiments.

Figure 6 presents the distribution of membership values associated with the fuzzy c-means clustering algorithm for selected datasets. The selected datasets exhibit the most pronounced differences between the removal of least-important and most-important features. The x-axis denotes the membership value to the cluster which had the highest membership value when no features were removed, and the y-axis denotes the density of the distribution. The blue line plots the density when all features are present. The orange line plots the membership values when the most-important feature, as selected by the SHAP-based feature importance method, is removed. The green line plots the membership values when the least-important feature is removed. The expected behavior is that the removal of the most-important feature should lead to a more noticeable shift in the distribution of the membership values, as compared to the removal of the least-important feature. This expectation is an extension from the assumption that suppressing the most important features will lead to more perceptible variations in the cluster structure as quantified through the membership values. Our observations reveal that, for most datasets, there is indeed a more noticeable shift in the distribution of membership values when removing the most-important feature compared to removing the least-important feature. This behavior holds for echocardiogram, ecoli, glass, hayes-roth, heart-statlog, new-thyroid, pima, vehicle, wine-quality-red and yeast. These observations reinforce the validity of the SHAP-based feature importance method in accurately distinguishing between features that result in more perceptible variations in the cluster structure, as quantified through the membership values.

5   Concluding Remarks

In this paper, we have proposed two post-hoc feature importance methods for prototype-based clustering. The first method uses the dimension-wise distances between cluster centers to determine the extent to which features help separate the clusters. In this approach, features with larger variance are more likely to report larger feature importance scores. The second method modifies the well-known SHAP method for computing feature importance in a supervised setting. The proposed variant determines the contribution of a feature based on changes in the membership degrees of data points to discovered clusters when that feature is removed from all possible permutations.

The experiments using real-world datasets demonstrated that both methods elicit the features responsible for the cluster structure. On a sizable portion of the datasets, both scoring methods report a similar behavior when using the fuzzy c-means method to discover the clusters. In the hard setting, the proposed SHAP variant seems to be a safer choice, although it is significantly slower compared to the PBFI algorithm. This difference in computing power is especially pronounced in datasets with a larger number of features and/or observations. Therefore, we would recommend the PBFI scoring method to save computational power for these larger datasets. For smaller datasets where computational efficiency is less of a concern, we recommend using both scoring methods to rank features and compare the results before making a decision.

The proposed methods suffer from some limitations that deserve being mentioned. Firstly, similarity to k-means and fuzzy c-means, their inner working might be influenced by high-variance features. Secondly, it can be argued whether the distance between each data point and the cluster centers is the best approach to determine the clustering changes after marginalizing a feature. Some interesting research avenues in that regard include expanding the proposed methods by considering internal and external validation measures [10,16] and measuring how the resulting feature relevance scores differ.

References

1. UC Irvine machine learning repository. https://archive-beta.ics.uci.edu/. Accessed 20 Nov 2022
2. Bezdek, J.C.: Pattern Recognition with Fuzzy Objective Function Algorithms. AAPR, Springer, Boston, MA (1981). https://doi.org/10.1007/978-1-4757-0450-1
3. Bobek, S., Kuk, M., Szelażek, M., Nalepa, G.J.: Enhancing cluster analysis with explainable AI and multidimensional cluster prototypes. IEEE Access **10**, 101556–101574 (2022)
4. Bora, D.J., Gupta, D., Kumar, A.: A comparative study between fuzzy clustering algorithm and hard clustering algorithm. arXiv preprint arXiv:1404.6059 (2014)
5. Dunn, J.C.: A fuzzy relative of the ISODATA process and its use in detecting compact well-separated clusters. J. Cybern. **3**(3), 32–57 (1973). https://doi.org/10.1080/01969727308546046
6. Frost, N., Moshkovitz, M., Rashtchian, C.: ExKMC: expanding explainable k-means clustering. arXiv preprint arXiv:2006.02399 (2020)
7. Gunning, D., Stefik, M., Choi, J., Miller, T., Stumpf, S., Yang, G.Z.: XAI-Explainable artificial intelligence. Sci. Robot. **4**(37), eaay7120 (2019)
8. Jain, A.K., Murty, M.N., Flynn, P.J.: Data clustering: a review. ACM Comput. Surv. **31**(3), 264–323 (1999)
9. Kauffmann, J., Esders, M., Ruff, L., Montavon, G., Samek, W., Müller, K.R.: From clustering to cluster explanations via neural networks. IEEE Transactions on Neural Networks and Learning Systems (2022)
10. Lei, Y., Bezdek, J.C., Romano, S., Vinh, N.X., Chan, J., Bailey, J.: Ground truth bias in external cluster validity indices. Pattern Recogn. **65**, 58–70 (2017)
11. Lundberg, S.M., Lee, S.I.: A unified approach to interpreting model predictions. In: Advances in Neural Information Processing Systems 30 (2017)
12. MacQueen, J.: Classification and analysis of multivariate observations. In: Proceedings of the 5th Berkeley Symposium on Mathematical Statistics and Probability, pp. 281–297 (1967)

13. Montavon, G., Kauffmann, J., Samek, W., Müller, K.-R.: Explaining the predictions of unsupervised learning models. In: Holzinger, A., Goebel, R., Fong, R., Moon, T., Müller, K.-R., Samek, W. (eds.) xxAI - Beyond Explainable AI: International Workshop, Held in Conjunction with ICML 2020, July 18, 2020, Vienna, Austria, Revised and Extended Papers, pp. 117–138. Springer, Cham (2022). https://doi.org/10.1007/978-3-031-04083-2_7
14. Oyewole, G.J., Thopil, G.A.: Data clustering: application and trends. Artif. Intell. Rev. **56**, 6439–6475 (2022). https://doi.org/10.1007/s10462-022-10325-y
15. Peng, X., Li, Y., Tsang, I.W., Zhu, H., Lv, J., Zhou, J.T.: XAI beyond classification: interpretable neural clustering. J. Mach. Learn. Res. **23**(6), 1–28 (2022)
16. Rendón, E., Abundez, I., Arizmendi, A., Quiroz, E.M.: Internal versus external cluster validation indexes. Int. J. Comput. Commun. **5**(1), 27–34 (2011)
17. Ribeiro, M.T., Singh, S., Guestrin, C.: " why should i trust you?" explaining the predictions of any classifier. In: Proceedings of the 22nd ACM SIGKDD International Conference on Knowledge Discovery and Data Mining, pp. 1135–1144 (2016)
18. Samek, W., Binder, A., Montavon, G., Lapuschkin, S., Müller, K.R.: Evaluating the visualization of what a deep neural network has learned. IEEE Trans. Neural Netw. Learn. Syst. **28**(11), 2660–2673 (2016)
19. Shapley, L.S.: Notes on the N-Person Game — I: Characteristic-Point Solutions of the Four-Person Game. RAND Corporation, Santa Monica, CA (1951). https://doi.org/10.7249/RM0656
20. Steinhaus, H., et al.: Sur la division des corps matériels en parties. Bull. Acad. Pol. Sci. **1**(804), 801 (1956)

Uncovering Manipulated Files Using Mathematical Natural Laws

Pedro Fernandes[1]([✉]), Séamus Ó Ciardhuáin[1], and Mário Antunes[2,3]

[1] Department of Information Technology, Technological University of the Shannon,
Moylish Park, Limerick V94 EC5T, Ireland
{Pedro.Fernandes,Seamus.OCiardhuain}@tus.ie
[2] School of Technology and Management, Polytechnic of Leiria, Leiria, Portugal
mario.antunes@ipleiria.pt
[3] INESC TEC, CRACS, R. Campo Alegre 1021/1055, 4169-007 Porto, Portugal

Abstract. The data exchange between different sectors of society has led to the development of electronic documents supported by different reading formats, namely portable PDF format. These documents have characteristics similar to those used in programming languages, allowing the incorporation of potentially malicious code, which makes them a vector for cyberattacks. Thus, detecting anomalies in digital documents, such as PDF files, has become crucial in several domains, such as finance, digital forensic analysis and law enforcement. Currently, detection methods are mostly based on machine learning and are characterised by being complex, slow and mainly inefficient in detecting zero-day attacks. This paper aims to propose a Benford Law (BL) based model to uncover manipulated PDF documents by analysing potential anomalies in the first digit extracted from the PDF document's characteristics.

The proposed model was evaluated using the CIC Evasive PDF-MAL2022 dataset, consisting of 1191 documents (278 benign and 918 malicious). To classify the PDF documents, based on BL, into malicious or benign documents, three statistical models were used in conjunction with the mean absolute deviation: the parametric Pearson and the non-parametric Spearman and Cramer-Von Mises models. The results show a maximum F1 score of 87.63% in detecting malicious documents using Pearson's model, demonstrating the suitability and effectiveness of applying Benford's Law in detecting anomalies in digital documents to maintain the accuracy and integrity of information and promoting trust in systems and institutions.

Keywords: Benford's law · Pearson's correlation · Tampered files · Anomaly detection · Forensic analysis

1 Introduction

As a result of the effects arising from aggressive digitalisation, supported by connectivity between different computer systems in various areas, such as economic, financial, health and military, there has been an exponential growth in

© Springer Nature Switzerland AG 2024
V. Vasconcelos et al. (Eds.): CIARP 2023, LNCS 14469, pp. 46–62, 2024.
https://doi.org/10.1007/978-3-031-49018-7_4

the number of attacks to which no one is immune. Recent studies point to a 42% increase in the first half of 2022 alone, compared to 2021, causing significant losses to businesses and governments in billions of dollars [27].

According to Oracle [1] attacks can be categorised as structural attacks, i.e. the structure of the document is changed (Return-Oriented Programming, ROP attacks) and attacks carried out by exploiting multimedia resources and applications, such as including elements like audio and video, making it possible to load and execute malicious code through this contents [8,32].

Digital documents are usually emailed, and users are encouraged to open the attachments. The proximity of digital documents to programming languages allows the incorporation of JavaScript code from the attachments or even in the email message itself, allowing it to activate its digital content. In this context, attackers try to find ways to include malicious code (malware) and trigger client-side attacks, allowing shellcode to be stored in memory and thus execute arbitrary code [23].

In this work, by a "natural law" we mean an observed statistical property of given data. Unlike machine learning (ML) or deep learning (DL) techniques, natural laws do not depend on the attack type but on the characteristics of the data being analysed. They have been used to detect economic fraud and fraudulent online payments. The most used natural law for this purpose has been Benford's law [5,18,25]. Some studies have had encouraging results in analysing Internet traffic behaviour with such laws [3].

Our aim is to apply Benford's law to detect malicious PDF files [11]. This law, also known as the law of the first digit, describes the frequency with which digits occur in numerical data. Deviations from what is expected may indicate malicious content within the file. Analysis of the frequency of digits in PDF file metadata is then used to classify documents as malicious or benign.

The objectives of this research are:

- To review the literature on the application of Benford's law in detecting anomalies in a computer network from PDF files with anomalies.
- To create a model that allows the extraction of the first digit from the features of PDF files to evaluate whether the files conform to Benford's law.
- To apply statistical models (Pearson, Spearman and Cramer-Von Mises) to evaluate the Benford's Law model.

The paper is organised as follows: Sect. 2 reviews related work. Section 3 studies the mathematical foundations of Benford's law. Section 4, gives a specific view on the dataset used. Section 5 discusses the method that allowed using Benford's law. Section 6 discusses the results obtained by applying the model. Section 7 concludes the study with several observations on using the model under investigation and specific procedures to be implemented in future work.

2 Related Work

Portable Document Format (PDF), developed by Adobe Systems Inc. and standardized as ISO 32000, is one of the most popular electronic document formats.

It provides facilities such as links, buttons, form fields, embedded audio and video, and electronic signing. It is highly portable [14,15,30].

PDF documents can contain embedded code, allowing attackers to include malware payloads, triggering client-side attacks by executing arbitrary code [23]. This can lead to the theft of information and the compromise of computer systems. PDF documents that carry malicious code are usually sent by email and executed when opened by unsuspecting or inexperienced users. For example, the May 2022 Snake keylogger malware was distributed via email, providing sensitive information to email attackers because the malicious content was executed on the client side [6,29]. As a result, software tools were developed to check whether PDF documents contained malicious content. Some of these tools, such as MalOffice, CWSanbox and MDScan, use machine learning approaches [10]. However, these methods are complex and probably ineffective against zero-day attacks.

A PDF document is a binary file that includes a collection of objects that can be accessed randomly using a cross-reference table, which reduces object access time. This type of access makes it possible to incrementally alter or update the PDF document, cancelling any modifications that may have occurred and thus allowing confidential information to be recovered. Figure 1 shows the structure of a PDF file, which consists of four parts: header, body, Xref table and trailer [20].

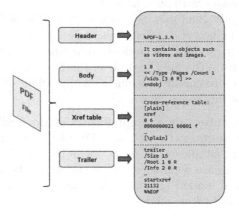

Fig. 1. Structure of a PDF file

The commonest cyber attacks on PDF documents include changes to the certification of the PDF document, the breaking of signatures and encryption, the presence of ransomware, shadow attacks and the existence of insecure features in PDF documents [8,28]. There are four types of objects in a PDF document: Page (/Contents), Annotation (/Link), Field (/AA), and Catalog (/OpenAction). Attackers develop JavaScript scripts embedded in documents and end up causing abnormal situations such as denial of service, infinite cycles and deflation bombs. In the detection and analysis of malicious PDF documents, the use of machine learning-based models has been gaining prominence. The essence of this type of tool relies on three fundamental pillars: pre-processing, feature extraction and learning, and classification.

In the pre-processing phase, machine learning-based methods are differentiated by static analysis, which results from direct code inspection, and dynamic analysis, which uses virtual environments to prevent the spread of infection. After the pre-processing phase, characteristics are extracted from PDF documents. These features are stored in a numeric vector, and their nature dictates the choice of the learning algorithm. Several techniques use Bayesian classifiers, SVM support vector machines and random forests [17, 21, 32].

3   Benford's Law Fundamentals

Benford's law empirically describes the frequency with which the first digits occur in a dataset. BL has been applied in various contexts with promising results, most notably the detection of financial fraud [19], anomalies in data reported by governments on specific diseases and election data [26].

The process developed for detecting manipulated photographs using Benford's Law is continued in the present work on malicious PDF documents. The difference between the two works lies in the extraction and the number of features analysed, which is reduced in PDF documents. The goal was to investigate the robustness of Benford's law using a set with few features.

Benford's law has been widely used in areas such as the detection of cyberattacks on power systems [24], in the detection of anomalies in a computer network using a set of features extracted from TCP streams [3,11] and more recently in building IDS for resource-constrained systems using the flow size difference of a network in conjunction with linear regression [31] and in detecting anomalous packets caused by DDoS attacks in network traffic [12].

BL has the following main requirements:

- It should be used on numerical data.
- Some studies point to using at least 500 observations [7]
- It can be applied to data with a huge variety in the number of digits.
- It uses logarithmic functions, so the zero digits must be restricted. In this case, an initial data reduction is necessary.
- The presence of decimal values and significant digits must be determined.

3.1   Benford's Law Statement

Several theories have emerged that attempt to explain the phenomenon of the law of the first digit, ranging from classical to more complex. They are usually differentiated by the nature of their study, whose demonstrations are based on the assumption of continuity, scale and basis invariance, or through static arguments [13].

The theory that best fits the explanation of Benford's law suggests that a probability domain must initially be found. Such a domain is obtained from a collection of subsets in $\mathbb{R}^+$, called a sigma-algebra defined by $\mathcal{A}$. Thus, the first significant digit of base 10 is 1 and is given by Eq. 1.

$$\{D_1 = 1\} := \cap_{n=-\infty}^{\infty} [1, 2) \times 10^n \tag{1}$$

Equation 1 is subsequently extended to the remaining digits, generalising into Eq. 2.

$$\{D_n = n\} := \cap_{n=-\infty}^{\infty} (a, b) \times 10^n \tag{2}$$

Considering each value of $i \in \mathbb{N}$, the function $D_i(d)$ represents each significant digit obtained from d. $\mathcal{A}$, will represent the collection of all significant digits obtained from the function $D_i(d)$, and the intended domain [13]. For example, if we want to get $D_1(\sqrt{2}) = 1$, $D_2(\sqrt{2}) = 4$ and so on. Equation 3 defines the general law that allows the probability of each significant digit to be calculated.

Theorem 1 (General law). *Let $k \in \mathbf{Z}$, $d_1 \in \{1, 2, 3, ..., 9\}$ and $d_j \in \{0, 1, 2, ..., 9\}$, $j = 2, ..., k$.*
Then

$$P(D_k = d_k) = \log \left(1 + \frac{1}{\sum_{i=1}^{k} d_i \times 10^{k-i}} \right) \tag{3}$$

From Eq. 3, Eq. 4 is derived, which allows the probability of the first significant digit to be calculated [2].

$$P(D_i(X)) = \log \left(1 + \frac{1}{d} \right), \text{ if } d = \{1, 2, 3, ..., 9\} \tag{4}$$

A proof of the General Theorem of Benford's Law (1) can be found in [34].

4   Dataset

The current approach uses Benford's law to detect anomalies in the features extracted from PDF's will be considered significant if they can distinguish which PDF documents can contain malicious or benign information.

Considering the nature of the features in a PDF document, i.e., whether numeric or text, their choice requires careful investigation due to the existing limitations of Benford's law. Assuming that BL follows a logarithmic distribution (4), one of BL's significant limitations is using zero digits. In the current approach, the feature selection is based on the public dataset CIC-Evasive-PDFMAL2022, hosted at UNB [16]. The PDF files were obtained from Contagio and VirusTotal and are summarised in Table 1.

After the first reduction, discussed in Sect. 5.1, the dataset has 10 025 files, of which 4 468 are benign, and 5 557 are malicious.

The features were extracted from the PDF documents using unsupervised machine learning techniques, which allowed the grouping of the data into two groups due to the similarity of the features. Thus, PDF files labelled malicious

Table 1. Dataset

Name	Malicious	Benign
Contagio	11 173	9 109
VirusTotal	20 000	–
Total	31 173	9 109

had an evasive set of malicious records, while those considered benign had no malicious records. Thus, 32 features were extracted, of which 12 are general features, which allow describing in a general way any PDF document, and 25 are structural features, which would enable describing PDF documents in terms of structure.

Figure 2 shows how PDF documents were categorised into malicious and benign.

Fig. 2. Flow diagram, which made it possible to categorise PDF files [22]

This study did not include documents that contained non-numeric features (e.g., the file name) or that had only absolute zeros or negative values. Table 2 shows some of the features of PDF documents not used in the original dataset.

Table 2. Some of the unused features.

ID	isEncry.	Emb. files	Text	Header	Images
aedaf3c5...	0	0	No	%PDF-1.3	0
c6453dd1...	0	0	Unclear	%PDF-1.4	−1
...	...	...	...	...	...

Table 3 shows the eight features extracted from the 32 main features appropriate for applying BL, which that does not have zero or negative values. Not using zero or negative values is related to the impossibility of the logarithmic function working with these values.

Eight features were selected from a preliminary analysis of the 32 features available in the dataset, subject to feature reduction (Subsect. 5.1).

Size, identified by S, ranges from malicious to benign due to the content embedded in the document, such as the number of images, multimedia files

Table 3. The eight features extracted after data reduction

General features	Structural features
PDF size	Xref length
Metadata size	Endobj
Pages	Endstream
Title characters	JS

or other PDF documents. Metadata, identified by M, where information about the PDF document is found. Usually, attackers tend to exploit this resource by including hidden illicit content. Pages identified by P tend to show fewer pages in malicious PDF documents. Some malicious PDF documents present blank pages. Xref length refers to the number and size of tables, where malformed tables indicate malicious content. Title characters, identified by T, are the number of characters in the title. Endobj, denoted by Eobj, is the keyword indicating the end of objects in the file. Endstream, identified by EStr, is the set of keywords showing the streams in the file. JS suggests the presence of Javascript code. Class is the tag classified as malicious or benign.

5   Benford's Law-Based Method

The proposed model is based on the analysis of the first digit extracted from the features of the PDF documents. An architecture was developed to make such a classification based on three main blocks: pre-processing, processing and analysis of results. Figure 3 illustrates the overall architecture that allowed BL to detect malicious files.

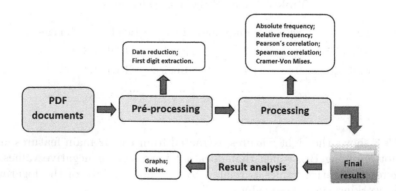

Fig. 3. Overall architecture of the proposed model

5.1 Pre-processing

Table 4, shows the content of some existing features for each PDF document before data reduction. Each line represents a PDF document.

Table 4. Features of some PDF documents present in the original dataset before data reduction

PDF document	Fine name	pdfsize	metadata size	pages	images	text	header	obj	...
1	9c39d6ca95dd38b12211ec0f34958e40e9fcbc7b6f3840252762d19ee847cce9	78	367	1	−2	unclear	%PDF-1.4	17	...
2	aedaf3c5428a2e3ba600c44b96ad78dfdf8ed76e7df129bcd8174d83b77a9c33	8	180	1	0	No	%PDF-1.3	10	...
3	c6453dd19503dee6d133018a266f0eabb40b6ab92ba08f921332b7a2b0f9625b	12	180	1	0	unclear	%PDF-1.4	10	...

The data reduction consisted of eliminating features containing zeros, negative numbers and non-numerical features, and was performed using Microsoft Excel. 1,191 PDF documents were obtained: 913 malicious PDFs and 278 benign ones, labelled malicious or benign. Two experiments were performed, one using the entire set of files, i.e. the 1,191 PDF files, and the second using a balanced dataset with 277 benign and 277 malicious files. The results can be seen in Sect. 6.1.

The data relating to extracting the first digit of each PDF file generates a feature vector, and each one is labelled, i.e. if the file is benign, the number 1 is assigned; otherwise, if the document is malicious, the number 0 is set. After reducing the data, Matlab extracted the first digit from each stream.

5.2 Processing

Figure 4 presents the processing phase. After the absolute frequency of each digit has been calculated from the data vector, each digit's relative frequency of occurrence is calculated, that is, the quotient between the frequency with which each digit occurs and the total frequency of digits present in the data set. This allows a graphical comparison with the empirical BL, enabling us to check whether the data agree with the BL. If not, we can conclude that outliers exist. This procedure allowed us to calculate the correlation between each digit's relative frequency of occurrence and the empirical Benford's law.

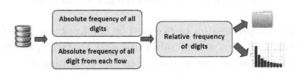

Fig. 4. Processing phase

The Fig. 5 compares the expected and observed values of the digits, where a strong distortion between all digits can be observed. Therefore, the first conclusion drawn from the graphic is the strong evidence of the existence of malicious PDF documents.

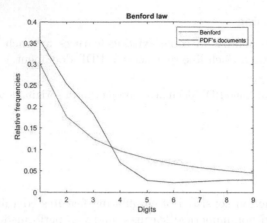

Fig. 5. Graphical comparison between Benford's Law and the relative frequency of occurrence of each digit when the data set is balanced, referring to the first experiment.

The second phase consists in counting the frequency of occurrence of the first digits of each PDF document and comparing it with the empirical frequency given by BL. As mentioned, the graphical visualisation does not allow us to conclude whether the PDF files are malicious.

The difficulty in obtaining a rigorous answer about the nature of a PDF document, that is, whether the document under analysis is considered malicious or benign, made it necessary to implement a set of statistical procedures to facilitate the classification of PDF documents. Since the data were appropriately tagged in the pre-processing phase, parametric and non-parametric statistical models allowed us to generate a binary classification, enabling a comparison with the original tags.

The statistical models used in the investigation were the parametric Pearson and non-parametric Spearman and the Cramer Von Mises models. In addition to allowing the calculation of the correlation between the relative frequencies of occurrence of each digit and Benford's empirical law, it enabled the determination of its p-value. A binary classification was generated from the p-value calculation and comparison with the desired significance level ($\alpha = 0.1, 0.001$, or 0.05).

Figure 6 shows the processing performed by the hypothesis tests. Each model allowed the generation of a set of labels related to the evaluation, indicating whether a given PDF document is malicious (0) or benign (1). The labels generated by the evaluation are stored in a separate .txt file for each model. Each .txt file is compared with the labels obtained in the pre-processing phase, generating results where the amount of malicious and benign PDF files are analysed, and the hit rate is calculated.

The choice of the Pearson, Spearman, and Cramer-Von Mises models considered the types of features extracted from the PDF files. The main features of these statistical models can be visualised in [9].

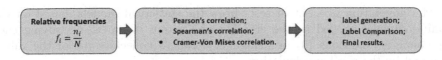

Fig. 6. Processing performed by the hypothesis tests

One of the main features of Pearson's model is the model's ability to quantify the strength of the linear relationship between two quantitative variables. Its use allowed us to determine the correlation between the frequency of occurrence of digits extracted from PDF documents and the empirical frequency of Benford's law.

A strong positive correlation between the variables suggests that we may be in the presence of possible anomalies indicating the presence of malicious code in PDF documents. Another essential feature of Pearson's model is its application to normally distributed variables. The presence of an outlier around digit 1 indicates that the frequency with which the first digit occurs is higher than that of the remaining digits.

Spearman's model was used to test whether there was an ordinal relationship between the two variables under investigation. Spearman's model is usually tested when distributions are abnormal or come from ordinal rankings (existence of an order).

The Cramer-Von Mises model allows one to check whether the marginal distribution of a set of independent and identically distributed random variables, $X_1, X_2, ..., X_n$ follows a given continuous distribution function, known as $F(x)$. The Cramer-Von Mises model is based on a square metric, which measures the distance between an empirical distribution (e.g., Benford's law) constructed around $X_1, X_2, ..., X_N$ and a specific non-negative cumulative function, defined at $[0, 1]$.

Applying the models described above, in conjunction with a set of well-defined metrics, resulted in the generation of labels that allowed us to classify the PDF document as malicious or benign. The metrics that allowed the data classification were based on a set of hypothesis tests that created a decision rule regarding the nature of the PDF files.

The value of the proof (the p-value) reflects whether a given outcome has a high probability of occurring, assuming the initial hypothesis is presented as accurate. In practice, hypothesis testing is unlikely to check whether a given result is true or false, so any value obtained will be subject to type I or II errors.

Usually, type I errors are mitigated by considering significance (α). This parameter gives the probability of committing a type I error when facing a true hypothesis in the data. For any p-value less than significance, the null hypothesis is rejected.

In these experiments, the infeasibility of such a procedure results in the impossibility of choosing which PDF files are considered relevant for applying the model based on Benford's Law.

The generation of a set of tests from the statistical models, based on the p-value calculation, led to the creation of a group of labels, classifying each document as malicious (0) or benign (1). Table 5 collects some of the values obtained by the p-value calculation.

Table 5. P-values based on Pearson, Spearman and Cramer-Von Mises models

PDF's Id	Pearson	Spearman	CVM
1	0000	0.0146	0.2417
2	0.0073	0.0397	0.0515
3	0.2188	0.0556	0.2432
..	...	...	...

5.3   Median Absolute Deviation

The Mean Absolute Deviation (MAD) calculates the average distance between the occurrence of each digit and the mean of that occurrence obtained in each stream. The MAD is a robust measure of variability that avoids the presence of outliers. Equation 5 defines the MAD.

$$MAD = Median\left(|X_i - median\left(X\right)|\right) \tag{5}$$

Observing the variability of the results obtained, the application of the MAD allowed the normalisation of the p-value. To do so, it was necessary to calculate an estimate of the normality parameter. Equation 6, shows the estimated parameter σ calculation.

$$\sigma = 1.253 * MAD\left(X, 0\right) \tag{6}$$

This was calculated using Matlab's MAD function.

5.4   Evaluation Metrics

Confusion matrix (or error matrix) was used to classify the model based on Benford's Law [4, 33].

True Positives (TP) refer to events where the model correctly predicted the existence of malicious files. In contrast, the True Negatives (TN) represent the events where the model correctly predicted benign files. False Positives (FP) and False Negatives (FN) refer to misclassified events, where benign files are classified as malicious and malicious files are classified as benign.

The correct p-value evaluation was calculated by the following metrics: precision, recall, F1-score and accuracy.

The metrics referred to are usually applied to machine learning-based classifiers. Applying such metrics in prediction based on statistical models is unusual, so their inclusion in this work facilitates future benchmarking with other learning and prediction models [9].

6   Results

This Section describes and discusses the experiments made with Benford's model.

6.1   Analysis of Results

Table 6 contains the results obtained after performing two experiments with 554 PDF documents (277 benign and 277 malicious documents) and 1191 PDF documents (278 benign and 913 malicious documents). The goal was to understand if there were fundamental differences in applying Benford's law to properly balanced datasets or if it could be applied to an unbalanced dataset. The values were obtained using the mean absolute deviation.

Table 6 also shows the values for False Positive (FP), False Negative (FN), True Positive (TP) and True Negative (TN), as well as the values for Precision (P), Recall (R), F1-Score (F1) and accuracy (A).

The best result is obtained using the Pearson model, with a significance level of 0.1. Comparing the two experiences performed, as well as the results obtained from the Spearman and Cramer-Von Mises models, it was with the Pearson model that the best results were obtained, with an F1 of 72.75% and 84.44% and with a precision of 60.14% and 81.19% for the first and second experiments respectively. Studies conducted without MAD have shown negligible effects, so using this technique in a parametric model is extremely useful. The non-parametric Cramer-Von Mises model obtained a high hit rate of true negatives. However, many false negatives (156) make the model unsuitable since most PDF files are considered benign when malicious. Considering that the mean absolute deviation is usually used in parametric models, we conclude that its use in a non-parametric model proves useless. The Pearson model is the most appropriate of the three models tested as long as it is used in conjunction with the mean absolute deviation.

Assuming that the initial hypothesis was based on a set of malicious files and that the test statistic rejected this hypothesis if the p-value was less than α, the Pearson model is the most reliable method for detecting malicious files. Looking at Table 6, the Cramer-Von Mises model was reliable in detecting benign files in the first experiment. Combining the two models (Pearson and the Cramer-Von Mises models) in classifying the files can be an added value. Given the results, it is essential to understand why there are so many false positives.

One possible cause is how the digits are distributed, and the relative frequency resulting from this distribution. Table 7 provides an overview of the relative frequency of the digit distribution for three malicious and three benign files, obtained from the experiments performed and referring to the behaviour of the application of Benford's law to benign and malicious documents, as well as the p-values obtained with and without using the median absolute deviation and the rank obtained given the p-values.

Using the mean absolute deviation, initial digits with a higher frequency of observations generally tend to classify files as malicious to the detriment of frequencies with a higher dispersion of values. Considering file three to file

Table 6. Results were obtained in the two experiments with MAD for each statistical model. The table shows the level of significance, as well as the number of documents classified as true positives, false positives and the accuracy of the model, among other parameters.

Statistical model	α	First experience (554 PDF files using MAD)								Second experience (1191 PDF files using MAD)							
		TP	TN	FP	FN	P	R	F1	A	TP	TN	FP	FN	P	R	F1	A
Pearson's	0.1	255	108	169	22	0.6014	0.9206	0.7275	0.6552	803	91	186	110	0.8119	0.8795	0.8444	0.7513
	0.05	274	30	247	3	0.5259	0.9892	0.6867	0.5487	889	50	227	24	0.7966	0.9737	0.8763	0.7891
Spearman's	0.1	127	137	140	150	0.4757	0.4585	0.4669	0.4765	471	102	175	442	0.7291	0.5159	0.6042	0.4815
	0.05	191	54	223	86	0.4614	0.6895	0.5528	0.4422	713	43	234	200	0.7529	0.7809	0.7667	0.6535
CVM	0.1	121	235	42	156	0.7423	0.4368	0.5500	0.6426	308	220	57	605	0.8438	0.3373	0.4820	0.4437
	0.05	194	155	122	83	0.6139	0.7004	0.6543	0.6300	532	146	131	381	0.8024	0.5827	0.6751	0.5697

Table 7. Comparison between the relative frequencies for three malicious documents and three benign documents based on the occurrence of each digit. Each document's p-value and MAD were calculated, as well as the decision obtained, being 0 if we are facing a malicious document and one if the document is benign.

	Malicious PDF files			Benign PDF files		
	1	2	3	1	2	3
1	0.5000	0.3750	0.1250	0.500	0.3750	0.1250
2	0.1250	0.1250	0.6250	0.2500	0.1250	0.5000
3	0.1250	0.1250	0.1250	m 0.2500	0.2500	0.2500
4	0.2500	0.2500	0	0	0.1250	0
5	0	0	0	0	0.2500	0
6	0	0	0.1250	0.2500	0	0
7	0	0.1250	0	0	0	0.1250
8	0	0	0	0	0	0
9	0	0	0	0	0.1250	0
P-value	0.0009	0.0073	0.2188	0.0601	0.4824	0.1998
MAD	0.1538	0.1164	0.0641	0.0946	0.3896	0.0451
Decision	0	0	1	1	0	1

four referring to malicious PDF files, the adopted model classified file three as benign and four as malicious. The frequency of observation of the digits in file 3 has greater dispersion of frequencies to the detriment of file four, where the concentration of frequencies is only in digits 1 and 2. This fact has an active influence on the calculation of the correlation with which each digit occurs with Benford's law. BL tells us that the frequency of occurrence of digit 1 is 30.10%, so the concentration of the event of digits around digits 1 and 2 will make the correlation relatively high, affecting the value obtained by the p-value, with consequences in the classification. From Table 6, we can draw some important conclusions. Balanced or unbalanced datasets impose no limitations on the model based on Benford's. On the other hand, by increasing the number of PDF files in the dataset, the model became more sensitive in detecting manipulated files.

Comparing the results of the two experiments with Spearman's model, we see an increase in the detection of the number of true positives from 34.47% to approximately 60%. This is visible in the remaining models, where the number of true positives (manipulated PDF files) increased from 35.01% to approximately 44.66% in the Cramer-Von Mises model and from 49.45% to 74.64% in the Pearson model. The results obtained show the importance of applying the mean absolute deviation.

The low number of PDF files to analyse (in the first experiment), the low number of features to analyse (both in the first and second experiments), and the fact that the dataset was balanced (first experiment) contributed significantly to the results being worse in the first experiment. After applying the MAD, there was an increase in the number of detections of manipulated PDF files, with false positives showing a marked decrease in all classification models, motivated by the increase in the dataset and the calculation of the average absolute deviation of the p-value in each PDF document with the average p-value. The Pearson model was the one that presented the best F1 score of the three classification models under analysis, with a result of 87.63%, registering an accuracy of 79.66%.

7   Conclusions and Future Work

The present research addresses several challenges anomaly detection faces in rapidly detecting malicious PDF documents through a low complexity and fast approach.

The application of Benford's law in this study does not require knowledge of the type of attack, but only that it can classify these anomalies and identify them as such.

Two experiments were conducted. The first experiment used 554 PDF files derived from the original dataset, with 277 manipulated PDF files and 277 benign PDF files (balanced dataset). The second experiment used the entire dataset with 1191 PDF files containing 913 manipulated PDF files and 278 benign PDF files (unbalanced dataset). One of the goals was to verify the robustness of the model against a balanced and unbalanced dataset to understand if the model was sufficiently robust in detecting manipulated files. It was found that Benford's law can be applied to unbalanced datasets. To increase the number of document classifications handled using the Pearson model, it is important to increase the number of files. If no further features can be extracted from the documents, it is important to use techniques to adjust the p-values obtained by applying the mean absolute deviation, where good results were found in classifying the manipulated PDF files.

The best result was obtained using the Pearson model with an F1-score of 87.63%, identifying almost all malicious PDF files but with some false positives. It was found that the high frequency of occurrence of digits around digits 1 and 2 influenced the results and, thus, the classifier, leading to false positives.

Benford's law is limited by its inapplicability to negative, and zero digits since its mathematical structure is based on logarithmic functions. This inapplicability may lead to possible information losses. In future work, it will be important to adapt the model based on Benford's law to avoid the existence of both zeros and negative digits and thus avoid possible information losses.

This study has demonstrated the existence of a robust, and enforceable method for detecting malicious PDF documents. Future research on the applicability of Benford's law and other natural laws in detecting abnormalities in PDF files should focus on implementing new methods that allow a classification with greater detail in addition to the application of MAD. Although MAD is a robust estimator and requires low computational resources, its applicability only to symmetric distributions and low Gaussian efficiency is a significant limitation. It is intended to investigate the applicability of other natural laws, such as Zipf's law and Stigler's law, in conjunction with new techniques, notably the S_n estimator and the Q_n estimator with the particularity of not being symmetric distribution estimators.

References

1. Adobe: Zero day malware threat prevention, July 2015. chrome-extension://efaidnbmnnnibpcajpcglclefindmkaj/ https://www.oracle.com/a/otn/docs/zero-day-malware-protection-brief.pdf
2. Arno Berger, T.P.H.: An Introduction to Benford's Law. Princeton University Press (2015). https://www.ebook.de/de/product/23323656/arno_berger_theodore_p_hill_an_introduction_to_benford_s_law.html
3. Arshadi, L., Jahangir, A.H.: Benford's law behavior of internet traffic. J. Netw. Comput. Appl. **40**, 194–205 (2014). https://doi.org/10.1016/j.jnca.2013.09.007
4. Caelen, O.: A Bayesian interpretation of the confusion matrix. Ann. Math. Artif. Intell. **81**(3–4), 429–450 (2017). https://doi.org/10.1007/s10472-017-9564-8
5. Cerioli, A., Barabesi, L., Cerasa, A., Menegatti, M., Perrotta, D.: Newcomb-Benford law and the detection of frauds in international trade. Proc. Natl. Acad. Sci. **116**(1), 106–115 (2018). https://doi.org/10.1073/pnas.1806617115
6. Check Point: Snake Keylogger (2022). https://research.checkpoint.com/2022/18th-july-threat-intelligence-report/
7. Collins, J.C.: Using excel and Benford's law to detect fraud, April 2017. https://www.journalofaccountancy.com/issues/2017/apr/excel-and-benfords-law-to-detect-fraud.html
8. Corum, A., Jenkins, D., Zheng, J.: Robust PDF malware detection with image visualization and processing techniques, June 2019. https://doi.org/10.1109/icdis.2019.00024
9. Ferreira, S., Antunes, M., Correia, M.E.: Exposing manipulated photos and videos in digital forensics analysis. J. Imaging **7**(7), 102 (2021). https://doi.org/10.3390/jimaging7070102
10. Gopaldinne, S.R., Kaur, H., Kaur, P., Kaur, G., Madhuri: Overview of PDF malware classifiers. In: 2021 2nd International Conference on Intelligent Engineering and Management (ICIEM). IEEE, April 2021. https://doi.org/10.1109/iciem51511.2021.9445341
11. Gottwalt, F., Waller, A., Liu, W.: Natural laws as a baseline for network anomaly detection. In: 2016 IEEE Trustcom/BigDataSE/ISPA, pp. 370–377 (2016). https://doi.org/10.1109/TrustCom.2016.0086

12. Hajdarevic, K., Pattinson, C., Besic, I.: Improving learning skills in detection of denial of service attacks with newcombe - Benford's law using interactive data extraction and analysis. TEM J., 527–534 (2022). https://doi.org/10.18421/tem112-05
13. Hill, T.P.: The significant-digit phenomenon **102**(4), 322–327. https://digitalcommons.calpoly.edu/cgi/viewcontent.cgi?article=1041&context=rgp_rsr
14. ISO: ISO 32000–1:2008document management - portable document format - part 1: Pdf 1.7, July 2008. https://www.iso.org/obp/ui/#iso:std:iso:32000:-1:ed-1:v1:en
15. ISO: ISO 32000–2:2020document management - portable document format - part 2: Pdf 2.0, December 2020. https://www.iso.org/standard/75839.html
16. Issakhani, M., Victor, P., Tekeoglu, A., Lashkari, A.: PDF malware detection based on stacking learning. In: Proceedings of the 8th International Conference on Information Systems Security and Privacy - Volume 1: ICISSP, pp. 562–570. INSTICC, SciTePress (2022). https://doi.org/10.5220/0010908400003120
17. Kang, A., Jeong, Y.S., Kim, S., Woo, J.: Malicious PDF detection model against adversarial attack built from benign PDF containing JavaScript **9**(22), 4764. https://doi.org/10.3390/app9224764
18. Kurien, K.L., Chikkamannur, A.: An ameliorated hybrid model for fraud detection based on tree based algorithms and Benford's law. In: 2020 Third International Conference on Advances in Electronics, Computers and Communications (ICAECC). IEEE, December 2020. https://doi.org/10.1109/icaecc50550.2020.9339471
19. Le, T., Lobo, G.J.: Audit quality inputs and financial statement conformity to Benford's law. J. Account. Audit. Finance **37**(3), 586–602 (2022). https://doi.org/10.1177/0148558X20930467
20. Mainka, C., Mladenov, V., Rohlmann, S.: Shadow attacks: hiding and replacing content in signed PDFs. In: Proceedings 2021 Network and Distributed System Security Symposium. Internet Society (2021). https://doi.org/10.14722/ndss.2021.24117
21. Maiorca, D., Biggio, B.: Digital investigation of pdf files: Unveiling traces of embedded malware. IEEE Secur. Priv. **17**(1), 63–71 (2019). https://doi.org/10.1109/MSEC.2018.2875879
22. Issakhani, M., Victor, P., Tekeoglu, A., Lashkari, A.H.: PDF malware detection based on stacking learning. In: The International Conference on Information Systems Security and Privacy, February 2022. https://www.unb.ca/cic/datasets/pdfmal-2022.html
23. Mavric, S.H.T., Yeo, C.K.: Online binary visualization for pdf documents. In: 2018 International Symposium on Consumer Technologies (ISCT). IEEE, May 2018. https://doi.org/10.1109/isce.2018.8408906
24. Milano, F., Gomez-Exposito, A.: Detection of cyber-attacks of power systems through Benford's law. IEEE Trans. Smart Grid **12**(3), 2741–2744 (2021). https://doi.org/10.1109/tsg.2020.3042897
25. Nigrini, M.J.: The patterns of the numbers used in occupational fraud schemes. Manag. Audit. J. **34**(5), 606–626 (2019). https://doi.org/10.1108/maj-11-2017-1717
26. Nunes, A., Inácio, H., Marques, R.P.: Benford's law and fraud detection in Portuguese enterprises. In: 2019 14th Iberian Conference on Information Systems and Technologies (CISTI), pp. 1–6 (2019). https://doi.org/10.23919/CISTI.2019.8760922

27. POINT, C.: Cyber attack trends: 2022 mid-year report, August 2022
28. Rohlmann, S., Mladenov, V., Mainka, C., Schwenk, J.: Breaking the specification: PDF certification, May 2021. https://doi.org/10.1109/sp40001.2021.00110
29. Schmitt, F., Gassen, J., Gerhards-Padilla, E.: PDF scrutinizer: detecting JavaScript-based attacks in PDF documents. In: 2012 Tenth Annual International Conference on Privacy, Security and Trust. IEEE, July 2012. https://doi.org/10.1109/pst.2012.6297926
30. Sergeev, A.V., Khorev, P.B.: Analysis of methods for hiding information in PDF documents and opportunities for their progress, March 2020. https://doi.org/10.1109/reepe49198.2020.9059117
31. Sethi, K., Kumar, R., Prajapati, N., Bera, P.: A lightweight intrusion detection system using Benford's law and network flow size difference. In: 2020 International Conference on COMmunication Systems & NETworkS (COMSNETS). IEEE, January 2020. https://doi.org/10.1109/comsnets48256.2020.9027422
32. Singh, P., Tapaswi, S., Gupta, S.: Malware detection in PDF and office documents: a survey. Inf. Secur. J. Global Perspect. **29**(3), 134–153 (2020). https://doi.org/10.1080/19393555.2020.1723747
33. Tharwat, A.: Classification assessment methods. Appl. Comput. Inform. **17**(1), 168–192 (2020). https://doi.org/10.1016/j.aci.2018.08.003
34. Wang, L., Ma, B.Q.: A concise proof of Benford's law. Fundamental Research (2023). https://doi.org/10.1016/j.fmre.2023.01.002. https://www.sciencedirect.com/science/article/pii/S2667325823000043

History Based Incremental Singular Value Decomposition for Background Initialization and Foreground Segmentation

Ibrahim Kajo[1]([✉]), Yassine Ruichek[1], and Nidal Kamel[2]

[1] UTBM, CIAD UMR 7533, 90010 Belfort, France
ibrahim.kajo@utbm.fr
[2] Faculty of Engineering and Computer Science, Vinuniversity, Hanoi, Vietnam

Abstract. Background initialization is an essential step for both hand-crafted and deep learning foreground segmentation approaches. In this paper, we propose a low-rank approximation algorithm that effectively handles the challenge caused by Stationary Foreground Objects (SFOs) on both offline and online bases. The proposed algorithm employs different incremental decomposition mechanisms that control the contribution of earlier and current frames in the overall covariance of the processed video. The proposed algorithm is able to identify the type of the detected SFO, whether it is an abandoned or removed object. Moreover, a background-updating mechanism is introduced to feed the proper background to learning models that are pretrained for foreground segmentation. The experimental results demonstrate the effectiveness of both proposed mechanisms: the SFO identification and the background initialization.

Keywords: background initialization · foreground segmentation · SFOs · incremental SVD

1 Introduction

Background/foreground extraction is a significant low-level processing stage that heavily affects the processing stages that follow it, such as motion detection, object tracking and object classification [1, 3]. The better the extracted background and foreground are, the better the outcome of the next processing mechanism is. Therefore, handling the challenges associated with such extraction has attracted the attention of many researchers, where challenges such as SFOs, camera jitter, dynamic background, and lighting changes, have been well addressed in the literature [4, 6]. However, the challenges caused by SFOs have gained more attention over the last decades, due to their significance in several image and video processing applications, such as anomaly detection, video surveillance, and crowd analytics. Dozens of effective SFOs detection techniques have been proposed in the literature [7–10]. Nonetheless, two main challenges associated with SFOs are not well addressed by the majority of the state-of-art detection techniques [6]: 1) the identification of the type of an SFO, 2) the background

© Springer Nature Switzerland AG 2024
V. Vasconcelos et al. (Eds.): CIARP 2023, LNCS 14469, pp. 63–75, 2024.
https://doi.org/10.1007/978-3-031-49018-7_5

initialization in the presence of an SFO. The first challenge arises when both abandoned objects and removed objects appear stationary for a specific period of time. Thus, identifying the status of an SFO as either an Abandoned Object (AO) or a Removed Objects (RO) becomes particularly problematic. Existing techniques often rely on analyzing the appearance characteristics, such as edges or color consistency, between the current image and the background model [11, 16]. However, several factors contribute to the difficulty of accurately classifying SFOs. Camouflage, occlusion, ghost effects, and lighting variations can lead to inaccurate classifications. Furthermore, computationally expensive methods that require continuous estimates of the background model can limit real-time applications.

On the other hand, the second challenge arises when the presence of SFOs negatively affects the initialization of the scene background, which is a crucial input for many recently introduced supervised deep learning algorithms. These algorithms aim to achieve more precise foreground segmentation compared to traditional hand-crafted approaches. Generally, the background is modelled using a simple median filtering mechanism for the first 50–100 frames which could be also updated later by recomputing the median again after a certain number of frames [17, 18]. However, the recent advance in the field of background initialization and the state-of-the-art approaches introduced in this field clearly show that median filtering based algorithms achieved poor performances in initializing an accurate background in the presence of several challenges including SFOs.

The contributions of this paper are as follows: 1) we propose an incremental singular value decomposition based SFO identification mechanism that can effectively classify SFOs into abandoned objects and removed objects. 2) we introduce a scheme that can initialize an accurate and robust background model in the presence of several types of SFOs. This background model helps to enhance the foreground detection performed by learning based approaches due to the importance of the initialized background as an input into their developed models.

2 Related Work

2.1 Identification of SFOs

Several SFOs detection techniques employ a post-processing stage with different procedures to handle the issue of SFO identification [11, 13]. This is achieved by analyzing the appearance (mainly edges) of both current and background images. Several empirical studies show that an SFO is identified as RO when the current image contains less edges than the background image and vice versa in the case of an AO. Such techniques are computationally expensive due to the need of having a point estimate of the background model every frame. Other approaches are proposed based on different assumptions such as the compatibility of a static object with its surrounding or the color inconsistency between a static object and its background surrounding [14, 16]. However, such assumptions do not hold true in the presence of several challenges such as camouflage and ghost effects.

It is worth mentioning that the incorporation of incremental Singular Value Decomposition (i-SVD) in detecting SFOs was initially introduced in [7, 19], where i-SVD based approaches were proposed to detect abandoned objects such as luggage and static boats. However, these proposed SFO detection techniques are limited to the detection of abandoned objects while the detection of removed objects was not addressed by these approaches. Furthermore, the identification mechanism required to distinguish between both SFOs (abandoned and removed) was not addressed nor discussed by these approaches.

2.2 Background Dependency

The initialized background is an essential input in the majority of the deep learning techniques proposed for Foreground Segmentation (FS). Several approaches simply extract the first frame of each image sequence and feed it into their networks as the reference background. Such a procedure simplifies the process of background initialization and does not take into account the dozens of challenges usually associated with such initialization. Because of its simplicity and widespread use in background subtraction, median filtering technique is also employed by many deep learning FS approaches to initialize the reference background required for their networks. Zheng et al. [20] utilized a simple median filtering to model the reference background of the image sequences where the background is visible at least 50% of the time. Likewise, Tezcan et al. [17] introduced a Convolution Neural Network (CNN) that, in addition to the current frame, accepts two background images initialized at different temporal scales. The first one, called "empty", is extracted manually as a static frame with no moving foregrounds while the second one, called "recent", is extracted by computing the median of the previous 100 frames. For better background initialization, several deep FS segmentation approaches decide to integrate the background initialization into the learning process of their proposed networks. To estimate a better background image in the presence of SFO, Patil et al. [21] proposed an architecture that consists of two temporal pooling layers, a spatial pooling layer and a group of learnable convolution filters, which together help suppress the effect of SFOs on the estimated background. Mandal et al. [22] presented a sequence of Maximum Multi-Spatial Receptive feature (MMSR) blocks that learn the background reference. For better adaptability, the maximum is computed among three responses extracted from three different levels of receptive fields. The learnable background is reinforced by the background estimated by computing the median of the previous images. Later, Mandal et al. [23] proposed another learnable background estimation block that contains a sequence of 3D average pooling with different levels of stride, in addition to 3D convolutions to encode the spatiotemporal patterns. The proposed block takes a sequence of history frames as input and generates a 2D feature map that represents the desired background.

3   Methodology

3.1   Notation and Preliminaries

Consider an RGB video as a four dimension tensor $\mathcal{X} \in \mathbb{R}^{m \times n \times c \times k}$, where m, n, c, and k refer to the number of rows, number of columns, number of color channels, and number of frames, respectively. For sake of simplicity, the video tensor is represented as a three dimension tensor $\mathcal{X} \in \mathbb{R}^{m \times n \times k}$, where the three single color channels are processed independently following the same procedure. The letters i, j, and t represent the indices of the three dimensions i.e., length m, width n, and time k, respectively. Thus, the frontal, lateral, and horizontal slices are indicated as $\mathcal{X}_{::t} \in \mathbb{R}^{m \times n}$, $\mathcal{X}_{:j:} \in \mathbb{R}^{k \times m}$, and $\mathcal{X}_{i::} \in \mathbb{R}^{k \times n}$, respectively. Considering $\mathcal{X}_{i::}$ as the slice of interest, the reduced-size (economic) SVD of a slice $\mathcal{X}_{i::}$ is computed as follows

$$\mathbf{U}^{\mathsf{T}} \mathcal{X}_{i::} \mathbf{V} = \mathbf{S} = diag(\sigma_1, \dots, \sigma_n) \in \mathbb{R}^{n \times n} \tag{1}$$

where $\sigma_1 \geq \sigma_2 \geq \dots \geq \sigma_n \geq 0$ are the singular values. The matrix of left singular vectors is defined as $\mathbf{U} = [\mathbf{u}_1 | \dots | \mathbf{u}_n] \in \mathbb{R}^{k \times n}$ and the matrix of right singular vectors is defined as $\mathbf{V} = [\mathbf{v}_1 | \dots | \mathbf{v}_n] \in \mathbb{R}^{n \times n}$. In certain cases, such as online application and when decomposing large matrices, a mechanism that updates the SVD components, rather than recomputing them from scratch, is employed. One of the widely used algorithms for SVD updating in the field of computer vision is the one proposed in [24], which allows for updating precomputed SVD components when new frames arrive.

3.2 Computation of Incremental SVD

As described in [24], the i-SVD accumulates y number of new frames $\mathcal{X}'_{i::} \in \mathbb{R}^{y \times n}$ and concatenate $\mathcal{X}_{i::}$ and $\mathcal{X}'_{i::}$ to construct $\widehat{\mathcal{X}}_{i::} \in \mathbb{R}^{k' \times n}$, where $k' = k + y$. The singular components of $\widehat{\mathcal{X}}_{i::}$ can be computed by updating the previously computed components i.e., $\mathbf{U}$, $\mathbf{S}$, and $\mathbf{V}$ as follows

$$\widehat{\mathcal{X}}_{i::} = [\mathcal{X}^T_{i::} | \mathcal{X}'^T_{i::}] = \acute{\mathbf{U}} \acute{\mathbf{S}} \acute{\mathbf{V}}^{\mathsf{T}} \tag{2}$$

where $\acute{\mathbf{U}} \in \mathbb{R}^{k' \times n}$, $\acute{\mathbf{S}} \in \mathbb{R}^{n \times n}$, and $\acute{\mathbf{V}} \in \mathbb{R}^{n \times n}$ are the updated matrices representing the left singular vectors, singular values, and right singular vectors, respectively. Taking into account the fact that $\mathbf{U}$ and $\mathbf{S}$ matrices are sufficient for both the update and reconstruction processes, the update computation steps of matrix $\mathbf{V}$ are omitted. Letting $\widetilde{\mathcal{X}}_{i::}$ be the component of $\mathcal{X}'_{i::}$ orthogonal to $\mathbf{U}$, the SVD of $\widehat{\mathcal{X}}_{i::}$ can be computed using the modified Sequential Karhunen–Loeve (SKL) algorithm [25] as follows

- Obtain $\widetilde{\mathcal{X}}_{i::}$ by computing the QR factorization of $[\mathcal{X}'_{i::} - \mathbf{U}\mathbf{U}^{\mathsf{T}}\mathcal{X}'_{i::}]$
- Construct the matrix $\mathbf{R} = \begin{bmatrix} f.\mathbf{S} & \mathbf{U}^{\mathsf{T}}\mathcal{X}'_{i::} \\ 0 & \widetilde{\mathcal{X}}_{i::}(\mathbf{U}\mathbf{U}^{\mathsf{T}}\mathcal{X}'_{i::}) \end{bmatrix}$
- Compute the SVD of matrix $\mathbf{R} = \acute{\mathbf{U}} \acute{\mathbf{S}} \acute{\mathbf{V}}^{\mathsf{T}}$.

where f is the forgetting factor, proposed to down-weigh the contribution of earlier frames. Its value ranges from 0 to 1, where $f = 1$ refers to no forgetting occurs.

3.3 History Based Incremental SVD (hi-SVD)

In several computer vision applications that highly consider the time dimension into their computations, more focus is given on the recent frames and less on the earlier ones. To achieve this purpose using i-SVD, the above mentioned forgetting factor mechanism is proposed to downgrade the contribution of the earlier frames to the overall covariance, which in turns reduces the effect of these frames on the current processing. For example, Ross et al. [25] incorporated such a forgetting mechanism into their incremental learning based visual tracking to tackle challenges such as tracking objects with changing appearance or under changeable lighting conditions. However, a challenging task such as SFOs detection draws attention to the fact that constructing a good background model in the presence of abandoned objects, requires more focus on the earlier frames. Therefore, we propose to upgrade the contribution of the earlier frames instead of downgrade it. Hence, we propose to incorporate a memory mechanism that can perform two memory based actions (forgetting and remembering) into the process of SFOs detection. A memory factor that can upgrade or downgrade the contribution of earlier frames to the overall covariance; is incorporated into the incremental eigenbasis update as follows

$$\mathbf{R} = \begin{bmatrix} f^{memory}.\mathbf{S} & \mathbf{U}^\mathsf{T}\acute{\mathcal{X}}_{i::} \\ 0 & \tilde{\mathcal{X}}_{i::}(\mathbf{U}\mathbf{U}^\mathsf{T}\acute{\mathcal{X}}_{i::}) \end{bmatrix} \tag{3}$$

where f^{memory} is the memory factor which is defined as follows

$$f^{memory} = \begin{cases} \alpha \in [0, 1[& \textit{if the target is to forget} \\ \beta \in]1, 2] & \textit{if the target is to remember} \end{cases} \tag{4}$$

Thus, to estimate a background model that is free of removed objects, more focus should be given on the current frames of processing. Therefore, the contribution of the earlier frames is downgraded by employing the memory factor $f^{memory} = \alpha$ where an ROs-free background slice can be constructed after successfully updating $\acute{\mathbf{U}}$ and $\acute{\mathbf{S}}$ matrices as follows

$$\widehat{\mathcal{B}}^{RO}_{t::} = \acute{\mathbf{u}}_1.(\acute{\mathbf{u}}_1^\mathsf{T}.\,\acute{\mathcal{X}}_{t::}) \tag{5}$$

where $\acute{\mathbf{u}}_1$ is the updated first singular vector which mainly represents the low-rank component with respect to pixel intensities (background). Subsequently, an ROs-free background image is estimated by computing the average of all frontal slices of tensor $\widehat{\mathcal{B}}^{RO}$ as follows

$$\mathbf{B}^{RO} = \frac{1}{k}\sum_{t=1}^{t=k}\left|\widehat{\mathcal{B}}^{RO}_{::t}\right| \tag{6}$$

On the other hand, to estimate an AOs-free background model, less focus on the current frames should be given, where incorporating earlier frames is more effective due to the absence of AOs in the scene. Hence, the contribution of the earlier frames to the overall covariance is upgraded using the memory factor $f^{memory} = \beta$, and an AOs-free background slice can be constructed as follows

$$\widehat{\mathcal{B}}^{AO}_{i::} = \acute{u}_1 \cdot (\acute{u}_1^T \cdot \mathcal{X}_{i::}) \tag{7}$$

An ROs-free background image can be estimated by computing the average of all frontal slices of tensor $\widehat{\mathcal{B}}^{AO}$ as follows

$$\mathbf{B}^{AO} = \frac{1}{k}\sum_{t=1}^{t=k}\left|\widehat{\mathcal{B}}^{AO}_{::t}\right| \tag{8}$$

Consequently, a map that localizes the SFOs, if any, regardless of their type, is estimated by computing the difference $\mathbf{D} = |\mathbf{B}^{RO} - \mathbf{B}^{AO}|$ between the two estimated backgrounds. The overall framework of our approach (hi-SVD) is shown in Fig. 1.

After properly localizing the SFOs, the challenge is how to identify the type of the detected SFOs whether they are removed or abandoned. It is crucial to emphasize that SFOs of a particular type consistently appear absent in the background image estimated for the other type. For example, if there is an abandoned object in the scene, the background image B^{AO} remains empty, while the abandoned object is clearly visible in the B^{RO} image. Based on this fact, the problem of localization can be solved by applying an object detector on two Regions of Interest (ROI): ROI^{RO}, and ROI^{AO}. These ROIs are extracted from $\mathbf{B}^{RO}$ and $\mathbf{B}^{AO}$ based on the localization result obtained from the map $\mathbf{D}$. For example, in the case of an abandoned object, an object should be detected in ROI^{RO} while no object will be detected in ROI^{AO}. Subsequently, the background image with no SFOs is selected as the best image that represents the scene background in the case of offline processing. The main steps of the proposed algorithm are presented in **Algorithm 1.** To achieve this identification, the widely recognized object detector, Mask R-CNN, is employed to validate the presence of a potential SFO corresponding to an existing object. On the other hand, the scene background can be continuously updated using the i-SVD algorithm, where the SFOs are detected as foregrounds regardless of their types. This leads to an updated version of the background at each incremental step. **Algorithm 2** lists the steps of the proposed online background initialization algorithm which provides multiple updated backgrounds. For instance, the estimated background prior to an object's removal should include that object, whereas the estimated background after its removal should be devoid of it. This ensures accurate foreground segmentation of AOs/ROs before or after they are abandoned or removed.

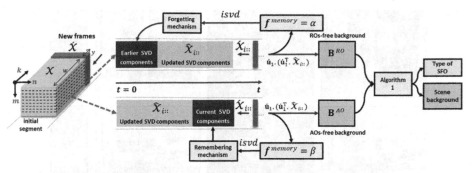

Fig. 1. Framework of *hi*-SVD for background initialization.

Algorithm 1 Identifying the type of SFO and updating the scene background

---

Input: $\mathbf{B}^{RO} \in \mathbb{R}^{m \times n}$, $\mathbf{B}^{AO} \in \mathbb{R}^{m \times n}$
Output: scene background $\mathbf{B}$, type of detected SFO.
Compute the difference map $\mathbf{D}$
if an SFO is detected **then**
    Crop the regions of interest $(\text{ROI}^{RO}, \text{ROI}^{AO})$ in $\mathbf{B}^{RO}$ and $\mathbf{B}^{AO}$
    Obtain the object detection results from both cropped ROIs
    if ROI^{RO} has an object **then**
        Identify the detection SFO as AO
        Use $\mathbf{B}^{AO}$ as the scene background
    else
        Identify the detection SFO as RO
        Use $\mathbf{B}^{RO}$ as the scene background
    end if
end if

Algorithm 2 Online background initialization for deep foreground segmentation

---

Input: Initial video segment $\mathcal{X} \in \mathbb{R}^{m \times n \times k}$
Output: $\mathbf{B}^{current}$
Initialization: $\mathbf{U}, \mathbf{S}$ extracted from $\mathcal{X}$
for $t = 1, 2, \ldots \infty$ **do**
  while $t < y$ **do**
    $\mathring{\mathcal{X}}_{::t} \leftarrow$ new frame
    $t \leftarrow t + 1$
  end while
  $\hat{\mathcal{X}} = [\mathcal{X} | \mathring{\mathcal{X}}] \in \mathbb{R}^{m \times n \times k}$
  $[\mathring{\mathbf{U}}, \mathring{\mathbf{S}}] = isvd\,(\hat{\mathcal{X}}, f^{memory})$
  Estimate the two scene backgrounds $\mathbf{B}^{RO}$, $\mathbf{B}^{AO}$ using Eq. (7) and Eq. (8)
  Identify the scene background $\mathbf{B}$ using **Algorithm 1**
  $\mathbf{B}^{current} = \mathbf{B}$
end for

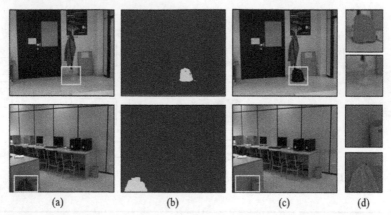

(a)   (b)   (c)   (d)

Fig. 2. SFOs type identification Examples: (a) RO background image, (b) difference map, (c) AO background image, and (d) object detection results.

4   Experimental Results

4.1   Datasets

There are several standard datasets that are proposed to evaluate the performance of SFO detection approaches. Two widely used datasets are employed in our experiments. These datasets not only provide ground truths of the scene backgrounds but also include ground truths for the foreground. This allows us to assess how the estimated background influences the learning process of a trainable model. The first dataset is CDnet2014 dataset [26] that contains a dedicated intermittent object motion category of 6 videos showing different SFOs such as objects removing away from the background and stopping object for a considerable time. The second dataset is known as LASIESTA [27] which consists of 20 indoor and outdoor videos. The indoor category has two videos of abandoned and removed object and two videos of temporally static persons whose clothes have the same color as the scene background.

4.2   SFO Status Identification Experiment

The first experiment focuses on the evaluation of the offline identification accuracy. The tested videos show 3 AO scenarios, 3 RO scenarios, and 4 scenarios involving objects that remain static for a period and then start moving. Our proposed algorithm exhibits robust and accurate performance in detecting the SFOs and categorizing them as AOs and ROs. Figure 2 shows two examples of SFO detection in the *IMB01* and *IMB02* sequences from LASIESTA dataset. Two background images are estimated in each example to construct the difference map and localize the ROIs, where the Mask R-CNN is employed to verify the presence of potential SFOs. Furthermore, Fig. 3 demonstrates two background initialization examples (*ICA02* from LASIESTA, and *streetlight* from CDnet2014) involving intermittent static objects. Despite the camouflage challenge presented in *ICA02* sequence, the proposed algorithm managed to accurately initialize the background model.

(a)   (b)   (c)

Fig. 3. Examples of background initializations in the case of intermittent SFOs: (a) AO background image, (b) RO background image, and (c) original frame

4.3 Foreground Segmentation Experiment

The second experiment concerns the evaluation of the impact of the background initialization stage on the performance of the approaches mainly proposed for foreground segmentation. To achieve this purpose, we test such impact on a Multi-Input GAN for Foreground Enhancement (FEMI-GAN) framework [30]. FEMI-GAN is designed to accept multiple inputs such as the original frame, initialized background, and initial foreground map as shown in Fig. 4. FEMI-GAN is tested following two different scenarios. In the first scenario, the background is estimated using median filtering and fed to the network. While, in the second scenario, the background is estimated employing **Algorithm 2** and fed to the network. Figure 5 shows the enhancement that the updated background brings to the final foreground segmentation results. Moreover, the estimated foreground results are compared with the results of several state of the art learning approaches following two testing protocols: seen and unseen videos. Table 1 lists the average F-score results of all tested learning approaches, including the proposed one with two different backgrounds (classical and updated). The numerical results clearly show the improvement in the performance of the tested GAN model in segmenting the foreground objects especially when dealing with unseen videos whose frames were not included in the training phase.

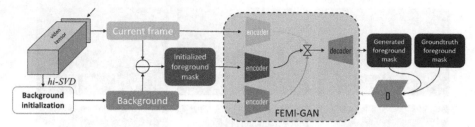

Fig. 4. Framework of the proposed foreground enhancement framework that accepts three inputs: current frame, initialized background, and initialized foreground. The output is the corrected foreground mask.

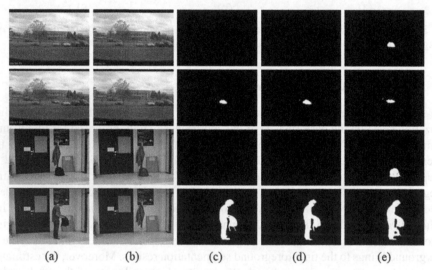

Fig. 5. Examples of the impact of our proposed algorithm on the foreground segmentation results obtained by FEMI-GAN: (a) original frame, (b) initialized backgrounds: 1st row and 3rd rows show the backgrounds when the SFOs are considered as part of the background while 2nd and 4th rows show the backgrounds after SFOs' removal (c) groundtruth, (d) estimated foreground in case of using the updated background, (e) estimated foreground in case of using a classical background.

Table 1. Average f-score values of several deep learning approaches tested using scene dependent and scene-independent scenarios.

Approach	CDnet2014		LASIESTA	
	Seen	Unseen	Seen	unseen
FgSegNet [28]	0.78	0.17	0.60	0.60
MSFS [29]	0.68	0.52	0.64	0.63
3DCD [23]	0.83	0.84	0.90	0.79
ChangeDet [22]	0.86	0.80	0.91	0.77
GAN-classical BG	0.87	0.80	0.90	0.80
GAN-proposed BG	**0.91**	**0.86**	**0.92**	**0.88**

5 Conclusion

In this paper, we propose a balancing mechanism that controls the contributions of both earlier and current frames to the overall SFOs segmentation process. Based on the SFO type, an appropriate procedure is followed to detect and identify such objects. The proposed mechanism helps extract accurate scene background that is free of SFOs regardless their type. To verify the effectiveness of the proposed algorithm, the performance of a deep learning based foreground segmentation is tested with our initialized background as a key input. The experimental results clearly show improved foreground segmentation performance, especially in the cases where abandoned and removed object scenarios are present in the scene.

References

1. Bouwmans, T., Porikli, F., Höferlin, B., Vacavant, A.: Background Modeling and Foreground Detection for Video Surveillance. Chapman and Hall/CRC, Boca Raton (2014)
2. Maddalena, L., Petrosino, A.: Towards benchmarking scene background initialization. In: Murino, V., Puppo, E., Sona, D., Cristani, M., Sansone, C. (eds.) ICIAP 2015. LNCS, vol. 9281, pp. 469–476. Springer, Cham (2015). https://doi.org/10.1007/978-3-319-23222-5_57
3. Bouwmans, T., Maddalena, L., Petrosino, A.: Scene background initialization: a taxonomy. Pattern Recognit Lett. **96**, 3–11 (2017)
4. Javed, S., Mahmood, A., Bouwmans, T., Jung, S.K.: Spatiotemporal low-rank modeling for complex scene background initialization. IEEE Trans. Circuits Syst. Video Technol. **28**, 1315–1329 (2018)
5. Kajo, I., Kamel, N., Ruichek, Y.: Self-motion-assisted tensor completion method for background initialization in complex video sequences. IEEE Trans. Image Process. **29**, 1915–1928 (2019)
6. Cuevas, C., Martínez, R., García, N.: Detection of stationary foreground objects: a survey. Comput. Vis. Image Underst. **152**, 41–57 (2016)
7. Kajo, I., Kamel, N., Ruichek, Y.: Incremental tensor-based completion method for detection of stationary foreground objects. IEEE Trans. Circuits Syst. Video Technol. **29**, 1325–1338 (2019)

8. Ingersoll, K., Niedfeldt, P.C., Beard, R.W.: Multiple target tracking and stationary object detection in video with Recursive-RANSAC and tracker-sensor feedback. In: 2015 International Conference on Unmanned Aircraft Systems (ICUAS), pp. 1320–1329. IEEE (2015)

9. Ortego, D., SanMiguel, J.C.: Multi-feature stationary foreground detection for crowded video-surveillance. In: 2014 IEEE International Conference on Image Processing (ICIP), pp. 2403–2407. IEEE (2014)

10. Lin, Y., Tong, Y., Cao, Y., Zhou, Y., Wang, S.: Visual-attention-based background modeling for detecting infrequently moving objects. IEEE Trans. Circuits Syst. Video Technol. **27**, 1208–1221 (2017)

11. Kim, J., Kang, B.: Nonparametric state machine with multiple features for abnormal object classification. In: IEEE Proceedings of 2014 11th IEEE International Conference on Advanced Video and Signal Based Surveillance (AVSS), pp. 199–203. Springer (2014)

12. Tian, Y., Feris, R.S., Liu, H., Hampapur, A., Sun, M.-T.: Robust detection of abandoned and removed objects in complex surveillance videos. IEEE Trans. Syst. Man Cybern. C. **41**, 565–576 (2011)

13. Thomaz, L.A., da Silva, A.F., da Silva, E.A.B., Netto, S.L., Bian, X., Krim, H.: Abandoned object detection using operator-space pursuit. In: IEEE Proceedings of IEEE International Conference on Image Processing (ICIP), Springer, pp. 1980–1984 (2015)

14. Muchtar, K., Lin, C.-Y., Kang, L.-W., Yeh, C.-H.: Abandoned object detection in complicated environments. In: IEEE Proceedings of 2013 Asia-Pacific Signal and Information Processing Association Annual Summit and Conference, pp. 1–6. Springer (2013)

15. Pan, J., Fan, Q., Pankanti, S.: Robust abandoned object detection using region-level analysis. In: IEEE Proceedings of 18th IEEE International Conference on Image Processing, pp. 3597–3600. Springer (2011)

16. Porikli, F., Ivanov, Y., Haga, T.: Robust abandoned object detection using dual foregrounds. EURASIP J. Adv. Signal Process. **2008** (2008)

17. Tezcan, M.O., Ishwar, P., Konrad, J.: BSUV-Net: a fully-convolutional neural network for background subtraction of unseen videos. In: IEEE, Proceedings of 2020 IEEE Winter Conference on Applications of Computer Vision (WACV), pp. 2763–2772. Springer (2020)

18. Tezcan, M.O., Ishwar, P., Konrad, J.: BSUV-Net 2.0: spatio-temporal data augmentations for video-agnostic supervised background subtraction. arXiv preprint arXiv:2101.09585 (2021)

19. Kajo, I., Kamel, N., Ruichek, Y.: Tensor-based approach for background-foreground separation in maritime sequences. IEEE Trans. Intell. Transport. Syst., 1–14 (2020)

20. Zheng, W., Wang, K., Wang, F.-Y.: A novel background subtraction algorithm based on parallel vision and Bayesian GANs. Neurocomputing **394**, 178–200 (2020)

21. Patil, P.W., Murala, S.: MSFgNet: a novel compact end-to-end deep network for moving object detection. IEEE Trans. Intell. Transport. Syst. **20**, 4066–4077 (2019)

22. Mandal, M., Vipparthi, S.K.: Scene independency matters: an empirical study of scene dependent and scene independent evaluation for CNN-based change detection. IEEE Trans. Intell. Transport. Syst., 1–14 (2020)

23. Mandal, M., Dhar, V., Mishra, A., Vipparthi, S.K., Abdel-Mottaleb, M.: 3DCD: scene independent end-to-end spatiotemporal feature learning framework for change detection in unseen videos. IEEE Trans. Image Process. **30**, 546–558 (2021)

24. Levey, A., Lindenbaum, M.: Sequential Karhunen-Loeve basis extraction and its application to images. IEEE Trans. Image Process. **9**, 1371–1374 (2000)

25. Ross, D.A., Lim, J., Lin, R.-S., Yang, M.-H.: Incremental learning for robust visual tracking. Int. J. Comput. Vis. **77**, 125–141 (2008)

26. Goyette, N., Jodoin, P.-M., Porikli, F., Konrad, J., Ishwar, P.: Changedetection.net: a new change detection benchmark dataset. In: 2012 IEEE Computer Society Conference on Computer Vision and Pattern Recognition Workshops, pp. 1–8. IEEE (2012)

27. Cuevas, C., Yáñez, E.M., García, N.: Labeled dataset for integral evaluation of moving object detection algorithms: LASIESTA. Comput. Vis. Image Underst. **152**, 103–117 (2016)
28. Lim, L.A., Keles, H.Y.: Foreground segmentation using convolutional neural networks for multiscale feature encoding. Pattern Recognit. Lett. **112**, 256–262 (2018)
29. Lim, L.A., Keles, H.Y.: Learning multi-scale features for foreground segmentation. Pattern Anal. Appl. **23**(3), 1369–1380 (2019). https://doi.org/10.1007/s10044-019-00845-9
30. Kajo, I., Kas, M., Ruichek, Y., Kamel, N.: Tensor based completion meets adversarial learning: a win–win solution for change detection on unseen videos. Comput. Vis. Image Underst. **226**, 103584 (2023)

Vehicle Re-Identification Based on Unsupervised Domain Adaptation by Incremental Generation of Pseudo-Labels

Paula Moral[✉][iD], Álvaro García-Martín[iD], and José M. Martínez[iD]

Video Processing and Understanding Lab, Universidad Autónoma de Madrid,
Madrid, Spain
{paula.moral,alvaro.garcia,josem.martinez}@uam.es

Abstract. The main goal of vehicle re-identification (ReID) is to associate the same vehicle identity in different cameras. This is a challenging
task due to variations in light, viewpoints or occlusions; in particular,
vehicles present a large intra-class variability and a small inter-class variability. In ReID, the samples in the test sets belong to identities that have
not been seen during training. To reduce the domain gap between train
and test sets, this work explores unsupervised domain adaptation (UDA)
generating automatically pseudo-labels from the testing data, which are
used to fine-tune the ReID models. Specifically, the pseudo-labels are
obtained by clustering using different hyperparameters and incrementally due to retraining the model a number of times per hyperparameter with the generated pseudo-labels. The ReID system is evaluated in
CityFlow-ReID-v2 dataset.

Keywords: Vehicle re-identification · Unsupervised Domain
Adaptation · Pseudo-labels · Deep learning · Image processing ·
Surveillance videos

1 Introduction

Vehicle Re-identification is a computer vision task that aims to retrieve images of
a given vehicle identity across non-overlapping cameras. In compliance with data
protection regulations, vehicle ReID datasets frequently anonymise the license
plates and people present in the images. Consequently, technologies should be
developed without relying on this information.

There are several applications in which the use of vehicle ReID can be useful,
such as Intelligent Transportation Systems (ITSs) [22] where it can be utilized
to assess the performance of dynamic traffic systems by estimating the flow of
circulation and travel times. In Urban Computing [25], it can help to calculate
origin-destination matrices. Additionally, it can be used in intelligent surveillance
to quickly detect, locate, and track target vehicles [8].

One of the main challenges in the ReID task [1, 11, 14, 26] is the large intra-
class variance and the small inter-class variability related to the significant

© Springer Nature Switzerland AG 2024
V. Vasconcelos et al. (Eds.): CIARP 2023, LNCS 14469, pp. 76–89, 2024.
https://doi.org/10.1007/978-3-031-49018-7_6

changes in the views of the same vehicle and the similarity of different identities sharing the same viewpoints and attributes [7].

In many practical situations, i.e., real-world scenarios, a common issue arises where the training and testing data belong to different domains [21]. An effective ReID model should still yield satisfactory accuracy on scenarios unseen during training. Unsupervised domain adaptation refers to the process of learning a model for the target domain using a fully annotated source dataset and an unlabeled target dataset [18]. For this study, the source domain will be the train set while the target will be the unannotated set of test.

This paper proposes an unsupervised domain adaptation by incremental generation of pseudo-labels. The main contributions are the following:

- An unsupervised domain adaptation in ReID task generating pseudo-labels to iteratively improve the baseline model.
- The study of the optimal configuration in terms of the maximum distance hyperparameter in the Density-Based Spatial Clustering of Applications with Noise (DBSCAN) method to generate pseudo-labels to fine-tune the model.
- The study of the optimal number of iterations of pseudo-label generation and fine-tuning to improve the performance.

2 Related Work

Numerous techniques for unsupervised domain adaptation in ReID have been introducing, aiming to maximize the potential of unlabeled images. These methods primarily address the mismatch in data distribution between domains and the absence of label information in the target domain using distribution aligning and clustering-based methods.

Distribution aligning methods are employed to reduce the domain gap between the source and target in a shared feature space. Most of the works have used Generative Adversarial Networks (GANs) based techniques to transfer source images into the target-domain style and train a model using the generated images [23, 26, 27].

Another unsupervised domain adaptation approach to learn a ReID model is through clustering-based adaptation. To compensate for the lack of labeled data, several studies [5, 9, 19] leverage unlabeled target data and adopt clustering methods to generate pseudo-labels. The core concept is to exploit the similarity of unlabeled samples by clustering features and producing pseudo-labels for supervised information. To enhance the quality of pseudo-labels, some studies aim to use the potential relational information of the data to improve the matching reliability. This is the case of [9], that includes the tracklet information and camera bias among images in the clustering process to generate more robust pseudo-labels.

3   Proposed Method

This section introduces the proposed unsupervised domain adaptation scheme for vehicle ReID using pseudo-labels. To address the domain gap issue between the source and target domains (in this case, the training and testing sets respectively), the proposed scheme employs a clustering algorithm to extract pseudo-labels from the test set. These pseudo-labels are then used to fine-tune the baseline models previously trained as ReID features extractors. In this case, the baseline models are the image-based models from our previous system [10]. This ReID system ensembles three image-based and one video-based model. It includes some post-processing steps in order to enhance the final results. As the pseudo-labels are generated for the image domain, the module of [10] related with video-path information is not going to be used as can be seen in Fig. 1.

The proposed unsupervised domain adaptation approach is based on generating pseudo-labels by modifying the hyperparameters used in the DBSCAN clustering algorithm and the number of iterations carried out.

Once the fine-tuning strategy is established, the study examines the impact of the cluster's density according to the maximum distance hyperparameter. Additionally, the study explores an incremental approach to generating pseudo-labels, which involves increasing the maximum distance value and then repeating the procedure for a variable number of iterations. The effect of this incremental approach is also analysed.

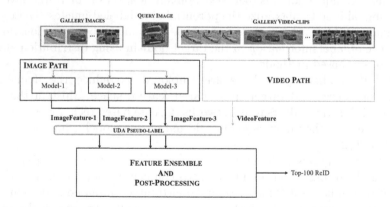

Fig. 1. ReID System scheme from our previous work [10]. The proposed work includes the UDA pseudo-label module and omitted the video-based model.

3.1   DBSCAN Pseudo-Labels

The DBSCAN [4] cluster algorithm is included in order to generate pseudo-labels. This method defines clusters as regions of high density separated by regions of low density. Core samples, which are located in areas of high density, are closer to each other based on a distance measure. The two hyperparameters that define the density of the clusters are the maximum distance and the minimal number of samples. The Eps-neighborhood determines how far apart two samples can be to be considered neighbours, i.e. the Eps is the maximum distance value for two neighbour samples. The minimal number of samples determines the number of samples in a neighborhood to be considered a cluster.

While the hyperparameter minimal number of samples primarily controls how tolerant the algorithm is towards noise, Eps is crucial to appropriately generate the clusters [13]: when is too small, the samples are not associated to a cluster, whilst when is too large, close clusters are merged into a single one. For this reason, this work is going to study the generation of pseudo-labels as clusters given by DBSCAN analyzing the impact of the cluster's density fixing the minimal number of samples and modifiying the Eps hyperparameter. The minimal number of samples will be fixed in this study according to [9, 24], and the influence of the maximum distance will be the object of analysis.

The system mentioned in Sect. 2 [9] includes an unsupervised domain adaptation method that generates pseudo-labels applying DBSCAN cluster method and improves their quality introducing the tracklet information and reducing the camera bias between images. Following [9], this work adopts the same strategies to reduce the aforementioned biases, calculating the distances among samples in order to apply the DBSCAN method.

The main difference between the proposed unsupervised domain adaptation method and the approach described in [9] lies in the generation of pseudo-labels. In the proposed method, the process of generating pseudo-labels is conducted iteratively and incrementally, taking into consideration the value of the Eps hyperparameter. On the other hand, in the work presented in [9], the generation of pseudo-labels is performed once, with the value of the Eps hyperparameter being set beforehand. This iterative nature of our method allows for a more dynamic and responsive adjustment of the pseudo-labels, potentially leading to improved performance in re-identification tasks.

3.2   Fine-Tuning

Let $X \subseteq \mathbb{R}^d$ be the input space and $Y \subseteq \mathbb{N}$ be the output label space. In the context of domain adaptation, the source domain will be denoted as S and the target domain as T [14].

To make the most of the generated pseudo-labels, a fine-tuning framework is introduced for re-identification that draws inspiration from domain adaptation techniques [12, 16, 17]. The proposed framework trains a classification model $\mathcal{G}(\cdot; \theta_{\mathcal{F}}, \theta_{\mathcal{C}})$, consisting of a feature extractor $\mathcal{F}$ and a classification head $\mathcal{C}$, being $\theta_{\mathcal{F}}$ and $\theta_{\mathcal{C}}$ their respective weights. To train $\mathcal{F}$ and $\mathcal{C}$ effectively, this method uses different learning rates. Specifically, a smaller learning rate is applied for $\mathcal{F}$ to prevent over-fitting of the source data weights to the target data.

To prevent the concept drifting of pseudo-labels [2], it is proposed a regularization technique that involves co-training with X_S and X_T. Specifically, the proposed method includes a common feature extractor $\mathcal{F}$ with an independent classification head $\mathcal{C}_i$ for each subset $i \in \mathcal{S}, \mathcal{T}$, which enables to effectively learn and generalize from both labeled and pseudo-labeled data.

The overall objective function is a combination of two terms: a classification loss on the labeled source data, and a classification loss on the pseudo-labeled target data. The description of the loss, a combination of Triplet loss and a Label Smooth Regularization (LSR), is described in [10]. The first term is straightforward and involves minimizing the classification loss between the ground-truth train labels y_s and the predictions made by the model on the labeled source data x_s. The second term involves the generated pseudo-labels $\hat{y}_t$ for the unlabeled target data x_t using the model's current predictions. By co-training on both types of data, the model can learn to effectively generalize to the target domain and prevent the concept drifting of the pseudo-labels. Formally, the overall objective function can be expressed as follows:

$$\min_{\theta_{\mathcal{F}}} \frac{1}{|X_{\mathcal{S}}|} \sum_{x_s} \mathcal{L}_{class}(x_s, y_s) + \frac{1}{|X_{\mathcal{T}}|} \sum_{x_t} \mathcal{L}_{class}(x_t, \hat{y}_t) \qquad (1)$$

The proposed regularization technique with co-training is designed to address the challenges of domain adaptation and improve the robustness of the model to changes in the target domain.

3.3   Unsupervised Domain Adaptation

The proposed approach is summarised in Algorithm 1. First, the baseline models (model-1, model-2 and model-3 from Fig. 1) are pre-trained independently on the source dataset. Then, a cycle is defined by each Eps value and each iteration. For each cycle, the images from the target dataset $X_{\mathcal{T}}$ are inferred to obtain the target features. Then, as is done in [9], a robust feature is obtained minimizing the camera bias and integrating all features of a tracklet, in order to obtain the distances between features applying a combination of Jaccard and Cosine distances.

After applying the DBSCAN algorithm, and generating the pseudo-labels of the target data, the classification heads of target and source domains to retrain the model $\mathcal{G}$ are defined. The modified model is used in the next cycle.

4   Experimental Validation

This section includes the description of the dataset, the parametrization involved and the experiments performed.

4.1   Dataset and Evaluation Environment

The vehicle ReID dataset used, CityFlow-ReID-v2 [11,15], is a combination of real-world and synthetic set of images. More in detail, CityFlow-ReID-v2 real-world images are captured by 46 cameras in a real-world traffic environment.

Algorithm 1. Pseudo-Labelling for Vehicle ReID.

```
1: for j in Eps do
2:     for i in iterations do
3:         if first iteration then
4:             Initialise F with weights trained on source domain
5:         else
6:             Initialise F with weights from previous iteration
7:         end if
8:         Feats_T = F(X_T)
9:         Feats_T = robustFeature( Feats_T )
10:        Dists_T = jaccard(Feats_T) * λ + cosDsist(Feats_T) * β
11:        Obtain pseudo-labels applying DBSCAN(Dists_T)
12:        Define C_T
13:        Define C_S
14:        Retrain G for a number of epochs
15:    end for
16: end for
```

It includes 880 vehicles identities in a total number of 85,058 images. This set of images is divided in a train set with 52,717 images of 440 vehicles and the remaining 31,238 images of 440 vehicles are for testing. The training set is captured by 40 cameras that overlap with the test set with the exception of 6 new cameras. The only annotations that the test set includes and can be used are the camera identities and the tracklet information, that are all the consecutive frames of the same vehicle recorded by one camera. VehicleX [20] expand the training data including 192,150 synthetic images generated by a 3D engine. The domain bias between real-world and synthetic domain is covered by VehicleX with the use of Similarity Preserving Generative Adversarial Network (SPGAN) [20].

CityFlow ReID-v2 [11,15] dataset is used to evaluate the performance. However, the test set does not provide ground-truth annotations and the online evaluator server for the AI City Challenge [11] only allows 50 uploads. To address these limitations, this work has opted to evaluate their performance in the subset of the test set that shares the same cameras and scenarios with the train set. This is because our previous work [10] obtained the annotations of these images and created an evaluation framework[1]. This specific subset with annotations includes 223 vehicles identities, having a total of 17,260 images.

4.2   Implementation Details

The resources used for the development of the proposed system are a NVIDIA GeForce GTX 1080Ti with GPU with 11 GB RAM and a Xeon Silver 4114 processor with 32 GB of RAM.

The image path includes three DenseNet121 [6] models pretrained on ImageNet [3] with the different training schemes explained in [10].

---

[1] http://www-vpu.eps.uam.es/publications/CityFlow-ReIDEvaluationFramework/.

The initial learning rate for the feature extractor $\mathcal{F}$ will be ten times smaller than the learning rate used for the classification head $\mathcal{C}_i$ for each subset $i \in \mathcal{S}, \mathcal{T}$.

Following [9,24], the minimal number of samples used as hyperparameter in the DBSCAN algorithm is fixed to 10. This method also includes a camera bias reduction between cross-camera images, and integrates all features of a tracklet to obtain a final feature. The aim is to improve the quality of the pseudo-labels, as this final feature will be the input to the DBSCAN method.

Finally, the proposed method studies the generation of incremental pseudo-labels according to the Eps hyperparameter. A cycle is defined by the generation of the pseudo-labels, and the following update of the specific model after applying fine-tuning in a total of 30 epochs applying the strategy described in Sect. 3.2.

Table 1. Results in the system porposed in [10] in terms of mAP.

	Feature-1	Feature-2	Feature-3
Evaluated at UDA pseudo-label			
Baseline	0.2215	0.2031	0.2350
GT	0.3029	0.2652	0.3731
[9] in the used scheme	0.1923	0.1964	0.2028
Our proposal	**0.2928**	**0.3266**	**0.3160**
Evaluated with post-processing			
Baseline	0.2935	0.2645	0.3028
GT	0.4899	0.4647	0.6355
[9] in the used scheme	0.1694	0.1526	0.1534
Our proposal	**0.3585**	**0.3647**	**0.3675**

4.3   Ablation Study for Eps-Neighborhood in DBSCAN

This section includes the experiments carried out to study the influence of Eps to reduce the domain gap between test and train sets. As mentioned in Sect. 3.1, the final performance of the fine-tuned models with the pseudo-labels is quite sensitive to the choice of the hyperparameter Eps in the DBSCAN cluster method. For all the experiments it is used the feature extractor model-1 as baseline that obtain a mAP of 0.2215 in the subset of the CityFlow-ReID-v2 test.

First, it is included the study of Eps hyperparameter in a range from 0.05 to 0.95 in steps of 0.05. Figure 2 shows the results of independently obtaining the pseudo-labels for each Eps value (blue points) and fine-tune the model vs iteratively fine-tune for each Eps in the range (orange curve). Option one involves obtaining each result by fine-tuning the baseline model with a specific Eps value. In contrast, option two begins the first cycle start with a Eps value of 0.05, obtaining the pseudo-labels from the baseline model, fine-tuning the model, and using the previously retrained model for the subsequent cycles with the next Eps value in the range. It can be seen that the performance of each independent Eps value does not achieve the result of generating incremental pseudo-labels, which for the last cycle for the Eps 0.95 obtains 0.2510 mAP. This implies an increase in the baseline result, from 0.2215 to 0.2510 mAP.

Then, Fig. 3 presents the study of various Eps ranges to iteratively generate pseudo-labels and fine-tune the model. As small values of Eps result in samples that do not belong to any cluster, as the Eps increases, a greater number of samples are pseudo-labeled. Consequently, sweeping the ranges from lower to higher Eps values generates a larger number of labeled samples. Each curve in Fig. 3 presents the same performance previously described; it increases as the network is retrained with the generated pseudo-labels. The only curve that does not follow this behaviour is the one for the range 0.85 to 0.95. This may be due to the fact that, for high Eps values, a large number of images are pseudo-labelled and many of the identities are grouped into a single cluster, leading to an increase in the noise of the pseudo-labels. Lastly, the range that obtain the higher mAP for the last cycle is the 0.55 to 0.95.

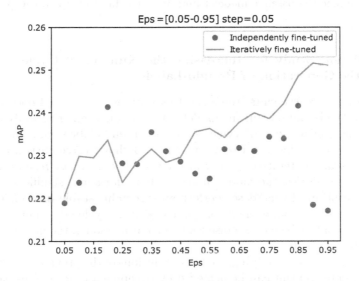

Fig. 2. Fine-tune the baseline model-1 individually for blue points and iteratively for orange curve; per each Maximum Distance (Eps) value between 0.05 to 0.95 with a step of 0.05. (Color figure online)

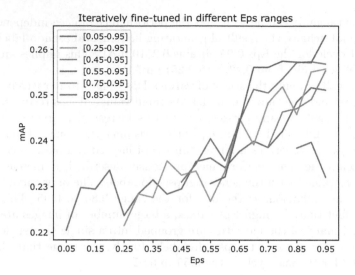

Fig. 3. Fine-tune the baseline model-1 iteratively for different Eps ranges with a step of 0.05.

4.4　Ablation Study for Increasing the Number of Cycles in the Generation of Pseudo-Labels

In the previous experiments, the same Eps value weis just used once to obtain the pseudo-labels and retrain the model. This section is going to study the effect of increasing the number of cycles (generate pseudo-labels and fine-tune the model) before changing the Eps value. Figure 4 shows three curves, each for a different Eps value, iteratively fine-tuned for 105 cycles. These results show that for the first few cycles, the model increases its performance significantly, but as the cycles continue to increase, performance remains stable or even begins to decrease (curve 0.4 from cycle 20 or curve 0.55 from cycle 60). In order to avoid generating pseudo-labels with noise that worsen the model, the number of cycles used will be set to 20.

The latest trial aims to integrate the conclusions of the previous results, which suggest an improvement can be achieved by sweeping a range of Eps values and iterating for a specific number of cycles per value. Figure 5 illustrates the three conducted experiments, starting with the one yielding the best results in Fig. 3 (Eps 0.55 to 0.95), and followed by two additional trials at both ends of the range (from 0.05 and from 0.85). 20 cycles are carried out before switching to the next Eps value. The best configuration (Eps in range from 0.55 to 0.95 with 20 cycles per Eps value) is used to fine-tune the model-2 and model-3 in Fig. 6.

Figure 7 shows two visual examples of the improvement derived from the proposal of this work. The example shows the results derived from the best configuration in model-1. For query-1, it can be observed that for the first cycle (upper row), the first error in the ranked result list appears at position twenty three, while for the last iteration (lower row), the first error does not appear until

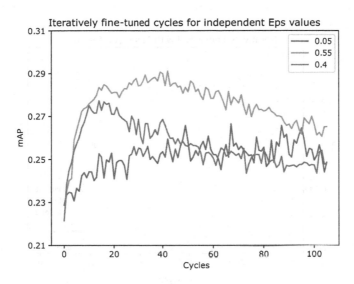

Fig. 4. Fine-tune the baseline model-1 iteratively during 105 cycles with three different Eps value per curve.

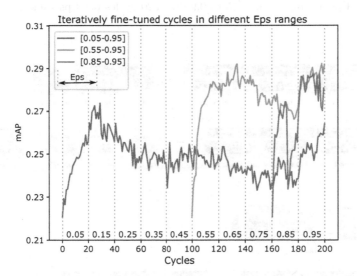

Fig. 5. Fine-tune the baseline model-1 iteratively for three different Eps ranges. For each range it iterates 20 cycles before change to the next Eps value.

position forty. For query-2, the results obtained after the first cycle (upper row) do not obtain any true match in the entire Top-100, while for the last iteration (lower row) the first error occurs for the rank-10.

All the experiments are evaluated at unsupervised domain adaptation pseudo-label (UDA pseudo-label) module level of Fig. 1. In Table 1 are showed

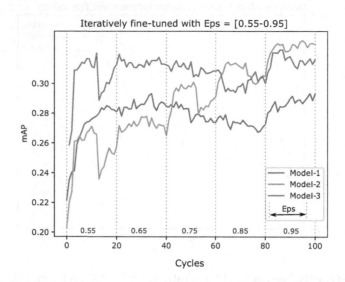

Fig. 6. Iteratively fine-tuned the baseline model-1, model-2 and model-3 with the best configuration (Eps in range from 0.55 to 0.95 with 20 cycles per Eps value).

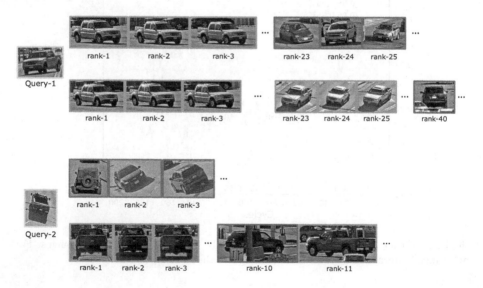

Fig. 7. Example of the visual results for the proposed approach applying the best configuration (Eps in range from 0.55 to 0.95 with 20 cycles per Eps value) in model-1. It shows the query in yellow, the true matches in green and the false matches in red. For each query, the upper row is the results obtained after the first cycle and for the lower row the results are for the last cycle. (Color figure online)

the results at unsupervised domain adaptation pseudo-label module and after the post-processing steps without the feature ensemble (in order to see independently in each feature). *Our proposal* follows the best configuration (20 cycles from 0.55 to 0.95 Eps range). The motivation of showing the results in the entire system is to check that the post-processing does not override the improvement given by the unsupervised domain adaptation module. It is also showed the *Baseline* results from which the experiments start and the Ground Truth (GT), that are the baseline models fine-tuned with the GT labels, in order to see the maximum in case the pseudo-labels generated were perfect. Then, the unsupervised domain adaptation proposed in [9] is included in the scheme of [10] and, as seen in the Table 1, performing just 3 epochs for just the value of Eps 0.55 worsens the baseline results.

A substantial improvement is observed in all three models, starting from baseline values of 0.2215, 0.2031, and 0.2350. At the UDA pseudo-label module, values of 0.2928, 0.3266 and 0.3160 were obtained, respectively, and this improvement is maintained after post-processing with values of 0.3585, 0.3647 and 0.3675.

5 Conclusions

This paper proposes an unsupervised domain adaptation by incremental generation of pseudo-labels. An ablation study of the different configurations has been included in the study of the influence of Eps hyperparameter of DBSCAN cluster algorithm and the number of cycles used to fine-tune the models. The best configuration improved the baseline results.

The proposed method is simple, and focuses on reducing the domain gap between test and train sets, generating incremental pseudo-labels optimally.

Future work includes the exploration and application of the unsupervised domain adaptation technique proposed in this work to other state-of-the-art re-identification systems,such as [9]. Moreover, the possible application of this technique to other related tasks, such as the objects or person re-identification, will be investigated.

Acknowledgements. This work is part of the tasks related to the HVD (PID2021-125051OB-I00) and the SEGA-CV (TED2021-131643A-I00) projects funded by the Ministerio de Ciencia e Innovación of the Spanish Government.

References

1. Capozzi, L., Pinto, J.R., Cardoso, J.S., Rebelo, A.: Optimizing person re-identification using generated attention masks. In: Tavares, J.M.R.S., Papa, J.P., González Hidalgo, M. (eds.) CIARP 2021. LNCS, vol. 12702, pp. 248–257. Springer, Cham (2021). https://doi.org/10.1007/978-3-030-93420-0_24
2. Cascante-Bonilla, P., Tan, F., Qi, Y., Ordonez, V.: Curriculum labeling: revisiting pseudo-labeling for semi-supervised learning. In: Proceedings of the AAAI Conference on Artificial Intelligence, vol. 35, pp. 6912–6920 (2021)

3. Deng, J., Dong, W., Socher, R., Li, L.J., Li, K., Fei-Fei, L.: ImageNet: a large-scale hierarchical image database. In: Proceedings of the IEEE/CVF Conference on Computer Vision and Pattern Recognition, pp. 248–255 (2009)
4. Ester, M., Kriegel, H.P., Sander, J., Xu, X., et al.: A density-based algorithm for discovering clusters in large spatial databases with noise. In: KDD, vol. 96, pp. 226–231 (1996)
5. Fan, H., Zheng, L., Yan, C., Yang, Y.: Unsupervised person re-identification: clustering and fine-tuning. ACM Trans. Multimed. Comput. Commun. Appl. **14**(4), 1–18 (2018)
6. Huang, G., Liu, Z., Maaten, L.V.D., Weinberger, K.Q.: Densely connected convolutional networks. In: Proceedings of the IEEE/CVF Conference on Computer Vision and Pattern Recognition, pp. 4700–4708 (2017)
7. Khan, S.D., Ullah, H.: A survey of advances in vision-based vehicle re-identification. Comput. Vis. Image Underst. **182**, 50–63 (2019). https://doi.org/10.1016/j.cviu. 2019.03.001
8. Liu, X., Liu, W., Ma, H., Fu, H.: Large-scale vehicle re-identification in urban surveillance videos. In: 2016 IEEE International Conference on Multimedia and Expo (ICME), pp. 1–6. IEEE (2016)
9. Luo, H., et al.: An empirical study of vehicle re-identification on the AI city challenge. In: Proceedings of the IEEE/CVF Conference on Computer Vision and Pattern Recognition, pp. 4095–4102 (2021)
10. Moral, P., García-Martín, Á., Martínez, J.M., Bescós, J.: Enhancing vehicle re-identification via synthetic training datasets and re-ranking based on video-clips information. Multimedia Tools Appl. **82**(24), 36815–36835 (2023). https://doi.org/ 10.1007/s11042-023-14511-0
11. Naphade, M., et al.: The 5th AI city challenge. In: Proceedings of the IEEE/CVF Conference on Computer Vision and Pattern Recognition, pp. 4263–4273 (2021)
12. Nowruzi, F.E., Kapoor, P., Kolhatkar, D., Hassanat, F.A., Laganiere, R., Rebut, J.: How much real data do we actually need: analyzing object detection performance using synthetic and real data. arXiv preprint arXiv:1907.07061 (2019)
13. Schubert, E., Sander, J., Ester, M., Kriegel, H.P., Xu, X.: DBSCAN revisited, revisited: Why and how you should (still) use DBSCAN. ACM Trans. Database Syst. **42**(3), 1–21 (2017). https://doi.org/10.1145/3068335
14. Song, L., et al.: Unsupervised domain adaptive re-identification: theory and practice. Pattern Recogn. **102**, 107173 (2020)
15. Tang, Z., et al.: CityFlow: a city-scale benchmark for multi-target multi-camera vehicle tracking and re-identification. In: Proceedings of the IEEE/CVF Conference on Computer Vision and Pattern Recognition, pp. 8797–8806 (2019)
16. Vu, T.H., Jain, H., Bucher, M., Cord, M., Pérez, P.: ADVENT: adversarial entropy minimization for domain adaptation in semantic segmentation. In: Proceedings of the IEEE/CVF Conference on Computer Vision and Pattern Recognition (2019)
17. Wang, H., Shen, T., Zhang, W., Duan, L.-Y., Mei, T.: Classes matter: a fine-grained adversarial approach to cross-domain semantic segmentation. In: Vedaldi, A., Bischof, H., Brox, T., Frahm, J.-M. (eds.) ECCV 2020. LNCS, vol. 12359, pp. 642–659. Springer, Cham (2020). https://doi.org/10.1007/978-3-030-58568-6_38
18. Wang, W., Shen, J., Xie, J., Cheng, M.M., Ling, H., Borji, A.: Revisiting video saliency prediction in the deep learning era. IEEE Trans. Pattern Anal. Mach. Intell. **43**(1), 220–237 (2019)
19. Wu, J., Liao, S., Wang, X., Yang, Y., Li, S.Z., et al.: Clustering and dynamic sampling based unsupervised domain adaptation for person re-identification. In:

2019 IEEE International Conference on Multimedia and Expo, pp. 886–891. IEEE (2019)

20. Yao, Y., Zheng, L., Yang, X., Naphade, M., Gedeon, T.: Simulating content consistent vehicle datasets with attribute descent. In: Vedaldi, A., Bischof, H., Brox, T., Frahm, J.-M. (eds.) ECCV 2020. LNCS, vol. 12351, pp. 775–791. Springer, Cham (2020). https://doi.org/10.1007/978-3-030-58539-6_46

21. Zhang, D., Han, J., Li, C., Wang, J., Li, X.: Detection of co-salient objects by looking deep and wide. Int. J. Comput. Vision 120(2), 215–232 (2016). https://doi.org/10.1007/s11263-016-0907-4

22. Zhang, J., Wang, F.Y., Wang, K., Lin, W.H., Xu, X., Chen, C.: Data-driven intelligent transportation systems: a survey. IEEE Trans. Intell. Transp. Syst. 12(4), 1624–1639 (2011)

23. Zhang, M., et al.: Unsupervised domain adaptation for person re-identification via heterogeneous graph alignment. In: Proceedings of the AAAI Conference on Artificial Intelligence, vol. 35, pp. 3360–3368 (2021)

24. Zhang, X., Cao, J., Shen, C., You, M.: Self-training with progressive augmentation for unsupervised cross-domain person re-identification. In: Proceedings of the IEEE/CVF International Conference on Computer Vision, pp. 8222–8231 (2019)

25. Zheng, Y., Capra, L., Wolfson, O., Yang, H.: Urban computing: concepts, methodologies, and applications. ACM Trans. Intell. Syst. Technol. 5(3), 1–55 (2014)

26. Zheng, Z., Yang, X., Yu, Z., Zheng, L., Yang, Y., Kautz, J.: Joint discriminative and generative learning for person re-identification. In: Proceedings of the IEEE/CVF Conference on Computer Vision and Pattern Recognition, pp. 2138–2147 (2019)

27. Zhong, Z., Zheng, L., Luo, Z., Li, S., Yang, Y.: Invariance matters: exemplar memory for domain adaptive person re-identification. In: Proceedings of the IEEE/CVF Conference on Computer Vision and Pattern Recognition, pp. 598–607 (2019)

How to Turn Your Camera into a Perfect Pinhole Model

Ivan De Boi[1]([⊠])[iD], Stuti Pathak[1][iD], Marina Oliveira[2][iD], and Rudi Penne[1][iD]

[1] InViLab, University of Antwerp, Groenenborgerlaan 171, 2020 Antwerp, Belgium
ivan.deboi@uantwerpen.be
[2] Institute of Systems and Robotics, Department of Electrical and Computer
Engineering, University of Coimbra, 3004-531 Coimbra, Portugal
http://www.invilab.be

Abstract. Camera calibration is a first and fundamental step in various
computer vision applications. Despite being an active field of research,
Zhang's method remains widely used for camera calibration due to its
implementation in popular toolboxes like MATLAB and OpenCV. However, this method initially assumes a pinhole model with oversimpli-
fied distortion models. In this work, we propose a novel approach that
involves a pre-processing step to remove distortions from images by
means of Gaussian processes. Our method does not need to assume
any distortion model and can be applied to severely warped images,
even in the case of multiple distortion sources, e.g., a fisheye image of
a curved mirror reflection. The Gaussian processes capture all distortions and camera imperfections, resulting in virtual images as though
taken by an ideal pinhole camera with square pixels. Furthermore, this
ideal GP-camera only needs one image of a square grid calibration pattern. This model allows for a serious upgrade of many algorithms and
applications that are designed in a pure projective geometry setting but
with a performance that is very sensitive to non-linear lens distortions.
We demonstrate the effectiveness of our method by simplifying Zhang's
calibration method, reducing the number of parameters and getting rid
of the distortion parameters and iterative optimization. We validate by
means of synthetic data and real world images. The contributions of this
work include the construction of a virtual ideal pinhole camera using
Gaussian processes, a simplified calibration method and lens distortion
removal.

Keywords: Pinhole camera · Zhang's method · Gaussian processes ·
Removing lens distortion

1 Introduction

Camera calibration is a vital first step in numerous computer vision applications,
ranging from photogrammetry [10] and depth estimation [13] to robotics [8] and

© Springer Nature Switzerland AG 2024
V. Vasconcelos et al. (Eds.): CIARP 2023, LNCS 14469, pp. 90–107, 2024.
https://doi.org/10.1007/978-3-031-49018-7_7

SLAM [4,24]. As such, it is still a very active field of research, resulting in a myriad of calibration techniques. [1,3,17,21,29]. In this work, we also use the term *camera* for systems such as multi-camera systems or catadioptric systems [6] which also include a mirror.

Several attempts have been made to unify the calibration procedures for different camera types and camera systems [7,11,18]. However, the most popular method in practice today is still Zhang's method [28], which is the basis for the camera calibration toolboxes of both MATLAB and OpenCV. This method assumes a pinhole model with additional lens distortions. The resulting calibration is a compromise between all intrinsic and extrinsic values, including the distortion model parameters. This approach has several drawbacks. First, the pinhole assumption is not applicable to non-central cameras. Second, the calibration is the result of a converging optimization process in which one parameter is adjusted in favour of another one to obtain a better optimum, without actual justification. Third, the proposed distortion models for radial and tangential distortion are in some cases oversimplifications, for instance when the distortion is not perfectly radially symmetric or when the centre of distortion is not at the principle point of the camera.

In this work, we propose a new approach in which we first perform a pre-processing step on the images to remove all distortions. Next, the undistorted images serve as input for a simplified version of Zhang's method for perfect pinhole cameras. By distortions, we mean all deformations resulting from lenses, camera hardware imperfections, faults in the calibration board and even noise. To capture these, we rely on Gaussian processes [20]. They are a non-parametric Bayesian regression technique that are very well suited to handle sparse noisy datasets.

The proposed method applies to a variety of 2D-cameras. For any such camera, we train a Gaussian process on the relationship from pixel coordinates of the corners detected in an image of a square grid pattern (e.g., a checkerboard) to a perfectly spread square lattice of virtual 2D points. Only one image of the calibration board is needed for this training. This lattice can be seen as the non-linear projection of the checkerboard corners in the original camera image to a virtual image plane, consisting of virtual pixels. The Gaussian process captures all possible distortions.

All future images can now be mapped to the same virtual image plane. As all distortions are removed, we are left with virtual images as though they were taken by an ideal pinhole camera. This method does not need to assume any distortion model and can be applied to severely warped images, even in the case of multiple distortion sources, e.g., a fisheye image of a curved mirror reflection.

The process of first taking images by the given camera followed by the Gaussian process mapping to this virtual image plane can be considered as acquiring images by a virtual camera, called the *GP-camera*, replacing the original camera. We will validate that the imaging by this GP-camera corresponds to a perspectivity (central projective transformation) from the 3D scene to the virtual image plane. In other words, we prove that the GP camera is a pinhole camera.

The main benefit of obtaining an ideal pinhole camera is that a lot of well-studied algorithms and applications can be employed on its images. These include pose estimation, depth estimation, epipolar geometry, shape from motion, 3D scene reconstruction, optical flow, externally calibrating multiple cameras and other 3D sensors, etc. Many of these assume a central projection. For a treatise on these topics, we refer to the industry standard book [7]. Our model allows for a serious upgrade of these algorithms and applications that are designed in a pure projective geometry setting. Their performance is very sensitive to non-linear lens distortions. In particular, the quality of calibration techniques that lean on sphere images is drastically improved when rectified images with square pixels are available [15,16,25].

In [19], Gaussian processes are also used to model lens distortions. However, they serve as a surrogate model for the function that captures the lens distortion. As such, they are still part of the iterative camera calibration process. Lens distortion based on one checkerboard pattern is proposed in [26]. However, they implement the Levenberg-Marquardt algorithm to find an optimal set of parameter values for their distortion models. Gaussian processes are non-parametric and as such do not depend on this. A deep learning variant of this can be found in [27].

The contributions of this works are:

- We explain how to construct a virtual ideal pinhole camera out of a variety of non-pinhole cameras using Gaussian processes.
- We show that our calibrated GP-camera using a simplified version of Zhang's method leads to more accurate measurements compared to the calibrated original camera using the general Zhang's method with iteration.
- We show how our method can be used to remove heavy distortions in images.

The rest of the paper is structured as follows: Sect. 2 provides some theoretical background of Gaussian processes. It explains the construction and operation of a virtual GP-camera, and describes how we will validate this pinhole model. In Sect. 3 we show the results and compare our method to the MATLAB implementation of Zhang's method. We discuss these results in Sect. 4. Finally, we formulate our conclusions in Sect. 5.

2   Methods

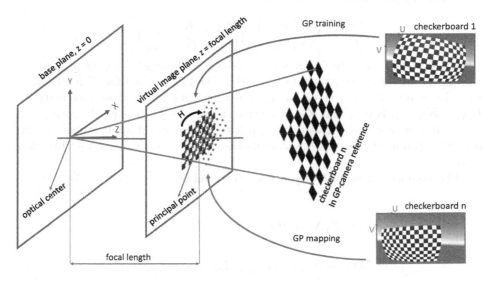

Fig. 1. An overview of our method. A GP model maps the found corners to the image plane, capturing all lens distortions and camera and checkerboard imperfections.

2.1   Gaussian Processes

This section includes a brief discussion of the most basic concepts of GPs. As a probabilistic machine learning technique, *Gaussian processes* (GPs) can be employed for numerous prediction tasks, especially in applications where uncertainty estimation plays a key role. They were originally designed as a probabilistic regression method, but appear to be useful as a machine learning technique keeping track of uncertainties. A Gaussian process can be defined as a continuous collection of random variables, any finite subset of which is normally distributed as a continuous multivariate distribution [20]. Consider a set of n observations $\mathcal{D} = \{X, \mathbf{y}\}$, where $X = \left[\mathbf{x}_1, \mathbf{x}_2, ..., \mathbf{x}_n\right]^T$ is an $n \times d$ input matrix and $\mathbf{y} = \left[y_1, y_2, ..., y_n\right]^T$ is a n-dimensional vector of scalar outputs. We are interested in the mapping $f : \mathbb{R}^d \to \mathbb{R}$,

$$y = f(\mathbf{x}) + \epsilon, \quad \epsilon \sim \mathcal{N}(0, \sigma_\epsilon^2), \tag{1}$$

where ϵ is the identically distributed observation noise.

A GP can be fully defined by its *mean* $m(\mathbf{x})$ and *covariance function* $k(\mathbf{x}, \mathbf{x}')$:

$$f(\mathbf{x}) \sim \mathcal{GP}(m(\mathbf{x}), k(\mathbf{x}, \mathbf{x}')) \tag{2}$$

where $\mathbf{x}$ and $\mathbf{x}'$ are two different inputs. It is common practice to normalise the data to zero mean [20]. By definition, a GP yields a distribution over functions that have a joint normal distribution,

$$\begin{bmatrix} \mathbf{f} \\ \mathbf{f}_* \end{bmatrix} \sim \mathcal{N}\left(\begin{bmatrix} m_X \\ m_{X_*} \end{bmatrix}, \begin{bmatrix} \mathbf{K}_{X,X} & \mathbf{K}_{X,X_*} \\ \mathbf{K}_{X_*,X} & \mathbf{K}_{X_*,X_*} \end{bmatrix} \right), \tag{3}$$

where X and X_* are the input vectors of the n observed training points and n_* the unobserved test points respectively. The mean value for $\mathbf{X}$ is given by m_X. Likewise, the mean value for X_* is given by m_{X_*}. The covariance matrices $\mathbf{K}_{X,X}$, $\mathbf{K}_{X_*,X_*}$, $\mathbf{K}_{X_*,X}$ and $\mathbf{K}_{X,X_*}$ are constructed by evaluating the covariance function k at their respective pairs of points. In real world applications, we are depending on noisy observations $\mathbf{y}$, as we don't have access to the latent function values.

The conditional predictive posterior distribution of the GP can be written as:

$$\mathbf{f}_* | X, X_*, \mathbf{y}, \boldsymbol{\theta}, \sigma_\epsilon^2 \sim \mathcal{N}\left(\mathbb{E}(\mathbf{f}_*), \mathbb{V}(\mathbf{f}_*) \right), \tag{4}$$

$$\mathbb{E}(\mathbf{f}_*) = m_{X_*} + \mathbf{K}_{X_*,X} \left[\mathbf{K}_{X,X} + \sigma_\epsilon^2 I \right]^{-1} \mathbf{f}, \tag{5}$$

$$\mathbb{V}(\mathbf{f}_*) = \mathbf{K}_{X_*,X_*} - \mathbf{K}_{X_*,X} \left[\mathbf{K}_{X,X} + \sigma_\epsilon^2 I \right]^{-1} \mathbf{K}_{X,X_*}. \tag{6}$$

In our work, we are mainly interested in $\mathbb{E}(\mathbf{f}_*)$, the expected value (or mean) of the function values at particular test points. The hyperparameters $\boldsymbol{\theta}$ are usually learned by maximising the log marginal likelihood. In our experiments we use L-BFGS, a quasi-Newton method described in [12]. There exists a large variety of covariance functions, also called kernels [5]. A common choice is the squared exponential kernel, which we also employ in our work. It is infinitely differentiable and thus yields smooth functions. This is a reasonable assumption to make in our context and has the following form:

$$k_{SE}(\mathbf{x}, \mathbf{x}') = \sigma_f^2 \exp\left(-\frac{|\mathbf{x} - \mathbf{x}'|^2}{2l^2} \right), \tag{7}$$

in which σ_f^2 is a height-scale factor and l the length-scale that determines the radius of influence of the training points. For the squared exponential kernel the hyperparameters are σ_f^2 and l.

2.2   Constructing an Ideal Pinhole Camera

Every pixel in an image taken by a camera corresponds to a ray of incoming light, which is a straight line. This means that all points on this straight line in 3D space are mapped to the same pixel. For a perfect *pinhole camera*, as described in [28], all these lines intersect in a central point called the optical centre. Consequently, for the pinhole model, taking images corresponds to a perspective projection, which is an important example of a projective map form $\mathbb{P}^3$ to $\mathbb{P}^2$. The image plane is perpendicular to the focal axis, which contains the optical centre and pointing in the direction in which the camera perceives the world. The distance between the optical centre and the image plane is called the focal length. In Fig. 1 we present a pinhole camera, where every checkerboard is projected to the image plane. More details are given in Appendix A.

A myriad of calibration techniques for (pinhole) cameras can be found in the literature. Due to its importance in computer vision, this is still a very active field of research. Here, we restrict ourselves to the methods most commonly used in practice, such as the camera calibration toolboxes in MATLAB and OpenCV. Their implementation is based on the well-known Zhang's method [28], which is based on the Direct Linear Transform (DLT) method, with the calibration points located in planar objects such as flat checkerboards. See [2,7] for a more in depth treatise. We provide an overview of Zhang's method in Appendix A.

To account for distortion, Zhang's method first assumes a perfect pinhole model with no distortion at all, and approximates the calibration matrix K by means of several homographies between checkerboard positions and the image plane. These homographies are used to determine the positions of these boards relative to the camera. These camera intrinsics and extrinsics serve as an initial guess for an iterative process in which the distortion model is integrated. A non-linear optimization process is then implemented to find, after convergence, values for both the intrinsics, extrinsics and the distortion model parameters.

Herein lies the main pitfall of this method. By iterating towards a convergence in the parameters values, there is a compromise between them. This means unjustly altering the value of one parameter in favour of another one. Both the pinhole and the distortion model might be an oversimplification of the underlying reality.

Now we will explain how we can construct an ideal pinhole camera by using Gaussian processes that first remove all factors that make the pinhole assumption invalid. What is left, is a virtual (ideal) pinhole camera for which a simplified Zhang's method can be used.

Using a fixed but arbitrary 2D-camera, physical or simulated, we take the image of **only one** checkerboard, or any planar square grid pattern. We introduce a local xy-reference frame on the board plane, with coordinate axes parallel to the grid lines, the unit equal to the square edges and the origin typically coincident with some grid corner. This board can be placed anywhere in 3D space, but for our purposes, it is best to position it in such a way that it fills up the entire image. Once this is done, we define a virtual image plane where the grid squares are the virtual pixels. We detect the grid corners in the image using any corner detection system, e.g., the MATLAB Camera Calibration Toolbox. We assign to every detected corner its local xy-coordinates on a virtual image plane. As a consequence, the original board image is mapped to a perfect square lattice of points in the virtual image plane.

These corresponding sets of data are used to train a Gaussian process model for a map between the uv-coordinates in the original image plane and the xy-

coordinates of the virtual image plane, explained in Sect. 3. In practice, we implement two independent scalar output Gaussian processes: one for the resulting x-coordinate and one for the y-coordinate.

In summary, for a given 2D-camera and the image of some spatially positioned square grid patterns, we have constructed a virtual GP-camera that obtains its images by mapping the real images to the virtual image plane by means of a Gaussian process.

Although the Gaussian map of this GP-camera was trained on a single checkerboard image, it apparently removes the distortion for any image of any spatial object. In Sect. 3 we investigate the images for many positions of the calibration board and observe the straightness of all the GP-images of the grid lines. In other words, the GP-camera maps every plane (board position) as a *collineation*, which must be a *homography* according to the fundamental theorem of projective geometry [22]. We conclude that the GP-camera images the world as a projective transformation.

Furthermore, this projective transformation is a *perspectivity* (central projection) since it can be described by the multiplication by a *projection matrix* $\mathbf{P}$ that can be decomposed as:

$$\mathbf{x} = \mathbf{PX} = \mathbf{K}[\mathbf{R} \mid \mathbf{t}]\mathbf{X}, \qquad (8)$$

where we work with homogeneous coordinates $\mathbf{x} = (x, y, 1)^T$ for points in the virtual image plane and $\mathbf{X} = (X, Y, Z, 1)^T$ spatial points.

In Sect. 3 we determine the calibration matrix $\mathbf{K}$ by a simplified version of Zhang's method, described in Appendix B. It is important to note that we need only three parameters for this. Since we are working on square pixels, there is no longer any skewness and the scale is the same in the $x-$ and y-direction:

$$\mathbf{K} = \begin{pmatrix} f & 0 & u_c \\ 0 & f & v_c \\ 0 & 0 & 1 \end{pmatrix}. \qquad (9)$$

The extrinsic component $[\mathbf{R} \mid \mathbf{t}]$ in Eq. 8 transforms the coordinates of the world reference frame to those in the GP-camera reference frame. The GP-origin is determined by the principal point (u_c, v_c) and the focal length f, which is measured in virtual pixel units. The X and Y axes of the latter are parallel to the corresponding axes of the virtual image plane as they appear on the checkerboard in its first position. The Z axis is perpendicular to this, yielding the equation $Z = f$ for the GP image plane.

We validate this pinhole model for the GP-camera by showing the almost perfect match of $\mathbf{K}$ in Eq. 9 as determined by the equations of Zhang's method for many board positions or homographies. In addition, we observe very small reprojection errors in the pinhole model of the GP-camera, using the relative positions of these boards as provided by the computed homographies (Sect. 3). Everything that makes our real world camera deviate from an ideal pinhole camera is captured by the Gaussian process model. This entails lens distortions, imperfections in the lens, camera, checkerboard and even noise in the image.

Although the GP-camera is a virtual camera, we have a geometric interpretation of its pinhole model. The centre of this virtual camera coincides with the theoretical optical centre of the physical camera, but the focal axis of the virtual camera is perpendicular to the square lattice board in the position of the first reference image. This lattice board is the virtual image plane, having world pixels equal to the square cells. The focal length is measured from the GP-centre to this virtual image plane in the same grid units. The images of the GP-camera are obtained by a central projection from this centre onto the virtual image plane (Fig. 1). These dimensions of the virtual camera are fixed once and for all from the moment the first picture is taken. However, they are linked to the physical camera, even if it is later on moved to a different position and rotation.

The calibration of the GP-camera by the simplified Zhang's method not only validates the pinhole model with square pixels, it also provides an interesting alternative to camera calibration, clearly outperforming state-of-the-art methods with respect to simplicity and accuracy (Sect. 3).

Table 1. The datasets.

Dataset	Nr of checkerboards	Nr of corners	Image resolution
Unity pinhole	30	15×9	3840×2160
Unity barrel	30	15×9	3840×2160
Unity pincushion	30	15×9	3840×2160
Webcam	11	15×9	2560×1440
Webcam with telelens	17	14×9	2560×1440
RealSense with mirror	26	14×9	971×871

2.3 The Datasets

We validate our findings on six diverse datasets of images of checkerboards. An overview is given in Table 1. The first three datasets are generated in a scene made in the Unity game engine software version (2020.2.5f1). These sets are based on the same scene, so they depict identical positions and rotations for the boards. The barrel distortion and pincushion distortion effect is obtained by the built-in post-processing package. The barrel distortion centre was placed in the centre of the image. The centre of the pincushion distortion is shifted to the left and bottom of the image.

The last three datasets are made with real world cameras. The Webcam is an Avalue 2k Webcam and the telelens is an Apexel Telelens x2. The RealSense is of the type D415. This can be seen as catadioptric system, as the camera is pointed at a mirror with unknown curvature. The centre is relatively flat while the edges show more spherical bending. There is no mathematical model to calibrate this system. In Fig. 2 we depict an example of one board out of every dataset in the first column.

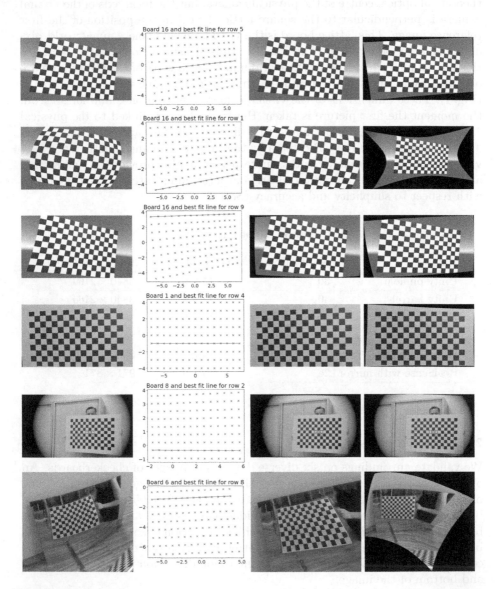

Fig. 2. Column 1: an example of one board of the datasets of Table 2. Column 2: the collinearity check on the result of the GP for that board. Column 3: undistorted image by MATLAB. Column 4: undistorted image by Gaussian processes (our method).

3   Results

We validate our findings with three assessments. First, we investigate how well the Gaussian processes predict collineations, meaning predictions for corners of a checkerboard result in rows and columns that are straight lines. Second, we calculate the reprojection error for found 3D coordinates of corners of the used checkerboards. Finally, we demonstrate the removal of distortion.

3.1   Collineation Assumption

In this section, we validate our method by showing that the mapping done by our GP-based algorithm from uv- to xy-coordinates is a projective transformation where any straight line fed into the said mapping remains straight. In other words, collinearity of points is preserved. We prove the aforementioned statement both visually (Fig. 2) and quantitatively (Table 2).

Let the grid formed by the corner points of a checkerboard have several rows and columns of multiple points each. We find the best fit lines through all of these individual rows and columns. Subsequently, we calculate the perpendicular distance of each point from the corresponding best fit line. The Root Mean Square (RMS) of these perpendicular distances for each row/column is divided by the distance between the end points of the respective row/column to obtain the unitless version of this RMS value. Finally, these scaled unitless RMS values are averaged across all the boards in each dataset to get the Average Root Mean Square Collinearity Error, which is abbreviated as GP CE in the third column of Table 2. For visual confirmation, we show one such best fit line for a given board in every dataset in the second column of Fig. 2.

It is worth mentioning that the positions of the corners on the virtual image plane should be interpreted by the projection of the virtual Gaussian process camera, whose location and rotation differs from the real (or Unity) camera. This depends on the image of the first checkerboard (see also Sect. 2.2). In other words, they are not just unwarpings of the original image.

3.2   Reprojection Error

In our method, we calculate a camera matrix $\mathbf{K}$ for our GP-camera such that for each board, the corresponding homography $\mathbf{H}$ has the form $\mathbf{H} \sim \mathbf{K}[\mathbf{r_1} \mid \mathbf{r_2} \mid \mathbf{t}]$. Furthermore, $\mathbf{K}$ has the form

$$\mathbf{K} = \begin{pmatrix} f & 0 & u_c \\ 0 & f & v_c \\ 0 & 0 & 1 \end{pmatrix}. \tag{10}$$

When solving the overdetermined linear system of equations (see Appendix B), we observe that we find a small value for the smallest singular values, meaning our method is in agreement with an algebraic point of view.

To interpret this from a geometrical point of view, we reproject the 3D coordinates of the corners to the image plane using the parameters found in the camera calibration process. Next, we compare those to the Gaussian process predictions and the found uv-coordinates of the corners for our method and the MATLAB Camera Calibration Toolbox (version R2023a) respectively. As done previously, we calculate the RMS error for these values and scale them by dividing by the distance between two corners, making them unitless. These errors are given in Table 2. From this table, we can see that our method outperforms the MATLAB Camera Calibration Toolbox, especially for datasets with severely warped images such as the Unity with severe barrel distortion, the Unity with eccentric pincushion distortion and the RealSense with curved mirror. This is due to the fact that the MATLAB Camera Calibration Toolbox is based on a model that oversimplifies these underlying realities.

Table 2. Collinearity Errors (CE) and Reprojection Errors (RE)

Dataset	GP CE ($\times 10^{-4}$)	GP RE	MATLAB RE
Unity pinhole	0.954	0.1197	0.1229
Unity barrel	1.334	0.1410	1.0315
Unity pincushion	1.062	0.1243	0.3879
Webcam	3.077	0.2478	0.2666
Webcam with telelens	3.903	0.2576	0.2667
RealSense with mirror	9.455	0.4404	0.5376

We demonstrate the *pinholeness* of our GP-camera further by visualising a grid of 10×10 lines that accompany a given set of pixels. For a pinhole camera, all lines intersect the optical centre, which is also the centre of the reference frame. This implies that all our pixels should correspond to straight lines going through the origin of the reference system. First, we define a 10×10 subset of pixels. We know $\mathbf{R_n}$ and $\mathbf{t_n}$ for each of the n checkerboards from the camera calibration. For each xy-coordinate of the 10×10 pixels, we calculate a corresponding 3D point on each checkerboard. Next, we group together coordinates of 3D points that belong to the same pixel. On these grouped points, we perform a least squares best fit line [9], including RANSAC. An alternative mechanical-inspired approach is given in [14]. We visualise these lines in Fig. 3. Notice how, for instance, the barrel distortion manifests itself as a warping of the grid of lines, while retaining the pinhole model.

3.3 Distortion Removal

The trained GP predicts a new location in virtual xy-coordinate frame for every pixel uv-coordinate frame of the original distorted image. Based on those virtual pixel values, we distil a new image. The predicted coordinates are non-integer numbers, which we round to an integer value. This rounding could imply that some pixels are left empty (black). This means no original pixel is mapped to that specific virtual pixel. We solve this issue by implementing a median filter on the surrounding pixels. Alternatives to this exist, but are outside the scope of this paper. The resulting undistorted images are shown in the last column of Fig. 2. Notice how our method is better equipped to handle distortions, as it is not limited to an underlying oversimplifying distortion model. The difference is most notable for severely warped checkerboards.

4 Discussion

The fact that collinearity is preserved when the GPs map uv-coordinates of found corners to the virtual image plane proves the projective transformation between the real world (including Unity) checkerboards and their virtual images. This means there is a homography between two virtual images of two checkerboards, validating our approach. All non-linearities are captured by the GPs. The decomposition of the projective transformation for every board contains the same **K**. We make use of this to calculate a line in 3D space for a given pixel. We observe that every line that corresponds to a pixel goes through the origin of the reference system of the GP-camera. This demonstrates the pinhole behaviour of our GP-camera.

An accurate corner detection algorithm is a crucial first step in our approach. Especially for the corners of the first checkerboard, on which the Gaussian processes are trained. Moreover, the checkerboard should consist of a sufficient amount of corners so that the distortion can be fully captured.

The reference system of the GP-camera is different from the real (or Unity) camera. This is the result of how the Gaussian process is trained. All uv-coordinates of corners are mapped to a regular square grid. This means the first board is perpendicular to the optical axis and all rows of points are horizontal. This can be seen in the last column of Fig. 2.

From these images we can also observe that for the severe barrel distortion and the eccentric pincushion distortion (row two and three), the MATLAB method fails. Our method is more flexible and can capture these. Notice how the undistortion in the lower left region is better in the last image of row three.

Our Gaussian process predictions decrease in reliability as we move away from the region of the corners in the image of the first board. For those points, the Gaussian processes fall back on their prior, accompanied by large uncertainties. The latter can be taken into account. We retained the pixel predictions with a large uncertainty in the last image of the last row of Fig. 2 for demonstration purposes. Notice that the results become meaningless far away from the first checkerboard where the GP was trained on.

A fallacy of working with Gaussian processes is that they sometimes smooth things out too much. Especially when working with a squared exponential kernel. This is an issue for example, when working with fisheye images, where a lot of corners are situated at the edge of the image and only a few are at the centre of

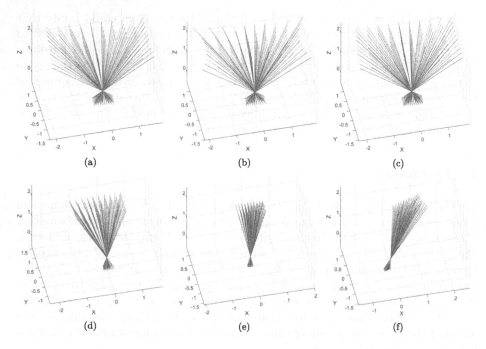

Fig. 3. Visualisation of the pinhole results for each of the six datasets. Only (a) and (d) are actual pinhole cameras, in Unity and the real world respectively.

the image. This data is in essence non-stationary. It varies more in some regions (the edges) than in others (the centre). A solution for this is to work with more corners, a different kernel or even active targets with Gray code instead of corners [23].

5   Conclusion

The aim of the present research was to construct a virtual ideal pinhole camera out of a given camera (including catadioptric systems with mirrors). We showed how this is possible by means of Gaussian processes, which capture everything that makes the camera deviate from an ideal pinhole model. This includes lens distortions and imperfections. Experiments confirmed that our approach results in a pinhole camera.

Further work is required to establish the benefit of our approach in real world camera calibration and compare them to other state of the art methods. We will address this in upcoming publications, as there exists a myriad of camera systems and likewise calibration procedures.

Our model allows for a serious upgrade of many algorithms and applications that are designed in a pure projective geometry setting but with a performance that is very sensitive to non-linear lens distortions.

Acknowledgements. Conceptualization, I.D.B. and R.P.; methodology, I.D.B., S.P. and R.P; software I.D.B.; validation, S.P.; formal analysis, I.D.B., S.P. and R.P; data curation, I.D.B. and M.O.; writing original draft preparation, I.D.B. and S.P.; writing review and editing, I.D.B., S.P., M.O. and R.P; supervision, R.P.; project administration, R.P.; funding acquisition, R.P.

Funding. The authors would like to acknowledge funding from the following PhD scholarships: BOF FFB200259, Antigoon ID 42339; UAntwerp-Faculty of Applied Engineering; 2020.06592.BD funded by FCT, Portugal and the Institute of Systems and Robotics - University of Coimbra, under project UIDB/0048/2020.

Appendix

A   Zhang's Method

The intrinsic camera matrix for a pinhole camera $\mathbf{K}$ can be written as

$$\mathbf{K} = \begin{pmatrix} fs_x & fs_\theta & u_c \\ 0 & fs_y & v_c \\ 0 & 0 & 1 \end{pmatrix}, \tag{11}$$

in which f is the focal length, s_x and s_y are sensor scale factors, s_θ is a skew factor and (u_c, v_c) is the coordinate of the image centre with respect to the image coordinate system. However, real world cameras and their lenses suffer from imperfections. This introduces all sorts of distortions, of which radial distortion is the most commonly implemented. Calibrating this non-ideal pinhole camera, means finding values for both $\mathbf{K}$ and $[\mathbf{R} \mid \mathbf{t}]$, and whichever distortion model is implemented.

Zhang's method is based on the images of checkerboards with known size and structure. For each position of the board, we construct a coordinate system where the $X-$ and Y-axis are on the board and the Z-axis is perpendicular to it. We assign all checkerboard corners a 3D coordinate in this system with a Z-component zero. This allows us to rewrite Eq. 8

$$\begin{pmatrix} x \\ y \\ 1 \end{pmatrix} = \mathbf{K}[\mathbf{R} \mid \mathbf{t}] \begin{pmatrix} X \\ Y \\ 0 \\ 1 \end{pmatrix} = \mathbf{K}[\mathbf{r_1} \mid \mathbf{r_2} \mid \mathbf{t}] \begin{pmatrix} X \\ Y \\ 1 \end{pmatrix}, \tag{12}$$

in which $\mathbf{r_1}$ and $\mathbf{r_2}$ are the first two columns of $\mathbf{R}$. This equation shows a 2D to 2D correspondence known as a homography. This means we can write

$$\begin{pmatrix} x \\ y \\ 1 \end{pmatrix} = \mathbf{H} \begin{pmatrix} X \\ Y \\ 1 \end{pmatrix}, \tag{13}$$

with $\mathbf{H}$ the 3×3 matrix that describes the homography. This matrix is only determined up to a scalar factor, so it has eight degrees of freedom. Each point correspondence yields two equations. Therefore, four point correspondences are needed to solve for $\mathbf{H}$. In practice, we work with several more points in an overdetermined system to compensate for noise in the measurements.

From these homographies, one for every position of the checkerboard, we estimate the camera intrinsics and extrinsic parameters. From Eq. 12 and 13, we can write a decomposition for $\mathbf{H}$, up to a multiple, as

$$\lambda\mathbf{H} = \lambda[\mathbf{h_1} \mid \mathbf{h_2} \mid \mathbf{h_3}] = \mathbf{K}[\mathbf{r_1} \mid \mathbf{r_2} \mid \mathbf{t}], \tag{14}$$

where λ is a scaling factor and $\mathbf{h_1}$, $\mathbf{h_2}$ and $\mathbf{h_3}$ are the columns of $\mathbf{H}$. We observe the following relationships:

$$\lambda\mathbf{K}^{-1}\mathbf{h_1} = \mathbf{r_1}, \tag{15}$$

$$\lambda\mathbf{K}^{-1}\mathbf{h_2} = \mathbf{r_2}. \tag{16}$$

Moreover, since $\mathbf{R}$ is a rotation matrix, it is orthonormal. This means $\mathbf{r_1}^T\mathbf{r_2} = 0$ and $\|\mathbf{r_1}\| = \|\mathbf{r_2}\|$. Combining these equations yields

$$\mathbf{h_1}^T\mathbf{K}^{-T}\mathbf{K}^{-1}\mathbf{h_2} = 0, \tag{17}$$

$$\mathbf{h_1}^T\mathbf{K}^{-T}\mathbf{K}^{-1}\mathbf{h_1} = \mathbf{h_2}^T\mathbf{K}^{-T}\mathbf{K}^{-1}\mathbf{h_2}. \tag{18}$$

These are now independent of the camera extrinsics.

We can write $\mathbf{K}^{-T}\mathbf{K}^{-1}$ as a new symmetric 3×3 matrix $\mathbf{B}$, alternatively by a 6-tuple $\mathbf{b}$. From Eqs. 17 and 18 we can write $\mathbf{Ab} = 0$, in which $\mathbf{A}$ is composed out of all known homography values of the previous step and $\mathbf{b}$ is the vector of six unknowns to solve for. For n checkerboards, and thus n homographies, we now have $2n$ equations. This means we need at least three checkerboard positions. Once $\mathbf{b}$ and thus $\mathbf{B}$ is found, we can calculate $\mathbf{K}$ via a Cholesky decomposition on $\mathbf{B}$. From $\mathbf{K}$, we know all camera intrinsics such as skewness, scale factor, focal length and principal point.

From Eqs. 15 and 16 we can determine $\mathbf{r_1}$ and $\mathbf{r_2}$. The scaling factor λ can be found by normalising $\mathbf{r_1}$ and $\mathbf{r_2}$ to unit length. Building on the orthogonality of the rotation matrix $\mathbf{R}$, we can write

$$\mathbf{r_3} = \mathbf{r_1} \times \mathbf{r_2}. \tag{19}$$

Lastly, we find

$$\mathbf{t} = \lambda\mathbf{K}^{-1}\mathbf{h_3}. \tag{20}$$

Up until this point, we have assumed an ideal pinhole camera model. MATLAB and OpenCV use this as a first step in an iterative process in which they introduce extra intrinsic camera parameters to account for image distortion. After convergence, a compromise is found for all camera parameters.

B   Simplified Zhang's Method

In this work, we construct an ideal virtual GP-camera. The Gaussian processes capture all distortions and other imperfections in a pre-processing step. This means that the images of the checkerboards on the virtual image plane are projections of a perfect checkerboard, up to noise. This allows us to simplify Zhang's method as follows.

First, since there is no skewness in the virtual image plane and all virtual pixels are squares, we can rewrite the intrinsic camera matrix $\mathbf{K}$ as

$$\mathbf{K} = \begin{pmatrix} f & 0 & u_c \\ 0 & f & v_c \\ 0 & 0 & 1 \end{pmatrix}. \tag{21}$$

This results in

$$\mathbf{B} = \mathbf{K}^{-T}\mathbf{K}^{-1} = \begin{pmatrix} \frac{1}{f^2} & 0 & \frac{-u_c}{f^2} \\ 0 & \frac{1}{f^2} & \frac{-v_c}{f^2} \\ 0 & 0 & \frac{u_c}{f^2} + \frac{v_c}{f^2} + 1 \end{pmatrix}. \tag{22}$$

The rest of the procedure is similar to Zhang's method. We combine Eqs. 17, 18 and 22 into the system $\mathbf{Ab} = 0$. The vector $\mathbf{b}$ is now the vector of four unknowns to solve for, instead of six. For n checkerboards, and thus n homographies, we still have $2n$ equations. This means we need at least two checkerboard positions to be able to solve this, instead of three. As before, more positions provide more equations, which are solved via Singular Value Decomposition (SVD). Notice that the form of Eq. 22 is such that we do not have to perform the Cholesky decomposition anymore.

Second, there is no need for a distortion model, nor for a converging iterative process. The camera calibration is reduced to a one-step analytical calculation.

References

1. Beardsley, P., Murray, D., Zisserman, A.: Camera calibration using multiple images. In: Sandini, G. (ed.) ECCV 1992. LNCS, vol. 588, pp. 312–320. Springer, Heidelberg (1992). https://doi.org/10.1007/3-540-55426-2_36
2. Burger, W.: Zhang's camera calibration algorithm: in-depth tutorial and implementation. HGB16-05 pp. 1–6 (2016)
3. Caprile, B., Torre, V.: Using vanishing points for camera calibration. Int. J. Comput. Vision 4(2), 127–139 (1990). https://doi.org/10.1007/BF00127813
4. Devernay, F., Faugeras, O.: Straight lines have to be straight. Mach. Vis. Appl. 13(1), 14–24 (2001). https://doi.org/10.1007/PL00013269
5. Duvenaud, D.K., College, P.: Automatic model construction with Gaussian processes. PhD thesis (2014). https://doi.org/10.17863/CAM.14087
6. Galan, M., Strojnik, M., Wang, Y.: Design method for compact, achromatic, high-performance, solid catadioptric system (SoCatS), from visible to IR. Opt. Express 27(1), 142–149 (2019)
7. Hartley, R.I., Zisserman, A.: Multiple View Geometry in Computer Vision. Cambridge University Press, ISBN: 0521540518, second edn. (2004)

8. Khan, A., Li, J.-P., Malik, A., Yusuf Khan, M.: Vision-based inceptive integration for robotic control. In: Wang, J., Reddy, G.R.M., Prasad, V.K., Reddy, V.S. (eds.) Soft Computing and Signal Processing. AISC, vol. 898, pp. 95–105. Springer, Singapore (2019). https://doi.org/10.1007/978-981-13-3393-4_11
9. Lesueur, V., Nozick, V.: Least square for Grassmann-Cayley agelbra in homogeneous coordinates. In: Huang, F., Sugimoto, A. (eds.) PSIVT 2013. LNCS, vol. 8334, pp. 133–144. Springer, Heidelberg (2014). https://doi.org/10.1007/978-3-642-53926-8_13
10. Li, Z., Yuxuan, L., Yangjie, S., Chaozhen, L., Haibin, A., Zhongli, F.: A review of developments in the theory and technology of three-dimensional reconstruction in digital aerial photogrammetry. Acta Geodaet. et Cartographica Sinica **51**(7), 1437 (2022)
11. Liao, K., et al.: Deep learning for camera calibration and beyond: a survey. arXiv preprint arXiv:2303.10559 (2023)
12. Liu, D.C., Nocedal, J.: On the limited memory BFGS method for large scale optimization. Math. Program. **45**(1–3), 503–528 (1989). https://doi.org/10.1007/BF01589116
13. Mertan, A., Duff, D.J., Unal, G.: Single image depth estimation: an overview. Digital Signal Process. **123**, 103441 (2022)
14. Penne, R.: A mechanical interpretation of least squares fitting in 3D. Bull. Belg. Math. Soc.-Simon Stevin **15**(1), 127–134 (2008)
15. Penne, R., Ribbens, B., Puttemans, S.: A new method for computing the principal point of an optical sensor by means of sphere images. In: Jawahar, C.V., Li, H., Mori, G., Schindler, K. (eds.) ACCV 2018. LNCS, vol. 11361, pp. 676–690. Springer, Cham (2019). https://doi.org/10.1007/978-3-030-20887-5_42
16. Penne, R., Ribbens, B., Roios, P.: An exact robust method to localize a known sphere by means of one image. Int. J. Comput. Vision **127**(8), 1012–1024 (2018). https://doi.org/10.1007/s11263-018-1139-6
17. Puig, L., Bermúdez, J., Sturm, P., Guerrero, J.J.: Calibration of omnidirectional cameras in practice: a comparison of methods. Comput. Vis. Image Underst. **116**(1), 120–137 (2012)
18. Ramalingam, S., Sturm, P.: A unifying model for camera calibration. IEEE Trans. Pattern Anal. Mach. Intell. **39**(7), 1309–1319 (2017). https://doi.org/10.1109/tpami.2016.2592904
19. Ranganathan, P., Olson, E.: Gaussian process for lens distortion modeling. In: 2012 IEEE/RSJ International Conference on Intelligent Robots and Systems, pp. 3620–3625 (2012). https://doi.org/10.1109/iros.2012.6385481
20. Rasmussen, C.E., Williams, C.K.I.: Gaussian processes for machine learning. The MIT Press (2006)
21. Raza, S.N., ur Rehman, H.R., Lee, S.G., Choi, G.S.: Artificial intelligence based camera calibration. In: 2019 15th International Wireless Communications and Mobile Computing Conference (IWCMC), pp. 1564–1569. IEEE (2019)
22. Sarath, B., Varadarajan, K.: Fundamental theorem of projective geometry. Comm. Algebra **12**(8), 937–952 (1984). https://doi.org/10.1080/00927878408823034
23. Sels, S., Ribbens, B., Vanlanduit, S., Penne, R.: Camera calibration using gray code. Sensors **19**(2), 246 (2019). https://doi.org/10.3390/s19020246, https://www.mdpi.com/1424-8220/19/2/246
24. Smith, P., Reid, I.D., Davison, A.J.: Real-time monocular SLAM with straight lines (2006)
25. Sun, J., Chen, X., Gong, Z., Liu, Z., Zhao, Y.: Accurate camera calibration with distortion models using sphere images. Opt. Laser Technol. **65**, 83–87 (2015)

26. Wu, Y., Jiang, S., Xu, Z., Zhu, S., Cao, D.: Lens distortion correction based on one chessboard pattern image. Front. Optoelectron. **8**(3), 319–328 (2015). https://doi.org/10.1007/s12200-015-0453-7
27. Zhang, Y., Zhao, X., Qian, D.: Learning-based framework for camera calibration with distortion correction and high precision feature detection. arXiv preprint arXiv:2202.00158 (2022)
28. Zhang, Z.: A flexible new technique for camera calibration. IEEE Trans. Pattern Anal. Mach. Intell. **22**(11), 1330–1334 (2000)
29. Zheng, Z., Xie, X., Yu, Y.: Image undistortion and stereo rectification based on central ray-pixel models. In: Artificial Intelligence and Robotics: 7th International Symposium, ISAIR 2022, Shanghai, China, October 21–23, 2022, Proceedings, Part II, pp. 40–55. Springer (2022). https://doi.org/10.1007/978-981-19-7943-9_4

Single Image HDR Synthesis
with Histogram Learning

Yi-Rung Lin[1]([✉]), Huei-Yung Lin[2], and Wen-Chieh Lin[3]

[1] Department of Electrical Engineering, National Chung Cheng University,
Chiayi 621, Taiwan
`glcopj1359@gmail.com`
[2] Department of Computer Science and Information Engineering,
National Taipei University of Technology, Taipei 106, Taiwan
`lin@ntut.edu.tw`
[3] Department of Computer Science and Information Engineering,
National Yang Ming Chiao Tung University, Hsinchu, Taiwan
`wclin@cs.nycu.edu.tw`

Abstract. High dynamic range imaging aims for a more accurate representation of the scene. It provides a large luminance coverage to yield the human perception range. In this paper, we present a technique to synthesize an HDR image from the LDR input. The proposed two-stage approach expands the dynamic range and predict its histogram with cumulative histogram learning. Histogram matching is then carried out to reallocate the pixel intensity. In the second stage, HDR images are constructed using reinforcement learning with pixel-wise rewards for local consistency adjustment. Experiments are conducted on HDR-Real and HDR-EYE datasets. The quantitative evaluation on HDR-VDP-2, PSNR, and SSIM have demonstrated the effectiveness compared to the state-of-the-art techniques.

Keywords: High Dynamic Range Image · Cumulative Histogram · Reinforcement Learning

1 Introduction

With the advances of imaging technologies, the photo quality has been improved significantly in the past few decades. The ultimate goal is to present the images with realistic scenery as been seen directly by human eyes. Although many aspects of visual sensing capabilities, such as image resolution, color accuracy and optical sharpness, have been enriched, the luminance range of current imaging devices compared to human vision system is still limited. This is mainly due to the adaptability of large illumination changes of visual perception, but only a fixed amount of light can be collected by a photosensitive element. Consequently, it is necessary to generate a high dynamic intensity range to increase the contrast of image contents for vivid scene representation.

© Springer Nature Switzerland AG 2024
V. Vasconcelos et al. (Eds.): CIARP 2023, LNCS 14469, pp. 108–122, 2024.
https://doi.org/10.1007/978-3-031-49018-7_8

The dynamic range of a digital image represents the ratio between the maximum and minimum measurable light intensities. A standard (or low) dynamic range image records 8-bit information per channel for each pixel. In general, it lacks of sufficient quantization levels to cover a wide brightness variation. To solve this problem, the idea of high dynamic range imaging (HDRI) is proposed, and the HDR rendering is synthesized by multiple image captures with various exposures [2]. Although this can effectively produce HDR images, the results could be affected by dynamic scene changes. Some later studies [7,20] focused on solving the problems of object shaking and artifacts when shooting bracketed images. Nevertheless, this method cannot be directly used for existing image data.

One approach to deal with this issue is to extend a single LDR (low dynamic range) image to its HDR counterpart. The so-called LDR2HDR technique utilizes inverse tone mapping operators for dynamic range expansion in the early research [10,15]. Recently, CNNs are also widely employed in LDR2HDR algorithms [4,9]. Eilertsen et al. proposed a fully convolutional hybrid dynamic range autoencoder to predict HDR values in saturated regions to recover the loss of details [3]. Santos et al. presented a feature masking mechanism to reduce the influences from saturated regions, and adopted a VGG-based perceptual loss function to synthesize images with visually pleasant textures [16]. Alternatively, a data-driven learning approach was proposed to reconstruct the information lost from the original images due to quantization, dynamic range clipping, tone mapping, or gamma correction [13]. Inspired by the LDR image formation pipeline, Liu et al. designed CNN architectures to reverse the steps to recover HDR images [11].

Recently, lightweight deep neural network models attract much attention due to the need of real-time processing capabilities. Wu et al. proposed to use an upsampling blocks in the decoder to alleviate artifacts while maintaining computation speed [19]. Guo and Jiang reformed the single image HDR problem with a camera pipeline modeling, and emphasized the lightweight DNN design for legacy LDR contents [6]. In [1], an inverse tone mapping model based on brightness adaptive kernel prediction was proposed. LDR images were convoluted with the adaptive kernels, and then re-weighted to derive the HDR results. To address the image super-resolution simultaneously, Kim et al. presented a joint image tone mapping and super-resolution framework using a multi-purpose CNN structure [8]. By decomposing the low and high frequency information, the proposed network is able to predict the missing high frequency detail while expanding the intensity range.

In this paper, we propose a technique for HDR synthesis with a single LDR image. The dynamic range of LDR images is expanded, and histogram learning is constructed to predict the brightness distribution of the HDR images. It is followed by reassigning the pixel intensities using histogram matching. The final HDR images are generated with regional adjustment using reinforcement learning. In the experiments, our method is compared with state-of-the-art techniques for performance evaluation. The results have illustrated that comparable HDR image quality can be achieved with the proposed low computation approach. The contributions of this paper are summarized as follows.

- An efficient two-stage LDR2HDR approach with histogram expansion and pixel allocation is proposed.
- A histogram learning method is proposed to predict the brightness distribution from a single LDR image.
- Reinforcement learning is incorporated for pixelwise adjustments of histogram-matched HDR images.

2 Method

The proposed single image HDR synthesis first applies pixelwise multiplication, image upsampling and downsampling to convert an input 8-bit LDR image to the one with 12-bit EDR (extended dynamic range). For an EDR image, the cumulative distribution function of intensity histogram is calculated. 16-bit HDR images are converted to the 12-bit representations, and used as the ground truth for EDR histogram reshaping. A deep neural network is utilized to learn the differnece between the cumulative histograms of EDR and HDR, and predict the target HDR histogram from the EDR image. The 12-bit EDR image with an intensity distribution update is obtained using the prediction as reference for histogram matching. It is then converted to a 16-bit HDR image with intensity scaling (multiplied by 2^4), followed by fine-tuning the intensity values using reinforcement learning with pixelwise rewards. The flow chart of our system as shown in Fig. 1.

2.1 LDR2EDR by Histogram and Resolution difference

To generate a 12-bit EDR representation from an LDR image, we first multiply the intensity of each pixel by 2^4. That is, the dynamic range is expanded from $[0, 255]$ to $[0, 4095]$. In order to fill up the empty bins in the histogram due to direct scaling, image upsampling with bicubic interpolation is carried out to derive a continuous histogram distribution. With resizing the image by $2^2 \times 2^2$, new pixel intensities are obtained through the weighted averages of neighboring 16 sample points. The weight of each sample is determined with the distance to the new pixel. We adopt the following interpolation weight kernel to perform the convolution operation:

$$W(\tau) = \begin{cases} (a+2)\,|\tau|^3 - (a+3)\,|\tau|^2 + 1, & |\tau| \leq 1 \\ a\,|\tau|^3 - 5a\,|\tau|^2 + 8a\,|\tau| - 4a, & \text{otherwise} \end{cases} \tag{1}$$

where a is -0.5 or -0.75. The pixel intensity $f(x, y)$ of the new image is then given by

$$f(x, y) = \sum_{i=0}^{3} \sum_{j=0}^{3} f(x_i, y_j) W(x - x_i) W(y - y_i) \tag{2}$$

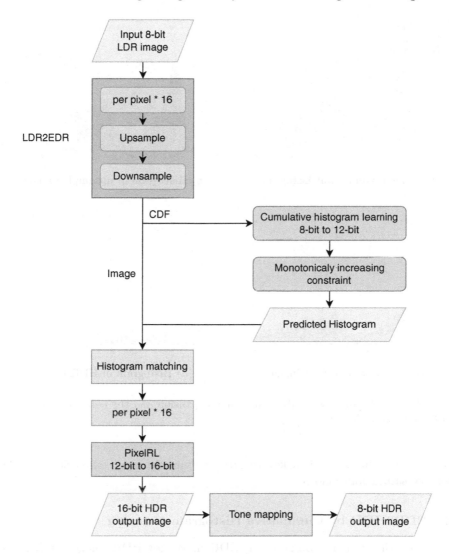

Fig. 1. The system flowchart of the proposed single image HDR synthesis technique.

With image upsampling, empty bins of the histogram are filled up but the image resolution is increased by 2^4 times. To restore the original image size, a downsampling operation is carried out with the pixel intensity average over a 4×4 region. This can reduce the image size while avoiding the appearance of differenceé pattern. After the proposed LDR2EDR computation, the image histogram is maintained as the original distribution and the image dynamic range is expanded to 2^{12}. As depicted in Fig. 2, the histogram obtained from LDR2EDR can eliminate the saturation appeared in the original image. Figure 2b shows that the histogram obtained from LDR2EDR can eliminate the saturation appeared

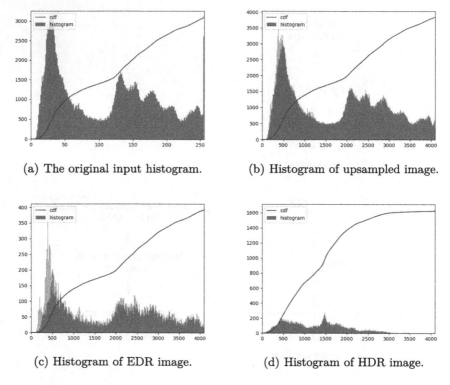

(a) The original input histogram.   (b) Histogram of upsampled image.

(c) Histogram of EDR image.   (d) Histogram of HDR image.

Fig. 2. LDR2EDR process for dynamic range expansion and the cumulative distribution for histogram learning.

in the original image. It is a key to further dynamic range expansions for the intensity saturation regions.

2.2 EDR2HDR by Cumulative Histogram Learning

To expand the dynamic range of an EDR image for HDR image synthesis, a cumulative histogram learning method based on deep neural network is proposed. The basic idea is to learn the pixel intensity distribution change from an ERD/HDR image pair. In addition to recovering the HDR luminance distribution from EDR, it also predicts the appropriate pixel intensity values. To learn the intensity distribution of high dynamic range images, we first convert 16-bit HDR images to their 12-bit representations as the ground truth for histogram learning. Figures 2c and 2d illustrate the histogram and its cumulative distribution of EDR and HDR images, respectively. It can be seen that a large amount of pixels are accumulated in the dark area, resulting in low contrast in the region. The histogram of HDR is relative flat and spreads a wider range of pixel intensity to provide better contrast and detail.

Depending on the scene environment and camera setting, the distributions of image histograms could be very different. Hence, it is not possible to formulate closed-form conversions between EDR and HDR. In this work, we propose a method to learn the histogram transformation using a deep neural network. Due to a great diversity of image histograms, learning from the intensity distributions directly is not able to provide stable data for network training. However, as depicted in Fig. 2, the cumulative distribution of the histogram (or cumulative histogram) has the properties of simplicity and monotonically increasing. It is also in a fixed range, convenient for data normalization, and suitable for model training. Nevertheless, the image intensity distribution can be restored from the predicted cumulative histogram.

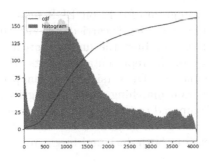 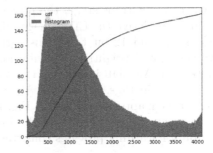

(a) After curve smoothing.   (b) Apply monotonic constraint.

Fig. 3. The cumulative histogram results of each stage.

Our proposed cumulative histogram learning architecture adopts a fully connected neural network with four hidden layers. There are 4096 neurons in both input and output layers to cover 2^{12} different intensity values. It contains 8192 neurons in each hidden layer, and rectified linear unit (ReLU) is used as the activation function. We add dropouts to last two hidden layers, and the dropout rate is set as 0.5. Although the model is relatively simple, it can effectively avoid the overfitting and improve the generalization ability. One issue associated with the network output is the necessity of monotonically increasing, which guarantees the nonnegative values appeared in the image histogram. To ensure this characteristic to be held, the output cumulative curves are processed by Savitzky-Golay filters for smoothing [17], and followed by adjustments to yield the monotonic constraint.

The cumulative curve smoothed by Savitzky-Golay filters and its image histogram are illustrated in Fig. 3a. Although slight jittering on the original curve prediction is mitigated, it still contains negative values in the histogram. To incorporate the monotonically increasing constraint, we enforce the first-order derivative of the curve to be non-negative at any points. For two consecutive bins, the cumulative value on the right is replaced by the one on the left if it is smaller. The processed curve is then used as the cumulative histogram, and converted back to the image histogram. Figure 3b shows the final cumulative distribution and image histogram from this cumulative learning stage.

The proposed network is trained by minimizing the mean square error (MSE) between the input and target cumulative histograms. We set the batch size to 16, train with 300 epochs, and store the best weights during training. The ADAM optimizer is adopted and the initial learning rate is given by 0.001. The input images are resized to 512×512 for training, so the cumulative histogram data are divided by 262,144 to normalized to the range of $[0, 1]$. This also improves the convergence speed of the model. We do not use any common data augmentation methods such as image rotation or horizontal flip since the histogram distribution will not change. However, it is possible to augment the data by adjusting the images with various image formation pipelines.

Up to the current stage, the predicted histogram of a 12-bit EDR image is obtained. To derive the EDR image, histogram matching is carried out with the LDR image and 12-bit target histogram. It is to adjust the image content to have the spatial brightness distribution close to the representation of its HDR counterpart. An LDR image and the corresponding HDR version (tone-mapped) are depicted in Figs. 4a and 4b. After histogram matching, the resulting EDR image (tone-mapped) is shown in Fig. 4c. Compared to the original LDR image, the contrast adjusted based on the target histogram generated by cumulative distribution learning provides consistent detail in both dark and bright regions.

(a)   (b)   (c)

Fig. 4. The histogram matching process: (a) LDR, (b) HDR, (c) result image of histogram matching with the histogram of ground truth image as the reference. (b) and (c) are tone-mapped.

2.3   Fine-Tuning with Reinforcement Learning

From histogram matching, the EDR and HDR images have a similar pixel intensity distribution. However, there might still lack some details due to the limited capabilities of histogram for image representation. To improve the regional pixel intensity consistency, the image is fine-tuned in the spatial domain using reinforcement learning. EDR images are first converted to an HDR representation by multiplying 2^4 to all pixel intensities for the subsequent fine-tuning. A multi-agent reinforcement learning architecture, PixelRL, is adopted for image adjustment [5]. PixelRL modifies the asynchronous advantage actor critic (A3C) architecture into a fully convolutional form. It takes each pixel as an agent, gradually modifies the image intensity through a predefined action set, and finds an optimal policy which minimizes the squared error between the resulting image and the ground truth.

Table 1. The action set used by the agents for training and testing.

	action	filter size	parameter
1	box filter	5×5	-
2	bilateral filter	5×5	$\sigma_c = 1.0, \sigma_S = 5.0$
3	bilateral filter	5×5	$\sigma_c = 0.1, \sigma_S = 5.0$
4	median filter	5×5	-
5	Gaussian filter	5×5	$\sigma = 1.5$
6	Gaussian filter	5×5	$\sigma = 0.5$
7	pixel intensity += 1	-	-
8	pixel intensity -= 1	-	-
9	pixel intensity += 10	-	-
10	pixel intensity -= 10	-	-
11	pixel intensity += 100	-	-
12	pixel intensity -= 100	-	-
13	do nothing	-	-

Chainer [18] and ChainerRL are used to build the network architecture for deep reinforcement learning. The 16-bit image converted from EDR is used as the input image I ($s^{(0)}$), where s is the state of the agent, and the HDR image is the ground truth I^{target}. During training and testing, the agents pick appropriate actions from the action set as tabulated in Table 1 for intensity adjustment. We include several classic image filters and value up/down in the action set. The reward is given by

$$r_i^t = (I_i^{target} - s_i^{(t)})^2 - (I_i^{target} - s_i^{(t+1)})^2 \qquad (3)$$

where I_i^{target} is the i-th pixel intensity of the ground truth. It calculates the squared error between the ground truth and the current image being decreased by the action $a_i^{(t)}$. As a result, maximizing the total reward is equivalent to minimizing the squared error between the final image $s^{t_{max}}$ and the ground truth I^{target}.

For the training hyperparameters, we use the batch size of 32, and perform data augmentation with 70×70 random crop, horizontal image flip, and random rotation. ADAM optimizer and polynomial learning rate policy are adopted in the training process. The initial learning rate is set as 10^{-3}, and decreased by $(episode/max\_episode)^{0.9}$ at each episode. We have the maximum episode, $max\_episode$, set as 3000, the length of each episode, t_max, set as 15, and store the weight every 20 episodes to select the best result.

3   Experiments

In the experiments, we use the HDR-Real dataset as training data [11], and the HDR-EYE dataset as testing data [14]. The image resolution is 512×512. There are 9,786 training image pairs and 46 testing image pairs in the original dataset. A subset is constructed by removing the extreme images with a large proportion of over-exposed and under-exposed regions to improve the learning performance. More specifically, if the number of pixels with the intensity value greater than 249 or less than 6 exceeds 25% of the total pixels, the image is removed from the subset. After the selection, there are 5,673 images in the HDR-Real training subset, Note that, during histogram learning, the complete set is used to extract the histogram information for training. But for the reinforcement learning, only the subset is used for training.

As illustrated in Fig. 5, our experiments are carried out in two stages. In the first stage, image histograms are extracted from the complete dataset for training. After the histogram of the high dynamic range image is predicted, histogram matching is performed on the EDR image to obtain a 12-bit image. In the second stage, the resulting 12-bit HDR image is multiplied by 2^4 on the pixel values, and then performs pixel-by-pixel adjustments with a reinforcement

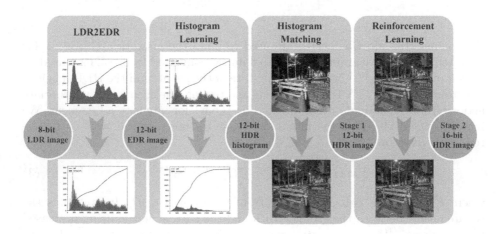

Fig. 5. The overview of the proposed two-stage image synthesis method.

(a) Input LDR image.   (b) Result image of stage.

(c) Result image of stage 2.   (d) Ground Truth HDR image

Fig. 6. The result images of the two stages. (b), (c) and (d) are tone-mapped representation.

learning framework to derive a 16-bit HDR image. The resulting images of the two stages are shown in Fig. 6.

For resulting HDR images, we evaluate using the QMOS score in HDR-VDP-2 [12]. QMOS ranges from 0 to 100, with a higher value indicating better image quality. 16-bit HDR are used as reference images, and the QMOS scores on the testing dataset for the two stages are 61.29 ± 2.71 and 61.54 ± 2.55, respectively. This indicates a 0.25 improvement in the QMOS score given by the second stage adjustment. In addition to the evaluation on an HDR domain, we also analyze the quality of tone-mapped images, and adopt the structural similarity index (SSIM) and PSNR as performance metrics for the comparison with tone-mapped ground truth images. Table 2 illustrates the evaluation results of original images and two output stages. It can be seen that both the PSNR and SSIM evaluation

(a) Input

(b) HDRCNN

(c) DrTMO (Debevec)

(d) ExpandNet

(e) SingleHDR+

(f) Ours

Fig. 7. The visual comparison of our technique with different algorithms.

metrics are improved significantly in the first stage. The results show that the proposed method can effectively learn the intensity distribution of the HDR, and redistribute the image intensity appropriately. In Stage 2, PSNR and SSIM are further improved by 0.273 and 0.013, respectively, from reinforcement learning based pixel-by-pixel adjustments.

Table 2. The PSNR and SSIM evaluation results of two output stages on tone-mapped images.

	PSNR(↑)	SSIM(↑)
Input image	14.218	0.560
Stage 1	19.314	0.711
Stage 2	19.587	0.724

We select four images from the HDR-EYE dataset for illustration. Figure 8 depicts the input LDR images, ground truth HDR, and our output HDR results. The images are converted to grayscale for better demonstration of the contrast enhancement. As shown in Fig. 8a, the original images are partially underexposed, resulting in the loss of detailed texture. Examples include tree trunks in the first image, stairs in the second image, building shadows in the third image, and shadows under parasols in the fourth image. All these difficult-to-identify details have been largely improved by the proposed technique, as shown in Fig. 8c. With the redistribution of pixel intensity based on the predicted HDR histogram, the pixels originally accumulated in the dark regions due to under-exposure can be allocated more properly. In the first result, the texture of the tree trunk becomes clearly visible, and the building has more vivid light and shadow changes. The second and third results are also improved for the under-exposure of stairs and leaves. In the last result, the texture details of the building can be seen clearly in the shadow, and the seating areas under the parasols are no longer dark.

In the last experiment, we compare our approach with recent CNN-based methods, including HDRCNN [3], DrTMO [4], ExpandNet [13], and SingleHDR [11]. For existing works, the pre-trained weights provided by the authors are used. The evaluation is carried out on the HDR-EYE dataset to construct HDR images. Table 3 shows the PSNR and SSIM scores on tone-mapped images. It indicates that our approach provides comparable results as obtained from the state-of-the-art methods. In the PSNR metric, we have better results than HDR-CNN and DrTMO. The proposed technique also outperforms HDRCNN in SSIM. Figure 7 shows the tone-mapped results of single image HDR synthesis obtained from different methods. The partially under-exposed regions are successfully restored in all images, but there are are some slight differences in hue and lightness. HDRCNN tends to variegate the image but also over-enhances the dark areas, which results in the loss of contrast. The output from DrTMO, on the other hand, has blurred and washed-out tones, and fails to faithfully represent

(a) LDR   (b) Ground truth HDR   (c) Ours

Fig. 8. The result images compared with the input LDR images and ground truth HDR images in tone-mapped grayscale representation.

the true colors of the scene. ExpandNet does not recover the locally underexposed regions well. As shown in Fig. 7f, the images obtained using our method have restored the details of the dark regions, retained the contrast, and provide vivid color information.

Table 3. The comparison of our method with other algorithms with PSNR and SSIM. 46 images are used for evaluation.

	PSNR($\uparrow$)	SSIM($\uparrow$)
HDRCNN	17.782	0.683
DrTMO (Mertens)	18.440	0.729
DrTMO (Debevec)	18.515	0.731
ExpandNet	20.387	0.729
SingleHDR	21.895	0.774
SingleHDR (refinement)	22.822	0.791
Ours	19.587	0.724

4 Conclusion

In this paper, we present a technique to generate HDR images from single LDR input. A cumulative histogram learning approach based on deep neural network is proposed for dynamic range expansion. The HDR intensity distribution is predicted from the LDR histogram, and the image pixels are reallocated by histogram matching. Finally, HDR images are constructed using reinforcement learning with pixel-wise rewards for local consistency adjustment. In the experiments, the evaluation with HDR-VDP-2, PSNR, and SSIM is carried on HDR-Real and HDR-EYE datasets.

Acknowledgments. The support of this work in part by the Ministry of Science and Technology of Taiwan under Grant MOST 106-2221-E-194-004 is gratefully acknowledged.

References

1. Cao, G., Zhou, F., Liu, K., Bozhi, L.: A brightness-adaptive kernel prediction network for inverse tone mapping. Neurocomputing **464**, 1–14 (2021)
2. Debevec, P.E., Malik, J.: Recovering high dynamic range radiance maps from photographs. In: ACM SIGGRAPH 2008 classes, pp. 1–10 (2008)
3. Eilertsen, G., Kronander, J., Denes, G., Mantiuk, R.K., Unger, J.: HDR image reconstruction from a single exposure using deep CNNs. ACM Trans. Graph. (TOG) **36**(6), 1–15 (2017)
4. Endo, Y., Kanamori, Y., Mitani, J.: Deep reverse tone mapping. ACM Trans. Graph. **36**(6) (2017). https://doi.org/10.1145/3130800.3130834
5. Furuta, R., Inoue, N., Yamasaki, T.: Pixelrl: fully convolutional network with reinforcement learning for image processing. IEEE Trans. Multimed. **22**(7), 1704–1719 (2019)
6. Guo, C., Jiang, X.: LHDR: HDR: reconstruction for legacy content using a lightweight DNN. In: Proceedings of the Asian Conference on Computer Vision, pp. 3155–3171 (2022)

7. Kalantari, N.K., Ramamoorthi, R., et al.: Deep high dynamic range imaging of dynamic scenes. ACM Trans. Graph. **36**(4), 144–1 (2017)
8. Kim, S.Y., Oh, J., Kim, M.: Deep SR-ITM: Joint learning of super-resolution and inverse tone-mapping for 4k uhd hdr applications. In: Proceedings of the IEEE/CVF International Conference on Computer Vision, pp. 3116–3125 (2019)
9. Lee, S., An, G.H., Kang, S.J.: Deep recursive HDRI: inverse tone mapping using generative adversarial networks. In: proceedings of the European Conference on Computer Vision (ECCV), pp. 596–611 (2018)
10. Lin, H.Y., Kao, C.C.: Hierarchical bit-plane slicing for high dynamic range image stereo matching. IEEE Trans. Instrum. Meas. **70**, 1–11 (2021)
11. Liu, Y.L., et al.: Single-image HDR reconstruction by learning to reverse the camera pipeline. In: Proceedings of the IEEE/CVF Conference on Computer Vision and Pattern Recognition, pp. 1651–1660 (2020)
12. Mantiuk, R., Kim, K.J., Rempel, A.G., Heidrich, W.: HDR-VDP-2: a calibrated visual metric for visibility and quality predictions in all luminance conditions. ACM Trans. Graph. (TOG) **30**(4), 1–14 (2011)
13. Marnerides, D., Bashford-Rogers, T., Hatchett, J., Debattista, K.: Expandnet: a deep convolutional neural network for high dynamic range expansion from low dynamic range content. In: Computer Graphics Forum. vol. 37, pp. 37–49. Wiley Online Library (2018)
14. Nemoto, H., Korshunov, P., Hanhart, P., Ebrahimi, T.: Visual attention in LDR and HDR images. In: 9th International Workshop on Video Processing and Quality Metrics for Consumer Electronics (VPQM). No. CONF (2015)
15. Rempel, A.G., et al.: LDR2HDR: on-the-fly reverse tone mapping of legacy video and photographs. ACM Trans. Graph. (TOG) **26**(3), 39-es (2007)
16. Santos, M.S., Ren, T.I., Kalantari, N.K.: Single image HDR reconstruction using a CNN with masked features and perceptual loss. arXiv preprint arXiv:2005.07335 (2020)
17. Savitzky, A., Golay, M.J.: Smoothing and differentiation of data by simplified least squares procedures. Anal. Chem. **36**(8), 1627–1639 (1964)
18. Tokui, S., et al.: Chainer: a deep learning framework for accelerating the research cycle. In: Proceedings of the 25th ACM SIGKDD International Conference on Knowledge Discovery & Data Mining, pp. 2002–2011 (2019)
19. Wu, G., Song, R., Zhang, M., Li, X., Rosin, P.L.: Litmnet: a deep CNN for efficient HDR image reconstruction from a single LDR image. Pattern Recogn. **127**, 108620 (2022)
20. Yan, Q., et al.: Attention-guided network for ghost-free high dynamic range imaging. In: Proceedings of the IEEE/CVF Conference on Computer Vision and Pattern Recognition, pp. 1751–1760 (2019)

But That's Not Why: Inference Adjustment by Interactive Prototype Revision

Michael Gerstenberger[1]([✉])[iD], Thomas Wiegand[1][iD], Peter Eisert[1,2][iD], and Sebastian Bosse[1][iD]

[1] Heinrich Hertz Institute, 10587 Berlin, Germany
michael.gerstenberger@hhi.fraunhofer.de
[2] Humboldt University, 10099 Berlin, Germany

Abstract. Prototypical part networks predict not only the class of an image but also explain why it was chosen. In some cases, however, the detected features do not relate to the depicted objects. This is especially relevant in prototypical part networks as prototypes are meant to code for high-level concepts such as semantic parts of objects. This raises the question how the inference of the networks can be improved. Here we suggest to enable the user to give hints and interactively correct the model's reasoning. It shows that even correct classifications can rely on unreasonable or spurious prototypes that result from confounding variables in a dataset. Hence, we propose simple yet effective interaction schemes for inference adjustment that enable the user to interactively revise the prototypes chosen by the model. Spurious prototypes can be removed or altered to become sensitive to object-features by the suggested mode of training. Interactive prototype revision allows machine learning naïve users to adjust the logic of reasoning and change the way prototypical part networks make a decision.

Keywords: Prototype Learning · human-AI-Interaction · interactive ML · deep neural networks · inference correction

1 Introduction

Human learning typically involves communication between individuals based on a shared understanding of symbolic mental representations [1]. However, neural networks are usually black box models. Consequently, the sub-symbolic processing in artificial neural networks has hindered analogous interactions between humans and machines.

A common strategy in interactive machine learning (IML) has therefore been additional training with modified data [2,3], e.g. by reusing samples with manually corrected annotations to refine an initial prediction [2]. Alternatively, a second network can be trained to predict the user feedback [3,4]. In both cases the concepts learned by the network remain hidden inside the black box. Only

© Springer Nature Switzerland AG 2024
V. Vasconcelos et al. (Eds.): CIARP 2023, LNCS 14469, pp. 123–132, 2024.
https://doi.org/10.1007/978-3-031-49018-7_9

recent developments in explainable AI allow for a better understanding of the models reasoning [5]. This enables the interaction on the conceptual level. In this paper, we build on recent advances in prototype-based learning that explains predictions using previously learned prototypes [5].

Prototype theory roots in cognitive science and suggests that human cognition relies on prototypes that structure conceptual spaces [6,7]. In such spaces similar concepts lie closer while there is a greater distance between dissimilar ones. Evidence suggests that such an order with strong association between similar features even exists in the human brain [8]. Leveraging this similarity of natural and artificial cognitive representation, we propose a novel approach to IML that allows for direct human-AI interaction on the conceptual level.

Our major contribution is an interactive procedure that allows to eliminate the effect of spurious features that do not relate to the objects classified by ProtoPNet including a quantitative evaluation. We apply and compare our approach to a simple removal of prototypes and illustrate the effect it has on the latent distribution. It shows that object-sensitivity can be achieved with few clicks abolishing the need to mask training images. While concurrent related work evaluated the possibility to remove artificially introduced artifacts we show quantitatively that object-sensitivity of virtually all prototypes can be achieved for a dataset where irrelevant features correlate naturally with the class concepts [9]. Moreover our approach yields good results even with fixed encoder weights reducing the computational cost and hence improving training speed during the interaction runs: It shows that inference of prototype networks can be adjusted solely by the interaction with concepts in latent space. In Sect. 2 we review the concept of prototype-based learning. Section 3 introduces our interaction scheme and shows how it can be used to adjust the inference. Section 4 presents experimental results showing how inference adjustment can be achieved.

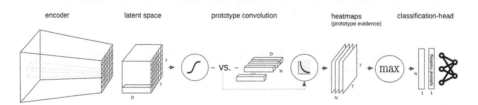

Fig. 1. A prototypical-part network: Suitable prototype vectors are learned during end-to-end training. The prediction relies on the detection of prototypes in feature space. This is achieved by prototype convolution that computes the inverted distance between the grid vectors and the prototypes.

2   Prototype-Based Learning

Prototype-based deep learning builds on the insight that similar features form compact clusters in latent space. Prototypes are understood as concepts that are central for a class [10].

In prototype based deep learning latent features are either represented by a single vector or a latent grid [5,11]. In prototypical part networks a prototype refers to (1) the prototype-vector and (2) the image patch of the image with the closest grid vectors in embedding space. Figure 1 shows the structure of the model of type ProtoPNet that was used here. Prototypes cluster feature space analogously to centroids in k-means. Prototype convolution computes the Euclidean distance between prototypes and feature-vectors. An evidence score is computed for each latent grid position (7×7) and prototype ($N = 2000$) as a function of the inverted distance. In prototype networks with Gaussian Prototype Layer the eucledean distance is weighted by the covariance matrix instead [12]. The maximum taken over each heatmap constitutes the input to the classification head. Prototypes are pushed to the closest feature vectors such that the latent encoding of an image patch coincides with the prototype-vector and corresponds to image patches from the training set [5]. Hence, they can be conceptually equated with image patches from the training set.

There is a growing body of work on prototype based deep learning. Typically, prototype based learning is used to classify images and explain what the relevant features are. Differences exist regarding the question whether or not negative reasoning is allowed or the way prototypes are fitted. Training can be end-to-end [5,11,13] or iterative [14,15]. Recent work also includes models that use configurations of prototypical parts for deformable prototypes [16] or hierarchical prototypes that follow predefined taxonomies and show how prototype based decision trees can be constructed [17,18]. Besides classification, prototype-based deep learning is used for explainable segmentation [12] or tracking of objects in videos, where prototypes are retrieved from previous frames and used as sparse memory-encodings for spatial attention [19]. Some approaches address data scarcity and have particularly been developed for few-shot problems making predictions for a new class with few labels ($n < 20$). Prototypes can be used for classification or semantic segmentation [15,20]. The benefits of deep prototype learning are the high accuracy and robustness its potentials for outlier detection [11], the ability to yield good predictions in few shot problems [15,20] and an increased interpretability that allows for intuitive interaction with the represented concepts [5,17].

3   Interactive Prototype Revision

We aim to adjust the inference of the model by having the user revise the prototypes. Here we compare two approaches. Inference can be adjusted (1) by masking the model's evidence for undesired prototypes or (2) by iterative prototype revision via repetitive user feedback and a custom loss-function: Prototypes are removed from areas of latent space that encode non-object patches as the user identifies antitypes i.e. vectors that code for features unrelated to the concepts represented by the prototypes (here: birds; Fig. 2).

Hence, the network acts like a student that consults the teacher to retrieve information about the learned prototypes. The user takes on the role of the

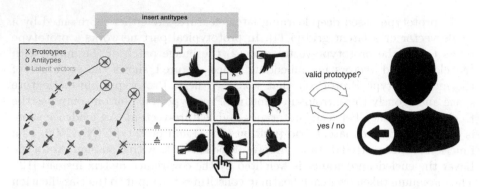

Fig. 2. Interactive prototype revision: The prediction of a class is based on the detection of its previously learned prototypes. Inference is adjusted as the user identifies antitypes that exert a repelling force on prototypes.

teacher and either accepts or rejects the prototypes presented by the student (Fig. 2).

For deselection without replacement, a mask is assembled to exclude prototypes from inference. To adjust for a decrease in accuracy, the final layer of the network is trained afterwards. Prototype revision with replacement can be achieved with a customized loss term *Refine* which relies on the L^2 norm of the prototypes p_j and the user-injected antitypes q_s.

The cost term measures the distance between p_j and q_s and is maximal if antitype and a prototype coincide. It thus encourages the prototypes to diverge from latent encodings of non-object patches.

$$\text{Refine} = \max_i \left[\log \left(\frac{d_i + 1}{d_i + \epsilon} \right) \right] \times \frac{1}{-\log(\epsilon)}, \text{ where } d_i = \|\mathbf{p}_j - \mathbf{q_s}\|_2^2$$

The antitype vectors are initialized with the prototypes that the user identifies as being of type non-object. The prototypes diverge from the antitypes that are inserted by the user due to the repelling term in the loss. Figure 2 illustrates this process.

A second term *Con* imposes a soft-constraint and ensures that prototypes stay in the hyper-cube that contains the latent vectors by penalizing any value i of each prototype-vector j that lies outside of the desired range.

$$\text{Con} = \sum_j \sum_i v_{i,j} \text{ where } v_{i,j} = \begin{cases} 1, & \text{if } (p_{i,j} > 1) \vee (p_{i,j} < 0) \\ 0, & \text{otherwise} \end{cases}$$

These terms are part of the final loss. It additionally contains the cross-entropy *CrsEnt*, the clustering cost *Clst* and the separation cost *Sep* as well as the *L1* term from the original loss of ProtoPNet.

$$L = \text{CrsEnt} + \lambda_1 \text{ Clst } + \lambda_2 \text{ Sep } + \lambda_2 \text{ L1 } + \lambda_3 \text{ Refine } + \lambda_4 \text{ Con}$$

Although prototypes diverge from the deselected feature vectors, some may move towards areas in latent space that encode other spurious features. Hence repeated training for several iterations is needed to remove these prototypes. In total around 350 clicks are necessary until convergence is achieved using the CUB dataset with 200 classes (Fig. 4; lower left). This is considerably less effort than the naïve approach of attributing pixelwise annotations to the 6200 training images, masking out irrelevant areas and training a standard model. The user repeatedly interacts with the model and successively explores areas in latent space that are covered by non-object patches. At the beginning of each repetition the prototypes are pushed to the closest feature vector to identify the prototypical image patches. In the next step the user is consulted to reject unreasonable prototypes.

1: $Q \leftarrow \emptyset$
2: **for** *repetition* $= 1, 2, \ldots N$ **do**
3:     $p_j \leftarrow \arg\min_{z \in Z_j} \|z - p_j\|$ ▷ Push prototypes
4:     $Q \leftarrow Q \cup non\text{-}object(P)$ ▷ Consult user
5:     **for** *epoch* $= 1, 2, \ldots, M$ **do**
6:         **for** *batch* of *Dataset* **do**
7:             *loss* $\leftarrow$ $L(y, net(batch), P, Q)$
8:             *net* $\leftarrow$ $SGD(net, loss)$ ▷ Move prototypes
9:         **end for**
10:     **end for**
11: **end for**

Alg. 1: Iterative prototype revision

The set of antitypes Q is united with the subset of the prototypes P that are identified as non-object prototypes and the network is trained using our deselection loss. Hence, the areas of latent space that encode spurious features are successively clustered by the interplay of user and model (Alg. 1).

4   Results

Our experiment includes three conditions. For the baseline condition we complete the first two stages of the original training schedule of ProtoPNet [5]. For that purpose we use the CUB200-2011 dataset. The model is trained for five epochs with the encoder weights fixed, then all layers and the prototype-vectors are trained together. Afterwards, we push the prototypes to the closest image patch of any training image of the prototype's class. Finally the classification head is trained to adjust for the changes and allow for the L1 term to converge [5].

We aim at a quantitative internal evaluation by dissecting different parts of the training schedule. Our experiment aims at comparing (1) the baseline condition, to (2) the presented interactive revision procedure (Alg. 1) and (3) the effect of a subsequent removal of prototypes without replacement by prototype masking. To enable reproducibility we simulate the user interaction computationally for named experiments. To this end we rely on pixelwise annotations included in the dataset and identify non-object prototypes by computing the image patch using the upscaling-rule [21]. If there is any overlap between the patch and with the object it is considered to be an object-prototype. In contrast denote prototypes have no overlap with the depicted objects non-object prototypes.

In the baseline condition between 157 and 251 of the 2000 prototypes were identified as non-object prototypes ($\bar{n}_{bg} = 206, \bar{n}_{fg} = 1794$). Figure 3b shows a random selection of non-object prototypes for a sample run. Here, an overlap of at least 75% exists only for 1077 prototypes. Analogously, Fig. 3c shows all prototypes for the class "Eastern Towhee". Red boxes indicate the location of the prototype. Blue areas indicate where the heatmap of prototype evidence is larger then 1% of its maximum.

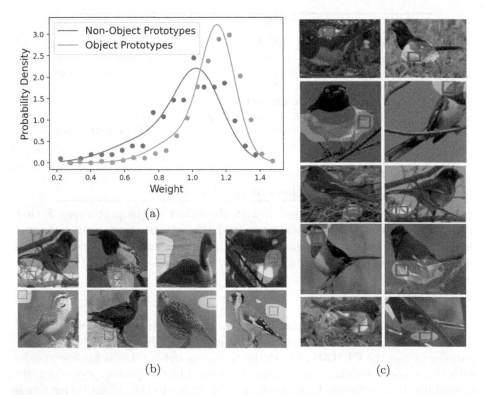

Fig. 3. The relevance of non-object prototypes: (a) The distribution of weights of object-prototypes and non-object prototypes in the final layer. (b) Examples for non-object prototypes (c) Prototypes of the class "Towhee". (Color figure online)

Figure 3a shows the distribution of the weights in the final layer for object and non-object prototypes in a representative run with a Gaussian fit for the PDF. Larger weights are more frequent for object prototypes with a peak in the PDF at $w = 1.2$. However, substantial weights exist also for non-object prototypes highlighting the relevance of the respective non-object prototypes.

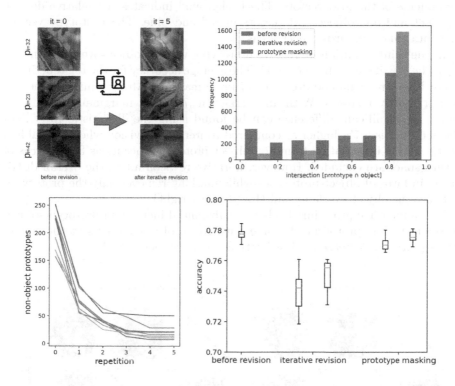

Fig. 4. Prototype revision: (a) Non-object prototypes become sensitive to object features upon iterative revision (upper row). The number of non-object prototypes during iterative revision (lower left). The change in accuracy for different schemes before and after fine-tuning of the last layer (lower right).

For condition two we apply our interactive procedure. It aims at moving prototypes in latent space to regions that encode object-related features. Inference adjustment is achieved by training the network with the presented deselection loss. After the first round between 38 and 77 prototypes revealed themselves as non-object prototypes. None of these prototypes covers the same image patch as before. The accuracy drops from .778 to an average of .732 and recovers to a level of .745 when the last layer is trained. The number of non-object prototypes converges to zero after five repetitions (Fig. 4, lower left) while the accuracy increases: The average accuracy is then slightly higher with .739 before and .749 after fine-tuning of the last layer (Fig. 4, lower right). The procedure yields the highest number of prototypes with near complete object-intersection as compared to prototype masking and the baseline condition (Fig. 4, upper panel). After iterative revision 1583 of the 2000 prototypes show an overlap of at least

75% with the object they represent: The prototypes become sensitive to object features (Fig. 4, upper left). Figure 5 shows the effect on the distribution of prototypes. Each subplot shows the 2D subspace defined by the specified principal components of the grid vectors. The background indicates the relative density of object and non-object grid vectors in red and blue. The prototype vectors (magenta) move towards object areas.

In the third condition we remove prototypes with no object-overlap by masking in the classification head (Fig. 1). The average accuracy for the ten repetitions is .778 before revision and drops to .77 upon masking. After training of the last layer it mostly recovers. With an average of .776 it is marginally lower then before. No significant difference can be found using the T-statistic ($p = .29$). The accuracy is 2.7% higher as compared to iterative revision where arguably a greater loss of information from non-object prototypes occurs as larger areas of feature space are excluded. However, iterative revision alters the overall distribution in favor of object-prototypes while masking removes only the prototypes that have no object-overlap at all (Fig. 4 upper right).

The interactive procedure leads to a substantial increase in prototype-object intersection: The prototypes become sensitive to object features at the cost of only a marginal decrease in classification accuracy (below 3%).

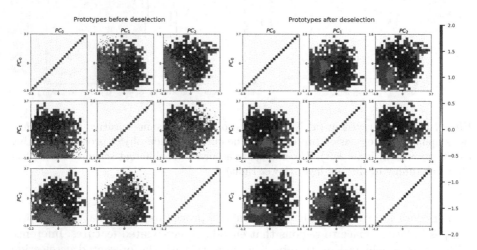

Fig. 5. Movement of prototypes upon iterative revision: The prototype vectors (magenta) in non-object areas (blue) move towards areas of object patches (red). (Color figure online)

5   Conclusion

We show that the inference of prototypical part models cannot only be understood but also adjusted interactively solely by the interaction with concepts in latent space without the need for encoder training. Prototypes can be refined such that the prediction is restricted to meaningful features. The resulting prototypes are highly sensitive to object-features while the effect of spurious features is eliminated. Removing the effect of antitypes leads to a marginal loss in accuracy (1.4% for masking and 2.9% for revision). With the presented interactive procedure we ensure that object features are used only for the prediction. As compared to the naïve approach of masking the objects in the 6200 training images manually it requires considerably less effort (only 350 clicks). This number could potentially be further reduced if only a subset of prototypes was evaluated in each round.

As different strategies for inference exist future works should focus on scenarios that allow for a systematic quantitative comparison. Future research could also address configurations of prototypes, for example by interactive learning of deformable parts [22]. Prototypes for the prediction of bounding boxes should be investigated. They could potentially help to model whole-part relationships in neural networks [23]. Active learning with prototypes may also leverage potentials for increased labeling efficiency [24]. Interactive prototype learning allows to alter the explanations of the model increasing trust in its predictions. As such it will arguably be a valuable tool for future applications of prototype networks.

Acknowledgements. This research has received funding from the German Federal Ministry for Economic Affairs and Climate Action as part of the NaLamKI project under Grant 01MK21003D.

References

1. Planer, R.: What is symbolic cognition? Topoi **40**, 1–12 (2021)
2. Ho, D.J., et al.: Deep interactive learning: an efficient labeling approach for deep learning-based osteosarcoma treatment response assessment. In: Martel, A.L., et al. (eds.) MICCAI 2020. LNCS, vol. 12265, pp. 540–549. Springer, Cham (2020). https://doi.org/10.1007/978-3-030-59722-1_52
3. Wang, G., Zuluaga, M.A., Li, W., et al.: DeepIGeoS: a deep interactive geodesic framework for medical image segmentation. IEEE Trans. Pattern Anal. Mach. Intell. **41**(7), 1559–1572 (2018)
4. Liao, X., Li, W., Xu, Q., et al.: Iteratively-refined interactive 3D medical image segmentation with multi-agent reinforcement learning. In: Proceedings of the IEEE/CVF Conference on Computer Vision and Pattern Recognition, pp. 9394–9402 (2020)
5. Chen, C., Li, O., Tao, D., Barnett, A., Rudin, C., Su, J.K.: This looks like that: deep learning for interpretable image recognition. In: Advances in Neural Information Processing Systems, vol. 32 (2019)
6. Geeraerts, D.: Prototype theory. In: Cognitive Linguistics: Basic Readings, pp. 141–166. De Gruyter Mouton (2008)
7. Balkenius, C., Gärdenfors, P.: Spaces in the brain: from neurons to meanings. Front. Psychol. **7**, 1820 (2016)
8. Bosking, W.H., Zhang, Y., Schofield, B., Fitzpatrick, D.: Orientation selectivity and the arrangement of horizontal connections in tree shrew striate cortex. J. Neurosci. **17**(6), 2112–2127 (1997)

9. Bontempelli, A., Teso, S., Tentori, K., Giunchiglia, F., Passerini, A., et al.: Concept-level debugging of part-prototype networks. In: Proceedings of the Eleventh International Conference on Learning Representations (ICLR 2023) (2023)

10. Bau, D., Zhu, J.-Y., Strobelt, H., Lapedriza, A., Zhou, B., Torralba, A.: Understanding the role of individual units in a deep neural network. Proc. Natl. Acad. Sci. **117**(48), 30071–30078 (2020)

11. Yang, H.-M., Zhang, X.-Y., Yin, F., Liu, C.-L.: Robust classification with convolutional prototype learning. In: Proceedings of the IEEE Conference on Computer Vision and Pattern Recognition, pp. 3474–3482 (2018)

12. Gerstenberger, M., Maaß, S., Eisert, P., Bosse, S.: A differentiable gaussian prototype layer for explainable segmentation, arXiv preprint arXiv:2306.14361 (2023)

13. Li, O., Liu, H., Chen, C., Rudin, C.: Deep learning for case-based reasoning through prototypes: a neural network that explains its predictions. In: Proceedings of the AAAI Conference on Artificial Intelligence, vol. 32 (2018)

14. Chong, P., Cheung, N.-M., Elovici, Y., Binder, A.: Towards scalable and unified example-based explanation and outlier detection. IEEE Trans. Image Process. **31**, 525–540 (2021)

15. Dong, N., Xing, E.P.: Few-shot semantic segmentation with prototype learning. In: BMVC, vol. 3 (2018)

16. Donnelly, J., Barnett, A.J., Chen, C.: Deformable protopnet: an interpretable image classifier using deformable prototypes. In: Proceedings of the IEEE/CVF Conference on Computer Vision and Pattern Recognition, pp. 10265–10275 (2022)

17. Hase, P., Chen, C., Li, O., Rudin, C.: Interpretable image recognition with hierarchical prototypes. In: Proceedings of the AAAI Conference on Human Computation and Crowdsourcing, vol. 7, pp. 32–40 (2019)

18. Nauta, M., van Bree, R., Seifert, C.: Neural prototype trees for interpretable fine-grained image recognition. In: Proceedings of the IEEE/CVF Conference on Computer Vision and Pattern Recognition, pp. 14933–14943 (2021)

19. Ke, L., Li, X., Danelljan, M., Tai, Y.-W., Tang, C.-K., Yu, F.: Prototypical cross-attention networks for multiple object tracking and segmentation. In: Thirty-Fifth Conference on Neural Information Processing Systems (2021)

20. Snell, J., Swersky, K., Zemel, R.S.: Prototypical networks for few-shot learning, arXiv preprint arXiv:1703.05175 (2017)

21. Wah, C., Branson, S., Welinder, P., Perona, P., Belongie, S.: The Caltech-UCSD Birds-200-2011 Dataset. California Institute of Technology, Technical report, CNS-TR-2011-001 (2011)

22. Branson, S., Perona, P., Belongie, S.: Strong supervision from weak annotation: interactive training of deformable part models. In: 2011 International Conference on Computer Vision, pp. 1832–1839. IEEE (2011)

23. Hinton, G.: How to represent part-whole hierarchies in a neural network, arXiv preprint arXiv:2102.12627 (2021)

24. Gal, Y., Islam, R., Ghahramani, Z.: Deep Bayesian active learning with image data. In: International Conference on Machine Learning, pp. 1183–1192. PMLR (2017)

Teaching Practices Analysis Through Audio Signal Processing

Braulio Ríos[1], Emilio Martínez[1], Diego Silvera[1], Pablo Cancela[1],
and Germán Capdehourat[1,2(✉)]

[1] Instituto de Ingeniería Eléctrica, Facultad de Ingeniería,
Universidad de la República, Montevideo, Uruguay
{braulio.rios,emartinez,dsilveracoeff,pcancela}@fing.edu.uy
[2] Ceibal, Montevideo, Uruguay
gcapdehourat@ceibal.edu.uy

Abstract. Remote teaching has been used successfully with the evolution of videoconference solutions and broadband internet availability. Even several years before the global COVID 19 pandemic, Ceibal used this approach for different educational programs in Uruguay. As in face-to-face lessons, teaching evaluation is a relevant task in this context, which requires many time and human resources for classroom observation. In this work we propose automatic tools for the analysis of teaching practices, taking advantage of the lessons recordings provided by the videoconference system. We show that it is possible to detect with a high level of accuracy, relevant lessons metrics for the analysis, such as the teacher talking time or the language usage in English lessons.

Keywords: teaching analysis · classroom activity detection · diarization · education · audio signal processing

1 Introduction

Classroom observation and teaching evaluation have historically been relevant activities in the field of education [9]. In this context, many efforts have been made to standardize lesson observation [6]. In all existing observation protocols, the effort required to apply them is still very high, mainly due to the time and human resources involved in the task to implement it on a large scale. To tackle this scalability limitation, in this work we propose and validate different automatic classroom observation tools to assist the analysis of remote teaching practices, based on the processing of the lessons recordings.

The increasing deployment of broadband internet access at schools enabled new ways of teaching, such as remote lessons through videoconference solutions. In Uruguay, even several years before the global COVID-19 pandemic, this technology was implemented for different educational programs [1]. Ceibal, an organization that provides technological support to the K-12 education system in Uruguay, for example used this approach to universalize the English lessons as

V. Vasconcelos et al. (Eds.): CIARP 2023, LNCS 14469, pp. 133–147, 2024.
https://doi.org/10.1007/978-3-031-49018-7_10

a second language at the primary education level. The main problem was the lack of local English teachers, which was solved in a joint work with the British Council [7], which provided the required teachers that are placed all over the world.

This innovative educational approach with remote teachers for English lessons was very successful. Thus, this methodology was also later extended for Computational Thinking courses. One of the key points of the remote English lessons program, that has been addressed from the very beginning, is the continuous quality monitoring process of the lessons and teachers involved. A group of education technicians, the so-called *quality managers*, attend every year to some lessons from different teachers, following a standardized observation protocol that allows them to review the different activities carried out. After each lesson observation, they write a report to give feedback and exchange ideas with the remote teachers. Their work contributes to the continuous improvement of pedagogical practices, which allows to identify strengths and weaknesses of the academic program and thus plan enhancements for the following year.

The limited number of quality managers, together with the great amount of time that an observation requires, only allows to monitor a reduced number of lessons throughout the year. Therefore, any automation that could be introduced to support their work, would have a great impact in the information available to analyze and improve the educational program. In this context, we propose different tools implemented with state-of-the-art audio processing techniques applied to the lessons recordings, that would be of great help in this regard.

As we show in the following sections, the results obtained validate the utility of the proposed tools for the automation of different relevant analyses of teaching practices observed during a lesson. In the next section, we review some previous work in the area related to classroom analysis. Then, in Sect. 3 we describe the dataset that was built for this work. Next, Sect. 4 focuses on classroom activity detection, addressing the problem of detecting whether the teacher or students are talking at each moment of the lesson. Additional tools are presented in Sect. 5, such as the identification of the spoken language during a lesson and the detection of key phrases, related to the particular content of the educational Unit that should be covered. Finally, the paper ends with Sect. 6, presenting the main conclusions and new lines of work that could be studied in further research.

2   Related Work

With the advances in machine learning, the automation of classroom activity detection have been addressed in several previous works. One of them is the Decibel Analysis for Research in Teaching (DART) [10], a simple approach based on the power of the audio signal. This method detects the lesson segments with only one voice, with more than one voice speaking simultaneously, and with no voices at all. The authors report an accuracy close to 90% for college classes.

More recent works are mostly based on deep learning techniques [4,13,14]. The typical approach is to first extract more powerful low level features rather than just the audio signal power, such as the cepstral coefficients of the Mel bands (MFCC). These features have proven to be very versatile and provide good results in a wide variety of applications, particularly in speech processing [8]. The next step is to train a neural network, defining suitable labels for the stated problem. Typical tags could be the same as in DART (no voice, single voice or multiple voices) or other higher level labels, such as the ones used in [13] to identify teaching practices (e.g. presenting, guiding or administration tasks), where they report 80% of accuracy for the two most common categories.

All the previous work found is based on supervised learning. To the best of our knowledge, this is the first paper to use diarization techniques to tackle the classroom activity detection problem. Speaker diarization responds to the question of *who spoke when* in an audio signal and deep learning techniques have significantly improved the performance of state-of-the-art algorithms for this purpose [11,15]. As we will see in more detail in Sect. 4, it is possible to directly apply unsupervised diarization for teaching analysis, thus avoiding the costly data labeling for training. Although the results obtained are worse than for the supervised approach, they could still be useful for various applications.

In addition to speaker analysis, in this paper we also propose to use state-of-the-art techniques for the identification of the spoken language throughout the lesson and the detection of key phrases. Language identification is a long date relevant problem in the audio processing community [5,16] and an important performance boost has been achieved with the recent Transformers based algorithms. Open source models such as Whisper [12], which serve for speech to text transcription, also has quite good performance in the language detection task, as we will show in Sect. 5. Moreover, we explore the usage of the lesson transcription for key phrases detection, something already studied in previous works [2,17]. This makes it possible to detect the lesson segments where certain specific content of each Unit is addressed, as well as to analyze the kind of feedback that teachers provide to the students.

None of the previous works found refer to the specific use case of remote teaching lessons. There are some particular characteristics of this context that pose specific challenges. On the one hand, the lessons to be analyzed are taught remotely, that is, the students are in a classroom with their local teacher, while the remote teacher guides the lesson through the videoconference system. This fact has an impact in the quality of the recordings, which are made with the microphones of the videoconference system itself, which are not high-fidelity equipment designed for further signal processing. On the other hand, the lessons dynamics in a primary school environment present considerably differences with respect to the classes taught in secondary or college level. This is a significant difference with other previous works and it also affects the recordings quality for the problem posed, since at the primary level it is more complex to control the behavior of students, and even more so in a lesson which is taught remotely.

3 Dataset Description

One of the most important aspects related to any machine learning algorithm development is the appropriate dataset collection and the corresponding data labeling. In this case, the raw data corresponds to the videoconference recordings of the lessons from the English and Computational Thinking courses managed by Ceibal. It is worth noting that both the teachers and the parents or the legal guardians of the students involved, gave their consent for the collection of the data for this research. Although the recordings include the videos of the lessons, only the audios were used for this work.

A carefully specified labeling protocol was defined in order to tag who is talking at each time in the recording of a lesson. Each tag includes its start and end times as well as a label. These labels indicate the presence of the teacher's voice (**teacher**), an individual student voice (**student**) or multiple overlapping student voices (**multiple**) such as during teamwork activities or due to answers in chorus. An example of the labels is illustrated in Fig. 1. The labeling protocol also included other higher level tags, such as classroom disorder, particular noises (crawling chairs, knocks on tables, noisy vehicles outside the classroom) and hushing utterances ("*shh*"). The latter were included to have more information about each particular lesson, and being able to relate performance drops observed with audio quality issues in the data.

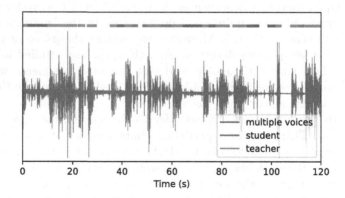

Fig. 1. Example of the labeling result along the raw audio waveform for a lesson.

It should be noted that the data labeling stage was essential for this work, due to the lack of public databases and the different educational context analysed in some previous works, such as university lessons. In order to ensure the quality of the labeling data, and also to minimize ambiguities in the defined protocol, two labeling stages were performed. In the first stage, the three people involved in the labeling process worked with the same six different recordings, which together reached a total of 4.5 h of teaching time. With the results of the first stage, the differences found between the labels were analyzed and discussed, and finally the labeling criteria were adjusted. Once the final protocol version was defined, the second labeling stage was carried out with all the available lessons recordings, reaching a total of 19 h of manually labeled audio data, complying with the aforementioned protocol. It is worth to mention that each hour of lesson recording requires approximately one and a half hour of manual labeling, being an important time consuming task to generate the dataset.

4   Classroom Activity Detection

One of the most important things to analyze teaching practices is to observe how does the teacher manage the spoken time during a lesson, which is in fact summarized in a lesson observation metric called *Teacher Talking Time* (TTT). Excessive use of speech by the teacher can lead to students not participating actively as expected during the lesson. In this section we evaluate different approaches to automatically distinguish the teacher's voice from students participation, through the analysis of the audio lesson recording. The basic output of this module is a list of time intervals, with their detected label, which is one of the three values {*teacher, multiple, student*} already shown in Fig. 1.

The problem posed is an audio classification problem, but it can also be tackled using an adapted diarization system [11,15]. So, two different approaches were implemented for comparison:

- **Unsupervised speaker diarization:** answers the question of *who spoke when?*, which means to discriminate all the different speakers in a conversation, and detect the exact segments in which each of them spoke. This is carried out without any prior information about the number of speakers or how their voices sound like.
- **Supervised audio classification:** given enough annotated data with a predefined set of labels {*teacher, multiple, student*}, a supervised model is trained to predict the label for each audio segment.

Comparing them is relevant because using a pre-trained diarization system does not require any custom training data, thus we consider it a simpler and less costly approach to solve this task. Diarization models are intended to be used on audios with new speakers, whose voices were not known during their training. They do not require fine-tuning over manually annotated data, like the supervised models do.

The unsupervised diarization system is the baseline to which we compare the supervised model. For the latter, we do need training data, which requires a large human effort to annotate the audio recordings. With enough training data, it is expected that the performance for the supervised approach turns out to be better. Thus, one of the key questions to answer is what is the minimum amount of training data needed to surpass the performance of the unsupervised diarization approach.

4.1   Training and Testing Data

As described in Sect. 3 the dataset considered for this work consists of 25 lessons of 45 min each, adding up a total of 19 h of audio recordings that were manually annotated following the defined protocol. Sections of the recordings with technical issues or that take place before the start of after the end of a lesson were discarded. All of them correspond to situations that would be difficult or impossible to manually annotate and can be considered as outliers and for that reason they were not included in the database. Only the annotated audio segments were used for training and testing, although the rest of the audio was not removed to preserve the context information.

The resulting useful annotated audio totaled 15 h of recorded lessons, and was split into 50% for training and 50% for testing. This amount of training data is more than enough for the supervised model as we will see next, so the test size was increased for better significance of the comparison results. The data is further divided into groups to analyze how the supervised model performance generalizes for novel teacher voices. For that purpose, the 25 recorded lessons were split into 5 groups, each with voices from 5 different teachers.

To evaluate the amount of annotated data needed, the training set was also divided into 5 splits, without taking into account teacher gender, which were added incrementally to assess model performance. The distribution of labels was approximately: 60% for *teacher*, 30% for *multiple* and 10% for *student*. The splits and groups were designed to keep that distribution as much as possible. This enables to train the models with incremental data in two ways: adding new lessons (i.e. novel teacher's voices), or adding more annotated time for the same lessons.

4.2   Unsupervised Diarization Approach

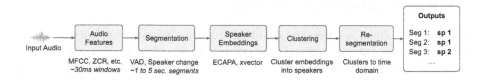

Fig. 2. Diagram of components of a typical diarization system.

Figure 2 shows a typical diarization pipeline, from the input audio waveform, to the output predictions, which consists of a list of segments with their assigned speaker. In our teaching analysis application, diarization does not solve the problem directly, since it does not indicate if any given speaker is the teacher or some of the students. However, we can assume that the teacher is always the most frequent speaker (which was the case in all the lessons recordings that were analyzed for this work). Thus, the speaker detected by the diarization scheme with the largest time across the lesson can be assigned to the teacher label. Please note that this heuristic may fail if the classroom context analyzed is different or if the diarization performance is below a certain quality threshold.

For the implementation, we used the pre-trained *speaker-diarization* pipeline from the *pyannote.audio* toolkit [3]. The module was adapted such that the output speaker with the largest amount of time during the lesson was assigned to label *teacher*. All other speakers were assigned to label *student*. Finally, any audio segment that is not silence neither assigned to *teacher* or *student*, is set to *multiple*.

4.3   Supervised Audio Classification

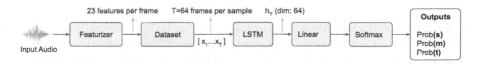

Fig. 3. Diagram of the LSTM-based pipeline for Classroom Activity Detection.

Figure 3 shows the basic blocks of the supervised audio classification pipeline. The key block is an LSTM network with two layers, where the hidden state of the first layer is used as the input to the second layer. The audio features were extracted using overlapping sliding windows that are 30 ms long and a hop of 15 ms. They consist of 12 MFCC coefficients as well as other audio specific features such as spectral flatness, centroid, bandwidth, contrast (7-dimensional vector), and signal power. The resulting feature vector has 23 components for each 30 ms audio frame and is normalized using the mean and variance from all the available training data.

The input to the LSTM network is a sequence of $T = 64$ consecutive normalized feature vectors. This sequence is considered as one audio sample (length $\simeq 1$ s) to be classified as *teacher*, *student* or *multiple*. The classification is done by selecting the largest score after the final layer, which maps the last hidden state h_T from the LSTM to a scores vector, using a fully-connected linear layer and a softmax block. The fact that the hidden state has dimension 64 like the sequence length T is only a coincidence.

4.4   Experiments and Results

Student classroom participations in primary schools are typically short and with lots of overlapping. Most of them are between 1 and 5 s, and the voices are not always clear. From the point of view of the quality managers, the goal is not to know the exact boundaries of each student participation. What they look for is a broader temporal picture of the different moments of the lesson and the participation level. Thus, discussing the optimal resolution with the education

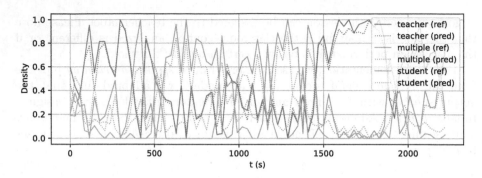

Fig. 4. Example output of the Classroom Activity Detection module, showing the results of the LSTM (supervised) model over 45 min of audio.

technicians, we decided that the most effective value for their purposes was to estimate each label's density on a moving window of 30 s. That is to say, to work with label moving averages of that time window, each of them indicating its fraction of speaking time during those 30 s.

Figure 4 shows an example of the density estimation for all labels during the 45 min of a whole lesson. The LSTM prediction is compared to the reference annotations. As we can see, this visualization allows a user to quickly find peaks of activity in any label, such as moments in which the teacher speaking predominates, or others of high student participation. Given the particular application, an additional metric to the well-known MAE (Mean Absolute Error) was considered. The problem with MAE, as an standard regression metric, is that it evaluates how close the curves are, and not necessarily if they follow the same general shape of peaks and valleys as the reference density. Thus, the other metric considered was the Pearson's correlation coefficient between the predicted and reference density values. Since the correlation is invariant to a vertical shift or scaling in the density estimation, MAE is a good complementary measure to check for systematic over/under estimation problems.

Figure 5 shows the comparison in terms of the correlation coefficient for all labels, and also in terms of MAE for label *teacher* only. In general, the diarization results show a lower accuracy and are less consistent for all the test groups. Test group 1 has very similar correlations between models in label *teacher* (approx. 90%), so to see in more detail what performance this value corresponds to, the estimated densities are shown for both models in Fig. 6. Despite the fact that the metrics values are similar and that the densities are actually comparable in terms of accuracy, the main errors in the predictions are in different times of the audio, which is understandable since the underlying considered techniques are very different.

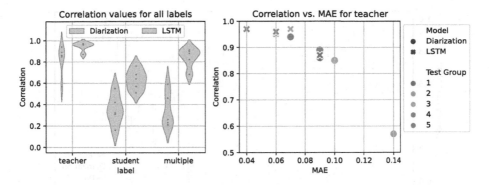

Fig. 5. Comparison between supervised/unsupervised models. **Left:** correlation values for all labels in all test groups (dots inside each violin plot). **Right:** Correlation vs. MAE only for label *teacher*, allowing to see groups individually. Note that test groups 2 and 5 are overlapping for the LSTM model.

For the labels *student* and *multiple*, the diarization system is poorly able to distinguish them with the simple heuristic discussed previously. The confusion is mostly between them, but does not affect the performance for the detection of label *teacher*, where the current approach of taking the most frequent speaker seems successful.

In addition to identifying the moments of teacher and students participation, quality managers also want to have the total time that the teacher or the students spoke, the aforementioned Teacher Talking Time (TTT). For this work, the total estimation error was measured on each of the five test groups, and is shown in Fig. 7. It can be seen that it is possible to estimate the TTT with errors below 5 min using both approaches. The supervised LSTM model is more precise in the total time for all labels, but it is important to note that for the *student* participation, the average total duration is only 4.5 min per lesson (only 10% of the time as mentioned before), so an overestimation of 5 min is very significant in this case.

It is also worth noting that the errors in labels *multiple* and *student* are complimentary, so considering them together as one single category reduces the error substantially. In fact, the error distribution becomes complimentary to the *teacher* label, for which the estimation is very precise. In summary, the total time estimation works well to estimate TTT and the overall student participation, but it should not be used to distinguish individual student participation from group work.

After the baseline performance comparison between both approaches, we want now to address the question of what is the amount of data needed to train the supervised classifier, such that it surpasses the performance of the unsupervised diarization pipeline, which does not require any training data. Figure 8 shows the performance improvement as new data groups are added to the training dataset of the LSTM classifier. The correlation is measured only on the test groups that were not present in the training sets, except for the last point which

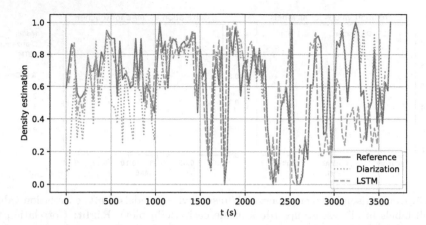

Fig. 6. Comparison of label *teacher* densities over **test group 1**. This is the group where the diarization and LSTM have the most similar metrics.

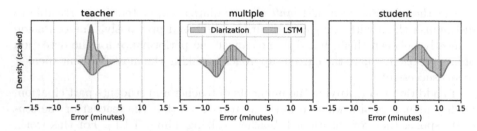

Fig. 7. Distribution of the total time estimation error for each label, for both models. The error is scaled to the duration of a standard lesson of 45 min. Error for the teacher time corresponds to the TTT.

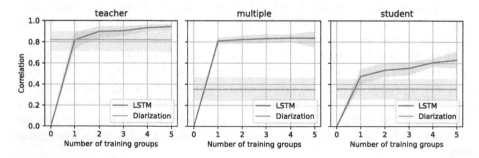

Fig. 8. Comparison between the unsupervised diarization model vs. the supervised LSTM approach, while adding new groups of audios (i.e., adding new lessons and teachers) to the supervised model. The colored spans around each curve represent the confidence interval of 95% of each metric, as measured on all 5 test groups.

includes all the five groups (but always keeping separate data to train and test). This means that except for the last data point, the classifier is being evaluated on new -unknown- teacher and student voices. These results could be used to estimate how the classifier will generalize and perform on new voices.

With respect to adding more annotated time from the same lessons, the detection of *teacher* and *multiple* labels reaches the plateau with the first training split (less than two hours of annotated data). The only label which keeps slightly improving in this case is the *student*, which makes sense considering the low proportion of samples that this label has. Hence we deduce that trying to annotate the whole lesson is not necessary, but instead the efforts should be focused on times with high student participation.

5   Additional Tools for English Lessons Analysis

For the particular case of English lessons, in addition to the speaker detection detailed in the previous section, some additional tools were developed to help in the analysis carried out by the quality managers. In the next sub-sections we present the approaches followed for language identification and key phrases detection. We end up this section introducing the user interface developed for the education quality managers, to support their lessons analysis work.

5.1   Language Detection

During English lessons, the best teacher practice is to encourage the use of English language as much as possible. Spanish usage (mother language in Uruguay) is only justified when they have technical issues due to the videoconference system. Thus, a language detection module is very helpful for the quality managers to analyze the English usage during the lesson. This tool enables to automatically alert when excessive Spanish usage is detected in a lesson.

As introduced in Sect. 2, the Whisper model [12] includes a language detection module prior to speech transcription. In this work, we exploit such functionality to detect the spoken language during each segment of the lesson audio recording. Before the language detection, the audio recording is segmented using a voice activity detection (VAD) module. For this purpose we used the same implementation from *pyannote.audio* included in the diarization pipeline shown in Sect. 4.

The prediction is based on a probability vector for the different possible languages. These probabilities are computed for each speech segment based on all the languages included in the model. Each value represents the algorithm confidence to assign each language to the audio segment. Thus, we decide the language spoken on each segment, as the one with the largest probability. Figure 9 illustrates the pipeline developed, where the raw audio recording is segmented with the VAD and each segment is classified as English or Spanish, according to the output probability vector for each speech audio segment. Based on the probabilities, it is also possible to define a threshold when both values are close to each other, in order to indicate borderline cases to be further analyzed by manual inspection.

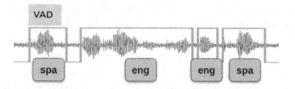

Fig. 9. Language detection example, where the blue signal is the raw audio waveform and the green line is the VAD output. Each audio segment is classified as English or Spanish, based on the maximum confidence values of the Whisper language model. (Color figure online)

For the evaluation of this language detection pipeline, another manual labeling was also necessary. A small dataset of four lessons was selected, totaling three hours of recorded audio labeled. Based on this data, the corresponding confusion matrices were generated (shown in Fig. 10), considering the amount of time for each language. The results show much better accuracy for English language (eng) detection, with almost perfect accuracy, unlike the one observed for the less used Spanish language (spa). In a deeper look at every single lesson recording, it was noticed that the latter improved when Spanish usage is greater. For example, the accuracy for the audio recording where more Spanish is spoken (19% of the total lesson) reaches 90%.

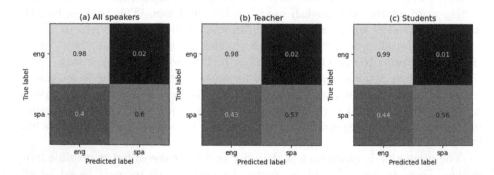

Fig. 10. Language detection confusion matrices for different speakers.

This observations verify that the proposed approach can still be effective for the desired application, which is to detect significant lesson segments where the language usage does not correspond to the best teaching practice of English usage. The performance drops detected, correspond to cases where very little

```
phrases = ['Is @ at @?', '@ is at @']
places = ['home', 'the park']
names = ['Julia', 'Charlie']
Is Julia at home?
Charlie is at the park.
```

Fig. 11. Phrase detection example with *wildcards*. Each "@" can only be replaced with a word belonging to one of the predefined lists.

Spanish usage occurs, so it is not a relevant situation for the teaching analysis application. Moreover, combining the result of this module with the one presented in the previous section, it is possible to identify if the language misusage corresponds to the teacher or the students.

5.2 Key Phrases Matching

Key phrases detection enables to find out if the expected grammar and vocabulary of the corresponding Unit is used during the lesson and to assess the number of times that are repeated. This is particularly important for English lessons, since each Unit has predefined learning exercises. Thus, one relevant thing that the quality managers seek to verify is if the appropriate vocabulary is trained during each lesson, analyzing the speech of both the teacher and the students.

The implementation of this module was also based on the Whisper model [12], but in this case using the speech-to-text output. In addition to the lesson transcription, another required input is the list of strings indicating the grammar and vocabulary that should be detected for a particular Unit. Finally, the goal is to search all the key phrases occurrences during the lesson, indicating for each of them the corresponding times in which they were detected. Since the transcription is not always perfect, a dynamic programming approach was implemented, using the Levenshtein distance to find the most similar matches.

As phrases may have variants, we integrated *wildcards* to deal with them, taking into account the minimum distance over all possible words for each wildcard. Figure 11 shows an example where two lists of places and characters have a predefined vocabulary that can be used. The purpose of wildcards is to allow flexibility in the vocabulary detection used in the lesson. The module detects the times at which the lesson transcription matches an entry from a set of predefined phrases defined by their grammar and vocabulary. The phrases are organized hierarchically so that each phrase corresponds to an educational Unit, allowing the module to produce a report indicating which and when the different Units have been covered in a lesson.

5.3 User Interface for Education Technicians

The final application requires the integration of all the previously described modules in a friendly user interface to be used by the quality managers. This is achieved through a simple web interface, where the education technicians select the lesson recording to be analyzed, enters some basic information about the teacher and the students group, in addition to the options of the required report (e.g. if language usage analysis is necessary or the word list for the key phrases detection).

Combining the results of the different modules (i.e. speaker identification, language usage and key phrases matching) a PDF report is automatically generated with all the relevant metrics summary for the particular lesson and the teaching practice observed. An output video is also generated, which enables to easily navigate through the recording, going directly to relevant excerpts for the quality managers, such as moments of high interaction with students, excessive use of Spanish or vocabulary usage associated with specific content from a course Unit. With the developed tool the teaching evaluation process is enhanced, supporting the quality managers with objective data which enables a faster and more detailed lesson analysis.

6   Conclusions and Further Work

In this work we present different machine learning modules integrated in a tool for education technicians who work on teaching practices analysis and evaluation. Based on lesson recordings for a particular remote teaching scenario, we generated a manually annotated dataset which serves both algorithm development and evaluation.

The first module deals with the speaker identification problem. The goal is to analyze how does the teacher manage the speaking time during the lesson and how much participation do the students have. For this purpose we compared two different approaches, an unsupervised diarization system versus a custom-trained LSTM network. The latter showed a better performance as expected, but we also validated that the diarization approach could be enough if we are only interested in the teacher speech.

Two additional modules were presented for English lessons. The first one uses language detection to identify excessive use of Spanish, while the other is focused on key phrases matching associated to particular Unit topics. All the modules were integrated into a web application, in order to help the quality managers with the teaching practices analysis and evaluation.

Further discussions with the quality managers, as soon as the tool is used more intensively, will bring novel requirements to be addressed. For example, we plan to add the detection of predefined recordings played during a lesson which indicate specific activities were covered, such as English listening exercises. We also expect to obtain new statistical information using the presented tools to contribute to research on education studies carried out within Ceibal.

Acknowledgements. This research was funded by Agencia Nacional de Investigación e Innovación (ANII) Uruguay, Grant Number FMV_1_2021_1_166660.

References

1. Banegas, D.L.: ELT through videoconferencing in primary schools in Uruguay: first steps. Innov. Lang. Learn. Teach. **7**(2), 179–188 (2013)
2. Blunt, P., Haskins, B.: A model for incorporating an automatic speech recognition system in a noisy educational environment. In: 2019 International Multidisciplinary Information Technology and Engineering Conference (IMITEC), pp. 1–7 (2019)
3. Bredin, H., et al.: pyannote.audio: neural building blocks for speaker diarization. In: IEEE ICASSP (2020)
4. Cosbey, R., Wusterbarth, A., Hutchinson, B.: Deep learning for classroom activity detection from audio. In: IEEE ICASSP, pp. 3727–3731 (2019)
5. Foil, J.: Language identification using noisy speech. In: IEEE ICASSP, vol. 11, pp. 861–864 (1986)
6. Guimarães, L.M., da Silva Lima, R.: A systematic literature review of classroom observation protocols and their adequacy for engineering education in active learning environments. Eur. J. Eng. Educ. **46**(6), 908–930 (2021)
7. Kaplan, G.: Innovations in Education: Remote teaching. British Council, London, UK (2019)
8. Martinez, J., Perez, H., Escamilla, E., Suzuki, M.M.: Speaker recognition using Mel frequency cepstral coefficients (MFCC) and vector quantization (VQ) techniques. In: 22nd International Conference on Electrical Communications and Computers, pp. 248–251 (2012)
9. Millman, J., Darling-Hammond, L.: The New Handbook of Teacher Evaluation: Assessing Elementary and Secondary School Teachers. Corwin Press Inc., SAGE Publications (1990)
10. Owens, M., Seidel, S., Wong, M., Tanner, K.: Classroom sound can be used to classify teaching practices in college science courses. PNAS Psychol. Cogn. Sci. **114**(12), 3035–3090 (2017)
11. Park, T.J., Kanda, N., Dimitriadis, D., Han, K.J., Watanabe, S., Narayanan, S.: A review of speaker diarization: recent advances with deep learning. Comput. Speech Lang. **72**, 101317 (2022)
12. Radford, A., Kim, J.W., Xu, T., Brockman, G., McLeavey, C., Sutskever, I.: Robust speech recognition via large-scale weak supervision. arXiv CoRR abs/2212.04356 (2022)
13. Schlotterbeck, D., Uribe, P., Araya, R., Jimenez, A., Caballero, D.: What classroom audio tells about teaching: a cost-effective approach for detection of teaching practices using spectral audio features. In: 11th LAK Conference, pp. 132–140 (2021)
14. Slyman, E., Daw, C., Skrabut, M., Usenko, A., Hutchinson, B.: Fine-grained classroom activity detection from audio with neural networks. arXiv CoRR abs/2107.14369 (2021)
15. Wang, Q., Downey, C., Wan, L., Mansfield, P.A., Moreno, I.L.: Speaker diarization with LSTM. In: IEEE ICASSP, pp. 5239–5243 (2018)
16. Zissman, M.A., Berkling, K.M.: Automatic language identification. Speech Commun. **35**(1), 115–124 (2001)
17. Zylich, B., Whitehill, J.: Noise-robust key-phrase detectors for automated classroom feedback. In: IEEE ICASSP, pp. 9215–9219 (2020)

Time Distributed Multiview Representation for Speech Emotion Recognition

Flavia Letícia de Mattos[1], Marcelo E. Pellenz[1(✉)], and Jr. Alceu de S. Britto[1,2]

[1] Graduate Program in Computer Science (PPGIa),
Pontifical Catholic University of Paraná (PUCPR), Curitiba, Brazil
{flavia.mattos,marcelo,alceu}@ppgia.pucpr.br
[2] State University of Ponta Grossa (UEPG), Ponta Grossa, Brazil

Abstract. In recent years, speech-emotion recognition (SER) techniques have gained importance, mainly in human-computer interaction studies and applications. This research area has different challenges, including developing new and efficient detection methods, efficient extraction of audio features, and time preprocessing strategies. This paper proposes a new multiview model to detect speech emotion in raw audio data. The proposed method uses mel-spectrogram features optimized from audio files and combines deep learning algorithms to improve the detection performance. This combination relied on the following algorithms: CNN (Convolutional Neural Network), VGG (Visual Geometry Group), ResNet (Residual neural network), and LSTM (Long Short-Term Memory). The role of the CNN algorithm is to extract the characteristics present in the images of the mel-spectrograms applied as input to the method. These characteristics are combined with the VGG and ResNet networks, which are pre-trained algorithms. Finally, the LSTM algorithm receives all this combined information to identify the predefined emotions. The proposed method was developed using the RAVDESS database and considering eight emotions. The results show an increase of up to 12% in accuracy compared to strategies in the literature that use raw data processing.

Keywords: speech · emotion recognition · mel-spectrogram · deep learning · CNN · ResNet · VGG · LSTM

1 Introduction

Nowadays, many applications and communication technologies apply machine learning strategies to analyze and interpret different aspects of [1] information. Emotions have become an object of research in different areas, such as psychology and other human behavior studies. This growth is because emotion is essential to human nature and behavior in society and culture. Recently, these studies have also attracted the attention of researchers in the Computing area, especially in

© Springer Nature Switzerland AG 2024
V. Vasconcelos et al. (Eds.): CIARP 2023, LNCS 14469, pp. 148–162, 2024.
https://doi.org/10.1007/978-3-031-49018-7_11

text processing, information retrieval, and human-computer interaction. These studies have some applications in different products, as some examples include smartphone virtual assistants and chatbots used in websites or [2] applications.

Many studies on emotion identification involve deep learning and can help us better understand human behavior. We currently have several examples, such as emotional chatbots. One of them is Loris [3], who listens to employees' concerns and then guides them on how to face daily challenges, whether in a relationship with an angry customer or a co-worker. These chatbots are intended to improve the efficiency of workflows and simplify communication. However, despite all the technological advances of recent years, speech analysis still presents many challenges. The difficulties are mainly related to the intonation and speed of the voice in which a person speaks with an application and how to identify a specific emotion [4] using this audio. Therefore, emotion recognition, analysis, or interpretation has gained even more prominence in several research areas and market branches.

Observing over two thousand Internet videos, a recent study by the University of Berkeley [5] identified 27 human emotions. These emotions include *fear, disgust, anger, romance, surprise* and *sadness*. Therefore, this study area is highly complex and presents many challenges. The related work in this area mainly focuses on the emotion classification of audio files extracted and processed by deep learning or machine learning algorithms. They consider spectral characteristics such as MFCC (Mel Frequency Cepstral Coefficients), spectrogram, or mathematical characteristics such as valence and excitation. In general, all methods proposed in the literature end up using the same types of features extracted from voice signals and similar approaches in terms of the type of deep learning technique, such as the CNN, ResNet, and VGG algorithms. In our study, we sought to explore two essential aspects: parameter settings when extracting spectral features and classifier combination techniques.

Therefore, this article proposes a time-distributed multiview method to identify emotions in raw audio data. The strategy explores the time-frequency characteristics of melspectrogram images, where the STFT parameters are previously optimized. It combines the results of three deep learning algorithms to improve overall performance, including a purpose-built CNN and two pre-trained networks, ResNet50 and VGG16. Furthermore, we explore time correlation using melspectrogram slices in an LSTM network. Through the proposed strategy, we aim to answer the following research questions:

Q1: Is the fusion of multiview representations provided by different CNN architectures considering temporal information a promising strategy to recognize audio emotions?

Q2: What is the impact of audio intensity in the proposed multiview representation-based model for emotion recognition?

Q3: What is the impact of LOSO protocol in the proposed multiview representation-based model?

These research questions will be revisited and addressed in the following sections. The results show that the proposed strategy can improve performance by 13% compared to other strategies using raw audio data. The rest of the paper is organized as follows: Section 2 presents a literature review of speech emotion recognition strategies. The proposed architecture for SER is presented in Sect. 3, where we detail all the processing steps and tools. The experimental results are discussed in Sect. 4. Discussions about the results are addressed in Sect. 5. Finally, the conclusions are drawn in Sect. 6.

2   Related Work

This section briefly reviews the literature, focusing on methods that use the RAVDESS (Ryerson Audio-Visual Database of Emotional Speech and Song) [11] database for speech emotion classification. We also investigate the main audio features explored by the different strategies.

In [6], the authors address emotion recognition based on two concepts: *valence* and *arousal*. Valence represents the aspect of the emotion, whether positive or negative. The arousal represents the strength of the mood. They are a psychological representation generally called activation-evaluation space. Based on these two concepts, we can represent emotions in a two-dimensional (2D) space divided into four quadrants. Depending on the value associated with the valence and arousal metrics in the 2D space, we can classify the type of emotion based on quadrant location. They first propose a discrete emotion classification method, called a *discrete model*, based on the audio's MFCC feature. This method involves a CNN with classical image classification architecture. A second method was proposed, called *dimensional model*. It considers the concept of valence and arousal, and it is a 2D quadrant predictor CNN model. They applied only four emotions in the method and a combination of the female and male gender. In the discrete model, the accuracy achieved was 50%, and with the dimensional model, the accuracy achieved was 76.2%.

In [7], the authors propose a deep-forward CNN for speech emotion recognition. The CNN architecture has nine layers, including seven convolutional layers and two fully connected layers. Additionally, they propose an audio preprocessing step, using a threshold for noise reduction and removing periods of silence. After this step, they generate the spectrogram images. The proposal was evaluated considering two databases, RAVDESS and IEMOCAP. The proposed CNN architecture achieved an accuracy of 70% when processing raw spectrogram images. The audio pre-processing stage for spectrogram cleaning introduces additional complexity and delay for real-time applications.

Another spectrogram information classification technique was proposed in [8]. The approach uses only one hidden layer neural network, composed of input, hidden, and output layers. The method employs reduced spectrogram images, which can lead to a loss of information about emotions. Four databases were evaluated: RML, TESS, EMODB, and RAVDESS. For validation with the application of the RAVDESS base, the work applied three scenarios: application of the complete base, application of the base with the audios recorded at low intensity, and application of the base with the audios recorded at high intensity. However, neutral emotion is not considered when using the database with high intensity. Within the scenarios presented, the highest value achieved with the RAVDESS database was 86.31%, using 5000 interactions and only seven emotions.

In [9] was proposed an algorithm called wide residual network blocks, which decreases the depth and increases the width of the residual networks. The algorithm employs a CNN-based model. The work applied two combined databases, RAVDESS and TESS, making it difficult to compare with the present work and related work. With this combination of bases, the accuracy reached 90%. However, it does not show results for separated databases.

In [10], the authors present an architecture based on CNN and LSTM, where the process starts with generating MFCC images, which are used as input in the classification network and LSTM. The difference in this process is the application of the LSTM algorithm twice with the same parameters preceded by a dropout layer. The author comments that this application of two layers is to help in the final classification of the method. They applied the RAVDESS database with music and speech. The music database has six emotions, and the speech database has eight emotions, so they only used six emotions in the two databases for comparison purposes. For the music database, the method reached 73.3%. For the speech database, 53.32%. Table 1 summarizes the main characteristics of the strategies presented in the related work. Our proposal aims to develop an improved method for detecting emotions by applying a multiview strategy based on combining classifiers to improve performance, as presented in the next section.

Table 1. Related Work Details

Ref	Audio Features	Types of Emotions	Number of Emotions	Database	Algorithms	DA	Split Ratio Training/Testing	Accuracy
[6]	Valence Arousal MFCC	A+H+N+S	4	RAVDESS	CNN-1D CNN-3D	No	70%/30%	76.20%
[7]	Spectrogram (128 × 128)	A+C+D+F+H +N+SA+SU	8	IEMOCAP RAVDESS	CNN-2D	No	80%/20%	70.00%
[8]	Spectrogram (150 × 66)	A+C+D+F+H +N+SA+SU	8	RML TESS EMODB RAVDESS	One Hidden Layer	No	80%/20% 85%/15%	86.31%
[9]	MFCC	A+C+D+F+H +N+SA+SU	8	RAVDESS +TESS	Wide Residual Network	Not Specified	70%/30%	90.00%
[10]	MFCC	N+C+H +SA+A+F	6	RAVDESS	CNN1D+LSTM	Not Specified	70%/30%	53.32%

A = Anger C = Calm D = Disgust F = Fear H = Happiness N = Neutral SA = Sadness SU = Surprise 1D = 1 Dimension 2D = 2 Dimensions 3D = 3 Dimensions DA = Data Augmentation

3   Proposed Strategy

We present an enhanced machine learning (ML)-based strategy for accurately identifying various speech emotions in raw audio data. Our primary objective is to classify eight distinct emotions using an optimized mel-spectrogram audio representation, employing a time-distributed multiview deep learning architecture. The developed architecture combines different pre-trained algorithms: CNN, VGG16, ResNet50, and LSTM. We considered the RAVDESS [11] database because it differentiates between genders, intonations, and emotions. In addition, the data files are indexed, helping to read and process the audio files. Furthermore, as discussed in Sect. 2, this database is widely used in the performance evaluation of many algorithms proposed in the literature.

3.1   Database Description

This study uses the RAVDESS database, which consists of two file models: audio and music. While the Speech Emotion Recognition (SER) area of study can be applied to both types of audio, our focus is specifically on analyzing audio files. The choice not to apply the musical basis in our method is due to the fact that the focus is on the emotions present in speech with the aid of audio recorded for this purpose. The audio database comprises a total of 1,440 files, with each actor contributing 60 recordings. There are 24 actors, including 12 male and 12 female actors. The files are available in **wav** format. File names are structured to indicate the actor/actress, the specific emotion, the stated phrase, and other relevant additional information. It is worth mentioning that the RAVDESS audio database covers a classification of eight distinct emotions. The database exhibits an uneven distribution of audio samples across emotions, resulting in an imbalance that affects analysis of results using metrics such as F1-Score. Among the 8 emotions in the database, the neutral emotion has fewer samples than the other emotions due to the lack of "high" intensity.

3.2   General Architecture

Figure 1 presents an overview of the general architecture. The implemented strategy combines the features extracted from the three algorithms: VGG16, ResNet50, and CNN. These integrated features are applied to the LSTM layers to guarantee the combination of different views over a time-distributed perspective. Explaining better the first steps of the method, in Step 1, we perform the extraction of audio features, specifically using the spectrogram with optimized parameters. Moving on to Step 2, we employ a pool of pre-trained classifiers. This fusion of algorithms establishes a multiview architecture, significantly improving classification results. Such an approach has been widely adopted in several applications and has consistently shown promising results.

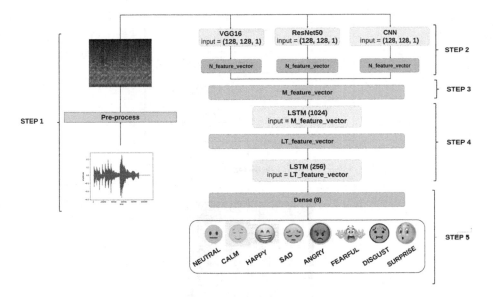

Fig. 1. Proposed end-to-end time-distributed multiview architecture.

3.3   Step 1 - Initial Procedures

This step includes defining the audio's reading parameters, data augmentation, and feature extraction. The audio reading consists of the sample's normalization using the z-score normalization procedure. Another step in audio reading is the setting of the *offset* parameter, which defines the beginning of the audio reading. The offset adjustment reduces periods of silence or containing non-relevant information at the beginning of database recordings. We tested three *offset* values: 0, 0.2, and 0.3 s.

Our approach is based on the audio spectrogram. The spectrogram includes time and frequency domain information whose resolution depends on the calculation parameters. We employ a post-processed version of the spectrogram called the melspectrogram. The spectrogram is obtained through the Short Time Fourier Transform (STFT) [12]. Depending on the STFT parameters, the time and frequency resolution of the spectral images is modified. Figure 2 presents an overview of the main parameters that affect the STFT and melspectrogram.

The audio files in the database were digitized using a sample rate of $S_r = 8$ kHz. It is important to highlight the main effect of these parameters for the spectral analysis. For a specific FFT window analysis (w_l), the frequency resolution of the spectrum is defined by the number of FFT points (N_{FFT}). When we use a smaller window size (w_l), we can more accurately identify audio frequency changes from one time window to another. The hop length (h_l) is the number of samples between the start times of adjacent frames. We implemented the proposed architecture using the **librosa** [13] package. This package provides a rich set of audio feature analysis and extraction functions. The melspectrogram is

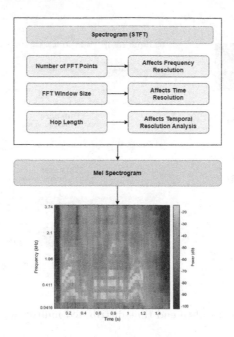

Fig. 2. Melspectrogram Generation Process

computed in Step 1. The generation of melspectrogram images is represented in Fig. 2. The audios of the RAVDESS database were recorded in a studio, without noise and interference. Therefore, in this step, we also introduce data augmentation by adding noise to the audio files, thus increasing the database and the SER complexity. This step also includes the slicing of the melspectrogram images. This procedure prepares the images for entry into the deep learning network for processing.

3.4 Steps 2 and 3 - Algorithms and Combiner

Step 2 aims to initialize and implement the pre-trained networks which compose the developed deep learning architecture. The proposed multiview architecture includes three views generated by the VGG16, ResNet50, and CNN algorithms. The choice of different algorithms aimed to guarantee a diversity of architectures in the development of the proposed strategy, therefore generating a diversity of data representations. The VGG16 algorithm was selected because it is frequently employed as a good feature extractor in the literature. The ResNet algorithm applies the concept of residual layers that are aggregated during processing. The proposed CNN guarantees a way to train a specific architecture while ensuring features' fusion with the other algorithms to feed the LSTM network in the next step. The three algorithms have the same input size to guarantee that the three views similarly process the images. In Step 3, we gather the output information from the three algorithms, where each output represents a specific view. After the independent processing of each algorithm for the previously generated images, all the output data are combined with the aid of a concatenation function. After this combination, we have a general result that will be applied in the next step.

3.5 Steps 4 and 5 - LSTM and Emotion Classification

In step 4, we implement the LSTM network in two layers. The first and second LSTM layers had dimensions of 1024 and 256, respectively. The output of the first LSTM is the input to the second LSTM layer. These two networks evaluate the features using two different time-distributed views. The second layer aims to ensure that at the end of step 4, the information will be processed and passed correctly to step 5, which receives clean information with a considerable size to classify the number of predefined classes, which is the number of emotions established by the method. These two layers were implemented to improve the processing of data generated by the classifiers and ensure that the classification is based on temporal correlation.

As mentioned, after processing the second LSTM layer, its output is applied as an input in step 5, which involves passing the information to a dense layer of predefined size, which aims to organize all past data and thus join with their probabilities of each class to present the general classification of the implemented strategy finally. This general classification involves the generation of the confusion matrix and the metrics table. As mentioned, the dense layer has a size parameter of 8 to obtain the classification for predefined emotions, and this quantity is the emotions available in the applied database.

4 Experimental Results

This section aims to present the results obtained with the proposed strategy. For these results, different experimental protocols were defined. The test and training database split applied in all experiments was 80% for training and 20% for testing.

The first approach (Experiment 1) used the entire database without any data filter, with all audio samples without division by genre or intensity. The objective is to evaluate the overall performance of the developed strategy. The second (Experiment 2) divided the database by audio intensity, which is high and low. These two audio intensity ratings are identified for each file within the database, so creating the filter from that information is easy. It is important to note that the *neutral* emotion does not have one of the intensities and therefore has a different number of samples in the database. The third approach (Experiment 3) ensures that the actors applied in the training phase are not present in the testing phase. For this purpose, we have used *LOSO* (Leave-One-Subject-Out). The database comprises 24 actors, and using such a scheme, at each iteration, the training phase is performed with 23 actors, leaving one actor out for testing. This procedure was repeated to consider all actors as testing. No additional filters were applied to the data in Experiment 3.

As mentioned in the previous section, some configuration parameters of the STFT algorithm and the CNN structure were investigated to determine the impact on the accuracy of the method. The choice of values for these parameters and the possible implications of the proposed method were also crucial in our study. Based on preliminary experiments, the best STFT parameters selected for the proposed method were $S_r = 8\,\mathrm{kHz}$, $N_{FFT} = 1024$, $w_l = 512$, $h_l = 256$ e $offset = 0.3$.

4.1 Experiment 1 - Results for All RAVDESS Database

In this experiment, we apply the entire database without any data filter, with all the audio samples, without splitting by genre or intensity. Table 2 shows the results of the method for *precision, recall,* and *F1-Score* metrics. Mathematically, the F1-Score is computed from *precision* and *recall* metrics and is the most used metric for unbalanced datasets. Table 3 presents the experiment's result in the format of a confusion matrix. The confusion matrix aims to visualize the general behavior of the method. The matrix compares the target values to those the machine learning model predicted. As discussed in Sect. 3, the RAVDESS database is not balanced. We can observe in Table 3 the differences among the classes.

Based on the F1-Score metric presented in Table 2, the three best-classified emotions, with values greater than 80%, are neutral, calm, and disgust. The other five emotions had performance greater than 70%. The analysis shows that the method performed well in classifying the eight emotions, achieving an accuracy of 83%. The database imbalance may influence the performance of specific emotions.

Table 2. Experiment 1 - Performance Metrics

Emotion	Precision	Recall	F1-Score
Neutral	0.85	0.92	0.88
Calm	0.92	0.76	0.83
Happy	0.73	0.69	0.71
Sad	0.71	0.82	0.76
Angry	0.88	0.61	0.72
Fearful	0.67	0.88	0.76
Disgust	0.89	0.86	0.88
Surprise	0.76	0.69	0.72
Accuracy			**0.83**

It is possible to verify that the three best emotions classified only two emotions continue to maintain their performance by the confusion matrix, with the

Table 3. Experiment 1 - Confusion Matrix

		Predicted Class								
		Neutral	Calm	Happy	Sad	Angry	Fearful	Disgust	Surprise	Total
True Class	Neutral	**34**	0	1	0	1	0	0	1	37
	Calm	0	**35**	1	3	0	6	1	0	46
	Happy	1	0	**27**	6	1	2	0	2	39
	Sad	0	0	3	**36**	0	4	1	0	44
	Angry	2	1	1	0	**14**	0	1	4	23
	Fearful	0	2	1	0	0	**30**	0	1	34
	Disgust	0	0	2	2	0	0	**25**	0	29
	Surprise	3	0	1	4	0	3	0	**25**	36
	Total	40	38	37	51	16	45	28	33	

highest values in the main diagonal, and we have the change of emotion from disgust to sad. This change in emotion concerning the confusion matrix analysis is due to the already mentioned fact that the baseline is unbalanced and emotions have a large difference in the number of test samples. Thus, the disgust emotion has fewer samples than the sadness emotion. However, the disgust emotion has better TP, TN, FN, and FP values than the sad emotion. These values directly influence the formulas of the metrics in Table 2, so we have this difference of the highest-ranked emotions.

4.2   Experiment 2 - Database Divided by Intensity

In this experiment, we investigate the performance of the multiview method when the audios were split by intensity. The RAVDESS database has two intensities: high and low. These intensity differences in the recording of emotions can influence the spectral characteristics of the audio. Table 4 shows the results for high intensity. The accuracy of the method improved regarding Experiment 1. However, considering the F1-Score metric, the emotion *calm* achieved 95%. Three emotions obtained a performance between 80% and 90%. The other four emotions remain above 70%.

Table 5 presents the results obtained for low intensity. In this case, the overall performance was reduced, and the accuracy was 80%. Only two emotions obtained an F1-Score higher than 80%: *happy* and *surprised*. The separation of the database by intensity demonstrates how the audio characteristics can affect the method's performance. This can lead to specific optimizations for each scenario, low or high intensity.

Although the neutral emotion lacks high intensity, it was included in the assessment to maintain the set of 8 emotions in the analysis. Comparing Tables 4 and 5, we can identify the effect of the architecture on this emotion when optimized for the classification of high and low intensities.

Table 4. Experiment 2 - Performance Metrics for High Intensity

Emotion	Precision	Recall	F1-Score
Neutral	0.79	0.71	0.74
Calm	0.90	0.98	0.95
Happy	0.79	0.81	0.79
Sad	0.80	0.77	0.70
Angry	0.82	0.74	0.75
Fearful	0.88	0.80	0.84
Disgust	0.88	0.86	0.89
Surprise	0.92	0.89	0.87
Accuracy			**0.85**

Table 5. Experiment 2 - Performance Metrics for Low Intensity

Emotion	Precision	Recall	F1-Score
Neutral	0.8	0.8	0.75
Calm	0.82	0.98	0.85
Happy	0.78	0.83	0.80
Sad	0.73	0.72	0.69
Angry	0.79	0.72	0.65
Fearful	0.78	0.73	0.75
Disgust	0.85	0.71	0.76
Surprise	0.82	0.86	0.84
Accuracy			**0.80**

4.3   Experiment 3 - LOSO Protocol

In this experiment, we apply the LOSO protocol in the context of the actors present in the database. This protocol evaluates the method's behavior without one of the 24 actors in the RAVDESS database. In this experimental protocol, an actor is excluded from the model during the training phase and only used in the test phase. This scenario is the most relevant from the point of view of practical applications, as it avoids any bias during network training and can perform well when we need to identify new actors' emotions. The results of this experimental protocol are shown in Fig. 3. The x-axis indicates which actor is excluded from the training process, and on the y-axis, we have the accuracy value obtained by the proposed strategy. The average accuracy obtained by the method was 79%. This value is quite relevant compared to the values obtained in previous experiments, as it also remains above 70%. In Fig. 3, even actors are women, and odd actors are men. Analyzing the metric results for the two sets of actors, the two sets reached the same average accuracy. Therefore, it is possible to conclude that the gender analysis had a low direct impact on the achieved performance, and the method behaved similarly. A significant result from this experimental protocol, which we can see in the figure, is that with the exclusion of some actors, the architecture performance significantly improved, reaching an accuracy of up to 88%. It means that these actors, when included in the training process, significantly degrade the method's performance. A possible reason for this degradation is that the audios of these actors have specific characteristics that affect the melspectrogram. Therefore, an important conclusion is that we can investigate and identify which characteristics are causing this problem to propose future improvements in the architecture and the audio pre-processing.

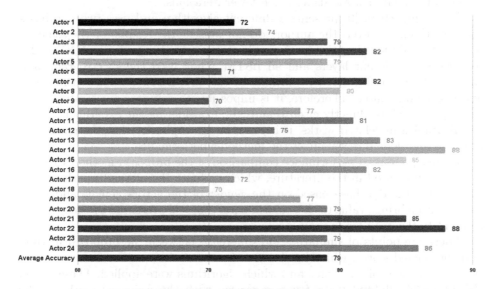

Fig. 3. Results for LOSO Protocol- Accuracy

5 Discussions

In [8], a different algorithm architecture was presented, using the audio spectrogram for feature extraction. The work presents three validation scenarios for classification using 2600, 2900, and 5000 epochs. This high number of interactions increases the computational complexity of the method. They also segment the audio database based on the audio intensity characteristic (low or high) for the performance analysis. We evaluated our proposed strategy under the same scenarios for a fair comparison. The results are shown in Table 6. The proposed approach reached a better accuracy with much fewer interactions using the low-intensity audio samples from the database. It is worth mentioning that they are different architectures. However, the number of interactions is an important

Table 6. Performance Comparison

Strategy	Audio Intensity	Epochs	Accuracy
Proposed Method	Low	150	80.00%
[8]	Low	2900	68.23%
Proposed Method	High	150	85.00%
[8]	High	5000	86.31%

point. For high-intensity audio samples, our proposed strategy reached a closer value to the related work, using much fewer iterations.

The approach in [9] presents a different algorithm and uses two databases in its method. However, the authors do not present the results for the separate databases, only for the combination. The authors in [10] present a CNN architecture with different layers and an LSTM. They used the spectrogram, and their analyzes employ the two formats of the RAVDESS database, the music and speech databases. Therefore, it is impossible to compare our strategy with those presented in directly citeb10 and [10]. Considering the experimental protocols applied in related works, it is essential to emphasize that the number of files and emotions directly affect the method's performance.

Table 7 presents a performance comparison of the proposed method with the strategies presented in the literature, which evaluate the method's performance by dividing the database based on the gender of the actors, female and male. This division brings value to the experimental protocol presented by the methods to identify the need for improvements or equality in processing.

There are points of similarity and difference between the proposed method and the related works, which involve the characteristics used for feature extraction, the number of emotions, and which algorithms were applied. These points should be highlighted for a fair comparison with the proposed method. The method in [6] has an approach similar to the present work in terms of algorithm and database but with other characteristics of the audio and classifying fewer emotions. The method in [7] was the basis for the proposed method since it uses spectral characteristics, a CNN architecture, and eight emotions for classification. Additionally, they must apply audio pre-processing to improve CNN performance.

Table 7. Performance Comparison

Strategy	Accuracy
Proposed Method	79.00%
[6]	76.20%
[7]	70.00%
[8]	75.50%

Consider the first research question, which was: *Is the fusion of multiview representations provided by different CNN architectures considering temporal information a promising strategy to recognize audio emotions?*. By analyzing the results previously presented, they emphasize the relevance of the implemented strategy and how the different representations improve the method. Therefore, we can conclude that the answer is yes.

The second research question was: *What is the impact of the audio intensity on the model based on the multiview representation proposed for the recognition of emotions?*. The answer is that it was possible to verify that the variation of the intensities directly interferes with the method's performance, as indicated by accuracy and F1-score metrics.

The third research question was: *What is the impact of the LOSO protocol on the model based on the proposed multiview representation?*. On average, the impact on the accuracy was minimal, and the performance was slightly inferior to the results obtained in Experiment 1. However, an important result was to identify the actors who degraded the method's performance. This observation makes room for future research to identify the characteristics that degrade the method, thereby enabling improvements in its architecture and audio preprocessing stages.

6   Conclusion

In this paper, we address the problem of speech emotion recognition using machine learning approaches. We focus on improving ML strategy using a combination of algorithms. We do not assume any audio preprocessing steps and consider classifying 8 emotions. Measurable results, such as precision and confusion matrix, helped to understand that the method is efficient and contributes to the SER research area using deep learning techniques. Currently we can consider that the work is in a position to add improvements to the area of study for emotion recognition with few raw processing steps and achieve relevant results. Therefore, in future steps, we plan to investigate the performance of the method using different databases and adjust the algorithm parameters to improve accuracy. We can also incorporate audio preprocessing steps to improve feature extraction and overall SER performance.

Acknowledgements. This study was financed in part by the Coordenação de Aperfeiçoamento de Pessoal de Nível Superior - Brasil (CAPES) - Finance Code 001. This work was partially supported by Conselho Nacional de Desenvolvimento Científico e Tecnológico - CNPq (Proc. 311065/2020-1).

References

1. Kulkarni, K., et al.: Automatic recognition of facial displays of unfelt emotions. IEEE Trans. Affect. Comput. **12**(2), 377–390 (2021). https://doi.org/10.1109/TAFFC.2018.2874996
2. Aleedy, M., Shaiba, H., Bezbradica, M.: Generating and analyzing chatbot responses using natural language processing. Int. J. Adv. Comput. Sci. Appl. (IJACSA) **10**(9) (2019). https://doi.org/10.14569/IJACSA.2019.0100910
3. Loris. www.loris.ai/company/
4. Das, A., Nair, K., Bandi, Y.: Emotion detection using natural language processing and ConvNets. In: Shukla, S., Gao, X.Z., Kureethara, J.V., Mishra, D. (eds.) Data Science and Security. LNNS, vol. 462, pp. 127–135. Springer, Singapore (2022). https://doi.org/10.1007/978-981-19-2211-4_11
5. Cowen, A.S., Keltner, D.: Self-report captures 27 distinct categories of emotion bridged by continuous gradients. Proc. Nat. Acad. Sci. USA **114**(38), E7900–E7909 (2017). https://doi.org/10.1073/pnas.1702247114. Epub 5 September 2017. PMID: 28874542. PMCID: PMC5617253
6. Rajak, R., Mall, R.: Emotion recognition from audio, dimensional and discrete categorization using CNNs. In: TENCON 2019–2019 IEEE Region 10 Conference (TENCON), Kochi, India, pp. 301–305 (2019). https://doi.org/10.1109/TENCON.2019.8929459
7. Mustaqeem, K.S.: A CNN-assisted enhanced audio signal processing for speech emotion recognition. Sensors **20**(1), 183 (2020). https://doi.org/10.3390/s20010183
8. Slimi, A., Hamroun, M., Zrigui, M., Nicolas, H.: Emotion recognition from speech using spectrograms and shallow neural networks. In: Proceedings of the 18th International Conference on Advances in Mobile Computing & Multimedia (MoMM 2020), pp. 35–39. Association for Computing Machinery, New York, NY, USA (2021)
9. Gupta, M., Chandra, S.: Speech emotion recognition using MFCC and wide residual network. In: 2021 Thirteenth International Conference on Contemporary Computing (IC3-2021), pp. 320–327. Association for Computing Machinery, New York, NY, USA (2021). https://doi.org/10.1145/3474124.3474171
10. Ayadi, S., Lachiri, Z.: A combined CNN-LSTM network for audio emotion recognition using speech and song attributs. In: 2022 6th International Conference on Advanced Technologies for Signal and Image Processing (ATSIP), Sfax, Tunisia, pp. 1–6 (2022). https://doi.org/10.1109/ATSIP55956.2022.9805924
11. Livingstone, S.R., Russo, F.A.: The Ryerson Audio-Visual Database of Emotional Speech and Song (RAVDESS) (2018)
12. Deckmann, S.M., Pomilio, J.A.: Analysis of discretized signals. in Electric Power Quality Assessment - UNICAMP (2020)
13. Raffel, C., Liang, D., Ellis, D.P.W., Nieto, O.: librosa: audio and music signal analysis in Python. In: Proceedings of the 14th Python in Science Conference (2015)

Detection of Covid-19 in Chest X-Ray Images Using Percolation Features and Hermite Polynomial Classification

Guilherme F. Roberto[1(✉)], Danilo C. Pereira[2], Alessandro S. Martins[2],
Thaína A. A. Tosta[3], Carlos Soares[1], Alessandra Lumini[4],
Guilherme B. Rozendo[4,5], Leandro A. Neves[5], and Marcelo Z. Nascimento[6]

[1] Faculty of Engineering, University of Porto (FEUP), Porto, Portugal
`guilhermefr@fe.up.pt`
[2] Federal Institute of Education, Science and Technology of Triângulo Mineiro (IFTM), Ituiutaba-MG, Brazil
[3] Science and Technology Institute, Federal University of São Paulo (UNIFESP), São José dos Campos-SP, Brazil
[4] Department of Computer Science and Engineering (DISI), University of Bologna, Cesena, Italy
[5] Department of Computer Science and Statistics (DCCE), São Paulo State University (UNESP), São José do Rio Preto-SP, Brazil
[6] Faculty of Computer Science (FACOM), Federal University of Uberlândia (UFU), Uberlândia-MG, Brazil

Abstract. Covid-19 is a serious disease caused by the Sars-CoV-2 virus that has been first reported in China at late 2019 and has rapidly spread around the world. As the virus affects mostly the lungs, chest X-rays are one of the safest and most accessible ways of diagnosing the infection. In this paper, we propose the use of an approach for detecting Covid-19 in chest X-ray images through the extraction and classification of local and global percolation-based features. The method was applied in two datasets: one containing 2,002 segmented samples split into two classes (Covid-19 and Healthy); and another containing 1,125 non-segmented samples split into three classes (Covid-19, Healthy and Pneumonia). The 48 obtained percolation features were given as input to six different classifiers and then AUC and accuracy values were evaluated. We employed the 10-fold cross-validation method and evaluated the lesion sub-types with binary and multiclass classification using the Hermite Polynomial classifier, which had never been employed in this context. This classifier provided the best overall results when compared to other five machine learning algorithms. These results based in the association of percolation features and Hermite polynomial can contribute to the detection of the lesions by supporting specialists in clinical practices.

Keywords: Percolation · Chest X-ray images · Covid-19 · Handcrafted features · Computer vision

© Springer Nature Switzerland AG 2024
V. Vasconcelos et al. (Eds.): CIARP 2023, LNCS 14469, pp. 163–177, 2024.
https://doi.org/10.1007/978-3-031-49018-7_12

1   Introduction

The novel coronavirus (Sars-CoV-2) has been first reported in Wuhan, at the Chinese province of Hubei at the end of the year 2019 and has rapidly spread to other countries and become a worldwide pandemic, having the World Health Organization (WHO) declare a global emergency [25]. The virus infects lung alveolar epithelial cells, causing the disease named Covid-19, wherein its initial clinical sign is often pneumonia, although gastrointestinal symptoms and asymptomatic infections have also been reported [29].

According to the Coronavirus Research Center of the Johns Hopkins University, over 545 million cases have already been reported worldwide, which have so far resulted in more than 6.88 million deaths [6]. The conventional diagnosis is made through a technique called reverse transcription-polymerase chain reaction (RT-PCR) [4]. However, this procedure relies on the existence of material kits, which can be costly, and there is also a contamination risk due to the proximity between medical staff and the patient at the moment of the exam [17]. Medical imaging techniques such as chest X-rays and CT-scans have also performed a major role in this pandemic, as the virus usually infects the lungs, the images obtained from these exams can be evaluated by radiologists in order to verify whether the patient is infected with Sars-CoV-2 and its severity. Besides being safe and easily accessible, signs of Covid-19 could be observed in chest X-ray images even in asymptomatic cases [1,5].

Since the start of the pandemic, the work required to evaluate all incoming chest X-ray images has become time-consuming and prone to inaccuracies, as misdiagnosis can happen due to lack of experience of some radiologists in not yet having deep knowledge of how a Sars-CoV-2 infection evolves [17]. Therefore, Artificial Intelligence tools, mostly based on deep learning, have been developed recently in order to reduce the workload of radiologists and provide support to the diagnosis of Covid-19 [9,20,27,28]. Due to good performances obtained in the detection of pneumonia in chest X-ray images, convolutional neural networks (CNN) have promptly been applied to the detection of Covid-19 as soon as the first chest X-ray datasets were available [9,17,20,27].

One of the earliest frameworks designed for the detection of Covid-19 was COVIDX-Net [9], which included seven different CNN models and was able to obtain accuracy values up to 90%. However, this framework had limited training since at the time of its publication, only 25 Covid-19 X-ray image samples were available. As the number of Covid-19 samples in public datasets began to grow, more complex solutions were proposed. In [20] a 17-layer CNN named DarkCovidNet was proposed. This model was able to obtain an accuracy of 98.08% for binary classification (Covid-19 vs. No findings) and 87.02% for multiclass classification (Covid-19 vs. No findings vs. Pneumonia), which was later improved to 99.27% by [27] after the application of the Fuzzy Color approach on the preprocessing stage and the use of two different CNN models: MobileNetV2 and SqueezeNet. Besides, other complex CNN models have been used for this task. In [12], InceptionV3, Xception and ResNeXt models have been evaluated on a dataset containing 6,432 samples of chest X-rays from Covid-19, Pneumonia and Healthy cases. After splitting the dataset into training (85%) and validation (15%) subsets the models were evaluated separately. The model that provided the highest accuracy was the Xception, with an value of 97.97%. Recently, a modified version of the DarkCovidNet which contains two extra convolutional layers was able to obtain an accuracy of 94.18% for multiclass classification [22], although the number of samples in the dataset has significantly increased from when the first DarkCovidNet model was published.

However, it can be noted that the majority of approaches proposed for detecting Covid-19 is purely based on deep learning techniques, and although CNN models became popular due to the high accuracy rates they have provided on several types of image datasets, they are also known to have a significant higher computational cost than more traditional techniques, which can provide features known as handcrafted. Few researchers have chosen this approach so far, and these have shown that handcrafted features can perform as well as learned features in the detection of Covid-19, as they are often good indicators of shape, size and regularity of structures present in medical images [13]. For instance, in [28], the authors presented a simple local binary pattern (LBP) extractor to obtain features from a chest X-ray dataset composed by 87 Covid-19 samples and 147 healthy samples. The most relevant features were selected using the ReliefF feature selector algorithm and an average accuracy of 99.55% was obtained using an SVM classifier and 10-fold cross validation. In [11], preprocessing multiresolution approaches such as shearlet transform were applied to the chest X-ray images prior to feature extraction. Then, entropy and energy features were obtained from the multiresolution images and given as input for an extreme learning machine (ELM) classifier. The average accuracy obtained was of 99.29%. Other classical handcrafted features such as Haralick's have been used in computer vision problems regarding the detection of Covid-19, providing accuracy values up to 98.04% [7,16]. Apart from the computational cost, researchers often choose handcrafted features as these often perform better on small datasets, as indicated in [19], wherein the authors obtained robust features from 260 images using four different feature extraction techniques. Then, stacked auto-encoder (SAE) and principal component analysis (PCA) were applied to reduce the number of features to 20. These features were given as input to a SVM classifier which provided an accuracy of 94.23%.

Although these works have presented great results, there seems to be a lack of information on how other types of handcrafted features could perform. Therefore, we have decided to investigate the performance of percolation features for detecting Covid-19 in chest X-ray images. These type of features are fractal-based and have recently been applied for the classification of colored histology images [23]. However, percolation features have not yet been applied for the classification of X-ray images, which are greyscale. In fact, fractal descriptors for classifying Covid-19 images have only been used so far in combination with CNN [8]. Besides feature extraction through percolation properties, we also propose the use of a polynomial classifier. Although several methods contribute in solving biomedical classification problems, the choice of the appropriate technique for this task depends on the classifier properties [21]. Polynomial classifiers have been showing promising results, especially when dealing with non-linearly separable data. This algorithm is a parameterized method that expands exponentially its polynomial basis according to the number of elements in the vector of characteristics and the degree of the function. One of the most recent polynomial classifiers available is the Hermite polynomial (HP), which consists on an expansion of the regular polynomial basis to a larger space in order to analyze complex contexts such as the one explored here [15].

We propose in this paper an approach for detecting Covid-19 in chest X-ray images based on local and global percolation features, using an adaptation of the method described in [23]. These features were then classified by the HP algorithm that allows the evaluation of data distribution of images from different classes. This is the first time wherein these features are used to describe properties of chest X-ray images. Another contribution of our work is that it defines the best association after applying well-known algorithms of artificial intelligence to evaluate the performance of the proposed approach.

The contributions of this paper can be summarised as follows: investigation of the behaviour of local and global percolation features for the feature extraction from chest X-ray images; association of an approach based on the HP algorithm and percolation features capable of evaluating classes in binary classifications and; a comparison of well-known algorithms of artificial intelligence with the proposed feature extraction approach to evaluate the classification performance.

This paper is organised as follows. In Sect. 2, we detail the methodologies used in the various stages of the proposed approach and the datasets used in this paper. We discuss the obtained results in Sect. 3 and we conclude the paper on Sect. 4.

2   Materials and Methods

2.1   Image Database

V7 Labs Dataset. The first analyzed dataset was made available by [14]. This dataset is composed of 2,002 chest X-ray images split into two classes (401 Covid-19 infections and 1,601 healthy lung images) with dimensions ranging from 148×122 to 4030×3133. Lung segmentation was performed by human-made annotations. Images containing graphic indicators such as arrows inside the lung area were disregarded. In Fig. 1, one example from each class and their respective segmented images are presented.

Ozturk et al. Dataset. The second analyzed dataset consists in a composition made by [20] from two other datasets: one containing 125 Covid-19 samples; and other composed of 500 pneumonia samples and 500 samples with no findings. Therefore, this datasets consists of a multiclass classification problem. There were no annotations available for lung segmentation in this dataset, so the images were used as they were originally available in the repository. Examples from each class of this dataset are shown in Fig. 2.

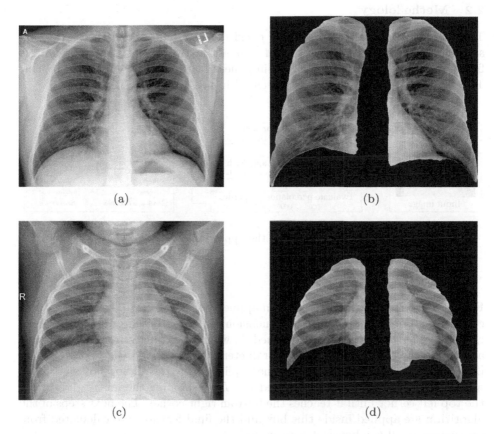

Fig. 1. Images from the v7 Labs dataset. (a) chest X-ray from an individual infected with Sars-CoV-2; (b) segmented version of the image in (a); (c) chest X-ray from an individual with a healthy lung; (d) segmented version of the image in (c).

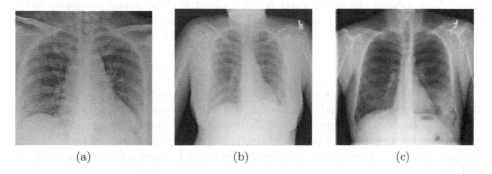

Fig. 2. Images from the Ozturk *et. al* dataset. (a) sample of a Covid-19 case; (b) sample of a healthy case; (c) sample of a pneumonia case.

2.2   Methodology

The approach proposed in this paper consists in applying the feature extraction algorithm available at [23] to obtain numeric features from the X-ray images from both datasets and then providing these features as input to a classifier. The overview of our approach is presented in Fig. 3.

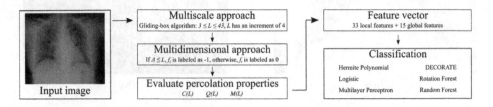

Fig. 3. Overview of the applied methodology.

Feature Extraction. The algorithm applied to obtain the percolation features consists in a multiscale and multidimensional method based on the analysis of the image under different scales valued L, wherein L ranges from 3 to 43 with an increment of 4, which are the parameters that provided the best results in [23]. This analysis is made through the application of the gliding-box algorithm, wherein for each iteration, a box sized $L \times L$ overlapping the image glides from the top left corner until it reaches the bottom right corner. The next steps of the algorithm are applied inside this box and the final features are calculated from an average of all the boxes of a given scale L.

Three percolation properties are verified: average number of clusters $(C(L))$; occurrence of percolation $(Q(L))$ and; average size of the largest cluster $(M(L))$. In the original paper, these properties were obtained after calculating the distance Δ between the RGB intensities of the center pixel f_c and any other given pixel f_i on each of the boxes sized $L \times L$. Since the chest X-ray images on both datasets are greyscale, we have adapted the distance calculation formula to the presented in Eq. 1:

$$\Delta = |f_i - f_c| .\qquad(1)$$

This procedure outputs a set of 33 local features based on the three described percolation properties, and 15 global features based on the metrics defined by area (A), skewness (S), area ratio (A_R), maximum point (Max) and scale of the maximum point (σ). The global features were obtained using the approach proposed by [2]. A summary of the 48 obtained features can be observed on Table 1.

Table 1. Summary of the 48 extracted features.

Local Features (33)			Global Features (15)		
$C(3)$	$Q(3)$	$M(3)$	$A(C)$	$A(Q)$	$A(M)$
$C(7)$	$Q(7)$	$M(7)$	$S(C)$	$S(Q)$	$S(M)$
$C(11)$	$Q(11)$	$M(11)$	$A_R(C)$	$A_R(Q)$	$A_R(M)$
$\vdots$	$\vdots$	$\vdots$	$Max(C)$	$Max(Q)$	$Max(M)$
$C(43)$	$Q(43)$	$M(43)$	$\sigma(C)$	$\sigma(Q)$	$\sigma(M)$

Classification. One of the main contributions of this paper consists on the evaluation of percolation features obtained from chest X-ray images when given as input to a polynomial classifier.

The polynomial classifier is a supervised method capable of non-linearly expanding an input feature vector into a higher dimension. This strategy allows one to obtain linear approximations in this space that is capable of labeling the input data for a desired output [24]. However, regular polynomial classifiers have a limited basis, which hinders the evaluation of the features.

Hermite polynomials (HP) are able to generate a complete orthogonal basis of Hilbert space that satisfies the orthogonality and completeness conditions of the family of elements of that space [10]. The orthogonality condition, Eq. (2), implies that any inner product between a pair of orthogonal polynomials, $P_m(x)_{n=0}^{\infty}$ and $P_n(x)_{n=0}^{\infty}$, of different degrees is equal to zero when the base function used $w(x)$ and interval $[a, b]$ are the same [3].

$$< P_m(x), P_n(x) >= \int_a^b P_m(x)P_n(x)w(x)dx = 0, \forall m \neq n. \tag{2}$$

According to [26], HP is orthogonal in the interval ($-\infty, \infty$) with respect to the base-function and enables gains at function approximation. Mathematically, HP can be defined as Eq. (3):

$$H_n(x) = (-1)^n e^{-x^2/2} \frac{d^n}{dx^n}[e^{-x^2/2}]. \tag{3}$$

A computationally efficient way to obtain the polynomial is to use the recurrence relation. In this case, only the first two terms are needed, so the others can be calculated iteratively. HP recurrence relation is expressed by Eq. (4) [30].

$$H_{n+1}(x) - xH_n(x) + nH_{n-1}(x) = 0, \quad n = 0, 1, 2, \dots \tag{4}$$

Finally, the polynomial classifier can be defined according to Eq. (5):

$$g(\mathbf{x}) = \mathbf{a}^T H_n(\mathbf{x}), \tag{5}$$

where $\mathbf{a}$, is the vector of coefficients at polynomial basis function, $H_n(\mathbf{x})$ is the base-function of Hermite and n corresponds to the order or degree of the polynomial function.

The polynomial classifier requires two steps for operation: training and testing. In the first stage, the training data vector is expanded through polynomial expansion to increase the separation of the different classes in the newly generated vector space. In the second part, a mapping of the expanded polynomial vectors to an optimal output sequence to minimize an objective criterion is performed. The mapped parameters represent the weights of the polynomial classifier which are known as class models [18].

In this paper, the 48 obtained features were given as input to a classifier in order to predict the class which the image represents. The order of the base-polynomial was set empirically and the best results were achieved with the $4th$ order and four features for class separation. In addition, the cross-validation method with 10-folds was employed, where 90% of the dataset was used for training and 10% on model prediction and evaluation.

Performance Evaluation. We evaluated the performance of our proposed approach in three stages. Firstly, we evaluated the features by applying the Mann-Whitney U test [15]. Then, the HP classifier was evaluated by comparing the performance of global and local features separately and then on a single feature vector. The performance of the HP algorithm was also compared with other five different classifiers that are based on three of the main supervised machine learning approaches: function-based, ensemble learning and tree-based. The chosen algorithms were logistic (LGT), multilayer perceptron (MLP), DECORATE (DEC), rotation forest (RoF) and random forest (RaF).

All classifications were performed in the software Weka v3.6.13 using the default parameters for the classifiers and a 10-fold cross validation approach. The metrics used for evaluation were the accuracy (ACC) as it is the most common metric and has been used in every related work previously presented; and the area under the ROC curve (AUC), since it is a more suitable metric to evaluate performance with imbalanced data.

All tests were performed on an Intel i5-9300H at 2.40GHz and 8GB of RAM. The percolation features were obtained using Matlab R2019a.

3   Results and Discussion

3.1   Feature Evaluation

In order to evaluate the significance of the features and how discriminating they are on the different tested scenarios, we applied the Mann-Whitney U test [15]. The results are shown in Fig. 4, with the empirical cumulative distribution function (CDF) for the p-values. For the v7 Labs dataset, all of the 15 global features had p-values close to zero, as seen on Fig. 4a. As for the local features, approximately 85% of these had p-values smaller than 0.1.

For the Ozturk *et al.* dataset, the two classification scenarios were evaluated separately. In the first one, presented in Fig. 4b, the local features have provided slightly more significant results, with p-values smaller than 0.2 for 80% of the features. On the other hand, in the second scenario presented in Fig. 4c, the smallest p-values were provided by the global features, as approximately 80% of these resulted in p-values close to zero. For better understanding how these features behave when given as input to the HP classifier, we performed the evaluations described in the following subsection.

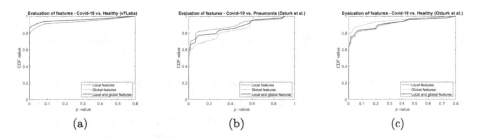

(a)   (b)   (c)

Fig. 4. The empirical cumulative distribution function of the p-values of the different feature sets: (a) v7Labs dataset; (b) Ozturk *et al.* dataset (Covid-19 vs. Pneumonia); (c) Ozturk *et al.* dataset (Covid-19 vs. Healthy).

3.2   Performance of the HP Classifier

We first evaluate the HP classifier using different orders for the polynomial function. This evaluation was made using the full set of global and local features in both Covid-19 vs. Healthy (CH), Covid-19 vs. Pneumonia (CP) and Covid-19 vs. Healthy vs. Pneumonia (CHP) scenarios. The results obtained for 1st (HPG1), 2nd (HPG2), 3rd (HPG3) and 4th (HPG4) orders are shown in Fig. 5.

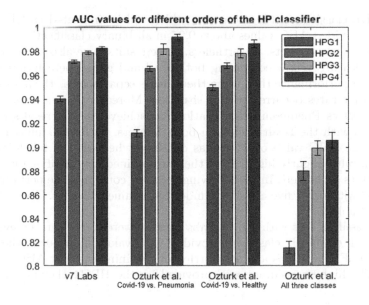

Fig. 5. Evaluation of the HP classifier with different orders.

Similarly to the results presented in [15], the 4th order for the polynomial function has presented the best results in all scenarios. Therefore, we chose the HPG4 for the following experiments. The tested scenarios were: Covid-19 vs. Healthy (v7 Labs dataset); Covid-19 vs. Pneumonia, Covid-19 vs. Healthy, and Covid-19 vs. Healthy vs. Pneumonia (Ozturk *et al.* dataset). The results obtained with AUC and ACC metrics and the standard deviation (SD) values are shown in Table 2.

Table 2. Evaluation of the HPG4 classifier using features obtained from the datasets.

Dataset	Groups	Features	AUC	SD	ACC	SD
v7 Labs	Covid-19 vs. Healthy	Local	0.9795	0.0011	97.09	0.0024
		Global	0.9712	0.0020	95.80	0.0031
		All	**0.9826**	0.0013	**97.89**	0.0015
Ozturk et al.	Covid-19 vs. Healthy	Local	0.9803	0.0047	96.83	0.0050
		Global	0.9734	0.0054	94.62	0.0055
		All	**0.9861**	0.0033	**97.93**	0.0036
	Covid-19 vs. Pneumonia	Local	0.9869	0.0026	97.86	0.0042
		Global	0.9718	0.0042	94.64	0.0067
		All	**0.9917**	0.0023	**98.37**	0.0042
	All classes	Local	0.8929	0.0093	91.31	0.0061
		Global	0.8674	0.0051	89.36	0.0044
		All	**0.9057**	0.0065	**92.13**	0.0048

The HP classifier in the 4th order, which will be indicated as HPG4 from now, has provided AUC values above 0.97 in all binary classification scenarios, considering both datasets. Nevertheless, these results have also shown that the combined feature vector containing both local and global percolation features has a better performance than using the features separately, as the highest AUC and accuracy rates occurred in this situation. Moreover, it can be noted that the Covid-19 vs Pneumonia classification has achieved better results than the comparison with the Healthy class in both datasets. As for the multiclass classification, an AUC value of 0.906 was obtained when use the full feature set, which is a value slightly higher than the one obtained for classifying global and local features separately. In the following tests, we compared the performance of the HPG4 with other five classifiers under these same scenarios. The results are shown in Table 3.

The classifications with the v7 Labs Dataset were the first to be evaluated. Although all of the six classifiers provided AUC values above 0.96, the HPG4, RoF and RaF algorithms are noteworthy for obtaining the best AUC and ACC values. The highest accuracy was provided by the HPG4 classifier using the

Table 3. Comparison among HPG4 and other five classifier regarding the classification of the combined feature set.

Dataset	Groups	Classifier	AUC	SD	ACC	SD
v7 Labs	Covid-19 vs. Healthy	HPG4	0.9826	0.0013	**97.89**	0.0015
		LGT	0.9777	0.0016	96.23	0.0024
		MLP	0.9832	0.0020	95.71	0.0046
		DEC	0.9834	0.0019	95.38	0.0036
		RoF	0.9884	0.0012	96.35	0.0034
		RaF	**0.9891**	0.0007	95.84	0.0013
Ozturk et al.	Covid-19 vs. Pneumonia	HPG4	**0.9917**	0.0023	**98.37**	0.0042
		LGT	0.9760	0.0043	94.43	0.0049
		MLP	0.9660	0.0040	92.70	0.0092
		DEC	0.9403	0.0097	91.48	0.0077
		RoF	0.9720	0.0055	94.78	0.0040
		RaF	0.9653	0.0036	93.00	0.0046
	Covid-19 vs. Healthy	HPG4	**0.9861**	0.0033	**97.93**	0.0036
		LGT	0.9746	0.0071	95.43	0.0040
		MLP	0.9581	0.0069	93.79	0.0078
		DEC	0.9689	0.0074	95.67	0.0063
		RoF	0.9770	0.0074	95.67	0.0063
		RaF	0.9787	0.0024	96.18	0.0026
	All classes	HPG4	**0.9057**	0.0065	**92.13**	0.0049
		LGT	0.8790	0.0058	89.68	0.0028
		MLP	0.8275	0.0102	86.42	0.0088
		DEC	0.8376	0.0089	87.03	0.0090
		RoF	0.8669	0.0056	88.84	0.0046
		RaF	0.8440	0.0068	87.65	0.0061

combination of global and local features and consisted in an ACC value of 97.89. The highest AUC value, which was 0.9891, was obtained by the RaF classifier.

The same experiments were also performed on the Ozturk *et al.* dataset. Three situations were evaluated: Covid-19 vs. Pneumonia classification; Covid-19 vs. Healthy classification; and Covid-19 vs. Pneumonia vs. Healthy classification. For the Covid-19 vs. Pneumonia comparison, the best results were also obtained using the HPG4 classifier, although the RaF algorithm did not perform as well as in the previous evaluated dataset, despite still having the second best performance in terms of AUC and ACC out of the six evaluated classifiers. In the Covid-19 vs. Healthy comparison, the HPG4 classifier has also provided the best AUC and ACC values, overcoming the obtained accuracy for the same class comparison in the v7 Labs dataset, which not only contains a greater number of samples, but these are also segmented.

The AUC and ACC values for the classification of the Covid-19 and Healthy groups were slightly smaller than the ones obtained for the Covid-19 vs. Pneumonia classification. However, we believe these results are influenced by the dataset composition, since the images for healthy and pneumonia samples and images for Covid-19 samples come from different sources, since other papers have reported high accuracy values for both comparisons [27,28]. Finally, for multiclass classification, the HPG4 classifier provided the best AUC and ACC results by a margin of at least 0.02 regarding the AUC. This may indicate a better adaptability for multiclass classification than the other five tested classifiers, although further analysis with different datasets and features are required.

We have also applied the non-parametric Friedman test in order to verify the average performance of each of the six tested classification algorithms considering the 48 features as input based on the AUC values. The classifiers that provided the highest AUC averages were the HPG4 (0.9665) and RoF (0.9511). The p-values for all pair-wise comparisons (Conover) are presented on Table 4, wherein statistically significant differences ($\alpha < 0.05$) are highlighted in bold. Although most of the p-values do not indicate a clear significant statistical difference among the classifiers, the best average ranking was obtained for HPG4, which reinforces its suitability for the classification of this type of features for both datasets.

Table 4. p-values obtained from the application of the Friedman test for comparing the performance of the six classifiers.

p-values	HPG4	LGT	MLP	DEC	RoF	RaF	Avg. Rank
HPG4	–	0.2268	**0.0235**	**0.0355**	0.5381	0.4140	2.0
LGT	0.2268	–	0.2268	0.3102	0.5381	0.6804	3.5
MLP	**0.0235**	0.2268	–	0.8365	0.0782	0.1136	5.0
DEC	**0.0355**	0.3102	0.8365	–	0.1136	0.1621	4.75
RoF	0.5381	0.5381	0.0782	0.1136	–	0.8365	2.75
RaF	0.4140	0.6804	0.1136	0.1621	0.8365	–	3.0

Overall, both local and global percolation features have shown to be a good discriminant factor in the classification of chest X-ray images regarding the detection of Covid-19, providing AUC values above 0.95 in all tested binary scenarios, considering both datasets. For the multiclass classification, the HPG4 classifier was the only one able to obtain AUC and ACC values above 0.9. Therefore, the HPG4 and RoF classifiers appear to be more suitable for these types of features, since the highest AUC and ACC values were obtained from them.

On Table 5, an overview of some of the published computer vision methods aimed at the detection of Covid-19 in chest X-ray images is presented. Our proposed method is on par with the other approaches available in the literature and we believe that percolation features can provide a relevant contribution to upcoming research regarding classification of chest X-ray images. Moreover, our approach is based on the extraction of handcrafted features and does not require the application of deep learning algorithms, thus generating a smaller computational demand than CNN methods, also due to considerably smaller dimension of the final feature vector.

Table 5. Overview of different approaches for detecting Covid-19 in chest X-ray images.

Method	Approach	Classes	Segmented	Samples	Features	ACC
[9]	DenseNet201	2	No	50	1,920	90.00%
[20]	DarkCovidNet	2	No	625	338	98.08%
		3		1,125		87.02%
[27]	Fuzzy color, MobileNetV2, SqueezeNet	3	No	458	1,357	99.27%
[28]	LBP, ReliefF, SVM	2	No	234	1,459	99.55%
[12]	Xception	3	No	6,432	2,048	97.97%
[11]	Shearlet transform, Entropy and Energy, ELM	2	No	561	98	99.29%
[7]	Haralick, Zernike features, SVM	4	No	6309	168	89.78%
[16]	GLCM, SVM ensembles	2	Yes	1,710	90	97.04%
[19]	Robust features, SAE, PCA, SVM	6	No	260	20	94.23%
[22]	Modified DarkCovidNet	2	No	4,572	507	99.53%
		3		10,190		94.18%
Proposed	Percolation features, HPG4	2	Yes	2,002	48	97.89%
		2	No	625		98.37%
		2		625		97.93%
		3		1,125		92.13%

4   Conclusion

In this paper, we proposed the use of local and global percolation features to represent chest X-ray images on the task of detecting infections by Covid-19. The method is based on [23] and was adapted to greyscale images. The HPG4 classifier was able to provide the highest ACC values on both tested datasets. We also verified that while global features were more significant for classifying images from the v7 Labs dataset, there was a more balanced contribution between the global and local features sets in classifying images from the Ozturk *et al.* dataset.

These obtained results are compatible with other methods that have been published since the beginning of the pandemic. We have shown that percolation based features can differentiate among Covid-19, Pneumonia and Healthy chest X-ray samples and encourage its application on different datasets and classification problems.

For future works, we suggest the application of this approach on new chest X-ray datasets containing more Covid-19 samples as well as the evaluation of these features on images from CT-scans. Other fractal measures such as lacunarity and fractal dimension could be evaluated as indicators of the presence of Covid-19 in chest X-ray images. Different parameters to obtain the features such as the maximum value of L and the type of distance could also be tested in order to improve the obtained results.

Funding. This study was financed in part by the Coordenação de Aperfeiçoamento de Pessoal de Nível Superior - Brasil (CAPES) - Finance Code 001, National Council for Scientific and Technological Development CNPq (Grants #132940/2019-1, #313643/2021-0 and #311404/2021-9), the State of Minas Gerais Research Foundation - FAPEMIG (Grants #APQ-00578-18 and #APQ-01129-21), the State of São Paulo Research Foundation - FAPESP (Grant #2022/03020-1). and the project NextGenAI - Center for Responsible AI (2022-C05i0102-02), supported by IAPMEI, and also by FCT plurianual funding for 2020-2023 of LIACC (UIDB/00027/2020 UIDP/00027/2020).

References

1. Bandirali, M., et al.: Chest radiograph findings in asymptomatic and minimally symptomatic quarantined patients in codogno, italy during covid-19 pandemic. Radiology **295**(3), E7–E7 (2020)
2. Căliman, A., Ivanovici, M.: Psoriasis image analysis using color lacunarity. In: 2012 13th International Conference on Optimization of Electrical and Electronic Equipment (OPTIM), pp. 1401–1406. IEEE (2012)
3. Chihara, T.S.: An introduction to orthogonal polynomials. Courier Corporation (2011)
4. Corman, V.M., et al.: Detection of 2019 novel coronavirus (2019-ncov) by real-time rt-pcr. Eurosurveillance **25**(3), 2000045 (2020)
5. Cozzi, D., et al.: Chest x-ray in new coronavirus disease 2019 (covid-19) infection: findings and correlation with clinical outcome. Radiol. Med. (Torino) **125**, 730–737 (2020)
6. Dong, E., Du, H., Gardner, L.: An interactive web-based dashboard to track covid-19 in real time. Lancet. Infect. Dis **20**(5), 533–534 (2020)
7. Gomes, J.C., et al.: Ikonos: an intelligent tool to support diagnosis of covid-19 by texture analysis of x-ray images. Res. Biomed. Eng. 1–14 (2020)
8. Hassantabar, S., Ahmadi, M., Sharifi, A.: Diagnosis and detection of infected tissue of covid-19 patients based on lung x-ray image using convolutional neural network approaches. Chaos, Solitons Fractals **140**, 110170 (2020)
9. Hemdan, E.E.D., Shouman, M.A., Karar, M.E.: Covidx-net: a framework of deep learning classifiers to diagnose covid-19 in x-ray images. arXiv preprint arXiv:2003.11055 (2020)
10. Hooshmand Moghaddam, V., Hamidzadeh, J.: New hermite orthogonal polynomial kernel and combined kernels in support vector machine classifier. Pattern Recogn. **60**, 921–935 (2016)
11. Ismael, A.M., Şengür, A.: The investigation of multiresolution approaches for chest x-ray image based covid-19 detection. Health Inform. Sci. Syst. **8**(1), 1–11 (2020)
12. Jain, R., Gupta, M., Taneja, S., Hemanth, D.J.: Deep learning based detection and analysis of covid-19 on chest x-ray images. Appl. Intell. **51**(3), 1690–1700 (2021)
13. Kurmi, Y., Chaurasia, V., Ganesh, N.: Tumor malignancy detection using histopathology imaging. J. Med. Imaging Radiation Sci. **50**(4), 514–528 (2019)

14. v7 Labs: Covid-19 x-ray dataset (2020). www.github.com/v7labs/covid-19-xray-dataset
15. Martins, A.S., et al.: A hermite polynomial algorithm for detection of lesions in lymphoma images. Pattern Anal. Appli. 1–13 (2020)
16. Mohammed, S., Alkinani, F., Hassan, Y.: Automatic computer aided diagnostic for covid-19 based on chest x-ray image and particle swarm intelligence. Inter. J. Intell. Eng. Syst. **13**(5), 63–73 (2020)
17. Nayak, S.R., Nayak, D.R., Sinha, U., Arora, V., Pachori, R.B.: Application of deep learning techniques for detection of covid-19 cases using chest x-ray images: a comprehensive study. Biomed. Signal Process. Control **64**, 102365 (2021)
18. Neves, L.A., et al.: Multi-scale lacunarity as an alternative to quantify and diagnose the behavior of prostate cancer. Expert Syst. Appl. **41**(11), 5017–5029 (2014)
19. Öztürk, Ş, Özkaya, U., Barstuğan, M.: Classification of coronavirus (covid-19) from x-ray and ct images using shrunken features. Int. J. Imaging Syst. Technol. **31**(1), 5–15 (2021)
20. Ozturk, T., Talo, M., Yildirim, E.A., Baloglu, U.B., Yildirim, O., Acharya, U.R.: Automated detection of covid-19 cases using deep neural networks with x-ray images. Comput. Biol. Med. **121**, 103792 (2020)
21. Padierna, L.C., Carpio, M., Rojas-Domínguez, A., Puga, H., Fraire, H.: A novel formulation of orthogonal polynomial kernel functions for svm classifiers: the gegenbauer family. Pattern Recogn. **84**, 211–225 (2018). https://doi.org/10.1016/j.patcog.2018.07.010, www.sciencedirect.com/science/article/pii/S0031320318302280
22. Redie, D.K., et al.: Diagnosis of covid-19 using chest x-ray images based on modified darkcovidnet model. Evolutionary Intell. 1–10 (2022)
23. Roberto, G.F., Nascimento, M.Z., Martins, A.S., Tosta, T.A., Faria, P.R., Neves, L.A.: Classification of breast and colorectal tumors based on percolation of color normalized images. Comput. Graph. **84**, 134–143 (2019)
24. Shanableh, T., Assaleh, K.: Feature modeling using polynomial classifiers and stepwise regression. Neurocomputing **73**(10–12), 1752–1759 (2010)
25. Sohrabi, C., et al.: World health organization declares global emergency: a review of the 2019 novel coronavirus (covid-19). Int. J. Surg. **76**, 71–76 (2020)
26. Thangavelu, S.: Hermite and laguerre semigroups: some recent developments. In: Seminaires et Congres (to appear) (2006)
27. Toğaçar, M., Ergen, B., Cömert, Z.: Covid-19 detection using deep learning models to exploit social mimic optimization and structured chest x-ray images using fuzzy color and stacking approaches. Comput. Biol. Med. **121**, 103805 (2020)
28. Tuncer, T., Dogan, S., Ozyurt, F.: An automated residual exemplar local binary pattern and iterative relieff based covid-19 detection method using chest x-ray image. Chemometr. Intell. Lab. Syst. **203**, 104054 (2020)
29. Velavan, T.P., Meyer, C.G.: The covid-19 epidemic. Tropical Med. Intern. Health **25**(3), 278 (2020)
30. Zanaty, E., Afifi, A.: Generalized hermite kernel function for support vector machine classifications. Int. J. Comput. Appl. **42**(8), 765–773 (2020)

Abandoned Object Detection Using Persistent Homology

Javier Lamar Leon[1(✉)], Raúl Alonso Baryolo[2], Edel Garcia Reyes[3],
Rocio Gonzalez Diaz[4], and Pedro Salgueiro[1]

[1] Centro ALGORITMI - VISTA Lab, Universidade de Évora, Evora, Portugal
{jlamarleon,pds}@uevora.pt
[2] Microsoft Corporation, Redmond, USA
[3] GEOCUBA Enterprise, Havana, Cuba
[4] Applied Math Department of School of Computer Engineering,
Campus Reina Mercedes, University of Seville, Seville, Spain
rogodi@us.es

Abstract. The automatic detection of suspicious abandoned objects has become a priority in video surveillance in the last years. Terrorist attacks, improperly parked vehicles, abandoned drug packages and many other events, endorse the interest in automating this task. It is challenge to detect such objects due to many issues present in public spaces for video-sequence process, like occlusions, illumination changes, crowded environments, etc. On the other hand, using deep learning can be difficult due to the fact that it is more successful in perceptual tasks and generally what are called system 1 tasks. In this work we propose to use topological features to describe the scenery objects. These features have been used in objects with dynamic shape and maintain the stability under perturbations. The objects (foreground) are the result of to apply a background subtraction algorithm. We propose the concept the *surveillance points*: set of points uniformly distributed on scene. Then we keep track of the changes in a cubic region centered at each *surveillance points*. For that, we construct a simplicial complex (topological space) from the k foreground frames. We obtain the topological features (using persistent homology) in the sub-complexes for each cubical-regions, which represents the activity around the *surveillance points*. Finally for each *surveillance points* we keep track of the changes of its associated topological signature in time, in order to detect the abandoned objects. The accuracy of our method is tested on PETS2006 database with promising results.

Keywords: abandoned objects detection · surveillance points · persistent homology · simplicial complex

1 Introduction

Abandoned objects detection concerns with the localization of objects left in the scene. Because of the amount of different possible objects, as well as their

V. Vasconcelos et al. (Eds.): CIARP 2023, LNCS 14469, pp. 178–188, 2024.
https://doi.org/10.1007/978-3-031-49018-7_13

diverse characteristics, there is not a universal definition of abandoned object. For example, a non-moving person may be an abandoned object if is dead, but may be a person waiting for something, a car parked may be abandoned or not, in the worst case it may be a pump car. The lack of an exact definition, together with the complexity of real scenarios (airports, streets, public spaces, etc.) makes the task of detecting abandoned objects a hard and open problem [20,21].

Despite the difficulty of giving a complete definition, some authors have given their own versions [1,5,17,20], always biased by the complexity of the scenario or other assumptions. In this work we don't give a definition, but follow the one in [20] which, despite having drawbacks, is the most general among the cited before.

There are two main approaches used to solve the problem up to now, one is based on tracking [5,7,15] and the other is based on background modeling and subtraction [16,20,22]. The first one suffers from many difficulties, like correspondence, splitting, merging, occlusion and others, besides of being difficulty to track objects in crowded scenes [22]. The second approaches have drawbacks too, like occlusions, but we follow the hypothesis that this direction is less error prone, and a good selection of a background subtraction algorithm according to the characteristics of the scene may give good classification rates.

Other approach is trying to model the social context of the scene in order to extract relations among groups of persons, and the relationship between abandoned object and its owner. This direction is recent and only one work introduces this novelties [6]. It is important to state that this approach seems promising, but is needed of more improvements yet.

In this work we use a proper implementation of the background subtraction algorithm in [2]. This algorithm is conceived to work in complex scenes with dynamic background, illumination changing, etc. The method provides a facility for selective updating the background model, which we use to learn the background of the scene for a fixed time. The authors of [2] demonstrate that their contribution outperforms the widely-used Mixture of Gaussians approach [18].

The shape of the foreground elements detected in the scene by background subtraction is usually variable. This is caused by changes in the pose, changes in the area because of varying distance from the camera, and in the case of non rigid objects the shape may change by stretching, shrinking, blending, etc. Moreover the shape of an object may vary because of errors inherited from the background subtraction step.

These difficulties guided us to use a descriptor like *persistent homology*, robust under deformations of the shape of foreground elements in the scene. In the other hand, in [9–13] use the topological features to describe the foreground items in video sequences. Furthermore, the authors in [9] show a mathematical develop to prove the topological feature stability under perturbations.

The rest of this work is organized as follow. In Sect. 2 we present the construction of the simplicial complex. In Sect. 3 are introduced the surveillance points and the simplicial complex associated to them. In Sect. 4 the filtration of a simplicial complex. In Sect. 5 an small introduction to persistent homology is given, and the topological signature of a surveillance point is introduced. Section 6 contains the method for detecting abandoned objects. In Sect. 7 are shown the experimental results. And finally we give conclusions and future directions of work.

2   From Background Subtraction to Simplicial Complex

In this section we present the transition from the background subtraction to simplicial complex, which is a topological representation of the last k frames from video sequence. In the first step, the foreground is segmented by applying an algorithm for background modeling and subtraction. A suitable algorithm may be selected according to the complexity of the scenario. Later k consecutive background images are extracted to construct the simplicial complex. In the Fig. 1 is shown an example for $k = 5$ consecutive images of the PETS2006 database and its background segmentation.

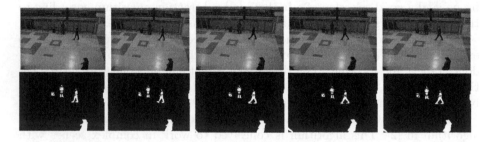

Fig. 1. Result of subtracting background in the PETS2006 database, in the first row is shown the scenario, in the second row is shown the background subtraction result of [2].

The next step consists of building a $3D$ binary digital image $I = (Z^3; B)$ (where $B \subset Z^3$ is the foreground). This is done by stacking k consecutive background subtracted images (the second row of Fig. 1). Later I is used to derive a cubical complex $Q(I)$, which consists of the unit cubes with vertices $V = \{(i; j; k); (i + 1; j; k); (i; j + 1; k); (i; j; k + 1); (i + 1; j + 1; k); (i + 1; j; k + 1); (i; j+1; k+1); (i+1; j+1; k+1)\}$ and all its faces (vertices, edges and squares), for all $V \subseteq B$. Finally, to obtain the simplicial complex $\partial K(I)$ we divide the squares of $Q(I)$ that are faces of exactly one cube in $Q(I)$ into two triangles, which together to their faces (vertices and edges), make up $\partial K(I)$ (see Fig. 2).

Figure 2 shows small connected components are obtained because of errors in the background segmentation step (see Fig. 1). *Persistent homology* helps us to gain invariance to these small errors in segmentation.

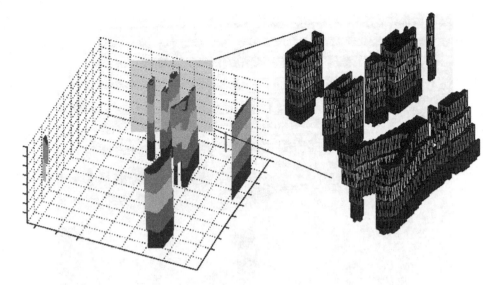

Fig. 2. Simplicial complex corresponding to the $k = 5$ background subtracted images shown in Fig. 1.

3 Surveillance Points

In this work, we analyze the video sequences as blocks of k images. Every k images we construct a simplicial complex following the procedure described in Sect. 2.

We fix a *region of interest* T in the video scene's, with dimension equal to size image or predefined $w_T \times h_T \times k$ (red rectangle in Fig. 3)

Let $i \in Z_+ \cup \{0\}$, stacking the frames between $i * k$ and $(i + 1) * k$ -1, we obtain the ith block Q^i of a video sequence to be analyzed. Let region of interest T with dimensions $w_T \times h_T$, then the ith block Q^i restricted to T is the *ith block of interest* Q^i_T, which has dimensions $w_T \times h_T \times k$. Later given a factor n^1 we divide the ith block of interest Q^i_T in smaller blocks of dimension $(w_T/n) \times (h_T/n) \times k$, which we call *surveillance blocks* (green cube in Fig. 3). The centroid of a surveillance block is called *surveillance point* (blue point in Fig. 3).

If $\partial K(I)^i$ is the simplicial complex introduced in Sect. 2, computed for the frames between $i * k$ and $(i + 1) * k$ -1 of a video sequence, it is embedded in the block Q^i_T. We can split $\partial K(I)^i$ in smaller simplicial complexes associated to surveillance points. Given a surveillance region R and the associated surveillance point p_R, the triangles in $\partial K(I)^i$ that have centroind inside R, form together to their faces the simplicial complex associated to p_R.

---

[1] We obtained good results when the abandoned objects have dimensions not greater than $(w_T/2n) \times (h_T/2n)$.

Fig. 3. Figure shows the *region of interest* T (red rectangle), surveillance block (green cube) and surveillance point (blue point).

4   Filtration

A filtration of a simplicial complex is an ordering of its simplices such that the faces of each simplice (vertices, edges, triangles or squared, etc.) appear before the simplice itself in the ordering. In this section we introduce the filtrations we use for the simplicial complex associated to a surveillance point.

Let $\partial K(I)_{(i,j,t)}$ be the simplicial complex associated to the surveillance point with coordinates (i,j,t). Two filtrations $\partial K(I)_{(i,j,t)min}$ and $\partial K(I)_{(i,j,t)max}$ of $\partial K(I)_{(i,j,t)}$ are obtained as follow:

1. For $\partial K(I)_{(i,j,t)min}$ the triangle t_1 appears before the triangle t_2 in the filtration if the gravity center $(i,j,t)_{t_1}$ of t_1 is nearer than the gravity center $(i,j,t)_{t_2}$ of t_2 from the surveillance point (i,j,t) according to L_∞ distance.
2. In the case of $\partial K(I)_{(i,j,t)max}$ the triangle t_1 appears before the triangle t_2 in the filtration if the gravity center $(i,j,t)_{t_1}$ of t_1 is farther than the gravity center $(i,j,t)_{t_2}$ of t_2 from the surveillance point (i,j,t) according to L_∞ distance.

In both cases before each triangle must be added its faces in any order, in case they were not added before.

5 Persistent Homology and Topological Signature

Persistent homology is an algebraic tool for measuring topological features of shapes and functions. It is built on top of homology, which is a topological invariant technique that captures the amount of components or 0-cycles, tunnels or 1-cycles, cavities or 2-cycles and similar in higher dimensions of a shape [8]. Small size features in persistent homology are often categorized as noise, while large size features describe topological properties of shapes [3].

Let $K = (\sigma_1, \sigma_2, \ldots, \sigma_k)$ be a filtration of a simplicial complex K, where $(\sigma_1, \sigma_2, \ldots, \sigma_k)$ is the list of sorted simplices. Suppose that the simplices of the filtration are added as they appear, if σ_i completes a $p-$cycle (p is the dimension of σ_i) when σ_i is added to $K_{i-1} = (\sigma_1, \ldots, \sigma_{i-1})$, then a $p-$homology class γ *born at time i*; otherwise, a $(p-1)-$homology class *dies at time i*. The difference between the birth and death time of a homology class is called its *persistence*, which quantifies the significance of a topological attribute. If γ never dies, we set its persistence to infinity. Drawing a point (i,j) for a $p-$homology class that borns at time i and dies at time j, we get the $p-$persistence diagram of the filtration, denoted as $Dgm_p(f)$ (Fig. 4). It represents a $p-$homology class by a point whose vertical distance to the diagonal is the persistence. Since always $i < j$, all points lie above the diagonal. For a detailed introduction see [4]).

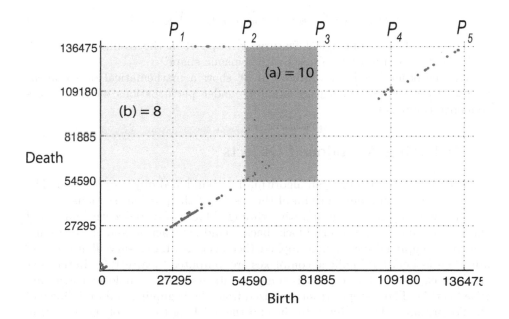

Fig. 4. An example of computation of the first element of a topological signature.

In this work, persistence diagrams are first computed for $\partial K(I)_{(i,j,t)_{min}}$ and $\partial K(I)_{(i,j,t)_{max}}$, given that (i,j,t) are the coordinates of a surveillance point

(Sect. 4). The persistence diagrams are explored according to a uniform sampling. More precisely, given a positive integer n, we compute the integer $h = \lfloor \frac{k}{n} \rfloor$ representing the width of the "window" we use to analyze the persistence diagram (Fig. 4), being k the number of simplices. Later, for $s = 0, \ldots, (n-1)$, the $p-$persistence diagram of $\partial K(I)_{(i,j,t)_{min}}$ (resp. $\partial K(I)_{(i,j,t)_{max}}$) is sampled to obtain (Fig. 4):

(a) Number of homology classes that born before $s \cdot h$ and persist or die after $s \cdot h$.
(b) Number of homology classes that born after $s \cdot h$ and before $(s+1) \cdot h$.

A vector of $2n$ entries is then formed containing (a) in entry $2s$ and (b) in $2s+1$; this way we obtain for a given dimension p, a vector for $\partial K(I)_{(i,j,t)_{min}}$ and the corresponding for $\partial K(I)_{(i,j,t)_{max}}$. Given that only $p = 0$ is used in this work, we have only two vectors for the surveillance point with coordinates (i, j, t), which we concatenate to obtain the *topological signature for a surveillance point*, denoted as $V_{(i,j,t)}$.

The topological signatures for two surveillance points, say $V_{(i_1,j_1,t_1)}$ and $V_{(i_2,j_2,t_2)}$, are compared in the experimental results provided in this work according to:

$$S(V_{(i_1,j_1,t_1)}, V_{(i_2,j_2,t_2)}) = \|V_{(i_1,j_1,t_1)} - V_{(i_2,j_2,t_2)}\|_1. \tag{1}$$

The use of this topological features is inspired in the results obtained in [9–13], shown be robust to objects with dynamic shape.

On the other hand, the authors in [9] show a mathematical development to prove the topological feature stability under perturbations on the images sequence (from video).

6   Detecting Abandoned Objects

In this section we introduce an algorithm for abandoned objects detection. The algorithm is nothing but the sum of the results of the previous sections.

The first step consists in iterating through blocks of k consecutive images of the video sequence. For every block, background subtraction and the construction of a simplicial complex is applied (See Sect. 2). Later surveillance points and their associated simplicial complexes are computed (See Sect. 3). In the next step, for each surveillance point is computed its associated topological signature (See Sect. 5). These steps are summarized from the beginning and until line 9 of the Algorithm 1. From step 9 to the end the finishing touches of the algorithm are given.

Let's focus in a surveillance point $p = (x_p, y_p, t_p)$ and its corresponding surveillance region R_p. In the 9th step of Algorithm 1 we compute the signature V_p associated to p. Using the signatures of points with the same spatial coordinates (x_p, y_p) and contiguous temporal coordinates, we apply an online k-mean

---

Algorithm 1: Detection of Abandoned Objects

---

input: video stream Sequence

1 **for** $i = 0$ **to** End(Sequence)$/k$ **do**

2   images ← Sequence $[i * k, (i + 1) * k)$;

3   background ← BackgroundSubtract(images);

4   simplicialComplex ← BuildComplex(background);

5   surveillancePts ← GetSurveillancePoints(i, *surveillanceRegion*);

6   **foreach** *point p in* surveillancePts **do**

7   complex(p) ← GetAssociatedComplex(p, simplicialComplex);

8   fmin, fmax ← GetFiltrations(complex(p));

9   signature(p) ← GetSignature(fmin, fmax);

10   classId ← GrowingKMean(signature(p));

11   **if** LastClass(p) *equal* classId **then**

12   RepeatLastClass(p) ← RepeatLastClass(p) + 1;

13   **if** RepeatLastClass(p) > *threshold* **then**

14   ABANDONED(SurveillanceRegion(p));

15   **end**

16   **end**

17   **else**

18   LastClass(p) ← classId;

19   RepeatLastClass(p) ← 1;

20   **end**

21   **end**

22 **end**

---

clustering approach [14] in order to decide if there is an abandoned object in the surveillance region R_p.

Suppose $(p_1, p_2, \ldots, p_n)$ are surveillance points such that $(x_{p_i}, y_{p_i}) = (x_{p_j}, y_{p_j})$ and suppose that there is no point p_k such that $t_{p_i} < t_{p_k} < t_{p_{i+1}}$. If all the topological signatures $(V_{p_1}, V_{p_2}, \ldots, V_{p_n})$ fall in the same class under the online k-mean clustering approach [14], we assume that the topology is repetitive in a surveillance region, which depicts the presence of an abandoned object. This is reflected in the steps from 10 to 16 of the algorithm. The value n corresponds to the variable *threshold* in the line 13 of the algorithm. If n contiguous point with the corresponding topological signature falling in the same class can not be found , then we reset the corresponding values, which is reflected from lines 17 to 20 of the algorithm.

7   Experimental Results

This work is evaluated using the Peet2006 database. The scene is a subway station where a man abandon a suitcase. This database has seven situations with increasing complexities according to the number of persons in the scene. The seventh situation is considered the most complex (the abandoned object is usually occluded)[2]. Each situation is taken by four cameras from different viewpoints. Each view is characterized by the size in pixels of the abandoned object, occlusion level, illumination variations, etc. The sequences are recorded with a resolution of 768 × 576 pixels, a frame rate of 25 frames per second, and compressed with JPEG. The Fig. 5 shows the four views from Peet2006 database.

Fig. 5. Views from the PETS2006 database.

For our experiments we took into account only the seventh situation, given it is the hardest. Furthermore, we reduced the resolution of the frames to 360 × 228. We use a own implementation from [2] to obtain the background subtractions. For our results we used 20 s as the time an object must be alone to be considered abandoned[3]. This value, together with $k = 5$ images per block (see Sect. 3), is enough to determine the value of the variable *threshold* introduced in the line 13 of the Algorithm 1.

Table 1 shows the results obtained in the four views. The column 5 and 6 represent the video frame number at which the alarm started (Alarm Start) and ended (Alarm End) respectively. The column 7 is the alarm continuous time (in seconds).

Furthermore, the performance of the systems is compared with other methods (see Table 2) using the Precision (see Eq. 2) and the Recall (see Eq. 3), which represent the percentage of true alarms and the percentage of detected events respectively.

Where T_P (true positive) represents abandoned object classified as abandoned by the system, F_N (false negative) represents a missed detection, F_P (false positive) corresponds to the classification of non-abandoned object as abandoned and T_N (true negative) stands for abandoned object classified as non-abandoned

$$Precision(\%) = \frac{T_P}{T_P + F_P} \tag{2}$$

$$Recall(\%) = \frac{T_P}{T_P + F_N} \tag{3}$$

---

[2] http://www.cvg.reading.ac.uk/PETS2006/data.html.
[3] According expert criterion for application area (airport, train station, etc) and empirical study from data sets.

Table 1. Results of our method in the four views of seventh situation of PETS2006.

Views	Abandoned Objects	T_P	F_P	Alarm Start (frame)	Alarm End (frame)	Abandon Time(sec)
View 1	1	1	0	2714	3397	27.32
View 2	1	1	0	3204	3514	12.4
View 3	1	1	0	1813	2478	26.6
View 4	1	1	0	1677	2232	22.2

Table 2. Precision and recall result on PETS2006.

Approach	Precision (%)	Recall (%)
[19]	100	85.7
Our approach	100	100

8   Conclusion and Future Works

Our method is capable of detecting the abandoned object in all the views of the most complex situation of PETS2006 database. From lateral view (view 3) we reach the major abandon time detection due to in this view the occlusion time is minimal. On the other hand in frontal view (view 2) the place of abandon is almost always occluded and it is far from the camera, which causes the lowest time of abandon. We implemented our method in Matlab reaching real time performance, which makes possible practical applications.

References

1. Beynon, M.D., Van Hook, D.J., Seibert, M., Peacock, A., Dudgeon, D.: Detecting abandoned packages in a multi-camera video surveillance system. In: Proceedings. IEEE Conference on Advanced Video and Signal Based Surveillance 2003, pp. 221–228. IEEE (2003)
2. Calderara, S., Melli, R., Prati, A., Cucchiara, R.: Reliable background suppression for complex scenes. In: Proceedings of the 4th ACM International Workshop on Video Surveillance and Sensor Networks pp. 211–214. ACM (2006)
3. Edelsbrunner, H., Harer, J.: Persistent homology-a survey. Contemporary Mathem. **453**, 257–282 (2008)
4. Edelsbrunner, H., Harer, J.: Computational Topology - an Introduction. American Mathematical Society (2010)
5. Ferrando, S., Gera, G., Regazzoni, C.: Classification of unattended and stolen objects in video-surveillance system. In: IEEE International Conference on Video and Signal Based Surveillance, 2006, AVSS 2006, pp. 21–21. IEEE (2006)
6. Ferryman, J., et al.: Robust abandoned object detection integrating wide area visual surveillance and social context. Pattern Recog. Lett. (2013)
7. Guler, S., Silverstein, J.A., Pushee, I.H.: Stationary objects in multiple object tracking. In: IEEE Conference on Advanced Video and Signal Based Surveillance, AVSS 2007, pp. 248–253. IEEE (2007)

8. Hatcher, A.: Algebraic topology (2002)
9. Lamar-Leon, J., Garcia-Reyes, E., Gonzalez-Diaz, R.: Topological signature for periodic motion recognition (2019)
10. Lamar-Leon, J., Baryolo, R.A., Garcia-Reyes, E., Gonzalez-Diaz, R.: Gait-based carried object detection using persistent homology. In: Bayro-Corrochano, E., Hancock, E. (eds.) CIARP 2014. LNCS, vol. 8827, pp. 836–843. Springer, Cham (2014). https://doi.org/10.1007/978-3-319-12568-8_101
11. Leon, J.L., Alonso, R., Reyes, E.G., Diaz, R.G.: Topological features for monitoring human activities at distance. In: Mazzeo, P.L., Spagnolo, P., Moeslund, T.B. (eds.) AMMDS 2014. LNCS, vol. 8703, pp. 40–51. Springer, Cham (2014). https://doi.org/10.1007/978-3-319-13323-2_4
12. Leon, J.L., Cerri, A., Reyes, E.G., Diaz, R.G.: Gait-based gender classification using persistent homology. In: Ruiz-Shulcloper, J., Sanniti di Baja, G. (eds.) CIARP 2013. LNCS, vol. 8259, pp. 366–373. Springer, Heidelberg (2013). https://doi.org/10.1007/978-3-642-41827-3_46
13. Lamar-León, J., García-Reyes, E.B., Gonzalez-Diaz, R.: Human gait identification using persistent homology. In: Alvarez, L., Mejail, M., Gomez, L., Jacobo, J. (eds.) CIARP 2012. LNCS, vol. 7441, pp. 244–251. Springer, Heidelberg (2012). https://doi.org/10.1007/978-3-642-33275-3_30
14. Parsons, S.: Introduction to Machine Learning by Ethem Alpaydin. MIT Press, 0–262-01211-1, 400 pp. Cambridge Univ Press (2005)
15. Martínez-del Rincón, J., Herrero-Jaraba, J.E., Gómez, J.R., Orrite-Urunuela, C.: Automatic left luggage detection and tracking using multi-camera ukf. In: Proceedings of the 9th IEEE International Workshop on Performance Evaluation in Tracking and Surveillance (PETS 2006), pp. 59–66. Citeseer (2006)
16. Sajith, K., Nair, K.R.: Abandoned or removed objects detection from surveillence video using codebook. In: International Journal of Engineering Research and Technology, vol. 2. ESRSA Publications (2013)
17. Spengler, M., Schiele, B.: Automatic detection and tracking of abandoned objects. In: Proceedings of the Joint IEEE International Workshop on Visual Surveillance and Performance Evaluation of Tracking and Surveillance. Citeseer (2003)
18. Stauffer, C., Grimson, W.E.L.: Adaptive background mixture models for real-time tracking. In: IEEE Computer Society Conference on Computer Vision and Pattern Recognition 1999, vol. 2. IEEE (1999)
19. Szwoch, G.: Extraction of stable foreground image regions for unattended luggage detection. Multimedia Tools Appli. **75**, 761–786 (2016)
20. Tian, Y., Feris, R.S., Liu, H., Hampapur, A., Sun, M.T.: Robust detection of abandoned and removed objects in complex surveillance videos. IEEE Trans. Syst. Man Cybern. Part C: Appli. Rev. **41**(5), 565–576 (2011)
21. Tripathi, R.K., Jalal, A.S., Agrawal, S.C.: Abandoned or removed object detection from visual surveillance: a review. Multimedia Tools Appli. **78**(6), 7585–7620 (2019)
22. Zin, T.T., Tin, P., Toriu, T., Hama, H.: A probability-based model for detecting abandoned objects in video surveillance systems. In: Proceedings of the World Congress on Engineering, vol. 2 (2012)

Interactive Segmentation
with Incremental Watershed Cuts

Quentin Lebon[1](✉), Josselin Lefèvre[1,2], Jean Cousty[1], and Benjamin Perret[1]

[1] LIGM, Univ Gustave Eiffel, CNRS, 77454 Marne-la-Vallée, France
`quentin.lebon@edu.esiee.fr`
[2] Thermo Fisher Scientific, Bordeaux, France

Abstract. In this article, we propose an incremental method for computing seeded watershed cuts for interactive image segmentation. We propose an algorithm based on the hierarchical image representation called the binary partition tree to compute a seeded watershed cut. We show that this algorithm fits perfectly in an interactive segmentation process by handling user interactions, seed addition or removal, in time linear with respect to the number of affected pixels. Run time comparisons with several state-of-the-art interactive and non-interactive watershed methods show that the proposed method can handle user interactions much faster than previous methods achieving significant speedup from 15 to 90, thus improving the user experience on large images.

Keywords: Interactive segmentation · watershed · binary partition tree · minimum spanning tree

1 Introduction

Image segmentation consists in partitioning an image into meaningful regions. A classical approach to this problem is the watershed (WS) where the image is seen as a topological relief and the regions corresponds to catchment basins: this method constitutes a fundamental stage of many image analysis workflows. The watershed segmentation is still widely used in today's learning era as a pre-processing or a post-processing (see e.g. [2,9,13,14,24]), achieving state-of-the-art results. The first approaches for WS on images considered an image as a vertex-weighted graph [3,23]. Nowadays, state-of-the-art WS methods are defined on edge-weighted graphs [7], allowing to characterize WS cuts as solutions to a global optimization problem related to Minimum Spanning Trees (MST). Further advanced WS are based on hierarchical representations describing how catchments basins are progressively merged into the most significant structures [16,17]. In this context, the authors of [18] have proposed an algorithm to compute a WS cut from a hierarchical representation called the Binary Partition Tree (BPT) [22].

One of the major drawbacks of WS is over-segmentation as each minimum of the image induces a catchment basin. To overcome this problem, weakly supervised WS segmentation through interactivity is a popular solution. By substituting the minima of the image by user-defined seeds, this procedure makes it

V. Vasconcelos et al. (Eds.): CIARP 2023, LNCS 14469, pp. 189–200, 2024.
https://doi.org/10.1007/978-3-031-49018-7_14

possible on the one hand to avoid over segmentation and on the other hand to introduce semantic information into the output segmentation.

But often, the segmentation has to be corrected by successively refining seeds *i.e.*, adding or removing seeds, until a result close to the desired segmentation is reached. Classical seeded WS algorithms cannot handle such incremental process and each update of the seeds requires to completely recompute the result on the whole image, even if a small number of pixels is affected by the change. This problem can seriously limit the speed of user interactions on large images or 3D volumes. This issue has been addressed in the framework of the Image Foresting Transform (IFT) [11] with the Differential Image Foresting Transform (DIFT) [10] a WS-based and fuzzy-connected segmentation method, whose response time for interactive segmentation is proportional to size of the modified regions of the scene. Interactivity has also been addressed in the context of object segmentation with hierarchical representations, especially for trees based on threshold decomposition *i.e.*, component trees [20] or tree of shapes [5,19].

In this work, we propose a seeded WS cut algorithm within the framework of BPTs. We show that this algorithm is particularly well suited for interactive segmentation, as it can handle user interactions in time linear with the number of affected pixels. The method is assessed on several images without using any GPU architecture where users were asked to interactively segment an object of interest by adding/removing seeds, we also assessed it on a more extensive dataset where random generated seeds are added. The run-time comparisons with state-of-the-art incremental and non-incremental methods show a significant advantage for the proposed method.

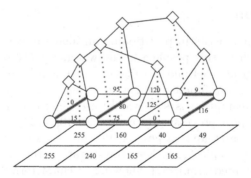

Fig. 1. The three working areas: grayscale image, graph, and BPT. The dashed red lines depict the bijection between non-leaf nodes and edges of the MST (in bold). (Color figure online)

2   Watershed Cuts

In this section, we briefly review the notion of a WS cut, recall its relation to minimum spanning trees and forests and highlight the important BPT datastructure that is used in the following. WS cuts are deeply related to MSTs [7]. In the semi-supervised case where an edge-weighted graph (representing the image) and a set of marked vertices are given, a WS cut can be obtained as the cut induced by a minimum spanning forest where each tree is rooted in one of the seeds [1]. The BPT is a hierarchical datastructure that allows one to represent a MST and to efficiently browse its edges. The BPT is obtained as a by-product of the efficient Kruskal's MST algorithm. Figure 1 illustrates the BPT of an edge-weighted graph and its relation to the MST of this graph (bold edges). In particular, it can be seen that the leaves of the BPT are mapped to the vertices of the MST and that each non-leaf node is a mapped to an edge of the MST (this mapping is represented by the dashed segments in Fig. 1). Intuitively, we can say that every non-leaf node of the BPT represents the addition of an edge to the MST during Kruskal's algorithm, the added edge being used to merge the two connected regions that contain the edges extremities. An efficient WS cut algorithm based on BPT is presented in [8,18] to handle the non-supervised case with no seed provided.

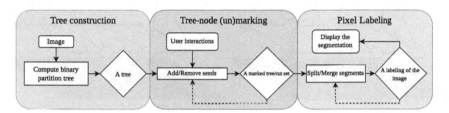

Fig. 2. Overview of our workflow of the interactive incremental watershed (WS) segmentation. (Color figure online)

3 Semi-supervised Watershed Cut Algorithm with Interactions

In this section, we present a novel WS cut algorithm that is able to (i) compute a seeded WS cut from user-provided seeds, and to (ii) efficiently update this WS cut from user's feedback given in the form of successive deletions and additions of seeds. The workflow of the method, presented in Fig. 2, comprises three main parts: **1)** Computation of an MST and an associated BPT; **2)** Identification of WS cut edges by browsing the BPT, taking into account the seeds provided by the user. These edges correspond to the edges that must be removed from the MST to obtain a minimum spanning forest rooted in the given seeds. **3)** Partitioning and labeling of the graph vertices according to the connected components (CC) of the forest resulting from step 2. This labeling is the resulting WS cut segmentation. Step 1 can be performed using the algorithm presented in [18]. Section 3.1 presents an efficient algorithm for step 2, Sect. 3.2 discusses the labeling involved at step 3, and Sect. 3.3 shows how to incrementally update the results of the workflow based on user's feedback.

3.1   Tree-Node Marking

This section is devoted to Algorithm 1 that identifies the WS edges from the BPT of the image for some user provided seeds. The algorithm proceeds seed by seed in an incremental manner. For each seed, taken in any order, one edge of the initial MST must be removed to cut the region of this seed from the regions of the current partition (obtained with the previous seeds). To this end, we search in the BPT for the lowest node that merges a region of the new seed with a region of the already processed seeds. In order to "prevent" this merging and to cut the region of the new seed from the rest of the partition, the MST edge associated to this node is tagged WS and added to the set of edges to be deleted from the MST.

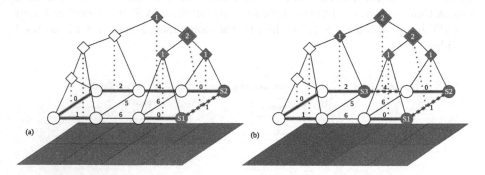

Fig. 3. Integer labels on non-leaf nodes represent the value of `visitCount`. (a) Initialization with S_1 and S_2 as seeds. (b) Update after the addition of S_3 as an additional seed. (Color figure online)

To make this idea practicable, we consider an array `visitCount` that marks each BPT node with the number of times that this node was visited during the successive searches. At the beginning, `visitCount` is initialized to 0 for each node. Then, for each leaf node of the tree corresponding to a seed to add, we browse its ancestors and update `visitCount` accordingly. Let `p` be a parent of the considered seeded node, if `visitCount[p]` $= 0$ then this value has to be incremented and the traversal continues by considering the parent of `p`. However, if `visitCount[p]` $= 1$, then node `p` has already been visited by a previous seed: it must then become a WS node separating 2 different seeds. We thus increment the value of `visitCount[p]` to 2 and the MST edge associated to `p` (denoted by `H.mstEdge[p]`) is added to the edge set WS. Note that the traversal stops when `visitCount[p]` $= 2$ as the separation induced by the addition of the new seed has been found. When a seed is added, only the nodes in the path from this seed to its closest WS node are visited. We can see this on Fig. 3(b), where adding seed S_3 results in browsing only three nodes. During a call or a succession of calls to Algorithm 1, each node is visited at most twice, leading to an overall linear-time complexity with respect to the number of vertices.

Algorithm 1: ADD SEED

Data: $\mathcal{H}$: a BPT, `seeds` : a set of seeds and `visitCount`;

Result: `ws` a set of edges to be removed and `visitCount` updated.

1 $ws \leftarrow \emptyset$

2 **foreach** *leaf n of seeds* **do**

3   **while** $n \neq \mathcal{H}.root$ **and** $visitCount[n] \neq 2$ **do**

4        $n := \mathcal{H}.parent[n]$

5        $visitCount[n] := visitCount[n] + 1$

6   **if** $visitCount[n] = 2$ **then**

7           $ws := ws \cup \{\mathcal{H}.mstEdge[p]\}$

3.2 Pixel Labeling

Once the WS nodes set is computed, we return to the image domain to compute the segmentation. First, we compute the minimum spanning forest representing the WS cut: for each added WS node, the corresponding WS edge is removed from the MST. Then we perform a labeling of the CCs of the forest with a simple Breadth First Search (BFS) algorithm.

3.3 Incremental Workflow

The workflow presented in the previous sections can be adapted to work in an incremental way, by considering the addition or removal of seeds and the differential update of the resulting WS cut and all intermediary structures.

Firstly, we adapt Algorithm 1 to obtain Algorithm 2 which accounts for seed removal. Such removal induces the merging of two regions and the disappearance of a WS node. The traversal procedure is the same as for adding seeds except that for each parent p, `visitCount[p]` is decremented and the parent browsing stops if `visitCount[p]` $= 1$ after decrementation. Indeed, if the value was decremented to 1, the current node is no longer be a WS node.

Regarding the labeling, Algorithm 2 can now restore WS edges which induces the merging of two CCs. That can be efficiently performed by constraining BFS to explore only the smallest CC associated with one extremity of a cut edge. The label of the larger CC is then spread on the smaller one. This can be done by keeping track of the size of each CC resulting in a linear time merging *w.r.t.* the number of vertices of the smallest CC. Note that when a WS edge is removed, we also only need to relabel the components at the two extremities of this edge. The split of a component thus also runs in time linear with the number of pixels in the affected component.

As a result, this incremental workflow enables updating of the segmentation in a time proportional to the number of pixels in the region affected by the seed refinement.

Algorithm 2: REMOVE SEED

    Data: $\mathcal{H}$: a BPT, `seeds` : a set of seeds and `visitCount`;
    Result: `ws` a set of edges to be added and `visitCount` updated.

1   $ws \leftarrow \emptyset$
2   **foreach** *leaf* n *of seeds* **do**
3   **while** $n \neq \mathcal{H}.root$ **and** $visitCount[n] \neq 1$ **do**
4        $n := \mathcal{H}.parent[n]$
5        $visitCount[n] := visitCount[n] - 1$
6   **if** $visitCount[n] = 1$ **then**
7           $ws := ws \cup \{\mathcal{H}.mstEdge[p]\}$

---

4 Experiments

We assessed the method with two different experiments. The first one confronts different methods with real user interactions while the second one relies on a publicly available dataset with a larger number of images and associated ground-truth segmentations, for which the interactions are simulated by randomly selecting markers from the ground truth.

4.1 Experiment with User Generated Seeds

For assessments, we confronted a user with a binary segmentation problem on three images from the INRIA Holidays dataset [12] and on one provided by ourselves. The images are all the same size: 2048 by 1536 pixels. To retrieve seeds, we ask the user to segment the images with an interactive tool: he could draw green seeds for the object of interest and red seeds for the background and if a mistake was made, he could remove seeds with an eraser tool. Each user interaction was recorded in batches of added/removed green/red seeds. The same sets of seeds were used for each tested method.

In this study, we consider two versions of the proposed method: (i) a non incremental version (denoted NIWS) where we first compute the BPT and then, at each user interaction, we completely recompute the WS edges (using only Algorithm 1) and the induced labeling; and (ii) an incremental version (denoted IWS) where at each interaction we update `visitCount` (Algorithms 1 and 2) and the labeling by considering only the added/removed seeds. We also consider three state-of-the-art implementations of seeded WS, namely OpenCV [4,15] (highly optimized library for image processing), Higra [21] (generic library for hierarchical graph analysis) and another incremental method: the Differential Image Foresting Transform (DIFT) [10]. IWS and NIWS have been tested with a C++ implementation available at https://github.com/lebonq/incremental_watershed. We used a C implementation of DIFT available at https://github.com/tvspina/ift-demo.

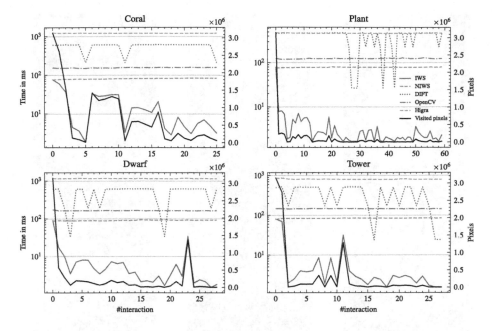

Fig. 4. Computation time of OpenCV (dot-dash blue), DIFT (dotted purple), Higra (dashed orange), NIWS (dashed green) and IWS (plain red) along an interactive segmentation session compared to the number of pixels to be updated (plain black). (Color figure online)

The results are presented in Fig. 4 and in Table 1. We see that the initialization cost (BPT creation) of both NIWS and IWS version of our method is quickly amortized during the segmentation process: IWS and NIWS have a much lower average execution time than other methods. In addition, we can see that the execution time of IWS is proportional to the number of pixels affected by seed updates as expected from the theoretical study. The upper bound is given by the

Table 1. Computation time (ms) of the methods. The second column corresponds to the initialization time. The third and fourth columns give the average and maximum computation time over all user interactions. The last column gives the total computation time (initialization plus all user interactions).

Method	Init	Average	Max	Accumulated
IWS	210.0	<u>9.2</u>	<u>89.3</u>	<u>514.1</u>
NIWS	210.0	82.2	94.8	3164.4
DIFT	<u>14.0</u>	478.4	622.8	15886.3
OpenCV	$\emptyset$	141.7	174.8	5089.4
Higra	$\emptyset$	829.1	1239.4	29012.0

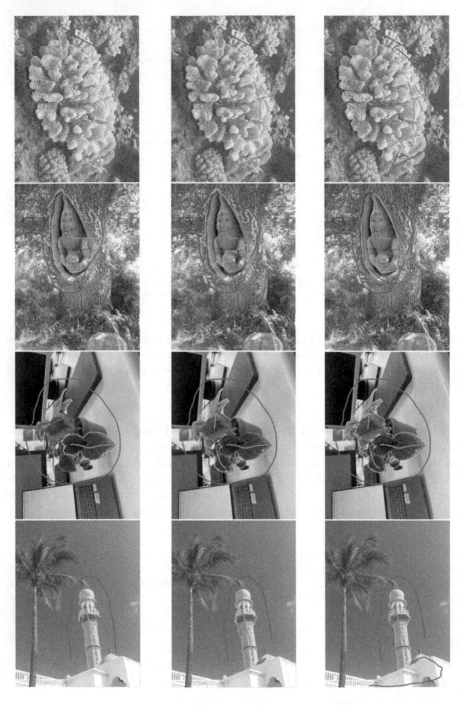

Fig. 5. Three stages in an interactive segmentation session. The red lines represent background seeds, the green lines represent object seeds, and the yellow lines represent the segmentation produced by the current set of seeds. (Color figure online)

first interaction, which labels all pixels (the first step is therefore equivalent to NIWS), and spikes in computation time occurs during mid- or end-interactions if the user updates seeds in a large CC, resulting in a significant number of pixel changes. Our results also indicate that there is no significant difference in computation time between adding or removing seeds.

4.2 Experiment with Randomly Generated Seeds

To assess the methods on a standard dataset with more samples, we chose the BIG dataset [6] which is composed of large natural images (from 2048 by 1600 to 5000 by 3600 pixels). We use all 150 images and pair each image with a series of 70 seeds. Seeds are divided into two balanced classes: object and background. Each seed is a ball of radius 11 centered on a randomly chosen pixel of the object (resp. background) mask eroded by a ball of radius 12 ensuring that the seed lies in the object (resp. background).

In this experiment, we consider the same methodology as in Sect. 4.1. During the iterative process, seeds are alternatively picked within the object and within the background. In this experiment, only addition of seeds is considered, never removal.

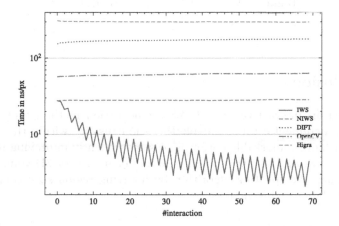

Fig. 6. Three stages in an interactive segmentation session. The red lines represent background seeds, the green lines represent object seeds, and the yellow lines represent the segmentation produced by the current set of seeds. (Color figure online)

The results are presented in Fig. 6 and Table 2. For each interaction, the computation time is the average over all 150 images. With this experiment, the tendency observed in Sect. 4.1 is confirmed: the execution times of other methods are constant over the iterations whereas our incremental method shows decreasing execution times over the iterations, as the number of pixels affected by the interaction decreases. Furthermore, we observe that our proposed incremental methods significantly improves the answer-time to user interactions compared to all other methods. We can see that the curve for the IWS has a "sawthooth" look. It can be easily explained as we alternately add seeds on the objects and on the background, and the size of the objects is often much smaller than that of the background. Up spikes correspond to seeds placed on the background, while down spikes correspond to seeds placed on the objects. This is due to the algorithm's dependence on the size of the regions to be updated.

Table 2. Computation times expressed in nanoseconds per pixels. The second column gives the initialization time of each method. The third and fourth columns give the average and maximum computation time over all interactions. The last column gives the total computation time, considering the sum over the 70 interactions and the initialization time.

Method	Init	Average	Max	Accumulated
IWS	88.1	6.8	27.6	563.6
NIWS	88.1	28.0	28.6	2051.3
DIFT	3.8	172.0	176.3	12044.9
OpenCV	∅	60.9	61.2	4262.0
Higra	∅	300.7	311.9	21045.3

5   Conclusion

We proposed an interactive seeded WS segmentation method that is compliant with an incremental process exploiting the causality within the interactive sessions to achieve remarkable performances, significantly improving responsiveness. In future works, we plan to test it on larger images or 3D volumes and to better optimize the splitting algorithm by inferring region size to ensure that the region to label is the smallest.

References

1. Allène, C., Audibert, J.Y., Couprie, M., Keriven, R.: Some links between extremum spanning forests, watersheds and min-cuts. Image Vis. Comput. **28**(10), 1460–1471 (2010)
2. Arbeláez, P., Maire, M., Fowlkes, C., Malik, J.: Contour detection and hierarchical image segmentation. IEEE Trans. Pattern Anal. Mach. Intell. **33**(5), 898–916 (2011). https://doi.org/10.1109/TPAMI.2010.161, conference Name: IEEE Transactions on Pattern Analysis and Machine Intelligence
3. Beucher, S., Meyer, F.: The morphological approach to segmentation: the watershed transformation. Math. Morphol. Image Process. **34**, 433–481 (1993). https://doi.org/10.1201/9781482277234-12

4. Bradski, G.: The OpenCV library. Dr. Dobb's J. Softw. Tools (2000)
5. Carlinet, E., Geraud, T.: Morphological object picking based on the color tree of shapes. In: 2015 IPTA, pp. 125–130. IEEE, Orleans, France, November 2015. https://doi.org/10.1109/IPTA.2015.7367111
6. Cheng, H.K., Chung, J., Tai, Y.W., Tang, C.K.: CascadePSP: toward class-agnostic and very high-resolution segmentation via global and local refinement. In: CVPR (2020)
7. Cousty, J., Bertrand, G., Najman, L., Couprie, M.: Watershed cuts: minimum spanning forests and the drop of water principle. IEEE TPAMI **31**(8), 1362–1374 (2009)
8. Cousty, J., Najman, L., Perret, B.: Constructive links between some morphological hierarchies on edge-weighted graphs. In: ISMM, pp. 86–97 (2013)
9. Eschweiler, D., Spina, T.V., Choudhury, R.C., Meyerowitz, E., Cunha, A., Stegmaier, J.: CNN-based preprocessing to optimize watershed-based cell segmentation in 3D confocal microscopy images. In: 2019 IEEE 16th International Symposium on Biomedical Imaging (ISBI 2019), pp. 223–227. IEEE, Piscataway, NJ, April 2019
10. Falcao, A., Bergo, F.: Interactive volume segmentation with differential image foresting transforms. IEEE Trans. Med. Imaging **23**(9), 1100–1108 (2004). https://doi.org/10.1109/TMI.2004.829335
11. Falcão, A., Stolfi, J., de Alencar Lotufo, R.: The image foresting transform: theory, algorithms, and applications. IEEE TPAMI **26**(1), 19–29 (2004). https://doi.org/10.1109/TPAMI.2004.1261076
12. Jegou, H., Douze, M., Schmid, C.: Hamming embedding and weak geometric consistency for large scale image search. In: Forsyth, D., Torr, P., Zisserman, A. (eds.) ECCV 2008. LNCS, vol. 5302, pp. 304–317. Springer, Heidelberg (2008). https://doi.org/10.1007/978-3-540-88682-2_24
13. Lux, F., Matula, P.: DIC image segmentation of dense cell populations by combining deep learning and watershed. In: 16th IEEE ISBI, pp. 236–239 (2019). https://doi.org/10.1109/ISBI.2019.8759594
14. Machairas, V., Faessel, M., Cardenas, D., Chabardes, T., Walter, T., Decencière, E.: Waterpixels. IEEE TIP **24**, 3707–3716 (2015). https://doi.org/10.1109/TIP.2015.2451011
15. Meyer, F.: Color image segmentation. In: 1992 International Conference on Image Processing and its Applications, pp. 303–306 (1992)
16. Meyer, F.: The watershed concept and its use in segmentation : a brief history, February 2012. http://arxiv.org/1202.0216, arXiv:1202.0216 [cs]
17. Najman, L., Schmitt, M.: Geodesic saliency of watershed contours and hierarchical segmentation. IEEE TPAMI **18**(12), 1163–1173 (1996). https://doi.org/10.1109/34.546254
18. Najman, L., Cousty, J., Perret, B.: Playing with kruskal: algorithms for morphological trees in edge-weighted graphs. In: ISMM, pp. 135–146 (2013)
19. Ngoc, M.Ô.V., Carlinet, E., Fabrizio, J., Géraud, T.: The Dahu graph-cut for interactive segmentation on 2D/3D images. Pattern Recogn. **136**, 109–207 (2023)
20. Passat, N., Naegel, B., Rousseau, F., Koob, M., Dietemann, J.L.: Interactive segmentation based on component-trees. Pattern Recogn. **44**(10), 2539–2554 (2011). https://doi.org/10.1016/j.patcog.2011.03.025, Semi-Supervised Learning for Visual Content Analysis and Understanding
21. Perret, B., Chierchia, G., Cousty, J., F. Guimarães, S., Kenmochi, Y., Najman, L.: Higra: hierarchical graph analysis. SoftwareX **10**, 100335 (2019). https://doi.org/10.1016/j.softx.2019.100335

22. Salembier, P., Garrido, L.: Binary partition tree as an efficient representation for image processing, segmentation, and information retrieval. TIP **9**(4), 561–576 (2000)
23. Vincent, L., Soille, P.: Watersheds in digital spaces: an efficient algorithm based on immersion simulations. IEEE TPAMI **13**(6), 583–598 (1991). https://doi.org/10.1109/34.87344
24. Wolf, S., Schott, L., Köthe, U., Hamprecht, F.: Learned watershed: end-to-end learning of seeded segmentation, September 2017. https://doi.org/10.48550/arXiv.1704.02249, arXiv:1704.02249 [cs]

Supervised Learning of Hierarchical Image Segmentation

Raphael Lapertot$^{(\boxtimes)}$ ⓘ, Giovanni Chierchia ⓘ, and Benjamin Perret ⓘ

LIGM, Univ Gustave Eiffel, CNRS, ESIEE Paris, 77454 Marne-la-Vallée, France
`raphael.lapertot@esiee.fr`

Abstract. We study the problem of predicting hierarchical image segmentations using supervised deep learning. While deep learning methods are now widely used as contour detectors, the lack of image datasets with hierarchical annotations has prevented researchers from explicitly training models to predict hierarchical contours. Image segmentation has been widely studied, but it is limited by only proposing a segmentation at a single scale. Hierarchical image segmentation solves this problem by proposing segmentation at multiple scales, capturing objects and structures at different levels of detail. However, this area of research appears to be less explored and therefore no hierarchical image segmentation dataset exists. In this paper, we provide a hierarchical adaptation of the Pascal-Part dataset [2], and use it to train a neural network for hierarchical image segmentation prediction. We demonstrate the efficiency of the proposed method through three benchmarks: the precision-recall and F-score benchmarks for boundary location, the level recovery fraction for assessing hierarchy quality, and the false discovery fraction. We show that our method successfully learns hierarchical boundaries in the correct order, and achieves better performance than the state-of-the-art model trained on single-scale segmentations.

Keywords: Image Segmentation · Supervised Learning · Ultrametric · Hierarchy · Graph

1 Introduction

Image segmentation is the process of partitioning an image into distinct regions, which simplifies the image by focusing on the structure of its objects. A characteristic of image segmentation (and of images in general) is the scale: in an image, the visible structure depends on the observation scale. The choice of the scale is crucial and strongly depends on the application. To overcome this limitation, one solution is to not choose a scale at all, by proposing several consistent segmentations at different scales (satisfying the principle of strong causality [7]),

---

This work is supported by the French ANR grant ANR-20-CE23-0019, and was granted access to the HPC resources of IDRIS under the allocation 2023-AD011013101R1 made by GENCI.

V. Vasconcelos et al. (Eds.): CIARP 2023, LNCS 14469, pp. 201–213, 2024.
https://doi.org/10.1007/978-3-031-49018-7_15

i.e., building a *hierarchy (of segmentations)*. In this case, the choice of the scale is not made during the segmentation, but *after* the segmentation, if even needed. In a hierarchy, an image is represented as a sequence of coarse to fine segmentations. Hierarchical segmentation also provides a more versatile and informative structure than traditional segmentation. It naturally allows multi-scale analysis, but also provides object hierarchy by capturing the relationships and dependencies between different segments. It is more flexibility by enabling users to adapt the segmentation output to suit their needs or application requirements. Finally, it can be useful in interactive scenarios, where a user can interact with the segmentation hierarchy to refine or adjust the segmentation results.

Hierarchies have long been used in computer vision as an intermediate representation to perform segmentation [1,5,6,13,15,18], or object detection and proposal [15,20]. Several works have been done to improve hierarchies using supervised learning techniques. In [16] the authors trained a cascade of edge classifiers based on classical human-designed features. Maninis *et al.* [9] trained a deep contour detector, the output of which is transformed into a hierarchy during post-processing ; they do not explicitly train their neural network for hierarchical segmentation, as they use classical image segmentation datasets with single scale annotations. More recently, Tao *et al.* [19] proposed a way to fuse segmentations at different scales using attention masks, but they do not predict a hierarchical segmentation. In general, while a variety of labeled datasets exist for image segmentation, this is not the case for hierarchical image segmentation, which is obviously a major problem for achieving supervised learning of hierarchical image segmentation.

The aim of this work is to train a neural network for hierarchical image segmentation. The contributions are threefold: (i) we build hierarchical segmentation ground truths for the Pascal-Part dataset, (ii) we propose a pipeline for supervised learning of a neural network that predicts hierarchies, and (iii) we define a benchmark to assess the quality of hierarchies.

Definitions. A *hierarchy* on an image is a sequence of partitions $P_1, ..., P_\ell$ of the image pixels, such that P_i is a *refinement* of P_{i-1}. Another possible representation of a hierarchy is the *ultrametric dissimilarity grid*, where the vertices are the pixels of the image, the edges represent the 4-adjacency relation between pixels, and the edge weights are a measure of dissimilarity satisfying the ultrametric property (a large dissimilarity means that the boundary represented by that edge persists along large scales). An ultrametric dissimilarity grid can be visualized as an image called an Ultrametric Contour Map (UCM), where interpixels are added to the original image to represent the grid edges; the size of an UCM is thus twice the size of the original image. The values of the interpixels are determined by the weights of the edges they represent (see Fig. 1).

2   Ultrametric Dataset

Our first contribution is to create a hierarchical dataset by transforming the existing annotations of the Pascal-Part dataset [2] into UCMs that enforce the principle of strong causality between objects and parts.

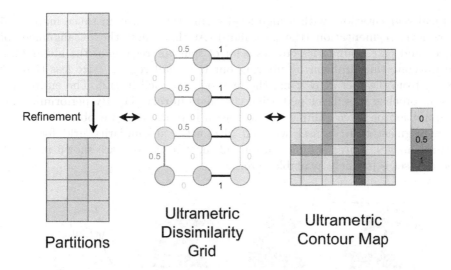

Fig. 1. A hierarchy of two partitions (left), ultrametric dissimilarity grid (middle) or Ultrametric Contour Map (right).

The Pascal-Part dataset extends the Pascal VOC 2010 dataset [4], which consists of 10103 natural images of different sizes, and annotations for different challenges such as classification, segmentation, and object detection. This dataset is a widely used dataset for supervised image segmentation learning and is challenging due to the complexity and diversity of its images. Pascal-Part provides an additional set of annotations for the Pascal VOC 2010 images, with segmentation masks for each instance of 20 classes of objects in each image, and segmentation masks for parts of these objects.

However, the segmentation masks of the Pascal-Part dataset have several limitations. First, some parts overlap with each other, and some parts sometimes completely cover other parts. Note that the objects, on the other hand, do not intersect with each other. Secondly, sometimes the contours of the parts do not match the contours of their object, being slightly off in the inner side of the object. Thirdly, there are parts of some objects that are behind non-annotated objects: sometimes the annotator *imagined* the continuation of the object behind. In the third sample of Fig. 2, a leg of the chair is imagined, even though it is hidden by the non-annotated stool. A consequence of these observations is that Pascal-Part annotations do not respect the principle of strong causality: they do not form hierarchies.

We now describe our method for constructing an ultrametric dataset from the Pascal-Part dataset. First, we build a high-level segmentation, the *instance* segmentation, by stacking the object masks. We then build a low-level segmentation, the *part* segmentation, by successively stacking the part masks on top of the instance segmentation. The parts are processed in an order that ensures that smaller parts are not covered by the larger ones. This results in a hierarchical segmentation, with a high-level segmentation (the instance map), and a low-level segmentation (the part map). At this stage, the misalignment of object and part boundaries creates a lot of spurious regions in the hierarchies. We mitigate this problem by filtering out the small regions (size less than 30 pixels) from the hierarchy using the method described in [12]. The entire processing pipeline was developed using the Higra library [11]. By performing this computation on each sample of the Pascal-Part dataset, we obtain a hierarchical segmentation dataset with an ultrametric dissimilarity grid for 10103 natural images. Three samples of our ultrametric dataset are shown in Fig. 2. The dataset is publicly available here.

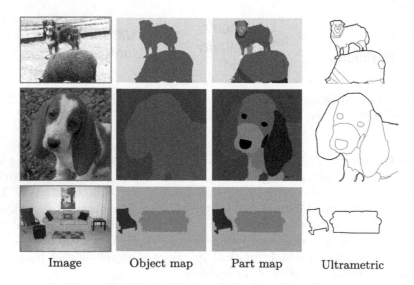

Image	Object map	Part map	Ultrametric

Fig. 2. Three samples from our Ultrametric Pascal-Part dataset, each consisting of an image, an object segmentation, a part segmentation, and the corresponding UCM.

The ultrametric dissimilarity grid will be the annotations that we will use in our method. Since the images are the same as the PASCAL Context [10], we can use the same dataset splits as used in COB [9]: *VOC train* refers to the official PASCAL Context train set, while the official PASCAL Context validation set is divided in two to create *VOC val* and *VOC test*. We have verified that this split maintains acceptable object class proportions. In the worst case (for the sofa class), there are still 19.2% of the total number of sofas in the *val* split, where we would expect 25%.

3   Model

Our second contribution is to train a neural network for hierarchical segmentation in a supervised manner, using the ultrametric dataset described in Sect. 2.

Specifically, we approach this as a classification problem on the edges of the 4-adjacency grid of the input image: each edge is classified as being a low, mid, or high-level edge in accordance with the ultrametric dataset annotations.

The central part of our pipeline is a U-Net [17] that outputs three dissimilarity grids, one for each level of the hierarchy. To predict such dissimilarity grids, we first predict in the pixel domain, and then compute the mean of neighboring pixels to obtain dissimilarity grid weights. In the last layer, a softmax activation is used so that the neural network predicts for each edge the probability of belonging to each of the three levels. To train the U-Net, we were inspired by the loss function used in [9,21]. Let Θ be the U-Net parameters, I the input image, and w the dissimilarity weights of the corresponding target ultrametric dissimilarity grid. Furthermore, let $\Lambda(w)$ be the set of unique values in the targets, which are 0 (low), 0.5 (medium), or 1 (high-level). Our balanced cross-entropy loss function is defined as

$$\mathcal{L}(\Theta, I, w) = \sum_{\lambda \in \Lambda(w)} -\beta_\lambda \sum_{e \in E_\lambda(w)} \log \mathbb{P}\big(e \mapsto \lambda \mid \Theta, I\big). \tag{1}$$

In this equation, $E_\lambda(w) = \{e \in E \mid w(e) = \lambda\}$ is the set of edges whose value is λ in the target w, where E denotes the set of all edges in the target. The parameter $\beta_\lambda = 1 - |E_\lambda(w)|/|E|$ mitigates the class imbalance. Finally, $\mathbb{P}(e \mapsto \lambda \mid \Theta, I)$ is the predicted probability that the edge e has value λ in the ultrametric dissimilarity grid, according to the U-Net with parameters Θ for the input image I. Note that this loss function has no hyperparameter.

Once training is complete, image segmentation requires the conversion of the predicted edge class probabilities into hierarchical regions. This is done by post-processing the network predictions: First, we compute a dissimilarity grid from the edge class probabilities:

$$(\forall e \in E) \quad \mathsf{predict}(I, e) = \sum_{\lambda \in \Lambda(w)} \lambda \, \mathbb{P}(e \mapsto \lambda \mid \Theta, I). \tag{2}$$

Second, we compute superpixels with a watershed cut on the dissimilarity grid. Then, we construct a hierarchy of superpixels with average linkage. Finally, we filter out the small regions (smaller than 30 pixels) from the hierarchy [12]. Our complete pipeline is shown in Fig. 3. It allows us to derive hierarchical segmentations for any input image.

4   Evaluation Metrics

Our third contribution is a benchmark for evaluating the quality of hierarchical segmentation using two evaluation metrics. First, we adapted the *Boundary Recovery Order by Hierarchy Level* benchmark proposed in HSA [8]. It originally reports the proportion of boundaries from each level of the target UCM that were recovered in the predicted UCM, as a function of the overall recall. We have adapted it by making it a function of the segmentation threshold $t \in [0, 1]$

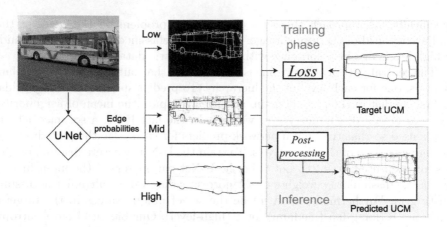

Fig. 3. Description of our method, for training and inference phase.

to see at what threshold the boundaries of each level are the most recovered. More formally, for a given predicted UCM U_{pred} and a given target UCM U_{tar}, we threshold U_{pred} at multiple thresholds t: for each edge e, $U_{pred_t}(e) = 0$ if $U_{pred}(e) < t$, and 1 otherwise. We then match U_{pred_t} with U_{tar}, and compute the Level Recovery Fraction (LRF) of each level $\lambda \in \Lambda(w)$:

$$LRF(\lambda, t) = \frac{|\{e \mid \text{match}_t(e) \text{ and } U_{tar}(e) = \lambda\}|}{|\{e \mid U_{tar}(e) = \lambda\}|}. \qquad (3)$$

where $\text{match}_t(e)$ is true if the edge e of U_{tar} matches an edge of U_{pred_t}. With a proper segmentation, the high-level boundaries will be recovered at a high threshold, and the mid-level boundaries will be recovered at a medium threshold. If the boundaries of each level are recovered simultaneously (i.e. the lines in the figure are similar), it means that the order of the target hierarchy is not reflected in the prediction.

Second, the False Discovery Fraction (FDF) is calculated as follows:

$$FDF(t) = \frac{|\{e \mid \text{not-match}_t(e) \text{ and } U_{pred_t}(e) = 1\}|}{|\{e \mid U_{pred_t}(e) = 1\}|}. \qquad (4)$$

where $\text{not-match}_t(e)$ is true if the edge e of U_{pred_t} does not match with any edge of U_{tar}. This measure, which is close to the false positives, calculates the proportion of boundaries that were detected but did not match with any level of the target UCM, at any segmentation threshold. With a good segmentation, this measure should be 0 for every threshold: this would mean that every boundary of the predicted UCM matched with a level of the target UCM.

Finally, we compute classical precision-recall curves for boundaries [1] and associated F-scores for the finest segmentation of the ground truth (this measure thus ignores the hierarchical nature of the ground-truth). We compute these three evaluation metrics on the test set *VOC test*.

5   Experiments

We evaluate the efficiency of our method through four questions: (i) Is it possible to infer the hierarchical levels of an image in the correct order? (ii) Are the boundaries placed correctly? (iii) Is the performance good for all classes? (iv) How do the models perform on a non-hierarchical dataset?

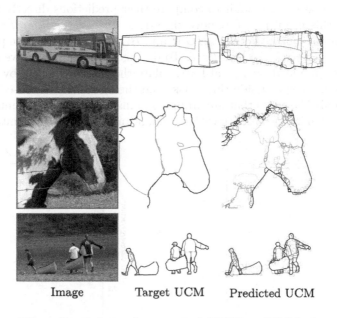

Image   Target UCM   Predicted UCM

Fig. 4. Predictions of our method *HGM* on *VOC test.*

To answer these questions, we trained several neural networks. For each of them, we use a U-Net with a ResNet50 backbone pre-trained on ImageNet. We also augment our training data with simple spatial and texture transformations such as slight rotations, Gaussian noise, and optical distortions. We also perform fine-tuning by freezing the encoder weights of our U-Net except for the last layer of the encoder, training it with a learning rate of $1e^{-4}$ for 30 epochs, then unfreezing the neural network completely, and training it with a learning rate of $1e^{-5}$ for 20 epochs. Finally, we use a learning rate scheduler that divides the learning rate by 3 if the loss on the validation set (*VOC val*) does not decrease for more than 5 epochs, with a batch size of 64. Since the images in the dataset have different dimensions, we had to use mini-batches of 1 and backpropagate every 64 samples. To optimize the neural networks, we used the loss function described in the previous section. We train a first neural network *Binary Grid-weight Model (Ultrametric Pascal-Part)* (BGM_{UPP}), which classifies edges as low-level or high-level boundaries, on the part segmentation of our ultrametric dataset. We train a second neural network *Hierarchical Grid-weight Model*

(HGM), which classifies edges as low-, mid-, or high-level boundaries, on our ultrametric dataset. Some of its predictions are shown in Fig. 4. Finally, we train a neural network (*Binary Grid-weight Model (PASCAL Context) BGM_{PC}*) that classifies edges as low or high-level boundaries, on the PASCAL Context dataset.

We now answer the first two questions by comparing BGM_{UPP}, HGM, COB [9], MCG [14], SCG [14] and *Quadtree* with the three metrics. Since the other methods' neural networks were trained on PASCAL Context and not Pascal-Part, it would be unfair to compare their predictions directly with ours. To mitigate this problem, we remove the edges *that are far from the objects of interest*, both for our predictions (HGM and BGM_{UPP}) and for the predictions of COB, MCG, SCG and *Quadtree*. To do this, for each image, we merge the object masks provided by Pascal-Part, dilate the resulting mask by 20 pixels, remove contour parts outside this mask, and then remove non-closed contours. This leaves only the edges that are around and inside the objects of interest. The results are shown in Fig. 5, along with the quantitative results. Let's interpret this

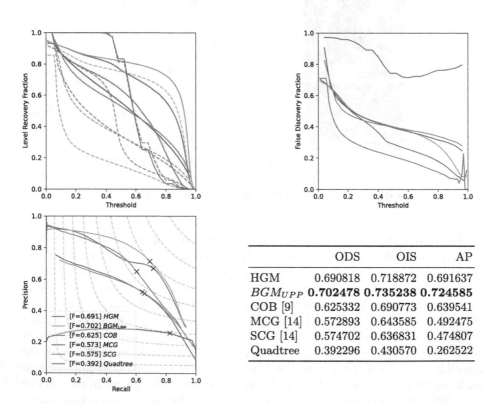

	ODS	OIS	AP
HGM	0.690818	0.718872	0.691637
BGM_{UPP}	**0.702478**	**0.735238**	**0.724585**
COB [9]	0.625332	0.690773	0.639541
MCG [14]	0.572893	0.643585	0.492475
SCG [14]	0.574702	0.636831	0.474807
Quadtree	0.392296	0.430570	0.262522

Fig. 5. Benchmarks our methods HGM (blue) and BGM_{UPP} (orange), on COB [9] (green), MCG [14] (red), SCG [14] (purple) and *Quadtree* (brown) on our ultra-metric dataset, with Level Recovery Fraction (top-left), solid lines for high-level, and dashed lines for mid-level), False Discovery Fraction (top-right), Precision-Recall Curves (bottom-left), and quantitative results (bottom-right). (Color figure online)

figure, starting with the Level Recovery Fraction on the top-left. As expected, for BGM_{UPP} (in orange), both high-level (solid lines) and mid-level (dashed lines) boundaries are recovered at the same high threshold (around 1). This is not the case for HGM (in blue) for which the mid-level boundaries are recovered at a medium threshold (around 0.5). This demonstrates the effectiveness of our method for learning hierarchical image segmentation in a supervised manner. COB, on the other hand, recovers mid-level boundaries at a low threshold, which is normal since their neural network has not been trained to detect them. They, as well as MCG and SCG, detect high-level boundaries almost linearly, whereas we would expect them to be mostly detected at a high threshold. $Quadtree$ recovers both mid-level and high-level boundaries indifferently and linearly at a medium threshold. Let's now focus on the False Discovery Fraction, on the top-right. Here, COB has the lowest amount of false detections, and the other methods except $Quadtree$ seem even. Finally, let's look at the Precision-Recall curves. BGM_{UPP} has the best F-score, followed closely by HGM, and then COB. Note the curve of HGM which, compared to BGM_{UPP}, has a plateau around medium threshold. This is due to the order of level recovery: high-level boundaries are recovered first, and when the medium level boundaries are finally recovered, they are accurately detected. All in all, HGM effectively recovers the levels in the right order while maintaining very good F-scores and FDF. The other methods do not recover the mid-level boundaries at the right threshold, and BGM_{UPP} achieves the best F-score on our ultrametric dataset.

We address the third question by comparing HGM to COB on our hierarchical dataset, class by class. To do this, for each class, we have cropped the predictions with a small margin around each instance of the class. We do the same on the target UCMs, as well as removing the boundaries of other classes, and compute the three evaluation metrics on them. Some results are shown in Fig. 6, and the full quantitative results are available at the end of the article (Fig. 8). Let's start with the Level Recovery Fraction: for most classes, the hierarchical order is reflected in HGM, except for *bicycles*, *bottles*, *plants*, and *trains* where it is more ambiguous. Regarding the False Discovery Fraction, there is basically no difference with the first experiment: COB has a lower False Discovery Fraction for every class. Finally, the Precision-Recall curves and the F-scores change significantly from class to class. HGM has better F-scores for some classes (*bus*, *car*, *cat*, *person*), COB is better for some other classes (*boat*, *bottle*, *chair*, *pottedplant*), and the F-scores are even for the remaining classes.

Finally, we answer the last question: how do we perform on a non-hierarchical dataset such as PASCAL Context, compared to other methods? To do this, we compare BGM_{PC}, COB, MCG, SCG and $Quadtree$ with the Precision-Recall, but not the Level Recovery Order as PASCAL Context is not hierarchical, nor the False Discovery Fraction as in a single-scale segmentation environment it is the false positive rate that is already reflected in the Precision-Recall curves. The results are shown in Fig. 7. Although COB still leads in terms of F-score, our simple neural network method remains competitive. This also proves that

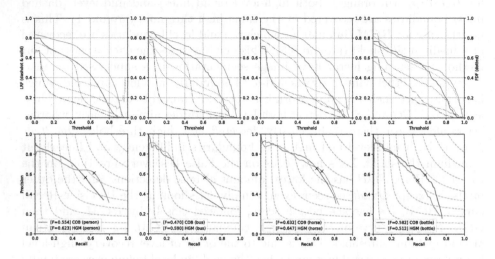

Fig. 6. Benchmarks on *COB* [9] (blue) and our method *HGM* (orange) on our ultra-metric dataset per class (person on the left, bus in the middle-left, horse in the middle-right, bottle on the right), with Level Recovery Fraction (top, solid and dash-dot lines), False Discovery Fraction (top, dotted lines) and Precision-Recall Curves (bottom). (Color figure online)

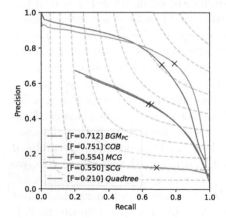

	ODS	OIS	AP
BGM_{PC}	0.711773	0.750888	0.749766
COB [9]	**0.750677**	**0.785024**	**0.773474**
MCG [14]	0.554028	0.614356	0.375386
SCG [14]	0.550186	0.608747	0.330745
Quadtree	0.210233	0.212235	0.126505

Fig. 7. Benchmark on the second experiment on PASCAL Context, with Precision-Recall Curves, and quantitative results.

	ODS	OIS	AP
COB (aeroplane)	0.680047	0.657341	0.686952
HGM (aeroplane)	0.658955	0.621932	0.670039
COB (bicycle)	0.640781	0.603702	0.561952
HGM (bicycle)	0.602852	0.575719	0.585744
COB (bird)	0.648532	0.560844	0.619821
HGM (bird)	0.621336	0.522969	0.586066
COB (boat)	0.482006	0.409506	0.359728
HGM (boat)	0.421522	0.379770	0.285375
COB (bottle)	0.581943	0.509849	0.514802
HGM (bottle)	0.512207	0.451764	0.454253
COB (bus)	0.469884	0.491188	0.462006
HGM (bus)	0.589534	0.562339	0.538801
COB (car)	0.448132	0.413176	0.386884
HGM (car)	0.550439	0.446858	0.455647
COB (cat)	0.574996	0.600807	0.582450
HGM (cat)	0.614230	0.623224	0.604625
COB (chair)	0.508438	0.507779	0.374837
HGM (chair)	0.444168	0.460078	0.325109
COB (cow)	0.623693	0.594259	0.607716
HGM (cow)	0.646403	0.588841	0.606057
COB (dog)	0.580688	0.592093	0.582632
HGM (dog)	0.618172	0.610517	0.589478
COB (horse)	0.631965	0.601856	0.606264
HGM (horse)	0.647019	0.606201	0.616302
COB (motorbike)	0.586627	0.570394	0.550539
HGM (motorbike)	0.545491	0.542184	0.525384
COB (person)	0.553968	0.518099	0.500073
HGM (person)	0.623382	0.547710	0.531034
COB (pottedplant)	0.541500	0.489925	0.462293
HGM (pottedplant)	0.456295	0.407224	0.340515
COB (sheep)	0.583776	0.515271	0.514862
HGM (sheep)	0.583537	0.500565	0.489362
COB (sofa)	0.421639	0.527803	0.279333
HGM (sofa)	0.440106	0.506891	0.340161
COB (train)	0.479617	0.519992	0.405943
HGM (train)	0.474488	0.495916	0.414802
COB (tvmonitor)	0.538754	0.572695	0.444587
HGM (tvmonitor)	0.507300	0.499316	0.379238

Fig. 8. Quantitative results on Ultrametric Pascal-Part, class by class

the performance improvement in the previous experiments is not due to the backbone models, but rather to the hierarchical structure.

6 Conclusion

Hierarchical image segmentation offers interesting advantages over traditional image segmentation methods by allowing for multi-scale analysis, capturing objects and structures at different levels of detail. In this paper, we described a comprehensive pipeline for supervised learning of hierarchical image segmentation. We performed an ultrametric adaptation of the Pascal-Part dataset, built a neural network to predict ultrametric dissimilarity grids, and trained it on the latter dataset. We also evaluated its performance, in terms of boundary localization, hierarchy order, and false discovery fraction, and demonstrated the effectiveness of our method for learning hierarchical image segmentation. We showed that the results vary significantly from class to class. We have shown that our method, although using a simple neural network, remains competitive for non-hierarchical image segmentation compared to more complex neural network architectures such as COB. In future work, we plan to incorporate continuous hierarchy optimization methods [3] to obtain an end-to-end supervised hierarchical segmentation method. Another interesting question would be the prediction of semantic information in the hierarchy of contours.

References

1. Arbelaez, P., Maire, M., Fowlkes, C., Malik, J.: Contour detection and hierarchical image segmentation. IEEE Trans. Pattern Anal. Mach. Intell. **33**(5), 898–916 (2010)
2. Chen, X., Mottaghi, R., Liu, X., Fidler, S., Urtasun, R., Yuille, A.: Detect what you can: detecting and representing objects using holistic models and body parts. In: Proceedings of the IEEE Conference on Computer Vision and Pattern Recognition (CVPR), June 2014
3. Chierchia, G., Perret, B.: Ultrametric fitting by gradient descent. In: Wallach, H., Larochelle, H., Beygelzimer, A., d' Alché-Buc, F., Fox, E., Garnett, R. (eds.) Advances in Neural Information Processing Systems, vol. 32. Curran Associates, Inc. (2019). www.proceedings.neurips.cc/paper/2019/file/b865367fc4c0845c0682bd466e6ebf4c-Paper.pdf
4. Everingham, M., Van Gool, L., Williams, C.K.I., Winn, J., Zisserman, A.: The pascal visual object classes (VOC) challenge. Int. J. Comput. Vision **88**(2), 303–338 (2010)
5. Funke, J., et al.: Large scale image segmentation with structured loss based deep learning for connectome reconstruction. IEEE Trans. Pattern Anal. Mach. Intell. **41**(7), 1669–1680 (2018)
6. Guigues, L., Cocquerez, J.P., Le Men, H.: Scale-sets image analysis. Int. J. Comput. Vision **68**, 289–317 (2006)
7. Koenderink, J.J.: The structure of images. Biol. Cybern. **50**(5), 363–370 (1984)
8. Maire, M., Yu, S.X., Perona, P.: Hierarchical scene annotation. In: British Machine Vision Conference (BMVC) (2013)

9. Maninis, K.K., Pont-Tuset, J., Arbeláez, P., Van Gool, L.: Convolutional oriented boundaries: from image segmentation to high-level tasks. IEEE Trans. Pattern Anal. Mach. Intell. **40**(4), 819–833 (2017)
10. Mottaghi, R., et al.: The role of context for object detection and semantic segmentation in the wild. In: Proceedings of the IEEE Conference on Computer Vision and Pattern Recognition, pp. 891–898 (2014)
11. Perret, B., Chierchia, G., Cousty, J., Guimarães, S.J.F., Kenmochi, Y., Najman, L.: Higra: hierarchical graph analysis. SoftwareX **10**, 100335 (2019)
12. Perret, B., Cousty, J., Guimarães, S.J.F., Kenmochi, Y., Najman, L.: Removing non-significant regions in hierarchical clustering and segmentation. Pattern Recogn. Lett. **128**, 433–439 (2019). https://doi.org/10.1016/j.patrec.2019.10.008
13. Perret, B., Cousty, J., Tankyevych, O., Talbot, H., Passat, N.: Directed connected operators: asymmetric hierarchies for image filtering and segmentation. IEEE Trans. Pattern Anal. Mach. Intell. **37**(6), 1162–1176 (2014)
14. Pont-Tuset, J., Arbeláez, P., Barron, J., Marques, F., Malik, J.: Multiscale combinatorial grouping for image segmentation and object proposal generation. In: arXiv:1503.00848, March 2015
15. Pont-Tuset, J., Arbelaez, P., Barron, J.T., Marques, F., Malik, J.: Multiscale combinatorial grouping for image segmentation and object proposal generation. IEEE Trans. Pattern Anal. Mach. Intell. **39**(1), 128–140 (2016)
16. Ren, Z., Shakhnarovich, G.: Image segmentation by cascaded region agglomeration. In: Proceedings of the IEEE Conference on Computer Vision and Pattern Recognition, pp. 2011–2018 (2013)
17. Ronneberger, O., Fischer, P., Brox, T.: U-net: convolutional networks for biomedical image segmentation. In: Navab, N., Hornegger, J., Wells, W.M., Frangi, A.F. (eds.) MICCAI 2015. LNCS, vol. 9351, pp. 234–241. Springer, Cham (2015). https://doi.org/10.1007/978-3-319-24574-4_28
18. Salembier, P., Garrido, L.: Binary partition tree as an efficient representation for image processing, segmentation, and information retrieval. IEEE Trans. Image Process. **9**(4), 561–576 (2000)
19. Tao, A., Sapra, K., Catanzaro, B.: Hierarchical multi-scale attention for semantic segmentation (2020). https://doi.org/10.48550/ARXIV.2005.10821, www.arxiv.org/abs/2005.10821
20. Uijlings, J.R., Van De Sande, K.E., Gevers, T., Smeulders, A.W.: Selective search for object recognition. Int. J. Comput. Vision **104**, 154–171 (2013)
21. Xie, S., Tu, Z.: Holistically-nested edge detection. In: Proceedings of the IEEE International Conference on Computer Vision, pp. 1395–1403 (2015)

Unveiling the Influence of Image Super-Resolution on Aerial Scene Classification

Mohamed Ramzy Ibrahim[1,2(✉)] 🆔, Robert Benavente[2] 🆔, Daniel Ponsa[2] 🆔, and Felipe Lumbreras[2] 🆔

[1] Computer Engineering Department, Arab Academy for Science, Technology and Maritime Transport, Alexandria, Egypt
m.ramzy@aast.edu
[2] Computer Vision Center and Computer Science Department, Universitat Autònoma de Barcelona, Barcelona, Spain
{mramzy,robert,daniel,felipe}@cvc.uab.es

Abstract. Deep learning has made significant advances in recent years, and as a result, it is now in a stage where it can achieve outstanding results in tasks requiring visual understanding of scenes. However, its performance tends to decline when dealing with low-quality images. The advent of super-resolution (SR) techniques has started to have an impact on the field of remote sensing by enabling the restoration of fine details and enhancing image quality, which could help to increase performance in other vision tasks. However, in previous works, contradictory results for scene visual understanding were achieved when SR techniques were applied. In this paper, we present an experimental study on the impact of SR on enhancing aerial scene classification. Through the analysis of different state-of-the-art SR algorithms, including traditional methods and deep learning-based approaches, we unveil the transformative potential of SR in overcoming the limitations of low-resolution (LR) aerial imagery. By enhancing spatial resolution, more fine details are captured, opening the door for an improvement in scene understanding. We also discuss the effect of different image scales on the quality of SR and its effect on aerial scene classification. Our experimental work demonstrates the significant impact of SR on enhancing aerial scene classification compared to LR images, opening new avenues for improved remote sensing applications.

Keywords: Super-resolution · Scene classification · Deep learning · Aerial images · Remote sensing

1 Introduction

Super-resolution (SR) techniques aim to generate a detailed and sharp high-resolution (HR) image from low-resolution (LR) images. The goal of SR is to improve the quality of images by enhancing the information they contain. This

V. Vasconcelos et al. (Eds.): CIARP 2023, LNCS 14469, pp. 214–228, 2024.
https://doi.org/10.1007/978-3-031-49018-7_16

is especially useful in some applications where obtaining HR images is difficult due to the environment (i.e., satellite imaging) or excessive costs (i.e., hardware for HR image acquisition is expensive) [15].

SR, as many other tasks in computer vision, has benefited from the rapid development of deep learning (DL), leading to an impressive improvement in performance in terms of image quality [2]. It seems plausible that improving the quality of input images could improve other subsequent high-level vision tasks and enhance their results. Indeed, some previous works have explored this hypothesis in low-visibility scenarios for object detection [24], and object classification [20]. The conclusions of these studies show that while the preprocessing approaches tested can enhance performance in some scenarios, in others the results can even degrade. Some other works have focused on the impact of SR on specific tasks, obtaining some contradictory results. SR improves object detection results [16], has a very limited favorable impact on object classification [26], but has no positive effect on human pose estimation [6]. Therefore, we can conclude that the impact of SR on other high-level vision tasks is not as clear as could be expected and is highly dependent on the specific task at hand. Among the different image classification problems, scene classification on aerial images could really benefit from an improvement in resolution. This task is particularly challenging due to two main factors. First, the way images are acquired, usually from a camera on an aircraft or satellite, implies that images often have very low resolution. This fact implies that important regions in the image have low detail, and valuable information from the complex spatial distribution of the scene is lost. Second, images are sometimes acquired in tough environmental conditions (rain, clouds, etc.). These factors cause some images from different classes to be very similar (i.e., inter-class similarity), while some images from the same class are quite different (i.e., intra-class difference). Examples of these problems are shown in Fig. 1. Hence, improving the image resolution of such images could allow the detection of fine-grained details that might facilitate accurate and reliable scene classification.

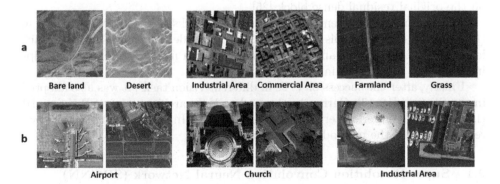

Fig. 1. Examples of (a) inter-class similarity, and (b) intra-class diversity. Images from AID [23], NWPU-RESISC45 [3], and RSSCN7 [27] datasets (left to right in each row).

In this paper, we present a comparative study to test the hypothesis that increasing image resolution can improve the results of aerial scene classification. The aim of this study is to complement previous works that studied the impact of SR on high-level tasks and to evaluate the impact of SR in the field of aerial scene classification. To achieve this goal, we have conducted experiments on (1) different SR models, from the traditional bicubic algorithm to convolutional neural networks (CNNs) and vision transformers (ViTs), (2) different state-of-the-art classifiers based on DL, (3) different benchmark datasets of aerial scenes with complex classes, and (4) simulated LR images with different scaling factors (×0.5 and ×0.25).

2   Super-Resolution

A number of SR methods have been proposed in the last few decades. Initially, traditional methods were based on interpolation, optimization, and learning. However, in terms of resolution and restoration quality, these methods have been superseded by DL approaches [4]. In recent years, different models based on CNNs have been applied to SR tasks. Dong *et al.* [5] proposed the first SR model based on CNN (SRCNN). It is a simple model trained to learn an end-to-end mapping between LR and HR image pairs. Later, Tai *et al.* [18] proposed a Deep Recursive Residual Convolution Neural Network (DRRCNN) consisting of 52 convolution layers with residual connections. They also proposed a memory network (MemNet) [19] based on a recursive and a gate units for explicitly mining persistent memory through an adaptive learning process.

Generative adversarial networks (GANs) have also been used in SR tasks. Ledig *et al.* [11] proposed SRGAN, a general GAN for SR, and Sajjadi *et al.* [14] presented EnhanceNet GAN, which uses automatic texture synthesis and perceptual loss to build realistic textures without focusing on ground truth pixels. Wang *et al.* [21] proposed the Enhanced Super-Resolution Generative adversarial network (ESRGAN) which includes a residual-in-residual dense block based on the original residual dense block [25].

Indeed, the development of residual dense blocks has shown a potential in the SR field. Following this idea, the 3DRRDB model [8] was designed to stack large number of features and bypass them between network layers which helps in generating a good HR image.

Finally, after the success of ViTs in different vision tasks, it was also adopted in the SR field [12,13]. Currently, the leading ViTs architecture in SR is SwinIR [12] that has a superior performance in image SR. In the following sections, we review in detail the methods used in our study.

2.1   Super-Resolution Convolution Neural Network (SRCNN)

SRCNN [5] was the first successful super-resolution CNN, and it pioneered many subsequent approaches. SRCNN structure is simple, consisting only of three convolution layers, each of which (except for the last layer) is followed by a rectified

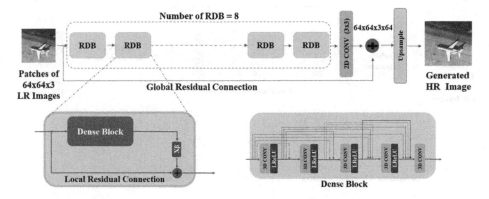

Fig. 2. Modified 3D Residual-in-residual dense block (m3DRRDB) with illustration of dense block.

linear unit (ReLU) non-linearity. The first convolution layer, referred to as feature extraction, is responsible for creating feature maps from the input images. The second convolution layer, known as non-linear mapping, is responsible for converting the feature maps into high-dimensional feature vectors. Finally, the last convolution layer combines the feature maps to produce the final HR image. Despite its simple architecture, SRCNN surpassed all classical SR techniques becoming a real breakthrough in the field.

2.2 Modified 3D Residual-in-Residual Dense Block (m3DRRDB)

The 3D residual-in-residual block model (3DRRDB) [8] was introduced for remote sensing SR and proposes a method to combine several LR images to generate a HR image. This scheme can be recast to estimate a HR multi-band (RGB) image from one LR RGB image by just substituting the 3D convolutions of its pipeline with 2D convolutions. We denote this modified model m3DRRDB.

The key values of this architecture are: (1) The usage of dense blocks that stack large amounts of 2D feature maps and establish maximum information flow between blocks, and (2) fusion of global and local residual connections with residual scaling that solves the problem of vanishing gradient and stabilize training. As shown in Fig. 2, the pipeline of m3DRRDB is composed of eight Residual Dense Blocks (RDBs) [25] with a convolution layer (CONV) with kernel of 3 × 3. The model has a global residual connection that connects the input to the output of the network before the upsampling layer. The importance of the global residual connection is to pass the information from the input to the output of the last RDB to avoid any gradient losses. In turn, as illustrated in lower part of Fig. 2, each RDB block consists of a dense block and a local residual connections that connect input of the dense block to its output after multiplying it by a residual scaling factor. Moreover, each dense block is composed of five convolutions (the first four convolutions are followed by ReLU) where the output of each CONV layer is fed as input to all subsequent CONV layers.

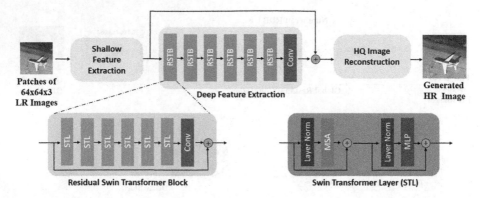

Fig. 3. SwinIR transformer architecture with illustration of RSTB block.

2.3 SwinIR Transformer

ViTs-based SwinIR [12] offers various advantages over CNN-based image restoration models which are: (1) Interactions between image content and attention weights that can be interpreted as spatially variable convolution which allows global interaction between contexts, and (2) the shifted window mechanism in SwinIR allows long-term dependencies modelling to avoid local interaction between image patches that happens in transformers.

As shown in Fig. 3, the pipeline of SwinIR starts with shallow feature extraction, followed by deep feature extraction, and ends with High-Quality (HQ) image reconstruction. The model has a skip connection that concatenates the inputs to deep features extraction phase to its output. In the shallow feature extraction phase, features are extracted from LR patches of size $64 \times 64 \times 3$ with a convolution layer, $CONV$, with kernel size 3×3 as shown in Eq. 1.

$$F_{SF} = CONV(LR),\tag{1}$$

where F_{SF} is the output of shallow feature phase.

For the deep feature extraction phase the Residual Swin Transformer blocks (RSTB) are used. This phase output, F_{DF}, is the output of six RSTB followed by $CONV$ layer with 3×3 kernel where intermediate feature is presented as $F_1, F_2, \ldots, F_6$. Equation 2 represents the complete scenario of deep feature extraction phase.

$$\begin{aligned} F_i &= RSTB_i(F_{i-1}), \quad i = 1, 2, \ldots, 6, \\ F_{DF} &= CONV(F_6), \end{aligned}\tag{2}$$

where $RSTB_i$ is the ith RSTB.

For the HQ image reconstruction phase can be noted as shown in Eq. 3

$$I_{HR} = H_{REC}(F_{SF} + F_{DF}),\tag{3}$$

where I_{HR} is the super-resolved image, and H_{REC} is the reconstruction module.

As shown in Fig. 3, the RSTB consists of six swin transformer layers (STL) followed by a CONV layer. A residual connection connects the input of the first STL to output of CONV layer to avoid gradient losses. Each STL is composed of 2 blocks. The first block is composed of a normalization Layer followed by a multi-headed self-attention (MSA) layer, with a residual connection between the input to the normalization layer and the output of the MSA layer. STLs are alternating between having a shifted window MSA or a regular window MSA. The second block is composed of a normalization layer followed by a multi-layer perceptron (MLP), with a residual connection between the input to normalization layer and the output of MLP.

3   Scene Classification

CNNs are widely used DL models for extracting high-quality representations and delivering robust results on different classification tasks [1,3]. From the pioneering Alexnet [10], several CNN models have been proposed for general object classification. Among them, VGG16 [17], and ResNet-101 [7] have been the most used CNNs for classification and also as a backbone for other tasks. For this reason, we have adopted these two models in our experiments on scene classification.

The VGG16 [17] architecture showed good performance, and this has been attributed to its small kernel size (3×3) and its number of trainable parameters, which aid in increasing the CNN's depth to extract more features [1]. The VGG16 model is made up of five groups. The first two groups consist of two convolution layers with ReLU activation function, and a single max pooling layer after the activation function. The last three groups include three convolution layers with ReLU activation function and a single max pooling layer. Finally three fully connected layers that also use a ReLU activation function. The last layer of the model is a softmax.

ResNet-101 [7] was proposed to solve the problem of vanishing gradients in previous deep CNNs. It is composed of 101 residual blocks, where each block consists of a pipeline of three convolutions. First, it starts with a convolution with multiple kernels, each of size 1×1, followed by ReLU activation function. Second, a convolution layer with multiple kernels, each of size 3×3, followed by ReLU activation function. Third, a convolution layer with multiple kernels, each of size 3×3. Finally, the input to the block is added to output of the third convolution and passes through ReLU activation function. The model ends with a fully connected layer for classification.

4   Experiments

To assess and quantify the impact of SR on aerial scene classification, we conducted two experiments applying different SR models before evaluating two CNN classifiers, namely VGG16 and ResNet-101, on three standard datasets of aerial images. In this section, we provide a general overview of the two experiments, explain the datasets used, and list the experimental settings.

Experiment 1: Ranking of SR Models. The goal of this experiment was to rank the different SR methods used in the classification experiments (see Experiment 2) and have an evaluation of them in terms of image quality. The models evaluated are the bicubic method (used as baseline in most SR works in the literature), SRCNN [5] (the first CNN proposed for SR), m3DRRDB [8] (a CNN model based on the idea of residual dense blocks), and SwinIR [12] (based on ViTs and the current state-of-the-art for SR). These models are evaluated using the Peak Signal-to-Noise Ratio (PSNR) and the Structural Similarity Index Measure (SSIM) which are the quantitative metrics mainly used to assess image quality in SR trials [22].

Experiment 2: Impact of SR on Aerial Scene Classification. In this experiment we asses the impact of the evaluated super-resolution (SR) methods on aerial scene classification across various low-resolution (LR) image scales and with different scene classification methodologies.

We utilized two well-known CNN-based classifiers, namely VGG16 [17] and ResNet-101 [7], and compared their performance on LR images at different scales to their performance on super-resolved images generated by different SR techniques. We also computed the classification results on HR original versions of the images to obtain the theoretical best possible result from the classifiers. A summary of the scenarios evaluated in this experiment is depicted in Fig. 4. The models were evaluated using four quantitative metrics [9]: accuracy, precision, recall, and F1-score. These metrics were computed for each class in each dataset and averaged across all classes.

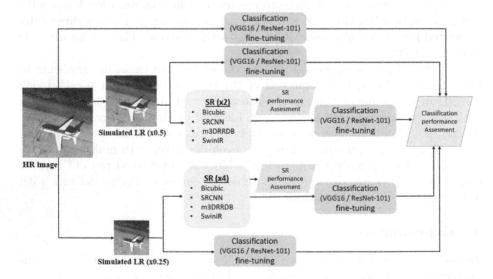

Fig. 4. Different experimental scenarios proposed on the three benchmark datasets.

4.1   Datasets

Experiments are conducted on three well-known benchmark aerial datasets: AID [23], NWPU-RESISC45 [3], and RSSCN7 [27]. Dataset details are listed in Table 1. The datasets are downscaled using the bicubic method to simulate LR images at scales of ×0.5 and ×0.25 for SR and scene classification experiments. The downsampling occured by adding blur to the HR images and after that downsampling them to the required scale.

Table 1. Comparison between the used datasets, AID [23] dataset, NWPU-RESISC45 [3] dataset, and RSSCN7 [27] dataset.

Datasets	Total Images	Scene Classes	No. of Scenes/Class	Spatial Resolution (m/pixel)	Image Sizes	Year
AID [23]	10,000	30	~220 to 420	8 to 0.5	600 × 600	2017
NWPU [3]	31,500	45	700	30 to 0.2	256 × 256	2017
RSSCN7 [27]	2,800	7	400	unspecified	400 × 400	2015

4.2   Experimental Settings

All the experiments were run on machines with a 3.80 GHz Core i7 processor, 32 GB of RAM, and an NVIDIA RTX 3090 with 24 GB of bandwidth. Moreover, the models are implemented using Python language and PyTorch DL framework. Each dataset used in the SR and classification experiments is divided into three portions as 78% for training, 2% for validation and 20% for testing.

Settings for Super-Resolution Methods. The training dataset is cropped to patches of size 64 and feed to the SR networks as batches of 16. The training data is augmented (on the fly) using random rotation, horizontal flip, and vertical flip before feeding it to SR networks. The SR models are trained for 600,000 iterations with learning rate of 0.0003 using a scheduler that decreases the learning rate to half at [250K, 400K, 450K, 475K] to avoid models overfitting during training. Adam optimizer is used with the parameters $\beta_1 = 0.9$ and $\beta_2 = 0.99$.

Settings for Classification Methods. The training data is batched into 32 samples and undergoes on-the-fly augmentation, including random horizontal and vertical flips, random rotations, and normalization before being fed into the classification network. Models are trained for 200 epochs with a learning rate of 0.001 using a cosine annealing scheduler and Adam optimizer with parameters $\beta_1 = 0.9$ and $\beta_2 = 0.99$. For VGG16, we employ a pretrained model from the Imagenet dataset and fine-tune the last three groups and three fully connected layers while freezing the first two groups. For ResNet-101, we utilize a pretrained ResNet-101 model from the Imagenet dataset and fine-tune the last 50 residual blocks while freezing the first 51 residual groups.

Table 2. Quantitative Results for different SR models on AID [23] dataset, NWPU-RESISC45 [3] dataset, and RSSCN7 [27] dataset, in terms of PSNR and SSIM.

Method	Scale	AID		NWPU		RSSCN7	
		PSNR	SSIM	PSNR	SSIM	PSNR	SSIM
Bicubic	×2	32.17	0.8623	29.81	0.8173	29.80	0.7923
SRCNN	×2	34.46	0.9011	31.86	0.8697	31.56	0.8385
m3DRRDB	×2	35.41	0.9152	32.67	0.8871	32.21	0.8538
SwinIR	×2	**36.91**	**0.9323**	**33.28**	**0.9217**	**33.18**	**0.8688**
Bicubic	×4	28.72	0.7358	26.88	0.6686	26.82	0.6323
SRCNN	×4	29.22	0.7443	27.42	0.6912	27.48	0.6712
m3DRRDB	×4	30.38	0.7779	28.19	0.7332	28.12	0.6823
SwinIR	×4	**31.49**	**0.8123**	**29.05**	**0.8992**	**28.92**	**0.6881**

5   Results and Discussion

In this section, we present the results of the experiments described in Sect. 4.

5.1   Experiment 1: Ranking of Super-Resolution Methods

Table 2 lists the results of Experiment 1. It arranges the SR models according to their quantitative results on the three benchmark datasets [3,23,27]. The results are organized according to the two upscaling factors (×2 and ×4). As can be seen in Table 2, the results obtained by the SR models agree with the previously reported results on other datasets [12]: SwinIR outperforms the other methods, obtaining the best PSNR and SSIM results at different scales on the three benchmark datasets, and the model based on residual dense blocks, m3DRRDB, overcomes SRCNN. The bicubic method obtained the lowest results in all the tested scenarios. Moreover, as could be expected, all the SR models get better results at scale (×2) than at scale (×4) in both PSNR and SSIM. Figure 5 shows qualitative results of the SR methods for ×4 scale on AID [23] dataset.

5.2   Experiment 2: Impact of Super-Resolution on Aerial Scene Classification

The results of Experiment 2, highlighting the impact of SR on aerial scene classification, are summarized in Tables 3 (results for VGG16) and 4 (results for ResNet-101). In each table, the four metrics (i.e., accuracy, precision, recall, and F1-score) are given for each evaluated scenario. For each dataset, we present the results on the LR images downsampled at two different scales, ×0.5 and ×0.25, and the corresponding results of the SR methods, with upsampling ×2 and ×4. For each dataset, we provide the results on the original HR images (i.e., before downsampling) as the theoretical best result achievable.

From the results in the tables, we can state that, overall, SR enhances the results on aerial scene classification. In the experiments, all the SR methods consistently yielded better classification performance in all the metrics compared to the results on LR images. Such improvement is achieved across all scales, classification methods, and datasets. Additionally, the scale of upsampling has an

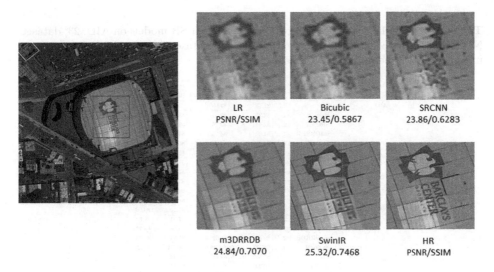

Fig. 5. Qualitative comparison of ×4 SR on AID [23] dataset.

Table 3. VGG16 classification results for different SR models on AID [23] dataset, NWPU-RESISC45 [3] dataset, and RSSCN7 [27] dataset in terms of mean accuracy, mean precision, mean recall, and mean F1-score.

Dataset	Method	Scale	Classification Metric			
			Accuracy (%)	Precision (%)	Recall (%)	F1-Score (%)
AID	LR	×0.5	74.20	74.88	74.14	74.51
	Bicubic	×2	75.45	76.28	75.23	75.75
	SRCNN	×2	77.85	78.38	77.42	77.90
	m3DRRDB	×2	78.40	78.86	77.89	78.37
	SwinIR	×2	**79.90**	**79.86**	**79.47**	**79.66**
	LR	×0.25	73.50	74.33	73.47	73.90
	Bicubic	×4	74.90	75.78	74.74	75.25
	SRCNN	×4	76.00	77.35	76.36	76.85
	m3DRRDB	×4	77.20	77.83	76.90	77.36
	SwinIR	×4	**77.90**	**78.44**	**77.47**	**77.95**
	HR	–	81.25	81.03	80.78	80.91
NWPU-RESISC45	LR	×0.5	85.89	86.70	86.93	86.86
	Bicubic	×2	86.14	87.00	87.11	87.06
	SRCNN	×2	87.60	88.16	88.48	88.32
	m3DRRDB	×2	88.30	88.52	89.02	88.77
	SwinIR	×2	**89.73**	**89.79**	**90.18**	**89.98**
	LR	×0.25	85.73	86.64	86.82	86.73
	Bicubic	×4	85.95	86.82	86.99	86.90
	SRCNN	×4	87.40	87.96	88.34	88.15
	m3DRRDB	×4	87.21	87.78	88.10	87.94
	SwinIR	×4	**88.44**	**88.66**	**89.11**	**88.88**
	HR	–	90.10	89.99	90.60	90.29
RSSCN7	LR	×0.5	73.75	74.18	73.84	74.01
	Bicubic	×2	74.46	74.95	74.56	74.75
	SRCNN	×2	75.71	76.16	75.85	76.00
	m3DRRDB	×2	76.07	76.51	76.19	76.35
	SwinIR	×2	**76.79**	**77.23**	**76.88**	**77.05**
	LR	×0.25	71.96	72.74	72.11	72.42
	Bicubic	×4	73.93	74.75	74.01	74.38
	SRCNN	×4	74.82	76.04	74.92	75.48
	m3DRRDB	×4	**75.71**	**77.06**	**75.83**	**76.44**
	SwinIR [12]	×4	75.18	76.44	75.28	75.86
	HR	–	78.21	79.28	78.27	78.77

Table 4. ResNet-101 classification results for different SR models on AID [23] dataset, NWPU-RESISC45 [3] dataset, and RSSCN7 [27] dataset in terms of mean accuracy, mean precision, mean recall, and mean F1-score.

Dataset	Method	Scale	Classification Metric			
			Accuracy (%)	Precision (%)	Recall (%)	F1-Score (%)
AID	LR	×0.5	81.75	81.95	81.53	81.74
	Bicubic	×2	82.25	82.55	82.04	82.30
	SRCNN	×2	83.45	83.76	83.28	83.52
	m3DRRDB	×2	84.60	84.75	84.35	84.55
	SwinIR	×2	**85.35**	**85.53**	**85.13**	**85.33**
	LR	×0.25	80.85	81.77	80.77	81.27
	Bicubic	×4	81.45	81.98	81.72	81.85
	SRCNN	×4	82.50	83.03	82.61	82.82
	m3DRRDB	×4	83.20	83.47	83.05	83.26
	SwinIR	×4	**84.80**	**85.01**	**84.58**	**84.80**
	HR	–	88.65	88.69	88.34	88.51
NWPU-RESISC45	LR	×0.5	86.68	86.42	87.10	86.76
	Bicubic	×2	87.49	87.27	87.81	87.54
	SRCNN	×2	88.71	88.46	89.08	88.77
	m3DRRDB	×2	89.35	89.02	89.74	89.38
	SwinIR	×2	**90.22**	**89.96**	**90.54**	**90.25**
	LR	×0.25	85.89	85.78	86.44	86.10
	Bicubic	×4	86.30	86.11	86.71	86.41
	SRCNN	×4	88.05	87.79	88.49	88.14
	m3DRRDB	×4	87.38	87.15	87.75	87.45
	SwinIR	×4	**88.86**	**88.55**	**89.15**	**88.85**
	HR	–	92.13	91.95	92.27	92.11
RSSCN7	LR	×0.5	82.68	83.21	82.68	82.94
	Bicubic	×2	83.75	84.40	83.75	84.07
	SRCNN	×2	84.11	84.64	84.11	84.37
	m3DRRDB	×2	84.82	85.21	84.82	85.02
	SwinIR	×2	**85.71**	**85.93**	**85.71**	**85.82**
	LR	×0.25	82.32	82.85	82.32	82.59
	Bicubic	×4	83.21	83.83	83.21	83.52
	SRCNN	×4	83.93	84.43	83.93	84.18
	m3DRRDB	×4	83.39	84.02	83.39	83.7
	SwinIR	×4	**84.82**	**85.21**	**84.82**	**85.02**
	HR	–	87.32	87.65	87.32	87.49

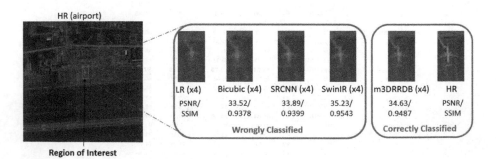

Fig. 6. Fine-tunned VGG16 missclassification example: Airport scene missclassified as Farm Land with illustration of degradation in RoI (note: PSNR/SSIM is for the full super-resolved image not the RoI region only).

impact on the classification metrics, with classifiers yielding better metrics when SR methods are applied to upsample at ×2 scale compared to the corresponding results at ×4 scale. This scale effect is consistent across all benchmark datasets.

Tables 3 and 4 also reveal that for the ×2 scale, SwinIR demonstrates the best performance as a SR method in terms of producing super-resolved images which obtain the highest classification results across all three benchmark datasets. This result is also observed at the ×4 scale for ResNet-101 (Table 4) for on the three datasets, and for VGG16 (Table 3) on the AID and NWPU-RESISC45 datasets. However, in the case of VGG16 on the RSSCN7 dataset, SwinIR super-resolved images do not achieve the best classification results, and m3DRRDB is the best SR method instead. Additionally, in the ×4 scale, SRCNN super-resolved images slightly outperform m3DRRDB super-resolved images in classification for NWPU-RESISC45 dataset using VGG16 (Table 3), and also for both NWPU-RESISC45 and RSSCN7 datasets using ResNet-101 (Table 4). Hence, we can observe that in some cases the ranking of SR methods obtained in terms of PSNR and SSIM in Experiment 1 is not kept when we evaluate them according to the improvement they provide on classification. This highlights the limitations of PSNR and SSIM as metrics for predicting the impact of SR methods on other tasks. Thus, from our results, we can conclude that achieving the highest PSNR and SSIM results does not guarantee the highest performance in aerial scene classification. This could be due to the fact that these metrics evaluate overall scene clarity rather than a specific region of interest (RoI) within the scene, which sometimes can be more relevant for a given task.

This possibility is illustrated in Fig. 6 which shows a missclassification example where the image represents class "airport". We hypothesize that inside the RoI (an aircraft that can be crucial to identify the class of the image) a lot of detail is lost in the LR image, and bicubic, SRCNN, and SwinIR can not recover it in the super-resolved images, which leads to wrongly classify the image as class "farm land". However, it is classified correctly as class "airport" with m3DRRDB which obtains a super-resolved image where details of the RoI are much clearer and more similar to the HR image. In this example, the full scene scored highest PSNR/SSIM of 35.23/0.9543 using SwinIR but it is missclassifed while it is correctly classified with m3DRRDB (second best PSNR/SSIM of 34.63/0.9487).

6 Conclusion

In this paper, we have studied the effect of different SR methods on aerial scene classification. We have first ranked different types of SR methods (traditional, CNN-based, and ViTs-based) in terms of image quality on two simulated LR image scales (×0.5 and ×0.25) from three benchmark datasets on aerial scene classification. Then, we have assessed how pre-processing images with SR techniques affects aerial scene classification by two well-known CNN classifiers (VGG16 and ResNet-101). This second experiment included aerial scene classification comparisons between simulated LR images at different scales (×0.5, ×0.25) and different super-resolved images obtained from the considered

SR methods. The extensive experimental work shows that SR methods consistently improve aerial scene classification compared to LR images for all the SR methods, all the scales, all the datasets, and all the classifiers tested.

Furthermore, we proved that the performance of a SR method in terms of PSNR and SSIM is not always directly related to the degree of improvement on aerial scene classification, especially when working with small LR scale images ($\times 0.25$). We draw the hypothesis that this result is due to the fact that PSNR and SSIM are metrics designed to measure the overall clarity of the image rather than that of specific RoIs, which can be determinants for a given task. In aerial scene classification, with the challenging inter-class similarity and intra-class diversity of aerial images, details of a certain RoI can be especially valuable for a correct classification. Moreover, we also hypothesize that sometimes SR of LR images with very small scales ($\times 0.25$) can amplify artifacts that affect the classification results. However, more experiments are needed to prove these hypothesis. As a future work, the presented study can be extended to test the effects of SR on other tasks such as image segmentation and object detection on specific datasets. Moreover, further studies on smaller scales ($\times 8$ and $\times 16$) are needed to test if the reported improvement on classification performance holds for lower resolutions, where the effects of artifacts and image degradation can be increasingly challenging.

Acknowledgment. This work is partially supported by Grant PID2021-128945NB-I00 funded by MCIN/AEI/10.13039/501100011033 and by "ERDF A way of making Europe". The authors acknowledge the support of the Generalitat de Catalunya CERCA Program to CVC's general activities, and the Departament de Recerca i Universitats from the Generalitat de Catalunya with reference 2021SGR01499.

References

1. Alom, M.Z., et al.: The history began from AlexNet: a comprehensive survey on deep learning approaches. arXiv (2018). http://arxiv.org/abs/1803.01164
2. Anwar, S., Khan, S., Barnes, N.: A deep journey into super-resolution: a survey. arXiv (2020). https://doi.org/10.48550/arXiv.1904.07523
3. Cheng, G., Han, J., Lu, X.: Remote sensing image scene classification: benchmark and state of the art. Proc. IEEE **105**(10), 1865–1883 (2017). https://doi.org/10.1109/JPROC.2017.2675998
4. Khoo, J.J.D., Lim, K.H., Phang, J.T.S.: A review on deep learning super resolution techniques. In: 2020 IEEE 8th Conference on Systems, Process and Control (ICSPC), pp. 134–139 (2020). https://doi.org/10.1109/ICSPC50992.2020.9305806
5. Dong, C., Loy, C.C., He, K., Tang, X.: Learning a deep convolutional network for image super-resolution. In: Fleet, D., Pajdla, T., Schiele, B., Tuytelaars, T. (eds.) ECCV 2014. LNCS, vol. 8692, pp. 184–199. Springer, Cham (2014). https://doi.org/10.1007/978-3-319-10593-2_13
6. Hardy, P., Dasmahapatra, S., Kim, H.: Can super resolution improve human pose estimation in low resolution scenarios? In: 17th International Conference on Computer Vision Theory and Applications, pp. 494–501 (2022). www.scitepress.org/Link.aspx?doi=10.5220/0010863700003124

7. He, K., Zhang, X., Ren, S., Sun, J.: Deep residual learning for image recognition. In: 2016 IEEE Conference on Computer Vision and Pattern Recognition (CVPR), pp. 770–778 (2016). https://doi.org/10.1109/CVPR.2016.90

8. Ibrahim, M.R., Benavente, R., Lumbreras, F., Ponsa, D.: 3DRRDB: super resolution of multiple remote sensing images using 3D residual in residual dense blocks. In: 2022 IEEE/CVF Conference on Computer Vision and Pattern Recognition Workshops (CVPRW), pp. 322–331 (2022). https://doi.org/10.1109/CVPRW56347.2022.00047

9. Ibrahim, M.R., Youssef, S.M., Fathalla, K.M.: Abnormality detection and intelligent severity assessment of human chest computed tomography scans using deep learning: a case study on SARS-COV-2 assessment. J. Ambient. Intell. Humaniz. Comput. 14(5), 5665–5688 (2023). https://doi.org/10.1007/s12652-021-03282-x

10. Krizhevsky, A., Sutskever, I., Hinton, G.E.: ImageNet classification with deep convolutional neural networks. In: Advances in Neural Information Processing Systems, vol. 25. Curran Associates, Inc. (2012). www.papers.nips.cc/paper_files/paper/2012/hash/c399862d3b9d6b76c8436e924a68c45b-Abstract.html

11. Ledig, C., et al.: Photo-realistic single image super-resolution using a generative adversarial network. In: 2017 IEEE Conference on Computer Vision and Pattern Recognition (CVPR), pp. 105–114 (2017). https://doi.org/10.1109/CVPR.2017.19

12. Liang, J., Cao, J., Sun, G., Zhang, K., Van Gool, L., Timofte, R.: SwinIR: image restoration using swin transformer. In: 2021 IEEE/CVF International Conference on Computer Vision Workshops (ICCVW), pp. 1833–1844 (2021). https://doi.org/10.1109/ICCVW54120.2021.00210

13. Lu, Z., et al.: Transformer for single image super-resolution. In: 2022 IEEE/CVF Conference on Computer Vision and Pattern Recognition Workshops (CVPRW), pp. 456–465 (2022). https://doi.org/10.1109/CVPRW56347.2022.00061

14. Sajjadi, M.S.M., Scholkopf, B., Hirsch, M.: EnhanceNet: single image super-resolution through automated texture synthesis. In: 2017 IEEE International Conference on Computer Vision (ICCV), pp. 4501–4510 (2017). https://doi.org/10.1109/ICCV.2017.481

15. Salvetti, F., Mazzia, V., Khaliq, A., Chiaberge, M.: Multi-image super resolution of remotely sensed images using residual attention deep neural networks. Remote Sens. 12(14), 2207 (2020). https://doi.org/10.3390/rs12142207

16. Shermeyer, J., Van Etten, A.: The effects of super-resolution on object detection performance in satellite imagery. In: 2019 IEEE/CVF Conference on Computer Vision and Pattern Recognition Workshops (CVPRW), pp. 1432–1441 (2019). https://doi.org/10.1109/CVPRW.2019.00184

17. Simonyan, K., Zisserman, A.: Very deep convolutional networks for large-scale image recognition. In: Bengio, Y., LeCun, Y. (eds.) 3rd International Conference on Learning Representations, ICLR 2015, San Diego, CA, USA, 7–9 May 2015 (2015). arxiv.org/abs/1409.1556

18. Tai, Y., Yang, J., Liu, X.: Image super-resolution via deep recursive residual network. In: 2017 IEEE Conference on Computer Vision and Pattern Recognition (CVPR), pp. 2790–2798 (2017). https://doi.org/10.1109/CVPR.2017.298

19. Tai, Y., Yang, J., Liu, X., Xu, C.: MemNet: a persistent memory network for image restoration. In: 2017 IEEE International Conference on Computer Vision (ICCV), pp. 4549–4557 (2017). https://doi.org/10.1109/ICCV.2017.486

20. Vidal, R.G., et al.: UG^2: a video benchmark for assessing the impact of image restoration and enhancement on automatic visual recognition. In: 2018 IEEE Winter Conference on Applications of Computer Vision (WACV), pp. 1597–1606 (2018). https://doi.org/10.1109/WACV.2018.00177

21. Wang, X., et al.: ESRGAN: enhanced super-resolution generative adversarial networks. In: Leal-Taixé, L., Roth, S. (eds.) ECCV 2018. LNCS, vol. 11133, pp. 63–79. Springer, Cham (2019). https://doi.org/10.1007/978-3-030-11021-5_5

22. Wang, Z., Bovik, A., Sheikh, H., Simoncelli, E.: Image quality assessment: from error visibility to structural similarity. IEEE Trans. Image Process. **13**(4), 600–612 (2004). https://doi.org/10.1109/TIP.2003.819861

23. Xia, G.S., et al.: AID: a benchmark data set for performance evaluation of aerial scene classification. IEEE Trans. Geosci. Remote Sens. **55**(7), 3965–3981 (2017). https://doi.org/10.1109/TGRS.2017.2685945

24. Yang, W., et al.: Advancing image understanding in poor visibility environments: a collective benchmark study. IEEE Trans. Image Process. **29**, 5737–5752 (2020). https://doi.org/10.1109/TIP.2020.2981922

25. Zhang, Y., et al.: Residual dense network for image super-resolution. In: 2018 IEEE/CVF Conference on Computer Vision and Pattern Recognition, pp. 2472–2481 (2018). https://doi.org/10.1109/CVPR.2018.00262

26. Zhou, L., Chen, G., Feng, M., Knoll, A.: Improving low-resolution image classification by super-resolution with enhancing high-frequency content. In: 2020 25th International Conference on Pattern Recognition (ICPR), pp. 1972–1978 (2021). https://doi.org/10.1109/ICPR48806.2021.9412876

27. Zou, Q., Ni, L., Zhang, T., Wang, Q.: Deep learning based feature selection for remote sensing scene classification. IEEE Geosci. Remote Sens. Lett. **12**(11), 2321–2325 (2015). https://doi.org/10.1109/LGRS.2015.2475299

Weeds Classification with Deep Learning: An Investigation Using CNN, Vision Transformers, Pyramid Vision Transformers, and Ensemble Strategy

Guilherme Botazzo Rozendo[1,4]($\boxtimes$) (ID), Guilherme Freire Roberto[2](ID),
Marcelo Zanchetta do Nascimento[3](ID), Leandro Alves Neves[4](ID),
and Alessandra Lumini[1](ID)

[1] Department of Computer Science and Engineering (DISI) - University of Bologna,
Bologna, Italy
guilherme.botazzo@unibo.it
[2] Faculty of Engineering, University of Porto (FEUP), Porto, Portugal
[3] Faculty of Computer Science (FACOM), Federal University of Uberlândia (UFU),
Uberlândia, Brazil
[4] Department of Computer Science and Statistics (DCCE),
São Paulo State University, São Paulo, Brazil

Abstract. Weeds are a significant threat to agricultural production.
Weed classification systems based on image analysis have offered innovative solutions to agricultural problems, with convolutional neural networks (CNNs) playing a pivotal role in this task. However, CNNs are
limited in their ability to capture global relationships in images due to
their localized convolutional operation. Vision Transformers (ViT) and
Pyramid Vision Transformers (PVT) have emerged as viable solutions to
overcome this limitation. Our study aims to determine the effectiveness
of CNN, PVT, and ViT in classifying weeds in image datasets. We also
examine if combining these methods in an ensemble can enhance classification performance. Our tests were conducted on significant agricultural
datasets, including DeepWeeds and CottonWeedID15. The results indicate that a maximum of 3 methods in an ensemble, with only 15 epochs
in training, can achieve high accuracy rates of up to 99.17%. This study
demonstrates that high accuracies can be achieved with ease of implementation and only a few epochs.

Keywords: Weeds classification · CNN · Pyramid Vision
Transformers · Vision transformers · Ensemble

1 Introduction

Ensuring food security and sustainability among the ever-growing global population requires a substantial increase in agricultural production. However, this
task is difficult due to numerous factors, including the proliferation of weeds,

© Springer Nature Switzerland AG 2024
V. Vasconcelos et al. (Eds.): CIARP 2023, LNCS 14469, pp. 229–243, 2024.
https://doi.org/10.1007/978-3-031-49018-7_17

which significantly affect crop yield and quality. Weeds compete with cultivated plants for water, sunlight, and nutrients, reducing crop productivity and economic losses for farmers [4]. Traditional weed control methods, such as manual labor or chemical herbicides, are often time-consuming, costly, and environmentally detrimental [4,9].

In recent years, with the advancements in computer vision and deep learning, weed detection and classification systems based on image analysis opened up new avenues for tackling agricultural challenges [4,10]. Weed image classification refers to categorizing images of plants into specific weed species or differentiating them from desirable crops. This task requires extracting meaningful features from the images and applying robust classification algorithms. Deep learning techniques, such as convolutional neural networks (CNNs), have proven highly effective in achieving outstanding accuracy levels in weed classification tasks [7,9, 12]. However, CNNs have inherent limitations in capturing global relationships in images due to their localized convolutional operations, and this led to the development of methods such as Vision Transformers (ViT) [3] and Pyramid Vision Transformers (PVT) [13].

ViTs have emerged as a recent paradigm shift in computer vision, leveraging the power of self-attention mechanisms to enable holistic image understanding and achieve state-of-the-art performance in various visual recognition tasks. Unlike CNNs, which rely on sequential convolutional layers, ViTs are built upon the Transformer architecture that operates on a patch-based representation of images [3]. Thus, the input image is divided into fixed-size patches, each represented by a learnable embedding vector. These patches are then linearly projected into a sequence of embeddings processed by multiple Transformer encoder layers. The self-attention mechanism allows the model to capture global dependencies and learn long-range relationships between patches.

PVT [13] is a model that combines the strengths of both CNNs and ViTs. It introduces a hierarchical approach that leverages a multiscale feature pyramid to capture images' local and global contextual information. The multiscale feature pyramid comprises multiple levels operating at different spatial resolutions. At each level, PVT employs a set of transformers to model the relationships between features, allowing it to capture fine-grained details and high-level semantic information. PVT also introduces novel attention mechanisms: local-global and global-local attention modules. The local-global attention module selectively attends to nearby and distant features, capturing both local details and global context. Conversely, the global-local attention module attends to the global representation while incorporating local information, enabling a comprehensive understanding of the image at different scales. Training PVT involves a combination of supervised and self-supervised learning, such as generative models, which can pre-train it on unlabeled data, enhancing its ability to capture robust and generalizable representations.

However, despite the relevance of the mentioned methods, it is still possible to improve the classification power of these techniques by combining them into an ensemble. Ensemble learning is a machine learning strategy that combines multiple models, known as base learners, to make more accurate predictions or decisions than any single model could achieve alone [8]. The idea behind ensemble learning is to leverage the collective intelligence of diverse models to improve overall performance. In ensemble learning, the base learners can be trained on the same dataset using different algorithms. Each base learner learns from the data and produces its prediction, which is combined to produce the final prediction. One way to combine the prediction is by applying the voting method. In this approach, each base learner gives a prediction, and the final prediction is determined by majority voting. Ensemble learning has several advantages: it can improve prediction accuracy, reduce overfitting, and increase the model's robustness to noisy data. It is particularly effective when the base learners are diverse and make uncorrelated errors [8].

Therefore, considering the relevance of the presented methods and the potential increase in the classification accuracy rate by combining them, we propose an investigation of the relevance of CNN, PVT, and ViT for classifying weeds in datasets of images. We also investigate an ensemble that merges these methods to verify if it can improve classification performance. To validate our approach, we tested it on significant agricultural datasets, such as DeepWeeds [9] and CottonWeedID15 [1]. This research makes four significant contributions:

1. A set of methods that effectively recognize and classify weeds with few epochs;
2. A comparative analysis of CNN, PVT, and ViT approaches in weed datasets;
3. The use of an ensemble strategy to boost the classification performance of individual models;
4. A study of the importance of CNN, PVT, ViT, and ensemble techniques in the newly released CottonWeedID15 dataset.

2   Related Work

Olsen et al. [9] applied the CNNs Inception-v3 and ResNet-50 on the DeepWeeds dataset using the transfer learning strategy. The CNNs were pre-trained on the ImageNet dataset, and their last fully connected layer with 1,000 neurons was replaced by a layer with nine fully-connected neurons. These models achieved an average classification accuracy of 95.1% and 95.7%.

Hu et al. [5] proposed a new deep learning architecture named Graph Weeds Net with the goal of classifying species from the DeepWeeds dataset. The model is based on multi-scale graph representations and uses the DenseNet-202 model as backbone. In general, the approach consists in splitting the input image in multiple-patches at different scales. Then a RNN cell applies graph concatenation and outputs the class with the highest prediction value. The method was able to provide an accuracy value of 98.1% using a 5-fold cross validation approach.

Huertas-Tato et al. [7] proposed an ensemble that combined the output probabilities of CNNs with the output probabilities of classic machine learning algorithms trained with statistical features. The result was a feature vector given as input to a set of classifiers. The ensemble was applied on the DeepWeeds dataset

[9] and provided an accuracy rate of 94.67% using DenseNet-201, the statistical features, and the K-nearest neighbors (kNN) classifier.

Saleem et al. [11] proposed a Faster Region-based CNN approach using the ResNet-101 model and anchor boxes, which consists on a multiscale analysis of different regions of the input image. The method was evaluated on the Deep-Weeds dataset using k-fold cross validation and provided a mean average precision of 96.02%.

Sharma and Vardhan [12] used a ViT model to classify crops and weeds using the transfer learning strategy. The model was pre-trained on the ImageNet-21k, fine-tuned on the ImageNet-1k, and evaluated on the DeepWeeds dataset. The authors used token representations at the end of every prediction to provide more accurate predictions. This strategy provided an accuracy of 98.54%.

Recently, a new weeds dataset consisting of 5,187 colored images from 15 types of weeds observed in cotton fields has been made available by [1]. In the same paper, the authors evaluated 35 deep learning models using transfer learning and Monte Carlo cross-validation in order to provide a benchmark for their new dataset. The best results were observed in models based on ResNet, ResNeXt and RepVGG architectures, which provided accuracy values of up to 98.93%. However, since this dataset is relatively new, there has not been much research directed to it and the few published papers have only applied well-known deep-learning approaches. Thus, the approaches explored with the results achieved here contribute significantly to the specialized literature on this topic.

3   Methodology

3.1   Datasets

The DeepWeeds [9] dataset (Fig. 1) comprises 17,509 labeled images of eight weed species from Australia selected due to their notoriety for invasiveness and damaging impact on rural areas. Table 1 shows the species and the number of samples. The negative class groups images of neighboring flora and backgrounds that did not contain the weed species of interest. The images were acquired through a ground-based weed control robot with high-resolution cameras and fast-acting solenoid sprayers that performed selective spot spraying of identified weed targets. The camera lens was 1 m from the ground, and the ground clearance underneath the robotic vehicle was 288 mm. The field of view of the optical system was 450 × 280 mm.

The CottonWeedID15 [1] dataset (Fig. 2) consists of 5187 color images of 15 weed species collected in the southern United States under natural light conditions and at varied weed growth stages. Table 2 shows each species and the number of samples. The images were collected from cotton fields using smartphones or hand-held digital cameras from different view angles, under natural field light conditions, at varying stages of weed growth, and at different locations across the cotton belt states (primarily in North Carolina and Mississippi).

Table 1. Distribution of classes in DeepWeeds dataset.

Species	Number of samples	Percentage
Chinee apple	1125	6.43%
Lantana	1064	6.08%
Parkinsonia	1031	5.89%
Parthenium	1022	5.84%
Prickly acacia	1062	6.07%
Rubber vine	1009	5.76%
Siam weed	1074	6.13%
Snake weed	1016	5.80%
Negatives	9106	52.01%

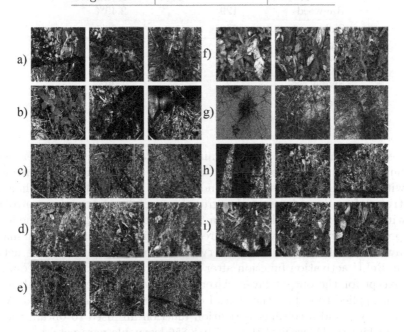

Fig. 1. Sample images from each class of the DeepWeeds dataset: a) Chinee apple, b) Lantana, c) Parkinsonia, d) Parthenium, e) Prickly acacia, f) Rubber vine, g) Siam weed, h) Snake weed and i) Negatives.

3.2   Models

For the CNN model, we used the DenseNet121 [6] available on Pytorch's Torchvision library[1]. DenseNet121 is a popular CNN used for image classification. Its architecture is densely connected, where each layer is connected to every other layer in a feed-forward manner. The network consists of multiple layers

---

[1] https://pytorch.org/vision/main/models/generated/torchvision.models.densenet121

Table 2. Distribution of classes in CottonWeedID15 dataset.

Species	Number of samples	Percentage
Carpetweeds	763	14,71%
Crabgrass	111	2,14%
Eclipta	254	4,90%
Goosegrass	216	4,16%
Morningglory	1115	21,50%
Nutsedge	273	5,26%
Palmer Amaranth	689	13,28%
Prickly Sida	129	2,49%
Purslane	450	8,68%
Ragweed	129	2,49%
Sicklepod	240	4,63%
Spotted Spurge	234	4,51%
Spurred Anoda	61	1,18%
Swinecress	72	1,39%
Waterhemp	451	8,69%

organized into dense blocks, each containing several convolutional layers with a fixed number of output feature maps. The dense blocks include a convolution layer with filter size 7×7, stride 2, and padding 3, a 3×3 max pooling layer with stride 2, four dense layers with a growth rate of 32, and another convolution layer with filter size 1×1 followed by a 2×2 average pooling layer. The feature maps produced by earlier layers are directly combined with the feature maps of later layers in each dense block, preserving the information flow. The network uses the ReLU activation function after each convolutional and fully connected layer, except for the output layer. After the last dense block, a global average pooling is applied to reduce the spatial dimensions of the feature maps. A fully connected layer and a softmax activation function are used to obtain the final class probabilities. DenseNet121 has 7,978,856 learnable parameters.

We used the second version of PVT (PVTv2) proposed by Wang et al. [14] and available on GitHub[2]. The PVT passes the input image through convolutional layers to capture low-level visual features and downsample its spatial resolution. Then, the model processes the images using a pyramid-style feature fusion strategy. It consists of multiple stages, each containing a set of transformer blocks with increasing spatial resolutions. Each block consists of multi-head self-attention layers and position-wise feed-forward networks. The self-attention layers capture relationships between patches, while the feed-forward networks provide non-linear transformations to the feature embeddings. At each stage, the previous stage's output is upsampled to match the current spatial resolu-

---

[2] https://github.com/whai362/PVT/tree/v2/classification

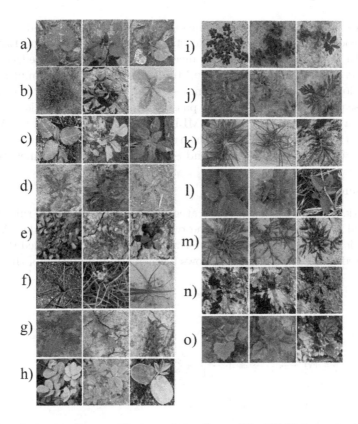

Fig. 2. Sample images from each class of the CottonWeedID15 dataset: a) Morning-glory, b) Carpetweeds, c) Palmer Amaranth, d) Waterhemp, e) Purslane, f) Nutsedge, g) Eclipta, h) Sicklepod, i) Spotted Spurge, j) Ragweed, k) Goosegrass, l) Prickly Sida, m) Crabgrass, n) Swinecress, o) Spurred Anoda.

tion. The upsampled features are then combined with the features from the previous stage to create a fused representation. The PVT employs a hierarchical self-attention mechanism that divides the input into multiple partitions and performs self-attention within each partition separately. After the transformer encoder, a classification head is applied to the final fused representation. The PVTv2 model adds three designs to reduce the computational complexity: a linear complexity attention layer, an overlapping patch embedding, and a convolutional feed-forward network. In this work, we used the PVTv2-B5 version with 82 million parameters.

For the ViT model, we used the one proposed by Dosovitskiy et al. [3], available on fast.ai's timm repository[3]. The model takes as input RGB images and divides them into non-overlapping 16×16 patches. Each patch is then linearly projected into a lower-dimensional embedding space, and position embeddings, encoding the spatial coordinates of the patches, are added to the patch embed-

---

[3] https://github.com/fastai/timmdocs/tree/master/

dings to incorporate positional information. The embeddings are fed into a transformer encoder consisting of multiple stacked transformer blocks. Each block consists of a multi-head self-attention mechanism and position-wise feed-forward networks. It allows the model to attend to different parts of the input sequence, capturing local and global dependencies. After the transformer encoder, the resulting sequence of embeddings is passed through a classification head that consists of a multi-layer perceptron (MLP). In this work, we used the ViT-base version that consists of 12 layers, the transformer's latent vector of size 768, MLP with 3072 neurons, 12 heads, and 86 million learnable parameters.

3.3 Ensemble

In this work, we used an ensemble strategy that involved the sum rule. We used the sigmoid function to normalize the output scores of each model. After this normalization process, we added the normalized scores and identified the highest score to predict the class. Figure 3 shows a schematic illustration of this process.

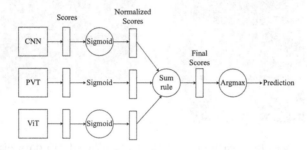

Fig. 3. Schematic summary of how works the ensemble strategy used in this work.

3.4 Evaluation

The k-fold cross-validation is the process of partitioning the dataset into k equally sized subsets (folds). One of the folds is considered the validation set, and the remaining k-1 folds are combined to form the training set. The model training is repeated k times, with each fold being the validation set once. The performance metric, such as accuracy, is calculated for each iteration, and the average of the accuracies of each fold calculates the final result. The k-fold cross-validation efficiently uses the data by utilizing all samples for training and validation, helping maximize the information extracted from the dataset. The repeated evaluation of different validation sets helps to mitigate overfitting issues. It also reduces bias, ensuring the model's performance is not influenced by the particular data distribution in a single train-test split.

The Monte Carlo cross-validation also involves splitting the data into folds. However, instead of using fixed or predetermined partitions, the dataset is randomly sampled multiple times to create different training and validation subsets. The random sampling process follows the Monte Carlo simulation approach, where each sample represents a different data configuration. The model is trained on each sample and evaluated on the corresponding validation set.

3.5   Training Configuration

We conducted this research through two experiments, one in the DeepWeeds dataset and the other in the CottonWeedID15 dataset. In the first experiment, we used the images from the DeepWeeds dataset at their original size of 256×256 pixels. The output layer of all methods was changed to support nine classes. Moreover, the evaluation method used was the stratified 5-fold cross-validation with a split of 60%-20%-20% for the training, validation, and test sets, according to the standard defined by Olsen et al. [9]. In the second experiment, all images from the CottonWeedID15 dataset were resized to 512×512 pixels, as done in work by Chen et al. [1]. We changed the output layer of the methods to hold 15 classes, and the evaluation method used was the 5-fold Monte Carlo cross-validation with a split of 65%-20%-15%. We used the accuracy metric to measure the performance of the methods in both experiments. All of the decisions set out above were made to be able to compare the classification performances with as little bias as possible. Furthermore, we also calculated the F1-score metric to provide better insight into the performance of the methods in the condition of unbalanced datasets.

In both experiments, classical data augmentation was performed in the training set. The chosen transformations were random rotation between -360 and $360°$, random horizontal flip with a probability of 50%, and color jitter with hue and saturation equal to 0.5. All the models were pre-trained on ImageNet [2] and fine-tuned on DeepWeeds and CottonWeedID15 datasets. We train all convolutional and classification layers of the models involved here for 15 epochs. We chose this value for the number of epochs because, through empirical experiments, we verified that there was little evolution of the methods around 10 to 15 epochs. Figure 4 shows examples of the evolution of accuracy in the validation groups of each fold over the epochs in the DeepWeeds dataset. It is possible to notice a stabilization of the accuracy around epoch 10. However, we chose 15 epochs to be able to define the weights that provide the peak of accuracy during the stabilization. For the DeepWeeds dataset, we used batches with size 32 for all models. For CottonWeedID15, we utilized different batch sizes for each model type to optimize memory usage. We set the batch size to 32 for CNN, 8 for ViT, and 4 for PVT, as these two models require more memory than CNNs.

Since the classes of the two datasets under study are highly unbalanced (Tables 1 and 2), we used the weighted loss function in training to avoid bias. The weights of each class were defined using the equation:

$$w_i = 1 - \frac{n_i}{N}, \tag{1}$$

where w_i is the weight of class i, n_i is the number of samples of class i and N is the total number of samples in the dataset. We used the cross entropy loss function and the stochastic gradient descent optimizer. Table 3 shows the summary of all parameters used in this work.

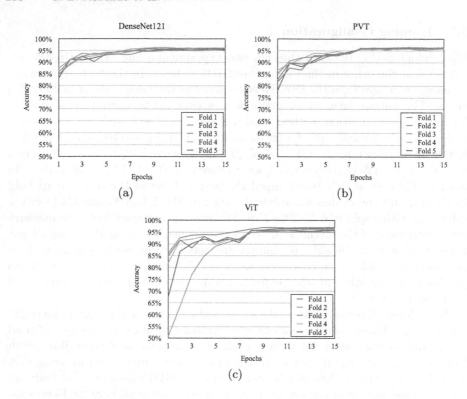

Fig. 4. Accuracy on each fold of the DeepWeeds validation set.

Table 3. Summary of settings used in training.

	Experiment 1	Experiment 2
Dataset	DeepWeeds	CottonWeedID15
Image size	256	512
Number of classes	9	15
Evaluation method	Stratified 5-fold cross-validation	5-fold Monte Carlo cross-validation
Split (train-val-test)	60%-20%-20%	65%-20%-15%
Data augmentation	Classical	Classical
Initial weights	ImageNet	ImageNet
Epochs	15	15
CNN batch size	32	32
PVT batch size	32	4
ViT batch size	32	8
Loss function	Weighted cross entropy	Weighted cross entropy
Optimizer	Stochastic gradient descent	Stochastic gradient descent

3.6   Execution Environment

The proposed method was implemented using Python 3.9.16 and the Pytorch 1.13.1 API. The experiments were performed on a computer with an 12th Generation Intel® Core™i7-12700, 2.10GHz, NVIDIA® GeForce RTX™3090 card, 64 GB of RAM and Windows operating system with 64-bit architecture.

4   Results and Discussion

Tables 4 shows the accuracy achieved with the models in the test sets of all folds of the DeepWeeds dataset (Experiment 1). After evaluating each method's performance, it was observed that they had similar accuracies, averaging around 96%. However, the ViT method had a slight edge over the others, with an average accuracy of 96.10%. When all three methods were combined in an ensemble (Ensemble 1), the accuracy improved in every fold, resulting in an average accuracy of 97.05%. We also conducted tests using ensembles combining the models two by two, identified as "Ensemble 2" (PVT + ViT), "Ensemble 3" (Densenet121 + PVT), and "Ensemble 4" (Densenet121 + ViT) in Table 4. However, the outcomes were inferior to those of Ensemble 1, indicating that all three methods played a role in achieving excellent results with Ensemble 1. When analyzing performance from the perspective of the F1-score metric (Table 5), it is possible to note that Ensemble 1 remains the best case among all those considered.

Table 4. Accuracy on each fold of the DeepWeeds test set.

	DenseNet121	PVT	ViT	Ensemble 1	Ensemble 2	Ensemble 3	Ensemble 4
Fold 1	95,81%	96,07%	96,64%	97,06%	**97.35%**	96.58%	96.78%
Fold 2	95,78%	95,92%	95,40%	**97,09%**	96.83%	96.57%	96.52%
Fold 3	95,09%	95,74%	96,57%	**96,74%**	96.63%	96.26%	96.00%
Fold 4	95,37%	96,31%	96,26%	**96,97%**	96.83%	96.51%	96.40%
Fold 5	96,20%	96,23%	95,63%	**97,57%**	96.97%	97,40%	97.06%
Average	95,65%	96,05%	96,10%	**97,09%**	96.97%	96.66%	96.55%

Table 5. Average F1-score on each fold of the DeepWeeds test set.

	DenseNet121	PVT	ViT	Ensemble 1	Ensemble 2	Ensemble 3	Ensemble 4
Fold 1	0,9475	0,9472	0,9569	**0,9931**	0,9649	0,9566	0,9598
Fold 2	0,9462	0,9481	0,9381	**0,9629**	0,9595	0,9565	0,9546
Fold 3	0,9395	0,9485	0,9570	**0,9600**	0,9585	0,9539	0,9516
Fold 4	0,9398	0,9538	0,9516	**0,9612**	0,9597	0,9553	0,9536
Fold 5	0,9531	0,9532	0,9438	**0,9696**	0,9607	0,9682	0,9630
Average	0,9452	0,9502	0,9495	**0,9694**	0,9607	0,9581	0,9565

The results of Experiment 2 on the CottonWeedID15 dataset are displayed in Tables 6 and 7. It is evident from the Table 6 that PVT and ViT models outperformed DenseNet121 in terms of classification accuracy. The average accuracies of PVT and ViT were almost 99%, precisely 98.77%, and 98.85%, respectively. In contrast, the DenseNet121 model yielded an average accuracy of 97.43%. It is possible to note that the average accuracy obtained from the Ensemble 1, that combines the threw models, was 98.63%, lower than the accuracy achieved by ViT alone. This drop in performance indicates that one of the methods in the ensemble introduced errors with a high degree of certainty. In Experiment 2, the best case was with Ensemble 2, which only included PVT and ViT methods. With Ensemble 2, we achieved an impressive average accuracy rate of 99.17%, surpassing the individual performances of the methods and the other ensembles. Regarding F1-score, although Ensemble 2 is not the best case in folds 2 and 3, it remains the best on average.

Table 6. Accuracy on each fold of the CottonWeedID15 test set.

	DenseNet121	PVT	ViT	Ensemble 1	Ensemble 2	Ensemble 3	Ensemble 4
Fold 1	97.54%	98.44%	98.83%	98.70%	**98.96%**	98.31%	98.44%
Fold 2	97.15%	98.96%	98.70%	98.18%	**99.09%**	98.05%	97.92%
Fold 3	98.44%	98.31%	98.44%	98.70%	**98.96%**	98.44%	98.70%
Fold 4	97.28%	98.83%	98.96%	98.70%	**99.35%**	98.31%	98.44%
Fold 5	96.76%	99.35%	99.35%	98.83%	**99.48%**	98.83%	98.18%
Average	97.43%	98.77%	98.85%	98.63%	**99.17%**	98.39%	98.34%

Table 7. Average F1-score on each fold of the CottonWeedID15 test set.

	DenseNet121	PVT	ViT	Ensemble 1	Ensemble 2	Ensemble 3	Ensemble 4
Fold 1	0,9457	0,9794	0,9774	0,9721	**0,9851**	0,9679	0,9655
Fold 2	0,9586	**0,9890**	0,9715	0,9691	0,9876	0,9680	0,9635
Fold 3	0,9768	0,9727	0,9695	**0,9812**	0,9811	0,9772	0,9809
Fold 4	0,9584	0,9867	0,9832	0,9773	**0,9927**	0,9712	0,9715
Fold 5	0,9464	0,9901	0,9873	0,9775	**0,9911**	0,9775	0,9617
Average	0,9572	0,9836	0,9778	0,9754	**0,9875**	0,9724	0,9686

Table 8 compares our proposed method to other works in the DeepWeeds dataset. The table includes information regarding the methods, evaluation techniques, the number of epochs, and accuracy. The works are arranged in ascending order by year of publication. In the table, CV stands for cross-validation. Sharma and Vardhan achieved the highest accuracy rate of 98.54%, followed by Hu et al. with 98.10% and our method with 97.09%. It is important to note that Sharma and Vardhan employed the hold-out evaluation method, utilizing 85% of the data for training and 15% for validation. Thus, the published accuracy is specific to the findings of the validation group. It is crucial to keep in mind that depending solely on the validation set's results and employing a hold-out method for evaluation could lead to an overestimation of the results.

It is essential to highlight that combining CNN, ViT, and PVT provided high performance, even with only a few training epochs (15 epochs). The results presented here indicate that achieving or surpassing state-of-the-art results can be done with few epochs and easy implementation, as these models are already available in popular APIs such as TensorFlow and PyTorch.

Table 8. Accuracy rates of the proposed method and related works on the DeepWeeds dataset.

	Accuracy	Method	Evaluation method (Split)	Epochs
Olsen et al. [9]	95.10%	Inception-v3	5-fold CV (60-20-20)	200
	95.70%	ResNet-50	5-fold CV (60-20-20)	200
Hu et al. [5]	95.30%	DenseNet202	5-fold CV (60-20-20)	-
	92.00%	InceptionResNet-V2	5-fold CV (60-20-20)	-
	97.40%	GWN (ResNet-50 backbone)	5-fold CV (60-20-20)	-
	98.10%	GWN (DenseNet202 backbone)	5-fold CV (60-20-20)	-
Huertas-Tato et al. [7]	94.67%	DenseNet-201 + statistical features + kNN	Hold-out (70-10-20)	1000
Saleem et al. [11]	96.02%	Faster Region-based ResNet-101	Stratified 5-fold CV (70-20-10)	-
Sharma and Vardhan [12]	98.54%	ViT + token representations	Hold-out (85-15)	30
Proposed method	97.09%	DenseNet121 + PVT + ViT	Stratified 5-fold CV (60-20-20)	15

Table 7 presents the top-10 accuracy rates published by Chen et al. [1] in the CottonWeedID15 dataset compared to our method. The ensemble proposed here, combining the PVT and ViT methods, provided the highest accuracy: 99.17%, followed by the RepVGG-B1 method with 98.90% and RepVGG-B2 with 98.88%. The noteworthy point is that the proposed ensemble had the best performance with only 15 epochs, 35 fewer epochs than the other methods.

An important observation is that the images of the CottonWeedID15 dataset were acquired by digital cameras and smartphones. Therefore, the excellent result (99.17% accuracy) provided by the proposed ensemble indicates the critical point that the acquisition, pattern recognition, and classification can be performed cheaply. However, it is worth pointing out that this dataset is highly unbalanced. Thus, despite the promising results, it is necessary to investigate the proposed method combined with data-balancing strategies (Table 9).

Table 9. Accuracy rates of the proposed method and related works on the Cotton-WeedID15 dataset.

	Accuracy	Method	Evaluation method (Split)	Epochs
Chen et al. [1]	98.93%	ResNeXt101	5-fold Monte Carlo CV (65-20-15)	50
	98.90%	RepVGG-B1	5-fold Monte Carlo CV (65-20-15)	50
	98.88%	RepVGG-B2	5-fold Monte Carlo CV (65-20-15)	50
	98.80%	ResNeXt50	5-fold Monte Carlo CV (65-20-15)	50
	98.70%	RepVGG-A2	5-fold Monte Carlo CV (65-20-15)	50
	98.65%	RepVGG-B0	5-fold Monte Carlo CV (65-20-15)	50
	98.35%	RepVGG-A1	5-fold Monte Carlo CV (65-20-15)	50
	98.24%	RepVGG-A0	5-fold Monte Carlo CV (65-20-15)	50
	98.02%	DPN68	5-fold Monte Carlo CV (65-20-15)	50
	97.97%	ResNet101	5-fold Monte Carlo CV (65-20-15)	50
Proposed method	99.17%	PVT + ViT	5-fold Monte Carlo CV (65-20-15)	15

5   Conclusions

In this work, we investigated the discriminatory potential of the PVT and ViT methods in weed datasets. We explore how an ensemble combining these two methods with CNN can improve the classification power. In the DeepWeeds dataset, the proposed method performance was comparable to state-of-the-art, providing an accuracy of 97.09%. In the CottonWeedID15 dataset, the results were even better. Our method reached an impressive accuracy rate of 99.17%, surpassing all published performances with this dataset. The results presented here were obtained with ensembles constituted with a maximum of 3 methods and trained with only 15 epochs, demonstrating that high accuracies can be achieved with few epochs and ease of implementation.

In future works, we intend to conduct new tests with different combinations of methods in the ensemble and different fusion techniques. Additional data augmentation techniques, such as generative adversarial networks, can also be explored to train models with even better generalization. We also understand that pre-processing techniques, such as segmentation and color normalization, can be applied to increase classification performance. Finally, the importance of exploring data balancing methods in these datasets is evident since they are highly unbalanced.

Acknowledgement. This study was carried out within the Agritech National Research Center and received funding from the European Union Next-GenerationEU (PIANO NAZIONALE DI RIPRESA E RESILIENZA (PNRR) - MISSIONE 4 COMPONENTE 2, INVESTIMENTO 1.4-D.D. 1032 17/06/2022, CN00000022). This manuscript reflects only the authors' views and opinions, neither the European Union nor the European Commission can be considered responsible for them. This work was also partially funded by the Coordenação de Aperfeiçoamento de Pessoal de Nível Superior - Brasil (CAPES) - Finance Code 001, and project NextGenAI - Center for

Responsible AI (2022-C05i0102-02), supported by IAPMEI, and also by FCT pluri-anual funding for 2020-2023 of LIACC (UIDB/00027/2020 UIDP/00027/2020). The authors gratefully acknowledge the financial support of National Council for Scientific and Technological Development - CNPq (Grants 311404/2021-9 and 313643/2021-0).

References

1. Chen, D., Lu, Y., Li, Z., Young, S.: Performance evaluation of deep transfer learning on multi-class identification of common weed species in cotton production systems. Comput. Electron. Agric. **198**, 107091 (2022)
2. Deng, J., Dong, W., Socher, R., Li, L.J., Li, K., Fei-Fei, L.: ImageNet: a large-scale hierarchical image database. In: 2009 IEEE Conference on Computer Vision and Pattern Recognition, pp. 248–255. IEEE (2009)
3. Dosovitskiy, A., et al.: An image is worth 16x16 words: transformers for image recognition at scale. arXiv preprint arXiv:2010.11929 (2020)
4. Hasan, A.M., Sohel, F., Diepeveen, D., Laga, H., Jones, M.G.: A survey of deep learning techniques for weed detection from images. Comput. Electron. Agric. **184**, 106067 (2021)
5. Hu, K., Coleman, G., Zeng, S., Wang, Z., Walsh, M.: Graph weeds net: a graph-based deep learning method for weed recognition. Comput. Electron. Agric. **174**, 105520 (2020)
6. Huang, G., Liu, Z., Van Der Maaten, L., Weinberger, K.Q.: Densely connected convolutional networks. In: Proceedings of the IEEE Conference on Computer Vision and Pattern Recognition, pp. 4700–4708 (2017)
7. Huertas-Tato, J., Martín, A., Fierrez, J., Camacho, D.: Fusing CNNs and statistical indicators to improve image classification. Inf. Fusion **79**, 174–187 (2022)
8. Mohammed, A., Kora, R.: A comprehensive review on ensemble deep learning: opportunities and challenges. J. King Saud Univ.-Comput. Inf. Sci. **35**, 757–774 (2023)
9. Olsen, A., et al.: DeepWeeds: a multiclass weed species image dataset for deep learning. Sci. Rep. **9**(1), 2058 (2019)
10. Rai, N., et al.: Applications of deep learning in precision weed management: a review. Comput. Electron. Agric. **206**, 107698 (2023)
11. Saleem, M.H., Potgieter, J., Arif, K.M.: Weed detection by faster RCNN model: an enhanced anchor box approach. Agronomy **12**(7), 1580 (2022)
12. Sharma, S., Vardhan, M.: Self-attention vision transformer with transfer learning for efficient crops and weeds classification. In: 2023 6th International Conference on Information Systems and Computer Networks (ISCON), pp. 1–6. IEEE (2023)
13. Wang, W., et al.: Pyramid vision transformer: a versatile backbone for dense prediction without convolutions. In: Proceedings of the IEEE/CVF International Conference on Computer Vision, pp. 568–578 (2021)
14. Wang, W., et al.: PVT v2: improved baselines with pyramid vision transformer. Comput. Vis. Media **8**(3), 415–424 (2022)

Leveraging Question Answering for Domain-Agnostic Information Extraction

Bruno Carlos Luís Ferreira(✉)(iD), Hugo Gonçalo Oliveira(iD),
and Catarina Silva(iD)

University of Coimbra, DEI, CISUC, Coimbra, Portugal
{brunof,hroliv,catarina}@dei.uc.pt

Abstract. Transformers gave a considerable boost to Natural Language Processing, but their application to specific scenarios still poses some practical issues. We present an approach for extracting information from technical documents on different domains, with minimal effort. It leverages on generic models for Question Answering and on questions formulated with target properties in mind. These are made to specific sections where the answer, then used as the value for the property, should reside. We further describe how this approach was applied to documents of two very different domains: toxicology and finance. For both, results extracted from a sample of documents were assessed by domain experts, who also provided feedback on the benefits of this approach. F-Scores of 0.73 and 0.90, respectively in the toxicological and financial domain, confirm the potential and flexibility of the approach suggesting that, while it cannot yet be fully automated and replace human work, it can support expert decisions, thus reducing time and manual effort.

Keywords: Information Extraction · Question Answering · Transformers · Toxicology Analysis · Exchange-Traded Fund Information

1 Introduction

The amount of data currently available to humans is unprecedented. However, data is often not in a structured form, and thus not ready for being exploited by users and applications. According to the International Data Corporation (IDC), unstructured data accounts for 95% of global data, with an estimate of the compound annual growth rate of 65% [1]. The main characteristics of unstructured data are: no schema, multiple formats, multiple sources and standardization [12,18], characteristics that prevent this data to be processed and analyzed via conventional data tools and methods[1].

Information Extraction (IE) from text is a subtask of Natural Language Processing (NLP) that aims precisely at extracting structured information, *e.g.*,

---

[1] https://www.ibm.com/cloud/blog/structured-vs-unstructured-data.

© Springer Nature Switzerland AG 2024
V. Vasconcelos et al. (Eds.): CIARP 2023, LNCS 14469, pp. 244–256, 2024.
https://doi.org/10.1007/978-3-031-49018-7_18

entities and their properties, from text. With an easier access to relevant information, businesses can unlock valuable insights, discover patterns, and make more informed decisions [7].

Recent developments in Artificial Intelligence (AI), specifically in NLP, are driving the future wave of data, which led to an increased interest in Large Language Models (LLMs) and their potential to meet the current surging challenges. Specifically, with the introduction of the Transformer Neural Network [17] and the consequent emergence of LLMs like Bidirectional Encoder Representations from Transformers (BERT) [8], Robustly Optimized BERT Pretraining Approach (RoBERTa) [11], T5 [15], GPT [5], and others which can be fine-tuned for a broad range of NLP tasks.

In this paper, we describe an approach for IE that relies on a transformer fine-tuned for extractive Question Answering (QA), and show its flexibility when applied to two different domains. This twofold approach extracts the values for a set of properties from given contexts. First, target documents are parsed, in order to identify the set of target properties to extract. Second, natural language questions are formulated for each of the target properties, also considering the structure of the documents. They are made to specific sections and the retrieved answer is used as their value. This is ideally done with models like BERT and RoBERTa because, unlike text generation models (*e.g.*, T5, GPT), in the former, answers correspond to sequences extracted directly from the given context. This minimizes hallucination and enables further explanation.

The potential of the approach is confirmed by its successful application to semi-structured documents of two different domains: toxicology and finance. For each case study, we detail how the practical effort and report on the results of manual evaluation by experts in each domain. F-Scores of 0.73 and 0.90 were achieved, respectively for the toxicological and for the financial domain, and we further conclude that manual work is indeed reduced, without losing full control on the process, which is not desired in such critical areas.

In the remainder of the paper, we overview some related work, describe the approach in more detail, and then report on the two case studies. We conclude with some final thoughts and future directions.

2 Related Work

Traditional approaches to IE fall into two major categories: the pipeline approach relies on tagging for identifying entities and then using Relation Extraction (RE) models for obtaining the relations between entity pairs; the joint approach combines the entity model and the relation model through different strategies, such as constraints or parameters sharing [4].

However, since the introduction of Transformers [17], several works proposed new approaches to IE, including Extractive QA. In this task, often applied to Machine Reading Comprehension (MRC), the model answers questions with spans of text directly extracted from a given textual context. For different domains, with different implementations and different levels of complexity, some reasons to use Extractive QA for IE are: (1) limited data for domain-specific documents where information needs to be retrieved from [14]; (2) features such as shorter answers are harder to learn from other models [2], *e.g.*, for Named Entity recognition (NER); or (3) generalization capability, as classification-based approaches cannot be generalized to new event types or argument roles without additional annotations [10].

Nguyen et al. [14] proposed a pre-trained BERT [8] model combined with a *Convolutional Neural Network* (CNN) for capturing the local context of a tag and an extracted sequence. Their approach has four main components: the representations of the input vectors of the tokens; BERT for learning hidden vectors for each token from the input tag and document; a convolutional layer for capturing the local context; and a *softmax layer* for predicting the value location. IE is reformulated as a *QA* task where the value is pulled from the document by querying the tag, *i.e.*, a list of required information is defined and represented as tags such as "Name of Institution" or "Deadline for Bidding".

Li et al. [10] formulate event extraction as a multi-turn QA approach. Event extraction is typically divided into two sub-tasks: trigger extraction and argument extraction. Possible approaches are broadly classified into two groups: (1) pipeline, where the extraction of triggers and arguments is performed in separate steps, and (2) joint approaches, where all sub-tasks are performed simultaneously in a joint learning fashion. Trigger identification is transformed into an extractive MRC problem where trigger words are identified from given sentences. Classification of triggers is formalized as a YES/NO QA problem that judges whether a possible trigger belongs to a given event type. Argument extraction is also solved via extractive MRC, where questions are constructed iteratively by a target event type and the corresponding argument roles.

Arici et al. [2] used QA for quantity extraction in a price-per-unit problem. They first predict the type of Unit of Measure, *e.g.*, volume, weight, or count, to formulate such as "*What is the total volume?*", then used to get all relevant answers. The approach divides event extraction into three sub-tasks: Trigger Identification, Trigger Classification, and Argument Extraction. These are modeled by a set of QA templates based on MRC [3], *i.e.*, questions are made on a specific context.

Our approach is inspired by the previous and applied to two new domains, toxicological and financial information extraction, where some encountered challenges are also present. These include the presence of limited data for their domain, as well as the variation in the nature and extent of expected responses.

3 Question Answering for Information Extraction

Our approach for IE relies on a model fine-tuned for extractive QA and suits especially well semi-structured documents, *i.e.*, written in natural language and organized in sections and subsections. If the documents follow a known template, assumptions can be made on the section where each property is mentioned. For each property, a natural language question is made using the target section as the context where the answer (*i.e.*, the value) is obtained from. Relying on a simple question avoids that specific patterns have to be handcrafted or learned for each property. Moreover, focusing on specific sections does not require that the documents are fully processed. In this section, we present the models used and a deeper explanation of the approach used.

3.1 Models and Questions

We explored multiple Transformer models for extractive QA, including BERT [8] and RoBERTa [11], fine-tuned in the Stanford Question Answering Dataset (SQuAD) [16]. Many available models were fine-tuned in this dataset, composed of paragraphs (contexts), questions about them, and extracted answers.

Such models learn to extract the answer to a question directly from a given context. Since SQuAD is not focused on a specific domain, resulting models should be able to answer questions on any domain without additional training, as long their format is not significantly different from those in SQuAD, where questions are based on the Six Ws (*Who, What, When, Where, Why, How*). This is why we formulate questions that are as straightforward as possible and start with one of the six Ws. For example, in order to extract species of animals, we rely on the straightforward question "*What is the species?*". When made to a context that includes the sentence "*The eye irritation potential of shampoo in rabbit eyes was not increased by the addition of ZPT*", models trained on SQuAD will retrieve the right answer, "*rabbit*".

3.2 Approach

The proposed approach, depicted in Fig. 1, leverages the capabilities of QA models for extracting information from semi-structured documents. These documents will follow different templates and, to support each, a correspondence has to be done, manually, between: properties to extract, sections where they should be found, and questions for obtaining their values. Focusing on sections enables to provide shorter contexts to the QA models and minimize noise. With aforementioned mapping available, the approach encompasses four steps.

1. **Structure Processing:** Sections of the document are identified by resorting to a table of contents, if available, or to regular expressions for finding section headers/titles. The exact process will depend on the type of document.
2. **Question Contextualization:** For each target property p, the corresponding section s_p is selected as the context for the QA model and a formulated natural language question q_p is selected in accordance from the mapping.
3. **Question Answering:** The QA model is prompted with question q using s_p as context, resulting in answer a_p, used as the value of property p.
4. **Output:** Answers are finally structured in property-value pairs, *i.e.*, (p, a_p).

In the end, for each property p_i, there should be a value, corresponding to a_{pi}. However, when target sections are longer than the maximum input sequence

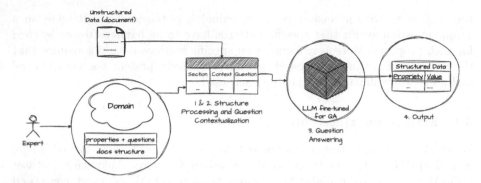

Fig. 1. Approach Workflow. The LLM fine-tuned for QA receives a section of the document and questions for each property to extract. The output is the answer to the question, *i.e.*, the information extracted from unstructured data.

length of the QA model, there might be more than one answer. This happens because, in order to consider the complete section, it has to be split into smaller overlapping fragments. These are iteratively traversed and may result in more than one answer. Here, different answers can be variations of the same, *e.g.* "OECD TG 414 (2001)" and "OECD TG 414". So, in order to remove duplicate answers, we rely on sentence similarity metrics, such as Recall-Oriented Understudy for Gisting Evaluation (ROUGE), Bi-Lingual Evaluation Understudy (BLEU), or the Damerau-Levenshtein Distance. Since all provided similar performance, we settled with ROUGE.

3.3 Visual Explainability on Evaluation

In order to evaluate the approach, we developed a web application with the objective of visually and quickly providing the evaluator the information extracted and an evaluation mechanism in order to obtain the evaluator's feedback.

The web application included a simple user interface for uploading supported PDFs. Once uploaded, the IE process would run and finally show all target properties and their extracted values. If required, these values could be further explained by showing the relevant context from the source document, with the value highlighted, marking the exact position on the context from where the information was extracted, visible in Fig. 2. The highlight component provides the evaluator a visual explanation of the results obtained. Even when incorrect, the highlight enables a quick identification of the incorrect result.

The expert would go through each extraction and check whether they were correct or not. In the latter case, they were also asked to provide the correct values[2].

---

[2] https://github.com/NLP-CISUC/QA-4-Domain-Agnostic-IE-Dataset.

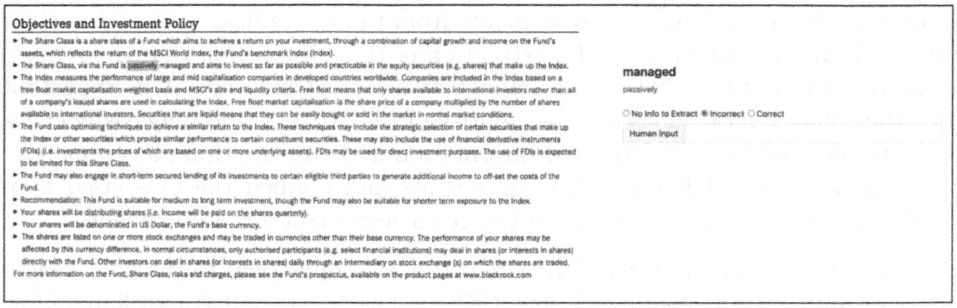

Fig. 2. Evaluation page of the web application. On the left, the context, and the information extracted highlighted. On the right, the information extracted and the evaluation mechanism used.

4 Case Studies

To confirm the flexibility of the QA-based approach for IE, we present case studies in two significantly different domains: toxicological analysis and finance. We report each practical application in detail, while describing the data used, the extracted information and its evaluation.

We experimented with different models, starting with BERT[3] and RoBERTa[4] fine-tuned for QA. For the first case study, ChemBERTa [6][5] and BioBERT [9][6] were also tested. These are all available from HuggingFace hub[7] and used with the **transformers** Python library. Since RoBERTa showed the best trade-off between results and computational resources in the first case study, we committed to its usage and present evaluation results for this model only.

Moreover, in both case studies, the documents used for evaluation were selected with the help of an expert on the respective domain, who also assessed the extractions through the web application developed for the purpose.

4.1 Application to Toxicology Analysis

The production of toxicological reports is a human-intensive task that involves collecting information of target chemical ingredients from various sources, including studies by different committees and organizations, often available in semi-structured documents. In order to take conclusions on the levels of toxicity of the target ingredient, such a report must address several questions.

This work is performed manually, by a chemical safety advisor, who spends hours searching for, comparing and labelling relevant information, to finally compile everything. It is thus a scenario that may benefit from some automation,

---

[3] `deepset/bert-base-cased-squad2`.
[4] `deepset/roberta-base-squad2`.
[5] `recobo/chemical-bert-uncased-squad2`.
[6] `dmis-lab/biobert-base-cased-v1.1-squad`.
[7] https://huggingface.co/.

namely in the collection of the documents from the relevant sources and in their automatic processing towards the extraction of the information that the reports need to address. Still, due to the impact of incorrect answers, it is important that the safety advisor does not lose control of the process, and may easily check the rationale behind the extracted information and edit anything incomplete.

This case study follows such requirements and applied the QA-based approach for extracting chemical properties from semi-structured documents.

Documents: Information was extracted from two references in the field: Scientific Committee on Consumer Safety (SCCS) Opinions[8] and Australian Industrial Chemicals Introduction Scheme (AICIS)[9] reports. These contain information on the physicochemical and toxicological properties of chemical substances. They are written in English and are organized in multiple sections, such as Irritation or Corrosivity, Skin Sensitisation, Acute Toxicity, Repeated Dose Toxicity, Mutagenicity, Carcinogenicity, Photo-induced Toxicity, and Safety Evaluation.

Properties/Questions: Together with the safety advisor (domain expert), we identified the fields required for each toxicological property, *i.e.*, keywords, which lead to the creation of a set of natural language questions using these keywords. Such questions, revealed in Table 1, are used for prompting the extractive QA models.

Table 1. Set of questions per property

Substance Property	Questions
Repeated Dose Toxicity	What is the NOAEL value? What is the Guideline?; What is the Study?
Acute Toxicity	What is the Guideline?; What is the Study?; What is the Species? What is the LD50?; What is the LC50?
Irritation	What is the Guideline?; What is the Study?; What is the Species?; What is the Concentration?; What is the Conclusion?
Mutagenicity	What is the Guideline?; What is the Study?; What is the Conclusion?
Skin Sensitization	What is the Guideline?; What is the Study?; What is the Conclusion?; What is the Concentration?
Carcinogenicity	What is the Species?; What is the Guideline?; What is the Study?; What is the Conclusion?
Photo-induced Toxicity	What is the Guideline?; What is the Study?; What is the Conclusion?; What is the Concentration?
Reproductive Toxicity	What is the Guideline?; What is the Study?; What is the Species?; What is the Conclusion?

Examples: Figures 3 and 4 illustrate the extraction process. They also show that text with relevant information can be present in multiple ways, such as a tabular-like format, or as a textual paragraph. However, this is not problematic because the QA models adapt well to both.

---

[8] https://health.ec.europa.eu/scientific-committees/scientific-committee-consumer-safety-sccs/sccs-opinions_en.

[9] https://www.industrialchemicals.gov.au/.

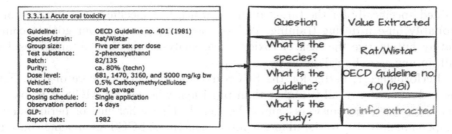

Fig. 3. Example of tabular section from a document, questions made and extracted values in the toxicology case study.

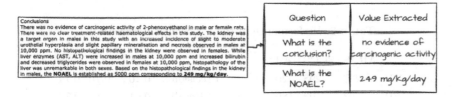

Fig. 4. Example of a document section (Conclusions), questions made and extracted values in the toxicology case study.

In Fig. 3 the study extracted value is "no info extracted", *i.e.*, an output generated in the case of when the model does not extract any value. That can be due to a recall error or due to the non-existence of information to extract.

Evaluation: Evaluation was based on a corpus of gold answers, created by the domain expert. It included 33 documents, 15 from SCCS and 18 from AICIS. From each, we capture multiple fields from eight different toxicological properties. In total, the corpus had 1,830 fields.

This allows for a quantitative evaluation with a confusion matrix based on comparing the results of the approach with the gold corpus. It uses the number of *True Positive (TP)*, *False Positive (FP)*, *False Negative (FN)* and *True Negative (TN)*, here defined as:

- *TP*: There is information in the document to be extracted, and the information extracted is correct;
- *FP*: Either there is no information in the document to be extracted, but some information was extracted; or there is information to be extracted but the extracted information is not correct;
- *FN*: There is information in the document to be extracted, but no information was extracted;
- *TN*: There is no information in the document to be extracted, and no information was extracted;

Table 2 summarizes the results obtained in the corpus of 33 documents. We can affirm that, although not perfect, overall results are globally positive. Average $F-Score$ was 0.73 with a RoBERTa model for QA which, we recall, was not trained specifically for the target domain, which has specific terminology, some probably unseen during training. We also note that it is not our goal to completely automatize the task of toxicological analysis and that obtained results will, in any case, be validated by the expert.

Performance is better for the SCCS documents, with F-Score of 0.79. For these documents, recall is higher than precision, whereas for AICIS it is lower, meaning that some answers cannot be found. This difference of performance happens mainly due to the different input structures. SCCS documents contain shorter sections, which require less or no additional splitting before QA. On the other hand, the sections of AICIS documents are generally longer, thus with more data, making extraction more challenging.

Table 2. Performance in the toxicological case study

	AICIS	SCCS	All			AICIS	SCCS	All
TP	222	298	520		Precision	0.70	0.76	0.74
FP	93	93	186		Recall	0.61	0.82	0.72
FN	140	66	206		Accuracy	0.72	0.68	0.75
TN	375	266	641		F-Score	0.66	0.79	0.73

4.2 Application to Finance

Exchange-Traded Fund (ETF)s are investment funds traded in stock exchanges. As with any investment, ETFs carry inherent risks that are associated with their particular asset class, investment strategy, sector and geographic focus[10]. With thousands of funds to choose from, investors require accurate information to make informed decisions. Following the changes in regulations put forward by Directive 2009/65/EC on UCITS IV[11] and Commission Regulation 583/2010[12], Key Investor Information Documents (KIID)s were introduced to provide investors with a concise and transparent overview of funds in a standardized format, with the explicit objective of increasing product comparability[13]. Still, extracting relevant information from KIIDs is both time-consuming and labor-intensive, requiring investors to manually analyze and search through the document to find the information they need.

An application of the QA approach in the extraction of critical fund information from KIIDs aims at further helping investors to gather the information they need more easily, hopefully resulting in more informed and faster investment decisions. Relevant information includes the objective of the funds, the type of

---

[10] https://www.blackrock.com/us/individual/education/ishares-etfs.
[11] https://eur-lex.europa.eu/legal-content/EN/ALL/?uri=celex%3A32009L0065.
[12] https://eur-lex.europa.eu/legal-content/EN/TXT/?uri=CELEX%3A32010R0583.
[13] https://www.blackrock.com/uk/solutions/key-investor-information-document.

management (active vs passive), the risk and return profile, and the charges or fees applied to the specific ETF.

Documents: Despite being a widespread product, ETFs have differences and are issued by different asset management firms, that charge different fee schemes [13]. KIIDs are human-written semi-structured reports, available in PDF. Despite being quite standardized in their content, there are still differences among KIIDs by different providers, and a massive comparison of content is not viable. In our study were used KIIDs from iShares[14] (hereafter IS), XTrackers[15] (XT), Amundi[16] (Am) and Vanguard[17] (Vg).

Properties/Questions: Each KIID is a multipage document with numerous paragraphs and phrases. As it happens to the toxicology documents, simply providing the complete document to the model is not feasible, due to size limitations. We thus rely on regular expressions for dividing the KIIDs into sections that pertain to the ETF, such as "Objectives and Investment Policy", "Risk and Reward Profile" and "Charges" (see Table 3). From each section, the values of six essential properties can be extracted with the following questions: "*What is the rate of the Share Class?*", "*How is the fund managed?*", "*What is the Entry Charge?*", "*What is the Ongoing Charge?*", "*What is the Performance fee?*", "*What is the Exit Charge?*".

Table 3. Information to extract from each KIID section

Section	Questions
Objectives and Investment Policy	How is the fund managed?
Risk and Reward Profile	What is the rate of the Share Class?
Charges	What is the Entry Charge?; What is the Ongoing Charge?; What is the Performance fee?; What is the Exit Charge?

Examples: As semi-structured documents, information in KIIDs can be formatted in multiple ways. Figures 5 and 6 illustrate parts of these documents with textual content, together with the set of questions and their extracted values after running the QA approach. In Fig. 6, we note an extraction that resulted in more text than the property value, which is not ideal.

Evaluation: A total of 38 KIIDs from the aforementioned ETF providers was used, comprising: 19 IS, 9 XT, 5 Am, and 5 Vg documents. These were selected with the help of a finance expert, having in mind the distribution of documents

---

[14] https://www.ishares.com/us.
[15] https://etf.dws.com/en-gb/.
[16] https://about.amundi.com/.
[17] https://investor.vanguard.com/investment-products/list/etfs.

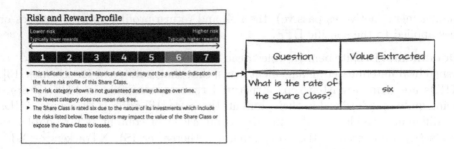

Fig. 5. Example of a document section (Risk and Reward Profile), questions made and extracted sections for the financial case study.

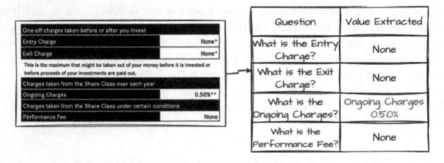

Fig. 6. Example of tabular text present in a document, questions made and extracted values for the financial case study.

they generally work on. The expert also assessed a total of 228 answers of the model for the previous documents, which resulted in a gold corpus used for quantitative evaluation. Results are summarized in Table 4.

Although the total number of extracted values is lower than for the application to toxicology, the overall performance is better, with an average F-Score of 0.90. This can be due to better contexts (sections) provided to the model, *i.e.*, smaller sections, and the vocabulary was not as technical as the toxicological domain. While confirming the straightforward adaptation of the QA approach to different domains, using the same model, it further shows that performance is variable between domains and, as in this case, it may be very competitive.

The best performance was for the IS and XT KIIDs, respectively, with an F-Score of 0.96 and 0.90. But performances for Am and Vg are still interesting, even if harmed by a lower recall. This lower performance of Am and Vg is due to: (i) the lower amount of documents used, leading to a greater impact of errors; (ii) issues in section detection in that type of documents or with the suitability of the questions used for the language of these documents.

Table 4. Performance in the financial case study

	AICIS	SCCS	All			AICIS	SCCS	All
TP	222	298	520	Precision		0.70	0.76	0.74
FP	93	93	186	Recall		0.61	0.82	0.72
FN	140	66	206	Accuracy		0.72	0.68	0.75
TN	375	266	641	F-Score		0.66	0.79	0.73

5   Conclusion

We confirmed that QA models can be leveraged for IE. A generic approach was first presented and its flexibility was demonstrated with practical applications. With minimal human effort, namely for mapping document sections with properties and formulating questions, information was successfully extracted from two different domains, where this kind of approach can be seen as novel. Due to specificities of the target documents and on the language they use, performance was better for finance than for toxicology, but results were still promising. Here, we highlight that the same RoBERTa model, trained in generic questions, could be used for both domains. We also see this kind of application as novel in the target domains and, specifically, documents used.

Performance on single domains could possibly benefit from domain-specific fine-tuning, but, for most domains, context-question-answer pairs are not available in a sufficient number. Another domain adaptation would involve pretraining the model in raw domain-specific data and then fine-tuning in SQuAD, as it happened to BioBERT and ChemBERTa for QA. However, this would make the approach no longer domain-agnostic, nor unsupervised.

Another word should be given on the recent trend of approaching several NLP tasks in a zero or few-shot scenario, by prompting Large Language Model (LLM)s like GPT3 [5]. This could indeed be an option, and should be explored in the future. Yet, we highlight that our approach requires less computational resources, is still unsupervised, and its answers are limited to sequences extracted from the given context. Which minimizes hallucination and enables human scrutinization, e.g., by looking and the answers in context.

In conclusion, the study on IE and the usage of QA models to extract information from natural language highlights the potential of this approach and provides a foundation for future research and applications in this area.

References

1. Adnan, K., Akbar, R.: Limitations of information extraction methods and techniques for heterogeneous unstructured big data. Int. J. Eng. Bus. Manag. **11**, 1847979019890771 (2019)
2. Arici, T., Kumar, K., Çeker, H., Saladi, A.S., Tutar, I.: Solving price per unit problem around the world: formulating fact extraction as question answering. arXiv preprint arXiv:2204.05555 (2022)

3. Baradaran, R., Ghiasi, R., Amirkhani, H.: A survey on machine reading comprehension systems. Nat. Lang. Eng. **28**(6), 683–732 (2022)
4. Bhutani, N., Suhara, Y., Tan, W., Halevy, A.Y., Jagadish, H.V.: Open information extraction from question-answer pairs. CoRR abs/1903.00172 (2019). http://arxiv.org/abs/1903.00172
5. Brown, T., et al.: Language models are few-shot learners. In: Advances in Neural Information Processing Systems, vol. 33, pp. 1877–1901. Curran Associates, Inc. (2020)
6. Chithrananda, S., Grand, G., Ramsundar, B.: ChemBERTa: large-scale self-supervised pretraining for molecular property prediction. CoRR abs/2010.09885 (2020). https://arxiv.org/abs/2010.09885
7. Dedić, N., Stanier, C.: Towards differentiating business intelligence, big data, data analytics and knowledge discovery. In: Piazolo, F., Geist, V., Brehm, L., Schmidt, R. (eds.) ERP Future 2016. LNBIP, vol. 285, pp. 114–122. Springer, Cham (2017). https://doi.org/10.1007/978-3-319-58801-8_10
8. Devlin, J., Chang, M.W., Lee, K., Toutanova, K.: BERT: pre-training of deep bidirectional transformers for language understanding. In: Proceedings of the 2019 Conference of the North American Chapter of the Association for Computational Linguistics: Human Language Technologies, Minneapolis, Minnesota (Volume 1: Long and Short Papers), pp. 4171–4186. ACL (2019)
9. Lee, J., et al.: BioBERT: a pre-trained biomedical language representation model for biomedical text mining. CoRR abs/1901.08746 (2019). http://arxiv.org/abs/1901.08746
10. Li, F., et al.: Event extraction as multi-turn question answering. In: Findings of the Association for Computational Linguistics: EMNLP 2020, pp. 829–838. ACL (2020)
11. Liu, Y., et al.: RoBERTa: a robustly optimized BERT pretraining approach. CoRR abs/1907.11692 (2019). http://arxiv.org/abs/1907.11692
12. Lomotey, R.K., Deters, R.: Topics and terms mining in unstructured data stores. In: 2013 IEEE 16th International Conference on Computational Science and Engineering, pp. 854–861 (2013). https://doi.org/10.1109/CSE.2013.129
13. McKinsey&Company: The great reset: North American asset management in 2022. Global Asset Management Practice (2022). https://mck.co/3lXfxVm
14. Nguyen, M.T., Le, D.T., Le, L.: Transformers-based information extraction with limited data for domain-specific business documents. Eng. Appl. Artif. Intell. **97**, 104100 (2021)
15. Raffel, C., et al.: Exploring the limits of transfer learning with a unified text-to-text transformer. J. Mach. Learn. Res. **21**(140), 1–67 (2020)
16. Rajpurkar, P., Jia, R., Liang, P.: Know what you don't know: unanswerable questions for SQuAD. In: Proceedings of 56th Annual Meeting of the Association for Computational Linguistics (Vol 2: Short Papers), pp. 784–789. ACL (2018)
17. Vaswani, A., et al.: Attention is all you need. In: Advances in Neural Information Processing Systems, vol. 30 (2017)
18. Wang, Y., Kung, L., Byrd, T.A.: Big data analytics: understanding its capabilities and potential benefits for healthcare organizations. Technol. Forecast. Soc. Chang. **126**, 3–13 (2018)

Towards a Robust Solution for the Supermarket Shelf Audit Problem: Obsolete Price Tags in Shelves

Emmanuel F. Morán$^{(\boxtimes)}$ (ID), Boris X. Vintimilla (ID), and Miguel A. Realpe (ID)

ESPOL Polytechnic University, Escuela Superior Politecnica del Litoral, ESPOL, CIDIS. Campus Gustavo Galindo Km. 30.5 Vía Perimetral, P.O. Box 09-01-5863 Guayaquil, Ecuador
{efmoran,boris.vintimilla,mrealpe}@espol.edu.ec

Abstract. Shelf auditing holds significant importance within the retail industry's industrial sector. It encompasses various processes carried out by human operators. This article aims to address the issue of identifying outdated price tags on shelves, bridging the gap of an automated shelf audit. Our proposal introduces a minimum viable process that effectively detects, recognizes, and locates price tags using computer vision and deep learning techniques. The outcomes of this study demonstrate the robustness of our approach in generating a comprehensive list of price tags on shelves, which can be subsequently compared with a database to identify and flag obsolete ones.

Keywords: Retail · Supermarket · Shelves Auditing · Deep Learning · Object Detection · Optical Character Recognition · Price Tags

1 Introduction

In a supermarket, shelves auditing can be defined as the process of comparing the current state of the shelves with the expected state according to its planogram [1]. Since the publication of the work by [4], the research in this topic has primarily focused on the object 'product' (localization and recognition) [5,6,8,9], but there is another object within this environment that deserves attention: the price tag. This object is crucial in the daily retail management [2,3] because it is the only object available to create a sense of order and give information about the products. However, to the best of our knowledge, it has not been used for shelves auditing solutions.

Some previous attempts have been made to extract information from product packaging in order to obtain the same data displayed on price tags [7], but some products do not have all the information in the front of the packing, and in terms of visibility, not all the packages are good enough to be easily read. The readability of the text on price tags is consistent, so there is little variation from one tag to another. In contrast, the readability of text on product packaging

© Springer Nature Switzerland AG 2024
V. Vasconcelos et al. (Eds.): CIARP 2023, LNCS 14469, pp. 257–271, 2024.
https://doi.org/10.1007/978-3-031-49018-7_19

Fig. 1. Examples of price tags in the shelves located in the bottom left side of the product shelving area.

varies widely. Besides, most retails use price tags not only to give a full product description, but as a delimiter between two products, as shown in Fig. 1, where the shelving space of each product is limited by its own price tag and the price tag of the adjacent product. For this, we can assure that the price tag is important not only for the information it contains, but also for its placement on the shelf.

It should be noted that the price tags come in different shapes and designs. They are normally personalized by each retail for their own needs, an example of price tag is shown in Fig. 1. The price tags display texts and images, from now and on called *items*, such as the description of the product and the barcode. These items give information about the product to the retail clients and employees. As an example, the barcode helps operators to quickly recognize products by using a barcode scanner included in a device for inventory. Other example of item is the price text, that helps clients to know the value that will be paid for the product. Others items could be percentage of discount (30% OFF in Fig. 1), product code (normally an internal code used only by the retail), QR (similar to barcode but usually can retrieve more information), and others. It is obvious to infer that each price tag could have different amounts and types of *items*, but it is clear that there are 3 *items* that always need to be on a price tag: Price, Description and Barcode.

This article will show a process to detect, recognize and locate price tags in a 2D space in order to create a report needed for the shelf auditing process: Obsolete Price Tags. For this, in Sect. 2 a dataset will be introduced, which is part of the one proposed in [1]. In Sect. 3 the proposed workflow will be explained, and finally in Sect. 4 the results obtained will be presented.

The paper presents the following contributions:

- A reliable methodology for detecting, localizing, and reading price tags on retail shelves, addressing a previously overlooked aspect in shelf auditing approaches.
- The publication of a dataset that includes all the necessary information to facilitate testing and evaluation of alternative approaches for the same process. available at https://t.ly/rkacT
- The outcome of this process serves as a validation tool for price tags on shelves, effectively reducing operational time for retailers.

2   Dataset Collection

As part of this research, a dataset was obtained using a robot which has three perception modules named *collectors*. Each collector is a Processing Unit (NUC from Intel) with two different types of cameras: an UHD Camera and a 3D Camera. This design allows the robot to gather multiples images in a single step. Further insights into this design can be found in [1], which provides a more detailed depiction and explanation about how it was proposed.

Since the robot uses an autonomous 2D navigation system programmed in ROS, it saves the relative location from where it captures the images for further processing. The data is acquired as the robot travels through the aisles of the store taking small steps to capture images in a steady position. Furthermore, since each aisle has two sides, a double tour should be done, one for each side of the aisle.

The information gathered by the robot at each step includes 3 types of data:

1. **RGB-UHD images:** Typical Red-Green-Blue images gathered with UHD or Ultra High Definition cameras. The resolution of these images is 3684 × 4912 pixels.
2. **Depth-images:** Depth images obtained with 3D cameras. This type of camera uses laser to measure the distance from the lens to objects in the scene. Thus, each pixel value represents the measured distance. The resolution of these images is 480 × 640 pixels.
3. **Positional Information:** An array of size 3 containing the X Y coordinates and YAW angle of the robot relative to the store map. From now and on this will be referred as *location*.

Fig. 2. Example of images taken in one robot's step by one collector; (left) RGB image, and (right) depth image, both images for the same scene.

It should be noted that the collector's cameras (UHD and Depth) rest in a structure created for them to be really near from each other, thus they point to the same scene, but with different field of views; for this, an alignment process to match each pixels from both cameras is used, and will be explain later.

A single execution of the robot acquisition process provides a large amount of data, so for this work a training dataset and a testing dataset were separated.

The training dataset is composed only of RGB-UHD Images. It has **851 images**, similar to the image shown in the left side of the Fig. 2, and collected ramdonly from different collectors in different steps of the robot. This dataset is used only for training the algorithms for detection in the workflow. It is emphasized that this randomly consist of price tags of only one design. This could lead to concerns about the model's ability to generalize to new designs. However, the present work relies on the power of deep convolutional networks to overcome this limitation. After being trained on a new dataset of price tags with a different design, the model should be able to detect price tags of that design with high accuracy.

The testing dataset has one side of two different aisles, named ZERO and ONE. Figure 2 shows an example of images acquired in the aisle ONE. The amount of images is mentioned in Table 1, here the steps means the number of stops the robot performed during the collection of images, thus more steps means a longer aisle, and for each step the location of the robot is also saved; The RGB-UHD and Depth images are three times the number of steps because the robot has three collectors.

Table 1. Testing dataset quantities

	Aisle ZERO	Aisle ONE
STEPS (Robot)	16	29
Locations	16	29
RGB Images	48	87
Depth Images	48	87

Both datasets were manually labeled. This means, for each RGB-UHD image an object detection labeling was performed with YOLO format ([class x y w h]). After that, each object of the RGB-UHD images (price tag) is extracted to create a dataset of price tags for detecting the *items* inside them. The testing dataset aggregates a location dataset of the aisle, the real texts of each price tag and a master database of all products in the aisle for evaluating the outputs.

This work focuses on processing the raw data in order to create a report that alerts the retailer which price tags in the sales area are obsolete. Figure 3 shows an example of the price tag we are going to work with, which includes price, description, barcode and product code.

In the case of the price tag's design presented in Fig. 3, only two of the visible items will be used, these are: Price and Description. Even though the barcode is one of the main items of a price tag, and it is present in this dataset, it will

Fig. 3. Example price tag in the dataset. Product description is in Spanish as the dataset is in that language.

not be used, mainly because the barcode of each price tag has a small dimension due to the way the dataset is obtained, and since it is problematic to read the lines of the barcode with cameras [10,11], it will be avoided. Nevertheless, it can be implemented since there are approaches of this type [12] but considering it requires more computation or even specialized hardware.

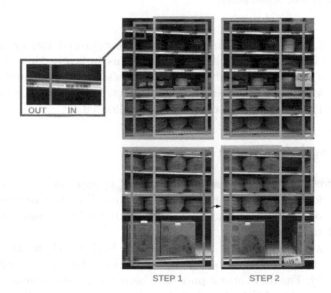

Fig. 4. RGB-UHD images acquired from 2 collectors and in 2 consecutive steps of the robot; (green boxes) example of a partial price tag removed by the filtering process. (Color figure online)

Given the nature of the dataset collection, there will be repeating shelf spaces. In Fig. 4, the redundancy between the collectors can be seen in the yellow boxes (vertical redundancy), as well as redundancy between the robot's steps in the purple boxes (horizontal redundancy). These redundancies allow us to affirm that there were no partial captures of the shelves. This also generates retakes

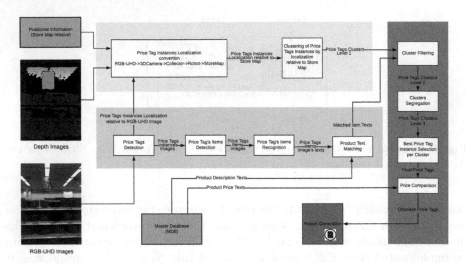

Fig. 5. Workflow of the proposed approach. Includes 3 Tracks: Location, Image-to-text and Selection. (Color figure online)

of the objects, that is, the same object can be captured two or more times, by different cameras and at different steps. This indicates that there will be redundancy of price tags, which favors the process to be presented later, since there will be several *instances* of the same object, from which the best one can be chosen and the information can be better validated.

3   Proposed Approach

The proposed solution can be seen in Fig. 5. It is important to note that this solution was developed based on the collected data. However, recommendations will be given on which blocks can be further extended to improve the solution for diverse data.

The inputs to the proposed solution are the RGB-UHD and Depth Images, as well as the Positional Information and the Master Database (MDB) (orange blocks in Fig. 5). The first three inputs have been extensively explained in Sect. 2. The fourth input, the MDB, is a database that contains the information of the products that should be in the store, including their updated prices. The MDB should be provided by the Retail, and it is expected to contain the Price and Description fields to be able to perform the necessary comparisons. Nevertheless, it is not limited to these two fields; it may also include other fields such as product code, barcode, or any other relevant information for further analysis.

The result of the proposed approach (green block) is a report that alerts retailers of price tags in the sales area that have obsolete or outdated price displayed.

The solution flow has two main tracks: Location (pink block) and Image-to-text (light blue block), that are joined in the Selection (red block) to finish the process.

3.1 Image-to-Text Track

This track uses deep learning and computer vision algorithms to detect, extract and read the price tags of all RGB-UHD images. Then, it uses the product descriptions from the MDB to compare and match the texts read with the product descriptions (which may contain minor errors). This track has two outputs, the first one is the location of the price tags instances in reference to the RGB-UHD image (used by the Location Track as input) and the second output is the texts of the items for each price tag (used by the Selection Track as input). Each block of this track will be explained in more detail below.

Price Tags Detection: This is the first block of the Image-to-text Track. It uses the RGB-UHD images as input and is responsible for detecting the location of price tags objects in the image. The output of this block is an array of the form [x1,y1,x2,y2,confidence,class], which will be called *detections*. The first four values of the array represent the top-left and bottom-right points of each detected object's bounding box, relative to the RGB-UHD image; the following values are the *confidence* and *class* of the object.For this, the deep convolutional neural network model YoloV5 [13] was used. It was trained using the training dataset consisting of 851 tagged images in Yolo format for price tag object detection. Using transfer learning techniques, it was possible to obtain a model that detects price tags with mAP and hits indicated in Table 2.

Table 2. Results for mean average precision (mAP) and hits of detections for Price Tag and Price Tag Item Detection

	Price Tag Detection	Price Tag Item Detection	
		Price	Description
mAP	96.43	95.17	91.78
Hits	100	99.89	97.93

The model is capable of detecting price tags even when they are partially visible in the image, typically located at the image borders. An instance of a partial price tag is depicted in Fig. 4, where the left price tag (zoomed within the green box) is cut in half. These detected partial objects serve no practical purpose and, in fact, pose an obstacle as incomplete information can lead to errors. To address this, a filter was developed to eliminate them by utilizing the detection centroid and determining whether it falls within the internal rectangle (indicated in red) in the RGB-UHD image. An example of this filter is shown in Fig. 4.

The detections that pass the filter will be called Price Tags Instances (PTI), since they are captures or instances of a real object with the camera and can be repeated as detailed above. The PTI will be used to crop the RGB-UHD image obtaining Price Tags Instances Images (PTI Images), used as input of the next block. Likewise, the location of each PTI, called Price Tags Instances Location (PTIL), is used as input for the Location Track.

Price Tags Items Detection: This block uses the output of the Price Tags Detection block: PTI Images. Similar to the previous one, this block produces an array of the form [x1,y1,x2,y2,confidence,class] which will also be called *detections*. The classes are Price and Description.

The same deep convolutional neural network model was used: YoloV5, trained on the data extracted from the training dataset (refer to Sect. 2). The confidence results of this model can be seen in the Table 2. It should be mentioned that most Prices are seen in the PTI Images, and Descriptions tends to have a lower mAP and hits due to the Price Tag design, since the Description is in the lower part of the Price Tag.

In a PTI Image, there can only be one instance of each item, that is, one Price and one Description. However, there were cases where more than one instance of the same object were detected, that is, two or more Prices or Descriptions on the same price tag. For this case, considering that the detections tend to overlap, a simple max filter over the confidence of the instances is done. At the end, for each PTI Image there will be only one Instances of each Item mentioned.

The detections that pass the filter will be call Price Tags Items Instances (PTII). The detections are used to create cropped images called Price Tags Items Instances Images (PTII Images), and will be used as input to the next block.

Table 3. Statistics on Price Tag Item Recognition texts confidences

Price Tag Item Recognition	Confidence			
	Mean	Std	Min	Max
Price	95.44	4.24	64.02	99.42
Description	87.10	13.69	0.0	97.55

Price Tags Items Recognition: This block uses the output of the previous block: PTII Images. This part of the track was not trained, instead an API [14] was used, as at the moment, it is considered a very laborious task to train a model to perform Optical Character Recognition (OCR) on this specific data. The obtained results were recorded in the Table 3. This will be used momentarily to collect data in order to generate a larger dataset big enough to get a robust model, in order to eliminate the need to pay for the usage of the API.

This block only contains the OCR algorithm due to the data that is being used (Price and Description), however, it could contain extra algorithms, such as QR readers or barcode readers, to be able to process other items of the image.

The output texts will be called Price Tags Items Instances Texts (PTII Texts) and will be used as input to the next block. As the response from the API could contain non-alphanumeric characters, a filter is used to eliminate them, thus the final result is an alphanumeric string.

Product Text Matching: This block uses the output of the previous block: PTII Texts. Specifically, it uses the Description item text to perform a match between the generated text and a text of the products in the MDB. The generated texts might contain small errors, in example: The PTII Text: *CAMISO BLANCA TAMONO XL* compared with the real text: *CAMISA BLANCA TAMANO XL*, smalls changes can be found. In order to achieve the matching, the cosine distance is used, which provide a numerical value to measure the distance between texts (if the distance is 0 the texts compared are exactly the same). The pairing method is simple, a matrix is created where the rows are represented by the PTII Texts and the columns by the products in the MDB. The matrix is filled with all the distances of PTII Texts vs. MDB Products. Then, for each row, the one with the lowest value is chosen to pair the text generated with a product description of the MDB. In this part, the top 1 match is being used, however, a top n matched products could be made to carry out a more in-depth analysis. Later, a cosine distance filter should be used to prevent using a large cosine distance. For this project we use 0.001 as threshold, meaning that if the lowest cosine distance obtain for the text is higher than the threshold, then no matching will be done as this text could have been mistakenly paired with a product.

Instead of creating a single matrix, it is recommended to estimate multiple matrices, each focusing on products belonging to the same category. This approach prevents situations where products from one category, such as detergent, are mistakenly associated with products from another category, like beverages.

The output texts will always be a product of the MDB that will be called Matched Items Texts (MIT). This output is the output of the Image-to-text Track, that is used jointly, with the output of the localization Track, as the input of the Selection Track.

3.2 Localization Track

This track consists of three inputs: Depth Images, location and PTIL. Its purpose is to generate Price Tags's clusters, wherein each cluster contains multiple PTIs. This output is referred to as Price Tags Clusters Level 1 (PTC L1). As the name suggests, it represents the initial level of grouping performed with the PTIL in a coordinate system relative to the store. This level serves as an initial approximation and may contain several grouping errors, as the location generated for each PTI is an estimation.

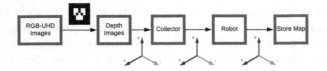

Fig. 6. Reference systems of the conversions done in the Localization Track.

Price Tags Instances Localization Conversion: This block plays a crucial role throughout the entire process. It takes the PTIL and subject it to multiple transformations across various reference systems. Each conversion introduces an inherent error. Figure 6 provides a visual representation of the reference systems involved, which consist of two types of transformations: one represented by an ARTAG and the other by a Cartesian plane.

The ARTAG represent a transformation from RGB-UHD images to Depth Images. These images are captured by each of the robot's collectors. The collector has the cameras in a structure designed so that they do not move, so it can be assumed that they are static and always at the same distance. Knowing this, some ARTAGs were used to automatically resolve the translation in the X and Y axes [19] from the RGB-UHD image to the depth image. Since the structure designed to hold the collectors is the same for all and the cameras are the same model, only one run is needed, making possible to transform one pixel of the RGB-UHD image to one pixel of the Depth image.

The Cartesian planes represent a transformation using position of robot's inside objects (cameras, lidar, etc.). This topic delves into robot navigation used in ROS [15] where every object has a referential position, normally referenced to the mobile base. In the case of the robot proposed in [1], each camera has a reference to its respective collector, and each collector has a reference to the robot. The native TF package in ROS [16], allows to keep track of these coordinates of each object relative to another and also enables changing the reference system of an observed or sensed point to one of the robot objects, that is, the location reference of an object obtained with the 3D camera can be transformed to the location reference of the collector, likewise it can also be passed to the robot by transitivity (multiplication of translation and rotation matrices). Finally, using the ROS package AMCL [17] an estimate of the position of a robot in an environment can be obtained using sensors such as lidars or 3D cameras. In the case of the robot, the environment is the Store Map, meaning that it has a reference to the Store Map at all times of its journey.

At the end of this block, the position of a point observed by the RGB-UHD camera in reference to the Store Map will be obtained. This is used for transforming the locations in the detections from the RGB-UHD image reference to the Store Map reference. Which will be in a [X Y Z] format, where X and Y represent the Store Map 2D plane and Z is the height.

Clustering of Price Tags Instances by Localization Relative to Store Map: This block processes the locations of the PTIs in reference to the Store Map in order to generate clusters. This grouping is done for a single side of an aisle. The algorithm used is DBSCAN, (Scikit-learn package implementation [18]). This algorithm allows data to be grouped without the need to indicate a cluster number, which favors this process since there is no way to determine in advance how many Price Tags will be in each aisle.

The algorithm parametrization is performed using the methodology in [20], which explains that the EPS (the maximum distance between two samples) can be obtained using Nearest Neighbors in the data and finding the Knee of the function. The second parameter is MIN_SAMPLES (number of samples in a neighborhood for a point to be considered as a core point) is stated in a value of 1 since a cluster could be only 1 PTI.

To evaluate this algorithm, the outputs (clusters) were passed over 2 scoring algorithms implemented in the Scikit-learn package. Rand Index (RI) [21] and Adjusted Rand Index (ARI) [22]. The mean score for the RI in the aisles of the testing dataset is 0.9925, while the ARI is 0.7694. In this case ARI uses cluster's order to evaluate, while RI do not. For this research, the RI is the metric selected over the ARI, since the clusters could not be grouped in the same order as the testing dataset.

The output of this block are clusters, which will be called Price Tags Clusters Level 1 (PTC L1), and which are then joined to the MITs (Image-to-text Track output) consolidating through the initial PTI label, to be used as input of the next Track.

3.3   Selection Track

This Track carries out the selection process of the best PTIs per cluster. That is, at the end of this Track, the redundancies of each object grouped in a cluster will be eliminated and there will be only a single PTI representing the object. For this, the clusters generated in the Location Track are filtered by recognition confidence of the PTII texts and then they are segregated to avoid the existence of different products in the same cluster. Finally, from each cluster, the best PTI is chosen using weights with the confidences of the PTII, obtaining a list of products which is then compared with the MDB to find those with different prices and report them.

Cluster Filtering: This is the first block of the Selection Track, its input is the PTC L1 joined with the MIT (outputs of the Location and Image-to-text Tracks respectively) to filter the PTIs according to the confidence of the PTII readings (when they matched, in the last block of the Image-to-text Track, the recognition confidence did not change). The choice of filters is dependent on the PTII selected for the process. In this article, Price and Description were used, thus the corresponding filters were chosen accordingly.

In the case of the Price, a common reading would be "$ 1.59" (see Fig. 3). For Price to be properly read, it must have: Exactly 2 decimal digits, and a decimal point separating the integer digits from the decimals. The dollar sign ($) is not considered. If the text follows this format, and the recognition confidence is over a threshold (in this study, greater than 0.9) then the PTI is not eliminated or filtered.

In the case of the Description, the text does not have a format to follow, so the selected filter was a threshold on its recognition confidence (in this study, greater than 0.75).

The recognition confidence threshold of the Description is lower than the Price's. This is due to the nature of the use of each type of PTII, in addition to the fact that the price is required to have greater confidence since it will be used for comparison at the end of the process and if it is wrong, it will cause false alerts.

The filtering occurs after creating the clusters, this way the DBSCAN algorithm has more data to automatically obtain the arguments it requires. If clustering were done after filtering, there would be less data, apart from the fact that the location estimate with respect to the Store Map is independent of the recognition confidence of each PTII, that is, the instance is real, but it could have had a tricky reading angle.

The output from this block are the same clusters but each one can have fewer instances. This output is called Price Tags Clusters Level 2 (PTC L2).

Clusters Segregation: This block uses the PTC L2 as input, which are groups of PTIs according to their proximity and which also went through a filter to eliminate PTIs that are not reliable enough in their PTII recognition confidence. By grouping them by their location, there could be PTIs of two or more different real Price Tags in the same cluster. In order to solve this problem, a segregation stage is carried out for the PTC L2. An example of this segregation can be seen in Fig. 7, where a cluster starts with 8 PTIs, which are actually instances of 3 different Price Tags: $1.99, $0.99 and $1.59. For this, cosine distance is calculated between the first PTI in the cluster and the others. PTIs with distance lower than a threshold (0.01 n this study) are separated into another cluster.

In the figure, the PTIs that are kept for the cluster are shown on the right side in the same level. The PTIs that are separated to create a new cluster are shown below in a new level. This segregation does not assure that all PTIs separated to the next cluster represent the same object. Therefore, the new cluster is put to segregation again. In the figure, it can be seen that the new cluster in level 2 has two different products (Price Tags with $1.59 and $0.99). So, the segregation is done again and it separates three PTIs to a new cluster. After a third segregation, no new cluster is created, then this process stops.

The output of this block is called Price Tags Clusters Level 3 (PTC L3), where no cluster have two or more products, only one product per cluster.

Best Price Tag Instance Selection per Cluster: This block uses PTC L3 as input, and selects the best PTI over the PTC L3. This is done by using a weighted confidence of the Price and Description, for this research, the weights

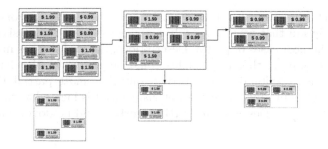

Fig. 7. Example of cluster segregation.

used were 0.75 for Price and 0.25 for Description. The price has more weight since it is more important for the final stages and the texts do not change like the description in the Product Text Matching block in the Image-to-text Track. For each cluster that has more that 1 PTI, this weighted confidence is calculated and the best one is selected over the others. The best PTI is called Final Price Tag Instance (FPTI) and is aggregated to a list called Final Price Tags (FPT).

Price Comparison: Having the FPTs as inputs, this block creates a list with the Price and Description. Then, the list is compared with the MDB, which has the updated price for all products. All the Descriptions in the FPT list are matched with the MDB's Descriptions leaving the comparison between the FPT's Prices and MDB's Prices ready for a one to one comparison. If the Price of the MDB and Price Tag are different, the FPT is then separated and aggregated to another list called: Obsolete Price Tags. In Table 4 can be seen the quantitative results of the comparison.

3.4   Report Generation

The last block represents the creation of the report. This report highlights the Price Tags in the sales area that have an obsolete or outdated price displayed. This report should be given to the operators at the beginning of the labor day for them to change the price tags as soon as possible.

4   Results

In this section quantitative results are shown. At the end of this work, the results where good enough to create a baseline pipeline. It should be mentioned that some errors came out, especially in the aisle ONE that is more extensive than aisle ZERO. Aisle ZERO only has one False Negative, that was produced due to partial text of a Description clustered correctly, but matched incorrectly with a similar product (specifically just a flavor change). For Aisle ONE, there were predicted products that were not in the ground truth, and there were also ground truth products that were not predicted. In most cases, these errors involved a partial Description that was incorrectly matched. A deeper analysis of the data showed that in most cases of partial Descriptions, the problem lies in a occlusion caused by the slope of the shelves (this could be unique for the dataset). This issue was further explored and it turned out that most partial Descriptions occurred when the product code was not visible as it is located at the same level as the last line of the Description. Further work should be done to address this particularity.

Table 4. Final results of aisle ZERO (left) and aisle ONE (right).

		Predictions Aisle ZERO				Predictions Aisle ONE	
		Negative	Positive			Negative	Positive
REAL	Negative	0	1	REAL	Negative	0	8
	Positive	0	45		Positive	6	94

5   Conclusions and Future Works

Retailers require a report indicating which Price Tags need updating. This work presents a way to obtain the report, automatically alerting that obsolete price tags in the shelves is presented with a success rate of 94.76% using only images captured by common cameras. The errors obtained lie mainly in the combination of the Price Tag design and a inclination of the place were them are placed, creating partial Descriptions that could not be correctly matched. A wider look in this errors could result in including another item in the process to correctly filter those Price Tags with occluded Descriptions.

Acknowledgements. This work has been partially supported by the ESPOL-CIDIS-11-2022 project and Tiendas Industriales Asociadas Sociedad Anonima (TIA S.A.). The authors would like to acknowledge TIA S.A., a leading grocery retailer in Ecuador, for providing access to an incredible environment for research and testing.

References

1. Moran, E., Vintimilla, B., Realpe, M.: Towards a robust solution for the supermarket shelf audit problem. In: Proceedings of the 18th International Joint Conference on Computer Vision, Imaging and Computer Graphics Theory and Applications - Volume 4: VISAPP, ISBN 978-989-758-634-7, ISSN 2184-4321, pp. 912–919 (2023)
2. Menon, R.V., Sigurdsson, V., Larsen, N.M., Fagerstrøm, A., Foxall, G.R.: Consumer attention to price in social commerce: eye tracking patterns in retail clothing. J. Bus. Res. **69**(11), 5008–5013 (2016)
3. Dutta, S., Bergen, M., Levy, D., Venable, R.: Menu costs, posted prices, and multiproduct retailers. J. Money, Credit, Bank. **31**(4), 683–703 (1999)

4. Goldman, E., et al.: Precise Detection in Densely Packed Scenes. (2019). https://doi.org/10.48550/ARXIV.1904.00853
5. Rubab, S., Khan, M.M., Ali, N., et al.: Hybrid approach for shelf monitoring and planogram compliance (hyb-smpc) in retails using deep learning and computer vision (2022). https://doi.org/10.1155/2022/4916818
6. Wei, Y., Tran, S., Xu, S., Kang, B., Springer, M.: Deep learning for retail product recognition: challenges and techniques (2020) https://doi.org/10.1155/2020/8875910
7. Chen, F., et al.: Unitail: Detecting, reading, and matching in retail scene (2022). https://doi.org/10.48550/ARXIV.2204.00298
8. Marder, M., Harary, S., Ribak, A., Tzur, Y., Alpert, S., Tzadok, A.: Using image analytics to monitor retail store shelves. IBM J. Res. Develop. **59**(2/3), 3:1–3:11 (2015). https://doi.org/10.1147/JRD.2015.2394513
9. Yilmazer, R., Birant, D.: Shelf auditing based on image classification using semi-supervised deep learning to increase on-shelf availability in grocery stores. Sensors **21**, 327 (2021). https://doi.org/10.3390/s21020327
10. Katuk, N., Ku-Mahamud, K.R., Zakaria, N.H.: A review of the current trends and future directions of camera barcode reading. J. Theor. Appl. Inform. Technol. **97**(8), 2268–2288 (2019). ISSN 1992–8645
11. Bantahar, M.A., Al-Gailani, S.A., Salem, A.A.: An automatic light control system for camera barcode reader. Springer, Singapore (2022). https://doi.org/10.1007/978-981-16-8129-5_25
12. Brylka, R., Schwanecke, U., Bierwirth, B.: Camera based barcode localization and decoding in real-world applications. In: 2020 International Conference on Omni-Layer Intelligent Systems (COINS) (2020). https://doi.org/10.1109/coins49042.2020.9191416
13. Jocher, G.: YOLOv5 by Ultralytics (Version 7.0) [Computer software] (2020). https://doi.org/10.5281/zenodo.3908559
14. Google Vision API. www.cloud.google.com/vision/docs/apis?hl=es-419
15. ROS Noetic. www.wiki.ros.org/noetic
16. ROS TF: Multi-coordinate frame Tracking over time. www.wiki.ros.org/tf
17. ROS AMCL: Probabilistic localization System. www.wiki.ros.org/amcl
18. DBSCAN: Density-Based Spatial Clustering of Applications with Noise. www.scikit-learn.org/stable/modules/generated/sklearn.cluster.DBSCAN.html
19. ROS AR_TAG_ALVAR: An open source AR tag tracking library. www.wiki.ros.org/ar_track_alvar
20. Rahmah, N., Sitanggang, I.S.: Determination of optimal epsilon (eps) value on DBSCAN algorithm to clustering data on peatland hotspots in Sumatra. IOP Conf. Ser.: Earth Environ. Sci. **31** 012012 (2016). https://doi.org/10.1088/1755-1315/31/1/012012
21. Rand Index Algorithm: Computes a similarity measure between two clusterings. www.scikit-learn.org/stable/modules/generated/sklearn.metrics.rand_score.html
22. Adjusted Rand Index Algorithm: Rand index adjusted for chance. www.scikit-learn.org/stable/modules/generated/sklearn.metrics.adjusted_rand_score.html

A Self-Organizing Map Clustering Approach to Support Territorial Zoning

Marcos A. S. da Silva[1]([✉]) [ID], Pedro V. de A. Barreto[1,2] [ID],
Leonardo N. Matos[2] [ID], Gastão F. Miranda Júnior[3] [ID],
Márcia H. G. Dompieri[4] [ID], Fábio R. de Moura[5] [ID], Fabrícia K. S. Resende[2] [ID],
Paulo Novais[6] [ID], and Pedro Oliveira[7] [ID]

[1] Embrapa Coastal Tablelands, 49025-370 Aracaju, SE, Brazil
`marcos.santos-silva@embrapa.br`
[2] Department of Computer Science, Federal University of Sergipe, São Cristóvão, SE, Brazil
{`pedro.araujo,leonardo`}`@dcomp.ufs.br`, fabricia_resende@academico.ufs.br
[3] Department of Mathematics, Federal University of Sergipe, São Cristóvão, SE, Brazil
`gastao@mat.ufs.br`
[4] Embrapa Territorial, 13070-115 Campinas, SP, Brazil
`marcia.dompieri@embrapa.br`
[5] Department of Economics, Federal University of Sergipe, São Cristóvão, SE, Brazil
[6] Department of Computing, Minho University, Braga, Portugal
`pjon@di.uminho.pt`
[7] ALGORITMI Centre/LASI Minho University, Braga, Portugal
`pedro.jose.oliveira@algoritmi.uminho.pt`

Abstract. This work aims to evaluate three strategies for analyzing clusters of ordinal categorical data (thematic maps) to support the territorial zoning of the Alto Taquari basin, MS/MT. We evaluated a model-based method, another based on the segmentation of the multi-way contingency table, and the last one based on the transformation of ordinal data into intervals and subsequent analysis of clusters from a proposed method of segmentation of the Self-Organizing Map after the neural network training process. The results showed the adequacy of the methods based on the Self-Organizen Map and the segmentation of the contingency table, as these techniques generated unimodal clusters with distinguishable groups.

Keywords: Alto Taquari basin · Ordinal categorical data · Spatial Analysis

Supported by National Council for Scientific and Technological Development - CNPq, Brazil, and by National Funds through the Portuguese funding agency, FCT - Fundação para a Ciência e a Tecnologia within project 2022.06822.PTDC. The work of Pedro Oliveira was also supported by the doctoral Grant PRT/BD/154311/2022 financed by the Portuguese Foundation for Science and Technology (FCT), and with funds from European Union, under MIT Portugal Program.

V. Vasconcelos et al. (Eds.): CIARP 2023, LNCS 14469, pp. 272–286, 2024.
https://doi.org/10.1007/978-3-031-49018-7_20

1 Introduction

The elaboration of public policies aimed at territorial development is becoming increasingly complex due to the various factors we must consider and mutually influence each other, such as climate, political choices, landscape, economic systems, migration, etc. [6]. One of the first steps in elaborating regional intervention policies is the analysis of homogeneous zones or zoning, which aims to identify areas with similar characteristics and, consequently, which should be submitted to different intervention regimes [16].

Territorial zoning can have different purposes (ecological, economic, risk prevention) and is conducted with the support of some spatial data management and analysis systems. Given the massive availability of spatial data, the complexity of the zoning process, and the urgency that specific applications demand, unsupervised zoning can play an essential role in elaborating territorial public policies [16].

Computer-assisted territorial zoning is elaborated from several superimposed layers of information. We consider one region a homogeneous zone if it presents similar characteristics for all the layers. We can define this homogeneity by choosing a similarity criterion (e.g., Euclidean distance) and a process for determining homogeneous areas (e.g., clustering algorithm).

This work investigates different forms to support automatic territorial zoning based on clustering thematic information layers with ordinal classes (ordinal categorical data). We evaluated three approaches to clustering ordinal data (thematic maps) to support the Alto Taquari basin territorial zoning. The first was based on the segmentation of the multi-way contingency table [7], the second applied a model-based clustering for ordinal data [3], and the last one conducted the multivariate ordinal data transforming it into numerical and used a proposed clustering method based on the segmentation of the Kohonen's Self-Organizing Map Artificial Neural Network [10].

The Sect. 2 briefly reviews territorial zoning with Self-Organizing Maps and clustering ordinal categorical data. The Sect. 3 unveils the case study, the *Alto Taquari* basin territorial zoning, and presents the three evaluated approaches. Section 4 shows the results and discussion, and the last section is dedicated to the conclusions.

2 Related Work

2.1 Zoning with Self-Organizing Maps

The capacity of Self-Organizing Maps to preserve the statistical properties of the data, including the proximity between observations, and its ability for quantization, topological ordering, and visualization impacted its application in several spatial analysis tasks, such as regionalization and zoning [1,12]. Recently, we can highlight several uses in zoning for ecosystem services [11,14,21], ecological-economic zoning [15], environmental zoning [18] and determination of adaptive zones [4].

Except [18], the other applications determine homogeneous zones from numerical data with low dimensionality and combine the use of ANN SOM with a clustering algorithm such as k-means [14] or hierarchical agglomerative [4,15,21]. In general, neural networks with few (up to 10^2) neurons predominate in this type of study, and it is possible to conduct studies with very small neural maps, such as a SOM 5×5 [4,21]. [18] proposed an approach for determining homogeneous zones from thematic maps (categorical data), transforming them to binary data, and adapting the ANN SOM to perform the clustering by segmenting the neural map without the aid of other clustering algorithms.

The literature shows that the SOM is scalable and adequate for non-linear data, enabling applying it to massive data analysis problems. They generally use Euclidean distance as the similarity measure between the numerical input vectors. Still, we can adapt this artificial neural network to tackle different datasets, such as networks, contingency tables, binary data, and graphs [10]. Liu et al. [11] compared the SOM against other automatic zoning methods, showing the unsupervised algorithm's robustness. Still, there is a lack of comparative studies (SOM vs. other) for zoning, zoning from ordinal categorical data (e.g., thematic maps), and clustering using SOM without the support of statistical clustering algorithms.

2.2   Clustering Ordinal Categorical Data

Ordinal categorical data is nominal categorical with a sense of order between each modality whose difference is not directly interpretable. It is common to obtain this information through surveys or other qualitative methods. Clustering this data type implies choosing a similarity measure considering the ordinal character or a distinction process between the probability distribution functions. However, there are cases where it is possible to convert ordinal to numerical, which allows using conventional clustering algorithms (e.g., k-means) by adopting a convenient dissimilarity measure (e.g., Euclidean distance) [2].

According to [13], in some situations, it is necessary to group the observations associated with ordinal data considering them as such, to allow greater interpretability of dissimilarities (e.g., Goodman & Kruskla γ coefficient), consistency during analysis, and universality of the clustering method (e.g., OrdCIAn-H hierarchical clustering) allowing the treatment of different scales.

This type of clustering can be non-parametric as proposed by [7], which does not use a distance metric between observations. Instead, the authors propose clustering the contingency table cells. Giordan & Diana [7] showed that this approach generates good results for data with low dimensionality compared to the FANNY and PAM methods [9].

Biernacki & Jacques [3] proposed a parametric approach based on modeling the data as a probability distribution function generated by a binary search algorithm. This approach has a solid mathematical foundation. It seeks to model the data to represent important characteristics for clustering, such as the presence of a single mode for each variable per cluster, decreasing probabilities for the other modalities around the mode, and the ability to distinguish between the models. However, the algorithm has a high computational cost as the number of classes increases.

Another way to cluster ordinal data is to transform it into interval numeric and apply a clustering algorithm. There are other ordinal data clustering algorithms, such as ROCK [8]. Still, in this study, we will limit ourselves to evaluating three different and possibly complementary clustering approaches: model-based [3], contingency table [7], and based on the transformation of ordinal data into numeric and subsequent use of the Self-Organizing Map as a clustering algorithm.

3   Material and Methods

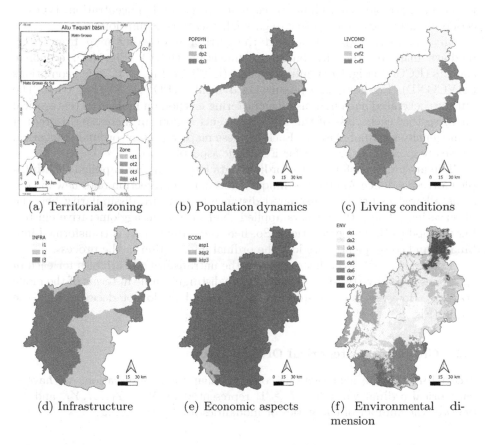

(a) Territorial zoning   (b) Population dynamics   (c) Living conditions

(d) Infrastructure   (e) Economic aspects   (f) Environmental dimension

Fig. 1. Territorial zoning of the *Alto Taquari* basin (a) obtained by [16] by applying a hierarchical agglomerative clustering over five ordinal data transformed into a binary: the population dynamics (b), the living conditions (c), the infrastructure (d), the economic aspects (e) and the environmental dimension (f).

3.1   The Alto Taquari Basin - MS/MT, Brazil

The Alto Taquari basin is found almost entirely in the northeast part of the MS, bordering the Pantanal, with a small region in the south part of the MT. The region crosses or includes 14 municipalities. According to [16], the delimited area comprises $28,046 km^2$, and they defined it from topographic maps on a scale of $1 : 250,000$ and digital images from the Landsat 5 satellite, TM sensor, mainly contour lines and drainage network (Fig. 1(a)).

This region was the subject of a comprehensive study that aimed to identify homogeneous zones to support public policies for the region [16]. The methodology consisted of combining pre-existing information (maps of vocation and environmental fragility) with a map of homogeneous zones for territorial planning, elaborated from categorical data of 37 fundamental indicators on the environment (e.g., geology, soil, climate), economy (e.g., land concentration, types of economic activities) and society (e.g., HDI, energy consumption).

Silva & Santos [16] elaborated the territorial ordering map from intermediate maps on the environmental dimension (ENV) (Fig. 1(f)), economic aspects (ECON) (Fig. 1(e)), infrastructure (INFRA) (Fig. 1(f)), living conditions (LIVCOND) (Fig. 1(c)) and population dynamics (POPDYN) (Fig. 1(b)). The authors generated each map using a clustering method on indicators (categorical data) so that each class of this map represents regions in increasing degrees of homogeneity (ordinal classes). Each of these maps represent ordinal categorical classes (e.g., asp1, asp2, asp3 for Economic aspects map).

The authors transformed the ordinal data into binary and applied the Multiple Correspondence Analysis technique. This data reprojection onto new dimensions allowed using the chi-squared distance to measure dissimilarity between observations. Then, the authors applied the hierarchical agglomerative clustering method to these reprojections. So, here the ordinal data is transformed into binary, and consequently, we lose the ordinal information in the process.

The evaluation of the histograms of the untransformed variables for each of the zones found by [16] shows that they are distinguishable in terms of the mode statistic. However, some histograms are bimodal or do not decay around the mode (Fig. 2).

3.2   Clustering Categorical Ordinal Data

Notation. To the next sections considers that for n observations we have J categorical ordinal variables, $J \geq 1$, represented by $Y = Y_1, \ldots, Y_J$, and N_j denotes the number of categories or modalities for Y_j, $j = 1, \ldots J$.

Clustering Based on Thresholding the Multi-way Contingency Table. The algorithm proposed in [7] is designed explicitly for ordinal categorical data and uses the multi-way contingency table generated from the data set as a starting point. Each cell on this table represents observations with the same characteristics, so they must be part of the same cluster. Besides, as we are dealing

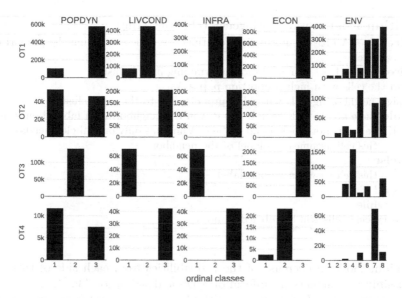

Fig. 2. Histogram for the reference territorial zoning (k=4), Fig. 1(a), according to [16].

with ordinal data, neighbor cells may imply some proximity between observations associated with each one. So, the authors use this idea of proximity between cells to propose a clustering algorithm that first considers the cell's density measured as a frequency or proportion and then considers the cell neighborhood to merge groups or associate non-labeled cells into a cluster.

According to the authors, the position of a cell C in a multi-way contingency table is given by $(C_1, \ldots, C_J)$, such that $C_j = 1, \ldots, N_j$, then the *neighborhood* of C is the set of cells whose coordinates are $(I_1, \ldots, I_J) \in Int(1) \times \cdots \times Int(J) \setminus (C_1, \ldots, C_J)$ where

$$Int(j) = \begin{cases} \{1, 2\}, & \text{if } C_j = 1 \\ \{N_j - 1, N_j\}, & \text{if } C_j = N_j \\ \{C_j - 1, C_j, C_j + 1\}, & \text{otherwise} \end{cases}$$

From this, we can describe the following algorithm [7, p. 1318]:

Model-Based Ordinal Clustering. As stated by [3], mixture models have become a well-established method for clustering due to their mathematical background for parameter estimation and model selection, capacity to generalize some geometric methods, and successful use in real-world situations. A key issue in this approach is defining an appropriate probability distribution function for ordinal data. It can be based on cumulative probabilities, on the constraining of a multinomial model to tackle the ordinality, on assuming that the ordinal data are the discretization of a continuous latent variable, or on the artificial construction of a model that exhibits some desired properties such as the presence

---

Algorithm 1. Multi-way contingency table clustering

Require: P — Multi-way normalized contingency table containing the proportions p
    for each cell
Require: $\lambda \in [0, 1]$ — Threshold for determining initial clusters
 1: Find the cells with high proportion p, $p \geq \lambda$
 2: Assign neighboring cells with high proportion p to the same cluster
 3: **while** there are no cells with $0 < p < \lambda$ in the neighboring of labeled ones **do**
 4:     **if** the cell is not labeled and $0 < p < \lambda$ and has only one labeled neighbor **then**
 5:         this cell assumes the label of the neighbor
 6:     **else**
 7:         this cell assumes a noise label
 8:     **end if**
 9: **end while**
10: Non-labeled cells will be labeled as noise

---

of a unique mode, a decrease of the probabilities from each side of this mode, the possibility to achieve a uniform or a Dirac distribution [3].

The proposition of [3] is based on the last strategy, and it assumes that the ordinal data-generating process results from a Binary Ordinal Search (BOS) algorithm that only uses ordinal information. This probabilistic model (Eq. 1) has two parameters, one related to the position (the mode of the distribution, μ) and the other associated with precision (prominence of the mode, π), that can be estimated by a Maximum Likelihood approach using an Expectation-Maximization algorithm. Then, the model is extended to perform a multivariate ordinal clustering based on these unimodal and univariate distributions (Eq. 2). We used the implementation of this method available at the *ordinalClust* R package, version 1.3.5.

$$p(Y_j, \mu, \pi) = \sum_{e_{N_j-1}, \dots, e_1} \prod_{i=1}^{N_j-1} p(e_{i+1} | e_i; \mu; \pi) p(e_1) \tag{1}$$

where Y_j is the jth-variable, N_j is the number of modalities in this variable, e_i represents an interval $e_i = \{b_i^-, \dots, b_i^+\} \subset \{1, \dots, N_j\}$ in the ith iteration of a binary search.

$$p(\mathbf{Y} | w_k = 1; \mu_\mathbf{k}, \pi_\mathbf{k}) = \prod_{h=1}^{J} p(Y_h; \mu_k^h; \pi_k^h) \tag{2}$$

where k represents the cluster, w_k is a variable, such that $w_k = 1$ if the observation belongs to cluster k, and $w_k = 0$ otherwise, $\mu_\mathbf{k} = (\mu_k^1, \dots, \mu_k^J)$ and $\pi_\mathbf{k} = (\pi_k^1, \dots, \pi_k^J)$.

Proposed Method Based on Self-Organizing Map. The proposed method transforms ordinal data into interval numerical data as its starting point so that the distance between two subsequent classes is the same for all variables. We transformed the ordinal data into numerical sequences (e.g., 1, 2, *and* 3 for the ECON map) to present the data to the SOM. After, we divided these values by the highest value of all maps (eight from the ENV map). Then, the input vector $\mathbf{Y}'$ has five components varying their values between 1/8 and 8/8.

The standard Self-Organizing Map defined by Teuvo Kohonen is an artificial neural network with unsupervised machine learning. The artificial neurons are represented by weight vectors, w, with the same dimension as the input data. They are organized in a two-dimensional grid, $N \times M$, with rectangular or hexagonal lattice that defines the neighborhood between neurons. The sequential or stochastic Machine Learning mechanism is iterative and can be divided into three phases. In the first phase, competitive, the input data are randomly presented to the neural network, and the neuron closest to the input vector according to the Euclidean distance is considered the Best Match Unit (BMU). In the second phase, cooperative, the neighboring neurons of the BMU are defined, which will also be updated in the third phase, adaptive, where each weight vector w of the BMU and its neighbors is updated according to the Eq. 3.

$$\mathbf{w}(t+1) = \mathbf{w}(t) + \alpha(t)h(t)(\mathbf{Y}'_i - \mathbf{w}(t)) \tag{3}$$

where t represents the iteration, $\mathbf{w}(t)$ is the neuron weight vector in the iteration t, $\alpha(t)$ is a small value representing the learning rate, $h(t)$ is a neighborhood function, and $\mathbf{Y}'_i$ an input data vector taken randomly.

At the end of the iterations, each input data will be associated with a single neuron, which can represent more than one input vector. SOM weights preserve the data's topology, meaning that neighboring neurons can represent nearby input vectors. This SOM feature allows clustering algorithms (e.g., k-means) on the neural network's weights as an indirect way to partition the input data [19].

However, it is possible to segment the SOM without the aid of traditional clustering algorithms using neural network internal information such as distance and neighborhood between the SOM weights, level of activation of neurons (number of input vectors associated with it), and data density between neurons. Costa & Netto [5] proposed a graph-based SOM partitioning model that uses all this information and automatically determines the number of clusters, which Silva et al. [18] successfully applied. This proposal has as a limiting factor, there are three hyperparameters that we must adjust to each data set. Silva & Costa [17] also proposed a graph-based method for segmenting the SOM, but using the density between neurons exclusively as a segmentation method using the Davies-Bouldin Validation Index (DBI). In this case, the algorithm automatically detects the number of clusters, we have a single hyperparameter, but we still do not have applications in real situations and do not use all the information available from SOM.

We propose a segmentation algorithm based on interpreting the SOM as an undirected graph, which uses all the information available after the machine learning process without needing hyperparameter adjustment. It is only necessary to define the desired number k of clusters (Algorithm 2). We implemented it in Python version 3.8 using the Minisom version 2.3.0 as the SOM solution [20].

Algorithm 2. SOM-based proposal

Require: $G = (V, E)$ — Graph of the trained SOM
Require: H — Neurons' activity level data
Require: D — Distance matrix between weights
Require: k — The number of desired clusters
 1: $T \leftarrow$ minimum spanning tree of G using D as edges' weights
 2: **for** each edge $(u, v) \in T$ **do**
 3:     $cost(u, v) \leftarrow DBI(u, v)$
 4: **end for**
 5: Prune the $k - 1$ edges in T with lesser costs
 6: Assign a cluster label to each set of connect nodes in T

Figure 3 shows the results of clustering some artificially and benchmark labeled data (spiral, gaussian, chainlink, and iris) using the k-means, DBSCAN, and the proposed method. We observed that the proposed method performs well for all four data sets. Different hyperparameters (number of neurons and grid lattice) were evaluated for the ANN SOM, using the accuracy measure ACC to choose the final configuration. The proposed method has a higher computational cost when compared to the k-means and DBSCAN methods. Still, it manages to be efficient in situations more appropriate for algorithms based on data partitioning, such as k-means (iris and Gaussian data sets) and density-based algorithms like DBSCAN (spiral and chainlink data sets).

3.3   Clustering Assessing

Considering that we will evaluate three very different forms of analysis of ordinal data groupings, we chose to assess the distinction between clusters based on analyzing the distribution of ordinal classes by variable and group. From the histograms, we evaluated the type of distribution per group (unimodal, bimodal, multimodal) and the distribution format (whether it decays around the mode or not).

4   Results and Discussion

We performed the model-based clustering for different values of the number of groups. However, the algorithm only converged for values equal to or less than four. For analysis, we have chosen to evaluate the result for $k = 4$ that generated the groups illustrated in the map in Fig. 4(a). The analysis of the histograms by each of the four clusters for the five ordinal variables considered showed that, although in most cases, we have unimodal distributions, it was not possible to

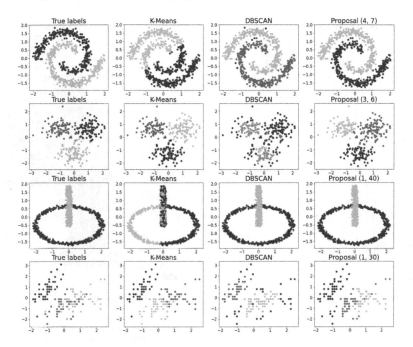

Fig. 3. In the first column on the left, we have the four labeled datasets (spiral, gaussian, chainlink, and iris data sets) used to evaluate the proposed method (last column), compared with the k-means methods in the second column and DBSCAN in the third column.

establish distinctions between the clusters through the analysis of the respective modes (Fig. 5).

We evaluated different values for the threshold for the clustering based on the contingency table, which varied between 0.007 and 0.082 in intervals of 0.005. The number of clusters generated ranged between eight for the lowest threshold and four for the highest value, and there was consistency between the different clusters, that is, without random variation between the clusters. Thus, we chose to analyze the partition with the largest number of clusters, eight, seeking to identify as many distinctions as possible, as seen in Fig. 4(b).

Figure 6 shows the histograms for each of the eight analyzed clusters, where we observe unimodal histograms with a clear distinction between the clusters, except for cluster 1 and the variable *ENV*, and for the variable *ECON* where no difference is observed between the groups as also shown in the model-based clustering histograms.

We evaluated different hyperparameters related to the ANN size, initial radius (sigma), and lattice of the neural grid (hexagonal or rectangular) for clustering the ordinal data transformed to numeric from the proposed post-training ANN SOM segmentation algorithm. To help choose the best result, we calculated the Davies-Bouldin clustering validation index, indicating the 6×5 hexagonal

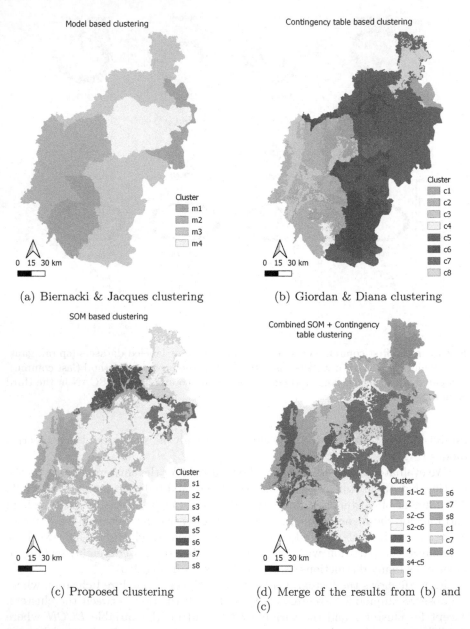

(a) Biernacki & Jacques clustering

(b) Giordan & Diana clustering

(c) Proposed clustering

(d) Merge of the results from (b) and (c)

Fig. 4. Maps in raster format obtained from the clustering of the five ordinal variables (POPDYN, LIVCOND, INFRA, ECON, and ENV) from the method proposed by Biernacki & Jacques (a), Giordan & Diana (b), and the method proposed in this article (c). In map (d) we have a merge of maps (b) and (c) in order to highlight their total coincidences (clusters 2, 3, 4, and 5), complementarities (clusters that represent regions differentiated by one and not by the other algorithm, clusters $s6$, $s7$, $s8$, $c1$, $c7$ and $c8$) and partial matches (when there are partial matches between them, clusters $s1 - c2$, $s2 - c5$, $s2 - c6$, and $s4 - c5$)

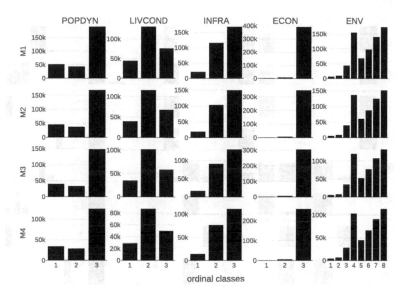

Fig. 5. Histogram for the model-based clustering method over the five ordinal categorical variables (POPDYN, LIVCOND, INFRA, ECON, and ENV) proposed by Biernacki & Jacques considering $k = 4$.

SOM ANN with sigma equal to 1.0 and partitioned it into six groups. After analyzing this 6×5 hexagonal ANN SOM for other values of k, we decided to analyze this same network partitioned into eight groups, resulting in the map in Fig. 4(c).

We observed that clustering based on the contingency table and the proposed method generated coincident partitions regarding spatial location, even if not entirely, and each strategy identified spatially distinct clusters. The analysis of the histograms from Fig. 7 for the clustering according to the proposed method shows that, as well as the method based on the contingency table, there is a substantial distinction between the clusters in terms of unimodal distributions, with emphasis on the variable *ECON* where the algorithm was able to distinguish, unlike the other techniques. In this clustering, the *ENV* variable shows better-defined distributions when compared to those generated by the contingency table method.

The model-based method was the only method that could have been more successful in partitioning the data to generate distinguishable groups from analyzing the distributions of the modalities. We should conduct further studies before discarding this approach for analyzing thematic maps with ordinal classes. This approach had a higher computational cost, followed by the proposed method, which depends mainly on the number of neurons in the ANN. The solution proposed by Giordan & Diana has the lowest computational cost.

The clustering method from the contingency table is deterministic. It depends solely on the definition of the threshold, which can be defined through trial

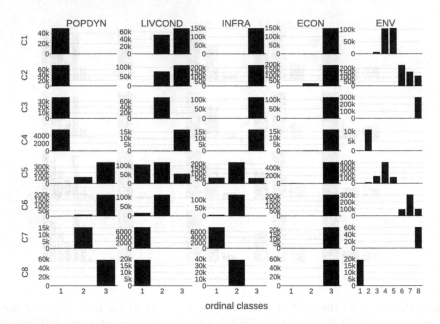

Fig. 6. Histogram for the multi-way contingency table clustering method over the five ordinal categorical variables (POPDYN, LIVCOND, INFRA, ECON, and ENV) proposed by Giordan & Diana considering $k = 8$ and a threshold equal to 0.007.

and error or by some other hyperparameter optimization method. The proposed method can be indicated for data with complex geometry. Still, it requires a more significant effort in choosing its hyperparameters, mainly the size of the neural network, which will determine the level of ability to separate the different intrinsic patterns in the dataset.

The environmental zoning process is complex and requires a multidisciplinary effort to establish the boundaries comprising the homogeneous zones. For the case of the Alto Taquari basin, the present work suggests, for example, the combined analysis of the partitions performed by the proposed method and the one based on the contingency table, as shown in Fig. 4(d). In this way, we could take advantage of the complementarities of each approach to have clusters with more significant distinctions.

The proposed method and the one based on the contingency table could partition the data into groups with characteristics of unimodal distributions, with the other classes decaying around the mode and quite distinguishable from each other. Both methods identified homogeneous areas with reduced area and, even so, distinguishable, as was the case of cluster $c4$ for clustering based on the contingency table and cluster $s5$ generated by the proposed algorithm. There is identification of regions with homogeneity detected by these two methods, emphasizing clusters 2, 3, 4, and 5, Fig. 4(d).

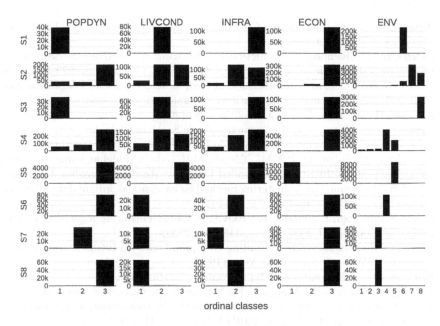

Fig. 7. Histogram for the SOM's segmentation clustering proposed method over the five ordinal categorical variables (POPDYN, LIVCOND, INFRA, ECON, and ENV) considering $k = 8$ and a and a 6×5 neural network.

5   Conclusions

We evaluated three approaches with different strategies for clustering ordinal data with many observations ($n = 448000$), low dimensionality ($d = 5$), and a few modalities (four variables with three and one with eight). The model-based clustering proposed by [3] did not achieve the objective of dividing the data into distinguishable groups, suggesting modifications in the algorithm's parameterization or the need to evaluate other model-based solutions for ordinal data. The proposed method and the one based on the contingency table showed good results regarding the distinction of groups. They showed the ability to identify coincident homogeneous regions but also complementary ones. The ordinal data clustering algorithm based on the contingency table is low computational cost, simple to understand, and requires the adjustment of a single parameter. However, we must evaluate its ability to separate ordinal data into distinguishable groups for data with higher dimensionality and complexity. The proposed method also deserves more exhaustive tests considering new datasets (ordinal and numerical) and comparisons with other non-parametric and parametric clustering methods.

References

1. Agarwal, P., Skupin, A. (eds.): Self-Organising Maps: Applications in Geographic Information Science. John Wiley and Sons, Chichester (2008)

2. Agresti, A.: Analysis of Ordinal Categorical Data. Wiley Series in Probability and Statistics. Wiley-Interscience, New York (2010)
3. Biernacki, C., Jacques, J.: Model-based clustering of multivariate ordinal data relying on a stochastic binary search algorithm. Stat. Comput. **26**, 929–943 (2016)
4. Bustos-Korts, D., et al.: Identification of environment types and adaptation zones with self-organizing maps; applications to sunflower multi-environment data in Europe. Theor. Appl. Genet. **135**, 2059–2082 (2022)
5. Costa, J.A.F., Netto, M.L.A.: Segmentação do SOM baseada em particionamento de grafos. In: VI Congresso Brasileiro de Redes Neurais, pp. 451–456 (2003)
6. Furtado, B.A., Sakowski, P.A.M., Tóvolli, M.H. (eds.): Modeling complex systems for public policies. Institute for Applied Economic Research, Brasília, DF (2015)
7. Giordan, M., Diana, G.: A clustering method for categorical ordinal data. Commun. Stat.-Theor. Methods **40**(7), 1315–1334 (2011)
8. Guha, S., Rastogi, R., Shims, K.: ROCK: a robust clustering algorithm for categorical attributes. Inf. Syst. **25**(5), 345–366 (2000)
9. Kaufman, L., Rousseeuw, P.J.: Finding Groups in Data: An Introduction to Cluster Analysis. J. Wiley & Sons, New York (1990)
10. Kohonen, T.: Self-Organizing Maps. Springer, Berlin (2001)
11. Liu, Y., Li, T., Zhao, W., Wang, S., Fu, B.: Landscape functional zoning at a county level based on ecosystem services bundle: methods, comparison and management indication. J. Environ. Manage. **249**(109315), 1–11 (2019)
12. Nikparvar, B., Thill, J.C.: Machine learning of spatial data. Int. J. Geo-Inform. **10**(600), 1–32 (2021). https://doi.org/10.3390/ijgi10090600
13. Podani, J.: Braun-Blanquet's legacy and data analysis in vegetation science. J. Veg. Sci. **17**, 113–117 (2006)
14. Pérez-Hoyos, A., Martínez, B., García-Haro, F.J., Álvaro Moreno, Gilabert, M.A.: Identification of ecosystem functional types from coarse resolution imagery using a self-organizing map approach: a case study for Spain. Remote Sens. **6**, 11391–11419 (2014)
15. Sadeck, L.W.R., de Lima, A.M.M., Adami, M.: Artificial neural network for ecological-economic zoning as a tool for spatial planning. Pesq. Agrop. Brasileira **52**(11), 1050–1062 (2022)
16. Silva, J.S.V., Santos, R.F.: Estratégia metodológica para zoneamento ambiental: a experiência aplicada na Bacia Hidrográfica do Rio Taquari. Embrapa Informática Agropecuária, Campinas, SP (2011)
17. Silva, L.A., Costa, J.A.F.: A graph partitioning approach to SOM clustering. In: 12th International Conference on Intelligent Data Engineering and Automated Learning (2011)
18. Silva, M.A.S.d., Maciel, R.J.S., Matos, L.N., Dompieri, M.H.G.: Automatic environmental zoning with self-organizing maps. MESE **4**(9), 872–881 (2018)
19. Silva, M.A.S.d., Matos, L.N., Santos, F.E.d.O., Dompieri, M.H.G., Moura, F.R.d.: Tracking the connection between Brazilian agricultural diversity and native vegetation change by a machine learning approach. IEEE Lat. Am. T. **20**(11), 2371–2380 (2022)
20. Vettigli, G.: Minisom: minimalistic and NumPy-based implementation of the Self Organizing Map (2018). https://github.com/JustGlowing/minisom/
21. Yan, Y., et al.: Exploring the applicability of self-organizing maps for ecosystem service zoning of the Guangdong-Hong Kong-Macao greater bay area. SPRS Int. J. Geo-Inf. **11**(481), 1–20 (2022)

Spatial-Temporal Graph Transformer for Surgical Skill Assessment in Simulation Sessions

Kevin Feghoul[1,2(✉)], Deise Santana Maia[2], Mehdi El Amrani[3], Mohamed Daoudi[2,4], and Ali Amad[1]

[1] Univ. Lille, Inserm, CHU Lille, UMR-S1172 LilNCog, 59000 Lille, France
kevin.feghoul@univ-lille.fr
[2] Univ. Lille, CNRS, Centrale Lille, UMR 9189 CRIStAL, 59000 Lille, France
[3] Department of Digestive Surgery and Transplantation, CHU Lille, PRESAGE, Univ. Lille, France
[4] IMT Nord Europe, Institut Mines-Télécom, Centre for Digital Systems, 59000 Lille, France

Abstract. Automatic surgical skill assessment has the capacity to bring a transformative shift in the assessment, development, and enhancement of surgical proficiency. It offers several advantages, including objectivity, precision, and real-time feedback. These benefits will greatly enhance the development of surgical skills for novice surgeons, enabling them to improve their abilities in a more effective and efficient manner. In this study, our primary objective was to explore the potential of hand skeleton dynamics as an effective means of evaluating surgical proficiency. Specifically, we aimed to discern between experienced surgeons and surgical residents by analyzing sequences of hand skeletons. To the best of our knowledge, this study represents a pioneering approach in using hand skeleton sequences for assessing surgical skills. To effectively capture the spatial-temporal correlations within sequences of hand skeletons for surgical skill assessment, we present STGFormer, a novel approach that combines the capabilities of Graph Convolutional Networks and Transformers. STGFormer is designed to learn advanced spatial-temporal representations and efficiently capture long-range dependencies. We evaluated our proposed approach on a dataset comprising experienced surgeons and surgical residents practicing surgical procedures in a simulated training environment. Our experimental results demonstrate that the proposed STGFormer outperforms all state-of-the-art models for the task of surgical skill assessment. More precisely, we achieve an accuracy of 83.29% and a weighted average F1-score of 81.41%. These results represent a significant improvement of 1.37% and 1.28% respectively when compared to the best state-of-the-art model.

Keywords: Graph Convolutional Networks · Transformer · Surgical Skill Assessment · Hand Skeleton · Simulation · Education

© Springer Nature Switzerland AG 2024
V. Vasconcelos et al. (Eds.): CIARP 2023, LNCS 14469, pp. 287–297, 2024.
https://doi.org/10.1007/978-3-031-49018-7_21

1   Introduction

Surgical skill assessment refers to the process of evaluating and measuring surgeon's technical proficiency and competence in executing surgical procedures. It delivers targeted feedback that enables efficient skill development through the provision of guidance, ultimately resulting in better patient treatment. Traditionally, evaluation has been performed by senior surgeons using both global and task-specific checklists [5,12]. However, classical surgical skill assessment checklists have several limitations, such as having a restricted scope, being prone to evaluator bias, lacking standardization, being a time-intensive and expensive process. Therefore, the development of automated tools to evaluate surgical skills is of significant interest. Collection and analysis of tool motion or video data can lead to an accurate assessment of the trainee's surgical proficiency. The proficiency can be quantified numerically through metrics such as the average OSATS score [12], or categorized into novice or expert levels, providing a clear and objective evaluation.

The conventional approach for automatically evaluating surgical proficiency relies on analyzing instrument motion, which can be obtained from various data sources such as video object tracking [14], video spatial-temporal features [24], and robotic kinematics [8,20]. Other techniques focus solely on utilizing video data. For instance, Funke et al. [3] proposed to use a Temporal Segment Network [19] by fine-tuning a pre-trained 3D Convolutional Neural Network on a stack of video frames. In [10], the authors proposed a unified multi-path framework for automatic video-based surgical skill assessment, taking into account various aspects of surgical skills, such as surgical tool usage, intraoperative event patterns, and other skill proxies. To capture the relationships between these factors, a path dependency module has been specially designed.

In recent years, Graph Convolutional Networks (GCNs) have become the de facto choice for modeling relational data due to their ability to capture both the local and global structure of graphs. This has resulted in GCNs achieving state-of-the-art performance in various tasks related to spatial-temporal data [6,17,21]. Similarly, Transformers [18] have revolutionized the field of natural language processing and have become the go-to method for various natural language processing (NLP) tasks. In addition to language-related applications, the Transformer architecture has also been applied to tasks beyond NLP, such as skeleton-based action recognition, and has produced outstanding results, as demonstrated in studies such as [13,16,23].

In this study, we explored the potential of using hand skeleton sequences for surgical skill assessment. Our framework offers several advantages, including (1) being lighter and easier to train than models that process entire video sequences, and (2) providing an affordable alternative to expensive robotic surgical systems that can provide kinematics data, since the hand skeleton can be extracted from affordable mobile phones. Additionally, hand skeleton detection is performed in real-time, which ensures its practicality and suitability for use in real-world scenarios. As far as our knowledge extends, this is the first attempt to use hand skeleton sequences for evaluating surgical proficiency. Considering the graph structure of the hand skeleton and the dynamic spatial-temporal patterns in sequences of hand movement, we propose the STGFormer framework that combines the strengths of spectral GCNs for learning spatial-temporal representations and Transformers for capturing long-range dependencies. Our framework has shown to outperform all existing state-of-the-art spatial-temporal skeleton-based deep learning models for surgical skill evaluation.

The contributions of this work are twofold and can be summarized as follows: (1) we propose to use sequences of hand skeleton for the task of surgical skill assessment. This approach offers several advantages, such as being non-invasive, objective, and extensible to operating rooms. Moreover, hand skeletons can be extracted from inexpensive devices, such as a smartphone. By analyzing hand dynamics, practitioners can gain valuable insights into their performance, which can be used for improvement and ultimately lead to better patient treatment; (2) we developed a new spatial-temporal model that learns the dynamic spatial-temporal correlations of hand skeletons. It consists of a spectral GCN for spatial-temporal feature learning follows by a Transformer encoder for capturing global temporal dependencies. This combination of proven techniques leads to the best prediction performances compared to existing state-of-the-art models.

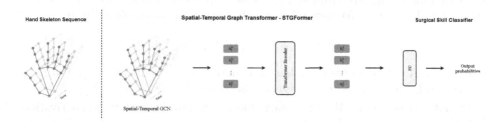

Fig. 1. Illustration of our STGFormer based surgical skill assessment framework, which is composed of two key components: Spatial-Temporal Graph Transformer and Surgical Skill Classifier.

2   Proposed Approach

This section introduces our STGFormer framework, which is illustred in Fig. 1. The framework consists of two essential components: (1) a spectral GCN responsible for learning spatial-temporal representation from hand skeleton sequences, and (2) a Transformer encoder designed to capture global temporal patterns.

2.1   Spectral Graph Convolutional Networks

In order to learn higher-level feature representations, we constructed a spatial-temporal graph and employed a spectral domain GCN.

Graph Construction. In this study, we constructed an undirected spatial-temporal graph $\mathcal{G} = (V, E)$ to obtain high-level representations of a hand skeleton sequence consisting of N joints over T frames. The set of nodes in the graph is represented by V, while E denotes the set of edges. The construction process is outlined as follows:

Nodes: the nodes in the graph consist of all joints in the sequence, expressed as $V = \{v_{ti} \mid t = 1, .., T, i = 1, .., N\}$. Each node v_{ti} is initialized with its 3D coordinate information. In this study, N is equal to 21.

Edges: the set of edges E is defined as the union of intra-skeleton connections, E_{intra}, and inter-frame connections, E_{inter}, in the graph, defined as follows:

$$E_{intra} = \{v_{ti}v_{tj} \mid (i, j) \in H, t \in \{1, .., T\}\} \tag{1}$$

$$E_{inter} = \{v_{ti}v_{(t+1)i} \mid i \in \{1, .., N\}, t \in \{1, .., T-1\}\} \tag{2}$$

In Eq. 2, H represents the set of naturally connected hand joints.

Graph Learning. We trained a spectral deep GCNs based on the previously constructed graph $\mathcal{G}$. We define the graph convolution operator as in [9]:

$$\widetilde{H}^{(l+1)} = \sigma(\widetilde{D}^{-\frac{1}{2}} \widetilde{A} \widetilde{D}^{-\frac{1}{2}} H^{(l)} W^{(l)}) \tag{3}$$

where $\widetilde{A} = A + I_n$ denotes the adjacency matrix of the undirected graph $\mathcal{G}$ with inserted self-connections, I_n represents the identity matrix, $\widetilde{D}_{ii} = \sum_j \widetilde{A}_{ij}$ is the diagonal degree matrix, $W^{(l)}$ is a learnable weight matrix, and $\sigma(.)$ an activation function. H^l represents the matrix of activations in the lth layer; $H^0 = X$, where X is the matrix of input node feature.

2.2   Transformer Encoder

In order to capture complex temporal patterns in the hand skeleton sequences, we feed the final high-level representation, previously extracted from the GCN, to a Transformer encoder. To be more specific, we concatenate the newly obtained representations of every joint j_{ti} into a vector h_t^0 for each frame. This concatenation process is illustrated in Eq. 4. Next, we combine the initial representations h_t^0 from all frames into a vector h^0 as depicted in Eq. 5. This vector h^0 serves as the input for a Transformer encoder.

$$h_t^0 = [j_{t1}, j_{t2}, .., j_{tN}] \tag{4}$$

$$h^0 = [h_1^0, h_2^0, .., h_T^0] \tag{5}$$

The Transformer is an advanced neural network architecture that relies on the self-attention mechanism, enabling the model to effectively process input sequences and generate predictions. Unlike traditional recurrent neural networks, which are limited by sequential processing, the Transformer can simultaneously attend to different parts of the input sequence, making it highly efficient at capturing long-term dependencies.

The self-attention mechanism in the Transformer calculates a weighted sum of the input sequence, with the weights being learned during the training process. This allows the model to assign importance to different positions in the sequence, focusing on the most relevant information for prediction. By considering the entire input sequence rather than just past representations, the Transformer can effectively capture contextual information and make accurate predictions.

A crucial component of the Transformer is the Multi-Head Attention (MHA) module. It enhances the model's ability to capture long-range dependencies and enables simultaneous attention across multiple representation subspaces at different positions. The MHA achieves this by utilizing Query-Key-Value (QKV) pairs. Each QKV triple is transformed into separate linear projections, and the scaled dot-product attention mechanism is applied. The scaled dot-product attention can be defined as follows:

$$Attention(Q, K, V) = softmax(\frac{QK^T}{\sqrt{d_k}})V \tag{6}$$

where $\frac{1}{\sqrt{d_k}}$ is used to counteract the vanishing gradient problem cause by the softmax function.

Each head of the MHA module is computed in parallel. The MHA module can be mathematically represented by the following equations:

$$MultiHead(Q, K, V) = Concat(head_1, ..., head_h)W^O \tag{7}$$

$$\text{with } head_i = Attention(QW_i^Q, KW_i^K, VW_i^V) \tag{8}$$

where $W_i^Q \in \mathbb{R}^{d_m \times d_k}, W_i^K \in \mathbb{R}^{d_m \times d_k}, W_i^V \in \mathbb{R}^{d_m \times d_v}, W_O \in \mathbb{R}^{hd_v \times d_m}$ represent the query, key, value, and output projection learnable weight matrices, respectively. h and d_m correspond to the number of heads and the output dimension of the encoder block. In this study, we choosed $d_k = d_v = d_m/h$.

2.3 Surgical Skill Classifier

After the forward pass through k-th Transformer encoder layer, the learned representation h^k, as shown in Eq. 9, is utilized as input to a fully connected neural network. This network is responsible for making predictions about whether the hand skeleton sequence is related to a senior surgeon or a surgical resident.

$$h^k = [h_1^k, h_2^k, .., h_T^k] \tag{9}$$

3   Experimental Results

This section presents the dataset collected for the surgical skill assessment task, as well as the results obtained using our proposed approach and several state-of-the-art deep learning-based models.

3.1   Dataset

Data Collection We gathered data from a total of 16 participants, consisting of 4 experienced surgeons and 12 surgical residents. The participants executed a circular cutting exercise using the VirtaMed medical simulator, as depicted in Fig. 2. The first step of the cutting exercise was to use a laparoscope, as illustrated in Fig. 3a, to enter the virtual environment and position the view at the correct location. Following that, the participants utilized an atraumatic grasper tool (Fig. 3b) to apply tension to the tissue and execute a precise cut along a circular incision between two lines using a pair of scissors (Fig. 3b). For each participant, we recorded their hand movements while they performed the exercise using a smartphone equipped with 4K recording capability.

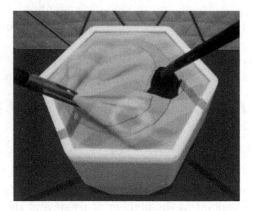

Fig. 2. Illustration of the circular cutting exercise using the VirtaMed simulator.

The circular cutting exercise was conducted in a simulated environment at the PRESAGE medical simulation center (Plateforme de Recherche et d'Enseignement par la Simulation pour l'Apprentissage des Attitudes et des GEstes), which is a department affiliated with the Faculty of Medicine at the University of Lille. This simulation center accurately replicates surgical training scenarios, making it an ideal setting for developing surgical skills. It is common for surgical novices to practice on medical simulator, where they perform tasks from curricula such as the Fundamentals of Laparoscopic Surgery (FLS) [15].

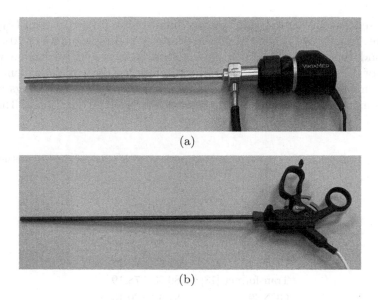

Fig. 3. (a) Laparoscope; (b) Atraumatic Grasper / Scissors.

FLS Program. The circular cutting exercise is an important part of the training of residents and is included in the FLS program. Initiated in 2004, the FLS program was designed to deliver standardized training for laparoscopic procedures and encompasses theoretical knowledge as well as practical skills. Surgeons often need to complete the FLS program to obtain certification in laparoscopic surgery.

The circular cutting exercise is a crucial component of the FLS program, along with few other simulation exercises, and holds significant importance within the training curriculum. Despite seeming straightforward, it remains an important aspect of the training curriculum. This exercise helps residents develop precise control over laparoscopic instruments, particularly scissors, and improves their hand-eye coordination. It also enables them to understand the tactile feedback and resistance encountered when cutting tissue using these instruments, and enhances their depth perception skills by accurately assessing the distance and thickness of simulated tissue.

3.2   Data Preprocessing

We used the method from [22] to extract the hand skeleton of both hands from the recorded videos of each individual. The hand landmark model outputs a set of 21 3D coordinates for each frame, based on the hand intra connectivity structure, as illustrated in Fig. 1. We opted to rely solely on right hand landmarks, as left hand detection was unreliable and played a minor role in this task. Indeed, the right hand was primarily responsible for the cutting, while the left hand primarily held the tissue with limited movement. Afterward, we normalized each hand skeleton sequence by subtracting the coordinate of the first wrist joint (v_{00}) from each joint. Finally, we generated non-overlapping sliding windows of 20 s, which correspond approximately to 600 data points. As a result, we have a varying number of data sequences for each subject, which are directly dependent on the time taken to complete the exercise, with a duration of the recordings ranging from 1 minute and 33 s to 6 min and 17 s, with an average duration of 3 min and 6 s.

Table 1. Surgical skill assessment: comparison with state-of-the-art methods.

Method	Acc	F1-score
SoCJ [2]	80.39	77.55
TCN [1]	80.08	78.25
LSTM [7]	81.21	79.36
DeepGRU [11]	81.42	79.48
Transformer [18]	80.53	78.19
GCN [9]	81.92	80.13
ST-GCN [21]	79.14	79.54
ASTGCN [6]	79.30	79.49
STGFormer (ours)	**83.29**	**81.41**

3.3   Results

Evaluation Framework. In line with the JIGSAWS [4] dataset, which is widely used as a benchmark for evaluating surgical skill assessment, our study also takes into account the surgeon's experience as a valuable indicator of surgical proficiency. In our case, given the existence of two distinct groups of practitioners, namely senior surgeons and surgical residents, we formulate the surgical skill assessment as a binary classification task.

Our evaluation strategy involved utilizing a subject-independent 6-fold cross-validation to enhance the robustness of our evaluation. This approach was necessary because the data sequences of hand movements from the same subjects are likely to exhibit correlations. To ensure fairness in distributing the limited number of surgeons in our dataset across each fold, we generated all possible combinations of two surgeons, resulting in a total of six combinations. This ensured that each surgeon had an equal presence in both the training and test sets.

Additionally, we ensured that surgical residents were evenly distributed across the six folds to maintain homogeneity. To assess the performance of our model, we employed accuracy as well as the weighted average F1-score. The inclusion of the F1-score allowed us to account for imbalanced class distributions within our dataset.

Surgical Skill Classification. We compared our approach with eight state-of-the-art models that we re-implemented. Our approach was compared to classical deep learning-based methods such as TCN [1], LSTM [7], DeepGRU [11], and Transformer [18], trained directly on sequences of raw hand landmarks. In addition, we compared our approach with state-of-the-art spatial-temporal graph-based models, including GCN [9], ST-GCN [21], and ASTGCN [6]. The ST-GCN consists of multiple spatial-temporal convolutional blocks, each of which includes two temporal gated convolution layers and one spatial graph convolution layer in the center. The ASTGCN consists of multiple blocks, each composed of a spatial-temporal attention mechanism and a spatial-temporal convolution that utilizes graph convolutions to capture spatial patterns and standard convolutions to describe temporal features simultaneously. We also compared our framework with a model trained on handcrafted features, namely the SoCJ descriptor [2], which extracts a descriptor from the hand skeleton based on its geometric shape. These features are then input into a LSTM model.

In Table 1, we presented the results of our STGFormer model and above mentioned state-of-the-art baselines. Our STGFormer achieves the best performance in terms of both evaluation metrics, achieving 83.29%, and 81.41% in terms of accuracy, and F1-score respectively, as shown in Table 1, which represent an improvement of 1.37% and 1.28% when compared to the best state-of-the-art model.

The SoCJ approach, which involves extracting spatial descriptors, exhibits the lowest F1-score among the evaluated methods. In addition, even when compared to a LSTM model trained directly on raw data, the SoCJ descriptor proves to be inefficient, highlighting the limitations of the descriptor extraction process for our particular task.

As part of our ablation study, we observed that STGFormer outperformed both the GCN and Transformer models by a significant margin, achieving an accuracy and F1-score improvement of at least 1.37% and 1.28% respectively. This outcome clearly demonstrates the effectiveness of combining graph-based and transformer-based approaches in the context of learning surgical skill evaluation. These results highlight the importance of incorporating spatial and temporal information for accurate and robust assessment of surgical skills.

Therefore, based on these findings, we can draw several conclusions regarding the effectiveness of using temporal data either individually or in combination with spatial data. Firstly, the superiority of STGFormer over the GCN and Transformer models suggests that leveraging both spatial and temporal information provides a more comprehensive understanding of surgical skill performance. By capturing the interplay between spatial relationships and temporal dynamics, STGFormer can extract more informative features, leading to improved accuracy and F1-score.

Secondly, the performance gap between STGFormer and the other models implies that solely relying on either spatial or temporal data may not be sufficient for accurate surgical skill assessment. Spatial information alone might not capture the dynamic nature of the surgical procedure, while temporal information alone might lack the contextual understanding provided by spatial relationships. Therefore, combining both spatial and temporal data, as done in STGFormer, proves to be crucial for achieving superior performance for the particular task of surgical skill assessment.

4   Conclusion

This study demonstrates the feasibility of utilizing hand skeleton sequences for accurate surgical skill assessment. The successful development of automated surgical skill assessment holds significant importance in training aspiring surgeons and enhancing their proficiency in performing safe interventions. In order to achieve this goal, we proposed a novel approach called STGFormer, which effectively captures spatial-temporal correlations and long-range dependencies in the hand skeleton sequences of practitioners as they perform tasks within a simulated environment. Extensive experiments were conducted on a dataset comprising both senior surgeons and surgical residents, and our STGFormer framework achieved an accuracy of 83.29% and a weighted average F1-score of 81.41%. These results strongly support the efficiency of our approach to accurately distinguish between senior surgeons and surgical residents, highlighting its potential as a valuable tool for evaluating surgical skills.

In a future study, we plan to extend our research by investigating a multimodal approach that combines hand skeleton sequences with RGB data to improve the accuracy of surgical skill assessment. This integration aims to leverage the complementary information provided by both modalities, further enhancing the robustness and effectiveness of our assessment framework.

References

1. Bai, S., Kolter, J.Z., Koltun, V.: An empirical evaluation of generic convolutional and recurrent networks for sequence modeling. arXiv preprint arXiv:1803.01271 (2018)
2. De Smedt, Q., Wannous, H., Vandeborre, J.P.: Skeleton-based dynamic hand gesture recognition. In: Proceedings of the IEEE Conference on Computer Vision and Pattern Recognition Workshops, pp. 1–9 (2016)
3. Funke, I., Mees, S.T., Weitz, J., Speidel, S.: Video-based surgical skill assessment using 3D convolutional neural networks. Int. J. Comput. Assist. Radiol. Surg. **14**, 1217–1225 (2019)
4. Gao, Y., et al.: JHU-ISI gesture and skill assessment working set (jigsaws): a surgical activity dataset for human motion modeling. In: MICCAI workshop: M2cai, vol. 3 (2014)
5. Goh, A.C., Goldfarb, D.W., Sander, J.C., Miles, B.J., Dunkin, B.J.: Global evaluative assessment of robotic skills: validation of a clinical assessment tool to measure robotic surgical skills. J. Urol. **187**(1), 247–252 (2012)
6. Guo, S., Lin, Y., Feng, N., Song, C., Wan, H.: Attention based spatial-temporal graph convolutional networks for traffic flow forecasting. In: Proceedings of the AAAI Conference on Artificial Intelligence, vol. 33, pp. 922–929 (2019)

7. Hochreiter, S., Schmidhuber, J.: Long short-term memory. Neural Comput. **9**(8), 1735–1780 (1997)
8. Ismail Fawaz, H., Forestier, G., Weber, J., Idoumghar, L., Muller, P.-A.: Evaluating surgical skills from kinematic data using convolutional neural networks. In: Frangi, A.F., Schnabel, J.A., Davatzikos, C., Alberola-López, C., Fichtinger, G. (eds.) MICCAI 2018. LNCS, vol. 11073, pp. 214–221. Springer, Cham (2018). https://doi.org/10.1007/978-3-030-00937-3_25
9. Kipf, T.N., Welling, M.: Semi-supervised classification with graph convolutional networks. arXiv preprint arXiv:1609.02907 (2016)
10. Liu, D., et al.: Towards unified surgical skill assessment. In: Proceedings of the IEEE/CVF Conference on Computer Vision and Pattern Recognition, pp. 9522–9531 (2021)
11. Maghoumi, M., LaViola, J.J.: DeepGRU: deep gesture recognition utility. In: Bebis, G., et al. (eds.) ISVC 2019, Part I. LNCS, vol. 11844, pp. 16–31. Springer, Cham (2019). https://doi.org/10.1007/978-3-030-33720-9_2
12. Martin, J., et al.: Objective structured assessment of technical skill (OSATS) for surgical residents. Br. J. Surg. **84**(2), 273–278 (1997)
13. Mazzia, V., Angarano, S., Salvetti, F., Angelini, F., Chiaberge, M.: Action transformer: a self-attention model for short-time pose-based human action recognition. Pattern Recogn. **124**, 108487 (2022)
14. Pérez-Escamirosa, F., et al.: Objective classification of psychomotor laparoscopic skills of surgeons based on three different approaches. Int. J. Comput. Assist. Radiol. Surg. **15**(1), 27–40 (2020)
15. Peters, J.H., et al.: Development and validation of a comprehensive program of education and assessment of the basic fundamentals of laparoscopic surgery. Surgery **135**(1), 21–27 (2004)
16. Plizzari, C., Cannici, M., Matteucci, M.: Spatial temporal transformer network for skeleton-based action recognition. In: Del Bimbo, A., et al. (eds.) ICPR 2021, Part III. LNCS, vol. 12663, pp. 694–701. Springer, Cham (2021). https://doi.org/10.1007/978-3-030-68796-0_50
17. Slama, R., Rabah, W., Wannous, H.: STR-GCN: dual spatial graph convolutional network and transformer graph encoder for 3D hand gesture recognition. In: IEEE FG, pp. 1–6 (2023)
18. Vaswani, A., et al.: Attention is all you need. In: Advances in Neural Information Processing Systems, vol. 30 (2017)
19. Wang, L., et al.: Temporal segment networks for action recognition in videos. IEEE Trans. Pattern Anal. Mach. Intell. **41**(11), 2740–2755 (2018)
20. Wang, Z., Majewicz Fey, A.: Deep learning with convolutional neural network for objective skill evaluation in robot-assisted surgery. Int. J. Comput. Assist. Radiol. Surg. **13**, 1959–1970 (2018)
21. Yu, B., Yin, H., Zhu, Z.: Spatio-temporal graph convolutional networks: a deep learning framework for traffic forecasting. arXiv preprint arXiv:1709.04875 (2017)
22. Zhang, F., et al.: Mediapipe hands: On-device real-time hand tracking. arXiv preprint arXiv:2006.10214 (2020)
23. Zhang, Y., Wu, B., Li, W., Duan, L., Gan, C.: STST: spatial-temporal specialized transformer for skeleton-based action recognition. In: Proceedings of the 29th ACM International Conference on Multimedia, pp. 3229–3237 (2021)
24. Zia, A., Sharma, Y., Bettadapura, V., Sarin, E.L., Essa, I.: Video and accelerometer-based motion analysis for automated surgical skills assessment. Int. J. Comput. Assist. Radiol. Surg. **13**, 443–455 (2018)

Deep Learning in the Identification
of Psoriatic Skin Lesions

Gabriel Silva Lima[1,3], Carolina Pires[2], Arlete Teresinha Beuren[1],
and Rui Pedro Lopes[3,4(✉)] (iD)

[1] Federal University of Technology - Parana, Santa Helena, Brazil
`gabriell.1997@alunos.utfpr.edu.br`, `arletebeuren@utfpr.edu.br`
[2] Institute of Biomedical Sciences Abel Salazar, University of Porto, Porto, Portugal
[3] Research Center in Digitalization and Intelligent Robotics (CeDRI),
Instituto Politécnico de Braganca, Bragança, Portugal
`rlopes@ipb.pt`
[4] Laboratório para a Sustentabilidade e Tecnologia em Regiões de Montanha
(SusTEC), Instituto Politécnico de Bragança, Bragança, Portugal

Abstract. Psoriasis is a dermatological lesion that manifests in several
regions of the body. Its late diagnosis can generate the aggravation of
the disease itself, as well as of the comorbidities associated with it. The
proposed work presents a computational system for image classification
in smartphones, through deep convolutional neural networks, to assist
the process of diagnosis of psoriasis.
The dataset and the classification algorithms used revealed that the classification of psoriasis lesions was most accurate with unsegmented and
unprocessed images, indicating that deep learning networks are able to do
a good feature selection. Smaller models have a lower accuracy, although
they are more adequate for environments with power and memory restrictions, such as smartphones.

Keywords: image processing · deep learning · psoriasis classification ·
mobile application

1 Introduction

Psoriasis is an immune-mediated, chronic and complex disease that mainly
affects skin, and is associated with others comorbidities such as rheumatological,
cardiovascular and psychiatric [11,12]. This condition has a considerable impact
on patients Quality of Life (QoL), that should be assessed on a holistic approach.

Psoriasis prevalence is estimated on more than 125 millions worldwide and
2,2% on US population [11]. There has also been reported significant geographic
and ethnic disparities on psoriasis incidence. Gerb et al. consider that both sexes
are equally affected by psoriasis [11], while other authors report a slightly higher
prevalence on women [12]. It is also less prevalent on paediatric group and ranges
from 0,5 to 2%. Recent retrospective cohorts have demonstrated an increase of
incidence through all age groups. This disease can begin at any age, but is usually

© Springer Nature Switzerland AG 2024
V. Vasconcelos et al. (Eds.): CIARP 2023, LNCS 14469, pp. 298–313, 2024.
https://doi.org/10.1007/978-3-031-49018-7_22

distributed by two age peaks, between 18 and 39 and between 50 and 69 years old.

Given the possible unfamiliarity with the disease, there may be a significant period from the first symptoms of psoriasis and the patients' consultation with the specialist to obtain the diagnosis.

This article addresses the development of a deep learning model, targeting smartphones based application, to allow taking and analysing pictures of human skin, providing a probability of psoriasis diagnose. We hope that this approach can help reduce the time of seeking medical opinion and early diagnosis.

The article os structure in seven sections, starting with this introduction. Section 2 provides the context and background, followed by the methodology, in Sect. 3. The results of the classification algorithms are presented in Sect. 4 and discussed in Sect. 5. Section 6 provides an introduction to the companion mobile application and the article ends with some final considerations.

2 Background and Literature Review

With the advent of deep learning and the remarkable impact on image processing, several applications and techniques are being developed in several areas, from interactive musical settings [3], innovative cognitive rehabilitation [21] as well as to assist medical professionals in the treatment and diagnosis of diseases. Song et al. developed a deep learning-based oral cancer image classification models, with the purpose of contributing to its early detection. The models are executed in a cost-effective smartphone, reaching an accuracy of 81% for distinguishing normal/benign lesions from clinically suspicious lesions [32]. They compared several MobileNet [14] and Inception [34] models ensuring that the model is kept as small and efficient as possible. In addition to the smartphone, it is also necessary an intraoral imaging probe, a light-emitting diode driver, and a rechargeable lithium battery for image acquisition.

Giavina-Bianchi et al. also focus on the development of a Computer-Aided Diagnosis (CAD) system to be used by Primary Care Physicians (PCP) on the diagnostic of skin cancers at early stages in the primary care attention [10]. The diagnosis system includes a protocol for image acquisition based on smartphone, for situations in which there are no dermoscopy devices available. The authors achieved 89% accuracy with ensemble models and, in addition to the classification model, the authors also included a VGG16 based model to classify the quality of the image taken with the smartphone. If the image is not "ideal" (as designated by the authors), a new picture must be taken.

Even if not resorting to smartphones, the identification of areas or regions of interest in medical imaging is important to diagnose or detect abnormal patterns. Jiang et al. seek to assess the current trends for tumor segmentation in images through deep learning techniques [19]. In this review, they focus on retrieving the trends in terms of deep learning architecture, topology, loss function and type of training as well as in the type of tumor approached in the literature. The source images are from CT (Computerized Tomography), PET (Positron

Emission Tomography) and MR (Magnetic Resonance) scans around a certain part of human body. According to the authors, there is a research trend focusing on 3D image segmentation, transfer learning, and model compression.

Jeong et al. did a similar study for Generative Adversarial Networks (GANs) [18]. Due to their characteristics, GANs are been increasingly used in medical applications, particularly to augment, balance, or improve classification and segmentation. Nevertheless, the wide use of this approach is still lacking the trust of the medical community, requiring more extensive verification.

Popescu et al. developed a systematic review of melanoma detection using artificial intelligence, especially neural network-based systems [26]. In the review, they checked for the use of different Neural Network (NN) families, including Resnet, Inception/GoogLeNet, U-Net, GAN, DenseNet, AlexNet, Xception, EfficientNet, VGG, NASNet, MobileNet, YOLO, FrNet, and Mask R_CNN. It is clear that the use of deep learning is prevalent. The authors also argue that the results are improved by using fusion or ensemble of deep neural networks.

Within skin diseases, skin cancer is one of the most threatening diseases worldwide so, Li et al. focus on a comprehensive review of works on deep learning for skin disease diagnosis [20]. In this context, the authors highlight several issues that should be resolved before deep learning can be extensively applied to real-life clinical scenarios of skin disease diagnosis, among others:

– Limited labeled skin disease data
– Unbalanced skin disease datasets
– Noisy data obtained from heterogeneous sources
– Lack of diversity among cases in existing skin disease datasets
– Missing of medical history and clinical meta-data of patients
– Explainability of deep learning methods
– Selection of deep neural networks for a specific skin disease diagnosis task

These issues are fundamental to increase the trust of the medical professionals as well as to increase the accuracy and overall usefulness of the systems.

In the approach of Shrivastava et al., a CAD dermatological system was proposed to classify skin images as healthy or psoriatic, using a base of 540 images [31]. From a unique resource space and a SVM (Support Vector Machine) classifier, the accuracy 99.81% in cross-validation was obtained. In another approach, Shrivastava et al. [30] proposes a diagnostic system for classifying healthy skin images and psoriasis lesions. Using a core SVM type two polynomial and color parameters for learning of 540 images, 99.94% accuracy was obtained, using the cross validation method with 10 divisions.

Velasco et al. used a MobileNet model with transfer learning to classify seven different types of dermatological diseases, one of them being psoriasis [36]. The model was trained from a dataset with 3406 images and achieved 94.4% accuracy. A mobile application was used to capture the images and display the results.

Dash et al. developed a framework that classifies an image as with psoriasis or without psoriasis [5]. Through a modified VGG16 model and an image base with 5000 examples in each class, the system achieved an average accuracy of 99.08%.

Zhao et al. developed an imaging classification system of nine dermatological lesions, one of them being psoriasis [37]. From an image base with approximately 8000 images, the authors tested the models DenseNet, InceptionV3, Inception-ResnetV2 and Xception. The InceptionV3 model performed better through the AUC (Area Under the Receiver Operating Characteristic (ROC) Curve) metric [2], with 0.981 ± 0.015.

In the research by Padilla et al., a skin classification system was developed to classify images among the classes psoriasis, atopic dermatitis and unknown [24]. Using the MobileNet architecture and a dataset composed of 6264 training images and 30 test images, the authors achieved an accuracy of 88% for atopic dermatitis and 90% for psoriasis.

Due to the growing efficiency of deep learning, several computational applications are been developed to assisting in the diagnosis of diseases performed by health professionals. Not many are designed for direct use by the patient, even if for increasing their awareness and attention toward the importance of early diagnosis. Thus, the development of such classification system is of great relevance.

3 Methodology

The main objetive of the work described in this article is the study and development of a personal recommendation tool, running in a smartphone, that can identify, with a good accuracy, psoriasis from pictures taken with the device's camera. With this in mind, the requirements include the ability to execute classification algorithms in the smartphone, with the restrictions in terms of quality of the picture, power consumption and classification model size.

3.1 Understanding the Problem

Psoriasis is an inflammatory and systemic disease that results from activation of innate and adaptive immune systems with release of pro-inflammatory cytokines (eg. tumor necrosis factor-α [TNF-α], interferon-gamma, interleukin-12/17/23) that damage multiple tissues and organs (Fig. 1).

In most cases, a specific trigger of the disease is not identified, but phenomena such as infection, trauma and stressful events have been considered [12]. However, once pathogenesis is initiated, leukocyte recruitment to the dermis and epidermis occurs and the mediators previously mentioned induce hyperproliferation of the keratinocytes, parakeratosis and high epidermal cell turnover rate, contributing to psoriasis typical lesions. Vascular dilatation of superficial blood vessels and abnormalities on conjunctival cytology are also driven by this process.

3.2 Deep Learning

Deep learning is an area in the field of machine learning that aims to build computer programs capable of acquiring new knowledge (supervised training) and

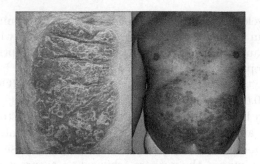

Fig. 1. Psoriasis lesion [35].

methods for organizing existing knowledge (unsupervised training), automatically improving its performance with experience (reinforcement learning) [23].

Methods that use deep learning aim to find a model through a collection of data used as example, and a procedure to guide learning through these examples. After the learning procedure a function is obtained which from a raw data input generates a representation relevant to the problem addressed [25].

Many of the deep learning approaches for image classification rely on Convolutional Neural Networls (CNNs), which are neural networks inspired by the visual cortex.

3.3   Dataset

Due to the difficulty in accessing dermatological image bases, because of privacy and rights of use, the images used in this work are composed of images obtained from the following dermatological platforms: International Psoriasis Council (IPC) [15], DermNetNZ [8], DermIS [16], DanDerm [1] and Hellenic atlas [7] (Table 1).

Table 1. Image sources and quantity.

Source	Number of images
International Psoriasis Council (IPC)	376
DermnetNZ	370
DermIS	193
DanDerm	170
Hellenic derm. atlas	97
Total	1206

Images of scalp, nails and genitalia were removed. The remaining images, in a total of 752, included different skin tones, lesion severity and location on the body, as well as different positions, angles and lighting conditions. Each image was annotated with two different labels: (i) psoriasis/non-psoriasis, and (ii) the bit mask corresponding to the skin lesion region. The split was 50%, meaning that half of the images did not have lesions and were composed of clean skin and skin with tatoos, freckles and others elements.

In order to compare and select the best approach, this dataset and the labels (label and mask) are used in the classification step in three scenarios: (i) the original images, without cropping or masking, (ii) the masks with the shape of the lesions, and (iii) the images multiplied by the masks, keeping only the lesion and the rest represented in black pixels (Fig. 2).

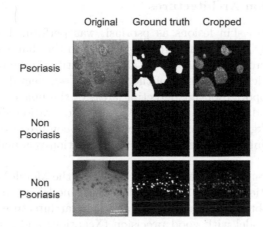

Fig. 2. Types of dataset used.

Deep learning for image classification is very sensitive to the number and variability of the training dataset. The current number of images is low, so, in order to maximize the performance of the models, data augmentation techniques were applied. Data augmentation involves changing the images to create a new images with sufficient changes to help the network to converge. The purpose is to be able to learn invariance, robustness, and reduce the impact of noise and interference. For that, several operations were performed, such as horizontal flip, vertical flip, rotation, color space changes and zoom (Table 2). These techniques generated a total of 18,368 images. It is worth mentioning that 150 images were kept unchanged, and used for testing.

In the training and validation phase, images were also preprocessed with CLAHE (Contrast Limited Adaptive Histogram Equalization) algorithm, to assess the importance of contrast [13]. The expectation was that with the optimization of the contrast of the images, the model would be able to increase its accuracy rate. CLAHE was implemented with the vision library OpenCV version 4.3.0. The grid size was defined as 8×8, generating 24 sub-regions, and a clip limit of 2.

Table 2. Data augmentation operations.

Operation	Proportion	Probability
Horizontal flip	1	70%
Vertical flip	1	70%
Rotation	90°, 180°, 270°	70%
Rotation	25°, −25°	70%
Random zoom	0.6	70%

3.4   Classification Architectures

The classification of skin lesions as psoriasis was performed with three main approaches, namely (i) with the full image, (ii) with the shape of the lesion only and (iii) with the cropped image. To be able to identify the shape, it is necessary to perform semantic segmentation of the image. So, segmentation is proposed in this work as a support tool, for detecting areas of the image that may contain psoriasis lesions and removing all the rest. Such an operation allows the removal of objects and backgrounds from an image that can confuse or disturb the model at the time of training for classification. Segmentation was performed with the U-Net model [27].

For classification, CNNs were used, specifically, the MobileNetV2 [28], Xception [4] and InceptionResnetV2 [33] models. Such models were selected to evaluate the results obtained from a small and light architecture (MobileNetV2), an intermediate model with good precision (Xception) and a more robust and heavy architecture (InceptionResnetV2).

All models use their original architectures, with the exception of the fully connected layer, which has been modified to perform binary classification (psoriases/non-psoriasis). The input layers have a $224 \times 224 \times 3$ dimension.

After the convolution and pooling layers the models apply a global mean pooling, calculating the average of each resource map produced and generating structured data. In the MobileNetV2 and Xception models this data is used directly as input of the 2 output neurons. In the InceptionResnetV2 model, a layer with 512 neurons activated by the Sigmoid function and a dropout of 40% is used before the 2 output neurons. The Softmax function is used as the final activation function.

The models were implemented in Tensorflow and Keras, which allowed the use of the initialized weights from the database Imagenet [6]. The weights are updated during the training, because the fine-tuning did not increase the performance of the models in the approach of this work.

3.5   Training

The training of the models was carried out in an AMD EPYC 7351 processor, with 16 GB RAM and a TITAN V GPU with 12 GB of memory. The machine learning libraries Tensorflow and Keras together with the CUDA 10.2 technology were used for parameter adjustments and training of the models.

The loss function was binary cross-entropy and the optimizador was Adam, with a learning rate of 0.0001. The networks have been trained for 30 epochs, the amount of steps per epoch is given by the ratio between the amount of samples from the dataset and the size of each batch created.

4 Results

A total of 15 tests were done, combining the three different models, the transformations in the dataset and the use of CLAHE (Fig. 3). Each combination aims to highlight the best architecture, algorithm or method to use.

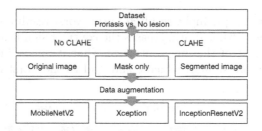

Fig. 3. Set of tests performed.

The sequence ilustrates the approached followed. First, the effect of CLAHE was assessed, followed by the type of input image and, finally, the architecture. After each step, only the best model, transformation or method is kept to the next step. In this way, when reaching the end of this test sequence, the best combination is obtained to achieve the project objective.

All the tests were made using the same evaluation metrics, namely: accuracy, loss rate, confusion matrix, sensitivity, specificity and precision. These were calculated with the machine learning library Scikit Learn version 0.23.1, and were based on the True Positive (TP), False Positive (FP), True Negative (TN) and False Negative (FN) counts (Table 3).

Accuracy correspond to the fraction of the correct classifications in all classifications. The confusion matrix is used to evaluate the performance of the classifier, by means of each pair of classes $C1, C2$, this metric verifies which proportion of documents $C1$ were incorrectly associated with $C2$ [22]. The Loss Rate refers to how far the classification is from the expected labels [9]. The Sensitivity points to the model's ability to classify positive samples. Specificity is used to calculate the capacity of the classifier to predict negative samples [29] and, finally, Precision is used to evaluate how certain the model is in the classification of positive samples in all samples that were labeled by the system as positive [17].

Table 3. Values used in the calculation of applied metrics.

Expression	description
TP	These are samples that are from the positive class and the model classifies them as belonging to the positive class
FP	These are samples that are not of the positive class and the model classifies as belonging to the positive class
TN	These are samples that are of the negative class and the model classifies as belonging to the negative class
FN	These are samples that are not of the negative class and the model classifies as belonging to the negative class

4.1 CLAHE

The use of CLAHE, surprisingly, resulted in the worst results, so it was removed from the next steps. Although it increases the contrast significantly, highlighting the lesions of the rest of the image, it seems that it makes images of psoriasis and other lesions more similar, which could lead to the worst results (Fig. 4).

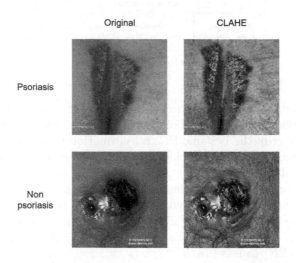

Fig. 4. Result of images after application of the CLAHE method.

4.2 Type of Input Image

To carry out the tests, 3 types of images were used:

- Original: formed by the original images,
- Mask: containing only the binary masks of the images (black/white),
- Cropped: formed by cropped images, where the lesions are highlighted and everything else that does not fit the lesion class is presented in black.

To create such datasets, the segmentation module was used to generate the masks and a subsequent AND operation was applied to cut the lesion regions of the images. These processing was done to test the efficiency of the models in images that do not contain the background, that is, objects or environments that are not part of the classification focus. In addition, the dataset that uses only binary masks was used to know if the model would be able to correctly classify images using only shape, size and location data.

Using the MobileNetV2, Xception and InceptionResnetV2 models, the 3 types of images were tested (Table 4). These tests were done before data augmentation.

Table 4. Test results with different dataset types.

ImagenModel	MobileNetV2	Xception	IncepResnetV2
Original	acc: 0,875	acc: 0,875	acc: 0,9
	loss: 0,6330	loss: 0,4587	loss: 0,56555
Mask	acc: 0,7	acc: 0,7500	acc: 0,8
	loss: 0,666	loss: 0,8622	loss: 0,71155
Cropped	acc: 0,85	acc: 0,8500	acc: 0,75
	loss: 0,2933	loss: 1,0534	loss: 0,6693

From the analysis of the data obtained from the tests, it is possible to observe that the original images had a better average performance in relation to accuracy and loss among the 3 image types in the 3 models, pointing out that segmented images are not effective in this approach. In this way, for the next test the original image dataset is used.

4.3 Data Augmentation

To test the influence of the number of images in the dataset and the operations used for data augmentation, a new test step was carried out, increasing the number of images according to the procedure described in Table 2. A total of 18,368 imagens were used for training and 150 (non-augmented) images were kept for testing.

A significant improvement was obtained, asserting the importance of a large number of training examples (Table 5).

5 Discussion

The best results were obtained with data augmentation in the MobileNetV2 model, in which 18,368 images were used for training. The models were trained for thirty epochs with 32 images per batch, a learning rate of 0.0001, and the

Table 5. Test results with new dataset and new data augmentation.

ImagenModel	MobileNetV2	Xception	IncepResnetV2
Original (no DA)	acc: 0,875	acc: 0,875	acc: 0,9
	loss: 0,6330	loss: 0,4587	loss: 0,56555
Original (DA)	acc: 0,9733	acc: 0,9666	acc: 0,9466
	loss: 0,0723	loss: 0,1266	loss: 0,1404

Table 6. List of results obtained in the classification.

Metrics	MobileNetV2	Xception	IncepResnetV2
Accuracy	97.33%	96.66%	94.66%
loss rate	7.23%	12.66%	14.05%
Sensitivity	96%	96%	92%
Specificity	98.66%	97.33%	97.33%
Precision	98.63%	97.29%	97.18%

Binary Cross Entropy loss function. Table 6 presents the results obtained in the tests for each model.

The MobileNetV2 model, although having the smallest structure and the smallest depth, obtained greater accuracy and lower loss rate. Thus, it is possible to conclude that larger and more robust models are not exactly the best option for this approach, with this image base, possibly by the limited amount of data.

Another relevant point is that the models had higher specificity value than sensitivity, which suggests a greater success in the classification of negative images, in this case images without lesion (Table 7).

Table 7. Confusion matrices of classification models.

	MobileNetV2		Xception		InceptionResNetV2	
	Psoriasis	Non-Psoriases	Psoriasis	Non-Psoriases	Psoriasis	Non-Psoriases
Psoriasis	72	3	72	3	69	6
Non-psoriases	1	74	2	73	2	73

The models are easier to classify images of non-psoriasis, in general the networks are missing in some predictions of images of psoriasis, classifying them as non-psoriasis, such a fact may be due to the similarity that exists between some images of other lesions and the psoriasis itself.

In order to understand what types of images model has difficulty classifying, the Fig. 5 displays images and their respective classifications, made by the MobileNetV2 model, which obtained better test performance.

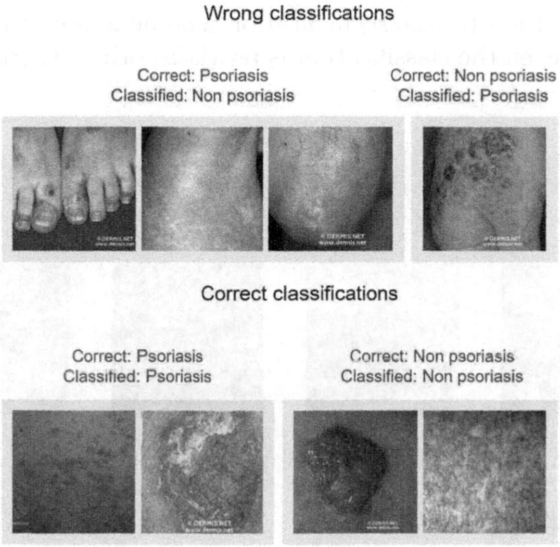

Fig. 5. Classifications performed by the MobileNetV2 model.

The first 4 images of Fig. 5 refer to the samples of false positive and false negative presented in the confusion matrix of the MobileNetV2 model. As can be seen, the model had difficulty classifying images of psoriasis with border without a clear definition and with not so reddish coloring. In the case of non-psoriasis images, the false positive was probably due to its coloration resembling the psoriasis images.

Despite the difficulties presented, in a general perspective the models obtained a satisfactory result, with the best performance presenting only 4 erroneous classifications and the worst model classifying 8 images in a wrong way, in a total of 150 images evaluated.

6   Mobile Application

After studying the behavior of the models and training process, a mobile application was developed to acquire the image and make a classification of the image taken. Its interfaces were kept as simple and intuitive as possible.

After clipping, the captured image is displayed on the main interface for user evaluation and the analysis button is enabled (Fig. 6(a)). Upon clicking the analysis button, a loading message is displayed until the classification is ready. The result is presented via two messages, one containing the classification, psoriasis or non-psoriasis, and the other presenting the probability of the classification (Fig. 6(b, c, d)). For better presentation of the information, the probability is mapped into 5 levels: very low, low, medium, high and very high, which are defined according to the accuracy from 0% to 100% returned by the classifier.

As can be seen in Fig. 6(c) the segmentation module is used to display the possible lesion area when the classification is psoriasis, otherwise the original image is displayed.

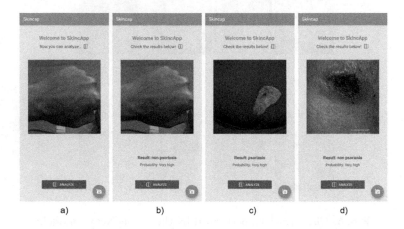

Fig. 6. Mobile application results display interface

7　Conclusion

Considering the diversity of existing dermatological lesions, correctly diagnose a specific dermatological disease is not a simple task. In this context, the work described in this article describes the phases and steps leading to a mobile app for classification photos of the skin as psoriasis or non-prosiasis. It is expected that the use of this tool by regular users can raise the awareness of the disease as well as providing conditions for seeking medical consultation and, thus, early diagnose.

The app is based on an image classification model based on deep learning architectures and, as such, a training dataset was created, composed of 752 clinical images of human skin, half of which with psoriasis and half of non-psoriasis.

The app has the possibility of classifying 150 unseen images with 97.33% accuracy, for which only 4 images were misclassified, which represents a good accuracy.

Nevertheless, the image dataset was still small, which required studying several approaches for increasing the accuracy. It is important to enlarge the datasets for a better performance of neural network models, as well as comparing the results presented with other deep learning techniques.

Acknowledgments. The authors are grateful to the Foundation for Science and Technology (FCT, Portugal) for financial support through national funds FCT/MCTES (PIDDAC) to CeDRI (UIDB/05757/2020 and UIDP/05757/2020) and SusTEC (LA/P/0007 /2021).

References

1. Atlas of Clinical Dermatology: Atlas of Dermatology (2022). https://www.danderm.dk/atlas/index.html
2. Bradley, A.P.: The use of the area under the ROC curve in the evaluation of machine learning algorithms. Pattern Recogn. **30**(7), 1145–1159 (1997)
3. Cardoso, M., Lopes, R.: Interactive musical setting with deep learning and object recognition. In: Proceedings of the 12th International Conference on Computer Supported Education, Prague, Czech Republic, pp. 663–667. SCITEPRESS - Science and Technology Publications (2020). https://doi.org/10.5220/0009856406630667. http://www.scitepress.org/DigitalLibrary/Link.aspx?doi=10.5220/0009856406630667
4. Chollet, F.: Xception: deep learning with depthwise separable convolutions. In: Proceedings of the IEEE Conference on Computer Vision and Pattern Recognition, pp. 1251–1258 (2017)
5. Dash, M., Londhe, N.D., Ghosh, S., Raj, R., Sonawane, R.S.: A cascaded deep convolution neural network based CADx system for psoriasis lesion segmentation and severity assessment. Appl. Soft Comput. **91**, 106–240 (2020)
6. Deng, J., Dong, W., Socher, R., Li, L.J., Li, K., Fei-Fei, L.: ImageNet: a large-scale hierarchical image database. In: 2009 IEEE Conference on Computer Vision and Pattern Recognition, pp. 248–255. IEEE (2009)
7. Dermatological Atlas: Home | Hellenic Dermatological Atlas - Over 2700 Dermatology pictures (2022). http://www.hellenicdermatlas.com/en/
8. DermNet NZ: Image library | DermNet NZ (2022). https://dermnetnz.org/image-library
9. Google for Developers: Machine Learning Crash Course (2020). https://developers.google.com/machine-learning/glossary
10. Giavina-Bianchi, M., et al.: Implementation of artificial intelligence algorithms for melanoma screening in a primary care setting. PLoS ONE **16**(9), e0257006 (2021). https://doi.org/10.1371/journal.pone.0257006
11. Greb, J.E., et al.: Psoriasis. Nat. Rev. Dis. Primers **2**(1), 16082 (2016). https://doi.org/10.1038/nrdp.2016.82. www.nature.com/articles/nrdp201682
12. Habashy, J.: Psoriasis: practice essentials, background, pathophysiology. Technical report, MedScape, November 2020. https://emedicine.medscape.com/article/1943419-overview
13. Heckbert, P.: Graphics Gems IV (IBM Version). Elsevier (1994)
14. Howard, A.G., et al.: MobileNets: efficient convolutional neural networks for mobile vision applications, April 2017. arXiv:1704.04861
15. International Psoriasis Council: IPC - Psoriasis Image Library (2022). https://www.psoriasiscouncil.org/imagelibrary.htm
16. ISIC: ISIC Archive (2022). https://www.isic-archive.com/
17. Japkowicz, N.: Why question machine learning evaluation methods. In: AAAI Workshop on Evaluation Methods for Machine Learning, pp. 6–11 (2006)
18. Jeong, J.J., Tariq, A., Adejumo, T., Trivedi, H., Gichoya, J.W., Banerjee, I.: Systematic review of generative adversarial networks (GANs) for medical image classification and segmentation. J. Digit. Imaging **35**, 137–152 (2022). https://doi.org/10.1007/s10278-021-00556-w
19. Jiang, H., Diao, Z., Yao, Y.D.: Deep learning techniques for tumor segmentation: a review. J. Supercomput. **78**(2), 1807–1851 (2022). https://doi.org/10.1007/s11227-021-03901-6

20. Li, H., Pan, Y., Zhao, J., Zhang, L.: Skin disease diagnosis with deep learning: a review. Neurocomputing **464**, 364–393 (2021). https://doi.org/10.1016/j.neucom.2021.08.096. https://www.linkinghub.elsevier.com/retrieve/pii/S0925231221012935

21. Lopes, R.P., et al.: Digital technologies for innovative mental health rehabilitation. Electronics **10**(18), 2260 (2021). https://doi.org/10.3390/electronics10182260. https://www.mdpi.com/2079-9292/10/18/2260

22. Manning, C.D., Schütze, H., Raghavan, P.: Introduction to Information Retrieval. Cambridge University Press, New York (2008)

23. Mitchell, T.M.: Machine Learning, 1st edn. McGraw-Hill, New York (1997)

24. Padilla, D., Yumang, A., Diaz, A.L., Inlong, G.: Differentiating atopic dermatitis and psoriasis chronic plaque using convolutional neural network MobileNet architecture. In: 2019 IEEE 11th International Conference on Humanoid, Nanotechnology, Information Technology, Communication and Control, Environment, and Management (HNICEM), pp. 1–6. IEEE (2019)

25. Ponti, M.A., da Costa, G.B.P.: Como funciona o Deep Learning. arXiv preprint arXiv:1806.07908 (2018)

26. Popescu, D., El-Khatib, M., El-Khatib, H., Ichim, L.: New trends in melanoma detection using neural networks: a systematic review. Sensors **22**(2), 496 (2022). https://doi.org/10.3390/s22020496. https://www.mdpi.com/1424-8220/22/2/496

27. Ronneberger, O., Fischer, P., Brox, T.: U-Net: convolutional networks for biomedical image segmentation. In: Navab, N., Hornegger, J., Wells, W.M., Frangi, A.F. (eds.) MICCAI 2015. LNCS, vol. 9351, pp. 234–241. Springer, Cham (2015). https://doi.org/10.1007/978-3-319-24574-4_28

28. Sandler, M., Howard, A., Zhu, M., Zhmoginov, A., Chen, L.C.: MobileNetV2: inverted residuals and linear bottlenecks. In: Proceedings of the IEEE Conference on Computer Vision and Pattern Recognition, pp. 4510–4520 (2018)

29. Shafiq, M., Yu, X., Bashir, A.K., Chaudhry, H.N., Wang, D.: A machine learning approach for feature selection traffic classification using security analysis. J. Supercomput. **74**(10), 4867–4892 (2018)

30. Shrivastava, V.K., Londhe, N.D., Sonawane, R.S., Suri, J.S.: Exploring the color feature power for psoriasis risk stratification and classification: a data mining paradigm. Comput. Biol. Med. **65**, 54–68 (2015)

31. Shrivastava, V.K., Londhe, N.D., Sonawane, R.S., Suri, J.S.: Reliable and accurate psoriasis disease classification in dermatology images using comprehensive feature space in machine learning paradigm. Exp. Syst. Appl. **42**(15–16), 6184–6195 (2015)

32. Song, B., et al.: Mobile-based oral cancer classification for point-of-care screening. J. Biomed. Opt. **26**(06), 065003 (2021). https://doi.org/10.1117/1.JBO.26.6.065003. https://www.spiedigitallibrary.org/journals/journal-of-biomedical-optics/volume-26/issue-06/065003/Mobile-based-oral-cancer-classification-for-point-of-care-screening/10.1117/1.JBO.26.6.065003.full

33. Szegedy, C., Ioffe, S., Vanhoucke, V., Alemi, A.: Inception-v4, inception-ResNet and the impact of residual connections on learning. arXiv preprint arXiv:1602.07261 (2016)

34. Szegedy, C., et al.: Going deeper with convolutions. In: 2015 IEEE Conference on Computer Vision and Pattern Recognition (CVPR), June 2015, pp. 1–9 (2015). ISSN 1063-6919. https://doi.org/10.1109/CVPR.2015.7298594

35. Torres, T., Sales, R., Vasconcelos, C., Selores, M.: Psoriasis and cardiovascular disease. Acta Med. Port. **26**(5), 601–607 (2013)

36. Velasco, J., et al.: A smartphone-based skin disease classification using MobileNet CNN. arXiv preprint arXiv:1911.07929 (2019)
37. Zhao, S., et al.: Smart identification of psoriasis by images using convolutional neural networks: a case study in China. J. Eur. Acad. Dermatol. Venereol. **34**(3), 518–524 (2020)

WildFruiP: Estimating Fruit Physicochemical Parameters from Images Captured in the Wild

Diogo J. Paulo[1,2](✉) [iD], Cláudia M. B. Neves[3,4] [iD],
Dulcineia Ferreira Wessel[3,4] [iD], and João C. Neves[1,2] [iD]

[1] University of Beira Interior, Covilhã, Portugal
[2] NOVA LINCS - NOVA Laboratory for Computer Science and Informatics,
Lisbon, Portugal
`diogo.paulo@ubi.pt`
[3] Polytechnic Institute of Viseu, Viseu, Portugal
[4] LAQV-REQUIMTE, Department of Chemistry,
University of Aveiro, Aveiro, Portugal

Abstract. The progress in computer vision has allowed the development of a diversity of precision agriculture systems, improving the efficiency and yield of several processes of farming. Among the different processes, crop monitoring has been extensively studied to decrease the resources consumed and increase the yield, where a myriad of computer vision strategies has been proposed for fruit analysis (e.g., fruit counting) or plant health estimation. Nevertheless, the problem of fruit ripeness estimation has received little attention, particularly when the fruits are still on the tree. As such, this paper introduces a strategy to estimate the maturation stage of fruits based on images acquired from handheld devices while the fruit is still on the tree. Our approach relies on an image segmentation strategy to crop and align fruit images, which a CNN subsequently processes to extract a compact visual descriptor of the fruit. A non-linear regression model is then used for learning a mapping between descriptors to a set of physicochemical parameters, acting as a proxy of the fruit maturation stage. The proposed method is robust to the variations in position, lighting, and complex backgrounds, being ideal for working in the wild with minimal image acquisition constraints. Source code is available at https://github.com/Diogo365/WildFruiP.

Keywords: Fruit Physicochemical Parameters · Computer Vision · Maturation Stage

1 Introduction

The agricultural industry is a vital sector of the global economy and plays a crucial role in the human food supply. Food production is constantly evolving, and technology has been a significant ally in this process. Vision-based systems

© Springer Nature Switzerland AG 2024
V. Vasconcelos et al. (Eds.): CIARP 2023, LNCS 14469, pp. 314–326, 2024.
https://doi.org/10.1007/978-3-031-49018-7_23

and artificial intelligence have enabled significant improvements in quality and productivity. In this context, this work aims to develop a computer vision strategy for automatically estimating fruit maturation stages from a single photo acquired by handheld devices. This will allow for a more accurate and efficient evaluation of the fruit production process, reducing resource waste and increasing agricultural sector productivity.

While computer vision systems have provided valuable information for crop monitoring, the assessment of fruit ripeness has received limited attention. Existing systems primarily rely on drone-based monitoring methods, which do not allow for the determination of the specific maturation stage of each fruit. Alternatively, some researchers have developed systems to classify the maturation stage of fruit after harvesting, which may not be particularly useful for farmers.

Considering the importance of determining fruit ripeness for optimal harvesting decisions, we introduce a method for determining the maturation stage of fruits using a single photo acquired from handheld devices while the fruit is still on the tree. The proposed method relies on an image segmentation model to crop and align the fruit, which a CNN subsequently analyses for extracting a visual descriptor of the fruit (illustrated in Fig. 1). The maturation stage is defined by a set of physicochemical parameters that are inferred from the visual descriptor using a regression model. To allow the learning of the image segmentation and regression model, we collected a dataset of 400 images of figs and prickly pears and their corresponding physicochemical parameters. To the best of our knowledge, this is the first dataset comprising both visual and physicochemical data, and we expect it to be of particular interest to the research community for carrying out studies of the relationship between the chemical properties of fruits and their visual appearance. The dataset used in this work is publicly available on https://github.com/Diogo365/WildFruiP.

Our main contributions in this work are as follows:

- We introduce a strategy for fruit ripeness estimation capable of operating in images acquired in the wild while the fruit is still in the tree.
- We assessed the performance of the proposed method in determining a set of physicochemical parameters of a fruit using a single image obtained in the visible light spectrum.
- To foster the research on the problem of fruit ripeness estimation from visual data, we introduce a dataset comprising 400 images from two fruit species and their respective physicochemical parameters, which serve as a proxy to the fruit maturation stage.

2   Related Work

2.1   Detection Methods

Object detection in images is a crucial task in computer vision, which had a tremendous progress in the last years due to the emergence of deep learning.

Several works have taken advantage of this progress for fruit detection. In [12], Parvathi *et al.* proposed an enhanced model of Faster R-CNN [10] for detecting coconuts in images with complex backgrounds to determine their ripeness. The performance of the model was evaluated on a dataset containing real-time images and images from the Google search engine. The results showed that the improved Faster R-CNN model achieved better detection performance compared to other object detectors such as SSD [7], YOLO [9], and R-FCN [2].

2.2   Segmentation Methods

Image segmentation is crucial for fruit image analysis, as it allows separating fruits from other parts of the image, such as leaves or background. Mask R-CNN [4] is an instance segmentation method that has proven effective in object segmentation tasks and has been extensively used for fruit analysis applications.

Siricharoen *et al.* [13] proposed a three-phase deep learning approach [13] to classify pineapple flavor based on visual appearance. First, a Mask R-CNN segmentation model was used for extracting pineapple features from the YCbCr color space. Then, a residual neural network pre-trained on COCO and ImageNet datasets was utilized for flavor classification. The authors concluded that their model successfully captured the correlation between pineapple visual appearance and flavor.

Ni *et al.* [8] developed an automated strategy for blueberry analysis. They employed a deep learning-based image segmentation method using the Mask R-CNN model to count blueberries and determine their ripeness. The results indicated variations among the cultivars, with 'Star' having the lowest blueberry count per cluster, 'Farthing' exhibiting less ripe fruits but compact clusters, and 'Meadowlark' showing looser clusters. The authors highlighted the need for objective methods to address fruit ripeness inconsistency caused by annotation inconsistencies in the trained model.

2.3   Methods for Estimating Fruit Ripeness in Images

Several strategies have been introduced to enable pre-harvest in-field assessment of fruit ripeness using handheld devices [6]. However, most approaches rely on the non-visible light spectrum, requiring thus dedicated hardware [11].

Regarding the approaches devised for visible light spectrum, most of them use CNNs for the estimation of fruit ripeness. Appe *et al.* [1] proposed a model for tomato ripeness estimation using transfer learning. They relied on the VGG16 architecture, where the top layer was replaced with a multilayer perceptron (MLP). The proposed model with fine-tuning exhibited improved effectiveness in tomato ripeness detection and classification. In another work, Sabzi *et al.* [12] developed an innovative strategy for estimating the pH value of oranges from three different varieties. A neural network was combined with the particle swarm optimization [5] to select the most discriminative features from a total of 452 features obtained directly from segmented orange images. This approach was able to rely on a subset of six features to obtain an accurate estimation of the pH values across different orange varieties.

In short, few approaches were devised for addressing the problem of fruit ripeness estimation from visual data, ranging from traditional extraction of hand-crafted features to deep-learning-based methods.

3   Proposed Method

The proposed approach can be broadly divided into three principal phases: the detection and segmentation of fruits in an image, the alignment and cropping of the fruit, and the determination of the physicochemical parameters of the fruit. The pipeline of this method is presented in Fig. 1.

3.1   Fruit Detection and Segmentation

This phase aims at removing the spurious information from the image keeping only the fruit region. Accordingly, the fruit is segmented automatically using the Mask R-CNN [4] allowing the prediction of a binary mask containing the pixels where exists a specific type of fruit. Considering the specificity of this task, the Mask R-CNN was fine-tuned on the proposed dataset, allowing thus it to generalize to the fruits targeted in this problem. To address the problem of multiple fruits in the image, we establish that the fruit to be analysed should be in the center of the image, and thus the remaining masks are discarded.

Figure 2 depicts the results obtained by applying Mask R-CNN to both figs and prickly pears.

3.2   Image Alignment

Considering that fruit orientation varies significantly in the images, it is particularly important to enforce a standard alignment to ease the learning of the fruit analysis model.

Considering the general shape of fruits, we propose to approximate their silhouette using an ellipse. Also, we concluded that the silhouette of the fruit can be modeled using the segmentation mask obtained from the previous phase.

Let M be the segmentation mask, and consider the general equation of the ellipse:

$$\frac{((x - x_0) \cos \theta + (y - y_0) \sin \theta)^2}{a^2} + \frac{(-(x - x_0) \sin \theta + (y - y_0) \cos \theta)^2}{b^2} = 1, \quad (1)$$

where (x_0, y_0) are the coordinates of the ellipse's center, a and b are the horizontal and vertical semi-axes, respectively and θ is the ellipse orientation, with $\theta \in \left[-\frac{\pi}{2}, \frac{\pi}{2}\right]$. The boundary of M is determined using the convex hull of the (x, y) points of M, and least square fitting [3] is used to determine x_0, y_0, a, b, and θ. The rotation angle θ is then used to rotate the original image and crop the fruit region based on the minimum bounding box containing the ellipse obtained. The results of the fruit alignment can be observed in Fig. 3.

3.3   Determination of Physicochemical Parameters

In the third phase, a CNN model is used to learn a visual descriptor which can encode the discriminative information regarding the physicochemical parameters of the fruit. A multi-layer perceptron is used as a regression model to infer the nine physicochemical parameters from the visual descriptors. The CNN and the regression model were trained in an end-to-end manner using the mean-squared error loss, and a k-fold cross-validation technique was adopted due to the reduced amount of training data.

4   Dataset

Considering the unavailability of public datasets comprising fruit images and their corresponding maturation stage or physicochemical parameters, we acquired 4 photos each from 60 figs and 40 prickly pears from local farmers. The fruits were subsequently harvested and analysed in the lab to extract 9 characteristics that are typically correlated with the maturation state of the fruit. The physical and chemical parameters obtained are listed in Table 1.

Table 1. Range of values for the physicochemical parameters used in this project.

Attribute	Range of Values
TSS (º Brix)	[13.5;19.0]
Hardness (N)	[1.1;61.4]
pH	[4.2;6.2]
mass (g)	[24.3;209.6]
L	[21.1;56.7]
a	[-16.4;23.2]
b	[2.6;26.8]
length (cm)	[59.6;110.0]
diameter (cm)	[39.0;62.9]

   To allow the development of a custom image segmentation model, we annotated the complete set of 400 images using the CVAT tool. An exemplar from each of the fruit species and its corresponding annotations can be observed in Fig. 4. To foster the research on the problem of estimating fruit ripeness from visual data, we make our dataset publicly available[1].

---

[1] https://github.com/Diogo365/WildFruiP.

5   Experiments

This section reports the performance of the proposed method for the problem of physicochemical parameter estimation from images of figs and prickly pears acquired using handheld devices. Tests are conducted using the aligned, and misaligned/cropped dataset using different neural networks. Also, we compare the proposed approach with a state-of-the-art method devised for inferring fruit physicochemical parameters from visual data.

5.1   Implementation Details

Detection and Segmentation. The backbone of the Mask R-CNN was a Residual Neural Network (ResNet), specifically the ResNet50 variant integrated into the PyTorch framework. Prior annotations were necessary for each fruit, including bounding boxes, labels, and masks to train the model. The data augmentation transformations were resizing, horizontal flipping, brightness and contrast adjustment. After defining the necessary transformations for data processing, the annotated initial dataset was split into 80% for training and 20% for testing. Finally, with the separated datasets and processed data, the model was trained for 50 epochs using the stochastic gradient descent optimizer with a learning rate of 0.001.

Determination of Physicochemical Parameters. The training data consisted of a set of images and their corresponding physicochemical parameters. Each parameter was normalized using a linear transformation estimated from the training data. A lightweight CNN architecture (ResNet18) was used for extracting 2048 dimensional visual descriptors from the aligned fruit images and a multi-layer perceptron was exploited for the estimation of nine parameters from the visual descriptors. The configurations used are presented in Table 2. All models were trained for a maximum of 100 epochs using the Early Stopping regularization technique and all of our experiments were conducted on PyTorch with NVIDIA GeForce RTX 3060 GPU and with Intel(R) Core(TM) i7-10700 CPU @ 2.90GHz. The inference times reported in Table 3 were obtained by executing the model on this hardware configuration.

Table 2. Configuration used for training the CNN.

Batch Size	16
Epochs	100
Learning Rate	0.001
Optimizer	Adam
Image Size	320×320

Table 3. Inference time and total size of the different models.

Models	Inference Time	Storage Size
MobileNetV2	10.1 ms	8.7 MB
ResNet-18	10.2 ms	89.9 MB

5.2   Metrics

To assess the performance of the proposed model, four metrics were employed, mean squared error (MSE), mean absolute error (MAE), mean absolute percentage error (MAPE), and the coefficient of determination (R^2). They are defined as follows:

$$MSE = \frac{1}{n} \sum_{i=1}^{n} (y_i - \hat{y}_i)^2, \tag{2}$$

$$MAE = \frac{1}{n} \sum_{i=1}^{n} |y_i - \hat{y}_i|, \tag{3}$$

$$MAPE = \frac{1}{n} \sum_{i=1}^{n} \left| \frac{y_i - \hat{y}_i}{y_i} \right| \times 100, \tag{4}$$

$$R^2 = 1 - \frac{\sum_{i=1}^{n}(y_i - \hat{y}_i)^2}{\sum_{i=1}^{n}(y_i - \bar{y})^2} \tag{5}$$

In these equations, n represents the total number of data points or observations in the evaluation set, y_i denotes the true value of the dependent variable for the ith observation, $\hat{y}_i$ represents the predicted value of the dependent variable for the ith observation, $\bar{y}$ denotes the mean value of the dependent variable across all observations.

Table 4. Performance of the proposed approach. The R^2 value (mean ± std) determined for both species denotes a strong predictive power for some physicochemical parameters. Also, the comparison with the approach of Sabzi *et al.* [12], evidences a clear improvement in all parameters.

Attributes		Sabzi *et al.* [12]		Proposed Method	
		Prickly Pears	Figs	Prickly Pears	Figs
TSS (º Brix)		−0.81±0.04	-	**0.18±0.07**	-
Hardness (N)		−2.66 ± 0.70	−1.13±0.66	**0.51±0.08**	**0.68±0.03**
pH		−1.26 ± 0.27	-	**0.13±0.11**	-
mass (g)		−0.83 ± 0.14	-	**0.22±0.11**	-
Color	L	−3.39 ± 0.84	−3.26±0.01	**0.25±0.09**	**0.75±0.04**
	a	−6.71 ± 0.03	−5.96±1.73	**0.83±0.03**	**0.87±0.01**
	b	−4.63 ± 0.71	−3.23±0.69	**0.42±0.08**	**0.79±0.03**
length		−0.17 ± 0.14	−0.04±0.08	**−0.03±0.08**	**0.40±0.03**
diameter		−0.65 ± 0.28	−0.49±0.34	**0.37±0.07**	**0.28±0.04**

5.3   Performance of the Proposed Approach

The proposed method was assessed in the evaluation split of both prickly pear and fig images using k-fold validation and repeating the training and evaluation process 10 times. The results are reported in Table 4.

The analysis of the results with respect to prickly pears shows a moderate correlation (refer to R^2) between some physicochemical parameters and the predictions of the network obtained from the fruit image. All parameters showed a positive correlation except for the length. Insufficient relevant information in the image might explain the lack of correlation with the prickly pear's length. Visual features such as shape, color, or texture are not informative about the length of a fruit. A strong predictive power was obtained for the 'a' parameter, hardness, and 'b' parameters, with correlations of 0.83, 0.51, and 0.42, respectively. The 'a' parameter represented the fruit chromaticity from green to red, which is strongly correlated with the fruit ripeness. Hardness also corresponds to ripeness, as riper figs are typically less firm. However, the sugar content, measured by the TSS (ºBrix) had a weak correlation possibly due to the dataset small size.

Table 5. Results obtained by the proposed model with the misaligned, and aligned datasets using prickly pears.

Attributes		R2		MSE		MAE		MAPE	
		Misaligned Image	Aligned Image	Misaligned Image	Aligned Image	Misaligned Image	Aligned Image	Misaligned Image	Aligned Image
TSS (º Brix)		−0.29 ± 0.42	**0.18 ± 0.07**	0.94 ± 0.25	**0.63 ± 0.04**	0.78 ± 0.10	**0.65 ± 0.03**	5.13 ± 0.64	**4.27 ± 0.16**
Hardness (N)		0.41 ± 0.13	**0.51 ± 0.08**	6.97 ± 1.46	**5.89 ± 0.95**	2.07 ± 0.21	**1.92 ± 0.19**	12.12 ± 1.15	**11.25 ± 1.43**
pH		−0.18 ± 0.25	**0.13 ± 0.11**	0.01 ± 0.00	**0.01 ± 0.00**	0.09 ± 0.01	**0.08 ± 0.00**	1.50 ± 0.14	**1.31 ± 0.06**
mass (g)		0.05 ± 0.15	**0.22 ± 0.11**	1022.83 ± 162.76	**854.54 ± 131.64**	25.10 ± 1.94	**23.06 ± 1.56**	21.95 ± 1.75	**20.50 ± 1.33**
Color	L	−0.18 ± 0.31	**0.25 ± 0.09**	17.61 ± 5.07	**10.55 ± 1.03**	3.20 ± 0.40	**2.45 ± 0.15**	6.71 ± 0.82	**5.14 ± 0.30**
	a	0.65 ± 0.11	**0.83 ± 0.03**	27.50 ± 8.16	**13.50 ± 2.39**	4.03 ± 0.46	**2.93 ± 0.24**	193.38 ± 54.40	**130.26 ± 21.16**
	b	0.15 ± 0.15	**0.42 ± 0.08**	5.02 ± 0.86	**3.43 ± 0.52**	1.79 ± 0.17	**1.44 ± 0.10**	8.39 ± 0.85	**6.80 ± 0.47**
length		−0.37 ± 0.21	**−0.03 ± 0.08**	127.65 ± 21.27	**97.24 ± 8.54**	8.92 ± 0.83	**7.80 ± 0.22**	11.02 ± 1.17	**9.72 ± 0.41**
diameter		0.08 ± 0.21	**0.37 ± 0.07**	28.70 ± 6.02	**19.96 ± 2.41**	4.22 ± 0.42	**3.57 ± 0.23**	8.34 ± 0.87	**7.03 ± 0.51**

Regarding the performance attained on figs, only six out of nine parameters were evaluated due to insufficient data for TSS (º Brix), pH, and mass. Nevertheless, our approach demonstrated a better aptitude for estimating physicochemical parameters in this fruit species (figs), likely due to the disparity in dataset sizes.

Regarding the comparison with the state-of-the-art, the method of Sabzi *et al.* [12] significantly underperformed when compared with our approach. The main justification for this difference is the fact that the method of Sabzi *et al.* [12] was originally intended to analyse fruit images in controlled scenarios (the method was devised for pH estimation of oranges in a uniform background). However, the images obtained when the fruits are still on the tree are inherently more challenging due to the varying pose, lighting, and complexity of the background.

5.4   Impact of Alignment Phase

In this experiment, the model was trained using misaligned/cropped, and aligned images for the nine physicochemical parameters.

Prickly Pears. Upon analyzing Table 5, it was observed that ablating the alignment of the images, as expected, led to worst results, making the difference between negative and positive values of R^2 as is the case of the TSS (ºBrix), pH and L parameters.

Table 6. Results obtained by the proposed model with the misaligned, and aligned datasets using figs.

Attributes		R2		MSE		MAE		MAPE	
		Misaligned Image	Aligned Image	Misaligned Image	Aligned Image	Misaligned Image	Aligned Image	Misaligned Image	Aligned Image
TSS (º Brix)		-	-	-	-	-	-	-	-
Hardness (N)		0.72 ± 0.05	0.68 ± 0.03	77.39 ± 12.30	88.20 ± 9.20	6.40 ± 0.78	6.76 ± 0.46	137.34 ± 29.17	138.02 ± 21.14
pH		-	-	-	-	-	-	-	-
mass (g)		-	-	-	-	-	-	-	-
Color	L	0.73 ± 0.06	0.75 ± 0.04	17.95 ± 4.04	16.55 ± 2.25	3.31 ± 0.37	3.17 ± 0.23	9.61 ± 0.99	9.18 ± 0.66
	a	0.87 ± 0.01	0.87 ± 0.01	11.20 ± 1.16	11.01 ± 1.20	2.63 ± 0.18	2.61 ± 0.16	84.93 ± 12.27	97.79 ± 8.78
	b	0.77 ± 0.03	0.79 ± 0.03	10.30 ± 1.12	9.39 ± 1.24	2.50 ± 0.15	2.40 ± 0.18	25.21 ± 1.34	23.89 ± 3.05
length		0.47 ± 0.04	0.40 ± 0.03	46.36 ± 3.48	53.68 ± 2.58	5.34 ± 0.19	5.87 ± 0.15	7.03 ± 0.32	7.75 ± 0.24
diameter		0.38 ± 0.08	0.28 ± 0.04	8.42 ± 0.96	9.47 ± 0.53	2.27 ± 0.12	2.38 ± 0.07	4.85 ± 0.26	5.12 ± 0.16

Figs. Regarding figs, upon analyzing Table 6 it was observed that the aligned dataset yielded slightly better results compared to the misaligned dataset. The aligned dataset led to improved training results for the color parameters, while the misaligned dataset performed better for shape features (diameter and length).

The diameter parameter proved challenging but outperformed the worst parameter in the prickly pear experiment.

5.5   Impact of Model Architecture

Considering that the proposed approach is planned to work in handheld devices with low computational resources, the proposed method is based on a lightweight architecture. To determine the best architecture for the problem, we compared the impact of the architecture on the performance of the proposed approach, as well as, on the inference time.

Therefore, for this experiment, we assessed the performance of our approach using two lightweight architectures: MobileNetV2 and ResNet18. The comparison of the model size and inference time of the different architectures is provided in Table 3, while Table 7 reports the performance of our approach along the different architectures.

It is interesting to observe that ResNet18 was able to consistently attain the best results over all parameters and simultaneously for both fruit species.

Despite its larger size, we claim that the superior predictive power obtained justifies its use in this problem. Also, it is important to note that the inference time is equivalent for both models.

5.6   Hard Samples

To further explain the obtained results, an additional test was conducted to identify figs where the proposed method significantly deviated from the correct physicochemical parameters using the MAE metric (less sensitive to outliers than MSE). Figure 5 shows the images of the two fruit species where the proposed approach had the largest MAE.

Table 7. Results obtained for the R^2 metric on prickly pears and figs utilizing the ResNet18 and MobileNetv2.

Attributes		Prickly Pears		Figs	
		MobileNetV2	ResNet18	MobileNetV2	ResNet18
TSS (º Brix)		0.15 ± 0.08	$\mathbf{0.18 \pm 0.07}$	-	-
Hardness (N)		0.49 ± 0.06	$\mathbf{0.51 \pm 0.08}$	0.62 ± 0.03	$\mathbf{0.68 \pm 0.03}$
pH		0.06 ± 0.10	$\mathbf{0.13 \pm 0.11}$	-	-
mass (g)		0.07 ± 0.07	$\mathbf{0.22 \pm 0.11}$	-	-
Color	L	0.08 ± 0.09	$\mathbf{0.25 \pm 0.09}$	0.64 ± 0.03	$\mathbf{0.75 \pm 0.04}$
	a	0.76 ± 0.04	$\mathbf{0.83 \pm 0.03}$	0.80 ± 0.02	$\mathbf{0.87 \pm 0.01}$
	b	0.28 ± 0.09	$\mathbf{0.42 \pm 0.08}$	0.68 ± 0.03	$\mathbf{0.79 \pm 0.03}$
length		-0.10 ± 0.10	$\mathbf{-0.03 \pm 0.08}$	0.24 ± 0.03	$\mathbf{0.40 \pm 0.03}$
diameter		0.26 ± 0.05	$\mathbf{0.37 \pm 0.07}$	0.16 ± 0.06	$\mathbf{0.28 \pm 0.04}$

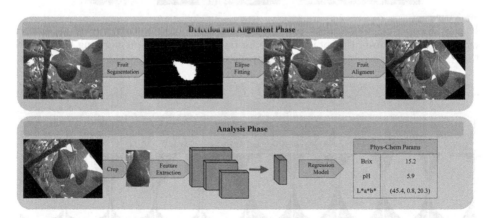

Fig. 1. Pipeline of the proposed approach. The fruit image is given to an image segmentation approach, which determines the fruit mask. Using the mask, a fitting process is performed to enclose the fruit in an ellipse, and the rotation angle of the ellipse is used to align the fruit in the image. Afterwards, the fruit is cropped using the bounding boxes also extracted from the segmentation mask. The cropped fruit is fed into a CNN for extracting a visual descriptor which is subsequently mapped to a set of physicochemical parameters through a regression model.

Several factors affect the performance of the model, including luminosity differences, blur, variations in fig shapes (length, diameter, and mass), and limitations in training due to a lack of examples of unripe figs.

Fig. 2. Fruit detection and segmentation. The Mask R-CNN was fine-tuned to provide a rough segmentation of the fruit allowing to discard irrelevant regions of the image in the analysis of the data. Even though the masks are not so accurate in the border, it is important to note that the accuracy of the segmentation mask is not crucial for the overall approach.

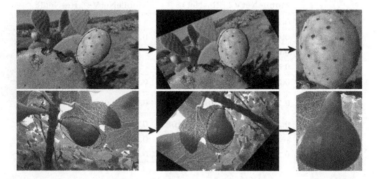

Fig. 3. Proposed alignment process. The fruit is approximated using an ellipse, which allows to obtain the rotation angle for image alignment and cropping the aligned fruit. The alignment process is depicted for the two fruit species considered in this study.

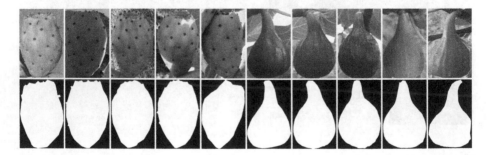

Fig. 4. Samples from the proposed dataset. Our dataset comprises 400 images of two fruit species and their corresponding physicochemical parameters. Also, we provide the location of fruit in the image using manually annotated bounding boxes and segmentation masks.

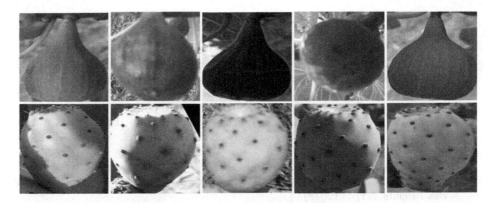

Fig. 5. Hard to predict samples. The five samples from the two fruit species that have the highest absolute error over the nine physicochemical parameters.

6   Conclusion and Future Work Prospects

In this work, we introduced an approach for estimating the maturation stage of fruit images acquired in the wild using handheld devices. The proposed approach relied on an innovative alignment strategy that increased the robustness to pose variations. Also, we introduced a novel dataset containing images with significant variations in lighting, and diversity of the background. The experimental validation of the proposed approach showed a strong correlation with some physicochemical parameters, which can serve as a proxy to determine the maturation stage of the fruits considered in this study. On the other hand, our approach was capable of remarkably surpassing a state-of-the-art approach specifically designed for fruit maturation estimation. To further validate the proposed method, we carried out several experiments, which showed that the alignment phase increased the performance of the method. Also, the analysis of the most challenging image samples evidenced that blur and brightness variation were the major causes of failure. In the future, we expect that our approach can be incorporated into a mobile application, providing farmers with an easy-to-use fruit ripeness estimation tool for efficient control and informed decision-making in agriculture.

Acknowledgements. This work was funded by the InovFarmer.MED project, which is part of the PRIMA Programme. Also, it was supported by NOVA LINCS (UIDB/04516/2020) with the financial support of FCT.IP.

References

1. Appe, S.R.N., Arulselvi, G., Balaji, G.N.: Tomato ripeness detection and classification using VGG based CNN models. Int. J. Intell. Syst. Appl. Eng. **11**(1), 296–302 (2023)

2. Dai, J., Li, Y., He, K., Sun, J.: R-FCN: object detection via region-based fully convolutional networks. In: Proceedings of the Advances in Neural Information Processing Systems (NIPS), vol. 29. Curran Associates, Inc. (2016)
3. Fitzgibbon, A., Pilu, M., Fisher, R.: Direct least square fitting of ellipses. IEEE Trans. Pattern Anal. Mach. Intell. (TPAMI) **21**(5), 476–480 (1999)
4. He, K., Gkioxari, G., Dollár, P., Girshick, R.: Mask R-CNN. In: Proceedings of the International Conference on Computer Vision (ICCV), pp. 2980–2988 (2017)
5. Kennedy, J., Eberhart, R.: Particle swarm optimization. In: Proceedings of the International Conference on Neural Networks (IJCNN), vol. 4, pp. 1942–1948 (1995)
6. Li, B., Lecourt, J., Bishop, G.: Advances in non-destructive early assessment of fruit ripeness towards defining optimal time of harvest and yield prediction - a review. Plants **7**(1), 3 (2018)
7. Liu, W., Anguelov, D., Erhan, D., Szegedy, C., Reed, S., Fu, C.-Y., Berg, A.C.: SSD: single shot multibox detector. In: Leibe, B., Matas, J., Sebe, N., Welling, M. (eds.) ECCV 2016. LNCS, vol. 9905, pp. 21–37. Springer, Cham (2016). https:// doi.org/10.1007/978-3-319-46448-0_2
8. Ni, X., Li, C., Jiang, H., Takeda, F.: Deep learning image segmentation and extraction of blueberry fruit traits associated with harvestability and yield. Horticult. Res. **7**, 110 (2020)
9. Redmon, J., Divvala, S., Girshick, R., Farhadi, A.: You only look once: unified, real-time object detection. In: Proceedings of the Conference on Computer Vision and Pattern Recognition (CVPR), pp. 779–788 (2016)
10. Ren, S., He, K., Girshick, R., Sun, J.: Faster R-CNN: towards real-time object detection with region proposal networks. IEEE Trans. Pattern Anal. Mach. Intell. **39**(6), 1137–1149 (2017)
11. Rizzo, M., Marcuzzo, M., Zangari, A., Gasparetto, A., Albarelli, A.: Fruit ripeness classification: a survey. Artif. Intell. Agric. **7**, 44–57 (2023)
12. Sabzi, S., Javadikia, H., Arribas, J.I.: A three-variety automatic and non-intrusive computer vision system for the estimation of orange fruit pH value. Measurement **152**, 107–298 (2020)
13. Siricharoen, P., Yomsatieankul, W., Bunsri, T.: Recognizing the sweet and sour taste of pineapple fruits using residual networks and green-relative color transformation attached with mask R-CNN. Postharvest Biol. Technol. **196**, 112–174 (2023)

Depression Detection Using Deep Learning and Natural Language Processing Techniques: A Comparative Study

Francisco Mesquita[1]($\boxtimes$) (ID), José Maurício[1] (ID), and Gonçalo Marques[2] (ID)

[1] Polytechnic of Coimbra, ISEC, Rua Pedro Nunes, 3030-199 Coimbra, Portugal
{a2018056868,a2018056151}@isec.pt
[2] Polytechnic of Coimbra, ESTGOH, Rua General Santos Costa, 3400-124 Oliveira dos Hospital, Portugal
goncalo.marques@estgoh.ipc.pt

Abstract. Depression is a frequently underestimated illness that significantly impacts a substantial number of individuals worldwide, making it a significant mental disorder. The world today lives fully connected, where more than half of the world's population uses social networks in their daily lives. If we interpret and understand the feelings associated with a social media post, we can detect potential depression cases before they reach a major state associated with consequences for the patient. This paper proposes the use of natural language processing (NLP) techniques to classify the sentiment associated with a post made on the Twitter social network. This sentiment can be non-depressive, neutral, or depressive. The authors collected and validated the data, and performed pre-processing and feature generation using TF-IDF and Word2Vec techniques. Various DL and ML models were evaluated on these features. The Extra Trees classifier combined with the TF-IDF technique emerged as the most successful combination for classifying potential depression sentiment in tweets, achieving an accuracy of 84.83%.

Keywords: Depression Detection · Sentiment Analysis · Natural Language Processing (NLP) · Twint · Antidepressant

1 Introduction

Depression is an increasingly common illness around the world and one that is likely to increase soon. Some of the symptoms of depression may be restlessness, irritability, impulsivity, anxiety, palpitations, sadness, loss of energy, a sense of hopelessness and many more [26]. According to World Health Organization (WHO), 3.8% of the population suffers from this disorder which means that approximately 280 million people around the world live with depression [1]. In Europe, 6.38% of the population suffers from depression, ranging from 2.58% in

V. Vasconcelos et al. (Eds.): CIARP 2023, LNCS 14469, pp. 327–342, 2024.
https://doi.org/10.1007/978-3-031-49018-7_24

the Czech Republic and 10.33% in Iceland [16]. It is often difficult to diagnose mental illness, however, the growth and globalization of social network usage can help to reduce the number of cases that go unnoticed. Social networks play a key role and have a direct correlation with depression as suggested by Yoon et al. [31]. Over the past few years, there has been an increase in the number of people interested in studying and using machine learning (ML) algorithms to create medical decision support systems [24]. This is due to the great evolution that has taken place in the industry in terms of computing power and the ever-increasing amount of data available [25].

The sentiment behind social media posts should be examined in order to diagnose depression as quickly and accurately as possible [6]. To achieve this, it needs a system capable of processing, analyzing, and deriving knowledge from a diverse and unstructured data set. One specific domain within the field of ML, particularly deep learning (DL), has the ability to accomplish this very task. Natural language processing (NLP) is capable of understanding human language and taking valuable information from it [32]. This has been a hot topic in research in recent years using data mining and ML techniques. The potential that these techniques could have for clinical use is very high as shown in the Ricard et al. study [23]. Consequently, it is proposed an ML approach to detect the sentiment associated with a tweet made on the Twitter social network. This could be the depressive, neutral, or non-depressive sentiment. Twitter was chosen because of its massive use, being the third most popular social network in the world. It also has a simple data model and an easily accessible API to collect data.

This paper objective is to create a predictive model capable of detecting a possible depressive feeling associated with a tweet. This will allow worldwide improvements in the way potential people at risk of depression are detected. The main contributions are: (i) Describe a full ML pipeline that covers collecting data, processing it, training a classification algorithm, and evaluating its performance, (ii) Comparative analysis of different feature generation techniques and classification algorithms, (iii) Compare the achieved results and proposed model with prior research works in the literature.

This paper has the following structure: Sect. 2 presents the summary of all similar papers in the literature, Sect. 3 is the methodology used, including information about the dataset, pre-processing techniques, label validation, exploratory data analysis, how we generate features with TF-IDF and Word2Vec, experimental setup and how we evaluate the models created. Section 4 shows all the results that we obtain with both ML and DL models. Section 5 presents the discussion where we state the findings of this study and make a comparison with what has been done previously in the literature. Section 6 are the conclusions we can draw from our work including contributions, limitations, and future work.

2   Related Work

In the last few years, several works have been proposed on how to automate the diagnosis of depression in a patient, to reduce the number of cases of depression that are increasing and reducing the number of suicides caused by major depression.

The study proposed by [7] used the public dataset Sentiment 140 which contains data without the presence of signals of depression. Adding to that data, they gathered a dataset with signals of depression, through the collecting data of Twitter with recourse to the Twint tool. The following keywords were used to pick tweets with the signals of depression: hopeless, lonely, antidepressant and depression. In pre-processing, the stop words, punctuation marks and hyperlinks were removed, and authors used Lemmatization to group different forms of the same word. After preprocessing the data, a Feature Generation was performed based on techniques such as Tokenization to separate the words in the text into a form that the machine understands. Valence Aware Dictionary and sEntimen-tReasoner (VADER) were also used to extract the polarity of the tweets to get the overall emotion of the text and finally Word2Vec was utilized to transform the text into word vectors.

After these data preparation processes, the dataset was divided into 60% for training and the rest was divided into validation and testing. They proceeded to classify the data using two types of approaches: a) a Long Short-Term Memory (LSTM) network; b) a hybrid CNN-LSTM model. In the first approach, they were able to obtain 90.33%, 91%, 91%, and 91% of Accuracy, F1-Score, Precision and Recall, respectively. On the other hand, in the second approach, they were able to get 91.35%, 91%, 92% and 91% on the same metrics, respectively.

Another study proposed by [9] achieved an improvement in these results using a combination of Word2Vec and DL models. It achieved 99.02% Accuracy, 99.04% Precision, 99.01% Recall and 99.02% F1-Score for the LSTM network. And, obtained 99.01% of Accuracy, 99.20% of Precision, 99.01% of Recall and 99.10% of F1-Score for the hybrid CNN+LSTM model.

Many works make use of feature extraction tools such as Bag-Of-Words (BOW), Tokenizer and TF-IDF models. In the study proposed by [29], the authors aimed to predict whether the person was not depressed, was half depressed, moderately depressed or severely depressed, so they used the unsupervised K-Means clustering algorithm to label the tweets. Decision Trees, Random Forest and Naive Bayes algorithms were used in this study for the classification of the tweets. The dataset was split into 80% for training and 20% for testing. In the end, they evaluated the performance of the algorithms through the classification metrics Accuracy, F1-Score, Recall, Precision and R-Score. The combination of TF-IDF models to generate features with the Random Forest algorithm stood out from this approach, having obtained 95% of Accuracy, 95% of Precision, 95% of Recall, 67% of R-Score and 95% of F1-Score.

The authors of [27] used a public dataset with 43000 tweets. For each tweet, a pre-processing was performed which consisted of removing non-alphabetic characters (e.g., HTML tags, punctuation, hashtags, numeric values, special characters, URLs), normalization the tweeters converting the text to lowercase, removing stop words (e.g., prepositions, conjunctions, and articles), and at last applied stemming. Since ML algorithms cannot process the raw text, TF-IDF was used to extract features to then be provided as input to the model. As a result, using

Multinomial Naive Bayes they achieved 72.97%, 74.58% and 75.04% accuracy, precision and recall, respectively.

Alsagri et al. [5] used almost the same pre-processing steps but the data was obtained through the Twitter API and was a much smaller amount, about 3000 tweets. It is also a differentiated approach as it tries to classify the user himself as depressive through the various tweets associated with him. Using TF-IDF it obtained 82.50% accuracy, 73.91% precision, 85% recall, 79% f1 score and finally, 77.50% AUC.

Kabir et al. [14] proposed a new topology to diagnose depression disease in Twitter messages called DEPTWEET and introduced a unique dataset labelled, with clinical validation, and for each label, a confidence score was assigned. The Twitter messages were retrieved using the Twint tool and the search keywords were defined based on the PHQ-9 questionnaire for depression. They classified each tweet as one of the four possible values: non-depressed, mildly depressed, moderately depressed, or severely depressed. As classifiers, the authors used Support Vector Machine (SVM), Bidirectional LSTM (BiLSTM), and two pre-trained transformer-based models: BERT and DistilBERT. ROC score was chosen as the evaluation metric and the best result was obtained using the transformer-based models, with the DistilBERT standing out and getting 78.88% in non-depressed, 74.72% in mildly, 78.79% in moderate and 86.60% in severe depression.

When compared to the literature, our work proposes a collection of tweets using the TWINT tool and the assignment of a label (POSITIVE, NEGATIVE, and NEUTRE) to the text through the sentence polarity score achieved by VADER. In addition, we will perform a manual validation of each phrase present in the data to ensure that the classes are assigned correctly, thus reducing the probability of error in the label assignment process done in the [14, 28] work. Word2Vec and TF-IDF were also used for feature generation. Finally, to classify the text, we use two DL architectures (LSTM and hybrid CNN+LSTM) and several ML algorithm.

3   Methodology

The methodology proposed in this article to detect depression consists of 3 steps: i) collecting data from Twitter using the Twint tool, cleaning and categorizing the collected tweets; ii) manual validation of the label assigned to each tweet and data augmentation; iii) generating features for each sentence using TF-IDF and Word2Vec to train ML and DL algorithms. As shown in Fig. 1.

3.1   Dataset

The data collection for this work consisted in acquiring Twitter data with signs of depression, using the Twint tool. The keywords lonely, depressed, frustrated, hopeless and antidepressant were used to obtain the phrases with signs of depression [7].

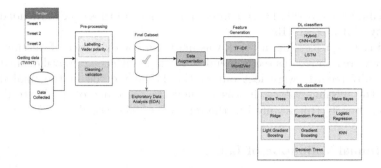

Fig. 1. Experimental setup.

Using Twint, we configure the search parameters such as the respective search keyword, the tweet limit, and the language in which the tweets are written. After that, the Twint tool will systematically extract the desired data from Twitter, aggregating tweets, user details, and even interaction metrics. At the end of this process, we have a dataset ready to be processed and analyzed.

All tweets were gathered on November 3, 2022. Twint tool starts by retrieving the latest tweets and continues to fetch older tweets until it reaches a stopping condition, such as a specified number of tweets or a certain time limit. In our case, it stopped when it reached the limit of 3500 collected tweets. Each word originated a dataset of tweets with 3500 records that were later converged to a final dataset with 17500 records.

3.2 Data Pre-processing

When collecting tweets for analysis, it is crucial to account for the presence of noise resulting from the limitations of the collection process, which is based on a single keyword. Where there is no control over the content of the tweets obtained. Therefore, pre-processing techniques were applied to the tweets. First, the sentences were normalized to convert them to lowercase. Next, all hyperlinks, hashtags, identifications of other users and emojis were removed. In addition, all stopwords were removed from the tweets collected through the stopwords function of the nltk library. These steps have been suggested in several previous works [11,15,22].

After the collected data had been cleaned, the VADER tool was used to categorize the tweets into Positive, Negative and Neutral. VADER is a tool widely used in sentiment analysis tasks due to its simplicity, effectiveness, and ability to handle domain-specific and colloquial language. It was introduced by Hutto et al. in 2014 [13] and has been employed in various applications, including social media monitoring, customer feedback analysis, and opinion mining [3,8,10,12,20]. To do the labelling of the tweets the compound score value generated by VADER was used as a base. It consisted in assigning the label Positive for compound score values greater than or equal to 0.05, the label Negative for compound score values less than or equal to -0.05 and values between -0.05 and 0.05 would be identified as Neutral. These compound values are those recommended in the article by Hutto et al. [13].

During pre-processing, a check was made for the existence of missing values, and they were removed as recommended by [21]. With the removal of the missing values, 1057 records were lost out of the 17500 records in the final dataset. Besides the missing values, it was verified the existence of duplicate data. As suggested by [30] this resulted in the removal of 955 records.

3.3   Manual Validation of Label

The manual validation process for the sentences in the collected dataset involved excluding sentences with fewer than three words after pre-processing. This was done because we cannot determine the sentiment from a sentence of less than 3 words [18]. This step resulted in 1520 sentences being eliminated.

After this validation, the resulting sentences were checked to ensure that the label assigned by the VADER algorithm was correct or not. In instances where there was ambiguity in interpreting the pre-processed sentence, we turn to the corresponding original sentence to better understand its meaning and decide. When incorrect labels were identified, appropriate corrections were made to ensure accuracy.

Furthermore, as part of this procedure, a check was performed to ensure that the collected sentences were relevant to the theme of the study. In cases where the semantics of the sentences were incorrect or they were in a different language, they were discarded. As a result, a total of 10,920 sentences were subjected to validation. Out of these, 8,519 sentences were removed from the dataset, leaving 2,512 sentences that were used for the study. This high number of discarded sentences is due to several factors: (i) many of the collected tweets did not fit the topic of depression, often consisting of reviews, opinions, quotes or other types of the text unrelated to a depressive feeling; (ii) despite the Twint settings, some tweets came in other languages and were therefore removed; (iii) a few tweets were ambiguous or even contradictory about the possible associated sentiment and we decided to discard them.

3.4   Exploratory Data Analysis

During the exploratory data analysis, the purpose was to examine the distribution of the assigned labels before and after manual validation and dataset augmentation. Figure 2 illustrates the class distribution prior to validation and augmentation, indicating an imbalance among the classes. From the graph analysis, it is evident that a class imbalance exists, as VADER tends to assign a strong Negative sentiment label to sentences containing negative keywords. However, Fig. 3 presents the class distribution after the dataset went through validation and augmentation, revealing that the data is now almost perfectly balanced.

Furthermore, we built three Word Clouds that allow us to visualize which words are present in the collected sentences. The Word Cloud with the words of the sentences that were considered positive and negative are shown in Fig. 4 and

Fig. 5, respectively. Finally, Fig. 6 shows the words of the sentences that were considered neutral.

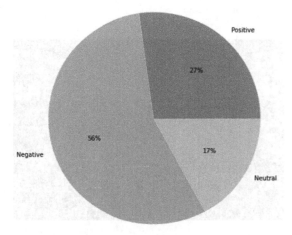

Fig. 2. Distribution of the classes before the manual validation and augmentation.

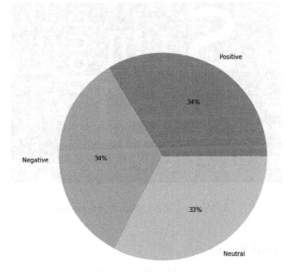

Fig. 3. Distribution of the classes after the manual validation and augmentation.

3.5 Feature Generation

The algorithms selected to generate features applying the algorithms described in the literature are presented in this section. Therefore, this work used the TF-IDF algorithm which consists of the vectorization of documents to calculate a score for each word based on its importance in the document and corpus [2].

Word2Vec algorithm allows representing all the words of the sentences extracted from Twitter in embeddings based on the similarity of the words [4]. This algorithm was configured with 300 for the embedding size and 10 for the window size.

Fig. 4. Word Cloud with label Positive.

Fig. 5. Word Cloud with label Negative.

3.6   Experimental Setup

For the development of this work, the Pycaret library in version 2.3.10, TensorFlow in version 2.11.0, Scikit-learn in version 1.2.0, NLKT in version 3.7 and Gensim in version 3.6.0 were used. In addition to these libraries, the nlpaug library was also used in version 1.1.11 to do data augmentation. To increase the number of instances and diversity when training the ML algorithms, data augmentation was used. The augmented sentences were generated by changing certain words by synonyms throughout the sentence, always maintaining their meaning and the associated sentiment. In addition, this step allowed the target to be balanced. From 2512 sentences we now have in the final dataset 5032 instances.

Therefore, after having augmented the dataset and the features generated by Word2Vec and TF-IDF, Several experiments have been done mixing ML and DL methods with the TF-IDF and Word2Vec feature selection techniques. ML algorithms were implemented using the Pycaret library properties and the top 10 algorithms with the best performance were selected. The result obtained by the models was achieved based on the parameters configured by default.

Fig. 6. Word Cloud with label Neutral.

On the other hand, the hybrid model and the LSTM network were built based on the Sequential API. The hybrid model consists of 1 Embedding layer, 2 Conv1D layers, 2 BatchNormalization layers, 2 MaxPooling1D layers, 1 LSTM layer, 4 Dense layers and 3 Dropout layers. The LSTM network consists of 1 Embedding layer, 1 LSTM layer, 1 Flatten layer, 5 Dense layers and 4 Dropout layers. Both DL models use SparseCategoricalCrossentropy as the Loss function and SGD as the optimizer, using a learning rate of 1.0e-04 during the experiment with TF-IDF and a learning rate 1.0e-02 for the experiment with Word2Vec. They were trained for 50 epochs with a batch size of 32. During training, an early stopping was applied with a patience of 5.

3.7 Evaluation

Before proceeding to the classification of the tweets, through the ML and DL algorithms, the dataset was divided into 70% for training and 30% for testing in the experiments performed [17].

The performance of the ML algorithms was evaluated through the classification metrics: Accuracy, F1-Score, Recall, Precision and AUC. On the other hand, the performance of the DL algorithms was assessed with the metrics: Accuracy, Loss, F1-Score, Recall and Precision. These metrics are calculated on the data of the testing dataset.

4   Results

Only the top outcomes from the experiments conducted will be provided in this section. Therefore, Table 1 has presented the top 10 best ML algorithms with the features generation being performed by the Word2Vec algorithm. Table 2 presents the top 10 best ML algorithms, where the TF-IDF algorithm was used to generate the features.

Table [3–4] are presented the best results obtained with the DL algorithms using TF-IDF and Word2Vec algorithms to generate features, respectively. In bold are marked the best result.

Table 1. Top 10 ML algorithms with Word2Vec.

Models	Accuracy	Precision	Recall	F1-Score	AUC
Extra Trees Classifier	**0.5775**	**0.5821**	**0.5775**	**0.5779**	**0.7430**
Extreme Gradient Boosting	0.5444	0.5448	0.5444	0.5439	0.7383
Random Forest	0.5417	0.5454	0.5417	0.5419	0.7279
Light Gradient Boosting	0.5411	0.5434	0.5411	0.5403	0.7281
Quadratic Discriminant Analysis	0.5391	0.5434	0.5391	0.5383	0.7172
Gradient Boosting	0.5060	0.5112	0.5060	0.5045	0.6980
K-Neighbors Classifier	0.4921	0.4953	0.4921	0.4920	0.6826
Linear Discriminant Analysis	0.4735	0.4748	0.4735	0.4730	0.6540
Ada Boost	0.4728	0.4738	0.4728	0.4687	0.6448
Naive Bayes	0.4477	0.4597	0.4477	0.4335	0.6298

Table 2. Top 10 ML algorithms with TF-IDF.

Models	Accuracy	Precision	Recall	F1-Score	AUC
Extra Trees Classifier	**0.8483**	**0.8501**	**0.8483**	**0.8487**	**0.9515**
SVM	0.8086	0.8232	0.8232	0.8231	0.0000
Ridge Classifier	0.8086	0.8087	0.8086	0.8086	0.0000
Random Forest	0.8060	0.8089	0.8060	0.8066	0.9292
Logistic Regression	0.7556	0.7562	0.7556	0.7557	0.8935
Naive Bayes	0.7417	0.7583	0.7417	0.7415	0.8088
Decision Tree	0.6960	0.6966	0.6966	0.6961	0.7719
Extreme Gradient Boosting	0.6927	0.6941	0.6927	0.6928	0.8564
Light Gradient Boosting Machine	0.6748	0.6756	0.6748	0.6750	0.8325
Gradient Boosting	0.6305	0.6339	0.6305	0.62975	0.8081

By analysing the results, is concluded that the Extra Trees Classifier algorithm combined with TF-IDF with 84.83%, 85.01%, 84.83% and 84.87% of Accuracy, Precision, Recall and F1-Score, respectively. It is the most accurate solution to predict whether a person has depression, through the sentences collected from

Twitter. However, it is observed that the SVM algorithm describes a near performance with 80.86%, 82.32%, 82.32% and 82.31% in the same metrics.

Based on these results, we can conclude that the DL algorithms do not demonstrate a good performance to predict depression in people through the gathered tweets. It has the combination of the hybrid model with Word2Vec demonstrating the higher result with: 52.05% of Accuracy, 59.25% of Precision, 52.15% of Recall, 51.92% of F1-Score and 1.0313 Loss.

Table 3. Results of the DL algorithms with TF-IDF.

Models	Accuracy	Loss	Precision	Recall	F1-Score
LSTM	**0.3305**	1.0986	**0.1102**	0.3333	**0.1656**
Hybrid model	0.3285	1.0986	0.1095	0.3333	0.1648

Table 4. Results of the DL algorithms with Word2Vec.

Models	Accuracy	Loss	Precision	Recall	F1-Score
Hybrid model	**0.5205**	**1.0313**	**0.5925**	**0.5215**	**0.5192**
LSTM	0.4642	1.0882	0.4968	0.4595	0.4031

5   Discussion

This comparative study between ML and DL algorithms uses different feature-generation techniques. By examining how people express themselves on social media, we can, with a certain degree of confidence, determine if they are feeling depressed or not. It serves as a benchmark for future research in sentiment analysis, offering a methodology that collects raw tweets and processes those sentences to predict the sentiment that person intends to express. In addition, it demonstrates the use of data augmentation tools associated with NLP problems.

Based on the findings of this study, it is clear that using the Extra Trees Classifier along with the TF-IDF feature generation technique achieves good prediction results. On the other hand, it is verified through the same results that the DL algorithms were inferior to the ML algorithms, in both combinations. Except that when using the Word2Vec algorithm, the hybrid model was able to obtain a better performance than algorithms such as Naive Bayes, Ada Boost, Gradient Boosting, K-Neighbors Classifier and Linear Discriminant Analysis. This can be explained by the low amount of data available and its high complexity [19]. Although this has been verified, we believe that DL techniques have the potential for better performance with a larger and more diverse dataset.

The best result presented in this study proves to be superior to the results obtained by a pre-trained model based on a transformer [14]. Where the authors collected data from Twitter, through the Twint tool and the validation of the sentences was done by a doctor, as well as the keywords selected were based on questionnaires previously performed. On the other hand, in studies that used deep learning algorithms the results obtained were significantly higher than those demonstrated in this work [7,9]. However, it was not possible to know the structures of the algorithms and their configurations to justify the results obtained.

In addition, most literature works used a larger dataset than the one used in this work. Either they used a public dataset such as sentiment 140, which is already labelled and in some cases medically validated, having a larger number of instances [27]. Or in other studies, data is collected following the same methodology as this study but is also used a validated public dataset to increase the number of tweets and to balance target [7,9]. Whereas, the authors of this study collected sentences from Twitter that were validated regarding the sentiment expressed in it by themselves.

The use of a two-step validation process, involving VADER initially and subsequent manually, enhances the accuracy of the sentiment associated with the phrase, thereby increasing confidence in the obtained results, despite the potential bias associated with the authors' interpretation. Also, it is possible to verify that the Extra Trees Classifier algorithm presents an AUC of 95.15%, which means that our ML algorithm has a good ability to distinguish between Positive, Negative and Neutral classes.

Table 5 presents a comparison between the proposed method and previous works in the literature.

When evaluating the insights provided by the literature, it remains unclear whether the algorithms' impressive performance translates into effective class distinction. We can take our study as an example, although the SVM and Ridge classifier algorithms exhibit strong predictive capabilities for sentiment analysis on tweets, a closer examination of the AUC value reveals their inability to effectively differentiate between the classes within the dataset.

This work shows that it is possible to classify sentences from Twitter according to the associated sentiment, doing so in sentences where there are many grammatical gaps, the vocabulary is not homogeneous and often there are both spelling and grammatical errors. According to the authors, utilizing raw data comprising colloquially written phrases that closely reflect real-life experiences enhances the classifier's performance and adds greater significance to the results. This approach is seen as more valuable compared to models trained on transformed phrases without spelling or grammatical errors, and homogeneous, failing to capture the authentic social media reality.

Table 5. Comparison of the results with the literature.

Ref	Model	Data	Labelling	Pre-processing	Performance
[7]	Hybrid CNN-LSTM	- Sentiment 140 non-depressive public data; - 1.6 million tweets; - Gathered depressive data with Twint which has no quantity information	- All non-depressive phrases came from public data; - Depressive sentences are the one gathered; with Twint; - No validation	- Remove hyperlinks, digits and stop words; - Text case change and Slang substitution; - Spell checking and lemmatization; - Feature generation with Word2Vec	Accuracy: 91.35%, Precision: 92%, Recall: 91% and F1-Score: 91%
[9]	Hybrid CNN-LSTM	- Random non-depressive tweets obtained from Kaggle; - Depressive tweets derived from Twint; - No information about quantity	- Random tweets from public data treated as not depressive; - Depressive data gathered with Twint; - No validation	- Remove links, images and URLs; - Remove punctuation and stop words; - Stemming and lemmatization; - Tokenization	Accuracy: 99.01%, Precision: 99.20%, Recall: 99.01% and F1-Score: 99.10%
[14]	DistilBERT	- Collected data using Twint; - final data contains 41191 tweets	- Human annotators; - Manual validation from expert psychologist	- WordPiece tokenizer	AUC: 79.75%
[29]	Random Forest	- Data collected from Twitter API, Kaggle and using Twint; - 16000 tweets where 8000 are negative and 8000 are positive	- Automatic annotation using K-means clustering	- Remove links and punctuation; - Feature extraction using Bag of Words TF-IDF and Tokenizer	Accuracy: 95%, Precision: 95%, Recall: 95% and F1-Score: 95%
[27]	Multinomial Naive Bayes	- Public data collected from Kaggle - 43000 tweets	- Already present on data; - No validation	- Emoji extraction and slung substitution; - Remove links, timestamp, digits, symbols, proper nouns and stop words; - Spelling correction and lemmatization; - Feature extraction with Bag of Words	Accuracy: 72.97%, Precision: 74.58% and Recall: 75.04%
[5]	Linear SVM	- Collected data manually and using Twitter API; - About 3000 tweets	- Manual human validation of the depressive tweets; - Non-depressive tweets were collected randomly and without validation	- Tokenization and Stemming; - Normalization: turn to lower case and remove links, emojis, symbols, mentions, retweets and punctuation; - Feature extraction using TF-IDF	Accuracy: 82.50%, Precision: 73.91%, Recall: 85%, F1-Score: 79% and AUC: 77.50%
-	Proposed method: Extra Trees	- Collected data manually using Twint - Data augmentation - 5032 tweets	- Two Step labelling; - Auto labelling with VADER; - Manual validation of each sentence sentiment	- Normalization: turn to lower case and remove links, hashtags, mentions, emojis and stopword; - Remove less than 3 words sentences; - Remove sentences that were not in English; - Feature extraction with both TF-IDF and Word2Vec	Accuracy: 84.83%, Precision: 85.01%, Recall: 84.83%, F1-Score: 84.87% and AUC: 95.15%

6   Conclusion

Depression is a highly common illness in our society, characterized by feelings of sadness, lack of interest, and potential psychological and physical harm. Individuals with depression tend to engage more with social media compared to those without the condition. Detecting the underlying emotions expressed in social media posts could aid in identifying and monitoring individuals who require mental health support, ultimately enhancing their well-being.

Throughout this work, a predictive model capable of predicting whether a given Twitter phrase has a negative, neutral, or positive sentiment was developed. The best DL model was the Hybrid combination (CNN + LSTM) with Word2Vec achieving a low accuracy value (52.05%). Overall, the best model created was the ML classifier Extra Trees combined with TF-IDF achieving 84.83% of accuracy. Therefore, it concludes that Extra Trees Classifier with TF-IDF is the best combination to predict a possible depressive feeling associated with a sentence.

Nevertheless, this work has several limitations. The dataset size is limited which can make it very difficult to create a model with the ability to generalize to external examples. Furthermore, future research will be needed to see if the model can maintain this performance on external data. Also, our validation of sentences will always have a bias associated with what may be the interpretation of reviewers, and which may not correspond to the real feeling behind the sentence. The use of Large Language Models (LLMs) can automate and improve the data labelling process leading to potentially better classifier performance.

On Future work, we will try to figure out how to generalize the model created to external data. The augmentation technique used can lead to a potential bias where despite considerably increasing the volume of data, we still have a low variance, which can affect the model's performance. To solve this, multiple different augmentation techniques and classifiers can also be analyzed to improve the results.

References

1. Depression.   https://www.who.int/news-room/fact-sheets/detail/depression. Accessed 26 Oct 2022
2. TF-IDF for Document Ranking from scratch in python on real world dataset. https://towardsdatascience.com/tf-idf-for-document-ranking-from-scratch-in-python-on-real-world-dataset-796d339a4089. Accessed 09 Jan 2023
3. Al-Garaady, J., Mahyoob, M.: Public sentiment analysis in social media on the SARS-CoV-2 vaccination using VADER lexicon polarity (2022)
4. Almeida, F., Xexéo, G.: Word embeddings: a survey (2019). https://doi.org/10.48550/arXiv.1901.09069
5. Alsagri, H.S., Ykhlef, M.: Machine learning-based approach for depression detection in twitter using content and activity features. IEICE Trans. Inf. Syst. **E103.D**(8), 1825–1832 (2020). https://doi.org/10.1587/transinf.2020EDP7023
6. Babu, N.V., Kanaga, E.G.M.: Sentiment analysis in social media data for depression detection using artificial intelligence: a review. SN Comput. Sci. **3**(1), 74 (2021). https://doi.org/10.1007/s42979-021-00958-1
7. Bhargava, C., Al, E.: Depression detection using sentiment analysis of tweets. Turk. J. Comput. Math. Educ. (TURCOMAT) **12**(11), 5411–5418 (2021)
8. Biswas, S., Ghosh, S.: Drug usage analysis by VADER sentiment analysis on leading countries. Mapana J. Sci. **21**(3) (2022)
9. Dessai, S., Usgaonkar, S.S.: Depression detection on social media using text mining. In: 2022 3rd International Conference for Emerging Technology (INCET), pp. 1–4 (2022). https://doi.org/10.1109/INCET54531.2022.9824931
10. Elbagir, S., Yang, J.: Sentiment analysis on twitter with Python's natural language toolkit and VADER sentiment analyzer. In: IAENG Transactions on Engineering Sciences, pp. 63–80. WORLD SCIENTIFIC (2019). https://doi.org/10.1142/9789811215094_0005
11. Gupta, B., Negi, M., Vishwakarma, K., Rawat, G., Badhani, P.: Study of twitter sentiment analysis using machine learning algorithms on Python. Int. J. Comput. Appl. **165**, 29–34 (2017). https://doi.org/10.5120/ijca2017914022
12. Hossain, M.S., Rahman, M.F.: Customer sentiment analysis and prediction of insurance products' reviews using machine learning approaches. FIIB Bus. Rev. (2022). https://doi.org/10.1177/23197145221115793
13. Hutto, C., Gilbert, E.: VADER: a parsimonious rule-based model for sentiment analysis of social media text. In: Proceedings of the International AAAI Conference on Web and Social Media, vol. 8, no. 1, pp. 216–225 (2014). https://doi.org/10.1609/icwsm.v8i1.14550

14. Kabir, M., et al.: DEPTWEET: a typology for social media texts to detect depression severities. Comput. Hum. Behav. **139**, 107503 (2023). https://doi.org/10.1016/j.chb.2022.107503

15. Kolchyna, O., Souza, T.T.P., Treleaven, P., Aste, T.: Twitter sentiment analysis: lexicon method, machine learning method and their combination (2015). https://doi.org/10.48550/arXiv.1507.00955

16. Arias-de La Torre, J., et al.: Prevalence and variability of current depressive disorder in 27 European countries: a population-based study. Lancet Publ. Health **6**(10), e729–e738 (2021). https://doi.org/10.1016/S2468-2667(21)00047-5

17. Macrohon, J.J.E., Villavicencio, C.N., Inbaraj, X.A., Jeng, J.H.: A semi-supervised approach to sentiment analysis of tweets during the 2022 Philippine presidential election. Information **13**(10), 484 (2022). https://doi.org/10.3390/info13100484

18. Mendon, S., Dutta, P., Behl, A., Lessmann, S.: A hybrid approach of machine learning and lexicons to sentiment analysis: enhanced insights from twitter data of natural disasters. Inf. Syst. Front. **23**(5), 1145–1168 (2021). https://doi.org/10.1007/s10796-021-10107-x

19. Najafabadi, M.M., Villanustre, F., Khoshgoftaar, T.M., Seliya, N., Wald, R., Muharemagic, E.: Deep learning applications and challenges in big data analytics. J. Big Data **2**(1), 1 (2015). https://doi.org/10.1186/s40537-014-0007-7

20. Newman, H., Joyner, D.: Sentiment analysis of student evaluations of teaching. In: Penstein Rosé, C., Martínez-Maldonado, R., Hoppe, H.U., Luckin, R., Mavrikis, M., Porayska-Pomsta, K., McLaren, B., du Boulay, B. (eds.) AIED 2018. LNCS (LNAI), vol. 10948, pp. 246–250. Springer, Cham (2018). https://doi.org/10.1007/978-3-319-93846-2_45

21. Prakash, T.N., Aloysius, A.: Data preprocessing in sentiment analysis using twitter data. Int. Educ. Appl. Res. J. **3**, 89–92 (2019)

22. Ramadhani, A.M., Goo, H.S.: Twitter sentiment analysis using deep learning methods. In: 2017 7th International Annual Engineering Seminar (InAES), pp. 1–4 (2017). https://doi.org/10.1109/INAES.2017.8068556

23. Ricard, B.J., Marsch, L.A., Crosier, B., Hassanpour, S.: Exploring the utility of community-generated social media content for detecting depression: an analytical study on Instagram. J. Med. Internet Res. **20**(12), e11817 (2018). https://doi.org/10.2196/11817

24. Shailaja, K., Seetharamulu, B., Jabbar, M.A.: Machine learning in healthcare: a review. In: 2018 Second International Conference on Electronics, Communication and Aerospace Technology (ICECA), pp. 910–914. IEEE (2018). https://doi.org/10.1109/ICECA.2018.8474918

25. Sidey-Gibbons, J.A.M., Sidey-Gibbons, C.J.: Machine learning in medicine: a practical introduction. BMC Med. Res. Methodol. **19**(1), 64 (2019). https://doi.org/10.1186/s12874-019-0681-4

26. Tiller, J.W.G.: Depression and anxiety. Med. J. Aust. **199**(S6), S28–S31 (2013). https://doi.org/10.5694/mja12.10628

27. tweets, Hemanthkumar, Latha: Depression detection with sentiment analysis of tweets. Turk. J. Comput. Math. Educ. (2019)

28. Wani, M.A., ELAffendi, M.A., Shakil, K.A., Imran, A.S., El-Latif, A.A.A.: Depression screening in humans with AI and deep learning techniques. IEEE Trans. Comput. Soc. Syst. (2022). https://doi.org/10.1109/TCSS.2022.3200213

29. Woods, C., Adedeji, M.: Classification of depression through social media posts using machine learning techniques. Univ. Ibadan J. Sci. Logics ICT Res. **7**(1), 19–28 (2021)

30. Yadav, N., Kudale, O., Rao, A., Gupta, S., Shitole, A.: Twitter sentiment analysis using supervised machine learning. In: Hemanth, J., Bestak, R., Chen, J.I.Z. (eds.) Intelligent Data Communication Technologies and Internet of Things. Lecture Notes on Data Engineering and Communications Technologies, pp. 631–642. Springer, Cham (2021). https://doi.org/10.1007/978-981-15-9509-7_51
31. Yoon, S., Kleinman, M., Mertz, J., Brannick, M.: Is social network site usage related to depression? A meta-analysis of Facebook-depression relations. J. Affect. Disord. **248**, 65–72 (2019). https://doi.org/10.1016/j.jad.2019.01.026
32. Zhou, B., Yang, G., Shi, Z., Ma, S.: Natural language processing for smart healthcare. IEEE Rev. Biomed. Eng., 1–17 (2022). https://doi.org/10.1109/RBME.2022.3210270

Impact of Synthetic Images on Morphing Attack Detection Using a Siamese Network

Juan Tapia$^{(\boxtimes)}$ and Christoph Busch

da/sec -Biometrics and Internet Security Research Group, Hochschule, Darmstadt,
Germany
{juan.tapia-farias,christoph.busch}@h-da.de

Abstract. This paper evaluated the impact of synthetic images on Morphing Attack Detection (MAD) using a Siamese network with a semi-hard-loss function. Intra and cross-dataset evaluations were performed to measure synthetic image generalisation capabilities using a cross-dataset for evaluation. Three different pre-trained networks were used as feature extractors from traditional MobileNetV2, MobileNetV3 and EfficientNetB0. Our results show that MAD trained on EfficientNetB0 from FERET, FRGCv2, and FRLL can reach a lower error rate in comparison with SOTA. Conversely, worse performances were reached when the system was trained only with synthetic images. A mixed approach (synthetic + digital) database may help to improve MAD and reduce the error rate. This fact shows that we still need to keep going with our effort to include synthetic images in the training process.

Keywords: Face Morphing · PAD · Biometrics

1 Introduction

Face Morphing can be understood as an algorithm to combine two o more look-alike facial images from one subject and an accomplice, who could apply for a valid passport exploiting the accomplice's identity. Morphing takes place in the enrolment process stage. The threat of morphing attacks is known for border crossing or identification control scenarios. Efforts to spot such attacks can be broadly divided into two types: (1) Single Image Morphing Attack Detection (S-MAD) techniques and Differential Morphing Attack Detection (D-MAD) methods.

S-MAD can detect both landmark-based morphing and synthetic GAN-based morphing methods [10,23] while integrating a variety of different feature extraction methods observing texture, shape, quality, residual noise, and others [21].

In an operational scenario, a manual passport or ID card inspection complements this automatic classification at border control gates to detect morphing attacks when a suspicious image is presented or the border police have received previous advice. Today, MAD is still an open challenge to develop generalisation capabilities. The following open issues can be determined according to the state-of-the-art: Cross-Dataset (CD), Cross-Morphing (CM), Leave-One-Out (LOO) evaluations, and a reduced number of bona fide images.

© Springer Nature Switzerland AG 2024
V. Vasconcelos et al. (Eds.): CIARP 2023, LNCS 14469, pp. 343–357, 2024.
https://doi.org/10.1007/978-3-031-49018-7_25

In order to train a robust MAD method (single or differential) is complicated due to the lack of a large-scale database available for this purpose. Only some existing databases allow us to create Morphing Attacks. It is essential to highlight that the image only can be used for the purpose informed in the consent form. Currently, most of the papers in the literature use FERET and FRGCv2 datasets to generate images with several morphing tools. Other open-access databases, such as FRLL and the London dataset, can be used directly with many Morphing Attacks. However, these databases have available only 204 subjects as bona fide images. 102 images are neutral expressions and the same subject (102 subjects) with a smiling expression. Overall, we have an unbalanced dataset.

On the other hand, capturing images from real Automatic Border Control (ABC) gates is a difficult task because of the equipment and the authorisation to perform this task in real scenarios. It is not trivial to capture sessions for many researchers.

This paper aims to complement previous approaches and analyse the impact of using purely synthetic images in a cross-dataset scenario. This means training with bona fide and morph face images (both from synthetic images) and evaluating in a benchmark test-set and assessment also in the state-of-the-art databases based on morph created from digital images and vice-versa. Those databases contain morphed images created using landmark and GANs-based morphing images. This cross-evaluation also considers different morphing tools and may help focus on the next steps of MAD. In summary, the main contributions of this paper are:

- This MA analysis complements the current state-of-the-art. An intra-dataset analysis was performed, training in synthetic face images (bona fide/morph) and testing purely on synthetic images.
- A cross-dataset evaluation was performed using purely synthetic images for training and evaluation on State-Of-The-Art (SOTA) digital databases such as FERET, FRGC, FRLL and others.
- A cross-dataset evaluation was performed using a mix-dataset (digital + synthetic) for training and evaluation on State-Of-The-Art (SOTA) digital databases such as FERET, FRGC, FRLL and others.
- Several Morphing tools were used in order to explore different qualities of morph images using a Siamese network based on a semi-hard triplet loss to improve the MAD accuracy.
- Our results outperform the SOTA on synthetic images but show that using only synthetic images as bona fice and morph is still not enough to detect real MAD based on digital images. Conversely, a system trained in SOTA or mixed can detect synthetic images with a low error rate.

The rest of the manuscript is organised as follows: Sect. 2 summarises the related works on MAD. The database description is explained in Sect. 3. The metrics are explained in Sect. 3.3. The experiment and results framework is then presented in Sect. 5. We conclude the article in Sect. 6.

2   Related Work

Due to legal and privacy issues, the use of face image data collected from the web is problematic for research purposes. Privacy regulations such as the GDPR assure individuals the right to withdraw their consent to use or store their private data, practically making the use of large face datasets difficult. This restriction and the difficulty of capturing images in a real-time process from ABC gates encourage the research community to develop new morphing images (bona fide and morphed) from Generative Adversarial Networks (GAN) to create synthetic face databases [9].

Damer et al. [9] raised the question: "can morphing attack detection (MAD) solutions be successfully developed based on synthetic data?". Towards that, it introduced the first synthetic-based MAD development dataset, the Synthetic Morphing Attack Detection Development dataset (SMDD).

Huber et al. [12] conducted a Competition on Face Morphing Attack Detection Based on Privacy-aware Synthetic Training Data. The competition was held at the International Joint Conference on Biometrics 2022. A new benchmark dataset was released called SYN-MAD-2022. The benchmark is based on the Face Research Lab London dataset (FRLL). In this competition, the best results according to ranking were reached by the MorphHRNet team obtained very good results. However, this team used a mixture of synthetic images with traditional FRLL, FERET and FRGC databases for evaluation. Then, the real impact of using only Synthetics images was not explored and tested for all the teams.

Related to the Siamese network on MAD, previous work has been proposed based on deep learning. Borgui et al. [6] proposed a differential morph attack detection based on a double Siamese architecture. The proposed framework consisted of two different modules, referred to as "Identity" and "Artefact" blocks, respectively, and each block was based on a Siamese network followed by a Multi-Layer Perceptron (MLP) that acts as fusion layers. Finally, a Fully Connected layer (FC) merges the features originating from the two modules and outputs the final score. Experimental results were obtained in three datasets: PMDB, MorphDB, and AMSL. This approach used a Contrastive loss.

Soleymani et al. [20] developed a novel differential morphing attack detection algorithm using a deep siamese network. The Siamese network takes image pairs as inputs and yields a score on the likelihood that the face images are from the same subject. They employ a pre-trained Inception ResNetv1 as the base network initialised with weights pre-trained on VGGFace2. The experiments are conducted on two separate morphed image datasets: VISAPP17 and MorGAN. This approach used a contrastive loss.

Chaudhary et al. [8] proposed a differential morph attack detection algorithm using an undecimated 2D Discrete Wavelet Transform (DWT). By decomposing an image to wavelet sub-bands, we can more clearly identify the morph artefacts hidden in the image domain in the spatial frequency domain.

3   Database

In order to measure the real impact of synthetic images, several morphing databases have been used. The description of each one is detailed as follows:

3.1   SYN-MAD-2022

This dataset was developed for the Competition on Face Morphing Attack Detection Based on Privacy-aware Synthetic Training Data (SYN-MAD) [12]. It is divided into a training set of 25k bona fide and 15k morphed images based on synthetic faces, which avoids using privacy-sensitive real-face images. Furthermore, the images bounding box and five facial landmark points are provided. The test set, named MAD evaluation benchmark database (MAD22), was created by the organisers as part of the competition, and it is publicly available[1].

The benchmark test set (SDD) contains 4,483 divided into 984 OpenCV, 1,000 FaceMorpher, 500 Webmorph, 1,000 MIPGAN-I, 999 MIPGAN-II morphed face images and 204 bona fide images from the FRLL dataset. It is essential to highlight that even using similar morphing tools in some cases, such as Face-Morpher, OpenCV and WebMorph, the resulting images are not the same as the original FRLL databases because the subjects were combined in a different order.

3.2   SOTA Databases

In this paper, four different databases of frontal faces images were also used: the Facial Recognition Technology (FERET) [15], the Face Recognition Grand Challenge (FRGCv2) [14], the Face Research London Lab (FRLL) [18] and AMSL database [1]. The morphed images in these datasets have been created using a morphing factor of 0.5, meaning both subject images contribute equally to the morphed image. The FRLL and AMSL morphed images have been generated with the following morphing tools: FaceMorpher, FaceFusion and WebMorpher based on landmarks and StyleGAN from FRLL without landmarks. The description of each database is explained as follows:

The FERET dataset is a subset of the Colour FERET Database, generated in the context of the Facial Recognition Technology program technically handled by the National Institute of Standards and Technology (NIST). It contains 569 bona fide face images.

The FRGCv2 dataset used in this work is a constrained subset of the second version of the Face Recognition Grand Challenge dataset. It contains 979 bona fide face images.

The FRLL dataset is a subset of the publicly available Face Research London Lab dataset. It contains 102 bona fide neutral and 102 smiling images. Three morphing algorithms were applied to obtain 1,222 morphs from the FaceMorpher

---

[1] https://github.com/marcohuber/SYN-MAD-2022.

algorithm, 1,222 morphs from the StyleGAN algorithm, and 1,222 morphs from the WebMorph algorithm [4].

The AMSL Face Morph Image Data Set is a collection of bona fide and morphed face images that can be used to evaluate the detection performance of MAD algorithms. The images are organised as follows: genuine-neutral with 102 genuine neutral face images, genuine-smiling with 102 genuine smiling face images and 2,175 morphing face images. Figure 1 shows a side-by-side image example of all the databases used in this work[2]. Table 1 shows a summary of all the databases used in this work.

Table 1. Summary databases

Database	Bona fide	Morph	Type	Tool	Notes
AMSL	204	2,175	Landmark	WebMorph	The same bona fide images as FRLL.
FRLL	204	1,222	Landmark	OpenCV	
		1,222	Landmark	FaceMorpher	
		1,222	Synthetics	StyleGAN2	
		1,222	Landmark	WebMorpher	The same bona fide images as AMSL.
FERET	529	529*3	Landmark	FaceFusion	(*) Print/Scan 300dpi, Print/Scan 600dpi, Digital-Resize (No P/S)
		529*3	Landmark	FaceMorpher	
		529*3	Landmark	OpenCV	
		529*3	Landmark	UBO-Morpher	
FRGC	979	979*3	Landmark	FaceFusion	(*)Print/Scan 300dpi, Print/Scan 600dpi, Digital-Resize (No P/S)
		979*3	Landmark	FaceMorpher	
		979*3	Landmark	OpenCV	
		979*3	Landmark	UBO-Morpher	
25.000	15.000 + 4,948*	25k	Synthetics	N/A	(*)Benchmark Test set contains: OpenCV, FaceMorpher, WebMorpher, MIPGAN-I, MIPGAN II.

According to the SOTA, different algorithms were used to create morph images. The following six morphing tools have been used:

- **FaceFusion** [2]: this proprietary mobile application developed by MOMENT generates realistic faces since morphing artefacts are almost invisible.
- **FaceMorpher** [3]: this open-source Python implementation relies on STASM, a facial feature finding the package, for landmark detection, but generated morphs show many artefacts which make them more recognisable.
- **OpenCV-Morpher** [7]: this open-source morphing algorithm is similar to the FaceMorpher method but uses Dlib to detect face landmarks. Again, some artefacts remain in generated morphs.
- **UBO-Morpher** [11]: The University of Bologna developed this algorithm. The resulting images are of high quality without artefacts in the background.

---

[2] All the image sources have been properly cited throughout the paper.

- **WebMorpher** [4]: this open-source morphing algorithm is a web-based version of Psychomorph with several additional functions. While WebMorph is optimised for averaging and transforming faces, you can delineate and average any image.
- **StyleGAN2** [13]: this open-source morphing algorithm by NVIDIA, No landmarks are used to create morph images.

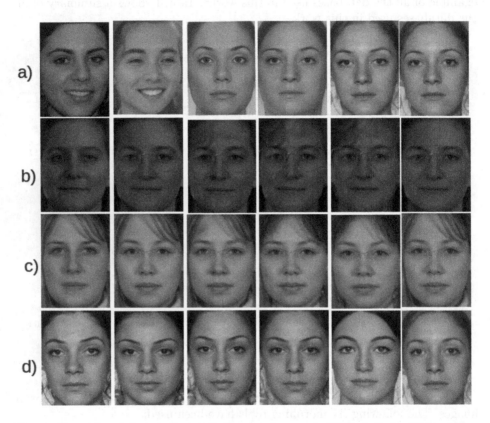

Fig. 1. Example of images from all the databases. a) SDD: Left to Right: Bona fide (Subject-1) synthetic, Morph Synthetics (Subject-2), MIPGAN-I, MIPGAN-II, OpenCV-Morpher, WebMorpher. B) FERET: Left to right: Bona fide (Subject-1), Bona fide (Subject-2), FaceMorpher, FaceFusion, OpenCV-Morpher, UBO-Morpher. C) FRGCv2: Left to right: Bona fide (Subject-1), Bona fide (Subject-2), FaceMorpher, FaceFusion, OpenCV-Morpher, UBO-Morpher. D) FRLL: Left to Right: Bona fide (Subject-1), Bona fide (Subject-2), AMSL, FaceMorpher, OpenCV-Morpher, and StyleGAN.

3.3 Metrics

The detection performance of the investigated S-MAD algorithms was measured according to ISO/IEC 30107-3[3] using the Equal Error Rate (EER), Bona fide Presentation Classification Error Rate (BPCER), and Attack Presentation Classification Error Rate (APCER) metric defined as (1) and (2).

$$BPCER = \frac{\sum_{i=1}^{N_{BF}} RES_i}{N_{BF}} \tag{1}$$

$$APCER = \frac{1}{N_{PAIS}} \sum_{i=1}^{N_{PAIS}} (1 - RES_i) \tag{2}$$

where N_{BF} is the number of bona fide presentations, N_{PAIS} is the number of morphing attacks for a given attack instrument species and RES_i is 1 if the system's response to the $i-th$ attack is classified as an attack and 0 if classified as bona fide. In this work, S-MAD performance is reported using the Equal Error Rate (EER), the point which corresponds to relevant security settings where the APCER is equal to BPCER. Also, two operational points are reported BPCER10 and BPCER20. The BPCER20 is the BPCER value obtained when the APCER is fixed at 5%, and BPCER10 (APCER at 10%).

4 Method

A Siamese network consists of two identical networks which can process different inputs. They are joined at their output layers based on the pre-trained networks by a unique function that calculates a metric between the embedding estimated by each network. Most of these networks used have been trained on ImageNet 1k weights. Because of that, we explored using 21k weights and also reduced the 21k to 1k weights. This kind of network is ideal for morphing attack detection because they are primarily designed to find similarities between two inputs. As a part of the process, a pre-computed template of four random bona fide images is processed in order to compare the embedding distances of the new input face image, which could be potentially morphed.

A contrastive loss is typically used to train a siamese network to distinguish between mated (bona fide) and non-mated (morphed) pairs. Traditionally, the contrastive loss function optimises two identical Convolutional Neural Networks outputs, each operating on a different input image and using a Euclidean distance measure or an SVM classifier to make the final decision. At the same time, contrastive representations have achieved state-of-the-art performance on visual recognition tasks and have been theoretically proven effective for binary classification. However, according to [22], the triple-loss function could separate the morphed images more effectively based on the semi-hard triplet loss function to separate several morphing tools used for the same pairs of subjects. Then, the

---

[3] https://www.iso.org/standard/79520.html.

contrastive loss is not able to separate this kind of image because we have images from the same subject with several morphing tools. Conversely, triplet-loss can explore and deal very well with easy, semi-hard and hard examples.

In order to extract the embedding from the bona fide and morphed images, ImageNet (1k and 21k) pre-trained general-purpose backbones such as MobileNetV2 [17], MobileNetV3 [16] and EfficientNetB0 [5] were used. Then, the network is retrained with a morphing database. Afterwards, the model is optimised by enforcing the triple loss function to measure the triplets relationships between both classes in a binary problem or N classes in a multi-class problem. Figure 2 show a Siameses Network.

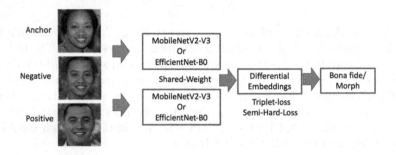

Fig. 2. Siamese network.

Triplet Loss. A triplet loss function was used to train a neural network to closely embed features of the same class while maximising the distance between embeddings of different classes. An anchor (bona fide) and one negative (potentially morphed), and one positive sample (bona fide or same class of anchor) are chosen to do this [19].

To formalise this requirement, a loss function is defined over triplets of embeddings:

- An anchor image (a) - For example, bona fide.
- A positive image of (p) the same class as the anchor.
- A negative image (n) of a different class - In our example, a morphed image.

For some distance (d) on the embedding space, the loss of a triplet (a, p, n) is defined as:

$$L_t = \max(d(a, p) - d(a, n) + margin, 0) \qquad (3)$$

We minimise this loss, which pushes $d(a, p)$ to 0 and $d(a, n)$ to be greater than $d(a, p) + margin$. As soon as n becomes an "easy negative", the loss becomes zero. We used a semi-hard triplet, which means that the negative is not closer to the anchor than the positive, but which still has a positive loss:

$$d(a, p) < d(a, n) < d(a, p) + margin \qquad (4)$$

5 Experiments and Results

Four experiments are conducted based on the Siamese network. A semi-hard triple loss function applied to morphing attack detection using MobileNetV2, MobileNetV3 and EfficientNetB0 was explored.

The morphing images were created based on OpenCV-Morphing tools for the SYN-MAD database. All the processes for pair selection and others were described in [12]. For the images from AMSL, FRGC, FERET, and FRLL databases, the faces were detected and aligned based on MTCNN [24]. For the SYN-MAD database, the coordinates of the faces are provided to crop the images.

For all the experiments, the optimisation function used is Adam; even were explored SGD and RMSprop, Adam reached the best results. A grid search was used to identify the best learning rate (lr) for each experiment. Further on that, 250 epochs, batch size of 128 on a 21GB-NVidia-GPU were used. An intra-dataset was performed in Experiment #1 using the different configurations. For Experiments #2 #3 and #4, a cross-dataset protocol was explored as explained as follows:

5.1 Exp1 - Train SYN-MAD/TEST SDD-Bechmark

For this experiment, three tests were conducted based on MobileNetV2 (ImagiNet-1k), MobileNetV3 (ImagiNet-21k) and EfficientNetB0 (ImagiNet-21k-1k). The SYN-MAD release's official training and test set was used for the process.

Test a) Both MobileNetV2 networks were trained from scratch using the ImagiNet weight from 1,000 classes and adapted to classify bona fide and morphed images.

Test b) Both MobileNetV3-small networks were trained from scratch using the ImagiNet weight from 21,000 classes and adapted to classify bona fide and morphed images.

Test c), Both EfficientNetB0 networks were trained from scratch using the ImagiNet weight from 21,000 classes but reduced to 1,000 as adapted to a binary problem to classify bona fide and morphed images.

For the tests a), b) and c), 23,000 bona fide and 13,000 morphed synthetic images were used to train. For the validation set, 4,000 (2,000 bona fide and 2,000 synthetics) images were used.

The test c) with EfficientNetB0 reached the best performance. Table 2 shows a summary with the best results for experiment 1. For the SDD competition, we include the best results reported for each dataset.

These results can be used as a reference for evaluation. Still, it is essential to highlight that some of the equipment use mixtures of synthetic and digital images in the evaluation process. Our method, based only on synthetic and the Siamese network, reached competitive results compared with the best result on similar E-CBAM-VCMI and MorphHRNet team conditions. Table 2 shows the summary results.

Table 2. Summary evaluation report based on the SYN-MAD and our proposal.

Database	IJCB 2022-SMDD			Our Proposal			Note
	EER (%)	BPCER10 (%)	BPCER20 (%)	EER (%)	BPCER10 (%)	BPCER20(%)	
Experiment 1							
OpenCV	27.54	46.68	33.03	22.06	33.59	49.80	(90037) Compared with E-CBAM-VCMI Train:SDD-Train-Set Test: Bench. Test-Set
FaceMorpher	41.20	92.80	62.80	2.00	0.00	0.00	
WebMorpher	30.60	86.80	46.80	26.39	46.14	55.80	
MIPGAN-I	32.50	84.90	60.20	11.76	13.23	22,07	
MIPGAN-II	25.93	64.66	37.34	8.82	8.83	10.78	
Experiment 2							
AMSL	N/A			3.70	2.02	3.29	(212642) Train:SDD-Train-Set Test: SOTA
OpenCV (FERET/FRGC)				21.55/19.25	41.39/21.45	63.53/22.15	
FaceMorpher (FERET/FRGC)				18.52/16.25	34.97/22.70	51.22/25.35	
FaceFusion (FERET/FRGC)				35.70/16.85	76.77/55.20	91.30/57.30	
UBO-Morpher (FERET/FRGC)				32.70/19.65	70.69/55.50	81.47/61.25	
Experiment 3							
OpenCV	N/A			16.66	28.92	54.51	(134122) Train: SOTA No synthetics images Test: Bench. Test-set
FaceMorpher				3.43	1.83	1.94	
WebMorpher				33.4	74.50	80.71	
MIPGAN-I				18.0	31.37	61.29	
MIPGAN-II				8.33	6.86	16.66	
Experiment 4							
OpenCV	5.69	1.96	1.47	6.37	1.90	0.84	Trained: Mix (synthetics+SOTA) Test: Bench. Test-set Compared with MorphHRNet
FaceMorpher	5.90	1.96	1.47	3.50	0.48	0.98	
WebMorpher	9.80	10.78	3.93	9.31	8.33	16.17	
MIPGAN-I	15.30	24.02	11.27	4.90	1.96	4.41	
MIPGAN-II	10.41	11.27	2.94	3.43	0.98	1.96	

Figure 3 show the T-SNE distribution of the SDD-test Benchmark dataset in a high dimensional space. The blue (star) and black (diamond) colours represent the validation data that belong to the SYN-MAD test dataset. All the images in the validation set are synthetics. The picture also shows that the SDD-test benchmark, even though most of them belong to the class morphs, was depicted in a different high-dimensional space.

Figure 4 show an individual analysis of the SDD benchmark datasets break for each morphing tool separately. Morphed images created by WebMorpher tools have been identified as the most difficult to detect in comparison with FaceMorpher, MIPGAN-I, MIPGAN-II, and OpenCV-Morpher.

5.2   Exp2 - Train SYN-MAD/TEST SOTA

The best results from experiment 1 were used for training the EfficientNetB0 again, but now only using the official training set of the SYN-MAD database release for the testing process was evaluated with SOTA: AMSL, FERET, FRGC and FRLL databases. 23,000 bona fide and 13,000 morphed synthetics images were used to train. For the validation set, 4,000 (2,000 bona fide and 2,000 synthetics) images were used. The best lr was $1e-4$. Table 2 shows a summary

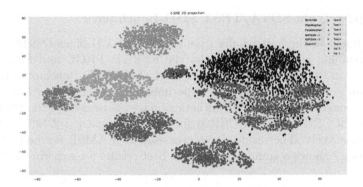

Fig. 3. T-SNE distibution of SDD-test benchmark. Each colour represents a morphing tool. Bona fide (test-0), WebMorpher (test-1), FaceMorpher (test-2), MIPGAN-I (test-3), MIPGAN-II (test-4), OpenCV-Morpher (test-5). Black represents the bona fide validation set (val-0), and Blue (start) represents the morph validation set (val-1) (Color figure online)

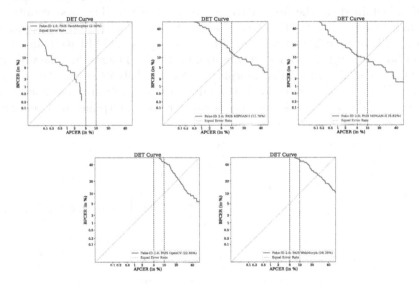

Fig. 4. Exp1: DET curves belong to each class of SDD-Test benchmark. Left to Right: FaceMorpher, MIPGAN-I, MIPGAN-II, OpenCV-Morpher, WebMorpher. Dotline indicates BPCER10 and BPCER20, respectively. The EER is reported in parentheses in percentages.

with the best results for experiment 2. This experiment shows that synthetic images can not reach generalisation capabilities, and the performance is worse when compared with digital images based on landmark morphing methods. The EER obtained was 29.12%. Table 2 shows the summary results.

5.3 Exp3 - Train SOTA/TEST SDD-Benchmark

For experiment 3, the best results from experiment 1 (EfficientNetB0) also was used for training again, but now used the FERET, FRGC and FRLL images for training and tested in the official test set of the SDD-Benchmark database. The lr was set up to $5e - 5$. This experiment can infer with a low error the synthetics images even when no synthetics images were used in the training set. The EER, BPCER10 and BPCER20 for SDD-test reached 3.49%, 1.96% and 2.94%, respectively. The results for a cross-set with AMSL reached 2.02% and 3.29%. Table 2 shows a summary with the best results for experiment 3.

5.4 Exp4 - Train Mix/TEST SDD-Benchmark

For Experiment 4, the synthetic images from the SYN-MAD, FERET, and FRGCv2 were mixed by increasing the number of images available for training. The SDD-test benchmark was used as a test set. This experiment allows us to compare our proposed method with the best results of the IJCB 2022 competition obtained by the MorphHRNet team (Table 2). The best results were obtained when at least 20% of the images belonged to digital faces (no synthetics). There were 23.000 synthetic face images plus 5,000 FERET and FRGC digital bona fide images and 13,000 morph synthetics images. For the validation set, 2,000 bona fide and 2,000 morphs were used. Table 2 shows the summary results. Figure 5 show an individual analysis of the train-mix database evaluated with SDD benchmark datasets break for each morphing tool separately.

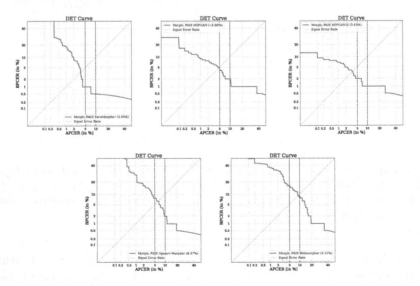

Fig. 5. Exp4: DET curves belong to each class of SDD-Test benchmark, trained on mix-database. Left to Right: FaceMorpher, MIPGAN-I, MIPGAN-II, OpenCV-Morpher, WebMorpher. Dot-line indicates BPCER10 and BPCER20, respectively. The EER is reported in parentheses in percentages.

Morphed images created by WebMorpher tools have been identified as the most difficult to detect in comparison with FaceMorpher, MIPGAN-I, MIPGAN-II, and OpenCV-Morpher. Figure 5 show the improvements reached when the 20% SOTA database is included in the pure synthetics database. The BPCER10 and BPCER20 for SDD-test reached 0.98% and 5.39%, respectively.

Figure 6 shows DET curves with the results when the test datasets are fully evaluated. This means all the morphed images belong to the benchmark dataset together as one morph class. We can observe that the performance of Exp-1, the SDD-test benchmark reached a higher EER of 14.22%. This result is valuable because it was obtained purely with synthetic images used as morph and bona fide. Also, the results show that controlling the quality of the synthetic images used for this process is necessary before creating the morph. Some images from SYN-MAD present relevant artefacts in some face areas. Then, these artefacts are also translated into morphed images. Exp-2 reached an EER of 3.49% which represents a third party when using only synthetic images. For Exp-4, which is based on a mixed dataset, the EER obtained was 5.13%.

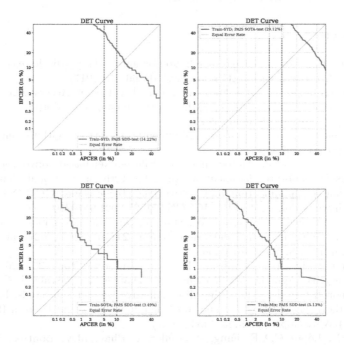

Fig. 6. DET curves. Left to Right: Exp1: Trained with SYN-MAD/test SDD-Benchmark. Exp2: Trained with SYN-MAD/test-SOTA. Exp3: Trained with SOTA/test SDD-Benchmark. Exp4: Trained with Mix database/test SDD-Benchmark. Dot-line indicates BPCER10 and BPCER20, respectively. The EER is reported in parentheses in percentages.

6   Conclusions

This work assesses the impact of using purely synthetic images to train MAD. Our proposal for a detection approach is based on a Siamese network with a semi-hard-loss function and outperforms the best results of the IJCB2021 competition. The results show that the performance based on only synthetic images, even reaching competitive results, cannot reach the best results and discards the digital images totally. The main lack regarding the SOTA database available was only an increased number of digital morphing tools despite bona fide and the fact that print/scan scenarios are one of the most realistic in border gate operation. However, developing this kind of database is time-consuming.

Acknowledgment. This work is supported by the European Union's Horizon 2020 research and innovation program under grant agreement No 883356 and the German Federal Ministry of Education and Research and the Hessen State Ministry for Higher Education, Research and the Arts within their joint support of the National Research Center for Applied Cybersecurity ATHENE.

References

1. DeBruine, I., Jones, B.: Face research lab London set. http://figshare.com/articles/dataset/Face_Research_Lab_London_Set/5047666/5 (2017)
2. FaceFusion. https://www.wearemoment.com/FaceFusion. Accessed 26 Jan 2023
3. FaceMorpher.   https://github.com/yaopang/FaceMorpher/tree/master/facemorpher. Accessed 26 Jan 2023
4. Webmorph morphing algorithm, implementation. https://github.com/debruine/webmorph. Edit 07 Dec 2021
5. Acharya, V., Ravi, V., Mohammad, N.: EfficientNet-based convolutional neural networks for malware classification. In: 2021 12th International Conference on Computing Communication and Networking Technologies (ICCCNT), pp. 1–6 (2021). https://doi.org/10.1109/ICCCNT51525.2021.9579750
6. Borghi, G., Pancisi, E., Ferrara, M., Maltoni, D.: A double siamese framework for differential morphing attack detection. Sensors **21**(10), 3466 (2021), https://www.mdpi.com/1424-8220/21/10/3466
7. Bradski, G.: The OpenCV Library. Dr. Dobb's J. Softw. Tools **25**, 120–123 (2000)
8. Chaudhary, B., Aghdaie, P., Soleymani, S., Dawson, J., Nasrabadi, N.M.: Differential morph face detection using discriminative wavelet sub-bands. In: Proceedings of the IEEE/CVF Conference on Computer Vision and Pattern Recognition (CVPR) Workshops, pp. 1425–1434 (2021)
9. Damer, N., López, C.A.F., Fang, M., Spiller, N., Pham, M.V., Boutros, F.: Privacy-friendly synthetic data for the development of face morphing attack detectors. In: 2022 IEEE/CVF Conference on Computer Vision and Pattern Recognition Workshops (CVPRW), pp. 1605–1616 (2022). https://doi.org/10.1109/CVPRW56347.2022.00167
10. Damer, N., Saladié, A.M., Braun, A., Kuijper, A.: Morgan: recognition vulnerability and attack detectability of face morphing attacks created by generative adversarial network. In: 2018 IEEE 9th International Conference on Biometrics Theory, Applications and Systems (BTAS), pp. 1–10 (2018). https://doi.org/10.1109/BTAS.2018.8698563

11. Ferrara, M., Franco, A., Maltoni, D.: Face demorphing. IEEE Trans. Inf. Forensics Secur. **13**(4), 1008–1017 (2018). https://doi.org/10.1109/TIFS.2017.2777340
12. Huber, M., et al.: SYN-MAD 2022: competition on face morphing attack detection based on privacy-aware synthetic training data. In: 2022 IEEE International Joint Conference on Biometrics (IJCB), pp. 1–10 (2022). https://doi.org/10.1109/IJCB54206.2022.10007950
13. Karras, T., Laine, S., Aittala, M., Hellsten, J., Lehtinen, J., Aila, T.: Analyzing and improving the image quality of stylegan (2020)
14. Phillips, P.J., Flynn, P.J., Scruggs, T., Bowyer, K.W., Chang, J., et al.: Overview of the Face Recognition Grand Challenge. In: Conference on Computer Vision and Pattern Recognition (CVPR), vol. 1, pp. 947–954 (2005)
15. Phillips, P.J., Wechsler, H., Huang, J., Rauss, P.J.: The FERET database and evaluation procedure for face-recognition algorithms. Image Vis. Comput. **16**(5), 295–306 (1998)
16. Qian, S., Ning, C., Hu, Y.: MobileNetV3 for image classification. In: 2021 IEEE 2nd International Conference on Big Data, Artificial Intelligence and Internet of Things Engineering (ICBAIE), pp. 490–497 (2021). https://doi.org/10.1109/ICBAIE52039.2021.9389905
17. Sandler, M., Howard, A., Zhu, M., Zhmoginov, A., Chen, L.C.: MobileNetV2: inverted residuals and linear bottlenecks. In: IEEE Conference on Computer Vision and Pattern Recognition, pp. 4510–4520 (2018)
18. Sarkar, E., Korshunov, P., Colbois, L., Marcel, S.: Are GAN-based morphs threatening face recognition? In: ICASSP 2022-2022 IEEE International Conference on Acoustics, Speech and Signal Processing (ICASSP), pp. 2959–2963 (2022)
19. Schroff, F., Kalenichenko, D., Philbin, J.: Facenet: a unified embedding for face recognition and clustering. In: CVPR, pp. 815–823. IEEE Computer Society (2015)
20. Soleymani, S., Chaudhary, B., Dabouei, A., Dawson, J., Nasrabadi, N.M.: Differential morphed face detection using deep Siamese networks. In: Del Bimbo, A., et al. (eds.) ICPR 2021. LNCS, vol. 12666, pp. 560–572. Springer, Cham (2021). https://doi.org/10.1007/978-3-030-68780-9_44
21. Tapia, J.E., Busch, C.: Single morphing attack detection using feature selection and visualization based on mutual information. IEEE Access **9**, 167628–167641 (2021)
22. Tapia, J., Schulz, D., Busch, C.: Single morphing attack detection using siamese network and few-shot learning. arXiv (2022). https://doi.org/10.48550/arXiv.2206.10969
23. Zhang, H., Venkatesh, S., Ramachandra, R., Raja, K., Damer, N., Busch, C.: MIPGAN-generating strong and high quality morphing attacks using identity prior driven GAN. IEEE Trans. Biometrics Behav. Identity Sci. **3**(3), 365–383 (2021). https://doi.org/10.1109/TBIOM.2021.3072349
24. Zhang, K., Zhang, Z., Li, Z., Qiao, Y.: Joint face detection and alignment using multitask cascaded convolutional networks. IEEE Signal Process. Lett. **23**(10), 1499–1503 (2016). https://doi.org/10.1109/LSP.2016.2603342

Face Image Quality Estimation on Presentation Attack Detection

Carlos Aravena[1], Diego Pasmiño[2], Juan Tapia[3(✉)], and Christoph Busch[2]

[1] IDVisionCenter, R&D Center, Santiago, Chile
[2] TOC Biometric, R&D Center, Santiago, Chile
[3] da/sec-Biometrics and Internet Security Research Group,
Hochschule Darmstadt, Germany
`juan.tapia-farias@h-da.de`

Abstract. Non-referential Face Image Quality Assessment (FIQA) methods have gained popularity as a pre-filtering step in Face Recognition (FR) systems. In most of them, the quality score is usually designed with face comparison in mind. However, a small amount of work has been done on measuring their impact and usefulness on Presentation Attack Detection (PAD). In this paper, we study the effect of quality assessment methods on filtering bona fide and attack samples, their impact on PAD systems, and how the performance of such systems is improved when training on a filtered (by quality) dataset. On a Vision Transformer PAD algorithm, a reduction of 20% of the training dataset by remoing lower-quality samples allowed us to improve the Bona fide Presentation Classification Error Rate (BPCER) by 3% in a cross-dataset test.

Keywords: Presentation Attack Detection · Face-PAD · FIQA

1 Introduction

Biometric and identity verification systems have numerous commercial and industrial applications in fields diverse such as access controls, video surveillance and user validation, among others. Among the techniques commonly used are fingerprint recognition, ID card validation, and iris and face biometric systems. [25], [20]. Due to the improvement in the quality and availability of capture devices (i.e. smartphones), and the increase in remote image processing, these applications are increasingly available to the general public in uncontrolled environments [26]. In this scenario, one of the most relevant challenges for the diffusion and viability of biometric systems is the problem of impersonation. In this case, a subject tries to present fraudulent evidence to the biometric recognition system to be authorised and to obtain resources within the system. This problem transversely affects companies, governments and individual users.

Facial PAD techniques on face biometric systems are used to overcome this challenge. The objective is to differentiate between a genuine biometric reading of a living subject's face (*bona fide*) and a fake one created by the attacker,

© Springer Nature Switzerland AG 2024
V. Vasconcelos et al. (Eds.): CIARP 2023, LNCS 14469, pp. 358–373, 2024.
https://doi.org/10.1007/978-3-031-49018-7_26

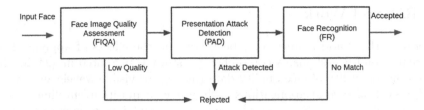

Fig. 1. Overview of a traditional remote face authentication system.

using, for example, a photo, video, mask or a different substitute for the face of an authorised subject. Presentation attacks (PA) tend to be the most common on authentication systems, especially in uncontrolled environments, because they do not necessarily require knowledge on the part of the attacker trying to be authorised. Active research is conducted on this topic due to the difficulty involved in designing an algorithm that generalises well to different sensors, environmental conditions and methods for identifying these attacks.

On many real-life remote authentication systems, the user submits an image of his/her face for registering or accessing. In case of accessing, the input face goes to a face recognition algorithm to find a match. This step is done after, or in parallel, with a PAD system to validate the face. Both actions usually follow a Face Image Quality Assessment (FIQA) that filters out faces unsuitable for recognition due to occlusion, inadequate illumination or heavily rotated faces. Many state-of-the-art face quality algorithms are explicitly designed with face recognition as a primary target, tuning their parameters to decrease the comparison error. Figure 1 shows an overview diagram of a typical remote authentication system.

In this paper, we study the influence of FIQA when applied in conjunction with state-of-the-art face PAD algorithms. In summary, the main contributions of this paper are:

- We show that a Vision Transformer PAD algorithm [14] performs better when presented with higher quality faces, even for attack samples, improving its discrimination performance at inference time.
- We show that applying FIQA pre-filtering on an image processing pipeline tends to benefit the selection of bona fide samples over low-quality attacks for most datasets tested.
- Training on a reduced dataset, by removing low-quality samples (of both bona fide and attacks) lowers the Equal Error Rate (EER) (up to 1%) and Bona fide Presentation Classification Error Rate (BPCER) (up to 4%) when using ViTranZFAS [14] as the PAD algorithm. A similar result is obtained on Single Side Domain Generalisation (SSDG) [17].

The article is organized as follows: Sect. 2 summarizes the related works on Presentation Attack Detection, Image Quality Assessment and database description. The methodology is explained in Sect. 3. The experiment descriptions and results are then presented in Sect. 4. Finally, the conclusion is reported in Sect. 5.

2   Relatwd Work

Several methods and studies have been done on the effects of applying FIQA algorithms on FR systems. A comprehensive study is presented in [24] in which the biometric utility of face quality data (mostly for visible wavelength images) is discussed on several applications such as face acquisition/enrollment, video frame selection, face detection filtering, conditional image enhancement and model selection, score fusion and PA avoidance. Several FIQA algorithms are summarised and evaluated, highlighting the current predominance of deep learning methods. Finally, future directions and challenges are presented on the topic, especially on the comparability of FIQA algorithms and standardisation.

2.1   Face PAD

The role of a PAD module within an FR system is to prevent unauthorised users from accessing illegally by posing as authorised users. This module checks if the user is real or fake. Generally, if the input image is found to be real, it will proceed to the facial recognition phase; otherwise, it will reject the attempt if it determines that the face is fake. Since the last few years, commercial FR systems are taking PAs into account as a major issue, seeking ISO 30107 normative compliance[1] and/or vendor certification such as iBeta[2] or NIST[3]. It is an important research topic that has gained relevance given the proliferation of FR systems and the technological ease with which physical and digital attacks can be carried out. In recent years, several articles have been published that comprehensively review state-of-the-art. At a general level, [20] is a complete book that explains the problem at the level of general biometrics. More specific reviews applied to face biometrics can be found at [19]. A comprehensive review of algorithms based on deep learning can be found at [31]. A new quality metric, based on vertical edge density, is presented in [28] that estimates pose variations on a database of 101 subjects, under different pose angles, illumination and distances captured on a smartphone device.

Single Side Domain Generalisation (SSDG). SSDG aims to improve the generalisation ability of a PAD system by learning a compact distribution of real faces' features and a dispersed distribution of the fake ones among domains. A combined feature generation is trained for both real and fake samples; then, the features are fed to a classifier that differentiates between bona fide and fake. Also, the features enter an asymmetric triplet loss computation to make the real samples compact and the fake ones separated by the origin domain. Lastly, the features of the real samples are used to train a discriminator classifier that aims to modify the feature extractor to learn more generic features across domains using a Gradient Reversal Layer (GRL).

---

[1] https://www.iso.org/standard/67381.html.
[2] https://www.ibeta.com/biometric-spoofing-pad-testing.
[3] https://pages.nist.gov/frvt/html/frvt_pad.html.

ViTranZFAS. This method uses a vision transformer architecture as its backbone, with the last layer replaced by a fully connected layer with one output node and fine-tunes using Binary Cross-Entropy loss (BCE). Our tested method replaces the transformer model with a hybrid model that uses a ResNet-26 Convolutional Neural Network (CNN) to compute the feature map that feeds the Transformer encoder.

MobileNetv3. MobileNetv3 [16] is a CNN tuned for smartphone CPUs. This CNN adds hard swish activation and squeeze-and-excitation modules, among other changes, to the previous model version, achieving similar PAD detection performance but a considerably faster performance for image classification. In this paper, we used the MobileNetv3 small architecture with pre-trained weights from Imagenet. We modified the net's last layer to be a two-class output instead of the original 1,000 classes.

2.2 Face Image Quality Assessment

Face Image Quality Assessment (FIQA) is the process of using face data as input to generate some kind of quality estimate as output. Models can be trained to automatically estimate the quality of a face using man-made scores of the input image (referential FIQA) or without any score as input (non-referential FIQA). In the latter case, the quality metric is usually designed with face comparison in mind; a high-quality face refers to the utility of that face for a FR task. In this paper, we evaluate three state-of-the-art non-referential FIQA methods and how they impact the process of PAD, even when the quality score is not directly related to this task.

MagFace. Face picture quality evaluation and FR are combined in Mag-Face [21]. The magnitude of the facial recognition feature vector directly relates to quality. Based on its magnitude, the authors' presented a loss function with adaptive margin and regularisation. The objective of this loss function is to move the challenging samples away from the class centre and the easily recognised examples toward it. Thus, during training, the face utility is inherently learnt by this loss function. In direct relation to the facial image utility, the magnitude of the feature vector is proportional to the cosine distance to its class centre. A large magnitude denotes a high face utility. Scores range from 0 (lower quality) to 40 (higher quality).

SER-FIQ. This method [27] relates the robustness of face representations with face quality. Face representations are chosen using sub-networks of randomly changed face models. High-quality face images should be more resilient and have less variability in face representations. A novel idea to quantify face quality based on an arbitrary face recognition model is developed to avoid the usage of unreliable quality labels. The resilience of a sample representation and, consequently, its quality is measured by examining the embedding variations produced by random sub-networks of a face model. Scores range from 0 (lower quality) to 1 (higher quality).

SDD-FIQA. Ou et al. [22] present an unsupervised FIQA technique that uses similarity distribution distance. The technique uses the Wasserstein Distance (WD) [6] between the intra-class and inter-class samples. The FR model uses ResNet-50 [15] trained on the MS1M database [10] to calculate the positive and negative sample distributions. A regression network for quality prediction is trained with Huber loss [22], using the WD metric as quality pseudo-labels. Scores range from 0 (lower quality) to 100 (higher quality).

2.3 Face Image Quality Applied to PAD

On FIQA applied to PAD, most works use quality as a means to reject fake faces directly. A software based on PAD is presented in [13] that may be applied to various biometric systems (iris, fingerprint and face) to identify several forms of fraudulent access attempts. The approach under consideration aims to increase the security of biometric recognition systems by incorporating PAD mechanisms in a quick and non-intrusive way using Image Quality Assessment (IQA). The method extracts 25 image-quality features from an image and can be applied in real-time to distinguish between real and fake biometric samples.

In order to discern between bona fide and fraudulent face appearances, Fourati et al. [12] provide a quick and unobtrusive PAD method based on IQA and motion cues. The results produced demonstrated superior performance to cutting-edge methods. The approach is particularly suited for real-time mobile apps since it considers both dependable robustness and minimal algorithm complexity.

Chang and Ye [7] present a face PAD method based on features for multi-scale perceptual picture quality evaluation. Specific hand-crafted texture features taken from facial photos are used for PAD. Generalised Gaussian density-based, asymmetric generalised Gaussian density-based, and top gradient similarity deviation features are the three main models into which the proposed features are classified. A total of 21 multiscale features are gathered for classification using a Support Vector Machine (SVM). Extensive tests on five benchmark databases and a novel dataset showed that the suggested framework is effective.

The same authors present in [30] a successful method for defending against face PAD based on perceptual picture quality evaluation features and multi-scale analysis. First, they show that a Blind Image Quality Evaluator (BIQE) [30] is capable of spotting spoofing attempts. Later, combine the BIQE with an IQA model called Effective Pixel Similarity Deviation (EPSD) to determine the standard deviation of the gradient magnitude similarity map by choosing effective pixels in the image. A multi-scale descriptor for categorisation is made up of a total of 21 features that were obtained from the BIQE and EPSD. Utilising three current benchmarks, Replay-Attack [8], CASIA [32], and UVAD [23], extensive research based on intra-class and cross-dataset methods were carried out, showing good performance compared with state-of-the-art methods.

An extendable multi cues integration framework for face PAD utilising a hierarchical neural network is proposed to increase the generalisability of face PAD systems [11]. This framework can combine picture quality cues with motion cues for liveness identification. A liveness feature based on image quality is created using Shearlet. In order to extract motion-based liveness features, dense optical flow is used. Different liveness features can be successfully integrated using a bottleneck feature fusion technique. Three public face PAD databases were used to evaluate the suggested methodology.

In [3], real biometric data are distinguished from data used in presentation/sensor PAs using non-reference image quality criteria. An experimental study demonstrates that bona fide versus fake iris, fingerprint, and facial data classification are possible with an average PAD detection performance of 90% based on a collection of 6 such measures. The target dataset, however, significantly impacts the optimal quality measure (combination) and classification setting, according to this research.

2.4 Datasets

In our experiments, training was done by merging several face PA datasets to improve generalisation as shown on Table 1. To enhance the variability of the dataset and increase the number of examples, we included a portion of two face databases that were not related to PA research: the Flickr Face database [18] and the UTK Faces database [34]. These datasets have images of regular faces, and print and screen attacks were created artificially using a texture transfer method developed in-house. This texture transfer method proposes automatically isolating the texture or artefacts in the capture process caused by the sensor-noise, frequency patterns or moire patterns without any training process. This way, we can quickly transfer this pattern to any new bona fide face image fast and transform it into a screen or printer attack promptly. These kinds of artefacts are inherent to the type of paper, camera sensors and scanners. In order to do that, we selected a palette of 50 solid colours. This palette was captured as raw images, which means that these colour palettes were printed on bond paper and high-quality glossy paper and then captured using five different smartphones (Samsung S5, Samsung S6, Samsung Galaxy Tab, Samsung A20, Xiaomi). The images were captured at several distances (ranging from 10 to 25 cm), keeping the paper focused on avoiding blurry faces. In total, we captured 5,000 single-colour images with different variations in illumination (day, afternoon and fluorescent light) and orientations (portrait and landscape) from the colour palette. More details can be found in [2]. We also selected a small number of samples[4] from the CelebA-Spoof dataset [33], for both bona fide and attacks, carefully selecting samples that resemble "selfie" images or had mostly frontal faces and attacks. This dataset also has "mask" attacks, most of them of low quality. We selected

---

[4] This file text will be available for reproducibility.

1,262 masks to be included in our dataset. Figure 2 shows some examples of these images.

Fig. 2. Right: Examples of mask attacks from the CelebA-Spoof and Left: Flicker-PAD Dataset. Top row: screen attacks. Bottom row: print attacks.

To have high-quality training samples and attacks, we also designed protocols to manually create our own PAs (both printed and screen). Original bona fide samples were taken from the Flickr dataset [18] (a different set than the one used for texture transfer) and from our proprietary dataset[5] of real-life selfies from users authenticating remotely with their smartphones. We called these datasets Flickr-PAD and Selfies-PAD, respectively. Figure 2 shows some attack samples from the Flickr PAD dataset.

Table 1. Dataset composition

Dataset	Bona fide	Print	Screen	Mask	Comments
CASIA-MFSD [35]	279	576	291	0	5 frames/video
CelebA-Spoof [33]	2,019	1,079	1,405	1,262	Man. sel. (frontal, selfie-like images)
UTK Faces [34] Text.Transfer	237	236	236	0	N/A
Flickr [18] Text.Transfer	1,580	1,580	1,579	0	N/A
Replay Attack Mobile [9]	2,726	3,295	1,119	0	10 frames/video (train+devel)
Flickr [18] - PAD	2,700	2,700	2,700	0	Proprietary dataset
Selfies - PAD	2,700	5,359	4,491	0	Proprietary dataset
Total	12,241	14,825	11,821	1,262	

Table 1 shows the composition of the full dataset. This full dataset has a total of 40,149 images, with 35,871 images were used for training purposes and 4,278 images were used as an intra-dataset test for all experiments in this paper, keeping the same class distribution. Performance of PAD algorithms sharply decreases on out-of-distribution data so we also tested on a cross-dataset testing

---

[5] This dataset is sequestered available only for evaluation.

partition using 2,830 images obtained from the OULU-NPU dataset [4]. This dataset is distributed as follows: 552 bona fide, 1,155 printed and 1,123 screen attacks. It is essential to highlight that this dataset was not used in training.

3   Method

In this paper, we study the influence of FIQA applied in conjunction with state-of-the-art face PAD algorithms. The following experiments were conducted:

- First, we study the distribution of FIQA scores on bona fide and attack samples from several public and proprietary PAD datasets. We evaluated the three FIQA methods described on Sect. 2.2.
- Second, the performance of PAD systems was evaluated when input faces were filtered by their score quality at inference time. We evaluated three deep learning-based Face PAD algorithms. Two of them present state-of-the-art results: SSDG and ViTranZFAS. The third one is a MobileNetv3 based network with a two-class output. We use this method to test the effects on simpler deep-learning networks.
- Finally, we assess the impact on PAD training by discarding low-quality bona fide and attack samples from the training dataset for all three PAD methods described above and all three FIQA methods of Sect. 2.2.

3.1   Training of PAD Algorithms

All three aforementioned PAD algorithms were implemented on PyTorch using the Kedro framework [1]. The same data augmentation scheme was used on all three algorithms using the albumentation library [5]. Input image size for all algorithms was 224×224 pixels. The training was done on a PC with an Intel i7-9700F CPU, 64 GB of RAM and dual NVIDIA GTX 2080Ti GPUS.

ViTranZFAS. Fine-tuning was performed from pre-loaded weights from the Timm package [29]. We used a fixed learning rate of $1e^{-4}$ and weight decay of $1e^{-6}$ with an Adam optimiser. As in the original paper, patch size was kept at 16×16 pixels. The training was done with batch size 64 over 100 epochs.

SSDG. The training was based on pre-trained ImageNet weights, using the following parameters: weight decay of $5e^{-4}$, the momentum of 0.9, and an initial learning rate of $1e^{-3}$ (first 150 epochs, then $1e^{-4}$). The λ parameters that control the balance between the adversarial loss, triplet loss and Cross-EntropyLoss were set to 1, 0.5 and 1, respectively. The training was done on 200 epochs with a batch size of 64.

MobileNetv3. We set an SGD optimiser with a momentum of 0.9, a learning rate of $5e^{-4}$, with Cross-EntropyLoss. The number of workers and batch size were both set to 32. The training was done over 150 epochs.

4 Experiments and Results

4.1 FIQA Effect on Filtering PAs

Quality scores were computed for all 4,278 images of the intra-dataset test for all three FIQA methods used. Then, all images on the lowest X% with X ranging from 0% to 95% (regardless of the class label) were discarded. Figure 3 shows the ratio of discarded images by class, normalised by the total of images of each class. Results show that bona fide samples are discarded at a lower rate than PAs, with masks being the easiest to filter out by quality. All three FIQA methods behaved similarly. Even if not designed as a PAD method per se, FIQA pre-filtering on an image processing pipeline tends to benefit the selection of bona fide samples over low-quality attacks.

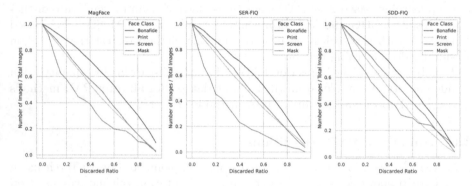

Fig. 3. Bona fide and PA filtering by quality scores for MagFace, SER-FIQ and SDD-FIQA.

Figure 4, shows a box plot of the quality score distribution for each dataset of the composing training dataset for MagFace, SER-FIQ and SDD- FIQA, respectively. As expected, attack presentations show a lower score on average than bona fide samples, with some important variation across datasets. It should be noted that samples, especially attacks, from the CelebA-Spoof dataset have lower scores than most other datasets. This is expected due to the in-the-wild nature of the images and because many of the PAs are very easily recognised, even for the selected fraction of the dataset we are using in this paper. This is unfortunate, given this dataset's massive size and adoption. Also, it can be seen that SER-FIQA scores are much less stable than the other two FIQA methods.

4.2 PAD Performance Versus Input Face Quality

In this experiment, we evaluated the impact on each PAD algorithm when varying the input test images according to their quality score. Images were discarded in steps of 5% from 0% to 55%, keeping the higher quality images for each class

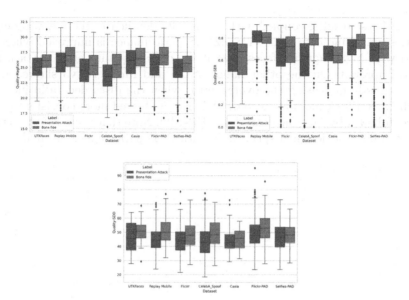

Fig. 4. Score statistics by original dataset when using MagFace, SER-FIQ and SDD-FIA.

(so class distribution remains intact across different discard ratios). The performance of each PAD algorithm was measured according to ISO 30107-3[6] using the BPCER and APCER metric defined as (1) and (2).

$$BPCER = \frac{\sum_{i=1}^{N_{BF}} RES_i}{N_{BF}} \qquad (1)$$

$$APCER = \frac{1}{N_{PAIS}} \sum_{i=1}^{N_{PAIS}} (1 - RES_i) \qquad (2)$$

where N_{BF} is the number of bona fide presentations, N_{PAIS} is the number of PAs for a given presentation species and RES_i is 1 if the system's response to the i-th attack is classified as an attack and 0 if classified as bona fide. A PAD performance can be reported as a single value of BPCER for a given APCER. For example, $BPCER_{20}$ is the BPCER value obtained when the APCER is fixed at 5%. $BPCER_{10}$ (APCER at 10%) and $BPCER_{100}$ (APCER at 1%) are also commonly used.

Figure 5 shows the $BPCER_{10}$ and $BPCER_{20}$ metrics on the cross-dataset test for VitranZFAS, SSDG and MobileNetv3. ViTranZFAS showed a lower error rate on both the intra-dataset (using SDD-FIQA) and cross-dataset (all three FIQA methods) when dealing with higher-quality faces. In the cross-dataset test, the BCPER decreases by 2%–4% when discarding 20% of the lower-quality

---

[6] https://www.iso.org/standard/67381.html.

images. These results, coupled with the results detailed in the following experiment, seem to indicate that ViTranZFAS performance is improved when presented with higher-quality faces even if the overall quality of the attack faces is also improved. On the other hand, SSDG and MobileNetv3 showed mixed results, mostly increasing the overall error when the discarded ratio increases. For all three FIQA methods tested, SDD-FIQA showed more consistent performance, followed by MagFace and then SER-FIQ. Note that these FIQA methods are not trained or designed specifically with PAD performance in mind.

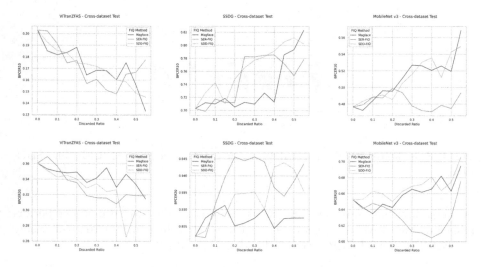

Fig. 5. PAD Performance using BPCER(%) versus Discarded Ratio of images done by Image Quality for the cross-dataset test.

Fig. 6 shows the higher and lower quality images for each FIQA method and their corresponding PAD scores for the VitranZFAS algorithm.

Fig. 6. Examples of the lower quality (3 images left) and higher quality images (3 images right) for each FIQA method: MagFace (left), SDD (center) and SER (right). The class and PAD score for the VitranZFAS method is shown.

4.3   FIQA Filtering of the Training Dataset

Face quality scores were computed for the training dataset with all three FIQA methods used in this paper. Then, a fraction of the images (Discarded Ratio) were discarded by sorting them and dropping images with the lowest quality score regardless of sample class (bona fide or attack). The training was done using the remaining dataset under the same conditions explained in Sect. 2.4.

Table 2 shows the metrics obtained in both the intra-dataset and cross-dataset tests when training with the full dataset and then training with the filtered dataset (EER corresponds to the error rate at the operating point where APCER equals BPCER).

Table 2. Model Performance versus train dataset Discarded Ratio (DR).

PAD Method	FIQA	DR	Intra-dataset test [%]			Cross-dataset test [%]		
			EER	$BPCER_{10}$	$BPCER_{20}$	EER	$BPCER_{10}$	$BPCER_{20}$
ViTranZFAS	-	0	3.22	0.69	2.39	13.85	20.29	36.1
	MagFace	0.2	2.00	0.31	0.62	13.04	**16.49**	**33.0**
	SER-FIQ	0.2	3.17	1.00	1.93	15.76	34.78	58.15
	SDD-FIQA	0.2	**1.85**	**0.08**	**0.54**	**12.90**	19.02	35.51
SSDG	-	0	1.93	0.39	0.69	20.10	70.29	92.21
	MagFace	0.2	2.25	0.38	1.08	18.61	64.05	87.68
	SER-FIQ	0.2	2.78	0.46	1.93	21.47	88.59	97.46
	SDD-FIQA	0.2	2.17	0.31	0.93	20.26	71.01	92.75
MobileNetv3	-	0	6.19	3.31	7.40	23.20	47.65	65.22
	MagFace	0.2	7.41	5.55	16.27	25.71	49.09	65.58
	SER-FIQ	0.2	7.32	5.09	19.66	22.17	49.46	67.75
	SDD-FIQA	0.2	7.32	5.24	10.17	25.45	55.62	69.75

ViTranZFAS showed an improvement of 1.5% on the EER of the intra-dataset test as well as improved BPCER when training with the MagFace and SDD-FIQA filtered dataset. On the cross-dataset test, improvement is even larger with an improvement close to 4% on $BPCER_{20}$. Note that training was done with 80% of the baseline dataset. This could be explained by the effect that can be seen over the scores of the cross-dataset test. A Kernel Density Estimation plot of the scores (Fig. 7) shows a much better separation of classes when training on the reduced dataset, compared to training on the full dataset for all three FIQ methods.

When training the SSDG algorithm, a small improvement is shown when training with MagFace on the cross-dataset test. Performance with other FIQA methods is worse than baseline training by a small margin. MobileNetv3 training did not show any improvements and cannot compensate for the loss of training images for any of the FIQA methods tested. MagFace showed the best improvements overall with any of the PAD methods. On the other hand, training with the dataset filtered by SER-FIQ did not lower the error metrics for any of the PAD methods tested.

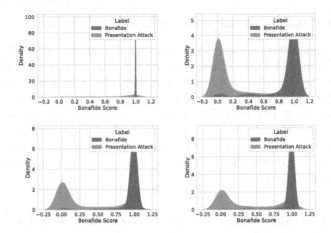

Fig. 7. Kernel Density Estimation Plots of Score Distributions for the ViTranZFAS PAD method in the cross-dataset test. (Top left) No discarded images. (Top Right) 20% discarded by MagFace. (Bottom Left) 20% discarded by SDD-FIQA. (Bottom Right) 20% discarded by SER-FIQA.

5   Conclusions and Future Work

In this paper, we evaluated the impact of face quality assessment methods on PAD systems. We showed that FIQA pre-filtering on an image processing pipeline benefits the selection of bona fide samples over low-quality attacks. Discarding 20% of the lower quality samples eliminates around 10% of bona fide but more than 20% of PA images, more if attacks are of lower quality. However, our suggestion to remove the 20% of the lower quality training samples should be analysed, considering the datasets used, the number of images available to train the algorithm, the quality of the pictures, attack complexity, the database diversity and the number of presentation attacks.

ViTranZFAS behaves better when presented with high-quality faces (even if they are PAs), improving its attack detection performance. Also, lower EER (up to 1%) and BPCER (up to 4%) are obtained when training ViTranZFAS with a filtered training dataset, dropping 20% of the lower quality samples. This also happens on SSDG when pre-filtering the dataset using MagFace.

Further study is needed on the design of pipelines that include FR, PAD and face quality. Score merging strategies and pipeline order need to be studied to yield better algorithmic performance and cost-effective solutions. Also, the computational efficiency of FIQA algorithms needs to be studied in real-life applications where cost and latency are of uttermost importance.

Finally, this study showed the importance of having more challenging high-quality datasets with carefully made PA images, so PAD algorithms can better learn from harder, more realistic spoofing attempts. There is a lack of publicly available datasets that have the variability, scale and level of crafting for attacks that are required to train PAD systems that can be applied to real applications in the wild with acceptable performance.

Acknowledgements. This work is supported by the European Union's Horizon 2020 research and innovation program under grant agreement No 883356 and the German Federal Ministry of Education and Research and the Hessen State Ministry for Higher Education, Research and the Arts within their joint support of the National Research Center for Applied Cybersecurity ATHENE.

References

1. Alam, S., et al.: (2021), https://github.com/kedro-org/kedro
2. Benalcazar, D., Tapia, J.E., Gonzalez, S., Busch, C.: Synthetic ID card image generation for improving presentation attack detection. arxiv:2211.00098 (2022)
3. Bhogal, A.P.S., Söllinger, D., Trung, P., Uhl, A.: Non-reference image quality assessment for biometric presentation attack detection. In: 5th Intl. Workshop on Biometrics and Forensics (IWBF), pp. 1–6 (2017)
4. Boulkenafet, Z., Komulainen, J., Li, L., Feng, X., Hadid, A.: OULU-NPU: a mobile face presentation attack database with real-world variations. In: 12th IEEE Intl. Conf. on Automatic Face Gesture Recognition (FG 2017), pp. 612–618 (2017)
5. Buslaev, A., Iglovikov, V.I., Khvedchenya, E., Parinov, A., Druzhinin, M., Kalinin, A.A.: Albumentations: fast and flexible image augmentations. Information **11**(2), 125 (2020)
6. Cai, Y., Lim, L.H.: Distances between probability distributions of different dimensions. IEEE Trans. on Inf. Theory **68**(6), 4020–4031 (2022)
7. Chang, H.H., Yeh, C.H.: Face anti-spoofing detection based on multi-scale image quality assessment. Image Vis. Comput. **121**, 104428 (2022)
8. Chingovska, I., Anjos, A., Marcel, S.: On the effectiveness of local binary patterns in face anti-spoofing. In: 2012 BIOSIG-Proceedings of the International Conference of Biometrics Special Interest Group (BIOSIG), pp. 1–7 (2012)
9. Costa-Pazo, A., Bhattacharjee, S., Vazquez-Fernandez, E., Marcel, S.: The replay-mobile face presentation-attack database. In: 2016 International Conference of the Biometrics Special Interest Group (BIOSIG), pp. 1–7 (2016)
10. Deng, J., Guo, J., Xue, N., Zafeiriou, S.: Arcface: additive angular margin loss for deep face recognition. In: Proceedings of the IEEE/CVF Conference on Computer Vision and Pattern Recognition (CVPR) (2019)
11. Feng, L., et al.: Integration of image quality and motion cues for face anti-spoofing: a neural network approach. J. Vis, Commn. Image Representation **38**, 451–460 (2016)
12. Fourati, E., Elloumi, W., Chetouani, A.: Anti-spoofing in face recognition-based biometric authentication using image quality assessment. Multimed. Tools Appl. **79**, 865–889 (2020)
13. Galbally, J., Marcel, S., Fierrez, J.: Image quality assessment for fake biometric detection: application to iris, fingerprint, and face recognition. IEEE Trans. Image Process. **23**(2), 710–724 (2014)
14. George, A., Marcel, S.: On the effectiveness of vision transformers for zero-shot face anti-spoofing. In: IEEE International Joint Conference on Biometrics (IJCB), pp. 1–8 (2021)

15. He, K., Zhang, X., Ren, S., Sun, J.: Deep residual learning for image recognition. In: Proceedings of the IEEE Conference on Computer Vision and Pattern Recognition (CVPR), pp. 770–778 (2016)
16. Howard, A., et al.: Searching for mobilenetv3 (2019), https://arxiv.org/abs/1905.02244
17. Jia, Y., Zhang, J., Shan, S., Chen, X.: Single-side domain generalization for face anti-spoofing. In: Proceedings of the IEEE/CVF Conference on Computer Vision and Pattern Recognition (CVPR) (2020)
18. Karras, T., Laine, S., Aila, T.: A style-based generator architecture for generative adversarial networks. In: Proceedings of the IEEE/CVF Conference on Computer Vision and Pattern Recognition, pp. 4396–4405 (2019)
19. Kumar, S., Singh, S., Kumar, J.: A comparative study on face spoofing attacks. In: International Conference on Computing, Communication and Automation (ICCCA), pp. 1104–1108 (2017)
20. Marcel, S., Nixon, M.S., Fierrez, J., Evans, N. (eds.): Handbook of Biometric Anti-Spoofing. ACVPR, Springer, Cham (2019). https://doi.org/10.1007/978-3-319-92627-8
21. Meng, Q., Zhao, S., Huang, Z., Zhou, F.: MagFace: a universal representation for face recognition and quality assessment. In: CVPR (2021)
22. Ou, F.Z., et al.: SDD-FIQA: Unsupervised face image quality assessment with similarity distribution distance. In: Proceedings of the IEEE/CVF Conference on Computer Vision and Pattern Recognition (CVPR), pp. 7666–7675 (2021)
23. Pinto, A., Schwartz, W.R., Pedrini, H., Rocha, A.d.R.: Using visual rhythms for detecting video-based facial spoof attacks. IEEE Trans. on Inf. Forensics Secur. **10**(5), 1025–1038 (2015)
24. Schlett, T., Rathgeb, C., Henniger, O., Galbally, J., Fierrez, J., Busch, C.: Face image quality assessment: a literature survey. ACM Comput. Surv. **54**, 1–49(2022)
25. Tapia, J.E., Gonzalez, S., Busch, C.: Iris liveness detection using a cascade of dedicated deep learning networks. IEEE Trans. Inf. Forensics Secur. **17**, 42–52 (2022)
26. Tapia, J.E., Valenzuela, A., Lara, R., Gomez-Barrero, M., Busch, C.: Selfie periocular verification using an efficient super-resolution approach. IEEE Access **10**, 67573–67589 (2022)
27. Terhörst, P., Kolf, J.N., Damer, N., Kirchbuchner, F., Kuijper, A.: SER-FIQ: unsupervised estimation of face image quality based on stochastic embedding robustness. In: Proceedings of the IEEE/CVF Conference on Computer Vision and Pattern Recognition CVPR, Seattle, WA, USA, June 13-19, 2020, pp. 5650–5659. IEEE (2020)
28. Wasnik, P., Raja, K.B., Ramachandra, R., Busch, C.: Assessing face image quality for smartphone based face recognition system. In: 2017 5th International Workshop on Biometrics and Forensics (IWBF), pp. 1–6 (2017)
29. Wightman, R.: Pytorch image models (2019)
30. Yeh, C.H., Chang, H.H.: Face liveness detection based on perceptual image quality assessment features with multi-scale analysis. In: 2018 IEEE Winter Conference on Applications of Computer Vision (WACV), pp. 49–56 (2018)
31. Yu, Z., Qin, Y., Li, X., Zhao, C., Lei, Z., Zhao, G.: Deep learning for face anti-spoofing: a survey. arXiv preprint arXiv:2106.14948 (2021)
32. Yu, Z., Wan, J., Qin, Y., Li, X., Li, S.Z., Zhao, G.: NAS-FAS: static-dynamic central difference network search for face anti-spoofing. In: TPAMI (2020)

33. Zhang, Y., et al.: CelebA-Spoof: large-scale face anti-spoofing dataset with rich annotations. In: Vedaldi, A., Bischof, H., Brox, T., Frahm, J.-M. (eds.) ECCV 2020. LNCS, vol. 12357, pp. 70–85. Springer, Cham (2020). https://doi.org/10. 1007/978-3-030-58610-2_5
34. Zhang, Z., Song, Y., Qi, H.: Age progression/regression by conditional adversarial autoencoder. In: Proceedings of the IEEE Conference on Computer Vision and Pattern recognition (CVPR). IEEE (2017)
35. Zhang, Z., Yan, J., Liu, S., Lei, Z., Yi, D., Li, S.Z.: A face antispoofing database with diverse attacks. In: 2012 5th IAPR International Conference on Biometrics (ICB), pp. 26–31 (2012)

Knowledge Distillation of Vision Transformers and Convolutional Networks to Predict Inflammatory Bowel Disease

José Maurício[1](✉)[iD] and Inês Domingues[1,2][iD]

[1] Instituto Politécnico de Coimbra, Instituto Superior de Engenharia, Rua Pedro Nunes - Quinta da Nora, 3030-199 Coimbra, Portugal
{a2018056151,ines.domingues}@isec.pt
[2] Centro de Investigação do Instituto Português de Oncologia do Porto (CI-IPOP): Grupo de Física Médica, Radiobiologia e Protecção Radiológica, Coimbra, Portugal

Abstract. Inflammatory bowel disease is a chronic disease of unknown cause that can affect the entire gastrointestinal tract, from the mouth to the anus. It is important for patients with this pathology that a good diagnosis is made as early as possible, so that the inflammation present in the mucosa intestinal is controlled and the most severe symptoms are reduced, thus offering the quality of life to people. Therefore, through this comparative study, we seek to find a way of automating the diagnosis of these patients during the endoscopic examination, reducing the subjectivity that is subject to the observation of a gastroenterologist, using six CNNs: AlexNet, ResNet50, VGG16, ResNet50-MobileNetV2 and Hybrid model. Also, five ViTs were used in this study: ViT-B/32, ViT-S/32, ViT-B/16, ViT-S/16 and R26+S/32. This comparison also consists in applying knowledge distillation to build simpler models, with fewer parameters, based on the learning of the pre-trained architectures on large volumes of data. It is concluded that in the ViTs framework, it is possible to reduce 25x the number of parameters by maintaining good performance and reducing the inference time by 5.32 s. For CNNs the results show that it is possible to reduce 107x the number of parameters, reducing consequently the inference time in 3.84 s.

Keywords: Inflammatory Bowel Disease · Knowledge Distillation · Vision Transformers · CNN · DeiT

1 Introduction

Crohn's disease and ulcerative colitis are two separate pathologies that are joined by inflammatory bowel disease, which is characterised by chronic inflammation throughout the intestinal tract. With 2.9 cases of ulcerative colitis per 100,000 residents per year and 2.4 cases of Crohn's disease per 100,000 residents per year in Portugal, it is estimated that it already affects 7,000 to 15,000 people there [8].

© Springer Nature Switzerland AG 2024
V. Vasconcelos et al. (Eds.): CIARP 2023, LNCS 14469, pp. 374–390, 2024.
https://doi.org/10.1007/978-3-031-49018-7_27

Crohn's disease affects any part of the gastrointestinal tract, from the mouth to the anus. The severity of the disease and its location determine the signs and symptoms, which have a wide spectrum of clinical presentations [11]. Ulcerative colitis is restricted to the large intestine and involves a superficial inflammatory process. This disease only affects the colon and rectum. Whereas Crohn's disease affects any part of the gastrointestinal tract. Normally, the inflammation starts in the rectum and can be continuously extended to the terminal ileum, where it is terminated by backwash ileitis [25].

The use of deep learning models to help the gastroenterologist in the diagnosis of patients with inflammatory bowel disease is important and, with time, more necessary. Not only in Portugal but worldwide the number of people who have this chronic disease is increasing and its diagnosis depends on the evaluation that the gastroenterologist makes of the endoscopic images, which makes this diagnosis subject to a great subjectivity [24]. Therefore, it is crucial to find tools that automate the diagnosis of the disease to prescribe the most suitable treatment for the patient.

These tools need to be fast in recognising the disease through endoscopic images since the patient is often under the anaesthesia effect. It is also important to produce light and low-complexity models to occupy few computational resources when deployed in medical systems. But that they guarantee the same accuracy in predicting inflammatory bowel disease. In this sense, this study seeks to make a comparative study, where six CNNs and five ViTs are used to process images collected using colonoscopy and video capsule endoscopic to recognise which type of inflammatory bowel disease is present: Crohn's disease and Ulcerative Colitis. Together with knowledge distillation to simpler models that guarantee similar performance [1] as pre-trained models on large volumes of data.

The organization of this paper is divided into four sections: Sect. 2 describes the created datasets for this study and the methodology implemented; Sect. 3 presents the quantitative results obtained by the teachers and student's models; Sect. 4 compare the results obtained in this study with the results of the other authors and critically assesses the findings made; and Sect. 5 summarises the findings and suggests some future research directions.

2   Methodology and Data

Aiming to facilitate the deployment of deep models in embedded systems to automate and improve the diagnosis of inflammatory bowel disease, this work suggests a methodology based on six phases.

In the first phase, the authors collected images of the inflammatory bowel disease (i.e. Ulcerative colitis and Crohn's disease). The videos captured during the patients' endoscopic examination were used in the second phase to extract frames. In the third phase, the images were pre-processed to fill in some undesired characteristics. In the fourth phase, different data augmentation techniques were applied to the training dataset. In the fifth phase, the convolutional networks and vision transformers were implemented and configured. In the sixth phase, the performance of the CNNs and ViTs was evaluated using six classification metrics, the Loss and the Inference time. Figure 1 illustrates the methodology proposed in this paper and in the following subsections will be explained.

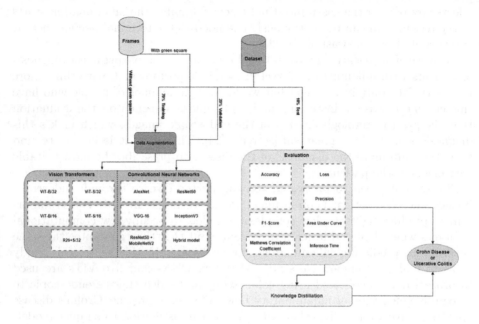

Fig. 1. Proposed methodology.

2.1 Dataset

Three datasets of images related to Crohn's disease and Ulcerative colitis were used to conduct this study, with the aim of combining the images from the two diseases into a single dataset. HyperKvasir [2,3] and LIMUC [23] datasets were used to get images of the Ulcerative colitis disease, while the CrohnIPI [7,32,33] dataset was selected to collect images of Crohn's disease.

Figure 2 shows an example of the images that exist in each collected dataset. Therefore, two different datasets were created:

– **Dataset 1**: this dataset compiles 850 images of ulcerative colitis from the HyperKvasir dataset and 1360 images of Crohn's disease from the CrohnIPI dataset. It also includes 64 video frames that were taken from the HyperKvasir dataset's labelled videos. Figure 3a shows the class distribution of this dataset;
– **Dataset 2**: this dataset consists of 446 instances of the LIMUC dataset, the instances are equally distributed by severity degree. That is, 25% of the total images referring to the severity degree were extracted from the original

dataset; 64 video frames were extracted from the labelled videos from the HyperKvasir dataset; 850 instances from the HyperKavsir dataset; and 1360 instances from the CrohnIPI dataset. Figure 3b shows the class distribution of this dataset.

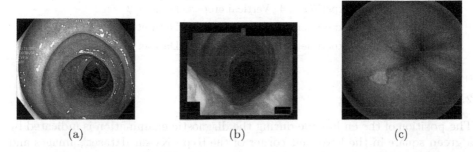

(a)   (b)   (c)

Fig. 2. Image Samples: (a) Ulcerative colitis image from the HyperKvasir dataset; (b) Ulcerative colitis image from the LIMUC dataset; (c) Crohn's disease image from the CrohnIPI dataset.

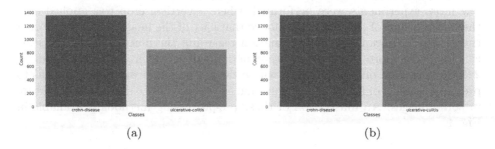

(a)   (b)

Fig. 3. Class distribution: (a) Dataset 1; (b) Dataset 2.

2.2   Experimental Setup

Tensorflow, version 2.8.0, and Tfimm, version 0.6.13, were utilised to carry out this investigation. The NVIDIA A100 GPU was used in conjunction with Google Colab as the programming environment for importing the libraries.

Based on the created datasets, a total of six experiments were realized, allowing different combinations. Experiments with odd numbers (1, 3, and 5) consisted in using dataset 1. Experiments with even numbers (2, 4, and 6) had the aim of balancing the classes and diversifying the images regarding class ulcerative colitis, for which was used the dataset 2. Table 1 summarizes the experiments realized, as well as pre-processing techniques (see Sect. 2.3) and dataset used.

Table 1. Configurations of the six performed experiments

Experiment	Pre-processing	Dataset
Experiment 1	Horizontal crop	Dataset 1
Experiment 2	Horizontal crop	Dataset 2
Experiment 3	Vertical crop	Dataset 1
Experiment 4	Vertical crop	Dataset 2
Experiment 5	Gaussian blur	Dataset 1
Experiment 6	Gaussian blur	Dataset 2

2.3   Pre-processing

The position of the endoscope during the diagnostic examination is indicated by a green square in the lower-left corner of the HyperKvasir dataset's images and frames extracted from a video. This might make it difficult to identify the specific type of inflammatory bowel disease depicted in the images given to CNN. When this kind of circumstance was confirmed, experiments were conducted where some transformations were applied.

One of the applied transformations was a horizontal crop across the entire width of the image to remove the green square in the lower left corner. Then a few more experiments were performed, where a vertical crop was applied over the entire height of the image to remove that part of the image. However, during these transformations, some information regarding the mucosa of the intestine is lost. Some more experiments were performed in which a Gaussian blur of 2/16 was applied only to the area where the green square was. The goal was to preserve as much of the image information as possible but to disguise that feature in some images. An illustration of these pre-processing methods is available in Fig. 4.

(a)   (b)   (c)   (d)

Fig. 4. Pre-processing examples: (a) Original image; (b) Horizontal crop; (c) Vertical crop; (d) Gaussian blur.

2.4   Data Augmentation

Before classifying the images, data augmentation was performed on the train set with horizontal and vertical Random Flip, Random Contrast with a factor of 0.15, Random Rotation with a factor of 0.2 and a Random Zoom with a portion of -0.2 for height and -0.3 for width.

2.5   Deep Learning Models

Three pre-trained CNN networks were selected for this study: ResNet50, VGG16, and InceptionV3. In addition, an ensemble model was built that combines two pre-trained CNN networks, a ResNet50 network and a MobileNetV2 network. All of these architectures have been pre-trained with the ImageNet dataset. Besides these architectures, two non-pretrained architectures were also implemented: (1) an AlexNet; and (2) a hybrid model that combines a CNN with an LSTM.

The architecture that combines the ResNet50 network and the MobileNetV2 network consists in an ensemble model, where the two architectures were combined. That is, the input images will be directed to the two architectures that will process these images without the last classification layer and the output of the two will be concatenated to be classified by a Dense layer with 2 neurons referring to class Crohn disease and ulcerative colitis. The construction of this model was inspired by [13,15].

The hybrid model built in this work was based on a similar architecture developed by other authors [26,30,34]. This architecture is based on: 8 Conv2D layers, 2 BatchNormalization layers, 4 MaxPooling2D layers, 1 Flatten layer, 7 Dense layers, 3 Dropout layers and 3 LSTM layers. The hybrid model is illustrated in Fig. 5. The forget gate present on the LSTM network will help the next layers remember the most important information and forget unnecessary information [30].

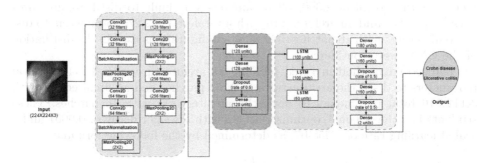

Fig. 5. The hybrid model's architecture [20].

Within the vision transformers, the following models were used: ViT-B/32, ViT-S/32, ViT-B/16, ViT-S/16 and R26+S/32. These models were pre-trained with the Imagenet21k dataset. The R26+S/32 model was created by Google and made available, as well as other versions very similar to this one, which consists of combining a ResNet network with a vision transformer [27]. That is, the first part of the model name (R26) represents the number of convolutional layers that were used from the ResNet network and the second part of the name (S/32) means that those layers extracted from the ResNet network were combined with a small vision transformer of 32 patches.

Using SparseCategoricalCrossentropy as the loss function, a batch size of 32, and Adam with a learning rate of 1E-05 as the optimizer, the models used to classify the pictures were trained for 200 epochs [16]. An EarlyStopping callback with patience of 5 was utilized during training to track the validity accuracy. The images were resized to 224×224 pixels in all experiments.

2.6   Knowledge Distillation

Deploying large architectures into medical systems to assist gastroenterologists in diagnosing the disease can be very complex and even impractical due to the time it takes to process the images, as well as the computational resources required to store these models. Therefore, this sub-section describes the process of distilling knowledge from core architectures to simpler and lighter architectures.

For the CNNs, a student with a total of 1,254,194 parameters was created, with the Adam optimizer using a learning rate of 1e-04 and the SparseCategoricalCrossentropy loss function. During distillation, the KLDivergence function was set as the loss function, with an alpha of 0.8 and a temperature of 105°C. For student training, 100 epochs and an EarlyStoping callback with the patience of 5 were set to monitor the validation accuracy.

In the case of ViTs, the distillation process was done in two different ways: i) using the temperature technique to scale the weights returned by the teacher's Softmax function, as shown in Fig. 6; ii) using the Data-Efficient Image Transformer (DeiT) technique. In the implementation of the first technique, an architecture with a total of 2,300,162 parameters was built to distil architectures with 32 patches and an architecture with a total of 3,442,274 parameters to distil architectures with 16 patches. The same configurations used in the distillation of the CNNs were considered.

The second method involved building an architecture with 3,554,884 total parameters to distil the architectures with 32 patches and an architecture with 3,112,516 total parameters to distil the architectures with 16 patches. These students had the AdamW optimizer set up with a weight decay of 0.0001 and a scaled learning rate of 6.25e-07, as determined by the following formula:

$$lr_{scaled} = \frac{lr}{512} \times batchsize$$

The learning rate (lr) is equal to 1.00e-05 and the batch size is 32. The student's loss function and the distillation function were both used to define the SparseCategoricalCrossentropy function. Using an EarlyStopping callback and a patience of 5, the students were trained for 100 epochs. Data augmentation

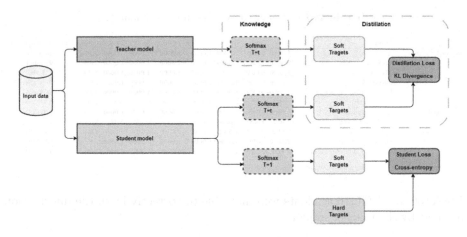

Fig. 6. Example of the Response-base knowledge, based on [14].

was also used in both of the ViTs' distillation methods and in CNNs' knowledge distillation [9,17].

2.7   Evaluation

The dataset used in all experiments was divided into three sets: training (70%), validating (20%) and testing (10%) [6]. As described in Sect. 2.1, the datasets include 64 video frames taken from the HyperKvasir dataset's labelled videos. Thirty of these 64 extracted frames present a green square in the lower left corner. All of these frames are in the validation dataset. The remaining frames from the extra video that were missing the green square were included in the training dataset.

Classification metrics were used to assess the performance of the CNNs and ViTs on the test dataset, as well as the performance of the students. These include Accuracy, Precision, Recall, F1-Score, Area Under Curve (AUC), Mathew's Correlation Coefficient (MCC), the inference time [5,29], and the Loss:

$$Loss = -\sum_{i=1}^{n} y_i \cdot \log \hat{y}_i$$

3   Results

This section shows the results obtained from all experiments. Table 2 and 3 show the results obtained by the CNNs and ViTs in experiment 1, respectively.

In the tables, it can be observed that both deep learning models managed to obtain a good performance in image classification and through knowledge distillation it was possible to reduce the inference time in most models, except in the AlexNet network. However, in this experiment, it is verified that the

Table 2. Results of the CNNs in experiment 1.

	Models	Acc	Loss	Recall	Precision	F1-Score	AUC	MCC	Inference time	Total of params
Main architectures	AlexNet	**1.0000**	0.0031	**1.0000**	**1.0000**	**1.0000**	**1.0000**	**1.0000**	**0.75 s**	46,760,706
	ResNet50	**1.0000**	0.0297	**1.0000**	**1.0000**	**1.0000**	**1.0000**	**1.0000**	3.65 s	23,591,810
	VGG16	**1.0000**	**0.0016**	**1.0000**	**1.0000**	**1.0000**	**1.0000**	**1.0000**	3.80 s	134,268,738
	InceptionV3	**1.0000**	0.0596	**1.0000**	**1.0000**	**1.0000**	**1.0000**	**1.0000**	2.95 s	21,806,882
	ResNet50+MobileNetV2	**1.0000**	0.0429	**1.0000**	**1.0000**	**1.0000**	**1.0000**	**1.0000**	4.20 s	25,852,354
	Hybrid model	0.9953	0.4380	0.9877	**1.0000**	0.9938	0.9938	0.9901	2.31 s	1,930,966
Knowledge Distillation	AlexNet student	**1.0000**	0.0026	**1.0000**	**1.0000**	**1.0000**	**1.0000**	**1.0000**	2.51 s	1,254,194
	ResNet50 student	0.3991	1.3687	**1.0000**	0.3876	0.5586	0.5152	0.1084	2.06 s	1,254,194
	VGG16 student	0.4272	2.6307	**1.0000**	0.3990	0.5704	0.5379	0.1739	1.61 s	1,254,194
	InceptionV3 student	**1.0000**	0.0702	**1.0000**	**1.0000**	**1.0000**	**1.0000**	**1.0000**	1.45 s	1,254,194
	ResNet50+MobileNetV2 student	0.9953	0.0476	**1.0000**	0.9878	0.9939	0.9962	0.9901	1.69 s	1,254,194

ResNet50 and VGG16 students were not able to correctly learn the information distilled by the teacher model.

Table 3. Results of the ViTs in experiment 1.

	Models	Acc	Loss	Recall	Precision	F1-Score	AUC	MCC	Inference time	Total of params
Main architectures	ViT-B/32	**1.0000**	0.0011	**1.0000**	**1.0000**	**1.0000**	**1.0000**	**1.0000**	3.07 s	87,417,602
	ViT-S/32	**1.0000**	**0.0007**	**1.0000**	**1.0000**	**1.0000**	**1.0000**	**1.0000**	2.63 s	22,475,138
	ViT-B/16	**1.0000**	0.0029	**1.0000**	**1.0000**	**1.0000**	**1.0000**	**1.0000**	5.25 s	86,417,594
	ViT-S/16	0.9953	0.0097	**1.0000**	0.9878	0.9939	0.9962	0.9901	3.26 s	21,590,402
	R26+S/32	0.9953	0.0169	**1.0000**	0.9878	0.9939	0.9962	0.9901	5.88 s	36,028,098
Knowledge Distillation	ViT-B/32 student	0.9624	0.1138	0.9012	0.9481	0.9506	0.9218	1.81 s	2,300,162
	ViT-S/32 student	0.9859	0.0251	**1.0000**	0.9643	0.9818	0.9886	0.9708	1.31 s	2,300,162
	ViT-B/16 student	0.9953	0.0132	**1.0000**	0.9878	0.9939	0.9962	0.9901	1.74 s	3,442,274
	ViT-S/16 student	0.9906	0.0147	0.9753	**1.0000**	0.9875	0.9877	0.9802	**0.91 s**	3,442,274
	R26+S/32 student	0.9953	0.0169	**1.0000**	0.9878	0.9939	0.9962	0.9901	1.42 s	2,300,162
DeiT	ViT-B/32 student	0.9671	0.0636	0.9506	0.9625	0.9565	0.9639	0.9302	0.93 s	3,554,884
	ViT-S/32 student	0.9671	0.0983	0.9136	**1.0000**	0.9548	0.9568	0.9314	**0.91 s**	3,554,884
	ViT-B/16 student	0.9577	0.1023	0.8889	**1.0000**	0.9412	0.9444	0.9122	1.03 s	3,112,516
	ViT-S/16 student	0.9624	0.0849	0.9012	**1.0000**	0.9481	0.9506	0.9218	1.01 s	3,112,516
	R26+S/32 student	0.9531	0.1419	0.8765	**1.0000**	0.9342	0.9383	0.9027	0.92 s	3,554,884

Tables [4–5] present the results of experiment 2 regarding CNN networks and ViT models, in the same order. Based on the obtained results, it is perceived by the value of the AUC metric, that the used architectures present a good ability to distinguish the images of the classes under study. Comparing the distillation results obtained by CNNs and ViTs, it can be concluded that the students of CNNs can perform better in classifying images of inflammatory bowel disease. When we compare the results between the two distillation techniques of the ViT models, we realize that the results worsened in the distillation through attention, having for example the model ViT-S/16 reduced 4.96% in Accuracy, 5.19% in AUC and 9.67% in MCC metric. Inference time was shorter using the DeiT distillation technique.

Tables [6–13] show the results obtained by the teacher and student models from experiments 3 to 6. The results in the tables express a behaviour similar to the one described during experiment 2, except that during experiment 3 the model R26+S/32 had a performance improvement with the distillation of knowledge through attention. The same was described by models ViT-B/32

Table 4. Results of the CNNs in experiment 2.

	Models	Acc	Loss	Recall	Precision	F1-Score	AUC	MCC	Inference time	Total of params
Main architectures	AlexNet	0.9858	0.0230	0.9852	0.9852	0.9852	0.9858	0.9716	**1.08 s**	46,760,706
	ResNet50	0.9965	0.0339	**1.0000**	0.9926	0.9963	0.9966	0.9929	2.68 s	23,591,810
	VGG16	0.9965	**0.0038**	0.9926	**1.0000**	0.9963	0.9963	0.9929	5.60 s	134,268,738
	InceptionV3	0.9929	0.0408	**1.0000**	0.9854	0.9926	0.9932	0.9859	3.96 s	21,806,882
	ResNet50+MobileNetV2	**1.0000**	0.0177	**1.0000**	**1.0000**	**1.0000**	**1.0000**	**1.0000**	4.18 s	25,852,354
	Hybrid model	0.9929	0.3326	**1.0000**	0.9854	0.9926	0.9932	0.9859	3.31 s	1,930,966
Knowledge Distillation	AlexNet student	0.9752	0.0998	0.9926	0.9571	0.9745	0.9759	0.9509	2.10 s	1,254,194
	ResNet50 student	0.9610	0.1058	0.9185	**1.0000**	0.9575	0.9593	0.9244	1.73 s	1,254,194
	VGG16 student	0.9929	0.1006	0.9926	0.9926	0.9926	0.9929	0.9858	1.76 s	1,254,194
	InceptionV3 student	0.9823	0.1043	**1.0000**	0.9643	0.9818	0.9830	0.9651	3.21 s	1,254,194
	ResNet50+MobileNetV2 student	0.9894	0.0738	0.9926	0.9853	0.9889	0.9895	0.9787	1.69 s	1,254,194

Table 5. Results of the ViTs in experiment 2.

	Models	Acc	Loss	Recall	Precision	F1-Score	AUC	MCC	Inference time	Total of params
Main architectures	ViT-B/32	0.9965	0.0116	0.9926	**1.0000**	0.9963	0.9963	0.9929	3.42 s	87,417,602
	ViT-S/32	0.9965	0.0070	**1.0000**	0.9926	0.9963	0.9966	0.9929	2.87 s	22,475,138
	ViT-B/16	**1.0000**	0.0004	**1.0000**	**1.0000**	**1.0000**	**1.0000**	**1.0000**	6.49 s	86,417,594
	ViT-S/16	**1.0000**	0.0065	**1.0000**	**1.0000**	**1.0000**	**1.0000**	**1.0000**	3.67 s	21,590,402
	R26+S/32	**1.0000**	0.0042	**1.0000**	**1.0000**	**1.0000**	**1.0000**	**1.0000**	6.45 s	36,028,098
Knowledge Distillation	ViT-B/32 student	0.9539	0.1103	0.9704	0.9357	0.9527	0.9546	0.9083	1.76 s	2,300,162
	ViT-S/32 student	0.9574	0.1615	0.9111	**1.0000**	0.9535	0.9556	0.9178	**1.00 s**	2,300,162
	ViT-B/16 student	0.9894	0.0413	0.9778	**1.0000**	0.9888	0.9889	0.9789	1.22 s	3,442,274
	ViT-S/16 student	0.9929	0.0376	**1.0000**	0.9854	0.9926	0.9932	0.9859	1.04 s	3,442,274
	R26+S/32 student	0.9504	0.2228	0.9926	0.9116	0.9504	0.9521	0.9042	1.06 s	2,300,162
DeiT	ViT-B/32 student	0.9362	0.1776	0.8963	0.9680	0.9308	0.9345	0.8739	1.08 s	3,554,884
	ViT-S/32 student	0.9468	0.1330	0.9111	0.9762	0.9425	0.9454	0.8950	1.10 s	3,554,884
	ViT-B/16 student	0.9397	0.2261	0.8815	0.9917	0.9333	0.9373	0.8837	1.16 s	3,112,516
	ViT-S/16 student	0.9433	0.2013	0.8963	0.9837	0.9380	0.9413	0.8892	1.16 s	3,112,516
	R26+S/32 student	0.9291	0.2303	0.8667	0.9832	0.9213	0.9265	0.8629	1.04 s	3,554,884

and ViT-S/32 in experiment 5. Also in the experiment, it is verified that the hybrid model of CNNs did not obtain a good performance. In addition, in each experiment using the same distillation technique and when comparing inference time between ViTs and CNNs, it was possible to reduce the inference time in the ViTs models substantially. Although the student models present a higher number of parameters than the CNNs' student model.

Table 6. Results of the CNNs in experiment 3.

	Models	Acc	Loss	Recall	Precision	F1-Score	AUC	MCC	Inference time	Total of params
Main architectures	AlexNet	**1.0000**	0.0024	**1.0000**	**1.0000**	**1.0000**	**1.0000**	**1.0000**	**0.78 s**	46,760,706
	ResNet50	**1.0000**	0.0376	**1.0000**	**1.0000**	**1.0000**	**1.0000**	**1.0000**	2.27 s	23,591,810
	VGG16	**1.0000**	0.0000	**1.0000**	**1.0000**	**1.0000**	**1.0000**	**1.0000**	3.56 s	134,268,738
	InceptionV3	0.9906	0.0644	**1.0000**	0.9759	0.9878	0.9924	0.9804	2.94 s	21,806,882
	ResNet50+MobileNetV2	**1.0000**	0.0338	**1.0000**	**1.0000**	**1.0000**	**1.0000**	**1.0000**	3.61 s	25,852,354
	Hybrid model	**1.0000**	0.3582	**1.0000**	**1.0000**	**1.0000**	**1.0000**	**1.0000**	2.43 s	1,930,966
Knowledge Distillation	AlexNet student	0.9906	0.0207	0.9753	**1.0000**	0.9875	0.9877	0.9802	2.15 s	1,254,194
	ResNet50 student	0.9859	0.0625	0.9877	0.9756	0.9816	0.9863	0.9702	1.72 s	1,254,194
	VGG16 student	**1.0000**	0.0006	**1.0000**	**1.0000**	**1.0000**	**1.0000**	**1.0000**	1.43 s	1,254,194
	InceptionV3 student	0.9906	0.0711	**1.0000**	0.9759	0.9878	0.9924	0.9804	1.70 s	1,254,194
	ResNet50+MobileNetV2 student	0.9906	0.0758	**1.0000**	0.9759	0.9878	0.9924	0.9804	1.90 s	1,254,194

Table 7. Results of the ViTs in experiment 3.

	Models	Acc	Loss	Recall	Precision	F1-Score	AUC	MCC	Inference time	Total of params
Main architectures	ViT-B/32	**1.0000**	**0.0003**	**1.0000**	**1.0000**	**1.0000**	**1.0000**	**1.0000**	3.21 s	87,417,602
	ViT-S/32	**1.0000**	0.0040	**1.0000**	**1.0000**	**1.0000**	**1.0000**	**1.0000**	2.91 s	22,475,138
	ViT-B/16	**1.0000**	0.0035	**1.0000**	**1.0000**	**1.0000**	**1.0000**	**1.0000**	5.62 s	86,417,594
	ViT-S/16	**1.0000**	0.0008	**1.0000**	**1.0000**	**1.0000**	**1.0000**	**1.0000**	3.33 s	21,590,402
	R26+S/32	**1.0000**	0.0033	**1.0000**	**1.0000**	**1.0000**	**1.0000**	**1.0000**	6.29 s	36,028,098
Knowledge Distillation	ViT-B/32 student	**0.9953**	0.0038	**0.9877**	**1.0000**	**0.9938**	**0.9938**	**0.9901**	1.70 s	2,300,162
	ViT-S/32 student	0.9906	0.0395	0.9753	**1.0000**	0.9875	0.9877	0.9802	1.58 s	2,300,162
	ViT-B/16 student	0.9906	0.0509	0.9753	**1.0000**	0.9875	0.9877	0.9802	1.15 s	3,442,274
	ViT-S/16 student	0.9906	0.0399	0.9753	**1.0000**	0.9875	0.9877	0.9802	1.65 s	3,442,274
	R26+S/32 student	0.9061	0.3198	0.7531	**1.0000**	0.8592	0.8765	0.8087	**0.97 s**	2,300,162
DeiT	ViT-B/32 student	0.9671	0.1054	0.9136	**1.0000**	0.9548	0.9568	0.9314	1.02 s	3,554,884
	ViT-S/32 student	0.9718	0.0793	0.9383	0.9870	0.9620	0.9653	0.9404	1.28 s	3,554,884
	ViT-B/16 student	0.9718	0.0738	0.9259	**1.0000**	0.9615	0.9630	0.9411	1.54 s	3,112,516
	ViT-S/16 student	0.9718	0.0964	0.9259	**1.0000**	0.9615	0.9630	0.9411	1.10 s	3,112,516
	R26+S/32 student	0.9671	0.0660	0.9383	0.9744	0.9560	0.9616	0.9302	1.03 s	3,554,884

Table 8. Results of the CNNs in experiment 4.

	Models	Acc	Loss	Recall	Precision	F1-Score	AUC	MCC	Inference time	Total of params
Main architectures	AlexNet	0.9858	0.0354	0.9926	0.9781	0.9853	0.9861	0.9717	**1.07 s**	46,760,706
	ResNet50	**1.0000**	0.0262	**1.0000**	**1.0000**	**1.0000**	**1.0000**	**1.0000**	3.70 s	23,591,810
	VGG16	**1.0000**	**0.0007**	**1.0000**	**1.0000**	**1.0000**	**1.0000**	**1.0000**	5.59 s	134,268,738
	InceptionV3	0.9965	0.0369	**1.0000**	0.9926	0.9963	0.9966	0.9929	3.87 s	21,806,882
	ResNet50+MobileNetV2	**1.0000**	0.0179	**1.0000**	**1.0000**	**1.0000**	**1.0000**	**1.0000**	4.96 s	25,852,354
	Hybrid model	0.9823	0.3591	**1.0000**	0.9643	0.9818	0.9830	0.9651	2.67 s	1,930,966
Knowledge Distillation	AlexNet student	0.9894	0.0467	0.9926	0.9853	0.9889	0.9895	0.9787	2.50 s	1,254,194
	ResNet50 student	0.9929	0.0754	**1.0000**	0.9854	0.9926	0.9932	0.9859	1.72 s	1,254,194
	VGG16 student	0.9929	0.0862	**1.0000**	0.9854	0.9926	0.9932	0.9859	1.88 s	1,254,194
	InceptionV3 student	0.9929	0.0422	**1.0000**	0.9854	0.9926	0.9932	0.9859	1.66 s	1,254,194
	ResNet50+MobileNetV2 student	0.9929	0.0592	**1.0000**	0.9854	0.9926	0.9932	0.9859	1.80 s	1,254,194

Table 9. Results of the ViTs in experiment 4.

	Models	Acc	Loss	Recall	Precision	F1-Score	AUC	MCC	Inference time	Total of params
Main architectures	ViT-B/32	**1.0000**	0.0028	**1.0000**	**1.0000**	**1.0000**	**1.0000**	**1.0000**	4.52 s	87,417,602
	ViT-S/32	0.9965	0.0051	**1.0000**	0.9926	0.9963	0.9966	0.9929	2.85 s	22,475,138
	ViT-B/16	**1.0000**	0.0019	**1.0000**	**1.0000**	**1.0000**	**1.0000**	**1.0000**	7.72 s	86,417,594
	ViT-S/16	**1.0000**	**0.0018**	**1.0000**	**1.0000**	**1.0000**	**1.0000**	**1.0000**	4.87 s	21,590,402
	R26+S/32	0.9787	0.0480	**1.0000**	0.9574	0.9783	0.9796	0.9583	8.11 s	36,028,098
Knowledge Distillation	ViT-B/32 student	0.9716	0.0697	0.9630	0.9774	0.9701	0.9713	0.9432	1.53 s	2,300,162
	ViT-S/32 student	0.8794	0.6815	0.9926	0.8024	0.8874	0.8841	0.7808	1.35 s	2,300,162
	ViT-B/16 student	0.9645	0.1487	0.9259	**1.0000**	0.9615	0.9630	0.9311	1.42 s	3,442,274
	ViT-S/16 student	0.9787	0.0924	0.9556	**1.0000**	0.9773	0.9778	0.9582	1.33 s	3,442,274
	R26+S/32 student	0.9645	0.0733	0.9333	0.9921	0.9618	0.9633	0.9303	1.56 s	2,300,162
DeiT	ViT-B/32 student	0.9574	0.1114	0.9333	0.9767	0.9545	0.9565	0.9154	1.14 s	3,554,884
	ViT-S/32 student	0.9468	0.1741	0.9037	0.9839	0.9421	0.9450	0.8958	**1.10 s**	3,554,884
	ViT-B/16 student	0.9433	0.2008	0.8963	0.9837	0.9380	0.9413	0.8892	1.37 s	3,112,516
	ViT-S/16 student	0.9397	0.1884	0.8963	0.9758	0.9344	0.9379	0.8815	1.31 s	3,112,516
	R26+S/32 student	0.9397	0.1539	0.9111	0.9609	0.9354	0.9385	0.8801	1.11 s	3,554,884

Table 10. Results of the CNNs in experiment 5.

	Models	Acc	Loss	Recall	Precision	F1-Score	AUC	MCC	Inference time	Total of params
Main architectures	AlexNet	**1.0000**	0.0007	**1.0000**	**1.0000**	**1.0000**	**1.0000**	**1.0000**	**0.74s**	46,760,706
	ResNet50	**1.0000**	0.0591	**1.0000**	**1.0000**	**1.0000**	**1.0000**	**1.0000**	2.58s	23,591,810
	VGG16	**1.0000**	**0.0002**	**1.0000**	**1.0000**	**1.0000**	**1.0000**	**1.0000**	3.65s	134,268,738
	InceptionV3	0.9953	0.0638	**1.0000**	0.9878	0.9939	0.9962	0.9901	3.48s	21,806,882
	ResNet50+MobileNetV2	**1.0000**	0.0356	**1.0000**	**1.0000**	**1.0000**	**1.0000**	**1.0000**	3.58s	25,852,354
	Hybrid model	0.4413	0.6797	**1.0000**	0.4050	0.5765	0.5492	0.1997	2.40s	1,930,966
Knowledge Distillation	AlexNet student	**1.0000**	0.0044	**1.0000**	**1.0000**	**1.0000**	**1.0000**	**1.0000**	2.15s	1,254,194
	ResNet50 student	**1.0000**	0.0448	**1.0000**	**1.0000**	**1.0000**	**1.0000**	**1.0000**	1.69s	1,254,194
	VGG16 student	0.9906	0.1358	**1.0000**	0.9759	0.9878	0.9924	0.9804	1.67s	1,254,194
	InceptionV3 student	0.9953	0.0457	0.9877	**1.0000**	0.9938	0.9938	0.9901	1.68s	1,254,194
	ResNet50+MobileNetV2 student	**1.0000**	0.0489	**1.0000**	**1.0000**	**1.0000**	**1.0000**	**1.0000**	2.22s	1,254,194

Table 11. Results of the ViTs in experiment 5.

	Models	Acc	Loss	Recall	Precision	F1-Score	AUC	MCC	Inference time	Total of params
Main architectures	ViT-B/32	**1.0000**	0.0024	**1.0000**	**1.0000**	**1.0000**	**1.0000**	**1.0000**	3.11s	87,417,602
	ViT-S/32	**1.0000**	0.0031	**1.0000**	**1.0000**	**1.0000**	**1.0000**	**1.0000**	2.70s	22,475,138
	ViT-B/16	**1.0000**	0.0012	**1.0000**	**1.0000**	**1.0000**	**1.0000**	**1.0000**	5.11s	86,417,594
	ViT-S/16	**1.0000**	0.0017	**1.0000**	**1.0000**	**1.0000**	**1.0000**	**1.0000**	3.26s	21,590,402
	R26+S/32	**1.0000**	**0.0007**	**1.0000**	**1.0000**	**1.0000**	**1.0000**	**1.0000**	5.86s	36,028,098
Knowledge Distillation	ViT-B/32 student	0.8779	0.4485	0.6790	**1.0000**	0.8088	0.8395	0.7532	2.66s	2,300,162
	ViT-S/32 student	0.9484	0.1526	0.8642	**1.0000**	0.9272	0.9321	0.8932	1.18s	2,300,162
	ViT-B/16 student	**1.0000**	0.0011	**1.0000**	**1.0000**	**1.0000**	**1.0000**	**1.0000**	1.76S	3,442,274
	ViT-S/16 student	**1.0000**	0.0035	**1.0000**	**1.0000**	**1.0000**	**1.0000**	**1.0000**	1.81s	3,442,274
	R26+S/32 student	0.8404	0.7652	0.5802	**1.0000**	0.7344	0.7901	0.6793	1.66s	2,300,162
DeiT	ViT-B/32 student	0.9484	0.1210	0.8765	0.9861	0.9281	0.9345	0.8918	**0.97s**	3,554,884
	ViT-S/32 student	0.9531	0.1131	0.8765	**1.0000**	0.9342	0.9383	0.9027	1.00s	3,554,884
	ViT-B/16 student	0.8779	0.3492	0.6790	**1.0000**	0.8088	0.8395	0.7532	1.53s	3,112,516
	ViT-S/16 student	0.9249	0.2402	0.8025	**1.0000**	0.8904	0.9012	0.8460	1.12s	3,112,516
	R26+S/32 student	0.9577	0.0981	0.9012	0.9865	0.9419	0.9468	0.9111	1.29s	3,554,884

Table 12. Results of the CNNs in experiment 6.

	Models	Acc	Loss	Recall	Precision	F1-Score	AUC	MCC	Inference time	Total of params
Main architectures	AlexNet	0.9858	0.0323	0.9778	0.9925	0.9851	0.9855	0.9717	**1.04s**	46,760,706
	ResNet50	**1.0000**	0.0104	**1.0000**	**1.0000**	**1.0000**	**1.0000**	**1.0000**	2.63s	23,591,810
	VGG16	**1.0000**	**0.0016**	**1.0000**	**1.0000**	**1.0000**	**1.0000**	**1.0000**	5.58s	134,268,738
	InceptionV3	**1.0000**	0.0271	**1.0000**	**1.0000**	**1.0000**	**1.0000**	**1.0000**	3.68s	21,806,882
	ResNet50+MobileNetV2	0.9965	0.0216	**1.0000**	0.9926	0.9963	0.9966	0.9929	4.27s	25,852,354
	Hybrid model	0.9787	0.3919	**1.0000**	0.9574	0.9783	0.9796	0.9583	2.72s	1,930,966
Knowledge Distillation	AlexNet student	0.9645	0.1238	0.9926	0.9371	0.9640	0.9657	0.9306	2.64s	1,254,194
	ResNet50 student	0.9787	0.0611	**1.0000**	0.9574	0.9783	0.9796	0.9583	1.74s	1,254,194
	VGG16 student	0.9858	0.3479	**1.0000**	0.9712	0.9854	0.9864	0.9720	3.20s	1,254,194
	InceptionV3 student	0.9752	0.0860	**1.0000**	0.9507	0.9747	0.9762	0.9515	1.76s	1,254,194
	ResNet50+MobileNetV2 student	0.9007	0.2484	**1.0000**	0.8282	0.9060	0.9048	0.8188	2.00s	1,254,194

Table 13. Results of the ViTs in experiment 6.

	Models	Acc	Loss	Recall	Precision	F1-Score	AUC	MCC	Inference time	Total of params
Main architectures	ViT-B/32	**1.0000**	0.0025	**1.0000**	**1.0000**	**1.0000**	**1.0000**	**1.0000**	3.46 s	87,417,602
	ViT-S/32	**1.0000**	0.0048	**1.0000**	**1.0000**	**1.0000**	**1.0000**	**1.0000**	4.00 s	22,475,138
	ViT-B/16	**1.0000**	0.0006	**1.0000**	**1.0000**	**1.0000**	**1.0000**	**1.0000**	7.26 s	86,417,594
	ViT-S/16	**1.0000**	0.0013	**1.0000**	**1.0000**	**1.0000**	**1.0000**	**1.0000**	4.80 s	21,590,402
	R26+S/32	0.9965	0.0195	**1.0000**	0.9926	0.9963	0.9966	0.9929	6.74 s	36,028,098
Knowledge Distillation	ViT-B/32 student	0.9681	0.1090	0.93333	1.0000	0.9655	0.9667	0.9378	1.45 s	2,300,162
	ViT-S/32 student	0.9787	0.0897	0.9926	0.9640	0.9781	0.9793	0.9578	1.44 s	2,300,162
	ViT-B/16 student	0.9716	0.0937	0.9778	0.9635	0.9706	0.9719	0.9433	1.94 s	3,442,274
	ViT-S/16 student	0.9929	0.0215	0.9926	0.9926	0.9929	0.9929	0.9858	1.44 s	3,442,274
	R26+S/32 student	0.9823	0.0506	0.9630	**1.0000**	0.9811	0.9815	0.9650	2.30 s	2,300,162
DeiT	ViT-B/32 student	0.9681	0.1108	0.9481	0.9846	0.9660	0.9673	0.9366	1.46 s	3,554,884
	ViT-S/32 student	0.9397	0.1967	0.8889	0.9836	0.9339	0.9376	0.8825	1.46 s	3,554,884
	ViT-B/16 student	0.9255	0.2187	0.8741	0.9672	0.9183	0.9234	0.8539	1.68 s	3,112,516
	ViT-S/16 student	0.9007	0.3263	0.8222	0.9652	0.8880	0.8975	0.8081	**1.20 s**	3,112,516
	R26+S/32 student	0.9327	0.1409	0.9037	0.9683	0.9349	0.9382	0.8807	1.49 s	3,554,884

4   Discussion

This study serves as a reference for hospitals and gastroenterology doctors to automate the process of diagnosing patients with inflammatory bowel disease. The AlexNet network in experiment 1 was the algorithm that achieved the shortest inference time (0.75 s), with a good performance: 100% Accuracy, 100% Recall, 100% Precision, 100% F1-Score, 100% AUC, 100% F1-Score and 0.0031 Loss. However, due to the self-attention mechanism and in the performance improvement that ViTs demonstrate over CNNs [22], ViT-S/16 with knowledge distillation could be a very promising alternative with: 99.06% accuracy, 97.53% recall, 100% precision, 98.75% f1-Score, 98.77% AUC, 98.02% MCC, 0.0147 Loss and took 0.91 s to infer the test results.

The models used in this study prove to be accurate in the diagnosis of inflammatory bowel disease, and through the analysis of the AUC and MCC metrics, the models show a good capacity in distinguish the classes, and the number of false positives and false negatives is very close to zero or even manages to be zero, respectively. The results obtained are superior to those presented by the authors [35] who used a ResNetXt-101 network. Furthermore, if we compare with the results presented in the same study corresponding to the accuracy of clinicians in diagnosing the disease, the accuracy of the models used manages to be much higher. However, the number of images used by the authors was higher than the one used in this study, as well as the images were collected using the same endoscopic tool.

The authors [30] in their study proposed a hybrid model, where they combine a CNN with an LSTM to perform the confirmation of mucosal healing in Crohn's disease. The best result achieved by the authors was 95.3% accuracy, 92.78% sensitivity, 94.6% specificity and an AUCROC of 0.98. In this paper, a hybrid model is also presented, where two more LSTMs were used concerning what was proposed by the authors, among other modifications. These changes represented a significant improvement in the results, with the best results being 100% accuracy, 100% recall, 100% precision and 100% AUC.

The literature has focused on the implementation of deep learning models in image processing of Crohn's disease, or ulcerative colitis disease. In this sense, the authors in the works [18,31] developed a Recurrent attention neural network in the detection of the type of lesion caused by inflammation in the mucosa of Crohn's disease. Although the study objective is different, the teacher and student models implemented in this work describe a performance improvement.

The implementation of knowledge distillation is one of the important contributions of this work so that it is possible to use these architectures in the diagnosis of IBD patients. Comparing the results between the different techniques used, applied to CNN networks and ViTs. It is concluded that knowledge distillation in ViTs allowed a greater reduction in inference time, preserving a good performance of the models. On the other hand, in ViTs it is still verified that in experiments 3 and 5 the hybrid model R26+S/32 obtained a performance improvement, using the distillation technique through the attention (DeiT) when compared with knowledge distillation. In experiment four, the same situation is verified in the ViT-S/32 model. The implementation of DeiT in this work represents a performance improvement when compared with the implementation in another study [12].

5   Conclusion

It is concluded with this work that ViT models and CNN networks show a good performance in the recognition of inflammatory bowel disease. It is considered that the knowledge distillation through attention in the ViTs models allowed in some experiments performed to obtain a lower inference time concerning the knowledge distillation through the logits. Furthermore, in some experiments, such as the R26+S/32 model in experiment 3, it was found that there was an improvement in the performance of the student model in the knowledge distillation through attention [28].

However, in this study, it is recognized as a limitation the number of images used to train the ViTs. Because these deep-learning models need a large number of images to be able to learn the information well [10]. Another limitation of this work is that the images were not collected based on the same endoscopic tool, which consequently the images are different and in the future may not demonstrate the same performance in processing colonoscopy images regarding Crohn's disease. The algorithms were not trained on images of normal bowels, because the number of images concerning normal intestines present on the CrohnIPI dataset was very small (fourteen images) which would cause an unbalance of the classes. Moreover, we could not assume the images with Mayo 0 as normal intestines, because these images are of patients with the diagnosis of ulcerative colitis inactive.

In future research, the interpretability of the deep learning models should be performed to understand which parts of the mucosa the models are using to predict inflammatory bowel disease [21]. This will ensure that the performance described by the algorithms corresponds to a correct prediction of the disease based on the inflammation caused. Another line of research lies in leveraging ordinal techniques [4,19] to predict the degree of severity according to the Mayo score of ulcerative colitis. Furthermore, it is important in the future to develop efforts to explore and implement deep learning models that process video from endoscopic examinations in real time and recognise the area affected by inflammation, predicting the type of inflammatory bowel disease.

References

1. Amorim, J.P., Domingues, I., Abreu, P.H., Santos, J.A.: Interpreting deep learning models for ordinal problems. In: European Symposium on Artificial Neural Networks (ESANN), pp. 373–378 (2018)
2. Borgli, H., Riegler, M., Thambawita, V., Jha, D., Hicks, S., Halvorsen, P.: The HyperKvasir Dataset. OSF (2019)
3. Borgli, H., et al.: HyperKvasir, a comprehensive multi-class image and video dataset for gastrointestinal endoscopy. Sci. Data **7**, 283 (2020)
4. Cardoso, J.S., Sousa, R., Domingues, I.: Ordinal data classification using kernel discriminant analysis: A comparison of three approaches. In: 11th International Conference on Machine Learning and Applications, vol. 1, pp. 473–477 (2012)
5. Chicco, D., Jurman, G.: The advantages of the Matthews correlation coefficient (MCC) over F1 score and accuracy in binary classification evaluation. BMC Genom. **21**, 6 (2020)
6. Chierici, M., et al.: Automatically detecting Crohn's disease and Ulcerative Colitis from endoscopic imaging. BMC Med. Inform. Decis. Mak. **22**, 300 (2022)
7. CrohnIPI. https://crohnipi.ls2n.fr/en/crohn-ipi-project/ (Accessed 21 Feb 2023)
8. Doença inflamatória do intestino — CUF. https://www.cuf.pt/saude-a-z/doenca-inflamatoria-do-intestino (Accessed 2 Nov 2022)
9. Das, D., Massa, H., Kulkarni, A., Rekatsinas, T.: An Empirical Analysis of the Impact of Data Augmentation on Knowledge Distillation (2020), arXiv Version: 2
10. Dosovitskiy, A., et al.: An image is worth 16x16 Words: transformers for image recognition at scale. In: International Conference on Learning Representations (2020)
11. Flynn, S., Eisenstein, S.: Inflammatory bowel disease presentation and diagnosis. Surg. Clin. North Am. **99**(6), 1051–1062 (2019)
12. Galdran, A., Carneiro, G., Ballester, M.A.G.: Convolutional Nets Versus Vision Transformers for Diabetic Foot Ulcer Classification. Diabetic Foot Ulcers Grand Challenge (2022)
13. Gamage, C., Wijesinghe, I., Chitraranjan, C., Perera, I.: GI-Net: anomalies classification in gastrointestinal tract through endoscopic imagery with deep learning. In: Moratuwa Engineering Research Conference (MERCon), pp. 66–71. IEEE, Moratuwa, Sri Lanka (Jul 2019)
14. Gou, J., Yu, B., Maybank, S.J., Tao, D.: Knowledge distillation: a survey. Int. J. Comput. Vision **129**(6), 1789–1819 (2021)
15. H. Kassani, S., Hosseinzadeh Kassani, P., Wesolowski, M., Schneider, K., Deters, R.: Classification of histopathological biopsy images using ensemble of deep learning networks. arXiv preprint (2019)
16. Khan, M.N., Hasan, M.A., Anwar, S.: Improving the robustness of object detection through a multi-camera-based fusion algorithm using fuzzy logic. Front. Artifi. Intell. **4**, 638951 (2021)

17. Li, W., Shao, S., Liu, W., Qiu, Z., Zhu, Z., Huan, W.: What role does data augmentation play in knowledge distillation? In: Computer Vision - ACCV 2022, LNCS. vol. 13842, pp. 507–525. Springer Nature Switzerland (2023). https://doi.org/10.1007/978-3-031-26284-5_31

18. Maissin, A., et al.: Multi-expert annotation of Crohn's disease images of the small bowel for automatic detection using a convolutional recurrent attention neural network. Endoscopy Int Open **09**, E1136–E1144 (2021)

19. Marques, F., Duarte, H., Santos, J.A., Domingues, I., Amorim, J.P., Abreu, P.H.: An iterative oversampling approach for ordinal classification. In: 34th ACM/SIGAPP Symposium on Applied Computing, pp. 771–774 (2019)

20. Maurício, J., Domingues, I.: Deep Neural Networks to distinguish between Crohn's disease and Ulcerative colitis. In: 11th Iberian Conference on Pattern Recognition and Image Analysis (IbPRIA) (2023)

21. Maurício, J., Domingues, I.: Interpretability of deep neural networks to diagnose inflammatory bowel disease. In: 29th Edition of the Portuguese Conference on Pattern Recognition (2023) (to appear)

22. Maurício, J., Domingues, I., Bernardino, J.: Comparing vision transformers and convolutional neural networks for image classification: a literature review. Appli. Sci. **13**(9) (2023)

23. Polat, G., Kani, H.T., Ergenc, I., Alahdab, Y.O., Temizel, A., Atug, O.: Labeled Images for Ulcerative Colitis (LIMUC) Dataset (2022)

24. Sairenji, T., Collins, K.L., Evans, D.V.: An update on inflammatory bowel disease. Primary Care: Clin. Office Pract. **44**, 673–692 (2017)

25. Seyedian, S.S., Nokhostin, F., Malamir, M.D.: A review of the diagnosis, prevention, and treatment methods of inflammatory bowel disease. J. Med. Life **12**, 113–122 (2019)

26. Shahzadi, I., Tang, T.B., Meriadeau, F., Quyyum, A.: CNN-LSTM: cascaded framework for brain tumour classification. In: IEEE-EMBS Conference on Biomedical Engineering and Sciences (IECBES), pp. 633–637 (2018)

27. Steiner, A., Kolesnikov, A., Zhai, X., Wightman, R., Uszkoreit, J., Beyer, L.: How to train your ViT? Data, Augmentation, and Regularization in Vision Transformers. arXiv Version: 2 (2021)

28. Touvron, H., Cord, M., Douze, M., Massa, F., Sablayrolles, A., Jégou, H.: Training data-efficient image transformers & distillation through attentio (2020), arXiv Version: 2

29. Turan, M., Durmus, F.: UC-NfNet: deep learning-enabled assessment of ulcerative colitis from colonoscopy images. Med. Image Anal. **82**, 102587 (2022)

30. Udristoiu, A.L., et al.: Deep learning algorithm for the confirmation of mucosal healing in crohn's disease, based on confocal laser endomicroscopy images. J. Gastrointestinal Liver Dis. **30**, 59–65 (2021)

31. Vallée, R., Coutrot, A., Normand, N., Mouchère, H.: Accurate small bowel lesions detection in wireless capsule endoscopy images using deep recurrent attention neural network. In: IEEE 21st Int WS on Multimedia Signal Proc (MMSP) (2019)

32. Vallée, R., Coutrot, A., Normand, N., Mouchère, H.: Influence of expertise on human and machine visual attention in a medical image classification task. In: European Conference on Visual Perception (2021)

33. Vallée, R., Maissin, A., Coutrot, A., Mouchère, H., Bourreille, A., Normand, N.: CrohnIPI: an endoscopic image database for the evaluation of automatic Crohn's disease lesions recognition algorithms. In: Medical Imaging: Biomedical Applications in Molecular, Structural, and Functional Imaging, p. 61. SPIE (2020)

34. Vankdothu, R., Hameed, M.A., Fatima, H.: A brain tumor identification and classification using deep learning based on CNN-LSTM method. Comput. Electr. Eng. **101**, 107960 (2022)
35. Wang, L., et al.: Development of a convolutional neural network-based colonoscopy image assessment model for differentiating crohn's disease and ulcerative colitis. Front. Med. **9**, 789862 (2022)

Analysis and Impact of Training Set Size in Cross-Subject Human Activity Recognition

Miguel Matey-Sanz[1]([⊠]) [iD], Joaquín Torres-Sospedra[2] [iD],
Alberto González-Pérez[1] [iD], Sven Casteleyn[1] [iD], and Carlos Granell[1] [iD]

[1] Institute of New Imaging Technologies, Universitat Jaume I, 12071 Castellón, Spain
{matey,alberto.gonzalez,sven.casteleyn,carlos.granell}@uji.es
[2] ALGORITMI Research Centre, University of Minho, 4800-058 Guimarães, Portugal
info@jtorr.es

Abstract. The ubiquity of consumer devices with sensing and computational capabilities, such as smartphones and smartwatches, has increased interest in their use in human activity recognition for healthcare monitoring applications, among others. When developing such a system, researchers rely on input data to train recognition models. In the absence of openly available datasets that meet the model requirements, researchers face a hard and time-consuming process to decide which sensing device to use or how much data needs to be collected. In this paper, we explore the effect of the amount of training data on the performance (i.e., classification accuracy and activity-wise F1-scores) of a CNN model by performing an incremental cross-subject evaluation using data collected from a consumer smartphone and smartwatch. Systematically studying the incremental inclusion of subject data from a set of 22 training subjects, the results show that the model's performance initially improves significantly with each addition, yet this improvement slows down the larger the number of included subjects. We compare the performance of models based on smartphone and smartwatch data. The latter option is significantly better with smaller sizes of training data, while the former outperforms with larger amounts of training data. In addition, gait-related activities show significantly better results with smartphone-collected data, while non-gait-related activities, such as *standing up* or *sitting down*, were better recognized with smartwatch-collected data.

Keywords: Human activity recognition · Training set size · Cross-subject evaluation · Smartphone · Smartwatch

1 Introduction

Human activity recognition (HAR) aims to determine human behaviour, actions, and activities from physical signals such as video- or inertial-based (i.e., accelerometers and gyroscopes) data. HAR has multiple application fields, from

© Springer Nature Switzerland AG 2024
V. Vasconcelos et al. (Eds.): CIARP 2023, LNCS 14469, pp. 391–405, 2024.
https://doi.org/10.1007/978-3-031-49018-7_28

security surveillance systems for abnormal or violent behaviour [17] to healthcare monitoring applications for fall risk detection [12]. In recent years, research interest in HAR has increased, driven by the availability and ubiquity of smartphone and smartwatch devices, equipped with inertial and other sensors [6].

HAR applications are traditionally based on classical machine learning techniques (e.g., decision trees, support vector machines, regression models, neural networks, etc.) [10] or deep learning techniques (e.g., convolutional neural networks, recurrent neural networks, etc.) [23]. To properly differentiate between target activities, these algorithms require large amounts of training data, which is critical for deep learning methods. Moreover, it is broadly accepted in the research community that the larger the amount of training data, the better the model will perform (i.e., "the more data the better").

Researchers have relied on public datasets for training HAR applications [5]. However, this strategy may have shortcomings. The available datasets may not be representative of the activities targeted by specific studies, data were collected using outdated sensors or devices, or the collection devices may have been placed in locations that are not suitable for the purpose of the study. Any of these factors force researchers to collect their own dataset, which is reportedly the most difficult and time-consuming task in HAR (i.e., recruitment of participants, set-up of devices and environment, segmentation and labelling of data, etc.) [3]. Researchers hereby face the challenge of deciding how much data they will need to meet the requirements of their study. By adhering to the rule of "the more data the better", they could end up collecting more data than necessary, thus needlessly investing resources in data collection.

In this paper, we aim to determine the appropriate tradeoff between recognition accuracy and training dataset size (i.e., number of subjects). To do so, we explore the existence of statistically significant improvements in the classification accuracy of a convolutional-based HAR model using an incremental cross-subject evaluation for five target activities (i.e., *seated, standing up, walking, turning* and *sitting down*). Performance improvements regarding the recognition of these individual target activities in terms of F1-score are also explored. Two independent devices, a smartphone and a smartwatch respectively placed in the pocket and the wrist, were used to collect data.

2   Related Work

Previous works have already explored the effect of the size of the training set on the performance of machine learning models in different domains, such as health [18] or remote sensing [20]. In the context of HAR, Yazdansepas et al. [25] compared the accuracy of six different classifiers using data from an ActiGraph GT3X+ placed in the hip and showed how the accuracy decreased when using the 50%, 25% and 10% of data from the original training set. A similar approach was used in [21], where three Colibri inertial measurement units placed in the arm, the chest and the ankle were used, showing that the accuracy quickly increases when the amount of training data is low, but it stabilizes with higher amounts of data. Equivalent results were obtained in [2], where the 3D accelerometer data from a smartphone placed in the waist was used.

One of the main drawbacks of the abovementioned works is that they apply random split (i.e., hold-out) to the dataset, being different data from the same subject in the training and test sets. This may generate data dependency between both sets and possible data leakage that may drive over-optimistic performance. Instead, a cross-subject evaluation is the recommended approach for evaluating HAR systems [7], where a user contributes data exclusively to the training set or to the testing set. This approach was used in [11], where half of the subjects were used to train the model and the rest to evaluate it using Kinect data. Training subjects were gradually added to the training set, showing a steady increase in accuracy until an apparent stabilization for 8–10 subjects. Yet, this reported stabilization could not be further confirmed, as no further data was available beyond 10 subjects. We also take a cross-subject evaluation approach, but expanding the number of subjects used by [11]. Compared to our work, previous works only analyzed the overall classification accuracy of the HAR model, ignoring how data addition affected activity-wise classification performance.

Table 1 compares the aforementioned related literature with the work presented in this paper. The main contribution of the present work is the statistical comparison of the impact of the training set size (in terms of number of subjects) in HAR with smartphones and smartwatches, analyzing the overall and activity-wise classification accuracy for both devices.

Table 1. Comparison with related works.

	Yazdansepas et al. [25]	Saez et al. [21]	Chen et al. [2]	Leightley et al. [11]	This work
Device	ActiGraph GT3X+	3 × Colibri	Smartphone	Kinect	Smartphone vs. Smartwatch
Body location	Hip	Arm, Chest Ankle	Waist	–	Pocket Wrist
Subjects	77 (18–64 yr)	8 (24–32 yr)	30 (19–48 yr) 31 (22–79 yr)	20	23 (23–66 yr)
Activities	25	12	6	10	5
Models*	SVM, RF, DT, NB, kNN, MLP	RF, MLP, LDA	CNN, LSTM	SVM, RF	CNN
Evaluation approach	Hold-out	Hold-out	Hold-out	Incremental cross-subject[1]	Incremental cross-subject[2]
Metrics	Overall accuracy	Overall accuracy	Overall accuracy	Overall accuracy Training time	Overall accuracy Activity-wise Training time
Statistical analyses	No	No	No	No	Yes

* SVM: Support Vector Machine; RF: Random Forest; DT: Decision Tree; NB: Naive Bayes; kNN: k-Nearest Neighbors; MLP: Multilayer Perceptron; LDA: Linear Discriminant Analysis; CNN: Convolutional Neural Network; LSTM: Long Short-Term Memory.
[1] Evaluation using up to 10 subjects for training and 10 for testing.
[2] Evaluation using Leaving-One-Subject-Out (up to 22 subjects for training and 1 for testing). See Sect. 3.4 for a detailed description.

3   Materials and Methods

This section contains a description of the data collection process, including information about the participants, used devices, processing techniques, and the data labelling procedure. We then describe the model architecture and the evaluation methodology employed to analyse performance improvements by using more training data. The collected datasets and code used for preprocessing, model training, and analytical processing are available as a reproducible package (including documentation, software dependencies, and versions) on Zenodo [14].

3.1   Data Collection

The data collection in this study was approved by the ethics committee of the Universitat Jaume I (reference No. CD/88/2022). We recruited 23 individuals (age range [23, 66]; mean age 44.3 ± 14.3; 13 male and 10 female) and asked them to participate in the data collection for this study. After obtaining their written informed consent, the participants were provided with a TicWatch Pro 3 GPS (WH12018) smartwatch and a Xiaomi Poco X3 Pro (M2102J20SG) smartphone, both equipped with an STMicroelectronics LSM6DSO IMU sensor[1] featuring a 3D accelerometer and a 3D gyroscope. The devices had a data collection application installed (an upgraded version of the one described in [15]) and configured to gather accelerometer and gyroscope samples at 100 Hz.

Participants were instructed to wear the smartwatch on the left wrist and the smartphone in the left pocket. No further instructions were given regarding the smartphone's orientation in the pocket. Then, the participants were told to execute 10 times a specific sequence of five unique activities: *seated* (on a chair), *standing up*, *walking* (three meters), *turning* (180°), *walking* (three meters), *turning* (180°) and *sitting down* (on the chair they stood up from). While the participants were performing the data collection, they were video-recorded with another smartphone to determine the ground truth.

Since three devices (i.e., smartwatch, smartphone, and video-recording smartphone) were involved in the data collection process, an internal clock synchronization mechanism had to be implemented [4]. We chose an NTP-based synchronization solution due to its simplicity and the only requirement of having an Internet connection [16]. Therefore, clock synchronization was implemented in the data collection application of both devices and the video recorder application using a variation of the solution described in [22] and Google's NTP server[2].

3.2   Data Processing

The 3D accelerometer and 3D gyroscope collected samples were processed offline. First, a temporal alignment process was executed using the collected data from each device to pair the accelerometer and gyroscope samples. Then, a Min-Max

---

[1]   https://www.st.com/en/mems-and-sensors/lsm6dso.html.
[2]   https://time.google.com.

scaling process was database-wise applied in order to rescale the samples into the $[-1,1]$ range. The Min-Max scaling to a certain range is defined as:

$$v' = \frac{v - min}{max - min} * (new\_max - new\_min) + new\_min, \tag{1}$$

where v is the original value, min and max are determined by the sensor's range (i.e., minimum and maximum values that the sensor is able to report), and new_min and new_max are -1 and 1.

Recorded videos of the participants were then manually analyzed to determine the start and end of each activity (i.e., bounds) and label the collected sensor data accordingly (i.e., ground truth). More concretely, Fig. 1 depicts the bounds extracted from the recorded videos and their associated activities on smartphone and smartwatch accelerometer samples. Table 2 shows the number of labelled samples collected for each activity and device.

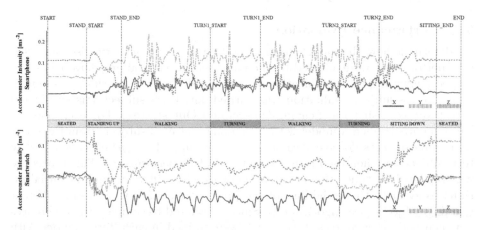

Fig. 1. Bounds and their associated activities of accelerometer data from the smartphone (top) and the smartwatch (bottom).

Table 2. Breakdown of labelled collected data.

Device	seated	standing up	walking	turning	sitting down	Total
Smartphone	32764	27303	115069	52209	31868	259213
Smartwatch	32025	27765	117126	53180	32457	262553

Finally, the sliding window partitioning technique was applied to the labelled data, which is used to build overlapping sets of raw data for further processing (e.g., feature extraction as input for a neural network). The window size was set to 50 individual measurements (≈ 0.5 s at 100 Hz sampling) due to the fact that short windows can obtain precise results in activity recognition [1,9], and 50% overlap has already shown successful results [23].

3.3 Model Architecture

The chosen architecture was the *Convolutional Neural Network* (CNN) due to its ability to identify temporal dependency on the raw (i.e., no need for feature extraction) accelerometer and gyroscope data in activity recognition applications while obtaining better results than recurrent models [23]. Therefore, following a trial and error approach, our CNN configuration was a low-weight network composed of a 1D convolutional layer of 64 filters of 1×10 size, a flattening layer, and a dense layer of 256 ReLU-activated neurons connected to a final 5 Softmax-activated neurons layer for activity classification. The *categorical crossentropy* loss function and the *Adam* optimizer were used during training, which was carried out during 100 epochs using a batch size of 20 instances.

3.4 Performance Evaluation

To determine how the model's overall performance and activity-wise classification performance were affected by the increasing size of training data, measured respectively using the classification *accuracy* and *F1-score* metrics [8], an incremental cross-subject evaluation was carried out. Each subject was considered a test subject using the Leaving-One-Subject-Out strategy. For each one, we subsequently trained a model with n randomly selected subjects from the set of remaining subjects, for every value of n in the range $[1, \ldots, nSubjects - 1]$. The test subject was then used to measure the model's performance. This procedure was repeated 10 times (for every value of n) with different random initializations to evaluate the variability of the proposed classifier when using different training subjects (see Algorithm 1). Since we wanted to compare the differences between both data sources, two models were trained in each iteration: one with smartphone data and the other with smartwatch data. In total, 5060 models were trained for each data source.

Algorithm 1. Incremental cross-subject evaluation

$nSubjects \leftarrow length(subjects)$
for $evalSubject$ in $subjects$ **do**
    for $n \leftarrow 1$ to $nSubjects - 1$ **do**
        for $repetition \leftarrow 1$ to 10 **do**
            $trainSubjects \leftarrow getRandomSubjects(n)$ ▷ $evalSubject$ never included
            $model \leftarrow trainModel(trainSubjects)$ ▷ For each data source
            $metrics \leftarrow evaluate(model, evalSubject)$ ▷ For each data source

---

Based on the performance metrics, descriptive statistics (i.e., mean, standard deviation, range) were computed to obtain insights about the model's classification accuracy and activity-wise F1-scores evolution, followed by Mann-Whitney

U statistical tests [13] to find significant differences between the models. The models were built and trained in Python 3.9 using TensorFlow 2.10 on an Intel Core i7-8700 and 8GB RAM Windows machine. Statistical analyses were executed using Pingouin 0.5.3 [24].

4   Results

In this section, we present and analyze the performance metrics obtained in the incremental cross-subject evaluation process described in Sect. 3.4.

Figure 2 depicts the mean accuracy, standard deviation and range (i.e., minimum and maximum accuracies) of the trained models for each value of n (i.e., subject's data used for training) using the collected data from the smartphone (sp) and smartwatch (sw). The training time of the models (i.e., mean, standard deviation and range) is also included.

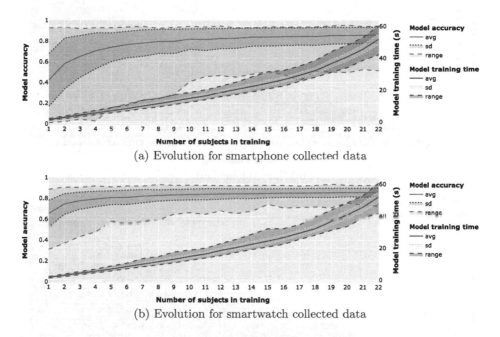

(a) Evolution for smartphone collected data

(b) Evolution for smartwatch collected data

Fig. 2. Model accuracy evolution and training time over the addition of training subjects' data collected from (a) the smartphone and (b) smartwatch.

Both figures show an ascending evolution in the classification accuracy over the increase of n. The sp models show an abrupt evolution mainly caused by the poor accuracy on $n < 4$, in contrast with the sw models, where a smoother evolution can be seen. Notwithstanding, the mean accuracy of the models seems to stabilize when $n > 8$, while the standard deviation appears to slightly improve.

The training time is linearly increased up to $n \leq 18$ when it starts to show a smooth exponential increase due to memory limitations.

To find statistical evidence of the improvement in the accuracies with higher n's, the non-parametric Mann-Whitney U-test (accuracies did not follow a normal distribution) was executed for each pair of n's. Figure 3 shows the results of this test for the sp and sw models, where as n increases, the classification accuracy improves, but not always significantly. For instance, in both types of models, for $n_1 \in [4, 10]$ other models trained with $n_2 = n_1 + 1$ subjects do not significantly improve the accuracy. Then, $n_1 > 10$ has an interesting pattern: significant improvements for consecutive n_1 are observed at a certain n_2. For example, in sp $n_1 \in [10, 11]$ are improved when $n_2 = 14$, or $n_1 \in [16, 18]$ are improved when $n_2 \in [21, 22]$. Similar results can be observed for the sw models.

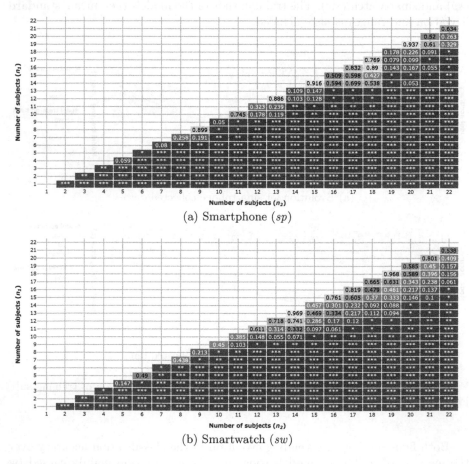

(a) Smartphone (sp)

(b) Smartwatch (sw)

Fig. 3. Mann-Whitney U-test results over the accuracy of the models trained with (a) smartphone and (b) smartwatch data. (*): p-value $< .05$, (**): p-value $< .01$, (***): p-value $< .001$. Results compare each n_1 with $n_2 > n_1$. Colour (darker with low p-values): green if $acc_{n_1} > acc_{n_2}$, red otherwise.

Regarding the activity-wise performance, Fig. 4 depicts the F1-score evolution for each of the five target activities for the *sp* and *sw* models. The observed trends are similar to those for overall classification accuracy: abrupt improvements occur for every activity in the *sp* models on low values of n, while smooth improvements are present in the *sw* models. It is noticeable that for the *seated*, *standing up* and *sitting down* activities, some models obtained very poor results (see *sp* ranges in Figs. 4a, 4b and 4e) with F1-scores 0 or close to 0.

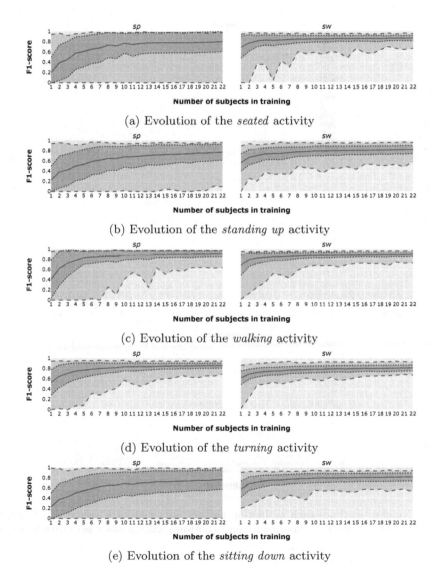

(a) Evolution of the *seated* activity

(b) Evolution of the *standing up* activity

(c) Evolution of the *walking* activity

(d) Evolution of the *turning* activity

(e) Evolution of the *sitting down* activity

Fig. 4. F1-score evolution of the five target activities for the models trained with smartphone (left) and smartwatch (right) data.

Further analysis of these underperforming cases where the F1-score was 0 (incorrect recognition 100% of the time) in the *sp* models for the three afore-mentioned activities is shown in Fig. 5. It can be observed that when $n < 10$, the underperforming cases are evenly distributed among all subjects. However, when $n \geq 10$, the subject *s06* monopolizes most of these cases. More specifically, *s06* is responsible for 97%, 91% and 95% of such cases in the activities *seated*, *standing up* and *sitting down*, respectively.

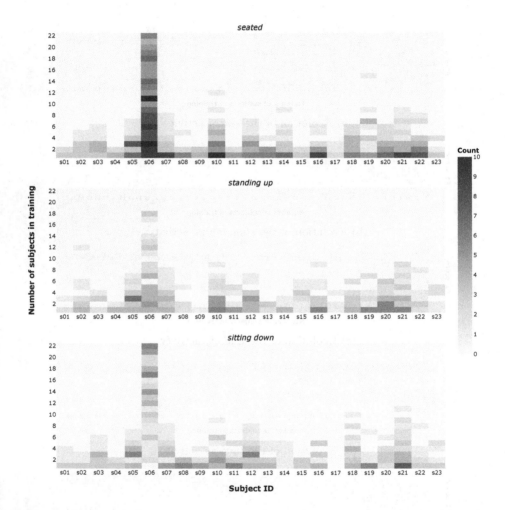

Fig. 5. Number of times each evaluation subject obtained an F1-score equal to 0 in the (top) *seated*, (middle) *standing up* and (bottom) *sitting down* activities in the *sp* models.

Visual inspection of Figs. 2 and 4 also indicate that with a reduced number of training subjects, the overall classification accuracy and the F1-score of all the activities in the *sw* models are superior to the *sp* models. When the number of training subjects increases, the overall mean accuracy appears to be similar for both models, despite a higher standard deviation and range in some of the *sp* models. For the individual activities, the *sw* models apparently perform better in the *seated, standing up* and *sitting down* activities, while the *sp* models seem superior for the *walking* and *turning* activities. The statistical analyses of Table 3 confirm the observed differences. Regarding the model classification accuracy, the *sw* models are superior for a small number of subjects (i.e., $n \leq 7$), while the *sp* models perform better for higher amounts of subjects (i.e., $n \geq 19$). Concerning the target activities, the *sw* models perform significantly better in the *seated, standing up* and *sitting down* activities no matter the n value, while the *sp* models performed better in the *walking* (for $n \geq 10$) and *turning* activities.

Table 3. Mann-Whitney U-test results comparing the mean accuracy and each activity F1-scores of the trained models with smartphone — M(sp) — and smartwatch — M(sw) — data for each number of training subjects (n). Differences are significant on *p-value* $< .05$.

n	Model accuracy			seated			standing up			walking			turning			sitting down		
	M(sp)	M(sw)	p-value	M(sp)	M(sw)	p-value	M(sp)	M(sw)	p-value	M(sp)	M(sw)	p-value	M(sp)	M(sw)	p-value	M(sp)	M(sw)	p-value
1	0.389	**0.670**	<.001	0.000	**0.772**	<.001	0.116	**0.588**	<.001	0.439	**0.740**	<.001	0.553	**0.642**	<.001	0.129	**0.579**	<.001
2	0.657	**0.766**	<.001	0.397	**0.831**	<.001	0.407	**0.690**	<.001	0.781	**0.836**	<.001	**0.708**	0.702	.677	0.315	**0.705**	<.001
3	0.701	**0.794**	<.001	0.561	**0.843**	<.001	0.454	**0.725**	<.001	0.811	**0.857**	<.001	**0.766**	0.724	.023	0.479	**0.731**	<.001
4	0.753	**0.812**	<.001	0.706	**0.862**	<.001	0.564	**0.760**	<.001	0.847	0.861	.057	**0.802**	0.745	<.001	0.599	**0.766**	<.001
5	0.772	**0.819**	<.001	0.757	**0.847**	<.001	0.618	**0.773**	<.001	0.865	0.876	.275	**0.812**	0.745	<.001	0.622	**0.778**	<.001
6	0.804	**0.824**	.012	0.784	**0.857**	<.001	0.695	**0.775**	<.001	0.877	0.878	.870	**0.823**	0.755	<.001	0.676	**0.778**	<.001
7	0.828	0.833	.063	0.798	**0.864**	<.001	0.704	**0.787**	<.001	0.886	0.884	.564	**0.835**	0.772	<.001	0.705	**0.796**	<.001
8	0.831	0.841	.253	0.824	**0.861**	<.001	0.737	**0.800**	<.001	0.892	0.885	.126	**0.831**	0.774	<.001	0.721	**0.807**	<.001
9	0.834	**0.843**	.041	0.805	**0.859**	<.001	0.742	**0.800**	<.001	0.895	0.893	.565	**0.836**	0.779	<.001	0.741	**0.815**	<.001
10	0.848	0.849	.736	0.820	**0.870**	<.001	0.747	**0.815**	<.001	**0.907**	0.896	.002	**0.851**	0.787	<.001	0.745	**0.808**	<.001
11	0.849	0.855	.502	0.818	**0.868**	<.001	0.754	**0.821**	<.001	**0.908**	0.897	.010	**0.850**	0.787	<.001	0.746	**0.824**	<.001
12	0.855	0.855	.905	0.831	**0.871**	<.001	0.758	**0.820**	<.001	**0.910**	0.901	.005	**0.856**	0.788	<.001	0.777	**0.822**	<.001
13	0.855	0.859	.918	0.836	**0.872**	<.001	0.764	**0.823**	<.001	**0.913**	0.900	.002	**0.855**	0.800	<.001	0.778	**0.820**	<.001
14	0.866	0.853	.070	0.836	**0.873**	<.001	0.786	**0.825**	<.001	**0.919**	0.902	<.001	**0.853**	0.792	<.001	0.796	**0.818**	.001
15	0.861	0.859	.352	0.839	**0.872**	<.001	0.777	**0.831**	<.001	**0.920**	0.899	<.001	**0.853**	0.800	<.001	0.784	**0.828**	<.001
16	0.869	0.857	.085	0.838	**0.871**	<.001	0.789	**0.824**	<.001	**0.919**	0.903	<.001	**0.857**	0.795	<.001	0.796	**0.825**	<.001
17	0.869	0.861	.147	0.835	**0.873**	<.001	0.794	**0.831**	<.001	**0.916**	0.905	<.001	**0.862**	0.804	<.001	0.793	**0.820**	.001
18	0.865	0.861	.334	0.845	**0.871**	<.001	0.790	**0.834**	<.001	**0.922**	0.905	<.001	**0.859**	0.801	<.001	0.786	**0.830**	<.001
19	**0.879**	0.862	.004	0.845	**0.875**	<.001	0.806	**0.830**	.001	**0.928**	0.904	<.001	**0.864**	0.796	<.001	0.802	**0.827**	.012
20	**0.877**	0.865	.015	0.843	**0.878**	<.001	0.808	**0.837**	<.001	**0.924**	0.904	<.001	**0.863**	0.802	<.001	0.810	**0.825**	.011
21	**0.878**	0.866	.009	0.849	**0.871**	<.001	0.808	**0.835**	.003	**0.926**	0.905	<.001	**0.863**	0.800	<.001	0.814	**0.834**	.007
22	**0.879**	0.867	.005	0.854	**0.874**	.002	0.816	**0.843**	.002	**0.927**	0.906	<.001	**0.864**	0.807	<.001	0.813	**0.838**	.010

5   Discussion

This study constitutes a medium-scale analysis of the effects of training set size compared to other works that included a significantly larger number of subjects [2,25] or activities [11,21,25]. In contrast to related works, we here compared the effect of the size of the training set using two different devices. Therefore, we only used one type of model (i.e., CNN) to exclude potential effects caused by the model characteristics. In addition, the compared devices (i.e., smartphone in the pocket and smartwatch in the wrist) were selected based on their feasibility of use in real-life setups, while other works employed specific devices [21, 25] or unnatural placements for the devices [2].

The presented results show that adding one new training subject is only significantly reflected in the overall classification accuracy of the CNN model employed in the study when $n < 4$. Significant improvements arise when adding several new subjects (from ≈ 3 if $n \leq 10$ to ≈ 5 if $n > 14$). This suggests that the larger the number of training subjects, the more additional training subjects are needed to achieve significant improvements. These results of the statistical analyses are in line with previous results obtained in studies that used hold-out evaluations [2, 21, 25] and one that used cross-subject evaluations [11]. More concretely, the latter showed a similar trend as the one reported in our work, i.e., quick increase with few subjects and signs of stabilization after 8 subjects, but only explored the evolution up to 10 subjects without statistical analysis.

In addition to the slowing performance gains observed, there is a disproportionate increase in the resources needed to collect the data (e.g. participant recruitment, data collection, processing, etc.) and an increase in the computational effort to train the model with more subjects. This raises the question of how much data to collect, or in other words, when to stop collecting more data. Ideally, researchers would perform the analyses described in Sect. 3.4 (i.e., consider each subject a test subject, train and evaluate a model 10 times for all subsets of n subjects) to check the performance evolution based on their own data. If the obtained performance metrics for a certain n are satisfactory, then no more data is necessary. Otherwise, the required number of additional subjects to obtain significant improvements can be estimated by observing the trends shown in Fig. 3-like plots, and researchers can determine if the effort is worth the benefit.

Regarding the type of device for collecting sensor data, a previous study showed that models trained with smartwatch data obtained better overall results than those trained with smartphone data [19]. In our study, we obtain similar results when the number of training subjects is small (i.e., $n \leq 7$). Nevertheless, with higher amounts of training subjects (i.e., $n \geq 19$) the *sp* models outperform the *sw* models. In addition, we analysed the activity-wise performance. Here, the *sw* models outperform *sp* models, particularly in non-gait-related activities (i.e., *seated*, *standing up* and *sitting down*), with F1-scores close to 0 in some models. Further exploration showed that a single subject was responsible for more than 90% of those F1-scores, which could possibly have been caused by alterations in the orientation and position of the smartphone (i.e., due to the particular shape and position of his/her pocket). It furthermore highlights the importance of the cross-subject evaluation to be aware of the weaknesses of the system (i.e., the model could fail against new subjects when considerable variations in test conditions are present, e.g. different phone orientation or motion due to specific clothing conditions, differences in pocket size and position, etc.). This issue is not reproduced in *sw* models, because the device position and orientation are "fixed" by design. Therefore, smartwatches are intrinsically more reliable to detect non-gait-related activities in real-world settings, due to the limited variability of their position and orientation, compared to smartphones.

When further examining the type of activities individually, the sp models outperform in gait-related activities (i.e., *walking* and *turning*). In the case of *turning* activity, sp models are better from a small number of training subjects, while for the *walking* activity, models require more training subjects ($n \geq 10$) to improve the sw models. These results can be explained due to different arm swing patterns among individuals, downgrading the performance of the sw models.

A disclaimer/limitation of this study is that the model architecture and hyperparameters were tuned in trial-error cross-validation using 14/4/5 subjects for training/validation/testing. Therefore, the chosen model hyperparameters were not specifically optimized for different training sizes (i.e., $n \neq 14$). In addition, only one type of model (CNN) has been used; future work should evaluate our method with other models.

6   Conclusion

HAR has gained interest in recent years, boosted by the proliferation of personal smart devices with embedded sensors, such as smartphones and smartwatches. However, the realisation of a HAR system often requires a time-consuming, labour-intensive data collection process (i.e., participants recruitment, environment set-up, data segmentation and labelling, etc.) to train the underlying models, and researchers usually try to obtain "as much data as possible".

In this study, we performed an incremental cross-subject evaluation where we explored statistically significant improvements in the performance of an activity classification model regarding its overall classification accuracy and activity-wise F1-scores while incrementing the amount of data used for training. This evaluation was executed using data collected from a smartphone and a smartwatch. We demonstrated that while performance metrics greatly improved when adding the data from a single subject for a low number of subjects ($n \leq 3$), the addition of an increasing number of subjects is required to obtain improvements when the training data is larger ($n > 14$). Therefore, an increasingly expensive (in terms of resources required) data collection process is needed, involving multiple subjects, to still significantly improve model performance as the training set increases. Researchers can replicate the evaluation process described in this article to determine if sufficient data has been collected (based on the desired accuracy), and observe the trend of increasing amounts of required subjects to obtain significant improvements as a cost-benefit analysis to decide on further data collection.

We also showed that non-gait-related activities were better recognized using smartwatch data (i.e., *seated, standing up* and *sitting down*), while gait-related activities (i.e., *walking* and *turning*) were better classified with smartphone data. Significant differences were observed in the overall classification accuracy between both data sources, where the smartwatch models performed better with a low number of training subjects ($n \leq 7$) while the smartphone models were superior with larger numbers of training subjects ($n \geq 19$). Future work could explore the improvement in the overall accuracy by fusing both data sources to exploit their respective strengths and reduce their individual weaknesses.

Acknowledgments. M. Matey-Sanz and A. González-Pérez are funded by the Spanish Ministry of Universities [grants FPU19/05352 and FPU17/03832]. This study was supported by project PID2020-120250RB-I00 (SyMptOMS-ET) funded by MCIN/AEI/10.13039/501100011033.

References

1. Banos, O., et al.: Window size impact in human activity recognition. Sensors **14**(4), 6474–6499 (2014). https://doi.org/10.3390/s140406474
2. Chen, H., et al.: Assessing impacts of data volume and data set balance in using deep learning approach to human activity recognition. In: IEEE International Conference on Bioinformatics and Biomedicine, pp. 1160–1165. IEEE (2017). https://doi.org/10.1109/BIBM.2017.8217821
3. Chen, W., et al.: Sensecollect: we need efficient ways to collect on-body sensor-based human activity data! Proc. ACM Interact. Mobile Wearable Ubiquitous Technol. **5**(3), 1–27 (2021). https://doi.org/10.1145/3478119
4. Coviello, G., Avitabile, G.: Multiple synchronized inertial measurement unit sensor boards platform for activity monitoring. IEEE Sens. J. **20**(15), 8771–8777 (2020). https://doi.org/10.1109/JSEN.2020.2982744
5. De-La-Hoz-Franco, E., Ariza-Colpas, P., Quero, J.M., Espinilla, M.: Sensor-based datasets for human activity recognition-a systematic review of literature. IEEE Access **6**, 59192–59210 (2018). https://doi.org/10.1109/ACCESS.2018.2873502
6. Demrozi, F., et al.: Human activity recognition using inertial, physiological and environmental sensors: a comprehensive survey. IEEE Access **8**, 210816–210836 (2020). https://doi.org/10.1109/ACCESS.2020.3037715
7. Gholamiangonabadi, D., Kiselov, N., Grolinger, K.: Deep neural networks for human activity recognition with wearable sensors: leave-one-subject-out cross-validation for model selection. IEEE Access **8**, 133982–133994 (2020). https://doi.org/10.1109/ACCESS.2020.3010715
8. Hossin, M., Sulaiman, M.N.: A review on evaluation metrics for data classification evaluations. International journal of data mining & knowledge management process **5**(2), 1 (2015). https://doi.org/10.5121/ijdkp.2015.5201
9. Jaén-Vargas, M., et al.: Effects of sliding window variation in the performance of acceleration-based human activity recognition using deep learning models. PeerJ Comput. Sci. **8**, e1052 (2022). https://doi.org/10.7717/peerj-cs.1052
10. Lara, O.D., Labrador, M.A.: A survey on human activity recognition using wearable sensors. IEEE Commun. Surv. Tutor. **15**(3), 1192–1209 (2012). https://doi.org/10.1109/SURV.2012.110112.00192
11. Leightley, D., Darby, J., Li, B., McPhee, J.S., Yap, M.H.: Human activity recognition for physical rehabilitation. In: IEEE International Conference on Systems, Man, and Cybernetics, pp. 261–266 (2013). https://doi.org/10.1109/SMC.2013.51
12. Li, H., Shrestha, A., Heidari, H., Le Kernec, J., Fioranelli, F.: Bi-lstm network for multimodal continuous human activity recognition and fall detection. IEEE Sens. J. **20**(3), 1191–1201 (2019). https://doi.org/10.1109/JSEN.2019.2946095

13. Mann, H.B., Whitney, D.R.: On a test of whether one of two random variables is stochastically larger than the other. The annals of mathematical statistics, pp. 50–60 (1947)
14. Matey-Sanz, M.: Reproducible Package for Analysis and Impact of Training Set Size in Cross-Subject Human Activity Recognition (Jul 2023). https://doi.org/10.5281/zenodo.8163542
15. Matey-Sanz, M., et al.: Instrumented timed up and go test using inertial sensors from consumer wearable devices. In: 20th International Conference on Artificial Intelligence in Medical, Proceedings, pp. 144–154. Springer (2022). https://doi.org/10.1007/978-3-031-09342-5_14
16. Mills, D.L.: Internet time synchronization: the network time protocol. IEEE Trans. Commun. **39**(10), 1482–1493 (1991). https://doi.org/10.1109/26.103043
17. Moënne-Loccoz, N., Brémond, F., Thonnat, M.: Recurrent bayesian network for the recognition of human behaviors from video. In: Crowley, J.L., Piater, J.H., Vincze, M., Paletta, L. (eds.) ICVS 2003. LNCS, vol. 2626, pp. 68–77. Springer, Heidelberg (2003). https://doi.org/10.1007/3-540-36592-3_7
18. Narayana, P.A., et al.: Deep-learning-based neural tissue segmentation of mri in multiple sclerosis: effect of training set size. J. Magn. Reson. Imaging **51**(5), 1487–1496 (2020). https://doi.org/10.1002/jmri.26959
19. Oluwalade., B., et al.: Human activity recognition using deep learning models on smartphones and smartwatches sensor data. In: Proc. of the 14th International Joint Conference on Biomedical Engineering Systems and Technologies, HEALTHINF, pp. 645–650. INSTICC, SciTePress (2021). https://doi.org/10.5220/0010325906450650
20. Ramezan, C.A., et al.: Effects of training set size on supervised machine-learning land-cover classification of large-area high-resolution remotely sensed data. Remote Sensing **13**(3), 368 (2021). https://doi.org/10.3390/rs13030368
21. Saez, Y., Baldominos, A., Isasi, P.: A comparison study of classifier algorithms for cross-person physical activity recognition. Sensors **17**(1), 66 (2016). https://doi.org/10.3390/s17010066
22. Sandha, S.S., et al.: Time awareness in deep learning-based multimodal fusion across smartphone platforms. In: IEEE/ACM Fifth International Conference on IoT Design and Implementation, pp. 149–156. IEEE (2020). https://doi.org/10.1109/IOTDI49375.2020.00022
23. Sansano, E., et al.: A study of deep neural networks for human activity recognition. Comput. Intell. **36**(3), 1113–1139 (2020). https://doi.org/10.1111/coin.12318
24. Vallat, R.: Pingouin: statistics in python. J. Open Source Soft. **3**(31), 1026 (2018). https://doi.org/10.21105/joss.01026
25. Yazdansepas, D., et al.: A multi-featured approach for wearable sensor-based human activity recognition. In: IEEE International Conference on Healthcare Informatics, pp. 423–431 (2016). https://doi.org/10.1109/ICHI.2016.81

Efficient Brazilian Sign Language Recognition: A Study on Mobile Devices

Vitor Lopes Fabris⬥, Felype de Castro Bastos⬥, Ana Claudia Akemi Matsuki de Faria⬥, José Victor Nogueira Alves da Silva⬥, Pedro Augusto Luiz⬥, Rafael Custódio Silva⬥, Renata De Paris⬥, and Claudio Filipi Gonçalves dos Santos(✉)⬥

Department of Software Applications - Eldorado Institute of Research, Campinas Alan Turing st, 13083-898 São Paulo, Brazil
claudio.santos@eldorado.org.br

Abstract. Automatic Sign Language Recognition (SLR) is a critical step in facilitating communication between deaf and hearing people. An interesting application of such a technology is a real-time mobile sign language translator since it could integrate both groups more easily. To this end,troduce a neBrazilian sign language (LIBRAS) recognition approach, the first for a mobile environment using an efficient 3D Convolutional Neural Network (CNN) to classify a sequence of frames extracted from a word being signaled in a video. Results show that our model is aproximately24 to 81 times faster than recent works in the field, and it is tested on a mobile device to understand the trade-off between performance and accuracy. Although slightly low accuracy, we have a significantly faster model at inference time and the beginning of something more relevant in the field, creating a discussion of future points of improvement to obtain an efficient real-time sign language system without greatly sacrificing accuracy in LIBRAS classification.

Keywords: 3D Convolutional Neural Networks · Computer Vision · Sign Language Recognition · Brazilian Sign Language · Mobile

1 Introduction

Sign language is a visual-gestural form of communication used by deaf individuals as a more accessible form of expression [8]. It is a rich and complex language with its grammar and syntax. Sign Language Recognition (SLR) plays a crucial role in facilitating communication between deaf and hearing individuals, as well as in bridging the language barrier that exists between them. For example, in Brazil, according to the 2010 demographic census [11], out of 190 million residents, 9.7 million suffer from some form of hearing impairment. And although Brazil has an extremely rich sign language, namely Brazilian Sign Language (LIBRAS), it is generally known that most Brazilians do not know it [20]. This language gap could be solved through the use of a computer application that

V. Vasconcelos et al. (Eds.): CIARP 2023, LNCS 14469, pp. 406–419, 2024.
https://doi.org/10.1007/978-3-031-49018-7_29

would automatically recognize the words being signaled and translate them into clear Portuguese.

Automatic recognition of sign languages, however, is not a trivial task. It requires a sufficiently robust system capable of learning important information from dynamic data (i.e. videos) coupled with a satisfactory learning pipeline that can capture this essential information from available data. Moreover, the underlying spatiotemporal features necessary for high generalization are influenced by many variables, such as signalers' fluency, physical characteristics (e.g. recording environment), and overall video quality [22].

In recent years, deep learning techniques, particularly Convolutional Neural Networks (CNNs), have shown remarkable success in various computer vision tasks, including image classification [30], object detection [21], and image segmentation [6]. However, the temporal nature of sign language requires the utilization of spatiotemporal information for accurate recognition. 3D Convolutional Neural Networks (3D CNNs) [14] have emerged as a powerful approach for capturing both spatial and temporal dependencies in video data. Extending the traditional 2D CNNs to 3D CNNs allows the latter to effectively process video sequences and exploit the inherent temporal dynamics present in sign language gestures.

To help Brazilians in their daily communication an application needs to be capable of translating signals in real-time. Otherwise, it risks losing semantic information connecting different signs and slowing down the pace of a conversation, unacceptable risks depending on the context, such as medical situations and emergencies. And despite the models mentioned earlier having great spatiotemporal feature extraction, working with video processing remains challenging, particularly in sign language. Among the challenges of working with this task, the following stand out:

- Differently from image classification, video processing has many points to consider, namely, computational resources, data volume, high-dimensional data, and temporal information.
- Given that our approach focuses mainly on the mobile environment by its huge flexibility and portability, some constraints have to be considered. Hardware limitation, inference time, model size, and power consumption are important variables to investigate.
- Lack of quality data is often a problem in deep learning approaches, and it is no different for sign language tasks, especially for LIBRAS, making the approach way more challenging.
- To mitigate sign language problems, a good understanding of the theme is necessary, given that sign language has characteristics that significantly differentiate it from spoken languages, and it is important to highlight that it is not a universal language each country has its own.

Although, by overcoming the above-mentioned challenges, we have the beginning of something more relevant in the field of sign language, where mobile phones can be used to communicate in LIBRAS, and consequently promote accessibility in social, academic, and work environments for deaf people who face this barrier daily.

In this paper, we propose the use of 3D CNNs for SLR, aiming to achieve high accuracy and robustness in mobile devices. We explore the application of 3D CNNs to capture the spatiotemporal features of sign language gestures, enabling the development of efficient and accurate systems for sign language interpretation.

The main contributions of this work are:

- Investigation of the effectiveness of 3D CNNs in SLR on a mobile device.
- First 3D CNN application for LIBRAS recognition based on a 3D MobileNet-V2 [16].
- Evaluation and comparison of the proposed approach with existing methods on benchmark Brazilian sign language datasets [23].

The remainder of this paper is organized in the following manner. Section 2 provides a review of related work in SLR and 3D CNNs. Section 3 presents the methodology and architecture of our proposed system, alongside the experimental setup, datasets, and evaluation metrics. The experimental results are presented in Sect. 4 and discussed in Sect. 5. Finally, Sect 6 concludes the paper and highlights potential future directions for research.

2   Related Works

It is understood that the application of a recognition system in a mobile setting can greatly benefit deaf individuals, especially in the context of a portable, scalable solution to smartphones, where it can be used individually and virtually anywhere. There are already two known examples of sign language translators for LIBRAS, namely Hand Talk[1] and VLibras[2]. However, they are based on Sign Language Production, which is the process of generating sign language videos from spoken language expressions. Unfortunately, as far as the authors could research, no studies regarding LIBRAS exist that propose a mobile solution with SLR and investigate its shortcomings, a serious step back in this field of research.

Nonetheless, some studies have been conducted for the automatic recognition of LIBRAS videos. Most of them used CNNs for the feature extraction and classification of RGB video samples [4,5,22,23], but others focused on different architectures, such as Temporal Convolutional Networks (TCNs) [22] and Support Vector Machines (SVMs) [20].

In [4], the authors propose a 3D-CNN model to classify the RGB samples. They developed a simple architecture, using only four convolutional layers, with max-pooling layers in-between for dimensionality reduction (although recent literature suggests that such operations might be redundant [26]), and two fully-connected ones at the end for a combination of extracted features and softmax

---

[1] https://www.handtalk.me/br/.
[2] https://www.gov.br/governodigital/pt-br/vlibras.

classification. They left signalers three and nine (known for having less samples than the rest) out of the training scheme, resulting in 995 videos total, and applied a frame summarization technique based on the Maximum Diversity Problem [17], leveraging only 10 frames at the end. The resulting frames are cut, resized to 224×224, and normalized in the range $[0, 1]$. Three data augmentation techniques were used: random temporal displacement in the frames, horizontal image mirroring, and random zoom between 5% and 15%.

In [5], a collection of 3D-CNNs were developed to classify signs from MINDS-Libras dataset [23]. To make the dataset more balanced, the authors used only 1,000 videos (10 signalers) from the total, and their training procedure employed the Leave-One-Out (LOO) technique with 10 folds. On the preprocessing step, they extracted video frames' features through a VGG16 [13] and used PCA + K-Means to cluster the features in 12 groups, retrieving the frames closest to their centroids, and dropping the preceding and leading ones. The chroma-key background of the videos was removed and video-wise horizontal mirroring, zooming, translation, and channel shuffling were performed for data augmentation purposes. Many other CNNs were developed by the authors, which are detailed and compared in Sect. 4.

As for the work reported in [23], a total of 10 experiments were conducted with the MINDS-Libras dataset. The baseline used for them, in which different techniques were aggregated with each experiment, is the same model developed in [4]. In total, the authors built six new 3D-CNN models, sequentially incrementing the original pipeline with new information, transformations, and augmentations. They developed three 2D-CNNs for LIBRAS signal recognition using spatial depth data, initially using only a recurrence matrix of the x, y position of the hands, elbow, wrists, hand tips and thumbs in reference to the position of the head, for the entire video. The last experiment consists of the combination of the best CNNs, both 2D and 3D.

In the PhD thesis [22], in which the entire process of capturing, preprocessing, and storing the MINDS-Libras videos is described, the author implements two different networks: a 3D-CNN trained on the RGB video information, and a TCN trained on hand-coordinates spatial information. The latter leverages all spatial information possible, such as coordinates of the signaler's hands, elbows, wrists, hand tips, and thumbs. As for the 3D-CNN, the data preprocessing pipeline was very similar to that of [4], involving the frame summarization technique through the Maximum Diversity Problem [17]. The data augmentation also followed the same steps, except for the removal of the temporal displacement in the frames. The model is a simple 4-convolutional layer network, with max-pooling layers in-between, and two fully-connected layers at the end for softmax classification.

The work in [20] differs from the usual 3D-CNN approaches. The authors decided to process the RGB video samples into a Gait Energy Image (GEI), which condenses the information of an entire video in a segmented, energy-based image using deep learning techniques [25,27]. They were able to reduce the dimensionality greatly with this approach, condensing an entire video in a 64×48 image. In their experiments, the CEFET LIBRAS dataset [9] was used to evaluate dimensionality reduction and data augmentation techniques alongside a classifier, as well as search for a good parameter space; in sequence, the chosen model was applied to LIBRAS-UFOP [2] and MINDS-Libras [23]. A total of five machine learning models were evaluated with varying hyperparameters and six possible preprocessing techniques, totaling 26 different experiments. The best combination when applied to the CEFET LIBRAS dataset is an SVM with SVD preprocessing of the GEIs. In Table 3, we include this architecture's results on MINDS-Libras.

The aforementioned works are quantitatively compared in Sect. 4 and discussed in Sect. 5.

3   Methodology

Figure 1 shows the development machine learning pipeline of the proposed application. Following this, we specify the collected data, the preprocessing applied along with augmentations used, and the model architecture. Furthermore, the mobile smartphone setting is explained and the conducted experiments are specified, as well as the hardware used for this study. The quality assessment is done in the following sections.

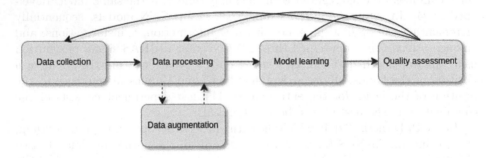

Fig. 1. Development flow of the LIBRAS recognition system

3.1   Dataset

Although the development of a sufficiently robust dataset, mapping enough words to allow a machine learning model to translate LIBRAS to Portuguese, has yet to be developed, some works have started to bridge the gap in this field of research. Of notorious mention to our work are the databases that deal with dynamic signaling, i.e. videos of LIBRAS signaling being performed under different conditions and speakers. The task that these databases bring forward is a more challenging one since the point of interest is not to understand only images of hand configurations, but to capture how a word would actually be signaled among speakers. Such databases could assist future research in different fields, e.g. Video Inpainting [15], Visual Question Answering [7], and Automatic Video Captioning [1].

As a dataset more convergent to our interests, the MINDS-Libras dataset [23] contains 1,158 1920 × 1080 RGB and RGBD video samples of 20 distinct words (see Table 1) being signaled five times by 12 different signalers (hearing and deaf), with varying levels of fluency. Some videos were not included due to recording issues. The background of the videos is a green screen, which provides a great way to artificially insert background complexity. It is publicly available[3].

Table 1. MINDS-Libras words [23]

Word	In Portuguese	# videos	Word	In Portuguese	# videos
To happen	Acontecer	60	To know	Conhecer	55
Student	Aluno	55	Mirror	Espelho	60
Yellow	Amarelo	55	Corner	Esquina	55
America	América	55	Son	Filho	56
To enjoy	Aproveitar	60	Apple	Maçã	60
Candy	Bala	60	Fear	Medo	55
Bank	Banco	60	Bad	Ruim	60
Bathroom	Banheiro	55	Frog	Sapo	60
Noise	Barulho	62	Vaccine	Vacina	60
Five	Cinco	55	Will	Vontade	60

In this study, we have decided to use the MINDS-Libras dataset for its interesting scope of chosen words, relative complexity in signaling fluency, and the possibility of artificially inserting noise in the form of different backgrounds - all of which can add to a recognition model's robustness in classification [5].

3.2   Data Preprocessing

To save time during training, we extracted all frames from each video in the dataset and stored only the images in separate folders for each signal. At this stage, no preprocessing was applied to the frames.

During training, the following sequence is applied on the fly to each of the examples used:

- Temporal augmentation: The first and last 15 frames were removed, as they had no semantic value for model training given that the signalers were in rest position, and after that, 30 frames were randomly selected from the remaining frames.
- Downscaling: All frames were rescaled to 224 × 224 pixels in order to fit into the model input.

---

[3] https://zenodo.org/record/2667329.

- Background replacement: Since the frames have a chroma key, background replacement was applied to insert noise into the samples, ensuring that the same background was chosen throughout the sequence.
- Normalization: The stack of frames was normalized using the Min-Max technique. Finally, the RGB data with dimensions (30, 224, 224, 3) was used for training.

The temporal augmentation and background replacement acted as data augmentations to regularize training and improve the model's generalization [28] without losing naturalness, usefulness on real-life scenarios and to make the dataset larger and more complex during training.

3.3 Proposed Model

For this study, the 3D MobileNet-V2 [16] was used. It introduces a way to enable the use of MobileNet-V2 [24] in a 3D context by modifying it to handle volumetric data, utilizing 3D convolutional layers.

MobileNet-V2 also introduces a width multiplier parameter (W_m), allowing the network to scale to different resource constraints [24]. This hyperparameter in the network construction adjusts the dimensionality of each layer, primarily in the number of channels, but intuitively as a reduction in the dimensionality of the activation space. Effectively, it reduces the number of parameters in a trade-off between accuracy and efficiency, and experimenting with different W_m values can lead us to understand the robustness intrinsic to the architecture and allow us to choose a more efficient approach at the cost of some accuracy points.

MobileNet-V2 is an efficient CNN architecture that improves upon MobileNet-V1 [10] by incorporating inverted residual blocks, enhancing non-linearity, gradient propagation, and parameter efficiency altogether. It achieves this through the use of depthwise separable convolutions, which split the standard convolution into depthwise and pointwise convolutions. The depthwise convolution applies a single filter per input channel, reducing computational complexity, while the pointwise convolution merges outputs to create new features. This approach significantly reduces parameters and computations compared to traditional convolutions.

3.4 Training Details

During the training of the model, the standard categorial cross-entropy loss was applied. The training process utilized a batch size of 8 and implemented the technique of accumulating gradient batches with a value of 4. This approach ensured efficient memory utilization and enhanced gradient estimation throughout the training.

AdamW optimizer [18] was employed with a learning rate of 0.0002 and a weight decay of 0.15. Additionally, a learning rate scheduler was implemented to dynamically adjust the learning rate by reducing it by a factor of 0.8 every 25 epochs in the absence of any improvement in the training loss. This approach enabled the model to fine-tune its learning rate based on the training progress, ensuring optimal convergence and preventing the model from getting stuck in suboptimal solutions.

To ensure the uniqueness and originality of our model, no pre-trained model was utilized in the training process.

3.5   Hardware Specifications

All the experiments were run on a computer equipped with an Intel Core i7-5930K processor, 32GB of RAM, and an NVIDIA Titan X graphics card with 12GB of GDDR5X video memory. All mobile experiments were conducted using an iPhone 12[4].

3.6   Experiments Conducted

Considering the experimental setup detailed previously, the conducted experiments are elaborated here. Initially, four 3D MobileNet-V2 networks, each with W_m equal to 0.10, 0.50, 0.75 or 1.00, were trained five times each. Several properties were extracted from these networks - number of parameters, time necessary to train them, inference time in a mobile setting, average and top accuracy - to compare the impact of the W_m parameter and understand the trade-off between generalization and efficiency in such a network (Table 2). It is worth noting that a network with $W_m = 2.0$ was also trained, but only once, since it showed a significant increase in the number of parameters and training time - we report this result without considering its accuracy as comparable to other accuracies.

After selecting the most accurate model, five rounds of inferences were performed on the entire dataset, measuring the average time taken to predict a single video in both GPU and iOS in comparison to other works in the field (Table 3). With this information, it is possible to understand whether the 3D MobileNet-V2's architecture can lead us to a proper real-time system for automatic SLR.

The accuracy metric is the only one considered since most works rely on it to evaluate their models. It provides a simple, yet informative way to assess the generalization of models. Moreover, the MINDS-Libras dataset does not have a standardized train/test partitioning, leaving each author to decide their own method to evaluate generalization. It is understood that testing on unforeseen data, with its unique characteristics, is a good way to test whether a model is capable of generalizing well to the domain. Hence, we partitioned the data by employing signalers 1 through 10 for training, and tested only on signalers 11 and 12.

---

[4] For information regarding the iPhone 12's processors, we refer the reader to https:// support.apple.com/kb/SP830.

4   Results

Considering the experiments detailed in the previous sections, the results are reported here. In Table 2, networks with different W_m values are compared. The training and mobile inference time refers to the model with the highest accuracy for each W_m. The best values for each column in this evaluation are highlighted.

Table 2. Comparison between different W_m values

W_m	# params	Train time	Mobile time		Accuracy	Top accuracy
			Processing	Inference		
0.10	**107k**	**57.4 min**	1.45 ± 0.32s	**1.58 ± 0.07s**	41.50 ± 9.24%	49.00%
0.50	775k	59.8 min		1.62 ± 0.09s	63.20 ± 2.17%	66.00%
0.75	1500k	66.0 min		1.62 ± 0.10s	65.20 ± 5.18%	70.50%
1.00	2400k	78.0 min		1.63 ± 0.06s	**68.50 ± 5.34%**	**75.00%**
2.00	9000k	240.0 min		1.70 ± 0.07s	74.00%	74.00%

The network with $W_m = 1.0$ is used as a comparison to other works using MINDS-Libras (Table 3), all described in Sect. 2. Such network was chosen because the mobile inference time does not present a huge increase, although it is slightly higher, and its training wielded a better accuracy in comparison to the other networks.

In Table 3, the proposed approach is compared with other works in the field regarding model accuracy and inference time on GPU and mobile. It is worth mentioning that [23], while reporting the inference time for the experiments developed, does not specify the hardware used for such, which hinders a truly parallel comparison between our proposed approach and theirs. Also, the different 3D-CNNs present in [23] are hereby described:

- (a) Their baseline without data augmentation;
- (b) Model (a) with test data normalization;
- (c) Model (b) with grayscale data instead of RGB;
- (d) Model (c) with Optical Flow [19] information;
- (e) Model (c) with Histograms of Oriented Gradients [3] information;
- (f) Model (c) with data augmentation.

The enumeration scheme for the models in [5] is the following:

- (1) A 3D-CNN with RGB and optical flow [19] information;
- (2) A 3D-CNN with RGB and synthetic depth maps, obtained by training a Pix2Pix Conditional Generative Adversarial Network (cGAN) [12] in RGB-D data and performing inference on MINDS-Libras videos;
- (3) A probabilistic ensemble of the previously described models (including their baseline);

- (4) A multi-stream 3D-CNN using hands, head, and distance map information, where each type of information is fed to its specific branch in the 6-layer feature extraction CNN (for the distance maps, only 2 layers) and concatenated before being passed to the fully-connected layers; and
- (5) A combination of all the above, using the multi-stream 3D-CNN with hands, head, distance, and speed maps, alongside the 3D-CNN with RGB and depth information.

Table 3. Results on MINDS-Libras dataset

Ref	Model	Inference time		Accuracy (%)
		GPU (ms)	Mobile (s)	
[5]	3D-CNN LOO	—	—	57.00 ± 12.00
[23]	3D-CNN (a)	270	—	59.70 ± 20.52
[5]	3D-CNN (1)	—	—	60.00 ± 9.00
[5]	3D-CNN (2)	—	—	64.00 ± 6.00
[5]	3D-CNN (3)	—	—	65.00 ± 8.00
[22]	TCN	—	—	67.76
[4]	3D-CNN	—	—	72.60
[20]	SVM + SVD	—	—	84.66 ± 1.78
[22]	3D-CNN	—	—	84.75
[23]	3D-CNN (e)	250	—	88.80 ± 3.14
[23]	3D-CNN (b)	280	—	89.90 ± 1.02
[23]	3D-CNN (d)	250	—	90.50 ± 1.55
[5]	3D-CNN (4)	—	—	90.00 ± 7.00
[23]	3D-CNN (c)	170	—	91.60 ± 1.57
[5]	3D-CNN (5)	—	—	91.00 ± 7.00
[23]	3D-CNN (f)	570	—	$\mathbf{93.30 \pm 1.69}$
Ours	3D MobileNet-V2	$\mathbf{6.94 \pm 2.27}$	$\mathbf{1.63 \pm 0.06}$	68.50 ± 5.34

In [5], the baseline developed was a 3-layer 3D-CNN, with 2 fully-connected layers on top, to process only RGB video frames, which achieved 96% accuracy without LOO - with this technique, it amounted to $57 \pm 12\%$, showing that the model was not capable of generalizing well.

5   Discussion

Our highest accuracy, the $68.50 \pm 5.34\%$ with $W_m = 1.0$, is yet to be improved further to better match the accuracies reported in other works [4,5,20,22,23]. Nevertheless, some architectures compared in the previous section utilize different data modalities for joint information embedding in their training, such as depth maps and various image descriptors [5,23]. This differs significantly from our RGB-only information approach, a necessity when considering portability to a wide range of smartphones available.

Furthermore, as can be seen in Table 2, although highly changing the W_m value for the networks, which led to an increase in the number of parameters, the training and inference time did not fluctuate much, except for the case where $W_m = 2.0$. It suggests that this architecture's efficiency in both attributes is quite intrinsic and stems directly from its implementation and special characteristics.

Its accuracy, however, does change. For $W_m = 0.1$, a considerable drop-off is observable, in which the highest obtained accuracy is only 49%, compared to the networks with $W_m \in \{0.5, 0.75, 1.0\}$. These networks, on the other hand, present a rather uniform average accuracy: all lie within the 60% range, even with quite distinct W_m values. This seems to indicate that, in terms of efficiency at least, the 3D MobileNet-V2 in this task does not benefit much from an increased W_m parameter, especially considering the training time for the last network reported. As for robustness, the accuracy obtained lacks sufficient generalization capacity to obtain an average of three quarters in the testing set, which might indicate a necessity to increase the W_m. However, different techniques, e.g. different prepro-cessing layers, ensembles, or multistream training [5], might benefit the network even further.

Many works reviewed in Sect. 2 use frame summarizing techniques [4,5,22,23] by optimizing certain variables, such as the approaches using the Maximum Diversity Problem [17]. This leads to a frame selection that's sufficiently infor-mative of the entire word being signaled and, disconsidering temporal data aug-mentations, always renders the same set of frames for training. Our approach, however, leverages a random frame selection within the desired range as its video summarization, which increases information variance during training and therefore leads to a high variance in final model accuracies, explaining the high standard-deviations observed in Table 2.

As for inference times, many observations can be made. Our best GPU time, the 6.94 ± 2.27ms, is more than sufficient for a real-time SLR system. Although the compared approaches are not far, ours represents close to a 24 times decrease in processing time when considering the fastest model (from 170 to 6.94 ms), or about 81 times with regards to the state-of-the-art (570 to 6.94 ms), outperform-ing all reported results in this dataset and rendering such real-time application more feasible.

The inference time on the iPhone 12, an average of 1.63 ± 0.06s, although not sufficient for real-time recognition, is still faster than all other approaches described and provides valuable information in an acceptable time. This value can be dropped further by implementing more efficient architectures and frame summarization techniques [5].

Finally, we consider the inevitable trade-off between accuracy and efficiency. As already discussed before, our best GPU time would be a great response time for the considered application. However, achieving good times in GPU inferences does not render such application as done. In fact, the true trouble arises when a mobile setting - necessary for a virtually omnipresent aid in communication between deaf and hearing people - is considered. The processing intricacies and resource constraints of this setting necessarily ensue a trade-off: bigger, more robust models will require higher processing times, while smaller, less generalizable models will achieve lower accuracies. Therefore, the scientific effort of developing a sufficiently fast and robust architecture for mobile real-time SLR is yet to be concluded, but the present work indicates an interesting path to follow.

6   Conclusion

In this study, we conducted a review of recent literature regarding the automatic recognition of LIBRAS signs, developed an approach using a 3D MobileNet-V2 [16], and tested it in a mobile smartphone environment using an iPhone 12. Our approach is the first developed for such a task and, notwithstanding the slightly low accuracy, we have a significantly faster model at inference time when compared to other approaches using MINDS-Libras dataset [23].

While our work presents a LIBRAS recognition solution that outperforms every other approach in recent literature in terms of efficiency, it is still possible to refine it to achieve a proper real-time solution for mobile devices. Different architectures with more efficient layers and connections between them can be leveraged, such as MobileOne [31] and its removal of residual connections in inference mode. Reducing the number of frames considered for training and inference could help too, since it reduces the input's dimensionality, requiring less computing power and time. Our results on GPU show a promising comparison to other works.

The training pipeline can benefit greatly from the application of different data modalities, as can be seen in [5,23], to increase the model's robustness. To reduce the high variance observed in the final accuracies, more deterministic video summarization techniques could be applied, as well as more data augmentations to increase training data, reducing learning variance.

Finally, for a meaningful use of such a LIBRAS recognition tool on smartphones, users need to be able to understand why the model is arriving at its classifications. To this end, eXplainable AI techniques can be applied to make these models more explainable, therefore making possible users more confident in its capability - and since the domain of language is an important one in our lives, for deaf and hearing people alike, a profound understanding of a model's classifications, such as the one required in the medical field [29], is necessary.

Acknowledgements. The authors are grateful to the Eldorado Research Institute.

References

1. Amaresh, M., Chitrakala, S.: Video captioning using deep learning: an overview of methods, datasets and metrics. In: 2019 International Conference on Communication and Signal Processing (ICCSP), pp. 0656–0661 (2019). https://doi.org/10.1109/ICCSP.2019.8698097

2. Cerna, L.R., Cardenas, E.E., Miranda, D.G., Menotti, D., Camara-Chavez, G.: A multimodal LIBRAS-UFOP Brazilian sign language dataset of minimal pairs using a microsoft kinect sensor. Expert Syst. Appl. **167**, 114179 (2021)
3. Dalal, N., Triggs, B.: Histograms of oriented gradients for human detection. In: 2005 IEEE Computer Society Conference on Computer Vision and Pattern Recognition (CVPR 2005), pp. 886–893 (2005). https://doi.org/10.1109/CVPR.2005.177
4. de Castro, G.Z., et al: Desenvolvimento de uma base de dados de sinais de libras para aprendizado de máquina: Estudo de caso com CNN 3D. In: Anais do 14Ž Simpósio Brasileiro de Automação Inteligente (2019). https://doi.org/10.17648/sbai-2019-111451
5. de Castro, G.Z., Guerra, R.R., Guimarães, F.G.: Automatic translation of sign language with multi-stream 3D CNN and generation of artificial depth maps. Expert Syst. Appl. **215**, 119394 (2023)
6. Dolz, J., Gopinath, K., Yuan, J., Lombaert, H., Desrosiers, C., Ayed, I.B.: Hyperdense-Net: a hyper-densely connected CNN for multi-modal image segmentation. IEEE Trans. Med. Imaging **38**(5), 1116–1126 (2018)
7. de Faria, A.C.A.M., et al: Visual question answering: a survey on techniques and common trends in recent literature (2023)
8. Gala, A.S.: A importância da libras para a comunidade surda. https://www.handtalk.me/br/blog/importancia-da-libras/. Accessed 5 July 2023
9. Gameiro, P.V., Passos, W.L., Araujo, G.M., de Lima, A.A., Gois, J.N., Corbo, A.R.: A Brazilian sign language video database for automatic recognition. In: 2020 Latin American Robotics Symposium (LARS), 2020 Brazilian Symposium on Robotics (SBR) and 2020 Workshop on Robotics in Education (WRE), pp. 1–6 (2020). https://doi.org/10.1109/LARS/SBR/WRE51543.2020.9307017
10. Howard, A.G., et al.: MobileNets: efficient convolutional neural networks for mobile vision applications. CoRR abs/1704.04861 (2017). https://arxiv.org/abs/1704.04861
11. IBGE: Demographic census of 2010. https://censo2010.ibge.gov.br/
12. Isola, P., Zhu, J.Y., Zhou, T., Efros, A.A.: Image-to-image translation with conditional adversarial networks. In: Proceedings of the IEEE Conference on Computer Vision and Pattern Recognition (CVPR) (2017)
13. Jadon, S., Jasim, M.: Unsupervised video summarization framework using keyframe extraction and video skimming. In: 2020 IEEE 5th International Conference on Computing Communication and Automation (ICCCA), pp. 140–145 (2020). https://doi.org/10.1109/ICCCA49541.2020.9250764
14. Ji, S., Xu, W., Yang, M., Yu, K.: 3D convolutional neural networks for human action recognition. IEEE Trans. Pattern Anal. Mach. Intell. **35**(1), 221–231 (2012)
15. Kim, D., Woo, S., Lee, J.Y., Kweon, I.S.: Deep video inpainting. In: 2019 IEEE/CVF Conference on Computer Vision and Pattern Recognition (CVPR), pp. 5785–5794 (2019). https://doi.org/10.1109/CVPR.2019.00594
16. Kopuklu, O., Kose, N., Gunduz, A., Rigoll, G.: Resource efficient 3D convolutional neural networks. In: Proceedings of the IEEE/CVF International Conference on Computer Vision (ICCV) Workshops (2019)
17. Kuo, C.C., Glover, F., Dhir, K.S.: Analyzing and modeling the maximum diversity problem by zero-one programming*. Decis. Sci. **24**(6), 1171–1185 (1993)
18. Loshchilov, I., Hutter, F.: Decoupled weight decay regularization. In: International Conference on Learning Representations (2019). https://openreview.net/forum?id=Bkg6RiCqY7

19. Lucas, B.D., Kanade, T.: An iterative image registration technique with an application to stereo vision. In: Proceedings of the 7th International Joint Conference on Artificial Intelligence - Vol 2, pp. 674–679. IJCAI'81, Morgan Kaufmann Publishers Inc., San Francisco, CA, USA (1981)

20. Passos, W.L., Araujo, G.M., Gois, J.N., de Lima, A.A.: A gait energy image-based system for brazilian sign language recognition. IEEE Trans. Circuits Syst. I: Regular Papers 68(11), 4761–4771 (2021). https://doi.org/10.1109/TCSI.2021.3091001

21. Redmon, J., Divvala, S., Girshick, R., Farhadi, A.: You only look once: unified, real-time object detection. In: Proceedings of the IEEE Conference on Computer Vision and Pattern Recognition, pp. 779–788 (2016)

22. Rezende, T.M.: Reconhecimento automático de sinais da Libras: desenvolvimento da base de dados MINDS-Libras e modelos de redes convolucionais. Phd thesis, Universidade Federal de Minas Gerais (2021). https://hdl.handle.net/1843/39785

23. Rezende, T.M., Almeida, S.G.M., Guimarães, F.G.: Development and validation of a Brazilian sign language database for human gesture recognition. Neural Comput. Appl. 33(16), 10449–10467 (2021)

24. Sandler, M., Howard, A., Zhu, M., Zhmoginov, A., Chen, L.C.: MobileNetV2: inverted residuals and linear bottlenecks. In: Proceedings of the IEEE Conference on Computer Vision and Pattern Recognition (CVPR) (2018)

25. dos Santos, Claudio Filipi Goncalves., Moreira, Thierry Pinheiro, Colombo, Danilo, Papa, João Paulo.: Does pooling really matter? An evaluation on gait recognition. In: Nyström, Ingela, Hernández Heredia, Yanio, Milián Núñez, Vladimir (eds.) CIARP 2019. LNCS, vol. 11896, pp. 751–760. Springer, Cham (2019). https://doi.org/10.1007/978-3-030-33904-3_71

26. Santos, Claudio Filipi Goncalves dos., Moreira, Thierry Pinheiro, Colombo, Danilo, Papa, João Paulo.: Does removing pooling layers from convolutional neural networks improve results? SN Comput. Sci. 1(5), 1–10 (2020). https://doi.org/10.1007/s42979-020-00295-9

27. Santos, C.F.G.d., et al.: Gait recognition based on deep learning: a survey. ACM Comput. Surv. 55(2) (2022). https://doi.org/10.1145/3490235

28. Santos, C.F.G.d., Papa, J.a.P.: Avoiding overfitting: a survey on regularization methods for convolutional neural networks. ACM Comput. Surv. 54(10s) (2022). https://doi.org/10.1145/3510413

29. da Silva, M.V.S., et al.: explainable artificial intelligence on medical images: a survey (2023)

30. Tan, M., Le, Q.: EfficientNet: rethinking model scaling for convolutional neural networks. In: International Conference on Machine Learning, pp. 6105–6114. PMLR (2019)

31. Vasu, P.K.A., Gabriel, J., Zhu, J., Tuzel, O., Ranjan, A.: MobileOne: an improved one millisecond mobile backbone. In: CVPR (2023). https://arxiv.org/abs/2206.04040

Presumably Correct Undersampling

Gonzalo Nápoles[1](✉) and Isel Grau[2,3]

[1] Department of Cognitive Science and Artificial Intelligence, Tilburg University,
Tilburg, The Netherlands
g.r.napoles@uvt.nl

[2] Information Systems, Eindhoven University of Technology, Eindhoven, The
Netherlands

[3] Eindhoven Artificial Intelligence Systems Institute, Eindhoven University of
Technology, Eindhoven, The Netherlands

Abstract. This paper presents a data pre-processing algorithm to tackle
class imbalance in classification problems by undersampling the majority class. It relies on a formalism termed Presumably Correct Decision
Sets aimed at isolating easy (presumably correct) and difficult (presumably incorrect) instances in a classification problem. The former are
instances with neighbors that largely share their class label, while the
latter have neighbors that mostly belong to a different decision class. The
proposed algorithm replaces the presumably correct instances belonging
to the majority decision class with prototypes, and it operates under the
assumption that removing these instances does not change the boundaries of the decision space. Note that this strategy opposes other methods
that remove pairs of instances from different classes that are each other's
closest neighbors. We argue that the training and test data should have
similar distribution and complexity and that making the decision classes
more separable in the training data would only increase the risks of
overfitting. The experiments show that our method improves the generalization capabilities of a baseline classifier, while outperforming other
undersampling algorithms reported in the literature.

Keywords: Pattern Classification · Class Imbalance ·
Undersampling · Presumably Correct Decision Sets

1 Introduction

Class imbalance is a prevailing challenge in pattern classification, where the distribution of classes is heavily skewed, leading to a scarcity of instances in the
minority class relative to the majority class. The class imbalance poses significant
difficulties for machine learning algorithms, as they tend to be biased towards
the majority class, resulting in suboptimal performance. In real-world scenarios,
such as fraud detection or disease diagnosis, where the minority class represents
critical instances of interest, accurate prediction becomes crucial. Several techniques have been proposed to address the issue of class imbalance in classification

© Springer Nature Switzerland AG 2024
V. Vasconcelos et al. (Eds.): CIARP 2023, LNCS 14469, pp. 420–433, 2024.
https://doi.org/10.1007/978-3-031-49018-7_30

problems by amending the dataset. These techniques can be broadly categorized into undersampling, oversampling, and hybrid approaches.

Undersampling methods aim to reduce the number of instances in the majority class. For example, this can be achieved through random undersampling (RU) [12], where randomly selected instances from the majority class are removed. Alternatively, informed variants of undersampling select instances based on specific criteria or clustering algorithms. For example, Cluster Centroid Undersampling (CCU) [12] utilizes k-means clustering to discover clusters of instances from the majority class and replaces the clusters of majority samples with the cluster centroids. Meanwhile, Edited Nearest Neighbors (ENN) [19] undersamples the majority class by removing samples that are not similar enough to their neighbors. Tomek Links Undersampling (TLU) method [18] identifies pairs of instances from different classes that are close to each other and removes the majority class instances. The Condensed Nearest Neighbor (CoNN) rule [8] is also used in the context of undersampling by iterating over the examples in the majority class and keeping them only if they cannot be classified correctly by the already selected instances. Due to the initial random selection, CoNN method is sensitive to noise. As a solution, One-Sided Selection (OSS) [10] uses Tomek Links to remove noisy instances after applying the CoNN rule. Other undersampling techniques include the NearMiss method [14], which selects majority class instances based on their proximity to minority class instances.

On the other hand, oversampling methods focus on increasing the number of instances in the minority class by replicating or generating synthetic samples [6]. Similar to random undersampling, random oversampling duplicates instances from the minority class randomly. In contrast, synthetic oversampling methods generate synthetic samples with different rules. The most prominent example of synthetic oversampling is the Synthetic Minority Over-sampling Technique (SMOTE) [4]. This method creates new instances by interpolating between the minority class instances and their neighbors. One drawback of SMOTE is that it can generate instances between the inliers and outliers of a class, effectively creating a suboptimal decision space. Several variants of SMOTE try to correct this problem by focusing on samples near the border of the optimal decision function or harder instances from the minority class [7,9,11,16].

Hybrid methods combine both undersampling and oversampling techniques to achieve a balanced dataset. For example, using SMOTE method for oversampling the minority class and then applying an undersampling technique to clean the data is the most common approach. In [2] the authors used Tomek Links, whereas in [1] they propose ENN for the undersampling step. Overall, oversampling methods do not cause data loss, but they are prone to lead to overfitting for the majority of machine learning approaches [3]. Undersampling methods, on the other side, use different heuristics to counteract the data loss from random undersampling. The main advantage of these methods is their ability to reduce the computational complexity of training machine learning models, leading to faster training times and improved efficiency, even when such reduction might not lead to improved algorithms' performance on the fresh data. For example, the increasingly popular SHAP (SHapley Additive exPlanations) method [13] for generating explanations for machine learning algorithms is one of the methods that benefits from operating with smaller datasets.

In this paper, we tackle the class imbalance problem in machine learning by undersampling the majority class. The proposed method is based on the introduced *presumably correct decision sets* [15], a set formalism to analyze uncertainty in pattern classification problems. In particular, it focuses on determining which instances are easy (presumably correct) or difficult (presumably incorrect). The former category consists of instances whose neighbors mostly share the same class label, whereas the latter includes instances whose neighbors predominantly belong to a different decision class. The proposed algorithm comprises two main steps. Firstly, we modify the definitions of presumably correct and incorrect instances by using a complete inclusion operator to ensure that *all* neighbors of presumably correct and incorrect instances are contained in the decision classes being analyzed. Secondly, we cluster the presumably correct instances associated with the majority class into groups and replace them with prototypes. Both steps ensure that the majority class is reduced while preserving the decision boundaries that define the classification problem, translating into reporting the same or improved prediction rates on unseen data.

This paper is organized as follows. Section 2 describes the fundamentals of the presumably correct decision sets, while Sect. 3 presents our method to undersample the majority class in classification problems. Section 4 conducts numerical simulations and performs a comparative analysis between our algorithmic proposal and state-of-the-art undersampling methods. Section 5 provides concluding remarks and some future research avenues.

2   Presumably Correct Decision Sets

Let $P = (X, F \cup y)$ be a structured classification problem where X is a non-empty finite set of instances, while F is a non-empty finite set of features describing these instances. The decision feature, denoted by $y \notin F$, gives the class label for each instance. For example, if $x \in X$ is associated with the i-th decision class, then we can denote the whole instances as a tuple (x, y_i) where $y_i \in Y$, with Y being the set of possible decision classes. The decision feature induces a partition $X = \{X_1, \ldots, X_i, \ldots, X_m\}$ of the universe where X_i contains instances labeled with the i-th decision class, with m being the number of decision classes. Note that this partition satisfies the conditions $\cup_i X_i = X$ and $\cap_i X_i = \emptyset$. Moreover, let us assume that $g : X \to Y$ is the ground-truth function such that $g(x) = y_i$ if and only if $x \in X_i$. This function returns the decision class of the instance as observed in the dataset. Similarly, the neighborhood function $f : X \to Y$ determines the decision class of an instance given its neighborhood, and several implementations of this function lead to different algorithms. The certainty degree of such assignment is given by a membership function $\mu : X \times Y \to [0, 1]$ where $\mu(x, i)$ denotes the extent to which x belongs to the i-th decision class as determined by the neighborhood function. In the remainder of this paper, we will refer to $\mu(x, i)$ as $\mu_i(x)$ for convenience.

The rationale of this method is as follows. Let us suppose an instance $x \in X_i$ belongs to the i-th decision class. If most of its neighbors also belong to the i-th decision class, then x is categorized as *presumably correct*. In contrast, if the majority of its neighbors are labeled with the l-th decision class (where $l \neq i$), then x is considered to be *presumably incorrect*, even when the ground truth indicates that it is labeled with the i-th decision class in the dataset. In this approach, the membership function quantifies the extent to which the instance resembles the instances in its neighborhood.

Equations (1) and (2) portray the presumably correct and incorrect sets, respectively, associated with the i-th decision class,

$$\beta\text{-}C_i = \{x \in X_i : x \in \beta\text{-}S_i \wedge g(x) = f(x)\} \tag{1}$$

$$\beta\text{-}I_i = \{x \in X_i : \exists l \in \{1, \dots, m\}, x \in \beta\text{-}S_{l \neq i} \wedge g(x) \neq f(x)\} \tag{2}$$

where $\beta\text{-}S_i$ is referred to as the β–strong region and contains instances with a *high* membership degree to the i-th decision class according to the neighborhood function. These instances are likely members of the target decision class since they are strong members of a neighborhood dominated by instances labeled with that class. Equation (3) mathematically formalizes this region,

$$\beta\text{-}S_i = \{x \in X : \mu_i(x) \geq \beta\} \tag{3}$$

such that $\alpha, \beta \in [0,1]$ is a parameter to be specified by the user. Setting β close to one will cause most instances to be excluded from the presumably correct/incorrect analysis, while setting β close to zero will cause most instances to be considered as strong members of their neighborhood.

The β-presumably correct region $\beta\text{-}C_i$ contains instances that must satisfy two primary properties. Firstly, these instances have membership degrees in the i-th strong region that meet the constraints imposed by the β parameter. Secondly, the decision classes determined by the function $f(x)$ for these instances align with those computed by the function $g(x)$ ($x \in X_i$ implies $g(x) = y_i$). Instances belonging to the strong region $\beta\text{-}S_{l \neq i}$ but having $f(x) \neq g(x)$ are considered presumably incorrect (assuming $f(x) = y_l$). Finally, instances that do not possess membership values fulfilling the membership constraint are neither labeled as presumably correct nor incorrect.

3 Presumably Correct Undersampling

In this section, we present the contributions of our paper, named the Presumably Correct Undersampling (PCU) method. This procedure starts with the assumption that there is a highly imbalanced classification problem with a clearly distinguished majority class. The intuition behind the PCU formalism is that we can safely replace instances that are considered presumably correct members of the majority decision class by prototypes. These instances do not define the decision boundaries of the classification problem, and their removal should not affect the discriminatory power of the data. Note that this assumption opposes the strategy used by the TLU method (Tomek, 1976), which removes pairs of instances from different classes that are each other's closest neighbors. By removing such instances, the TLU method improves class separability and addresses challenges arising from overlapping or misclassified instances. However, while this brings evident benefits during the classifier's training process, it may increase the difficulty of accurately classifying challenging instances when applying the classifier to unseen data. Similarly to the approach used by the TLU method, our algorithm could also remove the presumably incorrect instances belonging to the majority class. These instances are strong members of some other region, according to the neighborhood function, yet are labeled with the majority decision class. Therefore, they can be considered noisy instances rather than borderline instances, as they strongly belong to their respective neighborhoods. However, this paper focuses on the presumably correct instances only. Let us break the proposed algorithm into two main steps related to isolating the presumably correct instances and replacing them with prototypes.

Step 1. Determining the presumably correct and incorrect instances, as defined in Eqs. (1) and (2), requires defining a neighborhood function. In the original paper, Nápoles et al. [15] defined this function as the k-nearest neighbor classifier. Hence, an instance $x \in X$ will be included in β-C_i if $g(x) = i$, $x \in \beta$-S_i and $|\mathcal{N}_k(x) \cap X_i|/|X_i|$ is maximal for the i-th decision class, with $\mathcal{N}_k(x)$ being the k closest neighbors of x. The last condition might bring some issues for our algorithm since it is not required for the $\mathcal{N}_k(x)$ to be entirely included in X_i. In other words, x has some neighbors that do not belong to the i-th decision class. Let us redefine the concepts of presumably correct and incorrect sets based on a neighborhood function ensuring the full inclusion degree of $\mathcal{N}_k(x)$ in X_i. Hence, Eqs. (1) and (2) can be rewritten as follows:

$$\beta\text{-}C_i = \{x \in X_i : x \in \beta\text{-}S_i \wedge \mathcal{N}_k(x) \subseteq X_i\}, \tag{4}$$

$$\beta\text{-}I_i = \{x \in X_i : \exists l \in \{1, \ldots, m\}, x \in \beta\text{-}S_{l \neq i} \wedge \mathcal{N}_k(x) \subseteq X_l\}. \tag{5}$$

Moreover, the membership function needed to define β-S_i can be formalized in terms of the k-nearest neighbors as follows:

$$\mu_i(x) = \frac{\sum_{z \in \mathcal{N}_k(x)} \frac{1}{d(x,z)} \cdot \phi(g(x) = g(z))}{\sum_{z \in \mathcal{N}_k(x)} \phi(g(x) = g(z))} \tag{6}$$

such that

$$\phi(g(x) = g(z)) = \begin{cases} 1, & \text{if } g(x) = g(z) \\ 0, & \text{otherwise} \end{cases}. \tag{7}$$

Notice that simultaneously fulfilling that $\mathcal{N}_k(x) \subseteq X_i$ and $x \in \beta$-S_i is key in the proposed method. The former ensures that all neighbors of the instance being analyzed belong to the majority class, while the latter ensures that the instance itself is a strong member of its neighborhood. Both constraints lead to pure and cohesive presumably correct decision sets.

Step 2. Once we have isolated the presumably correct instances associated with the class to be undersampled, we can gather them into p clusters using any clustering method. This procedure requires specifying the number of clusters in advance, although it can be automatically estimated using a clustering validation measure such as the Davies-Bouldin Index or the Silhouette Coefficient. As a rule of thumb, the number of clusters can be estimated as $p = \lfloor (1-r) \cdot \lfloor \beta\text{-}C_i \rfloor \rfloor$, where $r \in [0,1]$ is the imbalance ratio computed as the number of minority instances divided by the number of majority instances. Finally, the presumably correct instances are removed from the dataset and replaced p prototypes obtained from the discovered clusters, meaning that there will be a prototype per cluster. The prototypes will be either the cluster centers themselves or the instances reporting the closest distances to their cluster centers.

Figure 1 depicts the intuition of our algorithm for a three-class classification problem where the first decision class is represented by 100 instances, and the other two classes are represented by 50 instances each. Therefore, the purpose is to undersample the majority class as much as possible without damaging the decision space characterizing the problem. The algorithm first finds the presumably correct instances and then clusters them into p groups according to their similarity. Note that we only analyze instances belonging to the presumably correct region associated with the majority class.

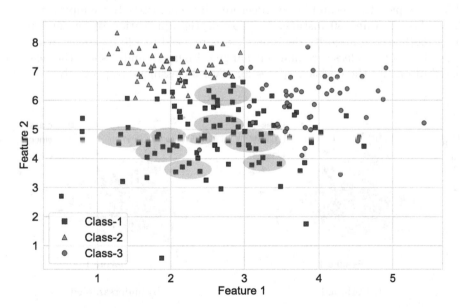

Fig. 1. Visual depiction of the proposed PCU algorithm for a toy example. Presumably correct instances belonging to the majority class are enclosed in colored circles. Note that these similarity classes or neighborhoods only contain instances that belong to the decision class to be undersampled.

Overall, the proposed PCU algorithm requires three parameters. The first parameter is the number of neighbors k used to build the neighborhood function $\mathcal{N}_k(x)$. The larger the number of neighbors, the fewer the number of presumably correct instances since finding pure neighborhoods will become more difficult. The second parameter, denoted as $\beta \in [0,1]$, specifies the minimum membership degree of an instance to be considered a strong member of its neighborhood. The last parameter concerns the reduction ratio $r \in [0,1]$, which indicates how many prototypes will be built from the isolated presumably correct instances. In short, larger values of this parameter translate into fewer prototypes. If $r = 1.0$ then all presumably correct instances will be deleted without building any prototypes. If $r = 0.0$ then each presumably correct instance will be deemed a prototype, thus inducing no modification to the dataset.

4   Experimental Simulations

In this section, we will conduct numerical simulations to validate the correctness of the proposed PCU method and contrast its performance against state-of-the-art algorithms devoted to undersampling the majority class.

First, let us illustrate the inner workings of our method in a two-dimensional dataset with three decision classes. In this problem, the majority class consists of 500 samples, the second largest decision class comprises 100 samples, and the third class contains 50 instances. Following the application of our undersampling method, the majority decision class is reduced to 130 instances, while the sizes of the other classes remain unchanged. Figures 2 and 3 show the instance distribution and the decision spaces for several classifiers, respectively.

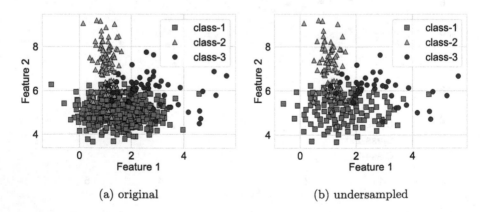

(a) original   (b) undersampled

Fig. 2. Imbalanced three-class classification problem described by two numerical features. In the original problem, the first decision class is represented by 500 instances, the second by 100, and the third by 50. After applying our algorithm, the majority decision class narrows down to 130 instances.

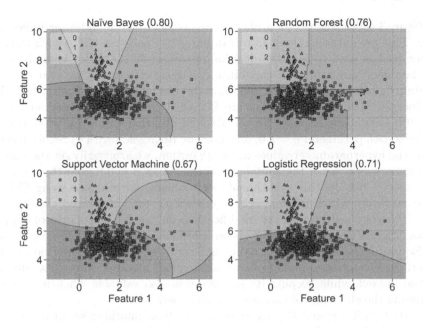

(a) original

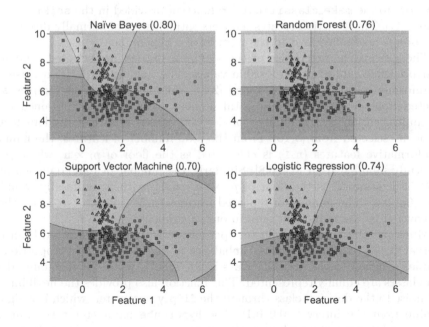

(b) undersampled

Fig. 3. Decision space of different classifiers for a three-class imbalanced problem before and after applying our algorithm. The number on the top of each sub-figure denotes Cohen's kappa score on the test set.

Figure 3 shows the decision of several classifiers before and after undersampling. This toy dataset is split into training (80%) and test (20%) such that the former part is used to plot the decision spaces while the latter is used to test the classifiers' generalization capabilities. It goes without saying that only the training data is undersampled and that the test data is kept untouched. The classifiers used in this simulation are Gaussian Naive Bayes, Random Forests with 100 estimators, Support Vector Machine classifier with a radial kernel and regularization parameter $c = 1.0$, and Logistic Regression with ℓ_2 regularization. As for the hyperparameter values of our algorithm, we arbitrarily set the number of neighbors as $k = 10$, the confidence threshold as $\beta = 0.8$, and the reduction ratio as $r = 0.8$. In this experiment, we evaluate performance using Cohen's kappa score [5], which is deemed a robust metric for imbalanced classification tasks. The results show that the modifications on the decision spaces of Support Vector Machine and Logistic Regression led to increased performance on the test set. Equally important is the fact that none of these classifiers reported any loss in performance after applying our undersampling method. Note that simplifying the dataset while keeping its properties is also valuable when it comes to improving the efficiency of machine learning methods.

Next, we will expand this experiment to a benchmarking set of imbalanced data. The initial step in our experimental methodology consists of generating 50 synthetic datasets with varying complexity levels. To generate these datasets, we resort to the `make_classification` function provided in the `scikit-learn` library [17]. This function creates datasets with clusters of normally distributed data points positioned around the vertices of a hypercube.

The general characteristics of the synthetic dataset are selected randomly from predefined intervals of possible values. The number of decision classes m is uniformly selected from the range [2,5], while the number of samples s is selected from [5000, 10000]. The total number of features n is obtained from the multiplication of the number of decision classes m with a random factor uniformly selected from [5,10]. From the total number of features, the number of informative features (n_1) is calculated as the floor of $p_i \times n$, where p_i is uniformly sampled from [0.4, 0.8]. Similarly, the number of redundant features (n_2) is determined as the floor of $p_r \times (n - n\_1)$, with p_r uniformly sampled from [0.2, 0.4]. The number of repeated features (n_3) is set to zero since these are covered by the redundant ones in our experiment.

Moreover, the number of clusters per class c is uniformly selected from the interval [1,5]. To introduce class imbalance in the dataset, the proportion of samples assigned to the majority class is set to 80%, while the remaining decision classes are equally represented. This function also provides the flexibility to add noise to the decision class through the `flip_y` parameter, which is assigned a value from the interval [0.0, 0.1]. The `hypercube` parameter takes random boolean values indicating whether the clusters are at the vertices of the hypercube or a random polytope. The dimension of the hypercube is determined by the number of informative features n_1. The length of the hypercube's sides is twice the value of the parameter `class_sep`, which is randomly selected from [1, 2] and controls the spread of the decision classes.

The second step in our experimental methodology is to contrast the performance of a baseline classifier before and after the undersampling process. The selected baseline classifier is a Random Forest using the default hyperparameter values as reported in the `scikit-learn` library [17]. This setting allows us to compare the proposed PCU method against well-established undersampling techniques, such as Random Undersampling (RU), Cluster Centroids Undersampling (CCU), Edited Nearest Neighbors (ENN), Tomek Links Undersampling (TLU), and One-Sided Selection (OSS). For RU and CCU the method removes the 80% of the majority class, while for ENN, TLU, and OSS, we specify that the undersampling is only performed on the majority class. In the case of our algorithm, we use the same hyperparameter values as described above.

Figure 4 shows Cohen's kappa scores (after performing 5-fold cross-validation) for all datasets before and after applying the undersampling methods. Besides coloring the scores to represent performance decrease (in red) or otherwise (in blue), we show the average score across all datasets reported for each method on the top of each violin plot. Our method reports the best performance overall, with the second smallest standard deviation after ENN. However, ENN is on par with TLU and OSS in terms of performance, while RU and CCU are more prone to lose performance after undersampling. It is worth mentioning that the average Cohen's kappa score associated with the original datasets is 0.68, thus confirming the added value of using the PCU method.

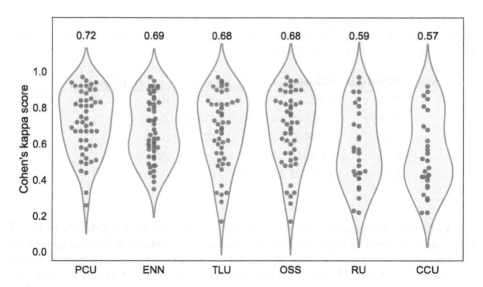

Fig. 4. Cohen's kappa scores (denoted as colored dots) computed by each algorithm for the synthetic datasets used for simulation. Red dots mean that the performance decreased compared to the baseline, while blue dots mean that the performance improved or remained unchanged. (Color figure online)

In order to test whether there are statistically significant differences in performance among the undersampling algorithms, we apply the non-parametric Friedman test. The null hypothesis H_0 for this test is that the algorithms result in negligible performance differences. The resulting p-value $= 0.0$ after rounding, meaning that there are statistically significant differences in the group of algorithms being compared. Subsequently, we use a Wilcoxon signed-rank test to determine whether the significant differences come from our undersampling method compared to each of the other methods. The null hypothesis H_0 asserts that there are no notable variances in the algorithm's performance among datasets, whereas the alternative hypothesis H_1 suggests the presence of significant differences. In addition, we use the Bonferroni-Holm post-hoc procedure to adjust the p-values produced by the Wilcoxon signed-rank test. This correction aims to control the family-wise error rate when performing a pairwise analysis by adjusting the p-values obtained in each comparison.

Table 1 portrays the results concerning the Wilcoxon test coupled with Bonferroni-Holm correction using the proposed PCU algorithm as the control method. In the table, R^- indicates the number of datasets for which PCU reports less performance than the compared algorithm, while R^+ gives the number of datasets for which the opposite behavior is observed. The corrected p-values computed by Bonferroni-Holm advocate for rejecting the null hypotheses in all cases for a significance level of 0.05 (corresponding to a 95% confidence interval). The fact that the proposed PCU method reports the largest R^+ values and that the null hypotheses are rejected in all pairwise comparisons allows us to conclude the superiority of our proposal for the generated datasets.

Table 1. Results concerning the Wilcoxon pairwise test with Bonferroni-Holm correction using PCU as the control algorithm.

Algorithm	Wilcoxon	R^-	R^+	Holm	Null Hypothesis
ENN	9.38E-04	11	33	9.38E-04	Reject
TLU	1.12E-06	2	32	3.37E-06	Reject
OSS	1.59E-06	2	31	3.37E-06	Reject
CCU	1.54E-11	3	47	7.72E-11	Reject
RU	6.54E-08	3	44	2.62E-07	Reject

As the last experiment, we show the reduction ratio for all datasets in Fig. 5 after applying undersampling. The area denotes the reduction percentage across all datasets (the larger the area, the smaller the undersampled dataset). Firstly, it is clear that the PCU method reduces a larger number of instances than the ENN algorithm without affecting the classifier's performance, as we concluded before. Secondly, although RU and CCU lead to the largest areas, they computed the worst prediction rates. Finally, TL and OSS barely modified the datasets, which explains their steady prediction rates.

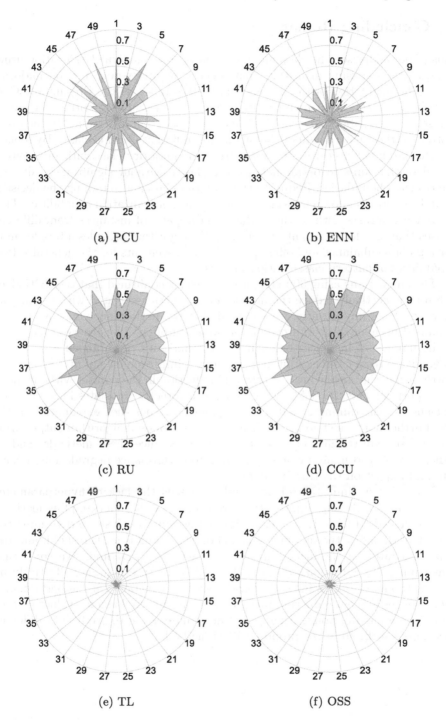

Fig. 5. Reduction ratio of instances that belong to the majority class reported by selected undersampling algorithms for each dataset. The numbers on the outer axis denote the indexes of synthetic datasets.

5   Concluding Remarks

This paper addressed the class imbalance problem in pattern classification using a data pre-processing algorithm that focuses on undersampling the majority class. The foundation of the proposed algorithm relies on the Presumably Correct Decision Sets, which differentiate between presumably correct instances (those with neighbors sharing the same class label), and presumably incorrect instances (those with neighbors belonging to a different decision class). By replacing the presumably correct instances from the majority class with prototypes, the algorithm aims to better balance the classes without significantly altering the decision boundaries. Reducing the majority class and retaining the decision boundaries would allow a classifier to increase its generalization capabilities. This approach contrasts with methods that remove pairs of instances from different classes that are closest neighbors. Overall, the algorithm provides a tool for mitigating class imbalance and offers potential improvements for the generalization capabilities of pattern classification models.

The simulations using 50 imbalanced problems confirmed that the PCU algorithm can effectively reduce the number of instances associated with the majority class. More importantly, we observed that the performance on unseen data remained unchanged or improved after undersampling. The observed reduction in the number of instances belonging to the majority class ranged from 8% to 60%, with certain cases exhibiting performance improvements of up to 28%. It is worth noting that achieving a substantial reduction in the dataset without sacrificing performance still holds great value, as it directly translates into improved efficiency when training classifiers or implementing post-hoc explanation methods. Furthermore, the findings indicate that the proposed algorithm outperforms existing state-of-the-art undersampling methods since these methods tend to either remove too many instances, leading to performance degradation or have minimal impact on the training data.

The primary limitation of our study concerns the lack of hyperparameter tuning since it will help improve the performance of all undersampling methods (ours included). In addition, conducting a sensitivity analysis to determine the impact of changing the number of neighbors, the confidence threshold, and the reduction ratio on the algorithm's performance is paramount. We conjecture that tuning the reduction ratio and the confidence threshold can effectively be done with binary search instead of grid search. For instance, if the performance decreases for a given reduction ratio, it will continue to do so for values greater than such a cutting point. Finally, it will be interesting to study how different classifiers benefit from the proposed PCU method.

References

1. Batista, G.E., et al.: Balancing training data for automated annotation of keywords: a case study. Wob **3**, 10–8 (2003)
2. Batista, G.E., Prati, R.C., Monard, M.C.: A study of the behavior of several methods for balancing machine learning training data. ACM SIGKDD Explorat. Newslett. **6**(1), 20–29 (2004)
3. Buda, M., Maki, A., Mazurowski, M.A.: A systematic study of the class imbalance problem in convolutional neural networks. Neural Netw. **106**, 249–259 (2018)
4. Chawla, N.V., Bowyer, K.W., Hall, L.O., Kegelmeyer, W.P.: Smote: synthetic minority over-sampling technique. J. Artif. Intell. Res. **16**, 321–357 (2002)
5. Cohen, J.: A coefficient of agreement for nominal scales. Educ. Psychol. Measur. **20**(1), 37–46 (1960)
6. Fernández, A., García, S., Galar, M., Prati, R.C., Krawczyk, B., Herrera, F.: Learning from imbalanced data sets. Springer, Cham (2018). https://doi.org/10.1007/978-3-319-98074-4
7. Han, H., Wang, W.-Y., Mao, B.-H.: Borderline-SMOTE: a new over-sampling method in imbalanced data sets learning. In: Huang, D.-S., Zhang, X.-P., Huang, G.-B. (eds.) ICIC 2005. LNCS, vol. 3644, pp. 878–887. Springer, Heidelberg (2005). https://doi.org/10.1007/11538059_91
8. Hart, P.: The condensed nearest neighbor rule (corresp.). IEEE Trans. Inf. Theory **14**(3), 515–516 (1968)
9. He, H., Bai, Y., Garcia, E.A., Li, S.: Adasyn: adaptive synthetic sampling approach for imbalanced learning. In: 2008 IEEE International Joint Conference on Neural Networks (IEEE World Congress on Computational Intelligence), pp. 1322–1328. IEEE (2008)
10. Kubat, M., et al.: Addressing the curse of imbalanced training sets: one-sided selection. In: ICML, vol. 97, p. 179. Citeseer (1997)
11. Last, F., Douzas, G., Bacao, F.: Oversampling for imbalanced learning based on k-means and smote. arXiv preprint arXiv:1711.00837 **2** (2017)
12. Lemaître, G., Nogueira, F., Aridas, C.K.: Imbalanced-learn: a python toolbox to tackle the curse of imbalanced datasets in machine learning. J. Mach. Learn. Res. **18**(17), 1–5 (2017)
13. Lundberg, S.M., Lee, S.I.: A unified approach to interpreting model predictions. Adv. Neural Inf. Process. Syst. **30** (2017)
14. Mani, I., Zhang, I.: KNN approach to unbalanced data distributions: a case study involving information extraction. In: Proceedings of Workshop on Learning from Imbalanced Datasets, vol. 126, pp. 1–7. ICML (2003)
15. Nápoles, G., Grau, I., Jastrzębska, A., Salgueiro, Y.: Presumably correct decision sets. Pattern Recognit. **141**, 109640 (2023)
16. Nguyen, H.M., Cooper, E.W., Kamei, K.: Borderline over-sampling for imbalanced data classification. Int. J. Knowl. Eng. Soft Data Paradigms **3**(1), 4–21 (2011)
17. Pedregosa, F., et al.: Scikit-learn: machine learning in python. J. Mach. Learn. Res. **12**, 2825–2830 (2011)
18. Tomek, I.: Two modifications of CNN. IEEE Trans. Syst. Man Cybernet. **6**, 769–772 (1976)
19. Wilson, D.L.: Asymptotic properties of nearest neighbor rules using edited data. IEEE Trans. Syst. Man Cybern. **3**, 408–421 (1972)

Leveraging Longitudinal Data
for Cardiomegaly and Change Detection
in Chest Radiography

Raquel Belo[1]($\boxtimes$)(iD), Joana Rocha[1,2](iD), and João Pedrosa[1,2](iD)

[1] Faculty of Engineering University of Porto, R. Dr. Roberto Frias s/n, 4200-465
Porto, Portugal
raquelmoraisbelo@gmail.com

[2] INESCTEC - Institute for Systems and Computer Engineering, Technology and
Science, Porto, Portugal

Abstract. Chest radiography has been widely used for automatic analysis through deep learning (DL) techniques. However, in the manual analysis of these scans, comparison with images at previous time points is commonly done, in order to establish a longitudinal reference. The usage of longitudinal information in automatic analysis is not a common practice, but it might provide relevant information for desired output. In this work, the application of longitudinal information for the detection of cardiomegaly and change in pairs of CXR images was studied. Multiple experiments were performed, where the inclusion of longitudinal information was done at the features level and at the input level. The impact of the alignment of the image pairs (through a developed method) was also studied. The usage of aligned images was revealed to improve the final mcs for both the detection of pathology and change, in comparison to a standard multi-label classifier baseline. The model that uses concatenated image features outperformed the remaining, with an Area Under the Receiver Operating Characteristics Curve (AUC) of 0.858 for change detection, and presenting an AUC of 0.897 for the detection of pathology, showing that pathology features can be used to predict more efficiently the comparison between images. In order to further improve the developed methods, data augmentation techniques were studied. These proved that increasing the representation of minority classes leads to higher noise in the dataset. It also showed that neglecting the temporal order of the images can be an advantageous augmentation technique in longitudinal change studies.

Keywords: Classification · Deep Learning · Longitudinal Image Pairs · Pathology · Thorax

1 Introduction

Chest X-Ray (CXR) imaging is a plain radiography technique that is commonly used to assess the thoracic area. Of all X-ray exams performed, 30–40% are CXR scans. This is because CXR scans are associated with fast acquisition times, low costs, and lower radiation exposure [11]. CXR images allow the visualization of

V. Vasconcelos et al. (Eds.): CIARP 2023, LNCS 14469, pp. 434–448, 2024.
https://doi.org/10.1007/978-3-031-49018-7_31

structures including the lungs and the heart. The analysis of such images is a laborious task, as there is a high volume of exams, displaying complex structures that may overlap. This factor led to the development of automated methods for CXR analysis, which intend to facilitate the job of radiologists and other medical professionals by, for instance, detecting and localizing abnormalities, or automatically generating reports. With the growth of such methods came the creation of multiple CXR datasets, such as ChestX-ray8 [15] and MIMIC-CXR [6], which can be used to train automatic abnormality detection systems. When the analysis is performed by human professionals, it is often done by comparing multiple scans from the same patient. It is important to look at images acquired at different time points simultaneously, so that a diagnosis can be done. This type of information is called longitudinal data, and consists of information from the same patient throughout time. The study of automated methods that employ longitudinal information to produce an output is, consequently, a field of high importance, as it allows a more realistic automation of the diagnosis process. However, the study of longitudinal information for automated analysis of CXR is still quite preliminary, since most systems normally consider a single input image at the time for analysis, isolating temporal information, rather than taking advantage of previously collected scans. With the rise of the COVID-19 pandemic, more studies that use sequential scans from the same patient arose, such as the ones in [3] and [12]. However, there is still much to uncover in this field, as the most common CXR datasets do not contain information specifically for longitudinal comparison, focusing on the abnormalities or findings in each individual image, and providing only the acquisition date or the age of the patient at the acquisition time, as for temporal information. In this work, the inclusion of longitudinal information was studied for the prediction of pathology and change (whether the abnormality remains or if it is no longer present) in a pair of CXR scans (information from the same patient in two different time points). The usage of longitudinal data may improve the performance of detection algorithms, by providing additional information. It may also increase the robustness and transparency of these methods, by providing a prediction with the reference of previous scans. Multiple experimental settings were designed, integrating this type of data in different manners in the used models, and allowing the study of its impact in the detection of pathology and in the comparison between scans. The used pathology in this study is cardiomegaly. Cardiomegaly is an abnormality that can be found in CXR, and it refers to the enlargement of the heart. Regarding the structure of this paper, initially, related work is exposed in Sect. 2. Then, the used methodology is described in Sect. 3, followed by the discussion of the results in Sect. 4. Finally, the conclusions can be found in Sect. 5. The main contributions of this paper are summarized below:

- Experimentation with the inclusion of longitudinal data at different points in a model for the prediction of pathology and change.
- Impact study of CXR alignment for the prediction of scans comparison.
- Proposal and comparison of a data augmentation technique for longitudinal problems.

2   Related Work

The majority of the developed automated methods for the analysis of CXR scans do not perform the comparison between exams, aiming at the analysis of a single image, as it is the most common form of image input and datasets are usually not equipped with longitudinal data for comparison. However, some studies have been developed regarding the utilization of longitudinal data for automated image analysis.

In [13], the focus is the detection of abnormalities on CXR scans and the detection of change in pathologies over sequential images. The abnormality detection is done through *Qure AI* (set of CNNs trained to identify certain diseases on frontal CXR). The used dataset is ChestX-ray8, where 874 scans were selected from. These images were annotated by two radiologists for four abnormalities. The image labels are compared between consecutive images, to assess the pathology change. The AUC reached by the method to detect change in the different pathologies ranges from 0.735 to 0.925, depending on the class. The results are compared with four test radiologists, which performed similarly or underperformed.

Another example can be found in [9], where the objective is to detect change in a lesion, when comparing two longitudinal images. A squeeze and excitation network (SENet) [5] is used to extract image features, which are used as local descriptors. A correlation score is computed for every possible local descriptor combination between the two images' feature maps, originating a geometric correlation map. Finally, a binary classifier is used to classify the sample as change or no change. The used dataset is a private CXR dataset from 5,472 patients. This method showed an AUC of 0.890, outperforming all the comparison methods, with an AUC of 0.780.

In [7] a model for tracking longitudinal relations between CXR (CheXRel-Net) is proposed. The used dataset is the Chest ImaGenome dataset [16]. A pre-trained Resnet-101 is used to extract global features, while a Graph Attention Network (GAT) [14] is used to extract local features, using the anatomical BB. GAT extracts inter and intra-image features, using an adjacency matrix that expresses these relationships. The global and local features are concatenated, and classification layers are added to provide the output. The output consists of the first image pathology labels, and in the second image an "improved" or "worsened" label. The final model outperformed the baseline models, which use only global or local information, presenting an accuracy of 0.680 (average the pathologies test accuracy).

In summary, a number of methods have been developed with the aim of utilizing sequential data in CXR analysis. Some of them focus on obtaining a better pathology detection or outcome prediction method by using longitudinal information, while others have the objective of predicting the difference between the images, which can be expressed as the presence of change or the presence of improvement or worsening in a pathology. In most of the methods, a DL model is used to extract features from the images, and in general, the inclusion of longitudinal information seems to improve the performance of the algorithms.

3 Methodology

3.1 Dataset

In this work, the ChestX-ray14 dataset was used. This dataset presents 14 labeled abnormalities, including atelectasis, cardiomegaly, effusion, infiltration, mass, nodule, pneumonia, pneumothorax, consolidation, edema, emphysema, fibrosis, pleural thickening, and hernia. ChestX-ray14 contains 112,120 frontal CXR images (size of $1,024 \times 1,024$ pixels) from 32,717 patients. Disease labels were generated from radiological reports using Natural Language Processing (NLP). Regarding this dataset, there are 13,302 patients with multiple images (patients with longitudinal information). In terms of image metadata, the dataset contains the patient age (at the time of acquisition) for each image as well as a follow-up number which refers to the sequential order of the scans of each patient. The minimum number of longitudinal frontal images found per patient is 2, and the maximum is 184. The median number of longitudinal frontal images per patient is 4.

Given the goal of introducing longitudinal information to predict pathology and change, a longitudinal dataset, based on the original one, was created. Here, only images of patients that have at least two scans are included, which are grouped into pairs of longitudinally acquired images. Each sample of the longitudinal dataset thus corresponds to two images with consecutive follow-up numbers, the first one being the oldest scan, and the second one (image pair) the most recent one. Thus, the original dataset was transformed into a longitudinal dataset composed of 81,315 samples.

Each image pair is associated with two labels: pathology detection and change detection.

- The **pathology detection** is related to the first image in the pair. Thus, an image was considered positive for the pathology if it has a positive label (1) for it, and negative (0) otherwise.
- The **change detection** is related to the comparison between the pathology labels in the two images in the pair. The presence of change is positive (1) if the pathology label is different between the two and negative (0) otherwise.

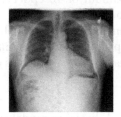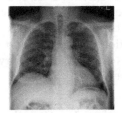

Fig. 1. Example of a longitudinal pair. The first image has a positive label for cardiomegaly, while its pair has a negative label for the abnormality. Thus, the change label is positive

In order to facilitate the interpretation of the results, as well to allow the development of multiple experiments and compare various methods, the performed experiments were based on an individual pathology: cardiomegaly. In Fig. 1 an example of an image pair is presented. This pair is part of the longitudinal dataset. In the longitudinal dataset, 1.75% of the samples are positive for both cardiomegaly and change. On the other hand, 1.69% are negative for cardiomegaly and positive for change, and 0.77% are positive for cardiomegaly and negative for change. 95.80% of the samples are negative for both cardiomegaly and change.

3.2   Experimental Settings

Different experiments were performed, aiming at the prediction of pathologic differences between longitudinal images. Four different experimental settings were designed, differing in the manner of integrating longitudinal information. In all of them, a ResNet-50 backbone was used, and the models were pre-trained with ImageNet [2].

In all the reported experiments, the images suffered the same pre-processing. The 3-channel images were resized to 256×256 pixels and normalized with the mean and standard deviation values of the ImageNet dataset. Data augmentation was used during training, by applying random affine transformations, including rotations from -5 to $5°$ and shear parallel to the x-axis, from -3 to $3°$. The used batch size was 8 in all training routines. The images were split into five folds, maintaining the original class distribution in each one and ensuring no patient overlap between folds. This class and patient distribution was preserved both in the original and longitudinal dataset. Three folds were used for training, while one was used for validation and one for testing.

The used loss function is the Binary Cross Entropy (BCE), commonly used in binary classification problems. The Adam optimizer [8] is used to compute the updated weights and biases. This optimizer was used with a defined learning rate of 10^{-4}. A scheduler was used to reduce the learning rate throughout training, which allows the model to reach the best possible performance. The validation loss is used as the controlling parameter, and the learning rate is reduced by a factor of 0.1 when the validation loss does not decrease for over 3 epochs. The training routine was extended for a maximum of 10 epochs, saving the model weights on the epoch that provided the best validation loss.

The described training conditions (including the model architecture, pretraining, loss function, scheduler, and optimizer) were maintained for all performed experiments unless stated otherwise. In Fig. 2 schemes of all experimental settings can be observed.

Experimental Setting 1 - Baseline. A baseline model was established for comparison with further experiments. This model focuses on a multi-label problem, predicting the 14 labeled pathologies in ChestX-ray14. This type of model was chosen as it allows the comparison of the performed experiments with the state-of-the-art, which often uses all available pathologies in a dataset. This baseline model was used to generate the probabilities of presence of each pathology, in each image of the test set. As in these experiments, individual labels were considered at a time, and only the results relative to cardiomegaly were kept. The change class was also predicted for the longitudinal test set. This was done by computing the absolute value of the difference of probability of pathology in a longitudinal pair, that is, $p_{change} = |p_{\text{pathology input image}} - p_{\text{pathology image pair}}|$.

Experimental Setting 2 - Features Model. In this experiment, the inclusion of longitudinal data was done at the image features level. This allows the assessment of the impact of pathology features for the longitudinal study. To do so, the encoder portion of the previously described multi-label model was used as a feature extractor. This was done by freezing all of its layers and removing the final fully connected layer. The deeper layers are used to generate the features for each image in a pair. The features are then concatenated (forming a features vector with double the length), before being fed to a new trainable dense classifier layer, which outputs the presence of the pathology in the input image and the change class for its pair. Training included 35 epochs, in this experimental setting, saving the model weights on the epoch that provided the best validation loss. The defined learning rate was 10^{-6}.

Experimental Setting 3 - Concatenated Model. This approach was developed in order to extract longitudinal features from the input level. In this experiment, both images from a longitudinal pair are used as the input to a model, aiming at the detection of both the presence of the pathology and the presence of change. The images in the pair are fed to the model after being concatenated, originating a 6-channel variable. The used model (ResNet-50 [4] backbone) was adapted to include a 6-channel input image, and return the two desired outputs. Training included 15 epochs, saving the model that provided the best validation loss. The defined learning rate was 10^{-6}.

Experimental Setting 4 - Concatenated Aligned Model. The previously described experiment (concatenated model) was repeated, but with image pairs that were aligned using a developed alignment method, which is further described in Sect. 3.3. Each image pair was aligned according to its reference, which is the first image in the pair.

3.3 Rigid CXR Alignment

An alignment algorithm was developed in order to align two CXR images. The alignment implies viewing the anatomical structures in the same regions of both images, thus helping to perceive any relevant differences between the pair and identify more easily the presence of a potential pathology. This developed method is mainly focused on the rigid alignment of the lungs, which includes rotation,

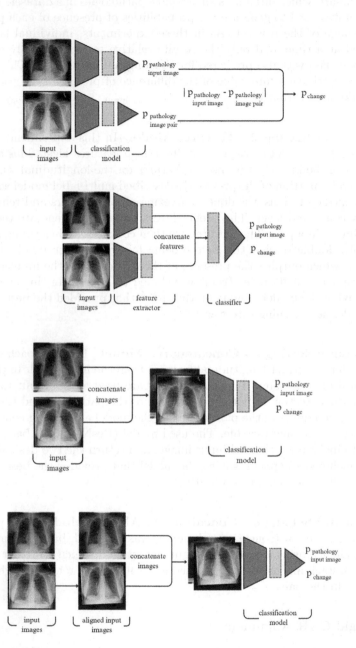

Fig. 2. Experimental settings schemes. From top to bottom: baseline model, features model, concatenated model, and concatenated aligned model

translation, and scaling. The features that are used to compute the transformations are lung segmentations and thoracic bounding boxes (BB). The lung segmentations were obtained using the work in [1]. For the thoracic BB generation, a localization algorithm in [10] was used.

In this algorithm, the lung masks are preprocessed and cleaned. Then, vertical and horizontal scaling parameters are computed, according to the reference image, followed by the computation of the final transformations. The final transformations to be applied are computed using Iterative Closest Point (ICP) algorithm [17], which is the main component of the alignment method. ICP is a classic technique for rigid image registration. This algorithm finds the rotation and translation parameters that better align two sets of points.

3.4   Longitudinal Data Augmentation

As previously described, the performed experiments have the objective of predicting both the presence of cardiomegaly (in the first image of the pair) and the change class between the two scans of the pair. The combination of the two exercises leads to four possible class combinations: [0,0], [1,0], [0,1] and [1,1], where the first label is relative to the presence of pathology, and the second one to the change class (for instance, [1,1] means that cardiomegaly is present in the first scan and not present in the second).

As previously mentioned, the cases that contain positive labels for pathology or change are far less common than doubly negative pairs. This fact could impede the learning process of longitudinal models as the low representation of the minority cases ([0,1], [1,0] and [1,1]) might hinder the learning of their representative features. Thus, as a possible solution, longitudinal data augmentation methods were explored.

Previously, the original dataset and the longitudinal dataset were described. The longitudinal dataset consists of pairs of consecutive scans from the same patient. However, as 9,189 patients have more than 2 images, multiple combinations can be done between the available scans. For instance, aside from combining two consecutive images in a pair, non-consecutive pairs can also be formed, either with a logical temporal order or an inverse temporal order. The datasets that can be formed by creating all possible combinations of images (independently of the time order) are hereinafter referred to as the **pseudolongitudinal datasets**. Different versions of pseudolongitudinal datasets were used in the augmentation studies, since they allow the usage of more minoritary pairs for training.

The fully pseudolongitudinal dataset (all possible pairs for each patient) yields a very high number of combinations in comparison with the longitudinal dataset. This fact would make the training process and the CXR alignment excessively computationally expensive, so, it was not used in this study. In alternative, other datasets derived from it were explored as less computationally expensive options. These datasets, that also aim at reducing the effect of class imbalance and leading to better final performance, are described in the following paragraphs.

- **Pseudolongitudinal Minoritary Dataset:** The first dataset consists of the longitudinal dataset (consecutive image pairs), to which the pseudolongitudinal combinations are added, but only if the formed sample belongs to a minority case ([1,1], [1,0] or [0,1]);
- **Pseudolongitudinal [1,0] Dataset:** Posteriorly, a similar dataset was generated, but where only the pseudolongitudinal samples with the [1,0] case were added to the longitudinal dataset. The objective of this experiment was introducing more minority samples in training, but without increasing the representation of the change class. As this class is by itself associated with a higher error (by being computed through the difference between two cardiomegaly labels), increasing its representation might hinder the model learning process;
- **Pseudolongitudinal < N Dataset:** Finally, another dataset was formed by using the longitudinal dataset and adding to it the pseudolongitudinal combinations formed by the patients with N or fewer images. This was done in order to generate an augmented dataset but without increasing too much number of samples. As for patients with more images, more combinations are possible, this restriction leads to the formation of multiple combinations only for patients with fewer images, while the remaining contribute with consecutive pairs only (longitudinal dataset). Experiments were performed for $N = 5$ and $N = 10$.

The pseudolongitudinal combinations were implemented only for the training and validation datasets. For testing, the original longitudinal dataset was used for all experiments.

In order to perform the initial experiments with the mentioned different datasets, the concatenated model (experimental setting 3, explained in Sect. 3.2) was trained again to predict the presence of cardiomegaly in the first image, and the presence of change in the image pair. This model was chosen since it is trained with longitudinal information from the input level, in opposition to the baseline and features models (experimental setting 1 and 2), which might be an advantage in inferring longitudinal features from the images. As the aligned concatenated model (experimental setting 4) depends on the previous alignment of the pairs, the concatenated model was chosen over it.

After determining the best longitudinal augmentation technique, the augmented dataset that provided the best final result for the concatenated model was used for training in the remaining experimental settings (2, the features model, and 4, the aligned concatenated model), maintaining the previously described conditions.

3.5   Evaluation and Metrics

The final results for each experiment were computed by applying the corresponding model to the longitudinal dataset test fold, containing 16,283 samples. The metrics used for evaluation are AUC, precision, recall, and accuracy.

4 Results and Discussion

4.1 Rigid CXR Alignment

This technique allowed the generation of aligned consecutive image pairs. As the ground truth of the alignment is not known, the evaluation of the method was done through the lung segmentations of the two images in the pair, which were used to compute the Dice Similarity Coefficient (DSC) score. The alignment of the pairs in the longitudinal dataset reached an average DSC of 0.895 ± 0.080.

The developed method was compared with a traditional rigid alignment solution, Scale-Invariant Feature Transform (SIFT). This comparison showed that the developed method is advantageous as it allows the alignment of all pairs by using anatomical features in the process, while SIFT fails at aligning a pair when the generated key points are too distinct. In a subset of 250 longitudinal pairs, SIFT failed at aligning 62 pairs. In this subset, the proposed method reached a DSC of 0.910 ± 0.063, while SIFT presented a DSC of 0.798 ± 0.205.

4.2 Pathology and Change Prediction

In Table 1, the results for the models can be seen. Regarding the detection of cardiomegaly, the baseline model (experimental setting 1) outperforms the remaining, with an AUC of 0.897. This is probably due to the fact that this model was trained with the original dataset, which contains more samples than the longitudinal version, and might have provided a better cardiomegaly detection performance. Concerning the detection of change, the features model outperforms the rest of the experiments, with an AUC of 0.858. The fact that the baseline model was not trained with longitudinal comparisons of the same patient might be the reason for its lower performance on change detection. The features model (experimental setting 2) performance in the pathology class is probably due to the usage of the same features as the baseline model (experimental setting 1) to reach the predictions. As the features from both images are used to compute the presence of cardiomegaly (and change), the similarity between these two cases was expected. Regarding the detection of change, the results show that the usage of features for cardiomegaly detection can be used to model the difference between two scans, predicting the label change when comparing the images in the pair.

In the concatenated model, the images are concatenated before feature extraction. The usage of all longitudinal information as input was thought to be an advantage, since the feature extraction could capture the difference between the scans from the beginning. However, this factor can be the main reason for the weaker results shown by this approach, as it might make the feature extraction process more difficult. It is clear however that the usage of aligned images in the concatenated model (experimental setting 4) improved the results. The alignment of the images leads to an alignment of the relevant structures in the scans and thus, it might facilitate the detection of features for comparison of the two images.

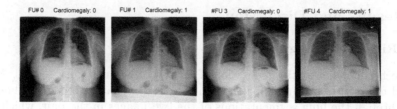

Fig. 3. Example of a case where similar images from the same patient display different ground truth labels for cardiomegaly

By looking at images from the same patient and corresponding ground truth labels, some situations where images appear practically equal, but have a different ground truth label can be found. In Fig. 3, an example of such a case is represented. It is important to acknowledge the existence of these cases, since in this dataset the labels were automatically generated from medical reports, and so, errors in the annotations can be present.

Table 1. Cardiomegaly and change detection results. Bold indicates the best result for each metric. *Card.* stands for Cardiomegaly. *Concat* stands for Concatenated

	1 - Baseline		2 - Features		3 - Concat.		4 - Concat. Aligned	
	Card.	Change	Card.	Change	Card.	Change	Card.	Change
AUC	**0.897**	0.824	0.893	**0.858**	0.833	0.795	0.868	0.820
Precision	0.096	**0.102**	**0.137**	0.088	0.079	0.09	0.078	0.076
Recall	**0.842**	0.716	0.734	**0.825**	0.734	0.679	0.803	0.818
Accuracy	0.799	**0.788**	**0.879**	0.722	0.782	0.769	0.762	0.678

4.3   Longitudinal Data Augmentation

The results obtained for the pseudolongitudinal datasets, all applied to the concatenated model (experimental setting 3) can be seen in Table 2. When comparing the model trained with the longitudinal dataset and the pseudolongitudinal minoritary dataset, the pathology detection metrics remain similar, and the change class metrics suffer from a small decrease. This fact is thought to be related to the noise associated with the change class. As previously explained, this class results from the combination of two cardiomegaly labels, thus, if one label is incorrect, this error will also be present in the change class. In the pseudolongitudinal dataset, as all possible combinations of images (from the same patient) are used, the noise associated with the change label is augmented. Thus, in this scenario, an incorrect cardiomegaly label can affect as many change annotations as the number of pairs the corresponding image is included in. For the pseudolongitudinal [1,0] dataset, as only the cases with negative change are considered, the noise associated with it is reduced. Thus, in this situation, both classes' metrics improve slightly, in comparison with both the longitudinal and pseudolongitudinal minoritary datasets. Regarding the pseudolongitudinal $<N$ datasets (for $N = 5$ and $N = 10$), both of them display a higher performance than the remaining experiments, with the pseudolongitudinal <10 dataset presenting the best AUC (0.863 for cardiomegaly and 0.827 for change). In these experiments, the representation of each case is similar to the longitudinal dataset, however, more samples are used for training. This fact is probably the reason these experiments provide the best results. The fact that the pseudolongitudinal <10 outperformed the pseudolongitudinal <5 dataset means that the usage of more combinations could be beneficial for predicting both the presence of cardiomegaly and change with higher performance.

As previously mentioned, the augmented pseudolongitudinal dataset experiments were also replicated on the remaining experimental settings. The chosen dataset was the pseudolongitudinal <5. This dataset was chosen over the pseudolongitudinal <10 dataset because there is only a slight performance improvement shown by the latter, and the concatenated aligned model (experimental setting 4) requires the previous alignment of each pair (with the first image is used as a reference). Thus, by using the pseudolongitudinal <5 dataset there is no need for aligning as many pairs as if the pseudolongitudinal <10 was used.

In Table 3 the results using the pseudolongitudinal <5 dataset for all experimental settings are shown. The features model (experimental setting 2) outperformed the remaining approaches, with an AUC of 0.896 for cardiomegaly and 0.863 for change. These results are similar to the ones presented for the longitudinal dataset and this experimental setting. Regarding the concatenated model (experimental setting 3), the results for the pseudolongitudinal dataset are better than the ones presented for the longitudinal dataset, with an increase of 2.2% for cardiomegaly and of 2.7% for change AUC. In the case of the concatenated aligned model (experimental setting 4), a very slight improvement is present, when comparing the pseudolongitudinal dataset with the longitudinal one. One possible hypothesis for explaining these results is the fact that training the models with more data (pseudolongitudinal augmentation) might facilitate the extraction of relevant features from the image pair. Also, as this augmentation leads to the combination of more temporally distant images, it is

Table 2. Results for the concatenated model (experimental setting 3) and the pseudolongitudinal dataset versions. Bold indicates the best result for each metric. *Card.* stands for Cardiomegaly. *Long.* stands for Longitudinal and *Psdlong.* stands for Pseudolongitudinal

	Long.		Psdlong. min.		Psdlong. [1,0]		Psdlong. < 5		Psdlong. <10	
	Card.	Change	Card.	Change	Card.	Change	Card.	Change	Card.	Change
AUC	0.833	0.795	0.830	0.775	0.849	0.808	0.855	0.822	**0.863**	**0.827**
Precision	0.079	0.090	0.062	0.058	0.082	0.079	**0.091**	0.090	0.088	**0.109**
Recall	0.734	0.679	**0.801**	**0.869**	0.746	0.777	0.736	0.741	0.781	0.691
Accuracy	0.782	0.769	0.695	0.547	0.789	0.701	**0.811**	0.752	0.796	**0.810**

possible that the change between them is more noticeable. The usage of these relevant features might lead to a better classifier. Thus, as the features model (experimental setting 2) already seemed to be extracting relevant features (for cardiomegaly, when using the longitudinal dataset), it is possible that the addition of new pairs did not result in improvement. As in this case, the change label is being predicted from the cardiomegaly features of both images, it is also possible that longitudinal features are not being further learned in this scenario. In the case of the concatenated model (experimental setting 3), the results using the longitudinal dataset suggest a weaker feature extraction. Thus, it is possible that the usage of more data (pseudolongitudal dataset) would lead to a more significant improvement in the final metrics. Regarding the concatenated aligned model (experimental setting 4), the usage of more information only led to a very slight improvement, thus, it is possible that the increase in information was overshadowed by the improvement provided by the alignment.

Table 3. Results for all experimental settings, using the pseudolongitudinal <5 dataset. Bold indicates the best result for each metric. *Card.* stands for Cardiomegaly. *Concat* stands for Concatenated.

	2 - Features		3 - Concat.		4 - Concat. Aligned	
	Card.	Change	Card.	Change	Card.	Change
AUC	**0.896**	**0.863**	0.855	0.822	0.876	0.829
Precision	**0.120**	0.091	**0.091**	0.090	0.105	0.087
Recall	**0.766**	0.825	0.736	0.741	0.746	0.772
Accuracy	**0.856**	0.730	0.811	**0.752**	0.837	0.735

5   Conclusions

Multiple experiments were carried out with the aim of integrating longitudinal data for the detection of pathology and change (comparison between the presence of pathology in two scans). This integration was performed at different levels. The best performance obtained for cardiomegaly detection was reached by the baseline model, with an AUC of 0.897, even though the features model reached a similar AUC of 0.893. Regarding the detection of change between the scans pair, the features model outperformed the remaining experiments, with an AUC of 0.858. The usage of the baseline model for feature extraction, allowing the computation of the presence of change, shows that longitudinal pathology data provides an advantage for the automatic comparison of exams. On the other hand, the usage of longitudinal pairs of images for training showed that the concatenation of the input images affects the prediction of the pathology and change. The usage of aligned images improved the cardiomegaly detection AUC by 3.5%, and the change detection AUC by 2.5%. The alignment of the structures probably facilitates the extraction of features relative to both images, which shows that it can be advantageous for longitudinal problems. Information from the two images in the pair is usually used to predict both the presence of pathology in the first image and the change in the pair. The use of non-consecutive longitudinal information as a data augmentation technique for the prediction of cardiomegaly and change in an image pair was explored. Experiments where the number of training samples was increased without affecting the ratio of each class seemed to improve the performance of the trained models. This shows that the usage of more image combinations during training can be used as an advantageous data augmentation technique, and it leads to the hypothesis that the higher the number of combinations used, the more notable the results will be. The noise associated with the change class (due to the accumulation of errors in the cardiomegaly label) proved to affect the performance of the model.

It is important to note that, in the performed experiments, only one of the pathology labels from ChestX-ray14 (cardiomegaly) was used. In future work, it should be a priority to validate the carried out experiments in the remaining abnormalities. It is also important to further validate these conclusions with more data and perform cross-validation. Few datasets include longitudinal information, which limits the available data for these type of studies. Furthermore, the possibility for the presence of wrong ground truth annotations was considered. This factor might be affecting the training and, thus, the formed conclusions. The usage of a higher number of pairs in the described data augmentation technique should also be assessed, in order to verify its advantage for acquiring better final metrics.

Acknowledgements. This work was funded by National Funds through the Portuguese funding agency, FCT - Portuguese Foundation for Science and Technology, within project LA/P/0063/2020. The work of J. Rocha was supported by the FCT grant contract *2020.06595.BD*.

References

1. Brioso, R.C., Pedrosa, J., Mendonça, A.M., Campilho, A.: Semi-supervised multi-structure segmentation in chest X-ray imaging. In: 2023 IEEE 36th International Symposium on Computer-Based Medical Systems (CBMS), pp. 814–820. IEEE (2023)
2. Deng, J., Dong, W., Socher, R., Li, L.J., Li, K., Fei-Fei, L.: ImageNet: a large-scale hierarchical image database. In: 2009 IEEE Conference on Computer Vision and Pattern Recognition, pp. 248–255 (2009)
3. Duanmu, H., et al.: Deep learning of longitudinal chest X-ray and clinical variables predicts duration on ventilator and mortality in COVID-19 patients. Biomed. Eng. Online **21**, 1–15 (2022)
4. He, K., Zhang, X., Ren, S., Sun, J.: Deep residual learning for image recognition. In: Proceedings of the IEEE Conference on Computer Vision and Pattern Recognition, pp. 770–778 (2016)
5. Hu, J., Shen, L., Sun, G.: Squeeze-and-excitation networks. In: 2018 IEEE/CVF Conference on Computer Vision and Pattern Recognition, pp. 7132–7141 (2018)

6. Johnson, A.E., et al.: MIMIC-CXR, a de-identified publicly available database of chest radiographs with free-text reports. Sci. Data **6**, 317 (2019)

7. Karwande, G., Mbakwe, A., Wu, J., Celi, L., Moradi, M., Lourentzou, I.: CheXRelNet: an anatomy-aware model for tracking longitudinal relationships between chest X-rays. In: Wang, L., Dou, Q., Fletcher, P.T., Speidel, S., Li, S. (eds.) MICCAI 2022. LNCS, vol. 13431, pp. 581–591. Springer, Cham (2022). https://doi.org/10.1007/978-3-031-16431-6_55

8. Kingma, D.P., Ba, J.: Adam: a method for stochastic optimization. In: Bengio, Y., LeCun, Y. (eds.) 3rd International Conference on Learning Representations, ICLR 2015, San Diego, CA, USA, 7–9 May 2015, Conference Track Proceedings (2015)

9. Oh, D.Y., Kim, J., Lee, K.J.: Longitudinal change detection on chest X-rays using geometric correlation maps. In: Shen, D., et al. (eds.) MICCAI 2019. LNCS, vol. 11769, pp. 748–756. Springer, Cham (2019). https://doi.org/10.1007/978-3-030-32226-7_83

10. Rocha, J., Pereira, S.C., Pedrosa, J., Campilho, A., Mendonşa, A.M.: Attention-driven spatial transformer network for abnormality detection in chest X-ray images. In: 2022 IEEE 35th International Symposium on Computer-Based Medical Systems (CBMS), pp. 252–257 (2022)

11. Schaefer-Prokop, C., Neitzel, U., Venema, H.W., Uffmann, M., Prokop, M.: Digital chest radiography: an update on modern technology, dose containment and control of image quality. Eur. Radiol. **18**, 1818–1830 (2008)

12. Shu, M., Bowen, R., Herrmann, C., Qi, G., Santacatterina, M., Zabih, R.: Deep survival analysis with longitudinal X-rays for COVID-19. In: 2021 IEEE/CVF International Conference on Computer Vision (ICCV), pp. 4026–4035. IEEE Computer Society, Los Alamitos, CA, USA (2021)

13. Singh, R., et al.: Deep learning in chest radiography: detection of findings and presence of change. PLOS ONE **13**, e0204155 (2018)

14. Velikovi, P., Cucurull, G., Casanova, A., Romero, A., Li, P., Bengio, Y.: Graph attention networks (2018)

15. Wang, X., Peng, Y., Lu, L., Lu, Z., Bagheri, M., Summers, R.M.: ChestX-ray8: hospital-scale chest X-ray database and benchmarks on weakly-supervised classification and localization of common thorax diseases. In: 2017 IEEE Conference on Computer Vision and Pattern Recognition (CVPR). IEEE (2017)

16. Wu, J., et al.: Chest ImaGenome dataset (2021)

17. Zhang, Z.: Iterative point matching for registration of free-form curves and surfaces. Int. J. Comput. Vision **13**(2), 119–152 (1994)

Self-supervised Monocular Depth Estimation on Unseen Synthetic Cameras

Cecilia Diana-Albelda[1]([✉])[ID], Juan Ignacio Bravo Pérez-Villar[1,2][ID],
Javier Montalvo[1][ID], Álvaro García-Martín[1][ID], and Jesús Bescós Cano[1][ID]

[1] Video Processing and Understanding Lab, Univ. Autónoma de Madrid,
28049 Madrid, Spain
{cecilia.diana,juanignacio.bravo,javier.montalvor}@estudiante.uam.es
{alvaro.garcia,j.bescos}@uam.es
[2] Deimos Space, 28760 Madrid, Spain
juan-ignacio.bravo@deimos-space.com

Abstract. Monocular depth estimation is a critical task in computer vision, and self-supervised deep learning methods have achieved remarkable results in recent years. However, these models often struggle on camera generalization, i.e. at sequences captured by unseen cameras. To address this challenge, we present a new public custom dataset created using the CARLA simulator [4], consisting of three video sequences recorded by five different cameras with varying focal distances. This dataset has been created due to the absence of public datasets containing identical sequences captured by different cameras. Additionally, it is proposed in this paper the use of adversarial training to improve the models' robustness to intrinsic camera parameter changes, enabling accurate depth estimation regardless of the recording camera. The results of our proposed architecture are compared with a baseline model, hence being evaluated the effectiveness of adversarial training and demonstrating its potential benefits both on our synthetic dataset and on the KITTI benchmark [8] as the reference dataset to evaluate depth estimation.

Keywords: Monocular Depth Estimation · Computer Vision · Self-Supervised Learning · Camera Generalization · Custom Synthetic Dataset · Adversarial Training

1 Introduction

Monocular depth estimation is the process in which a depth map is obtained from an RGB image. This is a fundamental task in computer vision, in particular for autonomous driving [9], computational medicine [14] and many others [1, 21].

Traditional methods typically use multi-view techniques through epipolar geometry or feature matching to perform depth estimation. However, these approaches have a major limitation in that they assume that the objects in the scene are rigid [10,31], which does not take into account deformable objects. Moreover, these methods cannot recover dense depth maps.

© Springer Nature Switzerland AG 2024
V. Vasconcelos et al. (Eds.): CIARP 2023, LNCS 14469, pp. 449–463, 2024.
https://doi.org/10.1007/978-3-031-49018-7_32

Deep learning (DL) is considered as a solution to overcome these limitations. Nevertheless, it requires training on very large datasets. To address this, self-supervised DL can be used to generate a supervisory signal from unlabeled data by exploiting the underlying structure on it through image reconstruction [9]. In addition, these models can be trained solely with monocular RGB sequences.

Self-supervised DL models for monocular depth estimation consist on performing feature extraction from a target frame and nearby ones. Then, a depth map from the target frame is predicted, as well as the relative pose changes from each pair of consecutive frames. Using this information, a reconstruction of the target frame is generated and used as the supervisory signal.

Even though these models show good performance on predicting depth maps from unlabeled data, they also show dependency to the intrinsic camera calibration parameters used during training, leading to a degradation of the results when applied to sequences recorded with different cameras [10,23].

To help overcome this limitation, we propose a method to increase the generalisation ability of the models to different cameras. We predict the camera that has captured each frame and we include it in the system in an adversarial manner, as we can see in Fig. 1. By doing this, we achieve a feature extraction process that encourages invariance to changes in the camera intrinsic parameters.

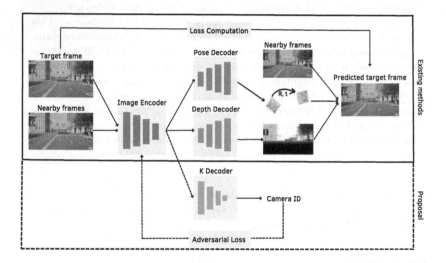

Fig. 1. Overview of self-supervised monocular depth estimation and the inclusion of adversarial training as our proposal. First, feature extraction is performed for a target frame and nearby ones (Image Encoder). Second, the pose change between each pair of consecutive frames is obtained (Pose Decoder), as well as the depth map of the target frame (Depth Decoder). Then, a reconstruction of the target frame is generated and used as the supervisory signal. Proposal: adding a classifier for the camera (K Decoder) in an adversarial manner to achieve features invariant to camera changes.

To support our research, we need a dataset containing identical sequences recorded by different cameras. As there is currently no public dataset that meets these characteristics, we have created a synthetic public dataset in which we can compare the same sequence taken by different cameras, thus allowing a fair comparison for the models.

2   Related Work

In this section, we are going to describe the concept of self-supervised monocular depth estimation and its main restrictions. Then, we particularise on the limitations introduced by the camera intrinsics.

2.1   Self-supervised Monocular Depth Estimation

Self-supervised monocular depth estimation allows inference of depth at the pixel level from images captured by a single camera, without the need of annotated data. It employs deep Convolutional Neural Networks (CNNs) to learn the relationship between input images and their corresponding depth maps. These models are trained in a self-supervised manner, comparing each image with a reconstruction of it as the supervisory signal.

2.2   Restrictions of Existing Methods

Restrictions for self-supervised monocular depth estimation methods can be categorized into three main areas: scale, photometric consistency, and camera model.

Scale. Scale consistency in depth estimation is assumed to ensure accurate results. However, it is inherently ambiguous in monocular depth estimation as these methods use a single image of the scene, thus posing a major challenge. Different approaches have been proposed to address this challenge, such as minimizing the difference between target and source depth [20], using inverse depth terms [28], or aligning depth estimations with sparse depth points [30].

Photometric Consistency. Self-supervised monocular depth estimation assumes static scenes for image reconstruction, where all pixels move relative to the camera's ego-motion. This assumption simplifies the depth estimation process but is limiting in dynamic scenes with moving objects or occlusions. Various techniques have been proposed to address this, as using a pose-explainability network [31], optical flow-based masking [12,17,27], depth difference measurements for occlusion handling [10], and weak supervision based on epipolar geometry [19].

Camera. Existing methods in this task assume constant intrinsic camera parameters, during both training and testing. While this is usually true in controlled environments, it often leads to a decrease in the generalization capability when using a different camera, thus limiting the applicability.

2.3   Approaches to Address Camera Restrictions

Several methods can be applied to deal with camera restrictions, which are mainly based on: adding camera information explicitly, continuous learning, fusion of multi-view geometry and DL or mixture of datasets.

Adding Camera Information Explicitly. One approach to address camera assumptions is by explicitly incorporating camera information into the learning process. This can be done: at the data level, by using synthetic and diverse datasets that simulate different camera parameters and situations [23]; and at the architecture level, by making changes to neural networks to adapt the convolution filters and introduce attention layers based on camera calibration [6]. These modifications improve the consistency and accuracy of depth estimation.

Continuous Learning. Continuous learning involves updating and adapting the model as new data is collected to overcome the limitations of static camera assumptions. This can be done, for example, by employing Bayesian inference and scene-independent geometric computations [13], which incorporate both optical flow and depth estimations. This approach enables fast adaptation to unseen scenes, but it requires online optimization and a big hardware support.

Fusion of Multi-View Geometry and DL. The fusion of multi-view geometry and DL [24] combines the spatial structure of the scene from multiple images with deep neural networks. By leveraging information from different viewpoints, more accurate and consistent depth estimations can be obtained. This approach uses traditional multi-view geometry methods to resolve camera variability issues and achieves end-to-end learning of 3D geometry.

Mixture of Datasets. The mixture of datasets strategy aims to address camera assumptions by combining diverse datasets, which can be mainly done by Multi-objective learning or through adversarial training. Multi-objective learning involves training the model using multiple datasets with variations in capture conditions, thus improving the generalization and accuracy of depth estimations [16]. Alternatively, adversarial training introduces a discriminator to identify the camera source to generate robust depth estimations across different cameras. This approach enables the system to learn invariant features and patterns with respect to the cameras, enhancing the generalization ability to changes in the intrinsic calibration parameters.

This paper focuses its research line on the use of adversarial training to mitigate the dependence of self-supervised monocular depth estimation models to changes in the intrinsic camera parameters.

3   Method

In order to have a reference of the validity of our work, we implement a baseline algorithm. Then, we describe our proposal, that is based on the inclusion of adversarial training to achieve feature invariance to the camera intrinsics.

3.1   Baseline Algorithm

Network Architecture. The architecture of this algorithm can be seen in Fig. 2. Firstly, there is an encoder which performs feature extraction. We follow the literature [9] and use a triplet of frames to ensure consistency: I_{t-1}, I_t and I_{t+1}. Subsequently, these features are used as input in two different decoders; one estimates the depth of I_t, D_t, and the other one predicts the pose changes between both pairs (I_t, I_{t-1}) and (I_t, I_{t+1}), representing them with transformation matrices, $T_{t \to t-1}$ and $T_{t \to t+1}$ respectively, where $T = [R|t]$, being R a rotation matrix and t a translation vector.

The encoder used is a standard Residual Neural Network, ResNet18 [11]. On the other hand, the Depth decoder is a U-Net [18] alike network, as it receives features at different resolution levels. Finally, the Pose decoder is a CNN that predicts rotation and translation from 2 feature vectors concatenated.

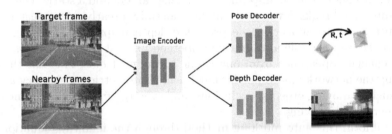

Fig. 2. Baseline Architecture. 'Image Encoder' performs feature extraction from a target frame (I_t) and two nearby ones (I_{t-1} and I_{t+1}). Then, from this features, 'Pose Decoder' predicts the relative pose change between each pair of consecutive frames, while 'Depth Decoder' estimates the depth map of I_t.

Bilinear Warping. We generate two reconstructions of the I_t frame, $\hat{I}_t$, one of them combining I_{t-1}, $T_{t \to t-1}$, the camera calibration matrix, K, and D_t, and the other one combining I_{t+1}, $T_{t \to t+1}$, K and D_t.

This computation is done according to the following formula [31]:

$$p_a \quad = \quad K T_{t \to a} D_t K^{-1} p_t, \tag{1}$$

being p_t the pixels of the target frame and p_a the pixels of the corresponding adjacent frame in each case. Intuitively, we first project p_t in the 3D world with

K^{-1}. However, as K projects the pixel up to a scale factor, we obtain this pixel in the 3D world through D_t. Then, we use the transformation matrix, $T_{t \to a}$, which contains the estimated translation vector t and rotation matrix R from I_t to I_a, to transform the pixel to the position of the adjacent frame. Finally, we use K to project it back to the camera dimension, thus obtaining p_a.

Loss Computation. The model computes the difference between the target frame and both reconstructions of it as the supervision signal.

This difference is computed by combining the L1 loss [29], the structural similarity index measure, SSIM, [25] and the smoothness loss [22]. This combination is done to ensure a better result both numerically and visually [25].

It is expressed as follows:

$$Loss \ = \ (1 - \alpha)L_1 + \alpha L_{ssim} + \beta L_{smooth}, \tag{2}$$

being $\alpha = 0.85$ and $\beta = 0.01$, as suggested in the literature [9].

Then, we average both losses, the one that extracts $\hat{I}_t$ from I_{t-1} and the one that uses I_{t+1} instead.

Auto-masking Stationary Pixels. Self-supervised monocular depth estimation operates under the assumption of a moving camera and a static scene. When this assumption breaks down, performance can suffer greatly, even causing 'holes' of infinite depth to appear in the predicted depth maps [15]. To mitigate this problem, we use a simple auto-masking method [9] that filters out pixels which do not change appearance from one frame to the next one. This has the effect of letting the network ignore pixels that break photometric consistency assumptions, such as moving objects, occlusions, and even to ignore whole frames when the camera stops moving.

We perform this auto-masking method through the following equation:

$$Loss^{i,j} \ = \ min\left(pe(I_t, I_{a \to t})^{i,j}, \ pe(I_t, I_a)^{i,j}\right), \tag{3}$$

where pe is the projection error defined by Eq. 2, $I_{a \to t}$ is the reconstruction of I_t from an adjacent frame I_a, and the terms i, j represent the pixel coordinates of the image.

We compare for each pixel of I_t what gives us the smallest error: using its corresponding value in a nearby frame or using its warped value. As this loss is a tensor of the same shape of the images, i.e. provides an error for each RGB pixel, we first compute the mean per channel and, then, global mean. Thus, a single error value per image is obtained.

3.2   Baseline Algorithm + Adversarial Training

In this section we describe our main contribution, which is based on adding robustness to the models so that the features extracted from each frame are invariant to changes in the intrinsic parameters of the camera that captured it. To do this, a K Decoder is added into the model architecture, as shown in Fig. 3, to classify the camera that has recorded each of the images. The proposed method involves training the architecture in an adversarial manner in which the Image Encoder acts as a feature generator and competes with the K Decoder as a discriminator, hence achieving a feature representation that remains independent of the camera model used. Apart from that, the same warping process is performed with respect to the baseline algorithm, as well as the same loss computation for the common part of the networks in Fig. 2.

The K Decoder is designed to be small, thus encouraging changes in the features. It consists of convolutional layers followed by LeakyReLU activation functions [26] and a final layer that generates the classifications.

As K is included in an adversarial manner in the algorithm, there are two different optimizers. The first one optimises the Image Encoder, Pose Decoder and Depth Decoder, whose aim is to obtain accurate depth and pose values while simultaneously maximizing the K Decoder loss. By doing this, the features extracted become indistinguishable between different cameras.

This optimization process is done according to the following equation:

$$L_{opt1} \quad = \quad (1 - \alpha)L_1 + \alpha L_{ssim} + \beta L_{smooth} - \gamma L_k, \quad (4)$$

being $\alpha = 0.85$ and $\beta = 0.01$, as suggested in the literature [9]. We take $\gamma = 0.001$ by experimental procedure, being L_k the Cross Entropy loss function.

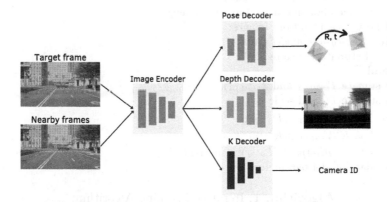

Fig. 3. Adversarial Training Architecture. The baseline algorithm for self-supervised monocular depth estimation contains an image encoder, a pose decoder and a depth decoder, as presented in Fig. 2. In this architecture there is an extra decoder, K, that is in charge of classifying the camera that recorded each frame. This decoder is included in an adversarial manner to achieve a feature extraction process invariant to changes in the camera parameters.

On the other hand, the second optimizer improves the prediction of the K decoder enhancing its ability to correctly identify the camera associated with each input image. Its objective is to optimize the network's proficiency in camera classification by:

$$L_{opt2} \quad = \quad L_k. \tag{5}$$

4   Experiments

4.1   Setup

Three models have been trained and evaluated: Base, Mult, and Multiseq-adv, the latter being our proposal.

Base model uses the baseline algorithm explained in Sect. 3.1 and has been trained with images of equal focal distance. Mult model also uses the Baseline architecture, but it has been trained with images of three different focal distances, hence capturing more camera variability. Finally, Multiseq-adv employs the Adversarial architecture proposed in Sect. 3.2 to enhance robustness to camera parameter changes, and it has been trained using the same images as Mult model, where three different focal distances are represented. The learning process is described in Algorithm 1.

Input : Dataset images, K
Output: Improved pose and depth estimators
foreach *iteration* **do**
    Randomly select I_t, I_{t-1} and I_{t+1};
    Introduce Gaussian noise into the images;
    Extract features from images;
    if *Multiseq-adv model* **then**
       | Estimate the camera ID;
    end
    Estimate $T_{t \to t-1}$ and $T_{t \to t+1}$;
    Estimate D_t and scale it;
    $p_{t-1}^{(1)} = K T_{t \to t-1} D_t K^{-1} p_t$;
    $p_{t+1}^{(2)} = K T_{t \to t+1} D_t K^{-1} p_t$;
    $Loss^{i,j} = min\,(pe(I_t, I_{a \to t})^{i,j}, pe(I_t, I_a)^{i,j})$;
    Update model parameters ;
end

Algorithm 1: Iterative Learning Algorithm.

We have trained all three models with Adam optimizer (lr $= 1e-5$), a batch size of 5, min_depth $= 0.1$ and max_depth $= 100$, for 300 epochs. These values have been selected according to the evaluation framework from [9].

Dataset. Our method requires a dataset in which the same sequence is available for cameras with different intrinsic parameters. However, since there is no public dataset that meets these conditions, we have generated a custom public dataset (available in [3]) that includes both RGB and depth images using CARLA simulator [4] through Unreal Engine 4 [7]. It comprises 6000 synthetic training images distributed across 3 video sequences, each captured by 5 different cameras with Field Of View (FOV) values of 40, 60, 80, 100, and 120. Therefore, we have 30000 images for the training phase.

On the other hand, there are 2 test sequences of 1200 images each, also captured by 5 different cameras, thus obtaining a total of 12000 test images.

It is important to note that Test Set 1 is captured in a complex urban environment with significant presence of buildings and traffic elements, while Test Set 2 takes place in a simpler rural environment, as we can see in Fig. 4. As a result, Test Set 1 can be considered more challenging due to its higher complexity, potentially yielding higher error values compared to Test Set 2.

Fig. 4. Images from Test Set 1 and Test Set 2 as urban and rural environments respectively, both belonging to the synthetic custom dataset created to enable comparisons of the same sequence captured by different cameras.

Evaluation Metrics. The metrics used to evaluate depth estimation are: Absolute Relative Error, RSE, RMSE, Log Scale Invariant RMSE [5] and Accuracy under a threshold [2].

4.2   Analysis of the Results

We validate that -1- models show a degradation of the performance when FOV value changes, and -2- adversarial training mitigates this effect. We evaluate our models with both test sequences of the custom created dataset in terms of depth estimation. Moreover, we also evaluate their performance on KITTI dataset [8].

Synthetic Dataset Results. Note that 'Cam' column in all figures and tables represents the camera used to record the test sequence that is being evaluated.

Table 1 shows the results of the models for each camera in Test Set 1.

Base model exhibits superior performance in Cam 1, suggesting its advantage in maintaining consistency between train and test images regarding intrinsic camera parameters. However, Base error values progressively rise as the camera changes, thus evidencing dependence on the parameters of the training camera.

Table 1. Results of the three models for depth estimation over Test Set 1. (KEY: Base model has been trained only with images taken from Cam 1; Mult model's training involves images from Cam 2, Cam 3 and Cam 4; Multiseq-adv model is our proposal, as it has been trained using images from Cam 2, Cam 3 and Cam 4 and it includes adversarial training for camera generalization).

Cam	FOV	Model	Abs Rel↓	Sq Rel↓	RMSE↓	RMSE log↓	$\delta < 1.25$↑	$\delta < 1.25^2$ ↑	$\delta < 1.25^3$ ↑
1	40	Base	**0.22**	**3.44**	**10.45**	0.27	**0.69**	**0.91**	**0.97**
		Mult	0.28	5.71	12.3	0.33	0.63	0.87	0.94
		Multiseq-adv	0.25	4.75	11.82	0.3	0.66	0.88	0.95
2	60	Base	0.26	**3.18**	**10.26**	0.3	0.6	**0.89**	**0.96**
		Mult	0.28	5.67	11.0	0.33	0.66	0.88	0.94
		Multiseq-adv	**0.24**	3.9	10.48	**0.29**	**0.67**	**0.89**	0.95
3	80	Base	0.3	**3.35**	10.42	0.35	0.51	0.83	**0.95**
		Mult	**0.26**	3.86	**9,47**	**0.31**	**0.69**	0.88	0.94
		Multiseq-adv	**0.26**	3.9	9.72	**0.31**	0.67	**0.89**	**0.95**
4	100	Base	0.37	3.79	10.62	0.41	0.43	0.74	0.89
		Mult	0.26	2.96	**8.46**	0.32	**0.68**	**0.87**	**0.94**
		Multiseq-adv	**0.24**	**2.83**	8.5	**0.31**	**0.68**	**0.87**	**0.94**
5	120	Base	0.39	3.64	11.45	0.48	0.33	0.66	0.84
		Mult	0.25	2.42	8.71	0.34	0.63	0.84	0.93
		Multiseq-adv	**0.23**	**2.32**	**8.63**	**0.32**	**0.65**	**0.86**	**0.94**

In contrast, Mult model shows a better consistency in the result as the FOV changes due to the higher variability in its training images. Nevertheless, Multiseq-adv outperforms the other two models in most error measures, also displaying enhanced generalization to the FOV value. In this case, error values remain consistent across different cameras, showcasing the benefits of adversarial training in improving the model's robustness to unseen cameras. Furthermore, Multiseq-adv achieves the best results on Cam 5, which is the only unknown camera for all the models.

Alternatively, Fig. 5 presents the depth predictions generated by the models for a specific RGB image captured by the lowest and highest FOV cameras.

As we can observe, Base model performs well on Cam 1 but shows considerable deterioration on Cam 5, resulting in blurred predictions. Alternatively, Mult model improves upon Base on Cam 5 but still exhibits some deterioration, specially on Cam 1. Finally, Multiseq-adv shows reduced blurring effects, outperforming both models for Cam 5 and also Mult model on Cam 1, accurately capturing nearby and further objects on unseen cameras.

Overall, after examining Table 1 and Fig. 5, Multiseq-adv model demonstrates the best performance in terms of depth prediction for Test Set 1.

Table 2 shows the error values obtained by each model in Test Set 2. In general, there is a significant reduction in errors compared to those achieved on Test Set 1, due to the lower complexity of this sequence.

Similar patterns to those from the Test Set 1 results are observed: Base performs the best on Cam 1, while Multiseq-adv outperforms the rest of the cases, showing lower errors on both seen and unseen cameras.

Cam	FOV	RGB	Ground truth	Base	Mult	Multiseq-adv
1	40	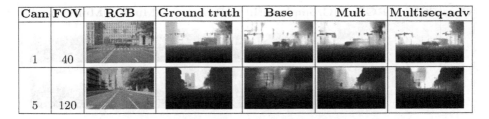				
5	120					

Fig. 5. Depth maps extracted from the predictions of the three models for a specific RGB image of Test Set 1. Note that darker pixels correspond to closer distances while brighter pixels are farther. (KEY: Base model has been trained only with images taken from Cam 1; Mult model's training involves images from Cam 2, Cam 3 and Cam 4; Multiseq-adv model is our proposal, as it has been trained using images from Cam 2, Cam 3 and Cam 4 and it includes adversarial training for camera generalization).

Table 2. Results of the three models for depth estimation over Test Set 2. (KEY: Base model has been trained only with images taken from Cam 1; Mult model's training involves images from Cam 2, Cam 3 and Cam 4; Multiseq-adv model is our proposal, as it has been trained using images from Cam 2, Cam 3 and Cam 4 and it includes adversarial training for camera generalization).

Cam	FOV	Model	Abs Rel↓	Sq Rel↓	RMSE↓	RMSE log↓	$\delta < 1.25$ ↑	$\delta < 1.25^2$ ↑	$\delta < 1.25^3$ ↑
		Base	**0.16**	**2.26**	**9.7**	**0.22**	0.74	**0.92**	**0.98**
1	40	Mult	0.21	4.15	11.69	0.26	0.75	0.89	0.96
		Multiseq-adv	0.18	3.19	10.95	0.23	**0.79**	**0.92**	**0.98**
		Base	0.18	**2.15**	10.02	0.26	0.72	0.89	0.96
2	60	Mult	0.18	3.47	10.44	0.26	0.78	0.89	0.96
		Multiseq-adv	**0.16**	2.98	**9.86**	**0.23**	**0.8**	**0.92**	**0.97**
		Base	0.2	2.57	10.81	0.34	0.7	0.85	0.91
3	80	Mult	0.17	2.53	8.69	0.25	0.79	0.91	**0.97**
		Multiseq-adv	**0.16**	**2.45**	**8.6**	**0.24**	**0.8**	**0.92**	**0.97**
		Base	0.22	2.93	11.56	0.42	0.67	0.81	0.88
4	100	Mult	0.2	2.46	8.44	0.3	0.74	0.86	0.92
		Multiseq-adv	**0.16**	**2.03**	**8.22**	**0.26**	**0.79**	**0.91**	**0.96**
		Base	0.25	3.9	13.47	0.54	0.62	0.77	0.83
5	120	Mult	0.21	2.36	9.29	0.34	0.71	0.84	0.9
		Multiseq-adv	**0.17**	**2.06**	**9.11**	**0.3**	**0.75**	**0.88**	**0.93**

Figure 6 displays the predictions of the models over an RGB image of Test Set 2. Base model performs well for Cam 1 but gets deteriorated results on Cam 5. Mult model shows a slight improvement but still struggles with generalization to unseen cameras, especially Cam 5. Finally, Multiseq-adv model provides the best depth predictions, even though some artifacts are observed in the top of the Cam 5 image, potentially caused by the presence of clouds.

KITTI Dataset Results. We have evaluated the monocular depth estimation capacity of the three models created using images from the KITTI dataset [8].

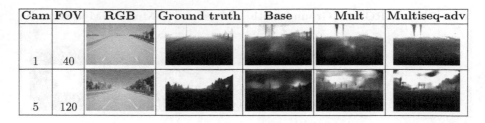

Cam	FOV	RGB	Ground truth	Base	Mult	Multiseq-adv
1	40					
5	120					

Fig. 6. Depth maps extracted from the predictions of the three models for a specific RGB image of Test Set 2. Note that darker pixels correspond to closer distances while brighter pixels are farther. (KEY: Base model has been trained only with images taken from Cam 1; Mult model's training involves images from Cam 2, Cam 3 and Cam 4; Multiseq-adv model is our proposal, as it has been trained using images from Cam 2, Cam 3 and Cam 4 and it includes adversarial training for camera generalization).

In this case, it should be noted that the model taken as a reference, Base, is a model that uses the Baseline algorithm detailed in Sect. 3.1 and that has been trained only with images taken by Cam 5, i.e. with FOV = 120.

This change in the training design for Base model is due to the fact that the KITTI dataset images have a focal length very close to 40, which is the one used by Cam 1 in the synthetic dataset, so using a model trained only with images from this camera would not be representative for this study.

Table 3. Results of the three models for depth estimation over KITTI evaluation dataset. (KEY: Base model has been trained only with images taken from Cam 5; Mult model's training involves images from Cam 2, Cam 3 and Cam 4; Multiseq-adv model is our proposal, as it has been trained using images from Cam 2, Cam 3 and Cam 4 and it includes adversarial training for camera generalization).

Model	Abs Rel↓	Sq Rel↓	RMSE↓	RMSE log↓	$\delta < 1.25$↑	$\delta < 1.25^2$ ↑	$\delta < 1.25^3$↑
Base	0.579	8.338	11.039	0.641	0.235	0.474	0.678
Mult	0.442	4.525	9.500	0.555	0.273	0.541	0.749
Multiseq-adv	**0.414**	**4.200**	**9.373**	**0.516**	**0.283**	**0.575**	**0.799**

As we can see in Table 3, the model that has been trained with images from a single camera, Base, gets significantly the worst results. On the other hand, Mult model improves considerably the error values, suggesting a higher generalisation capacity as it has been trained with three different cameras. Finally, Multiseq-adv model provides the best results, demonstrating the potential benefits of the use of adversarial training to mitigate the dependence of these models on the camera used during training.

On the other hand, Fig. 7 shows the visual predictions of the three models for an image of the KITTI evaluation set. As we can see, the depth map obtained with Base model is quite noisy, as we cannot intuit the structures present in the image. Mult model improves these results detecting parts of the car that is in front of the camera, although it still does not provide a good view of the depth information. Finally, Multiseq-adv model allows us to intuit the correct depth along the road, as well as the car and other structures in the frame.

Although Multiseq-adv model achieves the best predictions both numerically and visually, these is also a domain shift between the synthetic training images and the real evaluation ones. Nevertheless, it is demonstrated in this study that the inclusion of adversarial training is a potential improvement for the generalisation of models to unseen cameras in this task.

RGB	Base	Mult	Multiseq-adv

Fig. 7. Depth maps extracted from the predictions of the three models for a specific RGB image of KITTI Dataset. Note that the colors in the predictions of the models have been inverted for a better visualization i.e. closer pixels are whiter while further pixels are darker. (KEY: Base model has been trained only with images taken from Cam 5; Mult model's training involves images from Cam 2, Cam 3 and Cam 4; Multiseq-adv model is our proposal, as it has been trained using images from Cam 2, Cam 3 and Cam 4 and it includes adversarial training for camera generalization).

5 Conclusion/Discussion

In this work, the challenge of mitigating camera dependence on self-supervised monocular depth estimation models has been addressed. For this purpose, a synthetic public dataset has been created that allows a fair analysis of the effect that changes on the intrinsic camera parameters have on the models, which has been crucial for the development and evaluation of the project.

It has been observed evidence of degradation in the results when changing intrinsic camera parameters on models that are trained solely on one camera, highlighting the importance of achieving robustness of the models to these variations. Furthermore, the results obtained suggest improved performance when incorporating adversarial training into the model architecture.

In future work, we would like to evaluate the errors associated to the extracted pose values both on the created synthetic dataset and on real images.

Acknowledgments.. This work is part of the HVD (PID2021-125051OB-I00), IND2020/TIC-17515 and SEGA-CV (TED2021-131643A-I00) projects, funded by the Ministerio de Ciencia e Innovacion of the Spanish Government.

References

1. Antensteiner, D., Štolc, S., Huber-Mörk, R.: Depth estimation with light field and photometric stereo data using energy minimization. In: Beltrán-Castañón, C., Nyström, I., Famili, F. (eds.) CIARP 2016. LNCS, vol. 10125, pp. 175–183. Springer, Cham (2017). https://doi.org/10.1007/978-3-319-52277-7_22
2. Cadena, C., Latif, Y., Reid, I.D.: Measuring the performance of single image depth estimation methods. In: 2016 IEEE/RSJ International Conference on Intelligent Robots and Systems (IROS), pp. 4150–4157. IEEE (2016)
3. Diana, C., Bravo, J.I., J.M.l.G.J.B.: UNSYN-MF dataset: unified synthetic multiple FOV (2023). http://www-vpu.eps.uam.es/publications/UnSyn-MF_Dataset/
4. Dosovitskiy, A., Ros, G., Codevilla, F., Lopez, A., Koltun, V.: CARLA: an open urban driving simulator. In: Proceedings of the 1st Annual Conference on Robot Learning, vol. 78, pp. 1–16 (2017)
5. Eigen, D., Puhrsch, C., Fergus, R.: Depth map prediction from a single image using a multi-scale deep network. In: Advances in Neural Information Processing Systems, vol. 27 (2014)
6. Facil, J.M., Ummenhofer, B., Zhou, H., Montesano, L., Brox, T., Civera, J.: CamConvS: camera-aware multi-scale convolutions for single-view depth. In: Proceedings of the IEEE/CVF Conference on Computer Vision and Pattern Recognition, pp. 11826–11835 (2019)
7. Games, E.: Unreal engine 4 (2019). https://www.unrealengine.com
8. Geiger, A., Lenz, P., Urtasun, R.: Are we ready for autonomous driving? The KITTI vision benchmark suite. In: 2012 IEEE Conference on Computer Vision and Pattern Recognition, pp. 3354–3361. IEEE (2012)
9. Godard, C., Mac Aodha, O., Firman, M., Brostow, G.J.: Digging into self-supervised monocular depth estimation. In: Proceedings of the IEEE/CVF International Conference on Computer Vision, pp. 3828–3838 (2019)
10. Gordon, A., Li, H., Jonschkowski, R., Angelova, A.: Depth from videos in the wild: unsupervised monocular depth learning from unknown cameras. In: Proceedings of the IEEE/CVF International Conference on Computer Vision, pp. 8977–8986 (2019)
11. He, K., Zhang, X., Ren, S., Sun, J.: Deep residual learning for image recognition. In: Proceedings of the IEEE Conference on Computer Vision and Pattern Recognition, pp. 770–778 (2016)
12. Li, H., Gordon, A., Zhao, H., Casser, V., Angelova, A.: Unsupervised monocular depth learning in dynamic scenes. In: Conference on Robot Learning, pp. 1908–1917. PMLR (2021)
13. Li, S., Wu, X., Cao, Y., Zha, H.: Generalizing to the open world: deep visual odometry with online adaptation. In: Proceedings of the IEEE/CVF Conference on Computer Vision and Pattern Recognition, pp. 13184–13193 (2021)
14. Liu, X., et al.: Dense depth estimation in monocular endoscopy with self-supervised learning methods. IEEE Trans. Med. Imaging 39(5), 1438–1447 (2019)
15. Luo, C., et al.: Every pixel counts++: joint learning of geometry and motion with 3d holistic understanding. IEEE Trans. Pattern Anal. Mach. Intell. 42(10), 2624–2641 (2019)
16. Ranftl, R., Lasinger, K., Hafner, D., Schindler, K., Koltun, V.: Towards robust monocular depth estimation: mixing datasets for zero-shot cross-dataset transfer. IEEE Trans. Pattern Anal. Mach. Intell. 44(3), 1623–1637 (2020)

17. Ranjan, A., et al.: Competitive collaboration: joint unsupervised learning of depth, camera motion, optical flow and motion segmentation. In: Proceedings of the IEEE/CVF Conference on Computer Vision and Pattern Recognition, pp. 12240–12249 (2019)

18. Ronneberger, O., Fischer, P., Brox, T.: U-net: convolutional networks for biomedical image segmentation. In: Navab, N., Hornegger, J., Wells, W.M., Frangi, A.F. (eds.) MICCAI 2015. LNCS, vol. 9351, pp. 234–241. Springer, Cham (2015). https://doi.org/10.1007/978-3-319-24574-4_28

19. Shen, T., et al.: Beyond photometric loss for self-supervised ego-motion estimation. In: 2019 International Conference on Robotics and Automation (ICRA), pp. 6359–6365. IEEE (2019)

20. Tang, J., et al.: Self-supervised 3D keypoint learning for ego-motion estimation. In: Conference on Robot Learning, pp. 2085–2103. PMLR (2021)

21. Vieira, A.W., Nascimento, E.R., Oliveira, G.L., Liu, Z., Campos, M.F.M.: STOP: space-time occupancy patterns for 3D action recognition from depth map sequences. In: Alvarez, L., Mejail, M., Gomez, L., Jacobo, J. (eds.) CIARP 2012. LNCS, vol. 7441, pp. 252–259. Springer, Heidelberg (2012). https://doi.org/10.1007/978-3-642-33275-3_31

22. Wang, C., Buenaposada, J.M., Zhu, R., Lucey, S.: Learning depth from monocular videos using direct methods. In: Proceedings of the IEEE Conference on Computer Vision and Pattern Recognition, pp. 2022–2030 (2018)

23. Wang, W., Hu, Y., Scherer, S.: Tartanvo: a generalizable learning-based vo. In: Conference on Robot Learning, pp. 1761–1772. PMLR (2021)

24. Wang, Y., Luo, K., Chen, Z., Ju, L., Guan, T.: Deepfusion: a simple way to improve traditional multi-view stereo methods using deep learning. Knowl.-Based Syst. **221**, 106968 (2021)

25. Wang, Z., Bovik, A.C., Sheikh, H.R., Simoncelli, E.P.: Image quality assessment: from error visibility to structural similarity. IEEE Trans. Image Process. **13**(4), 600–612 (2004)

26. Xu, J., Li, Z., Du, B., Zhang, M., Liu, J.: Reluplex made more practical: Leaky ReLU. In: 2020 IEEE Symposium on Computers and communications (ISCC), pp. 1–7. IEEE (2020)

27. Yin, Z., Shi, J.: Geonet: unsupervised learning of dense depth, optical flow and camera pose. In: Proceedings of the IEEE Conference on Computer Vision and Pattern Recognition, pp. 1983–1992 (2018)

28. Zhan, H., Weerasekera, C.S., Garg, R., Reid, I.: Self-supervised learning for single view depth and surface normal estimation. In: 2019 International Conference on Robotics and Automation (ICRA), pp. 4811–4817. IEEE (2019)

29. Zhao, H., Gallo, O., Frosio, I., Kautz, J.: Loss functions for image restoration with neural networks. IEEE Trans. Comput. Imaging **3**(1), 47–57 (2016)

30. Zhao, W., Liu, S., Shu, Y., Liu, Y.J.: Towards better generalization: joint depth-pose learning without posenet. In: Proceedings of the IEEE/CVF Conference on Computer Vision and Pattern Recognition, pp. 9151–9161 (2020)

31. Zhou, T., Brown, M., Snavely, N., Lowe, D.G.: Unsupervised learning of depth and ego-motion from video. In: Proceedings of the IEEE Conference on Computer Vision and Pattern Recognition, pp. 1851–1858 (2017)

Novelty Detection in Human-Machine Interaction Through a Multimodal Approach

José Salas-Cáceres$^{(\boxtimes)}$ ⓘ, Javier Lorenzo-Navarro ⓘ, David Freire-Obregón ⓘ, and Modesto Castrillón-Santana ⓘ

Universidad de Las Palmas de Gran Canaria,
Instituto Universitario SIANI, Las Palmas, Spain
jose.salas@ulpgc.es

Abstract. As the interest in robots continues to grow across various domains, including healthcare, construction and education, it becomes crucial to prioritize improving user experience and fostering seamless interaction. These human-machine interactions (HMI) are often impersonal. Our proposal, built upon previous work in the field, aims to use biometric data of individuals to detect whether a person has been encountered before. Since many models depend on a threshold set, an optimization method using a genetic algorithm was proposed. The novelty detection is made through a multimodal approach using both voice and facial images from the individuals, although the unimodal approaches of just each single cue were also tested. To assess the effectiveness of the proposed system, we conducted comprehensive experiments on three diverse datasets, namely VoxCeleb, Mobio and AveRobot, each possessing distinct characteristics and complexities. By examining the impact of data quality on model performance, we gained valuable insights into the effectiveness of the proposed solution. Our approach outperformed several conventional novelty detection methods, yielding superior and therefore promising results.

Keywords: Novelty Detection · Human-Machine Interaction · Biometrics

1 Introduction

The interest in robots continues to rise over the years [19], and this growing fascination is well-founded. These machines have demonstrated a multitude of applications in various domains such as healthcare [21], construction [22], among others. Consequently, there is an increasing number of human-machine interactions (HMI) involving what are known as social robots [24]. Social robots are specifically designed to interact with humans and typically assist them in different tasks. Enhancing the user experience poses a challenge in creating more natural and personal interactions in this scenario. It has been observed that

© Springer Nature Switzerland AG 2024
V. Vasconcelos et al. (Eds.): CIARP 2023, LNCS 14469, pp. 464–479, 2024.
https://doi.org/10.1007/978-3-031-49018-7_33

Table 1. Modified extract from a table obtained from [5]

Task	Training Classes	Test Classes	Objective
Traditional Classifier	KKCs	KKCs	Classify data into one of the known classes.
Reject Option Classifier	KKCs	KKCs	Classify data and reject samples with low confidence.
Outlier Detection	KKCs and some KUCs samples	KKCs and KUCs	Detect outliers in the data.
Novelty Detection	KKCs	KKCs and UUCs	Differentiate between UUCs and KKCs.
Open-Set Classifier	KKCs	KKCs and UUCs	Identify samples belonging to known classes and categorize them correctly if they do belong.

people respond better to HMI if they are recognized by the robot, only if the interaction is after a previous encounter. Therefore, this work aims to develop a HMI model capable of detecting whether a person has been encountered before. This model would utilize biometric data of the individuals. By doing so, if these individuals run into the same robot again, the model would allow the robot to recognize them. To summarize, the goal is to design a novelty detection model for individuals based on biometrics and with the capability of efficiently and quickly expanding the database of enrolled identities.

2   Related Work

2.1   Terminology

Several terms will be used throughout this work in the context of novelty detection. First, we introduce the concept of Out of Distribution (OOD) data, which refers to data encountered during model exploitation that were not present in the training set. There are two types depending on their relationship with the original domain: novelties and anomalies. Novelties are related to the working domain, while anomalies are not. Next, a classification of the different types of classes based on their appearance in the training set and the knowledge about them is made, resulting in four categories [5,14]:

- **Known Known Class** (KKC): Refers to classes belonging to the known categories used to train the model.
- **Known Unknown Class** (KUC): Represent classes belonging to a class not in the KKCs but represented in the training set.
- **Unknown Unknown Class** (UUC): Denotes classes that belong to unknown categories and are not encountered during the training phase.
- **Unknown Known Class** (UKC): Indicates classes belonging to known categories but with no specific samples in the training set; instead, only another type of information is known.

Based on this classification, several tasks arise, presented in Table 1. Among these tasks, this work focuses on novelty detection. To achieve this, a multimodal approach will be employed, combining data to enhance the performance of the models. Specifically, facial images and voice recordings are going to be used. Due to the transient nature of the interaction, the amount of data taken from each person will be limited. As we said before, the database of enrolled persons has to be able to expand continuously. Therefore, selecting models poses a challenge, as some methods require a long time to train with the new data. This limitation will exclude models based on deep learning and neural networks with high training computational demands.

2.2 Existing Modeling Architectures

In the literature, multiple models have been proposed to address the task of novelty detection. Some will be mentioned here, especially those considered the most suitable for the selected scenario.

First, density-based models are considered, examining the spatial density across different regions to discern whether a sample is a novelty. Among these models, Density-Based Spatial Clustering of Applications with Noise (DBSCAN) [3] stands out as a clustering algorithm that groups samples based on proximity and classifies those located further of a certain distance as noise or, in our case, as a novelty. Another noteworthy approach is the Local Outlier Factor (LOF) [1], which leverages the notion of local density computed via distances to the K nearest neighbors. Shifting to classifier-based methods, various one-class classifiers are used in the literature. Below we employ the One-Class Support Vector Machine (OCSVM) [16]. Like other SVM-based models, this algorithm aims to maximize the margin between samples of distinct classes by using a hyperplane, being the one class of the binary classifier, identities in the dataset, ergo not novelties. Furthermore, Support Vector Data Description (SVDD) [20] has a similar principle but employs a hypersphere to enclose the classes instead of a hyperplane. Additionally, Isolation Forest (IF) [8] is often used in novelty detection. This algorithm, rooted in decision trees, strives to isolate samples. The underlying concept is that if a sample is quickly isolated, it is likely to be an outlier; conversely, if it is challenging to segregate from the rest, it is not an outlier. Lastly, we consider two models based on probability. Gaussian Mixture Model (GMM) fits Gaussian distributions to the available data. Alternatively,

Kernel Density Estimation (KDE) [6] estimates the density of a given set of points by aggregating different kernels, such as Gaussian or exponential distributions. Then, these probability-based models compute a probability of a sample belonging to the distribution made, setting a threshold is possible to differentiate between regular new samples and OOD data. There are other methods for novelty detection beyond those mentioned, including those based on reconstruction or deep learning [15,25]. It is also important to assert that many algorithms depend on setting up a threshold, limiting the number of ramifications for IF or the probability of belonging to a specific distribution like in KDE.

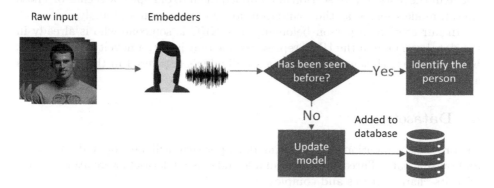

Fig. 1. Open-set classifier scheme.

3   Methodology

Figure 1 depicts the intended behavior of the proposal. Upon detecting a person, the robot captures his/her voice and face. This raw data then undergoes a preprocessing stage, being transformed into feature vectors. These vectors are then fed into trained models to determine if the person is known or unknown. The human-machine interface (HMI) proceeds uninterrupted if the person is recognized. However, if the person identity is unknown, the robot updates the database with the new identity.

As it was mentioned above, in our experiments we did not directly process the raw audio or image data. Instead, we employed a preprocessing step to convert the samples into fixed-dimensional numerical vectors known as embeddings. These embeddings were generated using specific neural networks called embedders, which were trained specifically for this task.

- For voice samples, we utilized the **X-Vector** network [18], which is trained to discriminate between different speakers. It was designed to convert audio of variable duration into fixed-dimensional vectors.

– For facial images, we employed the **FaceNet** network [17], which is trained to map facial images to a Euclidean space where distances reflect facial similarity. Similar to X-Vector, FaceNet generates fixed-dimensional embeddings.

By employing these dedicated embedders, we were able to transform the raw voice and facial image data into standardized and informative numerical representations. The two embedders used in our approach generate vectors of 512 elements each. In the multimodal approach, we concatenate these voice and face embeddings, resulting in a final feature vector of 1024 dimensions.

For our practical experiments, we selected a subset of those models described above in the related work section and applied them to our specific scenario. Those chosen models served as the foundation for our evaluation and analysis.

In our context, a person belonging to a KKC is someone who is already in the database. One in the UUC represents one that is not, a novelty. There is not KUC, however, that would encompass individuals registered in the database as a unknown.

4   Datasets

As previously mentioned, a multimodal approach will be adopted, requiring audiovisual data. Three distinct available audiovisual datasets were utilized, each with its characteristics and complexities.

First, let us describe AveRobot [9], a dataset specifically created with HMI in mind. The dataset consists of approximately 10-second videos where different individuals simulate interactions with a robot. These videos were recorded with eight different sensors in several indoor locations of a realistic and everyday environment, precisely the common spaces of a university building. Given those real life characteristics, the illumination conditions in these locations were not optimal, resulting in poor image quality, including noise, blurriness, and lighting issues. The audio quality suffers from a similar condition. The AveRobot dataset comprises samples from 111 individuals, most falling within 15 to 25 years. This dataset has been successfully used for multimodal user verification [4]. Another audiovisual dataset evaluated is VoxCeleb 1 [13], which consists of interviews with celebrities posted on Youtube. This dataset contains samples from 1251 celebrities from around the world. While the dataset exhibits large diversity, there is a predominant representation of males and native English speakers. Furthermore, owing to the data extraction source, the image and audio quality in VoxCeleb 1 are exceptionally high. Because of this, this dataset may not entirely represent the data one would encounter when attempting to integrate a model into a HMI environment. Lastly, the audiovisual dataset Mobio [7] was studied. This dataset was recorded using two mobile devices: one being a Nokia N93i mobile phone and the other being a standard 2008 MacBook laptop. The dataset consists of over 61 h of audiovisual data with 12 distinct sessions usually separated by several weeks. In total there are 192 unique audiovideo samples for each participant. This data was captured at 6 different sites over one and a half years with people speaking English. In this paper, we used the training and evaluation partitions, this two subsets have a total of 92 identities. The distinction in quality between the three datasets can be seen in Fig. 2.

(a) AveRobot samples [9].   (b) VoxCeleb samples [13].   (c) Mobio samples [7].

Fig. 2. Example images of the datasets.

Table 2. Characteristics of Data Subsets. Ids stands for identities.

Set	# known ids	# unknown ids	# images per id	# audios per id
Train	50	0	50	2
Validation	50	50	30	1
Test	50	50	30	1

5 Experiments

All experiments in our study followed a standardized structure, consisting of the following steps:

1. **Data loading:** Prior to conducting the experiments, three subsets of data were generated for each dataset described in Sect. 4: training, validation, and testing. The specific characteristics of each subset can be found in Table 2. It should be noted that for the experiments conducted with the Mobio dataset, 46 individuals were used per set, and 40 images were used per sample in the training set. This was due to limitations in the number of identities available in it.

2. **Model training:** After loading the data, each evaluated model was trained using the training set. Hyperparameter tuning was performed using a grid search with different combinations and leveraging the validation set. In some cases, we also tested different methods for calculating the threshold. It is important to note that during the training phase, there were no unknown samples. For threshold calculation, only elements from the training set, which all belonged to the Known Known Classes (KKCs), were used.

3. **Performance testing:** The model's performance was evaluated using the test set. Two separate tests were conducted: one to assess the model's ability to detect known samples and another to evaluate its capability in detecting novel samples. The final results were derived from a combination of the model's performance in both tests.

It is worth mentioning that for most models, we conducted a small search to identify the best possible hyperparameters. Furthermore, each model was tested using each dataset. In the following subsections, we provide a brief overview of the different experiments conducted in our study.

5.1   Distance-Based Experiment

The first experiment conducted is a variant of the Nearest Class Mean (NCM) algorithm [12], a classification algorithm that calculates a centroid for each class, and then the distances between each new sample and all the centroids are computed, and the sample is assigned to the class whose have the nearest centroid. This work has designed a modified version to adapt NCM for novelty detection.

The proposed experiment has k classes, where k is determined by the number of known individuals in the database at a specific moment. Each class is defined by n_i samples and a centroid $\bar{X}_i$, calculated using the formula 1.

$$\bar{X}_i = \frac{1}{n_i} \sum_{j=1}^{ni} \boldsymbol{X}_{ij} \tag{1}$$

When considering a new sample $\boldsymbol{X}_{new}$, it can either belong to one of the KKCs or be a novelty. This is determined by a threshold Thr_i, which can be calculated in various ways, always as a linear combination of some of the distances presented in the formulas: 2, 3, 4, 5, 6 and 7:

$$\boldsymbol{d_m} = \frac{1}{n_i} \sum_{j=1}^{ni} ||\boldsymbol{X}_{ij}, \bar{X}_i||_2 \tag{2}$$

$$\boldsymbol{d_M} = \max_{j=1,2,..,n_i} ||\boldsymbol{X}_{ij}, \bar{X}_i||_2 \tag{3}$$

$$\boldsymbol{D_m} = \frac{1}{k} \sum_{i=1}^{k} (\frac{1}{n_i} \sum_{j=1}^{ni} ||\boldsymbol{X}_{ij}, \bar{X}_i||_2) \tag{4}$$

$$\boldsymbol{D_M} = \max_{i=1,2,..,k} (\max_{j=1,2,..,n_i} ||\boldsymbol{X}_{ij}, \bar{X}_i||_2) \tag{5}$$

$$\boldsymbol{D_{mM}} = \frac{1}{k} \sum_{i=1}^{k} (\max_{j=1,2,..,n_i} ||\boldsymbol{X}_{ij}, \bar{X}_i||_2) \tag{6}$$

$$\boldsymbol{D_{Mm}} = \max_{i=1,2,..,k} (\frac{1}{n_i} \sum_{j=1}^{ni} ||\boldsymbol{X}_{ij}, \bar{X}_i||_2) \tag{7}$$

Thr_i forms a radius around $\bar{X}_i$, as shown in Fig. 3. According to expression 8, if the new sample falls within the influence zone of $\bar{X}_i$ it will be considered part of one of the known classes (KKC). If not, it will be considered a novelty (UUC).

$$\begin{cases} \boldsymbol{X}_{new} \in \text{KKC}, & \text{if } \exists i \text{ s.t. } ||\boldsymbol{X}_{new}, \bar{X}_i||_2 <= Thr_i \\ \boldsymbol{X}_{new} \notin \text{KKC}, & \text{if } ||\boldsymbol{X}_{new}, \bar{X}_i||_2 > Thr_i \forall i \end{cases} \tag{8}$$

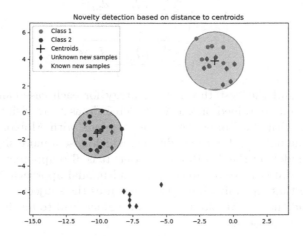

Fig. 3. Visual representation of the explained process for the NCM variant. The crosses depict the centroids, the diamonds represent the new samples, and the dots represent the existing samples in the databases. It is observed how the radii delimit the space that belongs to each class.

This experiment was conducted in five different cases, which differed in the type of data used, the centroid calculation process, or in the application of an additional step:

- **Unimodal:** To have a basis for comparison with the multimodal approach, the generated feature vectors from voice and images were separately used for analysis.
- **Multimodal:** The feature vectors from the unimodal cases were concatenated, forming a vector with twice the dimension of the original embeddings.
- **Multimodal variants:** Two variants were explored. One applies a dimensionality reduction technique, such as PCA, to the concatenated vectors. The second variant utilized GMM to calculate multiple centroids per class instead of a single centroid. These centroids corresponded to the mean positions of the Gaussian distributions that the GMM fitted to the data of each individual. Principal Component Analysis (PCA) was applied to achieve a 95% level of representation, resulting in a reduced dimensionality of 85 elements.

Various strategies for the Thr calculation were tested to find the one that gives the best results in each case. The different strategies are represented in Table 3, each σ the result of the formulas 9, 10 and 11.

$$\sigma_d = \sqrt{\frac{\sum_{j=1}^{n_i}(\boldsymbol{X}_{ij} - \bar{X}_i)^2}{n_i - 1}} \tag{9}$$

$$\sigma_m = \sqrt{\frac{\sum_{i=1}^{k}(\boldsymbol{d}_m - \bar{d}_m)^2}{k - 1}} \tag{10}$$

$$\sigma_M = \sqrt{\frac{\sum_{i=1}^{k}(\boldsymbol{d}_M - \bar{d}_M)^2}{k - 1}} \tag{11}$$

The results and the best threshold strategy for each case can be found in Table 4. The results obtained on the VoxCeleb 1 dataset are visibly better than those achieved with AveRobot, the same happens with Mobio. This can be attributed to the noise and data conditions, which pose a more significant challenge for the models in the AveRobot dataset. It is also apparent that the best results for each dataset are obtained from multimodal approaches. However, it is worth noting that the unimodal option that uses the subject's face also yields good results and that, in Mobio, use the voice alone lead to much better results that in the other datasets.

Table 3. Different methods used to calculate the threshold in the experiments

Threshold strategy	Equation
T_0	$Thr = 3 * \sigma_d$
T_1	$Thr = d_m + 3 * \sigma_d$
T_2	$Thr = d_M$
T_3	$Thr = d_M - 3 * \sigma_d$
T_4	$Thr = D_M$
T_5	$Thr = D_M + 3 * \sigma_M$
T_6	$Thr = D_M - \sigma_M$
T_7	$Thr = D_M - 3 * \sigma_d$
T_8	$Thr = D_{mM} + 3 * \sigma_M$
T_9	$Thr = D_{mM} + 3 * d_m$
T_{10}	$Thr = D_{Mm} + \sigma_m$
T_{11}	$Thr = D_{Mm} + 3 * \sigma_m$

Table 4. Results obtained for the three datasets in each one of the cases. Highlighted values correspond to best results for each set.

Experiment	AveRobot			VoxCeleb 1			Mobio		
	F1	Acc.	Thr	F1	Acc.	Thr	F1	Acc.	Thr
Face	0.8857	0.8953	T_3	0.9837	0.9837	T_2	0.9725	0.9728	T_3
Voice	0.0000	0.5000	T_0	0.5915	0.7100	T_5	0.9200	0.9130	T_5
Multimodal	0.8636	0.8557	T_{10}	**0.9977**	**0.9977**	T_{10}	0.9683	0.9674	T_6
Multi. PCA	**0.8982**	**0.8967**	T_5	0.9910	0.9910	T_1	0.9831	0.9833	T_9
Multi. GMM	0.3931	0.6223	T_0	0.9934	0.9933	T_5	**0.9856**	**0.9855**	T_5

5.2 Distribution and Density-Based Experiments

The second experiment involved applying the previously explained KDE (Kernel Density Estimation) algorithm. The novelty detection in this algorithm is also based on setting a threshold, in this case, on the score contributed by the model indicating the likelihood of a sample belonging to the distribution constructed with the train data. The threshold Thr will be the same for all the samples and will be calculated so that all training samples are always considered known. To achieve this, Thr is set at the 0th and 100th percentiles, ensuring that any new sample obtaining a score outside the original distribution of scores will be considered a novelty. The aforementioned score is calculated as the logarithm of the estimated density.

The density-based experiment utilizes a variant of the DBSCAN algorithm called Hierarchical DBSCAN (HDBSCAN) [2]. HDBSCAN applies the original DBSCAN algorithm with different radius values and integrates the results to find the most stable clustering [10]. The implementation used [11] can generate a score representing the probability of a new sample being OOD. This score is calculated using the GLOSH algorithm, a variant of the mentioned LOF that compares the density of the space where a sample is located with the density of the samples associated with it [2]. Similar to KDE, novelty detection sets a threshold Thr between the 0th and 100th percentiles.

The results obtained using KDE and HDBSCAN can be found in Table 5. Similar to the previous case, the performance achieved with VoxCeleb 1 and Mobio are significantly better than those achieved with AveRobot. Its worth noting that KDE performs better than HDBSCAN in those dataset with better sample quality (Mobio and VoxCeleb 1) but in AveRobot HDBSCAN is more effective, this is because HDBSCAN is designed to exhibit more robust behavior in noisy situations compared to KDE, which is more sensitive to the noise.

5.3 Non Threshold-Based Models

In addition to the threshold-based models mentioned above, we also evaluated some classification-based models that do not require any adjustment for threshold calculation. They rely on training with the available data. The results can

Table 5. Results obtained using KDE and HDBSCAN for the three datasets.

	AveRobot		VoxCeleb 1		Mobio	
Experiment	F1	Acc.	F1	Acc.	F1	Acc.
KDE	0.5811	0.6333	**0.9488**	**0.9513**	**0.9528**	**0.9522**
HDBSCAN	**0.7004**	**0.6703**	0.9105	0.9167	0.9094	0.9025

be seen in Table 6. We can notice that the results are much better in VoxCeleb 1 and Mobio. Additionally, these models do not adapt well to the specific problem a hand, as in almost all cases, accuracy higher than 60% is not achieved. An exception to this is seen in the OCSVM though, which achieve an accuracy above 70% in VoxCeleb 1 and Mobio. Another notable point is the substantial difference between the F1-Score and accuracy in some cases, such as KNN. This is due to a high disparity between Recall and Precision, indicating that either the model classified almost all new samples as known or all samples were considered novelties.

Table 6. Results using a variety of models in each dataset. Highlighted values correspond to best results.

	AveRobot		VoxCeleb 1		Mobio	
Model	F1	Acc.	F1	Acc.	F1	Acc.
OCSVM	0.6028	0.4773	**0.7540**	**0.7043**	**0.7816**	**0.7435**
LOF	0.5135	0.5207	0.2100	0.5587	0.4108	0.6290
IF	**0.6633**	**0.5043**	0.6709	0.5053	0.7475	0.6736
SVDD	0.6046	0.5100	0.5798	0.5077	0.6537	0.5540
KNN	0.1550	0.5420	0.1355	0.5363	0.2330	0.5659

6   Performance Optimization

In order to improve the performance of models that rely on a threshold value for determining novelty or known samples, the genetic algorithm (GA) was utilized, which is a bio-inspired heuristic optimization method [23]. A single-objective approach was adopted, where the accuracy of the models was maximized. To achieve this, the chromosome was encoded to explore various methods of calculating the threshold for each case and test previously unexplored combinations of hyperparameters. The threshold optimization was performed using train and validation subsets. Once the final performance was obtained, the configuration that yielded to the best results was tested on the validation set, resulting in the values presented in Table 8.

6.1 NCM-Based Algorithm

In this case, the chromosome consisted of six genes $[G_1, G_2, G_3, G_4, G_5, G_6]$, each limited to vary within the range of -5.0 to 5.0, being always a rational number. From these genes, Thr_i was calculated using the expression described in Eq. 12.

$$Th_i = G_1 * d_{mi} + G_2 * \sigma_d + G_3 * D_{mM} + G_4 * \sigma_M + G_5 * D_m + G_6 * \sigma_m \quad (12)$$

6.2 Distribution and Density-Based Approaches

For both models, the chromosome structure follows the same pattern, four gens $[G_1, G_2, G_3, G_4]$ where two are used for hyperparameter exploration and the other two for Thr calculation. The structure is shown in Table 7. The objective of this organization is twofold: firstly, to explore different configurations of hyperparameters and secondly, to vary the threshold location, all to improve the performance. The threshold will be calculated as indicated in Eq. 13, where L represents the limits obtained from the G_3 percentile and the $(100 - G_3)$ percentile and σ_{scores} is the standard deviation of the scores obtained from the training set.

Table 7. Codification of the chromosomes used in the GA for KDE and HDBSCAN.

KDE		
Gen	Hyperparameter	Search Space
G_1	kernels	$0 < G_1 < 3.$ $\quad G_1 \in \mathbb{N}$
G_2	bandwidth	$0 < G_2 < 1.$ $\quad G_2 \in \mathbb{R}$
G_3	Percentile	$95 < G_3 < 100.$ $\quad G_3 \in \mathbb{N}$
G_4	σ_{scores}	$-5 < G_4 < 1.$ $\quad G_2 \in \mathbb{R}$
HDBSCAN		
Gen	Hyperparameter	Search Space
G_1	min_cluster_size: 10	$2 < G_1 < 15.$ $\quad G_1 \in \mathbb{N}$
G_2	min_samples	$2 < G_2 < 15.$ $\quad G_2 \in \mathbb{N}$
G_3	Percentile	$95 < G_3 < 100.$ $\quad G_3 \in \mathbb{N}$
G_4	σ_{scores}	$-1 < G_4 < 5.$ $\quad G_2 \in \mathbb{R}$

$$Th = L_{G_3} \pm G_4 * \sigma_{scores} \quad (13)$$

In Table 8, a comparison of the results obtained by applying GA concerning the initial results is presented. Overall, there is an improvement, whether more or less significant, in the performance. Although all the results may not improve significantly, it is important to note that the thresholds shown in Table 3 were

Table 8. The results obtained by applying the GA. In each of the metrics △ represent the difference with the result achieved in the previous experiment.

AveRobot				
Experiment	F1-Score	△F1-Score	Accuracy	△Accuracy
NCM. Face	0.889	0.0033	0.8953	0
NCM. Voice	0.6095	0.6095	0.59	0.09
NCM. Multimodal	0.8966	0.033	0.8997	0.044
NCM. Multimodal. PCA	0.9046	0.0064	0.911	0.0143
NCM. Multimodal. GMM	0.8228	0.4297	0.8393	0.217
KDE	0.724	0.1429	0.723	0.0897
HDBSCAN	0.6983	−0.0021	0.6923	0.022
VoxCeleb 1				
Experiment	F1-Score	△F1-Score	Accuracy	△Accuracy
NCM. Face	0.9891	0.0054	0.989	0.0053
NCM. Voice	0.9184	0.3269	0.92	0.21
NCM. Multimodal	0.9973	−0.0004	0.9973	−0.0004
NCM. Multimodal. PCA	0.997	0.006	0.997	0.006
NCM. Multimodal. GMM	0.9973	0.0039	0.9973	0.004
KDE	0.9902	0.0414	0.9903	0.039
HDBSCAN	0.9065	−0.004	0.912	−0.0047
Mobio				
Experiment	F1-Score	△F1-Score	Accuracy	△Accuracy
NCM. Face	0.9803	0.0078	0.9804	0.0076
NCM. Voice	0.9247	0.0047	0.9239	0.0109
NCM. Multimodal	0.9942	0.0259	0.9942	0.0268
NCM. Multimodal. PCA	0.9772	−0.0059	0.9772	−0.0061
NCM. Multimodal. GMM	0.9902	0.0046	0.9902	0.0047
KDE	0.988	0.0352	0.988	0.0358
HDBSCAN	0.9273	0.0179	0.925	0.0225

obtained through trial and error, requiring multiple attempts and with no theoretical base. In contrast, using the GA to calculate these values requires minimal human intervention. Additionally, in the results of the experiment using only voice, a significant improvement is observed. This is likely because a good expression was not found in the trial-and-error process for setting the threshold. The same applies to the multimodal application of GMM in AveRobot.

7   Conclusions

Various techniques for novelty detection of individuals based on their biometrics have been developed throughout this work. The NCM-based algorithm has demonstrated the best performance in every dataset, specifically its multimodal application combining facial image and voice data. Furthermore, due to the high dimensionality of the problem, applying dimensionality reduction techniques such as PCA has been shown to decrease complexity without sacrificing performance.

Another essential aspect observed during the study is that threshold-based models, such as the mentioned implementation of NCM or KDE, are highly dependent on the proper adjustment of their hyperparameters. Therefore, it is considered good practice to use optimization methods to explore various combinations to find a practical expression. This is particularly important when considering that the optimal strategy for threshold calculation not only varies with the nature of the data but also with the dataset employed.

It is worth mentioning that the conditions under which the data is collected significantly impact the performance. This is evident in the apparent differences in results between the datasets used. Finally, distance-based or density-based models have been deemed the best option due to the data limitation and the desired training agility.

The logical next step following this work is to develop an Open-Set classifier that can not only successfully perform novelty detection but also classify the identified samples into their respective KKC. Additionally, despite the emphasis on agility mentioned earlier, exploring solutions based on deep learning, particularly those that fall under the Few-Shot learning paradigm, which requires only a small amount of training data, would be interesting.

Acknowledgments. This work is partially funded by the Spanish Ministry of Science and Innovation under project PID2021-122402OB-C22 and by the ACIISI-Gobierno de Canarias and European FEDER funds under project ULPGC Facilities Net and Grant EIS 2021 04, it is also supported by "Programa Investigo" refference code 32/39/2022-0923131539 of Servicio Canario de Empleo. "Fondos del Plan de Recuperación, Transformación y Resiliencia - Next Generation EU".

References

1. Breunig, M.M., Kriegel, H.P., Ng, R.T., Sander, J.: LoF: identifying density-based local outliers. SIGMOD Rec. **29**(2), 93–104 (2000). https://doi.org/10.1145/335191.335388
2. Campello, R.J.G.B., Moulavi, D., Zimek, A., Sander, J.: Hierarchical density estimates for data clustering, visualization, and outlier detection. ACM Trans. Knowl. Discov. Data **10**(1) (2015). https://doi.org/10.1145/2733381
3. Ester, M., Kriegel, H.P., Sander, J., Xu, X.: A density-based algorithm for discovering clusters in large spatial databases with noise. In: Proceedings of the Second International Conference on Knowledge Discovery and Data Mining. KDD'96, pp. 226–231. AAAI Press (1996)

4. Freire-Obregón, D., Rosales-Santana, K., Marín-Reyes, P.A., Penate-Sanchez, A., Lorenzo-Navarro, J., Castrillón-Santana, M.: Improving user verification in human-robot interaction from audio or image inputs through sample quality assessment. Pattern Recogn. Lett. **149**, 179–184 (2021). https://doi.org/10.1016/j.patrec.2021.06.014

5. Geng, C., Huang, S., Chen, S.: Recent advances in open set recognition: a survey. CoRR abs/1811.08581 (2018)

6. Hu, W., Gao, J., Li, B., Wu, O., Du, J., Maybank, S.: Anomaly detection using local kernel density estimation and context-based regression. IEEE Trans. Knowl. Data Eng. **32**(2), 218–233 (2020). https://doi.org/10.1109/TKDE.2018.2882404

7. Khoury, E., El Shafey, L., McCool, C., Günther, M., Marcel, S.: Bi-modal biometric authentication on mobile phones in challenging conditions. Image Vision Comput. 1147–1160 (2014). https://doi.org/10.1016/j.imavis.2013.10.001

8. Liu, F.T., Ting, K.M., Zhou, Z.H.: Isolation forest. In: 2008 Eighth IEEE International Conference on Data Mining, pp. 413–422 (2008). https://doi.org/10.1109/ICDM.2008.17

9. Marras, M., Marín-Reyes, P.A., Navarro, J.J.L., Santana, M.F.C., Fenu, G.: Averobot: an audio-visual dataset for people re-identification and verification in human-robot interaction. ICPRAM (Setúbal) (2019). https://doi.org/10.5220/0007690902550265

10. McInnes, L., Healy, J.: Accelerated hierarchical density based clustering. In: 2017 IEEE International Conference on Data Mining Workshops (ICDMW). IEEE, November 2017. https://doi.org/10.1109/icdmw.2017.12

11. McInnes, L., Healy, J., Astels, S.: HDBScan: hierarchical density based clustering. J. Open Source Softw. **2**(11), 205 (2017)

12. Mensink, T., Verbeek, J., Perronnin, F., Csurka, G.: Distance-based image classification: generalizing to new classes at near-zero cost. IEEE Trans. Pattern Anal. Mach. Intell. **35**(11), 2624–2637 (2013). https://doi.org/10.1109/TPAMI.2013.83

13. Nagrani, A., Chung, J.S., Zisserman, A.: Voxceleb: a large-scale speaker identification dataset. In: INTERSPEECH (2017)

14. Salehi, M., Mirzaei, H., Hendrycks, D., Li, Y., Rohban, M.H., Sabokrou, M.: A unified survey on anomaly, novelty, open-set, and out-of-distribution detection: solutions and future challenges. CoRR abs/2110.14051 (2021)

15. Schlegl, T., Seeböck, P., Waldstein, S.M., Schmidt-Erfurth, U., Langs, G.: Unsupervised anomaly detection with generative adversarial networks to guide marker discovery. CoRR abs/1703.05921 (2017)

16. Schölkopf, B., Williamson, R.C., Smola, A., Shawe-Taylor, J., Platt, J.: Support vector method for novelty detection. In: Solla, S., Leen, T., Müller, K. (eds.) Advances in Neural Information Processing Systems, vol. 12. MIT Press (1999)

17. Schroff, F., Kalenichenko, D., Philbin, J.: Facenet: a unified embedding for face recognition and clustering. In: 2015 IEEE Conference on Computer Vision and Pattern Recognition (CVPR), pp. 815–823 (2015). https://doi.org/10.1109/CVPR.2015.7298682

18. Snyder, D., Garcia-Romero, D., Sell, G., Povey, D., Khudanpur, S.: X-vectors: robust DNN embeddings for speaker recognition. In: 2018 IEEE International Conference on Acoustics, Speech and Signal Processing (ICASSP), pp. 5329–5333 (2018). https://doi.org/10.1109/ICASSP.2018.8461375

19. Stock-Homburg, R.: Survey of emotions in human-robot interactions: perspectives from robotic psychology on 20 years of research. Int. J. Soc. Robot. **14**(2), 389–411 (2022)

20. Tax, D.M., Duin, R.P.: Support vector data description. Mach. Learn. **54**(1), 45–66 (2004)
21. Uluer, P., Kose, H., Gumuslu, E., Barkana, D.E.: Experience with an affective robot assistant for children with hearing disabilities. Int. J. Soc. Robot. **15**(4), 643–660 (2023)
22. Wang, X., Liang, C.J., Menassa, C.C., Kamat, V.R.: Interactive and immersive process-level digital twin for collaborative human-robot construction work. J. Comput. Civ. Eng. **35**(6), 04021023 (2021)
23. Whitley, D.: A genetic algorithm tutorial. Stat. Comput. **4**(2), 65–85 (1994)
24. Youssef, K., Said, S., Alkork, S., Beyrouthy, T.: A survey on recent advances in social robotics. Robotics **11**(4) (2022). https://doi.org/10.3390/robotics11040075
25. Zhou, C., Paffenroth, R.C.: Anomaly detection with robust deep autoencoders. In: Proceedings of the 23rd ACM SIGKDD International Conference on Knowledge Discovery and Data Mining. KDD '17, pp. 665–674. Association for Computing Machinery, New York, NY, USA (2017). https://doi.org/10.1145/3097983.3098052

Filtering Safe Temporal Motifs
in Dynamic Graphs for Dissemination
Purposes

Carolina Jerônimo[1,2], Simon Malinowski[1,3], Zenilton K. G. Patrocínio
Jr.[2], Guillaume Gravier[3], and Silvio Jamil F. Guimarães[2(✉)]

[1] Université de Rennes 1, Rennes, France
carolinajeronimo@gmail.com
[2] PUC Minas, Belo Horizonte, Brazil
zenilton@pucminnas.br, sjamil@pucminas.br
[3] Université de Rennes 1, CNRS, INRIA / IRISA, Rennes, France
{simon.malinowski,guillaume.gravier}@irisa.fr

Abstract. In this paper, we address the challenges posed by dynamic
networks in various domains, such as bioinformatics, social network analysis, and computer vision, where relationships between entities are represented by temporal graphs that respect a temporal order. To understand
the structure and functionality of such systems, we focus on small subgraph patterns, called motifs, which play a crucial role in understanding
dissemination processes in dynamic networks that can be a spread of
fake news, infectious diseases or computer viruses. To address this, we
propose a novel approach called temporal motif filtering for classifying
dissemination processes in labeled temporal graphs. Our approach identifies and examines key temporal subgraph patterns, contributing significantly to our understanding of dynamic networks. To further enhance
classification performance, we combined directed line transformations
with temporal motif removal. Additionally, we integrate filtering motifs,
directed edge transformations, and transitive edge reduction. Experimental results demonstrate that our proposed approaches consistently
improve classification accuracy across various datasets and tasks. These
findings hold the potential to unlock deeper insights into diverse domains
and enable the development of more accurate and efficient strategies to
address challenges related to spreading process in dynamic environments.
Our work significantly contributes to the field of temporal graph analysis
and classification, opening up new avenues for advancing our understanding and utilization of dynamic networks.

The authors thank the Pontifícia Universidade Católica de Minas Gerais – PUC-Minas,
Coordenação de Aperfeiçoamento de Pessoal de Nível Superior – CAPES – (Grant
COFECUB 88887.191730/2018-00, Grant PROAP 88887.842889/2023-00 – PUC/MG
and Finance Code 001), the Conselho Nacional de Desenvolvimento Científico e Tecnológico – CNPq (Grants 407242/2021-0, 306573/2022-9) and Fundação de Apoio à
Pesquisa do Estado de Minas Gerais – FAPEMIG (Grant APQ-01079-23), PUC Minas
and INRIA under the project *Learning on graph-based hierarchical methods for image
and multimedia data.*

© Springer Nature Switzerland AG 2024
V. Vasconcelos et al. (Eds.): CIARP 2023, LNCS 14469, pp. 480–493, 2024.
https://doi.org/10.1007/978-3-031-49018-7_34

Keywords: Temporal motifs · temporal graphs · temporal graph classification

1   Introduction

Graphs are used to represent linked data in various domains such as social network [2,3], disease analysis [9,22,25], bioinformatics [24] and computer vision [19]. Modeling problems in terms of graphs allows for a deep understanding of the relationship between all elements. In social networks, for example, we can estimate how much a part of the network influences another [23]. However, many systems are not static since the links between objects may dynamically change over time. This kinds of systems may be modelled as temporal graphs, also called dynamic, evolving, or time-varying graphs. Applications involving temporal graphs were first introduced in 1997 [8]. Furthermore, temporal graphs have been used in numerous domains, such as communication networks [15,16,20], biological systems in cell and microbiology, neural networks [11,18], and economics [5,13].

The analysis of temporal graphs has led to the emergence of temporal motifs, which extend the concept of small subgraph patterns, commonly referred to as graphlets, to incorporate temporal properties. Temporal motifs capture not only the structural characteristics of subgraphs but also the chronological ordering of the edges, enabling a deeper understanding of the temporal dynamics within networks. Building upon the seminal work of previous researchers, [17] introduced the notion of temporal network motifs as elementary units of temporal networks and developed efficient algorithms for counting these motifs. That work revolutionized the field by providing a general methodology for quantifying temporal motifs, achieving significant speed improvements compared to existing methods. Inspired by the advancements in temporal motif analysis, [14] introduced the temporal graphlet kernel for classifying dissemination processes in labeled temporal graphs. The kernel represents labeled temporal graphs in the feature space of temporal graphlets, which are small subgraphs distinguished by their structure, time-dependent node labels, and the chronological order of edges.

While [14,17] have significantly advanced the understanding and analysis of temporal graphs, our work builds upon their foundations by filtering these motifs that are small subgraphs that can safely be removed without changing neither the overall graph structure nor disseminated information. The removal of these safe motifs leads to a reduction in the graph size and provides potential benefits in terms of time computational efficiency. In addition to proposing this new concept, we perform a classification task, aiming to discriminate between dissemination processes. Here, we also present a comprehensive evaluation of our approach, comparing it to existing state-of-the-art methods. The results demonstrate that our approach consistently achieves competitive accuracies while offering a significant graph size reduction. Moreover, according to our statistical analysis, made by *t-test*, to compare the classification task taking into account the original graph and filtered one, both results as statistically equivalent.

Another approach to reduce the graph size without changing its structure, in terms of reachability involves applying the *Transitive Reduction* (TR) algorithm [1]. In a directed graph, having two or more different paths between two vertices, known as the transitive relation, can introduce redundancy or ambiguity, making it challenging to interpret the graph's behavior. The TR algorithm yields a subgraph with the same reachability relation as the original graph but with the minimum number of edges. It preserves all paths between nodes in the original graph while eliminating unnecessary edges that can be inferred from the remaining edges. This process streamlines the graph representation, enhancing its clarity and reducing complexity. TR is only applied for *Directed acyclic graphs* (DAGs) and many real-world graphs are not DAGs such as social networks or transportation networks. This motivates our decision to use TR for temporal graphs. When working with temporal graphs, since one cannot come back in time, it is possible to avoid cycles and be able to use the TR algorithm. And, as it will be shown, avoiding redundancy can improve the accuracy of classification tasks. Our major objectives are summarized as follows: (i) a comprehensive analysis of motif removal regarding a dissemination process; (ii) the impact of this removal when performing a classification task with two different approaches, using a temporal kernel and a mapping process to use static kernel; and (iii) a combination of motif removal with transitive reduction for improving classification accuracy.

This work is organized as follows. In Sect. 2, some related works are described. Section 3 presents the main concepts related to graph theory that are necessary for understanding the proposed methods for filtering temporal graph's motifs, which are described in Sect. 4 and for transitive reduction in Sect. 5. An evaluation is given in Sect. 6. And finally, some conclusions are drawn in Sect. 7.

2   Related Work

In [17], the authors introduce the notion of motifs, which are small subgraph patterns that occur frequently in networks. Traditional motif analysis considers static networks; however, temporal networks demand a more sophisticated approach due to the inclusion of time-dependent interactions. To address this, they defines temporal motifs as induced subgraphs that preserve the temporal ordering of edges and present a methodology for quantifying them in temporal networks. These motifs capture both the network's structural information and the temporal dynamics. In [17], the authors also discuss the challenges of counting temporal motifs efficiently. In [10], it expanded the scope of motifs in temporal networks to hypergraphs. A hypergraph is a generalization of a traditional graph that allows edges to connect more than two nodes. The authors emphasize that hypergraph motifs offer a powerful tool for understanding relationships and interactions that goes beyond pairwise connections. Both works [10,17] propose algorithms for counting motif occurrences in graphs or hypergraph that helped us to create our algorithm, but nothing related to the use of those motifs for analysis or filtering involving a learning task.

A graphlet can be considered as a subset of motifs that emphasize the local connectivity within a network focusing on connected subgraphs of a fixed size. In [14], they introduce the concept of temporal graphlets and proposed a temporal graphlet kernel, which is a similarity measure that quantifies the resemblance between temporal graphlet features. To evaluate the effectiveness of the proposed approach, they conduct experiments on real-world datasets involving various dissemination processes, such as information spreading and disease propagation. While the mentioned works focused on motifs in temporal networks, hypergraphs, or evolving networks, our work takes a unique perspective by investigating the removal of these motifs in terms of their impact and their performance. With this, we can (i) simplify their structure, (ii) reduce the noise or irrelevant information, (iii) improve interpretability, (iv) facilitate the extraction of meaningful insights, (v) potentially improve computational efficiency; and (vi) in a context of classification task removing these edges, the classification model can focus on the most discriminative features, leading to more accurate and reliable classification.

3   Fundamental Concepts

In this section, we provide formal definitions of temporal graphs and temporal motifs. This work uses the definition proposed in [15] considering temporal graphs with edges existing at specific integral points in it and node labels and in [17] for temporal motifs.

3.1   Temporal Graph

Let $G = (V, E, l')$ be a labeled temporal graph in which V is a finite set of vertices, E is a finite set of directed temporal edges $e = (u, v, t)$ with u and v in V, $u \neq v$, the availability time (or time stamp) $t \in \mathbb{N}$, and $l' : V \cup T \mapsto \Sigma$, a labeling function that assigns a label to each vertex at each time step $t \in T = 1, \ldots, t_{\max} + 1$ with $t_{\max}$ being the largest timestamp of any $e \in E$.

For a temporal graph, the number of edges is not polynomially bounded by the number of vertices. A *temporal walk* of length k is an alternating sequence of vertices and temporal edges $(v_1, e_1 = (v_1, v_2, t_1), v_2, \cdots, e_k = (v_k, v_{k+1}, t_k), v_{k+1})$ such that $t_i < t_{i+1}$ for $1 \leq i < k$. Moreover, for a temporal walk, the waiting time at vertex v_i with $1 < i \leq k$ is $t_i - (t_{i-1} + 1)$.

3.2   Temporal Motifs

Temporal motifs are generalizations of small subgraph patterns, *i.e.* graphlets, that incorporate temporal properties like the chronological ordering of the edges.

A k-node, ℓ-edge, δ-temporal graphlet is a sequence of ℓ temporal edges, $g = ((u_1, v_1, t_1), (u_2, v_2, t_2), \ldots, (u_\ell, v_\ell, t_\ell))$ that satisfies the following conditions: (i) the edges are chronologically ordered, *i.e.*, $t_1 < t_2 < \ldots < t_\ell$, (ii) the time span between the first and last edges is within a δ time interval, *i.e.*, $t_\ell - t_1 \leq \delta$, and (iii) the induced static graph is connected and has k nodes. This definition is an extended definition of motifs that have the same properties, then we are going to refer temporal motifs as $M_{k,\ell}$ δ-temporal motif.

3.3 Transitive Reduction

Let $G = (V, E)$ be a directed acyclic graph with vertex set V and edge set E. A (directed) path from vertex u to vertex v in a graph G is a walk from u to v without repetition of vertices. The TR of G is another directed graph $G' = (V, E')$ such that:

- G' has the same vertex set V as G;
- For every pair of distinct vertices $u, v \in V$, if there is a directed path from u to v in G, then there is also a directed path from u to v in G';
- G' has the minimum number of edges among all graphs satisfying the first two conditions.

To define E', we first define the relation R on V as follows: for $u, v \in V$, uRv if and only if there is a directed path from u to v in G. Then, the transitive closure of R is the smallest transitive relation R' on V such that $R \subseteq R'$. We can define the edge set $E' = \{(u, v) \in E : u \neg R'v\}$. In other words, E' contains only the edges in E that are not necessary for maintaining the reachability relation in G. The resulting graph $G' = (V, E')$ is the transitive reduction of G [1]. The transitive reduction is well-defined only for DAGs.

The complexity of the transitive reduction algorithm depends on the size and structure of the input graph. Generally, there are several optimizations and one of them achieve $\mathcal{O}(|V|^2 \log |V|)$ for sparse graphs, in which $|V|$ is the number of vertices in the graph [21].

4 Filtering Motifs on Temporal Graphs

Here, our major goal is to study the pattern of small motifs regarding a classification task about the dissemination of a disease or of information in temporal graphs. We use a binary label alphabet Σ to represent the status of nodes in the temporal graph. The labels encode whether a node is infected or susceptible, or whether a node has received or not received certain information. The authors of [17] identified 36 non-equivalent temporal unlabeled motifs for $k \in \{2, 3\}$ nodes and $\ell = 3$ edges.

If we work with $M_{2,1}$ and $M_{2,2}$ we have only one unlabeled motif for the first configuration and two for the second one, see Fig. 1a. The motifs contain the chronological order of the edges but not the actual time stamps, which would be too restrictive, so in Fig. 1a the values $1, 2$ represent two time instants such that the first one is lower than the second one (whatever those time stamps). When considering node labeling, the number of distinct graphlets increases. According to [14], the number of distinct labels of ℓ-edge is $L^{2\ell}$ in which $L = |\Sigma|$. Then, for $k = 2$ nodes and $\ell = 1$ edge, there are 4 labeled motifs, as shown in Fig. 1b.

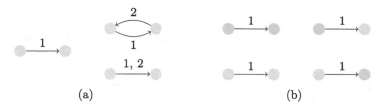

Fig. 1. Example of motifs of size (i) $k = 2$ nodes and $\ell = 1$ edge; (ii) and $k = 2$ nodes and $\ell = 2$ edges (a) and labeled motifs of $k = 2$ nodes and $\ell = 1$ for $\Sigma = 2$ (b).

Definition 1. *Let $f(v)$ be the time at which the label of vertex v in the function l' changed, represented as:*

$$f(v) = argmin\{t \in T : l'(v,t) \neq l'(v, t-1)\} \tag{1}$$

Since the dissemination of a disease or an information (fake news, for instance) is usually a mainly local process [15], here, we are going to check each pairwise vertices in a smaller motif, *i.e.* we will work only with $M_{2,1}$ motifs. Regarding the definition in Sect. 3, a 2-node, 1-edge, δ-temporal motif is a sequence of 1 edge, $M = (u_1, v_1, t_1)$ such that the induced static graph from the edges is connected and has 2 nodes. Using Eq. 1, as in our dictionary if a vertex starts to be infected it will stay infected forever, the label is going to change only once. Let α be the first changing time of u, $\alpha = f(u)$ while $\beta = f(v)$.

Our analysis starts from regarding the chronological order of α, t, β. For $M_{2,1}$ motifs, we have $(k+\ell)! = 6$ different combinations for the infected label, regarding the time ordering of α, t, β, as show in Table 1.

Table 1. Safety removal configuration

		$\alpha \xrightarrow{t} \beta$		
$\alpha < \beta < t$	Safe	$\alpha < t < \beta$	No
$\beta < \alpha < t$	Safe	$t < \alpha < \beta$	Unsafe
$\beta < t < \alpha$	Unsafe	$t < \beta < \alpha$	Unsafe

In our dissemination study, the vertices are persons and the edges represent interaction between them. Each one was infected in different time instances. The safe labels, as shown in Table 1 are the labeled motifs that are safe to remove. The notion of safe to remove indicates that this edge cannot be responsible of the dissemination (of disease or information). So it seems safe to remove it

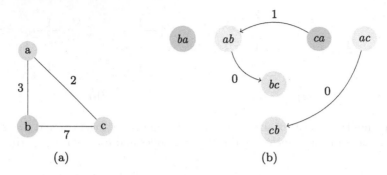

Fig. 2. Example of directed line transformation. The walk $(n_{ca}^2, n_{ab}^3, n_{bc}^7)$ of length 2 in (b) corresponds to the temporal walk $(c,(c, a, 2), a,(a, b, 3), b,(b, c, 7), c)$ of length 2 in the temporal graph (a) [15].

from the graph without loosing information about the dissemination process. For instance, if a person 1 (infected on day 1) has contact on day 3 with a person 2 (infected on day 2), person 1 can not be responsible of the infection of person 2 (they have been in contact after both were infected). So such an edge might be removed from the graph. Other cases are called unsafe, as such edges might be responsible of the dissemination in the graph. For instance, if a person 1 has contact with person 2 on day 1, and they show to be infected on day 2 and 3 respectively, then, there is a possibility that person 1 passed the infection for person 2. In that case, it is not safe to remove such an edge as it might bring information about the dissemination process. The *no* safety removal corresponds to a classical dissemination process and hence cannot be removed. This classical dissemination is when a person gets infected after having been in contact with an infected person.

To remove safe edges, for each vertex $O(|V|)$, there is an inner loop that iterates over adjacent temporal edges, which can take up to $O(|E|)$ time in the worst case if the graph is fully connected. Overall, the complexity of the algorithm can be considered as $O(|V| \cdot |E|)$, assuming constant time for comparisons and removals. Filtering safe temporal motifs enable to reduce the size of the graph. In our application case, it is a preprocessing step that enables to fasten the next steps (including classification one). In the next section, we will explain another way of reducing the size of the graph by filtering transitive edges.

5   Filtering Transitive Edges

Transitive edges can be removed using an algorithm called TR. This algorithm can be applied only to directed acyclic graphs and not on temporal ones. For our purpose, the temporal graph needs first to be mapped to a static one.

In [15], the authors introduced several methodologies to convert temporal graphs into static representations. Among these methods, the *Directed line graph expansion* (DL) demonstrated superior performance in fully capturing the temporal information.

Each temporal edge is represented by two vertices and a temporal walk can be performed, since a walk in the *DL* graph is related to the temporal walks in the original temporal graph having the same label sequence, being able to model waiting times and keeping the temporal information. Following [15], DL can be defined as follows: given a temporal graph $(\mathbf{G}, l)$, the directed line graph expansion $DL(\mathbf{G}, l) = (\mathbf{G}', l')$ in which $\mathbf{G}' = (V', E')$ is the directed graph, where every temporal edge $(\{u, v\}, t)$ is represented by two vertices $n_{\overrightarrow{uv}}^{t}$ and $n_{\overrightarrow{vu}}^{t}$ and there is an edge from $n_{\overrightarrow{uv}}^{t}$ to $n_{\overrightarrow{xy}}^{s}$ if $v = x$ and $t < s$. For each vertex $n_{\overrightarrow{uv}}^{t}$, the label $l'\left(n_{\overrightarrow{uv}}^{t}\right) = (l(u, t), l(v, t+1))$ is set. Figure 2 shows an example of the directed line transformation. The edge value is related to waiting time: if the previous node edges are consecutive (e.g., edges 2 and 3 in Fig. 2a), then there is no waiting time between edges, represented by value 1, otherwise, 0 is used. TR is only applied to DAGs. Proposition 1 shows that the conversion of a temporal graph to DL leads to a DAG, then we are able to apply the TR algorithm to deal with the size of edges.

Proposition 1. *Let $(\mathbf{G}, l)$ be a temporal graph. The resulting $DL(\mathbf{G}, l)$ is a Directed acyclic graph (DAG).*

Proof. Suppose there exists a directed cycle in $DL(\mathbf{G}, l)$. Let $n_{\overrightarrow{uv}}^{t}$ be a vertex on the cycle with the earliest timestamp t. Let $n_{\overrightarrow{xy}}^{s}$ be the last vertex of the cycle that connects $n_{\overrightarrow{uv}}^{t}$. Since there is a path from $n_{\overrightarrow{xy}}^{s}$ to $n_{\overrightarrow{uv}}^{t}$, we know that $y = u$ and $s > t$. However, $n_{\overrightarrow{uv}}^{t}$ represents a temporal edge (u, v, t), and $n_{\overrightarrow{xy}}^{s}$ represents a temporal edge (x, y, s). Therefore, $u = y = x$ and $t > s$. But this contradicts the fact that $n_{\overrightarrow{uv}}^{t}$ has the earliest timestamp on the cycle. Therefore, there are no directed cycles in $DL(\mathbf{G}, l)$, and $DL(\mathbf{G}, l)$ is a DAG.

Hence, TR can be applied just after the DL conversion. The combination of filtering motif on temporal graph, conversion to DL and application of TR algorithm leads in a less complex graph (as we will see in Sect. 6), making it easier for the classifier to identify the patterns and features that are truly relevant to the classification task. Using the two different filtering methods presented in this paper (Sects. 4 and 5), we can derive three strategies for our task of classification of temporal graphs:

TGF: the temporal graph is filtered as explained in Sect. 4

TGF-DL: same as **TGF** but the resultant graph is then converted to a static graph using DL conversion

TGF-DL-TR: same as **TGF-DL** but the resultant graph is then filtered by the method explained in Sect. 5

Note that these strategies are then followed by the classification step using either a graph kernel (either graphlet or Weisfeiler-Lehman kernel) for **TGF-DL** and **TGF-DL-TR**, or a temporal graphlet kernel (**TGL**) for **TGF**. In the next section, we will evaluate the performance of these strategies for classification tasks and compare them to a classical approach.

6   Experiments

We evaluate the impact of filtering temporal graphs and compare the effectiveness and efficiency to the baselines provided in [12,14].

6.1   Experimental Setup

Databases. We utilized 6 databases sourced from TUDataset [12] for our study. This benchmark provides temporal graph classification databases derived from *Tumblr, Dblp, Facebook* as well as contacts between students at MIT, in a *Highschool*, and visitors at the *Infectious* exhibition, for dissemination process study. For each database, a dissemination process simulation was done, in which nodes are infected at different time-stamps, providing two classification tasks for each database, see [14] for the description of each database. Since the size of the database for each task is the same (what changes is the node label), we use the same database name and then separated the values in *Task 1* and *Task 2*. The first task involves discriminating between temporal graphs with vertex labels resulting from a dissemination process and those without. To accomplish this, the authors ran a *Susceptible-infected* (SI) simulation [4], with fixed parameters on half of the data set and used it as the first class. The second class was made up of the remaining graphs. For each graph in the second class, the authors counted the number of infected vertices, reset the labels, and then randomly infected a number of vertices at a random time. The second task involves discriminating between temporal graphs that differ in the dissemination process itself. For this task, the authors ran the simulation with different parameters for each of the two subsets. For both subsets, $I = 0.5$ (initial infection rate), but for the first subset, the authors set $p = 0.2$ (infection probability), and for the second subset, $p = 0.8$. The simulation runs repeatedly until at least $|V| \times I$ vertices are infected or no more infections are possible, for example, if a graph has 100 vertices and the initial infection rate is set to 0.5, then initially 50 vertices are infected. In order to stop the SI simulation, at least $50 \times 0.5 = 25$ vertices (i.e., 50% of the total number of vertices) need to be infected. The simulation continues until either this condition is met or no more infections are possible.

Kernel Instances. As a baseline for temporal kernel, we use the wedge-based temporal graphlet kernel [14], since temporal wedges possess the ability to effectively capture dissemination patterns, and they can be easily quantified. As a baseline for static graphs we use two well known kernels, the 3-node Graphlet (GL) and the Weisfeiler-Lehman subtree (WL). The source code is provided by [12].

Implementation Details. We used the C-SVM implementation of LIBSVM [6] to determine the classification accuracies. We performed 10-fold cross-validation to select the C parameter from the range of $10^{-3}, 10^{-2}, \ldots, 10^2, 10^3$ on the training folds. We repeated the 10-fold cross-validation ten times to obtain the average accuracies and standard deviations. The DL method proposed in [15] is used to convert temporal graph to line graph. A NetworkX implementation of a transitive reduction will be used in **TGF-DL-TR** method [7].

6.2 Evaluation and Discussion

First we want to evaluate the classification when safe edges are removed. The results obtained after removing safe edges (**TGF** method) are presented in Table 2. Compared to **Baseline TG**, it is evident that the accuracy of our dissemination analysis was preserved, as evidenced by the maintenance of accuracy values (with standard deviation) for most databases. This confirms that the edges are safe to remove and barely modify the information extracted from the graph. Notably, in the case of the MIT database in *Task 2*, the removal of safe edges led to an improvement in accuracy, further supporting the contention that these edges are dispensable for our analysis.

Table 2. Classification accuracies in percent and standard deviation for *Task 1* and *Task 2*.

	Databases											
	Highschool		Infectious		Tumblr		Dblp		Facebook		MIT	
	Task 1	Task 2	Task 1	Task 2	Task 1	Task 2	Task 1	Task 2	Task 1	Task 2	Task 1	Task 2
Baseline TG												
TGL	98.27±0.4	91.33±1.5	97.00±0.6	88.20±1.3	92.60±1.1	77.74±0.9	98.31±0.4	83.41±0.3	94.84±0.4	75.46±0.7	92.26±1.3	63.07±4.3
TGF												
TGL	97.66±0.6	91.66±1.1	97.45±0.5	87.20±2.0	92.14±1.2	79.57±0.9	98.27±0.2	83.10±0.7	95.07±0.2	76.41±0.4	93.35±2.1	75.55±2.3
TGF-DL												
GL	95.33±0.4	**92.44**±0.9	96.55±0.9	**87.75**±1.5	89.97±1.0	**79.97**±0.7	97.24±0.3	80.55±0.2	92.56±0.3	74.10±0.4	92.61±2.2	**79.33**±3.0
WL	97.27±0.5	91.38±0.6	**98.50**±0.3	78.75±1.8	**94.33**±0.6	76.82±2.1	**98.31**±0.2	77.20±0.8	**96.43**±0.3	**80.96**±0.8	89.06±2.8	49.45±5.0
TGF-DL-TR												
GL	96.61±0.7	**94.83**±1.4	97.05±0.6	83.15±1.5	93.15±0.6	**81.58**±0.6	97.04±0.2	80.72±0.5	94.26±0.1	75.05±0.7	**95.66**±1.0	**89.53**±1.7
WL	96.72±0.5	**94.83**±0.8	**97.70**±0.5	83.45±2.2	**93.49**±0.7	**80.75**±1.1	**99.24**±0.2	79.12±0.9	**96.08**±0.3	**83.60**±0.8	91.40±2.3	74.91±3.4

To assess the statistical significance of our findings, we conducted a *t-test* on all 10-fold cross-validation sets mentioned in *Experimental Setup*, comparing the **Baseline TG** with the reduction process, **TGF**. To ensure a comprehensive understanding, we have employed two different *t-test* analysis. Firstly, we conducted individual *t-tests* for each database, generating a set of six p-values for each task. Secondly, we aggregated all the databases and performed a global *t-test*, resulting in a single p-value. By presenting both sets of results, we seek to provide a comprehensive perspective on the similarity and consistency of the methods across diverse contexts. The p-value globally was 0.580044, so we cannot confidently say that there is a significant difference between the two groups at the 95% confidence level. In Table 3 is shown the *t-test* for each database. The *t-test* results indicate that for databases MIT and *Dblp* in *Task 2*, there is a highly statistically significant difference in accuracies ($p < 0.05$). If we take a look on accuracy table, the MIT database is the one for which accuracy was improved.

Table 3. Results of *t-tests* for **Baseline TG** and **TGF**.

	Datasets											
	Highschool		Infectious		Tumblr		Dblp		Facebook		MIT	
	Task 1	Task 2	Task 1	Task 2	Task 1	Task 2	Task 1	Task 2	Task 1	Task 2	Task 1	Task 2
P-value	0.041	0.872	0.610	0.546	0.573	0.179	0.479	7.1×10^{-9}	0.916	0.079	0.021	0.00001
CI95%	$[0.02, 0.8]$	$[-1.5, 1.3]$	$[-0.4, 0.7]$	$[-1.1, 2.0]$	$[-0.7, 1.3]$	$[-1.5, 0.3]$	$[-0.2, 0.5]$	$[2.9, 4.1]$	$[-0.3, 0.4]$	$[-0.08, 1.2]$	$[-3.2, -0.3]$	$[-15.5, -7.4]$

Complete results comparing the three proposed methods with the baseline are given in Table 2. The directed line transformation explicitly incorporates temporal information by adding directed edges, which might capture temporal dependencies in the data. We propose a combined preprocessing approach, referred to as **TGF-DL**, which integrates the directed line transformation with temporal motif removal. Our experimental results demonstrate that the **TGF-DL** approach consistently outperforms the baseline method, representing as bold, preserving accuracy for most databases while significantly improving classification performance in certain cases. The combination of these preprocessing steps synergistically enhances the graph's representation, enabling classifiers to leverage relevant temporal patterns and reduce noise, ultimately leading to more accurate and efficient classification outcomes. These findings highlight the importance of tailored preprocessing strategies for temporal graph analysis and shed light on the potential benefits of incorporating temporal dependencies in graph-based classification tasks.

Another proposed approach, denoted as **TGF-DL-TR**, incorporates the benefits of filtering motifs, the directed edge transformation and transitive edge reduction. The Directed Line transformation enriches the graph structure by explicitly capturing temporal dependencies through directed edges as we saw before, while Transitive Reduction streamlines the graph by removing transitive edges, further enhancing the discriminative power of the representation. Our experimental results demonstrate significant accuracy improvements over the baseline method across various databases and tasks (in bold). Notably, in the case of the MIT database for *Task 2*, the accuracy substantially increased of more than 25% in accuracy. The Facebook database also showed a substantial improvement, increasing 8% in accuracy with the Weisfeiler-Lehman kernel (WL). Moreover, for the Tumblr database, the GL method achieved an accuracy boost from 77.74% to 81.58%. Additionally, in *Task 2* for the Highschool database the accuracy surge from 91.33% to 94.83%. These results underscore the effectiveness of the combined preprocessing steps, resulting in remarkable percentage improvements and enhancing the classifiers ability to capture relevant temporal dependencies, ultimately improving the overall classification performance.

In addition, we conducted an investigation into the reduction in graph size, quantified in terms of edge quantity, resulting from the application of **TGF** compared to the baseline represented. Regarding Fig. 3, it is evident that the **TGF** preprocessing resulted in a considerable reduction in graph size, with the MIT database experiencing a reduction of more than half in terms of edge quantity.

This reduction in graph size is advantageous for preprocessing, leading to faster computation and a less complex graph representation.

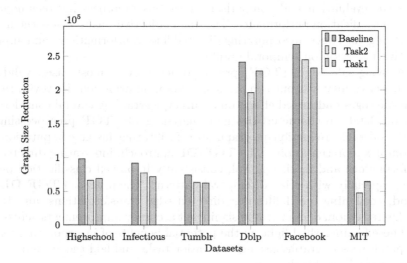

Fig. 3. Graph Size Reduction of *Task 1* and *Task 2* compared to the baseline.

We performed a classification removing the unsafe edges to evaluate the result (see Table 4). In some cases, removing unsafe edges may lead to a loss of accuracy, as seen in the Highschool and Facebook databases (in bold). However, in other cases, such as the MIT database, removing unsafe edges can actually lead to an improvement in accuracy. This highlights the importance of considering the database characteristics and the nature of the analysis task when deciding whether to remove unsafe edges. It may not be a one-size-fits-all approach, and the decision should be made based on empirical results and the specific goals of the analysis.

Table 4. Classification accuracy in percent and standard deviation for the both tasks removing the unsafe edges.

	Datasets											
	Highschool		Infectious		Tumblr		Dblp		Facebook		MIT	
	Task 1	Task 2	Task 1	Task 2	Task 1	Task 2	Task 1	Task 2	Task 1	Task 2	Task 1	Task 2
Baseline												
TGL	98.27±0.4	91.33±1.5	97.00±0.6	88.20±1.3	92.60±1.1	77.74±0.9	98.31±0.4	83.41±0.3	94.84±0.4	75.46±0.7	92.26±1.3	63.07±4.3
TGF												
TGL	**96.55**±0.5	92.50±0.8	97.60±0.5	89.60±1.2	92.89±0.7	77.12±1.4	97.37±0.2	**76.82**±3.4	94.71±0.1	**70.83**±0.6	91.55±1.8	76.82±1.4

7   Conclusion

In this work, we conducted an analysis of temporal graph classification, focusing on the evaluation and comparison of various preprocessing techniques to enhance classification performance. The main objective was to assess the impact of removing safe edges, incorporating directed line transformations, and applying transitive reduction on temporal graphs.

Removing safe edges (**TGF**) preserved accuracy in most cases, validating their dispensability without compromising graph information. However, removing unsafe edges had mixed effects on accuracy, warranting careful consideration based on database characteristics and analysis goals. **TGF** preprocessing led to a considerable reduction in graph size, facilitating faster computation and less complex representations. The **TGF-DL** approach, integrating directed line transformations and motif removal, consistently improved classification performance, especially with critical temporal patterns. The proposed **TGF-DL-TR** method, combining motif filtering, directed edge transformations, and transitive edge reduction, demonstrated significant accuracy improvements across data sets. These findings shed light on the potential benefits of incorporating temporal dependencies in graph-based classification tasks and laid the groundwork for further research in this domain.

References

1. Aho, A.V., Garey, M.R., Ullman, J.D.: The transitive reduction of a directed graph. SIAM J. Comput. **1**(2), 131–137 (1972)
2. Akhtar, N., Ahamad, M.V.: Graph tools for social network analysis. In: Research Anthology on Digital Transformation, Organizational Change, and the Impact of Remote Work, pp. 485–500. IGI Global (2021)
3. Amara, A., Hadj Taieb, M.A., Ben Aouicha, M.: Multilingual topic modeling for tracking COVID-19 trends based on Facebook data analysis. Appl. Intell. **51**(5), 3052–3073 (2021)
4. Bai, Y., Yang, B., Lin, L., Herrera, J.L., Du, Z., Holme, P.: Optimizing sentinel surveillance in temporal network epidemiology. Sci. Rep. **7**(1), 4804 (2017)
5. Barunik, J., Ellington, M., et al.: Dynamic networks in large financial and economic systems. arXiv preprint arXiv:2007.07842 (2020)
6. Chang, C.C., Lin, C.J.: LIBSVM: a library for support vector machines. ACM Trans. Intell. Syst. Technol. (TIST) **2**(3), 1–27 (2011)
7. Hagberg, A.A., Schult, D., Swart, P.: Networkx (2008). https://networkx.github.io/. Accessed May 8 2023
8. Harary, F., Gupta, G.: Dynamic graph models. Math. Comput. Model. **25**(7), 79–87 (1997)
9. Karaivanov, A.: A social network model of COVID-19. PLoS ONE **15**(10), e0240878 (2020)
10. Lee, G., Ko, J., Shin, K.: Hypergraph motifs: concepts, algorithms, and discoveries. arXiv preprint arXiv:2003.01853 (2020)
11. Meng, X., Li, W., Peng, X., Li, Y., Li, M.: Protein interaction networks: centrality, modularity, dynamics, and applications. Front. Comp. Sci. **15**, 1–17 (2021)

12. Morris, C., Kriege, N.M., Bause, F., Kersting, K., Mutzel, P., Neumann, M.: Tudataset: a collection of benchmark datasets for learning with graphs. arXiv preprint arXiv:2007.08663 (2020)
13. Nonejad, N.: An overview of dynamic model averaging techniques in time-series econometrics. J. Econ. Surv. **35**(2), 566–614 (2021)
14. Oettershagen, L., Kriege, N.M., Jordan, C., Mutzel, P.: A temporal graphlet kernel for classifying dissemination in evolving networks. arXiv preprint arXiv:2209.07332 (2022)
15. Oettershagen, L., Kriege, N.M., Morris, C., Mutzel, P.: Classifying dissemination processes in temporal graphs. Big Data **8**(5), 363–378 (2020)
16. Ozcan, S., Astekin, M., Shashidhar, N.K., Zhou, B.: Centrality and scalability analysis on distributed graph of large-scale e-mail dataset for digital forensics. In: 2020 IEEE International Conference on Big Data (Big Data), pp. 2318–2327. IEEE (2020)
17. Paranjape, A., Benson, A.R., Leskovec, J.: Motifs in temporal networks. In: Proceedings of the tenth ACM International Conference on Web Search and Data Mining, pp. 601–610 (2017)
18. Paulevé, L., Kolčák, J., Chatain, T., Haar, S.: Reconciling qualitative, abstract, and scalable modeling of biological networks. Nat. Commun. **11**(1), 4256 (2020)
19. Pradhyumna, P., Shreya, G., et al.: Graph neural network (GNN) in image and video understanding using deep learning for computer vision applications. In: 2021 Second International Conference on Electronics and Sustainable Communication Systems (ICESC), pp. 1183–1189. IEEE (2021)
20. Tadić, B.: Dynamics of directed graphs: the world-wide web. Phys. A **293**(1–2), 273–284 (2001)
21. Tang, X., Zhou, J., Qiu, Y., Liu, X., Shi, Y., Zhao, J.: One edge at a time: a novel approach towards efficient transitive reduction computation on DAGs. IEEE Access **8**, 38010–38022 (2020)
22. Wang, J., et al.: scGNN is a novel graph neural network framework for single-cell RNA-Seq analyses. Nat. Commun. **12**(1), 1–11 (2021)
23. Yang, S.: Networks: An Introduction by Mej Newman, 720 p. Oxford University Press, Oxford (2013)
24. Zhang, X.M., Liang, L., Liu, L., Tang, M.J.: Graph neural networks and their current applications in bioinformatics. Front. Genet. **12**, 690049 (2021)
25. Zhu, Y., Ma, J., Yuan, C., Zhu, X.: Interpretable learning based dynamic graph convolutional networks for Alzheimer's disease analysis. Inf. Fusion **77**, 53–61 (2022)

Graph-Based Feature Learning
from Image Markers

Isabela Borlido Barcelos[1] , Leonardo de Melo João[2] ,
Zenilton K. G. Patrocínio Jr.[1] , Ewa Kijak[3] , Alexandre X. Falcão[2] ,
and Silvio J. F. Guimarães[1(✉)]

[1] ImScience, Pontifical Catholic University of Minas Gerais, Belo Horizonte, Brazil
isabela_borlido@hotmail.com, {zenilton,sjamil}@pucminas.br
[2] LIDS, University of Campinas, Campinas, Brazil
leomelo168@gmail.com, afalcao@ic.unicamp.br
[3] Université de Rennes 1, CNRS, INRIA/IRISA, Rennes, France
ewa.kijak@irisa.fr

Abstract. Deep learning methods have achieved impressive results for
object detection, but they usually require powerful GPUs and large
annotated datasets. In contrast, there is a lack of explainable networks in the literature. For instance, Feature Learning from Image
Markers (FLIM) is a feature extraction strategy for lightweight CNNs
without backpropagation that requires only a few training images.
In this work, we extend FLIM for general image graph modeling,
allowing it for a non-strict kernel shape and taking advantage of the
adjacency relation between nodes to extract feature vectors based on
neighbors features. To produce saliency maps by combining learned
features, we proposed a User-Guided Decoder (UGD) that does not
require training and is suitable for any FLIM-based strategy. Our
results indicate that the proposed Graph-based FLIM, named GFLIM,
not only outperforms FLIM but also produces competitive detections
with deep models, even having an architecture thousands of times
smaller in the number of parameters. Our code is publicly available at
https://github.com/IMScience-PPGINF-PucMinas/GFLIM.

Keywords: Graph-based feature learning · Image markers ·
Lightweight CNN · Object detection

The authors thank the Pontifícia Universidade Católica de Minas Gerais – PUC-
Minas, Coordenação de Aperfeiçoamento de Pessoal de Nível Superior – CAPES
– (Grant COFECUB 88887.191730/2018-00, Grant PROAP 88887.842889/2023-00
– PUC/MG and Finance Code 001), the Conselho Nacional de Desenvolvimento
Científico e Tecnológico – CNPq (Grants 303808/2018-7, 407242/2021-0, 306573/2022-
9) and Fundação de Apoio à Pesquisa do Estado de Minas Gerais – FAPEMIG (Grant
APQ-01079-23), PUC Minas and INRIA under the project *Learning on graph-based
hierarchical methods for image and multimedia data.*

V. Vasconcelos et al. (Eds.): CIARP 2023, LNCS 14469, pp. 494–509, 2024.
https://doi.org/10.1007/978-3-031-49018-7_35

1 Introduction

In recent years, deep learning models have played an essential role in several areas of knowledge, usually achieving state-of-the-art results. However, they typically require large annotated datasets and powerful computational resources to train millions of parameters, becoming unsuitable for tasks with hardware constraints or small datasets. In object detection, some deep models achieve impressive results, whereas most suffer from the same limitations. To solve object detection, bounding boxes can be extracted from binarized saliency maps [8,18] or be a direct result of bounding box regressors [1,16,17]. For instance, U^2Net [8] is an encoder-decoder model that extracts multiscale features using U-shaped blocks. Transformer-based models also may produce saliency maps to perform detection, such as SelfReformer [18], which extracts global and local context features and avoids losing fine features by using an up/downsampling method. Both U^2Net and SelfReformer require large annotated datasets and powerful hardware for training. In contrast, DETReg [1] is an end-to-end object detection model based on a transformer that predicts a fixed number of bounding boxes based on image priors. Although demanding less annotated data, DETReg has millions of parameters to estimate, requiring costly hardware.

Despite having several deep learning proposals suitable for object detection, there is a lack of explainable models in the literature. Recently, [11] proposed a lightweight CNN learned through a methodology named Feature Learning from Image Markers (FLIM). FLIM learns kernels for its convolutional layers from user-drawn markers in regions with different image classes, providing user control over the training process. Also, FLIM trains CNN layers without back-propagation and requires only a few training images. Recent works demonstrated FLIM's advantages for different tasks in Image Processing [3,4,10]. Despite the good results provided by FLIM, it works on a pixel level, making it more sensitive to noise, and it uses a very strict kernel shape (for instance, a square).

In order to cope with these two drawbacks in the FLIM strategy, we propose, in this work, a more general framework based on graph analysis. Thus, here, we extend FLIM to a graph-based feature extraction by using, for instance, a superpixel graph. Moreover, we demonstrate the effectiveness of our proposal with superpixel graphs against FLIM results. The proposed Graph-based FLIM

(a) FLIM$_3$ (b) DETReg   (c) SR   (d) U^2Net   (e) GFLIM$_2$

Fig. 1. Example of *helminth eggs* detection. Blue and red bounding boxes indicate the detections and groundtruth, respectively, and their intersections are in green. (Color figure online)

(GFLIM) maintains FLIM's advantages, allowing feature extraction from general image graph modeling. Our proposal computes superpixel graphs from images and propagates user-drawn markers to the graph nodes. Instead of computing kernels based on the spatial information around the pixel markers, GFLIM takes advantage of the adjacency relation between nodes and computes feature vectors based on neighbors' features. For a superpixel graph, one may compute a pixel-superpixel map (label map) from superpixel segmentation to transform images into graphs and vice-versa. Like FLIM, GFLIM is trained layer-by-layer, and the user may select kernels that best represent the desired image patterns. The pixel-superpixel maps can be used to compute activation images from the kernels' weights of GFLIM, allowing a manual kernel selection based on their visualization. To combine the features produced by GFLIM, we propose a user-guided decoder that takes advantage of the user's knowledge from kernel selection. Our results demonstrated the effectiveness of our proposal using only four images and user-drawn scribbles for training. With only three layers and selecting ten kernels per layer, GFLIM achieved competitive results compared to deep learning models. In Fig. 1, we illustrate an example of helminth egg detection in optical microscopy images.

Therefore, the major contributions of this paper may be summarized as follows: (i) a strategy to extend FLIM for graphs, independent of the graph modeling; (ii) a user-guided decoder for FLIM-based strategies that use prior knowledge concerning the foreground kernels; and (iii) a novel flyweight graph-based CNN strategy for object detection that can achieve competitive results using simple architectures thousands lesser than deep learning strategies.

This paper is organized as follows. Section 2 presents important works in object detection and FLIM-based strategies. Subsequently, Sect. 3 describes our framework for feature learning based on graph analysis. Section 4 presents the experimental setup, measures, and results. Finally, conclusions and future works are presented in Sect. 5.

2   Related Works

Single-class object detection tasks are often solved either by bounding-box regression or saliency estimation. The former category estimates minimum bounding boxes for all objects of interest, while the latter assigns an object probability to each image pixel. In both categories, deep-learning-based approaches achieve state-of-the-art performance. For such, the models are often pre-trained in large datasets [5,6,13] and fine-tuned for specific applications.

For bounding-box regression, current object detectors use self-supervised approaches to create weak labels for ImageNet and train a backbone more suitable for object detection [1,7,16,17], requiring less annotated data to adapt the backbone for object-detection tasks. More recently, the Detection with Transformer using Region priors (DETReg) [1] achieved state-of-the-art results in one of the most popular datasets for Object Detection (MSCoco). Using Selective Search [12] for creating the bounding boxes, DETReg proposes learning a class-agnostic model (*i.e.* a model that can detect objects disregarding its class) whose classification head that can be adapted to other classes in a few-shot learning manner. Despite its few-shot learning capabilities and high-quality results, DETReg has over 40 million parameters, requiring expensive GPUs to execute fast, and it outputs a fixed number of bounding boxes, which requires filtering.

Saliency estimation methods highlight objects that stand out in a scene according to specific salient criteria, which may be learned using deep learning approaches [14]. Because saliency is more related to the visual relationship between the objects of the foreground and the background, the models are already class-agnostic. Among the recent approaches, SelfReformer (SR) [18] has achieved the highest scores in most saliency estimation benchmarks. SelfReformer uses a Pyramid Vision Transformer [15] and Pixel Shuffle [9] for improving long-range information dependency and to keep fine segmentation details. Considering CNN approaches, UNet [8] achieved state-of-the-art performance when considering models that can be trained using a small number of training images. U^{2}Net explores multi-scale information using a multi-level U-shaped network, where each of the network's encoder blocks are smaller U-shaped networks. Nevertheless, considering the parasite egg detection application, both saliency estimation methods also require expensive GPUs to execute in a viable time.

As an alternative for learning more explainable models with few training images, Feature Learning by Image Markers (FLIM) was proposed [11] as a methodology that allows convolutional feature extractors to be learned from scratch with very few training images and without backpropagation. In FLIM, a user annotates images with markers (*e.g.*, scribbles) that are used by FLIM as an attention mechanism for learning a sequence of convolutional kernels that will compose the feature extractor of a CNN. So far, FLIM has been explored as a feature extractor for graph-based segmentation [11]; for classification tasks as an end-to-end network with Fully Connected Layers as the classifier [4]; and recently, FLIM-based networks were extended to medical image segmentation [3,10] for identifying COVID-19 positive exams in CT images of the thorax, and brain tumor segmentation in MR images. However, to the best of our knowledge, no FLIM-based network operates in graphs or has been proposed for object detection.

3   Graph-Based Feature Learning from Image Markers

In this paper, we propose a lightweight Graph-based CNN with little supervision that generalizes the Feature Learning from Image Markers (FLIM) for superpixel graphs. The proposed Graph-based FLIM (GFLIM) explores superpixel features, providing higher-level features than image pixels and reducing information redundancy. The feature extraction in GFLIM is also independent of the graph modeling and is not restricted to superpixel graphs. GFLIM follows a similar pipeline to FLIM. For training, GFLIM estimates the kernels layer-by-layer (Fig. 2). Also, if a layer requires retraining, only subsequent layers are affected. Likewise, one can add or remove layers without affecting the previous ones. First, the user provides foreground/background markers for a small

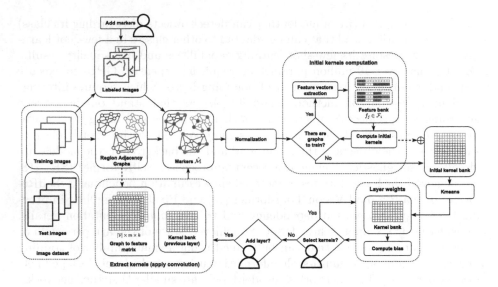

Fig. 2. GFLIM training diagram.

set of training images. Then, our proposal computes a graph for each marked image using a superpixel method (*e.g.*, DISF [2]). Afterward, a set of initial kernels is computed for each graph based on the k most similar neighbors of the marked vertices. In GFLIM, representative kernels are computed using a clustering strategy (*e.g.*, using K-means or DBSCAN). Then, the user can select the most suitable kernels by visualizing their object probability maps (saliency maps). Finally, whether another layer is desired, GFLIM extracts features from the previous layer by applying the learned kernels on the previous layer's inputs. In the following, GFLIM training steps are explained in detail. Afterward, we present our decoder strategy for object detection.

3.1 Superpixel Graphs

An image I can be defined as a pair $\langle \mathcal{P}, \mathbf{F} \rangle$ in which $\mathcal{P} \subset \mathbb{Z}^2$ is the set of *space elements* (*i.e.*, spels) and $\mathbf{F}$ maps every $p \in \mathcal{P}$ to its *feature vector* $\mathbf{F}(p) \in \mathbb{R}^m$, given $m \in \mathbb{N}^*$, here the *spels* can be defined as a set of feature vectors $I = \langle \boldsymbol{q}_1, \ldots, \boldsymbol{q}_N \rangle$, where $\boldsymbol{q}_t \in \mathbb{R}^m$ and $m > 0$ is the number of channels in I in some arbitrary feature space (*e.g.*, the CIELAB color space). One may segment I into $k \in \mathbb{N}^*$ regions using a function $\mathbf{S}(I, k)$ (*e.g.*, a superpixel segmentation method), resulting in a partitioning $\mathcal{P} = \{P_1, \ldots, P_k\}$, where $\bigcup_{t=0}^k P_t = I$ and $P_t \cap P_{t'} = \emptyset$ for all $1 \leq t < t' \leq k$. A segmentation map $L = \{t \mid \boldsymbol{q} \in P_t \forall \boldsymbol{q} \in I\}$ may be computed. An image segmentation $\mathbf{S}(I, k)$ may be interpreted as a graph $G = (\mathcal{V}, \mathcal{B}, \psi)$ with k vertices, where $\mathcal{V} = \{\boldsymbol{v}_1, ..., \boldsymbol{v}_k\}$ is the vertex set, $\psi : \mathcal{P} \to \mathcal{V}$ assigns a feature vector (*e.g.*, the mean color) to each vertex, and $\mathcal{B}$ is the superpixel *adjacency relation* $\mathcal{B} = \{(\boldsymbol{v}_x, \boldsymbol{v}_y) \in \mathcal{V} \times \mathcal{V} | \exists \boldsymbol{p} \in P_x, \boldsymbol{q} \in P_y, \|\boldsymbol{p} - \boldsymbol{q}\| \leq \rho\}$.

Let an image I, whose pixels may be associated with one of $c > 0$ classes, and a graph G from its segmentation $\mathbf{S}(I,k)$ (*e.g.*, using DISF [2]). In GFLIM, the user provides a set $\mathcal{L} \subseteq I$ of markers, whose labels are given by $\lambda : \mathcal{L} \to \{1, 2, ..., c\}$. One may extend the markers $\mathcal{L}$ and its labeling function λ for a set of vertices $\mathcal{V}$ as $\mathcal{M} = \{v_t \in \mathcal{V}|q \in P_t, q \in \mathcal{L}\}$, labeled by $\Lambda : (\mathcal{V}, \lambda) \to [1, c]$ (Eq. 1), in which $C_{tj} = \{p \in P_t | p \in \mathcal{L}, \lambda(p) = j\}$ corresponds to the pixels $p \in P_t$ that belongs to the image markers set $\mathcal{L}$ with label j.

$$\Lambda(v_t, \lambda) = \operatorname{argmax}_{j \in [1,c]} \left\{ |C_{tj}| \right\} \tag{1}$$

Like FLIM, GFLIM requires a marker-based normalization to avoid learning a bias for kernels. Let the training graphs $\langle G_1, G_2, ..., G_n \rangle$, in which $G_i = (\mathcal{V}_i, \mathcal{B}_i, \psi_i)$, and the markers $\hat{\mathcal{M}} = \langle \mathcal{M}_1, \mathcal{M}_2, ... \mathcal{M}_n \rangle$, where $\mathcal{M}_i$ has the markers of $\mathcal{V}_i$ and $1 \le i \le n$. For each G_i, $\mathcal{V}_i$ is normalized by computing $v = \frac{v - \mu(\hat{\mathcal{M}})}{\sigma(\hat{\mathcal{M}})}$ for all $v \in \mathcal{V}_i$, in which $\mu(\hat{\mathcal{M}})$ and $\sigma(\hat{\mathcal{M}})$ are the mean and standard deviation, respectively, of $\hat{\mathcal{M}}$.

3.2   Kernel Computation

A CNN kernel with size $r \times r$ usually operates on image patches with the same size. In graphs, such spatial relation is absent, but an adjacency relation may contain useful information. To compute kernels, GFLIM extracts features from superpixel graphs based on graph-based markers $\mathcal{M}$ and a labeling function Λ. Despite the user drawing markers on images, GFLIM's inputs and outputs are graphs. Although not preserving the spatial information of I, the adjacency relation $\mathcal{B}$ preserves neighboring information, and their ordering based on a weight function $w : \mathcal{V} \times \mathcal{V} \to \mathbb{R}$ (*e.g.*, the Euclidean distance) allows extracting similar kernels for similar neighborhood patterns.

To extract features from G, GFLIM performs a neighbor selection in a graph G guided by the adjacency relation $\mathcal{B}$ and by sets of markers $\mathcal{M}$. Let $\mathcal{V} \in \mathbb{R}^{k \times m}$ and a set $X_s \subseteq \mathcal{V}$, in which $s \in \mathcal{M}$. Let $\mathcal{N}$ the neighbors of X_s as $\mathcal{N} = \{v \mid (u, v) \in \mathcal{B}, u \in X_s, v \in \mathcal{V} \backslash X_s\}$ and $X_s \cap \mathcal{N} = \emptyset$. One may vectorize X_s as x_s. For each $s \in \mathcal{M}$, a feature vector $x_s \in \mathbb{R}^{m \times \gamma}$ is computed by selecting

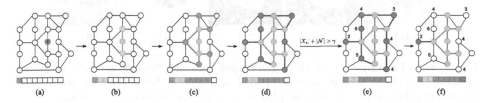

Fig. 3. Neighbors selection example. Given a marked pixel $s \in \mathcal{M}_i$ and $\gamma = 9$ (a). While $\mathcal{N}$ has lesser (or equal) neighbors than desired (b-c), include in X_s the nodes $v \in \mathcal{N}$ in a non-decreasing order according to $w(s, v)$. When $|X_s| + |\mathcal{N}| > \gamma$, select the $\gamma - |X_i|$ nodes in $\mathcal{N}$ with smaller values of w (f).

from $\mathcal{V}$ the $\gamma > 0$ nearest neighbors of s according to the following procedure. Starting with $X_s = \{s\}$, the nodes $v \in \mathcal{N}$ are inserted in X_s in non-decreasing order according to $w(s, v)$. Then, $\mathcal{N}$ is recomputed and this process continues until $|X_s| = \gamma$. Figure 3 demonstrates the neighbors' selection procedure with $\gamma = 9$. Let b_t the number of feature vectors $\boldsymbol{x}_s$, in which $s \in \mathcal{M}$ and $t = \Lambda(s)$. One may compute a *feature bank* $f_t \in \mathbb{R}^{b_t \times \gamma}$, where $\mathcal{F} = \langle f_1, f_2, ..., f_c \rangle$, f_t is composed of vectors $\boldsymbol{x}_s$ for $s \in \mathcal{M}$, in which $t = \Lambda(s)$.

Let α, β, and Γ be the number of desired features per label, features per graph and final kernels, respectively, and a graph G_i in a training set $\mathcal{G} = \langle G_1, ..., G_n \rangle$, from which a feature bank $\mathcal{F}_i = \langle f_1, ..., f_c \rangle$ may be computed. To control the number of kernels, remove redundancy, and find the patterns that best represent the user-marker features, GFLIM performs clustering operations (*e.g.*, K-means or DBSCAN). First, it computes $\hat{f}_t \in \hat{\mathcal{F}}_i$ by selecting α centroids using a clustering operation for each $f_t \in \mathcal{F}_i$. Then, it computes an initial kernel bank $\mathbf{F}$ by selecting β centroids from each $\hat{\mathcal{F}}_i$ with another clustering operation. Finally, a kernel bank $\mathcal{K}$ is computed by selecting the Γ centroids of the clusters in $\mathbf{F}$.

The GFLIM's layer weights are represented by the kernel bank $\mathcal{K}$ and the user may select those that best represent the image classes, finishing that layer training process. To perform kernel selection, the user may select those that activate regions of the same class by visualizing them. Figure 4 presents examples of image activations produced by GFLIM. Ideally, the selected kernels must have strong activations (Fig. 4d) and activate only foreground or background regions — *i.e.*, if there are activation values greater than 0 in foreground regions, they must be 0 in background regions. However, similar patterns may produce kernels that activate in both, foreground and background regions (Fig. 4c). Also, since the image objects/background may contain distinct features (*i.e.*, heterogeneous regions) within an image or between images, one may require more kernels per class to represent the image patterns. Through visualization, the user may choose kernels that best represent different image patterns (*e.g.*, Fig. 4e). Although it

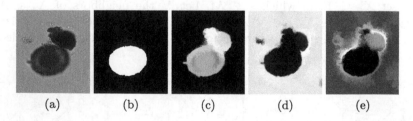

(a)   (b)   (c)   (d)   (e)

Fig. 4. Kernel example, in which brighter pixels indicate stronger activations. Let (a) an image and (b) its ground truth, (c) kernels activated by foreground and background regions may be discarded. Although (d) kernels strongly activated by background (or foreground) regions may be computed, the user may consider including (e) kernels activated by other background patterns.

has a manual kernel selection procedure, GFLIM may take around $1\,s^1$ to train a layer and less than it to perform kernel selection.

For $\mathcal{V}_i = \langle v_1, v_2, \ldots, v_k, \rangle$ and $v_j \in \mathbb{R}^m$, one may interpret $\mathcal{V}_i$ as a matrix $\mathcal{V}_i \in \mathbb{R}^{k \times m}$, in which $1 \le j \le k$ and v_j is the jth row. The previously mentioned neighbor selection process for each $v \in \mathcal{V}_i$ may result in a combination of neighboring vertex features, which can be interpreted as a matrix $\mathcal{V}_i' \in \mathbb{R}^{k \times m\gamma}$. Let the learned kernels $\mathcal{K} \in \mathbb{R}^{m\gamma \times \Gamma}$. The product between $\mathcal{V}_i'$ and $\mathcal{K}$ produce a matrix $\mathcal{S}_i \in \mathbb{R}^{k \times \Gamma}$. By using the adjacency relation $\mathcal{B}_i$, a graph $\hat{G}_i = (\hat{\mathcal{V}}_i, \mathcal{B}_i)$ may be computed, where the jth row in $\mathcal{S}_i$ is the jth vertex in $\hat{\mathcal{V}}_i$. Let $v_{x,y} \in \hat{\mathcal{V}}_i$ be the value at the xth row and yth column in a matrix $\hat{\mathcal{V}}_i$. From a graph $\hat{G}_i$ and a segmentation map L_i, one may compute Γ activation images $\langle \hat{I}_{i,1}, \ldots, \hat{I}_{i,\Gamma} \rangle$, in which each $\hat{I}_{i,j}$ is a grayscale image and $\hat{I}_{i,j}(p) = \{v_{r,j} \in \hat{\mathcal{V}}_i \mid r \in L_i(p)\}$. Whether the user desires to add more layers, a new training process will start with $\hat{G}_i$ as input — $i.e.$, the new layer input is the previous layer output. After training a GFLIM model, one may define an architecture whose layers are composed of the learned kernels followed by an activation function ($e.g.$, ReLU) and a pooling operation ($e.g.$, max pooling).

3.3 User-Guided Decoder for Object Detection

GFLIM outputs a graph from which activation images (saliency maps) $\hat{I}_i = \langle \hat{I}_{i,1}, \ldots, \hat{I}_{i,\Gamma} \rangle$ may be computed using a segmentation map L_i and the original image I_i. Let an activation map $\hat{I}_{i,j}$ with pixel intensities $\hat{I}_{i,j}(p) \in [0, 1]$. For object detection, one may apply a function $\Psi : \hat{I}_i(p) \to \hat{T}(p) \in \{0, 1\}$. During GFLIM's training, the user may select kernels that best represent the foreground or background features. In this work, we propose a User-Guided Decoder (UGD)

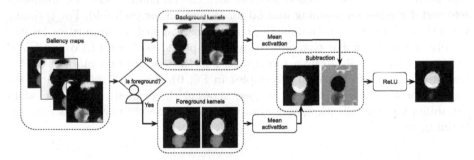

Fig. 5. Proposed User-Guided Decoder (UGD). From a set of activation maps — in which brighter pixels indicate strong activations, and the user information about which maps activate foreground regions, the average activation removes noised predictions. Then, the combination of both maps with negative importance for the background map can filter weak predictions, producing a saliency map.

(Fig. 5) that takes advantage of the user's prior knowledge, requiring a minimum effort, since the decision of which kernels are for object or background was already made during kernel selection in training. Let $T_O \subseteq \hat{I}_i$ be the kernels selected for foreground — consequently, $\hat{I}_i \backslash T_O$ for background. The kernels in $\hat{I}_i$ may activate distinct regions and have some noise (wrong activations). Let a saliency map X, a thresholding function $\delta(X, \tau)$ that assigns 1 when $X(p) > \tau$ and 0 otherwise, and $\mu(X) = \sum_{x_j \in X} x_j / |X|$ the mean saliency map of X.

To amortize wrong activations and maintain the strong ones, one may compute the mean object and background saliencies as $\mu(T_O)$ and $\mu(\hat{I}_i \backslash T_O)$, respectively. Finally, the most probable object regions may be obtained by removing $\mu(\hat{I}_i \backslash T_O)$ from $\mu(T_O)$ and thresholding the resulting image. Equation 2 presents the UDG function. For detection, a binarized image is computed using a thresholding function δ with a threshold $\tau \in [0, 1]$, and a minimum bounding box may be estimated from each connected component of it. Therefore, after training a GFLIM model, the detection process is automatic.

$$\Psi(T_O, \hat{I}_i) = \Big\{ \max\{t_o - t_b, 0\} | t_o \in \mu(T_O), t_b \in \mu(\hat{I}_i \backslash T_O) \Big\} \tag{2}$$

4   Experiments

In this section, we present our experimental setup and results. To evaluate the effectiveness of GFLIM, we compared it with the image-based proposal FLIM in three architectures, along with two other saliency estimation methods, U^2Net and SR, and a bounding box estimation method, DETReg. For the dataset, we selected a private one, Parasites, with 77 optical microscopy images of *helminth eggs*. We randomly divided the dataset into 30% (*i.e.*, 23 images) for the test set and performed a 3-fold cross-validation with the remaining ones. We manually selected 4 images for training and 50 for validation for each fold. For training, we selected images with challenging characteristics, such as impurities located closely or having similar characteristics to the parasites. For GFLIM and FLIM training, we manually draw a foreground and two background markers on the training images of each split (examples in Fig. 6). The test set evaluates the model's performance in unseen images, allowing the evaluation of the model's capability for generalization, and cross-validation is useful to avoid bias from the training set.

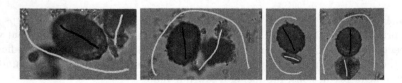

Fig. 6. User-drawn markers on training images of split 3, in which blue and red scribbles mark, respectively, foreground and background regions. (Color figure online)

4.1   Architecture Parameters

We evaluated three architectures for FLIM and GFLIM, with three layers each, and called them $FLIM_1$, $FLIM_2$, and $FLIM_3$, and $GFLIM_1$, $GFLIM_2$, and $GFLIM_3$. We set related architecture parameters for FLIM and GFLIM, such that for a layer with kernels of size γ in GFLIM (*i.e.*, the number of neighbors in neighbors selection), an architecture with $\sqrt{\gamma} \times \sqrt{\gamma}$ kernels was defined for FLIM (same for max pooling). In GFLIM and FLIM, we set centroids quantities as $\alpha = 5$, $\beta = 10000$, $\Gamma = 64$, and performed a ReLU and max pooling operation after convolution in all layers of all architectures. For GFLIM, the first two layers of all architectures have kernels' size 9 and a max pooling of 9. The third layer of $GFLIM_1$ and $GFLIM_2$ have kernels' sizes of 9 and 25, respectively, with 9 for max pooling in both. Finally, $GFLIM_3$ has kernels' size and max pooling 25 on its third layer. For FLIM, the first two layers of all architectures have kernels' size 3×3 and a max pooling of 3×3. At the third layer, $FLIM_1$ and $FLIM_2$ have kernels' sizes of 3×3 and 5×5, respectively, with 3×3 for max pooling in both. Finally, $FLIM_3$ has kernels' size and max pooling 5×5 on its third layer. During training, we manually selected 10 kernels (5 of foreground and 5 of background) for each layer and architecture in GFLIM and FLIM. To compute graphs from the image dataset, we performed a superpixel segmentation with DISF [2], producing 1000 superpixels per image with an initial grid sampling of 10000 seeds. Superpixel graphs are then computed from the superpixel segmentations using $\rho = \sqrt{2}$ and ψ as the mean color.

4.2   Bounding Boxes Computation

To combine the activation maps from GFLIM and FLIM's output, we used the proposed decoder and a set of foreground kernel indexes. For a fair comparison, the chosen binarization and detection parameters were optimized for each of the methods. Since GFLIM may produce higher activations, we set τ as 150, 180, and 200 for $arch_1$, $arch_2$, and $arch_3$, respectively, for GFLIM. For FLIM, we set τ as 100. Like, GFLIM and FLIM, SR and U^2Net produce saliency maps. For those, we also binarized the output with $\tau = 100$. For a binarized image, we computed the minimum bounding boxes for the image's connected components and filtered those with less than 4% of the image's size for all evaluated methods. We also filtered the bounding boxes produced by DETReg, with a non-maximum suppression with IoU 0.1 followed by a filtering of bounding boxes, removing those with less than 10% of the image's size.

4.3   Evaluation Measures

We evaluated GFLIM with the object-detection measures: Precision (Pre), Recall (Rec), F-measure (F-M), Mean Average Precision (μAP), and Precision-Recall curve (PR). Let B_I a set of bounding boxes in an image I, B_G a set of bounding boxes on the ground truth of I, and $IoU(b_I, B_G) = \max_{b_G \in B} \left\{ \frac{b_I \cap b_G}{b_I \cup b_G} \right\}$ the Insection over Union between b_I and B_G. For $t \in [0, 1]$, the precision $Pre^t = \frac{TP^t}{TP^t + FP^t}$

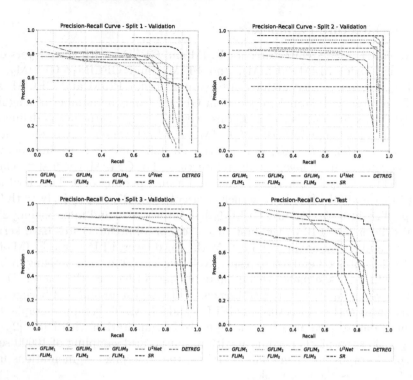

Fig. 7. Precision-recall curves for validation sets of splits 1 to 3 and the test set.

and recall $\mathrm{Rec}^t = \frac{TP^t}{TP^t + FN^t}$ for B_I, respectively, in which the true positives TP^t and the FP^t are the number of bounding boxes $b_I \in B_I$ with $\mathrm{IoU}(b_I, B_G) > t$ and $\mathrm{IoU}(b_I, B_G) \leq t$, respectively, and the false negatives FN^t is the number of bounding boxes $b_G \in B_G$, in which $\mathrm{IoU}(b_I, B_G) \leq t$.

Precision (Pre) measures the fraction of correct predictions (i.e., whose IoU $> t$) concerning the model's predictions, while Recall (Rec) measures the fraction of correct predictions concerning ground truth predictions. Considering $t = \{0.5, 0.75\}$, the F-Measure for object detection combine precision and recall as $\mathrm{F\text{-}M}^t = \frac{2\mathrm{Pre}^t \times \mathrm{Rec}^t}{\mathrm{Pre}^t + \mathrm{Rec}^t}$.

For a threshold $t \in T = \{0.5, 0.55, 0.6, 0.65, 0.7, 0.75, 0.8, 0.85, 0.9, 0.95\}$, one may compute (AP^t) as the mean precision of a dataset considering a fixed t. By considering AP^t the area under the curve for a fixed threshold t, the Precision-Recall curve (PR) may be computed considering all $t \in T$. Also, the Mean Average Precision ($\mu\mathrm{AP}$) is the mean of precision Pre^t for all $t \in T$.

4.4   Results

This section presents the quantitative and qualitative results. For all methods, we evaluated the test set using the split whose model has the highest $\mu\mathrm{AP}$ (Table 1), which is split 3 for U^{2}Net, SelfReformer, DETReg, and FLIM's archi-

Table 1. Results of validation and test sets of GFLIM and compared methods. The three best results for each evaluation measure for each split and the test set are highlighted in blue, green, and red, respectively.

Split 1	$\mu Pre^{0.5}$	$\mu Rec^{0.5}$	$\mu F\text{-}M^{0.5}$	$\mu AP^{0.5}$	$\mu Pre^{0.75}$	$\mu Rec^{0.75}$	$\mu F\text{-}M^{0.75}$	$\mu AP^{0.75}$	μAP
$FLIM_1$	0.815	0.863	0.853	0.843	0.611	0.647	0.640	0.627	0.472
$FLIM_2$	0.763	0.882	0.856	0.863	0.695	0.804	0.779	0.784	0.626
$FLIM_3$	0.776	0.882	0.859	0.863	0.586	0.667	0.649	0.647	0.562
U^2Net	**0.941**	0.941	**0.941**	0.922	**0.941**	0.941	**0.941**	**0.922**	**0.821**
SelfReformer	0.868	0.902	0.895	0.882	0.830	0.863	0.856	0.843	0.701
DETReg	0.576	0.961	0.848	0.941	0.529	0.882	0.779	0.863	0.639
$GFLIM_1$	0.754	0.843	0.824	0.824	0.649	0.725	0.709	0.706	0.663
$GFLIM_2$	0.808	0.824	0.820	0.804	0.712	0.725	0.723	0.706	0.587
$GFLIM_3$	0.878	0.843	0.850	0.824	0.714	0.686	0.692	0.667	0.554
Split 2	$\mu Pre^{0.5}$	$\mu Rec^{0.5}$	$\mu F\text{-}M^{0.5}$	$\mu AP^{0.5}$	$\mu Pre^{0.75}$	$\mu Rec^{0.75}$	$\mu F\text{-}M^{0.75}$	$\mu AP^{0.75}$	μAP
$FLIM_1$	0.836	0.902	0.888	0.882	0.727	0.784	0.772	0.765	0.630
$FLIM_2$	0.842	0.941	0.920	0.922	0.789	0.882	0.862	0.863	0.718
$FLIM_3$	0.793	0.902	0.878	0.882	0.741	0.843	0.821	0.824	0.680
U^2Net	0.960	0.941	0.945	0.922	0.920	0.902	0.906	0.882	0.778
SelfReformer	**0.961**	**0.961**	**0.961**	**0.941**	**0.961**	**0.961**	**0.961**	**0.941**	**0.808**
DETReg	0.533	**0.961**	0.828	0.693	0.511	0.922	0.794	0.664	0.584
$GFLIM_1$	0.857	0.941	0.923	0.922	0.821	0.902	0.885	0.882	0.770
$GFLIM_2$	0.925	**0.961**	0.953	**0.941**	0.887	0.922	0.914	0.902	**0.819**
$GFLIM_3$	0.904	0.922	0.918	0.902	0.865	0.882	0.879	0.863	0.755
Split 3	$\mu Pre^{0.5}$	$\mu Rec^{0.5}$	$\mu F\text{-}M^{0.5}$	$\mu AP^{0.5}$	$\mu Pre^{0.75}$	$\mu Rec^{0.75}$	$\mu F\text{-}M^{0.75}$	$\mu AP^{0.75}$	μAP
$FLIM_1$	0.789	0.882	0.862	0.863	0.754	0.843	0.824	0.824	0.722
$FLIM_2$	0.789	0.882	0.862	0.863	0.754	0.843	0.824	0.824	0.719
$FLIM_3$	0.793	0.902	0.878	0.882	0.759	0.863	0.840	0.843	0.780
U^2Net	**0.961**	**0.961**	**0.961**	**0.941**	**0.961**	**0.961**	**0.961**	**0.941**	0.852
SelfReformer	0.925	**0.961**	0.953	**0.941**	0.906	0.941	0.934	0.922	**0.865**
DETReg	0.495	**0.961**	0.809	**0.941**	0.485	0.941	0.792	0.922	0.787
$GFLIM_1$	0.842	0.941	0.920	0.922	0.789	0.882	0.862	0.863	0.756
$GFLIM_2$	0.891	**0.961**	0.946	**0.941**	0.855	0.922	0.907	0.902	0.816
$GFLIM_3$	0.906	0.941	0.934	0.922	0.811	0.843	0.837	0.824	0.704
Test	$\mu Pre^{0.5}$	$\mu Rec^{0.5}$	$\mu F\text{-}M^{0.5}$	$\mu AP^{0.5}$	$\mu Pre^{0.75}$	$\mu Rec^{0.75}$	$\mu F\text{-}M^{0.75}$	$\mu AP^{0.75}$	μAP
$FLIM_1$	0.704	0.760	0.748	0.431	0.630	0.680	0.669	0.296	0.237
$FLIM_2$	0.029	0.040	0.037	0.004	0.000	0.000	0.000	0.000	0.001
$FLIM_3$	0.724	0.840	0.814	0.552	0.655	0.760	0.736	0.431	0.378
U^2Net	0.840	0.840	0.840	0.800	0.760	0.760	0.760	0.452	0.452
SelfReformer	0.920	**0.920**	0.920	0.743	**0.840**	**0.840**	**0.840**	**0.574**	**0.567**
DETReg	0.429	0.840	0.705	0.304	0.408	0.800	0.671	0.256	0.235
$GFLIM_1$	0.769	0.800	0.794	0.540	0.654	0.680	0.675	0.311	0.270
$GFLIM_2$	**0.957**	0.880	0.894	**0.840**	0.739	0.680	0.691	0.329	0.382
$GFLIM_3$	**0.957**	0.880	0.894	**0.840**	0.783	0.720	0.732	0.367	0.421

tectures, and split 2 for GFLIM's architectures. Figure 7 presents the precision-recall curves for validation sets of splits 1 to 3 and the test set. As one may see, GFLIM's models achieved consistently better results than FLIM. Also, despite GFLIM requiring much less hardware than the deep-based models, it achieves greater precision than DETReg. DETReg generates a fixed number of 30 bounding boxes, and despite filtering most of these boxes in our evaluation, many false positives remain. $GFLIM_2$ and $GFLIM_3$ had similar results to SR and U^2Net, especially in split 2. For the test set, only the DETReg had similar performance among the test set and the splits, while the SR obtained the best results. However, all models performed worse in the test set than the splits. Despite having worse results in the test set, GFLIM was competitive with U^2Net in this set.

Table 1 presents the quantitative results for all methods in all splits and the test set. As one may see, GFLIM outperforms FLIM in almost all measures,

Table 2. Comparison among the number of parameters (#Param.) for different models. The architectures used for FLIM have the same number of parameters as GFLIM.

Model	DETReg	U²Net	SR	**GFLIM₁**	**GFLIM₂**	**GFLIM₃**
#Param.	39,847,265	44,009,869	91,585,457	2,100	3,700	3,700

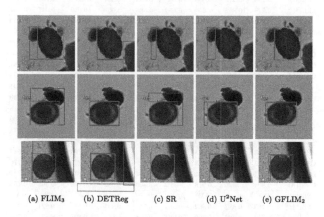

(a) FLIM₃   (b) DETReg   (c) SR   (d) U²Net   (e) GFLIM₂

Fig. 8. Qualitative comparison. Blue, red, and green bounding boxes indicate the detections, ground truth, and their intersections, respectively. (Color figure online)

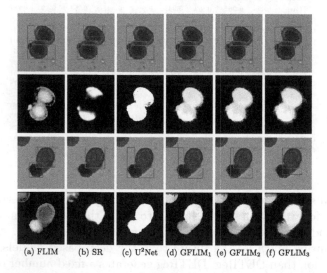

(a) FLIM   (b) SR   (c) U²Net   (d) GFLIM₁   (e) GFLIM₂   (f) GFLIM₃

Fig. 9. Failure cases in GFLIM. The detections (first and third row) have blue and red bounding boxes indicating the detected region and the ground truth, respectively, and their intersections are in green. The second and fourth rows have the activation maps of the first and third rows, respectively, in which brighter pixels are higher activations. (Color figure online)

especially on splits 2 and 3. Also, $FLIM_2$ had the worst result due to some of its foreground kernels having strong background activation in most of the images in the test set. $GFLIM_2$ and $GFLIM_3$ had similar results on splits 2 and 3 compared to the deep models. On the test set, $GFLIM_2$ and $GFLIM_3$ outperformed U^2Net when considering 0.5 of IoU ($\mu Pre^{0.5}$, $\mu Rec^{0.5}$, and $\mu AP^{0.5}$). Table 2 presents a comparison of the number of parameters of each model. Our proposal not only produces better results than FLIM but can also produce competitive detections with deep models, even having an architecture thousands of times smaller in the number of parameters.

Figures 8 and 9 present the qualitative results, in which the former contains correct GFLIM detections, while the second contains failure cases. As one may see, both DETReg and FLIM tend to generate false positives in parasite-like regions (first row of Fig. 8). Furthermore, DETReg erroneously detects image features distinct from those of the training set, such as the homogeneous white region and the black region, in the third row of Fig. 8. In contrast, FLIM, SR, and U^2Net tend to detect impurity regions at the parasite boundary (first and second rows of Fig. 8). Although the failure cases in GFLIM indicate a similar difficulty (third row of Fig. 9), their saliency maps (fourth row of Fig. 9) suggest that a threshold could solve this problem. In our evaluation, we binarized the saliency maps of all methods (except DETReg) with optimized values. However, a fixed threshold may not be an optimal solution. Nearby objects are also challenging for our proposal, as they tend to be detected as a single object (first row of Fig. 9). Finally, GFLIM maps tend to have higher activations (visually brighter) in regions with parasites than FLIM's. Although this also occurs in some impurities similar to the parasite, the impurities' saliency values in GFLIM are lower (visually darker) than the parasite, which does not occur in FLIM.

5   Conclusion

In this work, we proposed a lightweight CNN for feature extraction that does not require backpropagation or a large annotated dataset. The Graph-based Feature Learning from Image Markers (GFLIM) inputs and outputs graphs and its training process is independent of the graph modeling. Although graph properties may interfere with the network, such as digraphs and disconnected graphs, our proposal presents a generalization that can be applied to other graph models, not just superpixels or images. To produce saliency maps by combining learned features, we proposed a User-Guided Decoder (UGD) that does not require training and is suitable for any FLIM-based strategy — and any other strategy that can inform which kernels mostly activate foreground pixels. We evaluated GFLIM against deep-learning models from the state-of-the-art for object detection using UDG. The results indicate that our proposal not only generalizes FLIM but also is capable of extracting visually better features from graphs with competitive results. In future works, we intend to explore the graph structure with other architectures and graph modeling.

References

1. Bar, A., et al.: Detreg: unsupervised pretraining with region priors for object detection. arXiv preprint arXiv:2106.04550 (2021)
2. Belém, F.C., Guimarães, S.J.F., Falcão, A.X.: Superpixel segmentation using dynamic and iterative spanning forest. IEEE Signal Process. Lett. **27**, 1440–1444 (2020). https://doi.org/10.1109/LSP.2020.3015433
3. Cerqueira, M.A., Sprenger, F., Teixeira, B.C., Falcão, A.X.: Building brain tumor segmentation networks with user-assisted filter estimation and selection. In: 18th International Symposium on Medical Information Processing and Analysis, vol. 12567, pp. 202–211. SPIE (2023)
4. De Souza, I.E., Falcão, A.X.: Learning cnn filters from user-drawn image markers for coconut-tree image classification. IEEE Geoscience and Remote Sensing Letters (2020)
5. Deng, J., Dong, W., Socher, R., Li, L.J., Li, K., Fei-Fei, L.: Imagenet: a large-scale hierarchical image database. In: 2009 IEEE Conference on Computer Vision and Pattern Recognition, pp. 248–255. IEEE (2009)
6. Lin, T.-Y., et al.: Microsoft COCO: common objects in context. In: Fleet, D., Pajdla, T., Schiele, B., Tuytelaars, T. (eds.) ECCV 2014. LNCS, vol. 8693, pp. 740–755. Springer, Cham (2014). https://doi.org/10.1007/978-3-319-10602-1_48
7. O Pinheiro, P.O., Almahairi, A., Benmalek, R., Golemo, F., Courville, A.C.: Unsupervised learning of dense visual representations. Advances in Neural Information Processing Systems (2020)
8. Qin, X., Zhang, Z., Huang, C., Dehghan, M., Zaiane, O.R., Jagersand, M.: U2-net: going deeper with nested u-structure for salient object detection. Pattern Recognition (2020)
9. Shi, W., et al.: Real-time single image and video super-resolution using an efficient sub-pixel convolutional neural network. In: Proceedings of the IEEE Conference on Computer Vision and Pattern Recognition, pp. 1874–1883 (2016)
10. Sousa, A.M., Reis, F., Zerbini, R., Comba, J.L., Falcão, A.X.: Cnn filter learning from drawn markers for the detection of suggestive signs of covid-19 in ct images. In: 2021 43rd Annual International Conference of the IEEE Engineering in Medicine & Biology Society (EMBC), pp. 3169–3172. IEEE (2021)
11. de Souza, I.E., Benato, B.C., Falcão, A.X.: Feature learning from image markers for object delineation. In: 2020 33rd SIBGRAPI Conference on Graphics, Patterns and Images (SIBGRAPI), pp. 116–123. IEEE (2020)
12. Uijlings, J.R., Van De Sande, K.E., Gevers, T., Smeulders, A.W.: Selective search for object recognition. Int. J. Comput. Vis. (2013)
13. Wang, L., et al.: Learning to detect salient objects with image-level supervision. In: Proceedings of the IEEE Conference on Computer Vision and Pattern Recognition, pp. 136–145 (2017)
14. Wang, W., Lai, Q., Fu, H., Shen, J., Ling, H., Yang, R.: Salient object detection in the deep learning era: an in-depth survey. IEEE Trans. Pattern Anal. Mach. Intell. (2021)
15. Wang, W., et al.: Pyramid vision transformer: a versatile backbone for dense prediction without convolutions. In: Proceedings of the IEEE/CVF International Conference on Computer Vision, pp. 568–578 (2021)
16. Wei, F., Gao, Y., Wu, Z., Hu, H., Lin, S.: Aligning pretraining for detection via object-level contrastive learning. Advances in Neural Information Processing Systems (2021)

17. Yang, C., Wu, Z., Zhou, B., Lin, S.: Instance localization for self-supervised detection pretraining. In: Proceedings of the IEEE/CVF Conference on Computer Vision and Pattern Recognition, pp. 3987–3996 (2021)
18. Yun, Y.K., Lin, W.: Selfreformer: Self-refined network with transformer for salient object detection. arXiv preprint arXiv:2205.11283 (2022)

Seabream Freshness Classification Using Vision Transformers

João Pedro Rodrigues[1,2]([✉]), Osvaldo Rocha Pacheco[3,4], and Paulo Lobato Correia[1,2]

[1] Instituto Superior Técnico, Av. Rovisco Pais 1, 1049-001 Lisboa, Portugal
joao.pedro.carvalho.rodrigues@tecnico.ulisboa.pt,
http://www.tecnico.ulisboa.pt, http://www.it.pt
[2] Instituto de Telecomunicações, Av. Rovisco Pais 1, 1049-001 Lisboa, Portugal
[3] Universidade de Aveiro, Campus Universitário de Santiago, 3810-193 Aveiro, Portugal
https://www.ua.pt, http://www.ieeta.pt
[4] Instituto de Engenharia Electrónica e Telemática de Aveiro, Campus Universitário de Santiago, 3810-193 Aveiro, Portugal
http://www.tecnico.ulisboa.pt, http://www.it.pt

Abstract. Many different cultures and countries have fish as a central piece in their diet, particularly in coastal countries such as Portugal, with the fishery and aquaculture sectors playing an increasingly important role in the provision of food and nutrition. As a consequence, fish-freshness evaluation is very important, although so far it has relied on human judgement, which may not be the most reliable at times.

This paper proposes an automated non-invasive system for fish-freshness classification, which takes fish images as input, as well as a seabream fish image dataset.

The dataset will be made publicly available for academic and scientific purposes with the publication of this paper. The dataset includes meta-data, such as manually generated segmentation masks corresponding to the fish eye and body regions, as well as the time since capture.

For fish-freshness classification four freshness levels are considered: very-fresh, fresh, not-fresh and spoiled. The proposed system starts with an image segmentation stage, with the goal of automatically segmenting the fish eye region, followed by freshness classification based on the eye characteristics. The system employs transformers, for the first time in fish-freshness classification, both in the segmentation process with the Segformer and in feature extraction and freshness classification, using the Vision Transformer (ViT).

Encouraging results have been obtained, with the automatic fish eye region segmentation reaching a detection rate of 98.77%, an accuracy of 96.28% and a value of the Intersection over Union (IoU) metric of 85.7%. The adopted ViT classification model, using a 5-fold cross-validation strategy, achieved a final classification accuracy of 80.8% and an F1 score of 81.0%, despite the relatively small dataset available for training purposes.

Supported by organizations 1, 2, 3 and 4.

V. Vasconcelos et al. (Eds.): CIARP 2023, LNCS 14469, pp. 510–525, 2024.
https://doi.org/10.1007/978-3-031-49018-7_36

Keywords: Image Classification · Fish-freshness estimation · Image Segmentation · Feature Extraction · Vision Transformer

1   Introduction

The global fisheries and aquaculture sectors have an increasingly important role in the provision of people's food and nutrition. In 2020, the total production of fisheries and aquaculture reached a record 214 million metric tonnes of mostly aquatic animals (178 million ton) but also of algae (36 million ton). This is mostly due to the growth of aquaculture worldwide, notably in Asia. These sectors make up for a huge market with a total revenue of 281.5 billion dollars [1].

Portugal is a country with a great heritage in marine activities with fish taking a major role in its cultural and social life. According to the Food and Agriculture Organization of the United Nations (FAO)'s statistical report of the year 2020 [2], Portugal was the sixth country with the largest fish and seafood consumption per capita in the world and the largest of the European Union (EU), with 56.84 kg consumed by year, well above the global average of 20.3 kg, making it of great economic importance.

One of the fish species that Portuguese people consume the most is the seabream (*Sparus Aurata*), due to its flavour, relative cheap price and abundance, since it is the most cultivated fish species in the aquaculture sector of Portugal, with 3091 ton cultivated in the year of 2021 [3].

Since the seabream takes such a prominent role both in the Portuguese aquaculture sector and consumer's nutrition, an automated process for its freshness estimation is very desirable, particularly if it is not invasive and if it does not require to touch the fish for this estimation.

This paper proposes a novel fish-freshness classification system, using seabream images as input and for the first time using vision transformers for fish-freshness estimation. Moreover, recognizing the difficulty of benchmarking against previously published research, a publicly available dataset of seabream images is proposed, reflecting the fish-freshness degradation along time. It's also accompanied by metadata, which includes segmentation masks for the fish eye and body regions.

2   State-of-the-Art Review

Fish-freshness classification systems based on image analysis have been reported at least since 2009 [4]. These systems can be grouped into two main categories: (i) traditional approaches, in which researchers select the image characteristics considered relevant for fish-freshness classification; and, (ii) deep learning approaches, in which the selected model learns which features are relevant to perform the classification. Previous works which employ both these categories are briefly discussed in the following subsections.

2.1 Traditional Systems

Lalabadi *et al.* developed a fish-freshness classification system employing the gill and eye region of rainbow trouts (*Oncorhynchus Mykiss*) [5]. The eyes and gills of the fishes were segmented using Otsu's thresholding method [6], extracting 20 color components from the corresponding segmented images in the Red Green Blue (RGB), Hue Saturation Value (HSV) and L*a*b color spaces. The extracted components underwent statistical calculations like the geometric and arithmetic means (among others), resulting in a total number of 54 features. The classification process employed both an Artificial Neural Network (ANN) and Support Vector Machines (SVM), classifying the fish into 5 different classes - 1, 3, 5, 7 and 9 days since capture. The ANNs with best results had a topology of 54-7-5 (54 input color components, 7 hidden layer nodes and 5 output classes) for the eye images, with a classification accuracy of 84 percent (21 out of 25 samples correctly predicted) and 54-24-5 for the gill images, with an accuracy of 96 percent (24 out of 25 samples correctly predicted). A proprietary dataset was used to report these results.

Tomás Rosário *et al.* in [7], developed a system to perform seabream freshness classification into 3 classes: fresh, moderately fresh or not fresh. After an automatic segmentation of the fish eye region, classification relied on colourfulness and brightness features, due to their fairly consistent decrease along with fish storage time. Using these features, the classification used a 5-fold cross validation method, employing a k-Nearest Neighbors (kNN) model. The model's hyperparameters were fine tuned to optimize performance, reaching a final classification accuracy of 92.5%, but the system was only able to successfully segment the eye region in 36% of the images.

2.2 Deep Learning Based Systems

Taheri-Garavand *et al.* proposed a deep learning based approach for freshness classification of the common carp (*Cyprinus Carpio*) [8]. Features were extracted using a Convolutional Neural Network (CNN) model, an adaptation of the VGG-16 network [9], paired with a transfer learning approach, with pre-training on the ImageNet dataset [10]. The classifier used as input a feature array of 8192 features (flattened $4 \times 4 \times 512$ output from the feature extraction model); followed by dense layers of 256, 128 and 64 neurons with rectified linear unit (ReLU) activation functions, interspersed with dropout layers of dropout rate of 60%; the output layer contained 4 neurons, corresponding to 4 classes based on the elapsed time since the catch: most fresh (days 1–2), fresh (days 3–4), fairly fresh (days 5–7), and spoiled (days 8–14), with softmax as its activation function. The final classification accuracy was of 98.2%, out of a private dataset containing 56 test samples.

Prasetyo *et al.* compared the performance of state-of-the-art convolutional neural networks (CNNs) for eye region feature extraction and freshness classification of milkfish (*Chanos Chanos*) [11]. Three freshness classes were considered: very fresh (1st day), fresh (2nd-5th day) and not fresh (6th day). The

CNNs employed were: ResNet-50 and ResNet101 [12]; MobileNet V1 [13] and MobileNet V2 [14]; DenseNet 121 and DenseNet 169 [15]; Xception [16]; and NASNet Mobile [17]. The DenseNet 121 and NasNet mobile networks produced the best results, reaching a classification accuracy of 65.9% on a dataset of 234 total samples with a partition of 60% for the training set and 20% for validation and test sets.

Rayan *et al.* used a VGG-16 architecture for feature extraction, combining it with a bi-directional Long Short Term Memory (LSTM), for freshness classification of nile tilapia fish (*Oreochromis niloticus*) [18]. The classifier performed binary classification with the classes: fresh, for capture time up to 2 days; and non-fresh, for capture time larger than 2 days (and up to 7). The reported accuracy of the model was 98% out of 1000 total test samples.

Anas *et al.* [19], used tiny YOLOv2 [20] for both eye region detection and freshness classification of three different fish species: short mackerel (*Rastrelliger*), mackerel tuna (*Euthynnus Affinis*) and milkfish, each being classified into three freshness levels, based on the elapsed time since capture: good quality, for capture up to 2 h; medium quality, for capture between 2 and 5 h; and poor quality, after 5 h since capture. The fish were not refrigerated during the image acquisition. Classification accuracy of the test set (600 samples) was of 57.5%, with the system being unable to detect the eye region for 25% of the fish.

2.3   Discussion

The fish-freshness classification methods reported in the literature show some limitations that the present work tries to address:

- Poor eye region segmentation results - this is a critical step when relying in the eye region for classification. This limitation was reported in the works of Rosário *et al.* [7], and Anas *et al.* [19];
- Limited number of freshness classes considered - for instance the work of Rayan *et al.* performs only a binary classification, which can lead to an insufficient detail of the freshness assessment;
- Low fish-freshness accuracy for some of the reported work - this is the case of the works by Anas *et al.* [19] and Prasetyo *et al.*, showing margin for improvement;
- Nonexistence of a common dataset, publicly available - such a dataset would allow to compare the merits of the different proposals using the exact same conditions. Besides using private datasets, some of the discussed works have used small datasets, limiting in particular the size of the test set and possibly inflating the results, as is the case of the works of Taheri-Garavand *et al.* [8] and Lalabadi *et al* [5].

From the literature review about fish-freshness estimation it was observed that classification has typically followed either classic computer vision methods or deep learning strategies relying on CNNs. This raises another challenge for this work, notably to explore the emerging transformer models [21] for image classification. Although typically the use of transformers may require more data and resources for training purposes, they have recently achieved improved results in many vision tasks such as object detection and classification [22], notably when using the ViT [23], which has shown to be more robust against object occlusions and image perturbations when compared to state-of-the-art CNNs [24].

3   Seabream Freshness Dataset

A good quality image dataset is crucial for training and testing image-based classification systems, such as the fish-freshness classification system being proposed in this paper. Moreover, a limitation of the datasets used by state-of-the-art works is that they are not publicly available, besides typically including a limited number of samples. The available fish datasets, in platforms such as Kaggle, Hugging Face or GitHub, typically only contain images of living fish, which are not suited for a study about the degradation of fish-freshness.

This paper presents a new dataset, made publicly available for scientific and academic purposes in [25]. It is comprised of 4319 images of 40 seabream fish, containing information of the time elapsed since the capture, as well as segmentation masks for the eye and body regions. The dataset images reflect the visual characteristics of the seabream throughout its decomposition process, and the images were captured from different angles and under varying lighting conditions. All the images were submitted to an annotation process, which consisted of the manual identification and labelling of the eye and body regions of the fish, using the online platform Segments.ai [26] for the annotation.

3.1   Image Acquisition

Image acquisition for the seabream freshness dataset was done in two stages, the first was coordinated by Renato Pinto in 2020 [7]; and the second coordinated by João Rodrigues in 2023.

In 2020, images were taken in the Universidade de Aveiro's laboratory with for a duration of 7 days (168 h). In the first 4 days, 9 acquisition sessions took place, corresponding to 0, 17, 24, 41, 48, 65, 72, 89, and 96 h since the capture; the final session took place on the 7th day - 168 h. Each session includes 6 images of each fish, captured from various angles, with and without focus of the eye region, with 2 different types of lighting: neutral, direct lighting and a yellow-saturated, indirect lighting.

In 2023, the image acquisition process had a duration of 14 days, which roughly corresponds to the shelf-life of vacuum-packed seabream fillets (12 days according to [27]) and the threshold of acceptability quality of the fish when stored in ice (11 days according to [28]). A total of 22 acquisition sessions were performed, with a periodicity of 12 h until the 9th day of storage, and of 24 h in the last 5 days, due to changes in the fish's appearance being more noticeable, unlikely to be observed in retail, due to the fish's advanced decomposition during this stage.

Each session performed in 2023 contains 8 photographs of each fish - 4 of each side of the fish from various angles, favoring either the anterior or posterior axis of the fish, as well as the ventral or dorsal axis, either focusing in the fish eye region or the whole body. Each session also contains images of groups of 4 fish, with the fish chosen for the group image being changed for each session. The images were captured with fluorescent and direct lighting, with the fish on top of a brown wooden desk. During the handling of the fish, rubber gloves were used.

Some dataset image examples are presented in Fig. 1.

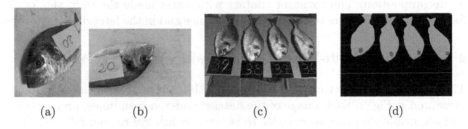

(a) (b) (c) (d)

Fig. 1. Seabream fish-freshness dataset: examples of images acquired in 2020 using a smartphone (a) and a digital camera (b); and example of a group image acquired in 2023 (c) and its respective annotation (d).

3.2 Equipment

The Dataset images were captured using 2 smartphone cameras and a digital camera.

The smartphone cameras employed were the Samsung's SM-N975F (quad-camera, 12.2 MP model, f/1.5 F-stop, 1/100 exposure time, ISO-125 speed and 4mm focal length) in 2020, and the SM-A226B (f/1.8 F-stop, 1/24 exposure time, ISO-250 speed and 5 mm focal length) in 2023, which produced 4032 × 3024 and 4000 × 3000 pixel images, respectively. The SM-N975F camera was used to capture images with neutral colors of fish 1 to 24, whereas the SM-A226B was used to capture all the images of fish 25 to 40.

The digital camera used, in 2020, was a Canon EOS 800D (f/4.5 F-stop, 1/80 exposure time, ISO100 speed and 20mm focal length) for capturing the images of fish 1 to 24, which display yellow-saturated colors, with a spatial resolution of 6000 × 4000 pixels.

3.3 Storage

During the 2020 acquisition campain, the fish were stored in the Universidade de Aveiro's laborathory refrigerators, which allowed for a highly controlled environment, keeping all 24 fishes at the same temperature throughout the 7-day storage period.

For the 2023 image acquisition campain, fish were kept in a regular fridge, stored into: 2 drawers at the bottom, a top and a middle shelf. This caused the decaying process of the fish to have slightly different characteristics. Fish 35 to 40, stored on the top shelf, were closer to the cooling source keeping them fresh for longer compared to fish 27 to 34, placed in the middle shelf. Fish stored on the shelves were gently watered after each image acquisition session, to make them look less dry. Fish 25 and 26 were stored in plastic drawers and were in contact with a shallow layer of water - the fish side in contact with the water remained the same throughout the decomposition process, in order to study the impact of constant contact with water on the appearance of the fish throughout its decomposition. The constant contact with water made the right side of the fish have a softer and more lacklustre appearance and in the later days of storage.

4   Proposed Fish-Freshness Classification System

The high-level architecture of the proposed fish-freshness classification system is presented in Fig. 2. Each image of the dataset undergoes an image preprocessing and segmentation process, in order to isolate the fish eye region; followed by an augmentation process to generate more data, to be used for training the feature extraction and classification modules. A visual scheme of the system is displayed in the following figure - Fig. 2.

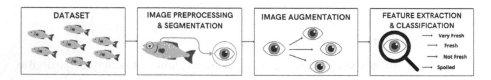

Fig. 2. Proposed fish-freshness classification system architecture.

4.1   Image Preprocessing and Segmentation

The dataset images must undergo the process of image segmentation in order to isolate the eye region of the fish, so that the most prominent features can be extracted in order to classify its freshness. The adopted methodology relies on the characteristics of the fish eye appearance, as these can be acquired without touching the fish, unlike alternatives that might require fish manipulation, for instance the observation of the appearance of the gills.

Prior to segmentation, the images were resized from their original size (4000 $\times$ 6000 pixels when captured with the digital camera, and 3000 $\times$ 4000 pixels for the smartphone cameras) to 512 $\times$ 512 pixels. This dimension provides a good compromise between the computational cost of the training of the model and the preservation of the original image's details, considering that the eye region of the images will still have a minimum diameter of 30 pixels.

An image normalization step is also applied, after resizing, by applying the following equation:

$$[R \quad G \quad B]_{norm} = \frac{[R \quad G \quad B] - \mu}{\sigma} \tag{1}$$

where $[R \quad G \quad B]$ corresponds to the R, G or B values of each pixel in the image, μ corresponds to the mean value of each color component and σ corresponds to its standard deviation.

The preprocessed images are used as input to the segmentation model. The selected model for segmentation of the eye and body regions of the fish was the Segformer model [29], as it provides state-of-the-art segmentation performance, with fewer parameters than alternative models such as DeepLabV3 [30] or SETR [31], among others. The Segformer divides the input image of size H (height) × W (width) into patches of size 4 × 4. These patches are used as input to an encoder to obtain multi-level features at $1/4, 1/8, 1/16, 1/32$ the size of the original image resolution. The encoded patches are passed to a Multilayer Perceptron (MLP) to predict the desired fish body and fish eye segmentation masks at a resolution of $H/4 \times W/4$.

The chosen Segformer encoder was the Mix Transformer Encoder (MiT)-b0 [29] as it leads to the less computationally expensive model, while achieving a good performance on a set of preliminary tests. This encoder is pre-trained on the ImageNet-1k dataset, which contains 1000 object classes and 1,281,167 training images, providing a good starting point in terms of segmentation model weights, to be later refined using transfer learning.

For evaluation of the segmentation model, the accuracy and the Intersection over Union (IoU) performance metrics were considered. Although both being useful for the evaluation of the model's performance, the mean IoU may provide a more reliable segmentation evaluation, as it measures the overlap between the automatically segmented area and the ground truth, whereas the accuracy only accounts for the amount of pixels correctly identified as belonging to the desired object, erroneously returning a perfect accuracy value for a segmentation mask spanning the complete image area, for example.

4.2   Image Augmentation

The training of transformer models for classification requires a significant amount of data, ideally far superior to the relative small dimension of the proposed seabream freshness dataset. To overcome this limitation data augmentation processes were applied in this work. Augmentation was applied to the segmented fish eye images, creating an augmented dataset comprised of 25914 segmented eye images.

The data augmentation included adjustments of saturation, contrast and brightness. Geometric augmentation processes, although quite useful in many classification tasks didn't have much application on this particular task, due to the eye's circular nature and geometry.

Saturation adjustments were done in the HSV color space, multiplying the saturation (S) channel by a scaling factor of 0.5 or 6, to allow the less saturated smartphone images to look more similar to the images taken by the digital camera, and vice-versa.

The contrast adjustment was done in each of the R, G and B color channels, applying the following operation, with the goal of making important regions of the eye, such the pupil and iris, easier to discern:

$$x_{contrast} = (x - \mu) \times \alpha_{contrast} + \mu \tag{2}$$

where μ is the arithmetic mean and $\alpha_{contrast}$ is the contrast factor. The contrast factor can take values of 2 or 6, to accentuate the contours of the images.

Brightness adjustment consisted in adding a constant, $\delta_{brightness}$, to the R, G and B components of the image. The $\delta_{brightness}$ takes a value of 0.3, to compensate the brightness difference observed in the dataset images, which results mostly from the different image acquisition angles.

Examples of augmented images are included in Fig. 3.

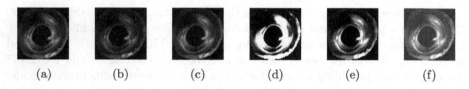

(a)   (b)   (c)   (d)   (e)   (f)

Fig. 3. Examples augmented images: original (a), saturation augmentation with factors 6 (b) and 0.5 (c), contrast augmentation with $\alpha_{contrast}$ value of 6 (d) and 2 (e), and brightness augmentation with $\delta_{brightness}$ value of 0.3 (f).

4.3   Feature Extraction and Classification

The seambream freshness dataset has information about the time since capture associated with every image. Taking into account the appearance and organoleptic observation of the fish decay along time, as well as previous works related to fish freshness such as [32], this work considers 4 fish-freshness classes:

- **Very fresh** - until 48 h since the fish capture;
- **Fresh** - more than 48 h and less 120 h;
- **Not Fresh** - more than 120 and less than 240 h;
- **Spoiled** - more than 240 h since the fish capture.

To perform automatic feature extraction and classification of fish-freshness based on the analysis of the eye region, the visual transformer (ViT) [23] was selected, considering its favourable performance, attaining state-of-the-art results on most recognition benchmarks and classification tasks, while involving a relatively low computational cost for model training, due to its smaller amount of parameters compared to other state-of-the-art models.

The ViT follows the original transformer architecture [21] and its attention mechanism but, in order to handle 2D images, an input image with original dimension of $H \times W \times C$, is flattened into 2D patches with $p^2 \times C$ dimension, where p^2 is the resolution of each image patch. A 1-D representation of each patch is created, including positional information, to create the patch embedding. The ViT's patches are typically non-overlapping and may have different resolutions, e.g., 16×16, 32×32, 64×64. For the present work a resolution of 16×16 was selected, to prevent the loss of local information of particular regions of the eye, such as its ridges and its main components (iris, pupil, etc.).

Transfer learning was used for training the ViT model, thus contributing to improve its feature extraction and classification performance. A ViT model pre-trained on the ImageNet-21k (14 million images, 21,843 classes) and fine-tuned on ImageNet 2012 (1 million images, 1,000 classes) was used as the staring point for ViT model refinement.

Given the relatively small size of the available dataset, and the need of a large amount of data to train the ViT [22], a 5-fold cross-validation strategy was adopted. This helps to prevent over-fitting by testing the ViT model on a chosen dataset partition while still retaining a sizeable amount of data to be used for training. The dataset was thus split into 5 folds, each containing images of different fish, with 4 folds used for training and 1 for testing. This way the training set consisted of approximately 20500 images, i.e. approximately 80% of the augmented dataset. The test set does not include the augmented images, thus being composed of 700 to 800 image for the different folds, which although not a very large number still exceeds the test sets reported in previous fish-freshness classification papers. Each test set includes 5 fish whose images were acquired during the 2020 campaign and 3 fish from the 2023 campaign, to have a better representation of the different image acquisition conditions covered in the seabream freshness dataset. The 3 fish from the 2023 campaign were also chosen taking into account their storage location in the fridge (drawer, top and middle shelves), to account for the different impacts of storage on the decomposition.

The system was evaluated using the accuracy and F1 score metrics in each of the cross validation test folds, and then averaging the results for reporting the overall results.

5   Results and Discussion

The solution presented in this work estimates fish-freshness in a non-intrusive way, by performing automated analysis of images containing seabream fish. Freshness analysis is based on the eye region characteristics, so the first set of results presented is related to the fish eye segmentation performance. Then results of using the ViT model for freshness classification are presented and discussed.

5.1   Segmentation Model

The Segformer segmentation model was trained using a model pretrained on the ImageNet-1k dataset: the MiT-b0 encoder checkpoint ("nvidia/mit-b0") was used for this purpose, available from the transformers library mentioned in the original article [29]. It used AdamW [33] as its optimizer with learning rate 6e-5 and as its optimization metrics are the loss, mean accuracy and mean Intersection over Union (IoU).

The test set corresponded to fish 1, 10, 20, 25, 30, 35 and the group photo of fish 29, 30, 31 and 32, for a total of 706 images; the remaining 3599 images of the dataset were used for training; corresponding to a 83.6/16.4 percent split.

The total training process took 1451 steps in batches of size 2, for a total elapsed time of approximately 1 h and 45 min, on a PC with a 11th Gen Intel Core i7-1165G7 @ 2.80 GHz CPU and a Intel Iris Xe GPU.

The Segformer model converged after 100 training steps, achieving a mean accuracy of 91.7%, showing that the "nvidia/mit-b0" checkpoint already contained much of the information the model needed for the segmentation. The final mean accuracy score was **97.37**, corresponding to an accuracy of 96.9% accuracy of the body region and **96.28%** accuracy of the eye region. The final mean IoU score was **94.53%**, corresponding to a score of 98.8% IoU for the body and **85.7%** for the eye region.

For a visual analysis of the model's performance, inference was ran on several images, both in images from the dataset and images available online. The images were chosen, based on different qualities, such as lighting, angle of the photo, freshness of the fish (in the case of the dataset images), and whether they were individual or group photos.

The results were positive, with the model having no difficulty at detecting and segmenting both the eye and body regions, presenting a detection rate of **98.77%** (53 failed detections for a total of 4319 images). The cases where the model was unsuccessful in detecting the eye region correspond to images where the fish's eye was very small and facing away from the camera. Examples of the model's prediction of dataset images are shown in Fig. 4.

(a)   (b)   (c)   (d)

Fig. 4. Segmentation model predictions on different images of the dataset. (a) through (c) demonstrate correct predictions of the eye region in singular and group photos in different angles and types of lighting. (d), shows an example where the model was unable to detect the eye region.

In order to test the model's universality in the prediction of different species fish, the model was tested on species that are very regularly consumed in Portugal; such as the black scabbard (*Aphanopus carbo*), sardines (*Sardina pilchardus*) and mackerel (*Trachurus trachurus*). It also performed well in these instances, as shown in Fig. 5.

(a)   (b)   (c)   (d)

Fig. 5. Segmentation model predictions of images of different species of fish. (a) displays a prediction on an image of a black scabbard, (b) a prediction on a group of sardines, (c) a prediction on a group of mixed fish and (d) on a big group of mackerel fish.

5.2   Feature Extraction and Classification Model

The feature extraction and classification was done through a 5-fold cross-validation method. Each fold had the following fish as the test sets (the training set corresponds to the remaining fish from the database not used in the test set):

– Fold 1 - fish 1, 2, 3, 4, 5, 25, 27 and 35;
– Fold 2 - fish 6, 7, 8, 9, 10, 26, 28 and 36;
– Fold 3 - fish 11, 12, 13, 14, 15, 29, 30 and 37;
– Fold 4 - fish 16, 17, 18, 19, 20, 31, 32 and 38;
– Fold 5 - fish 21, 22, 23, 24, 33, 34, 39 and 40.

Each fold was trained using a starting checkpoint which is pre-trained on ImageNet-21k (14 million images, 21,843 classes) and fine-tuned on ImageNet 2012 (1 million images, 1,000 classes). Both the pre-training and fine-tuning of the checkpoint used an input image with a resolution of 224 × 224 pixels.

The model was trained using the Google Colab platform, with the Tesla 4 (T4) GPU. Each fold trained for approximately 1500 steps, using a batch size of 32, which translated in close to 2 h and 30 min of training. In each fold, the model was trained until the 400th step without a validation set, for a faster training process; from step 400 on wards, it was evaluated every 50 steps, with the testing dataset.

As mentioned previously, the metrics used for performance evaluation were the accuracy and F1 scores. Table 1 shows the scores for each fold, as well as the average scores for the whole cross-validation method.

Table 1. Top 1 percent scores for the performance metrics - accuracy and F1 score.

Fold #	Accuracy (%)	F1 Score (%)
1	83.44	82.56
2	79.80	81.79
3	83.11	82.57
4	79.72	79.79
5	77.98	78.38
Average	**80.81**	**81.02**

In order to have a better understanding on which classes the model had more/less difficulty correctly classifying, a confusion matrix was performed on fold number 1. The confusion matrix is represented in Table 2.

Table 2. Confusion matrix of fold 1.

Predicted\Truth	Very Fresh	Fresh	Not Fresh	Spoiled
Very Fresh	**225**	46	1	0
Fresh	12	**150**	31	0
Not Fresh	1	21	**137**	15
Spoiled	0	0	12	**69**

By analysis of the confusion matrix for fold 1, it can be observed that the model only misclassifies adjacent classes (except on 2 occasions which can be considered as outliers). It had the most difficulty discerning between the fresh and not fresh classes (52 misclassifications) and the very fresh and fresh classes (58 misclassifications). This can be explained due to the similarities in appearance in adjacent sessions, for instance, the fish are considered as very fresh if 48 h have passed since capture and as fresh if 63 h have passed, which corresponds to a 15 h interval which may not present significant changes in the eye's appearance.

6   Conclusions and Future Work

The classification system showed promising results performing the eye region detection and segmentation in a reliable manner of several fish species, marking an improvement upon previous works. In regards to the classification of the freshness, the system achieved accuracies averaging 80 percent, with misclassifications only in adjacent classes (fresh as not fresh, for example), making a fair assessment of the seabream fish's freshness, and proving it can be a reliable non-invasive system to universally classify fish-freshness.

The developed database also proves to be the first of its kind, containing the time since the fish was caught, providing knowledge of each fish's freshness and also containing image masks, which annotate the eye and body region of the fish.

In future efforts, the model should be able to reach higher classification accuracy scores even when discerning adjacent classes; comparisons between the accuracy of classification of different images of the database, based on their characteristics, such as lighting, eye size, should also be considered; the model should also be trained with other species of fish, in order to provide a more universal classification of fish-freshness.

Acknowledgements. This work is partly funded by FCT/MEC under the project UID/50008/2020.

References

1. FAO, The State of World Fisheries and Aquaculture (SOFIA), Rome, Italy (2022). https://doi.org/10.4060/cc0461cn
2. United Nations Food and Agricultural Organization (FAO). (2020) Fish, Seafood-Food Supply Quantity (Kg/Capita/Yr) (FAO, 2020). http://www.fao.org/faostat/en/#data/FBS. Accessed 2 May 2023
3. Cultivated fish in Aquaculture: total and main species. Source: PORDATA. https://www.pordata.pt/db/portugal/ambiente+de+consulta/tabela. Accessed 2 May 2023
4. Muhamad, F., Hashim, H., Jarmin, R., Ahmad, A.: Fish freshness classification based on image processing and fuzzy logic. In: Proceedings of the 8th WSEAS International Conference on Circuits, Systems, Electronics, Control, pp. 109–115 (2009)
5. Lalabadi, H.M., Sadeghi, M., Mireei, S.A.: Fish freshness categorization from eyes and gills color features using multi-class artificial neural network and support vector machines. Aquacult. Eng. **90**, 102076 (2020)
6. Otsu, N.: A threshold selection method from gray-level histograms. IEEE Trans. Syst. Man Cybern. **9**(1), 62–66 (1979)
7. Rosario, T., Correia, P.L., Pacheco, O.: Image-based fish freshness estimation. In: RECPAD Portuguese Conference on Pattern Recognition (2022)
8. Taheri-Garavand, A., Nasiri, A., Banan, A., Zhang, Y.-D.: Smart deep learning-based approach for non-destructive freshness diagnosis of common carp fish. J. Food Eng. **278**, 109930 (2020)
9. Simonyan, K., Zisserman, A.: Very deep convolutional networks for large-scale image recognition, arXiv preprint arXiv:1409.1556 (2015)
10. ImageNet Homepage. https://www.image-net.org/. Accessed 13 June 2023
11. Prasetyo, E., Purbaningtyas, R., Adityo, R.D.: Performance evaluation of pre-trained convolutional neural network for milkfish freshness classification. In: 2020 6th Information Technology International Seminar (ITIS), pp. 30–34. IEEE (2020)
12. He, K., Zhang, X., Ren, S., Sun, J.: Deep residual learning for image recognition. In: Proceedings of the IEEE Conference on Computer Vision and Pattern Recognition, pp. 770–778 (2016)
13. Howard, A.G., et al.: Mobilenets: efficient convolutional neural networks for mobile vision applications, CoRR, vol. abs/1704.04861 (2017)

14. Sandler, M., Howard, A.G., Zhu, M., Zhmoginov, A., Chen, L.: Inverted residuals and linear bottlenecks: mobile networks for classification, detection and segmentation, CoRR, vol. abs/1801.04381 (2018)

15. Huang, G., Liu, Z., Van Der Maaten, L., Weinberger, K.Q.: Densely connected convolutional networks. In: Proceedings of the IEEE Conference on Computer Vision and Pattern Recognition, pp. 4700–4708 (2017)

16. Chollet, F.: Xception: deep learning with depthwise separable convolutions. In: Proceedings of the IEEE Conference on Computer Vision and Pattern Recognition, pp. 1251–1258 (2017)

17. Zoph, B., Vasudevan, V., Shlens, J., Le, Q.V.: Learning transferable architectures for scalable image recognition. In: Proceedings of the IEEE Conference on Computer Vision and Pattern Recognition, pp. 8697–8710 (2018)

18. Rayan, M.A., Rahim, A., Rahman, M.A., Marjan, M.A., Ali, U.M.E.: Fish freshness classification using combined deep learning model. In: 2021 International Conference on Automation, Control and Mechatronics for Industry 4.0 (ACMI), pp. 1–5. IEEE (2021)

19. Anas, D., Jaya, I., et al.: Design and implementation of fish freshness detection algorithm using deep learning. In: IOP Conference Series: Earth and Environmental Science, vol. 944, no. 1, p. 012007. IOP Publishing (2021)

20. Redmon, J., Divvala, S., Girshick, R., Farhadi, A.: You only look once: unified, real-time object detection. In: Proceedings of the IEEE Conference on Computer Vision and Pattern Recognition, pp. 779–788 (2016)

21. Vaswani, A., et al.: Attention is all you need. In: Advances in Neural Information Processing Systems, vol. 30 (2017)

22. Carion, N., Massa, F., Synnaeve, G., Usunier, N., Kirillov, A., Zagoruyko, S.: End-to-end object detection with transformers. In: Vedaldi, A., Bischof, H., Brox, T., Frahm, J.-M. (eds.) ECCV 2020. LNCS, vol. 12346, pp. 213–229. Springer, Cham (2020). https://doi.org/10.1007/978-3-030-58452-8_13

23. Dosovitskiy, A., et al.: An image is worth 16x16 words: transformers for image recognition at scale. arXiv preprint arXiv:2010.11929 (2020)

24. Naseer, M.M., Ranasinghe, K., Khan, S.H., Hayat, M., Shahbaz Khan, F., Yang, M.H.: Intriguing properties of vision transformers. Adv. Neural. Inf. Process. Syst. **34**, 23296–23308 (2021)

25. Seabream Database Homepage. http://www.img.lx.it.pt/SeabreamDB. Accessed 27 June 2023

26. Segments.ai Homepage. https://www.segments.ai/. Accessed 7 May 2023

27. Khemir, M., Besbes, N., Khemis, I.B., Bella, C.D., Monaco, D.L., Sadok, S.: Determination of shelf-life of vacuum-packed sea bream (sparus aurata) fillets using chitosan-microparticles-coating. CyTA J. Food **18**(1), 51–60 (2020)

28. Erkan, N., Alakavuk, D., Tosun, S., Ozden, O.: Gutted effect on quality and shelf-life of sea bream stored in ice, vol. 86, pp. 105–110 (2006)

29. Xie, E., Wang, W., Yu, Z., Anandkumar, A., Alvarez, J.M., Luo, P.: SegFormer: simple and efficient design for semantic segmentation with transformers. Adv. Neural. Inf. Process. Syst. **34**, 12077–12090 (2021)

30. Chen, L.C., Papandreou, G., Schroff, F., Adam, H.: Rethinking atrous convolution for semantic image segmentation. arXiv preprint arXiv:1706.05587 (2017)

31. Zheng, S., et al.: Rethinking semantic segmentation from a sequence-to-sequence perspective with transformers. In: Proceedings of the IEEE/CVF Conference on Computer Vision and Pattern Recognition, pp. 6881–6890 (2021)

32. Huss, H.H.: Fresh fish-quality and quality changes: a training manual prepared for the FAO/DANIDA Training Program on Fish Technology and Quality Control, no. 29. Food & Agriculture Org. (1988)
33. Loshchilov, I., Hutter, F.: Decoupled weight decay regularization. arXiv preprint arXiv:1711.05101 (2017)

Explaining Semantic Text Similarity in Knowledge Graphs

Rafael Berlanga$^{(\boxtimes)}$ and Mario Soriano

Departament de Llenguatges i Sistemes Informàtics, Universitat Jaume I,
Castellón de la Plana, Spain
berlanga@uji.es, magregor@uji.es

Abstract. In this paper we explore the application of text similarity
for building text-rich knowledge graphs, where nodes describe concepts
that relate semantically to each other. Semantic text similarity is a
basic task in natural language processing (NLP) that aims at measuring
the semantic relatedness of two texts. Transformer-based encoders like
BERT combined with techniques like contrastive learning are currently
the state-of-the-art methods in the literature. However, these methods
act as black boxes where the similarity score between two texts cannot
be directly explained from their components (e.g., words or sentences).
In this work, we propose a method for similarity explainability for texts
that are semantically connected to each other in a knowledge graph. To
demonstrate the usefulness of this method, we use the Agenda 2030 which
consists of a graph of sustainable development goals (SDGs), their subgoals and the indicators proposed for their achievement. Experiments
carried out on this dataset show that the proposed explanations not
only provide us with explanations about the computed similarity score
but also they allow us to improve the accuracy of the predicted links
between concepts.

Keywords: Semantic Text Similarity · Transformers · eXplainable
AI · Knowledge Graphs

1 Introduction

Semantic text similarity (STS) is a basic task in natural language processing
(NLP) that aims at measuring the semantic relatedness of two texts. This
task has been widely studied and applied in more complex tasks like ques-
tion/answering, conversational systems, and semantic information retrieval systems. With the disruption of large language models (LLM) like chatGPT, STS is
being massively used to retrieve relevant chunks of texts for answering questions
[4]. Currently the main trend in STS consists of adaptations of the transformer-
based encoders like BERT (Bidirectional Encoder Representations from Transformers) [3]. There are two main approaches, namely: the supervised ones, which
train a BERT model to classify pairs of sentences as similar or not, and the unsupervised ones that use a pre-trained model of BERT to generate a metric space

© Springer Nature Switzerland AG 2024
V. Vasconcelos et al. (Eds.): CIARP 2023, LNCS 14469, pp. 526–539, 2024.
https://doi.org/10.1007/978-3-031-49018-7_37

for comparing sentences. Supervised approaches have many drawbacks. Firstly, they only fit well with examples similar to the training dataset. Secondly, they have severe limitations with respect to the size of the input sequences as the model needs to see both sequences in the input. Thus, the current trend is to use a pre-trained model of BERT to generate sentence embeddings which can be compared to perform other downstream tasks like semantic search or concept linking. These sentence embeddings can be also generated by more powerful LLMs, which is the current trend in the industry for semantic search.

Unfortunately, encoders act as black boxes and therefore we cannot know which elements of the sentences make them similar or dissimilar. In other words, we cannot justify the final similarity score. This fact can be dangerous in some applications as STS encoders are trained on benchmarks with little or no relation to the domains where we finally apply them [14]. Therefore, similarity results can be completely wrong and we are not aware of it. Furthermore, in the experiments we will see other unexpected factors that can affect the similarity of two sentences, like the way words are tokenized to serve as input of these encoders.

Explainable AI [8] (XAI) emerged as a field whose main aim is to provide methods and techniques to create interpretable models that allow explaining the decisions made by AI methods. Formerly applied to machine learning methods like support vector machines, they were highly demanded in the area of deep learning where millions of hidden parameters take part in the decision making. In the area of natural language processing (NLP), the great success of transformers-based models like BERT or GPT-3, makes it necessary to provide explanations that help developers and end-users to understand how these models deal with different NLP tasks.

In this paper we first revise the main methods related to XAI for STS methods based on transformers. Then, we propose a method to generate graph-based explanations for textual similarity. The proposed method is evaluated over a knowledge graph (KG) about sustainable goals and indicators, which are taken from the Agenda 2030 report. It is worth mentioning that this scenario has no relation to the STS benchmarks used for training text encoders, so that we can measure their performance as a zero-shot problem. Our hypothesis is that concepts related in the KG should show higher similarity scores than the concepts that are not explicitly related.

2   Related Work

Although text similarity has a long history within the area of information retrieval (IR), it has not been until recently that text similarity has been able to deal with latent semantic features of texts. The proposal of the first methods based on the distributional hypothesis, mainly Word2Vec [7] and GloVe [9], showed that neural-like encoders were able to uncover semantic patterns in the text words that enhanced text similarity metrics. Basically, all these methods produce dense vectors for words (called embeddings) so that similar words are placed close to each other in the resulting metric space. In this context, a document embedding just consists of aggregating somehow (e.g., mean, sum, minmax, concatenation, etc.) the embeddings of its words. Although these methods deal well with the long-tail problem they were not able to deal with the ambiguity of words and the context of words in general. Another issue of these methods is that they suffer from the out-of-vocabulary (OOV) problem, as words never seen cannot be assigned to any vector. Transformers, and more specifically its encoding part (e.g., BERT), accounts for these limitations. Firstly, the embedding of a word depends on its context (i.e.- surrounding words) and therefore different senses of the same word will have different representations. Secondly, encoders like BERT tokenize words into smaller pieces (tokens) so that they can cover new words, alleviating thus the OOV problem. As a matter of fact, transformer-based encoders have become the state of the art in the STS task.

Sentence-BERT [10] and its derived library sentence-transformers[1] is being widely used by the academia and the industry to perform a great variety of tasks like text/image retrieval, ontology alignment and concept linking among others. Basically, this method consists of an encoder that has been trained with pairs of similar and dissimilar sentences. Well-known datasets such as the STS-n benchmarks [2] have been traditionally used to evaluate semantic text similarity models. There are several ways to train such an encoder. Usually, the same BERT model is fine-tuned by duplicating the models and defining a head layer that is associated with a loss function based on the cosine of the output embeddings of these siamese models. Then the fine-tuned BERT model is simply used as an encoder of sentences. Similarly, SetFit [15] was trained on a richer benchmark and can be used as a few-shot learner where more specific examples can be provided. Other popular models for sentence encoding derive from large language models like GPT3 [1], where text embeddings can be obtained for measuring semantic similarity. In this paper we are mainly focusing on non-supervised approaches (or zero-shot learners).

XAI methods for transformers are scarce and they are mainly based on attribution-based methods developed for CNN models for image classification. Basically, these methods calculate the level of attribution or salency of the input features (pixels) for each decision. Other classical approaches like LIME (Local Interpretable Model-agnostic Explanations) [11] and SHAP (SHapley Additive exPlanations) [5] have also been applied to deep learning models but they do not explain well the complex interactions produced in the hidden connections of these neural networks. More recently, the work presented in [6] provides a new focus where graphs and clustering are used to build more accurate explanations of the output of BERT-based models. Our work is based on the conceptual background provided by this work, but adapted to the problem of applying XAI to explain a simple task of concept linking in knowledge graphs.

The main challenge of XAI in transformers resides in explaining how the self-attention mechanism affects the saliency of tokens and words when combined in the output to make the final decision. In the case of the STS task, we need to

---

[1] https://sbert.net.

explain how the sentence embeddings generated by sentence-transformers affect the decision of considering them similar or not.

3   Preliminary Definitions

In this section we introduce the background of our approach. We will mainly focus on the extraction of interesting properties from BERT encoders that can be used to build explanations of similarity scores.

3.1   Knowledge Graphs

In this paper we start from a given knowledge graph (KG) consisting of a set of concepts c_i which have associated a textual description, denoted as D_i. Concepts are then related through properties that connect semantically pairs of concepts. Formally, the KG is represented with a set of triples of the form (c_i, p, c_j) where c_i and c_j are concepts and p is the property connecting the concepts. Usual properties used in a KG are the *is_a* and *part_of* relationships. In this work, we restrict ourselves to a very simple structure of KG, which consists of a tree of concepts which can be stratified in different conceptual layers. As earlier mentioned, our hypothesis is that two connected concepts should show a higher textual similarity score on their descriptions than unconnected ones.

3.2   Saliency Scores

In this section we describe a method to calculate the saliency score of each word in a text when compared to another text. If the textual similarity is high, the most relevant words participating in the match should present a higher saliency score. Notice that, contrary to other attribution-like methods like IGrad [13], the task of comparing two text embeddings is considered an unsupervised task with no reference classes. Instead, we need to simulate the process in which tokens of each sentence pay attention to each other in the context of a final similarity score [6].

In order to identify salient words, we have adopted the method proposed in [6], where a metric for *token saliency* ($saliency_i \in \mathbb{R}$) is proposed. This metric is intended to provide higher scores to the tokens that pay more attention to each other according to the self-attention mechanism used in all transformer models.

The STS problem is just defined as measuring the similarity of two texts p_1 and p_2. When using a transformer like BERT, texts are tokenized into sequences of tokens, namely q_1 and q_2 respectively. Each text word can be reconstructed by joining its corresponding tokens of the input.

The saliency score of the i-th token in the sentence p_2 when it is compared to the sentence p_1, can be derived from the gradients of the first attention layer as shown in Fig. 1.

The calculation of the saliency score is then obtained with the following formula:

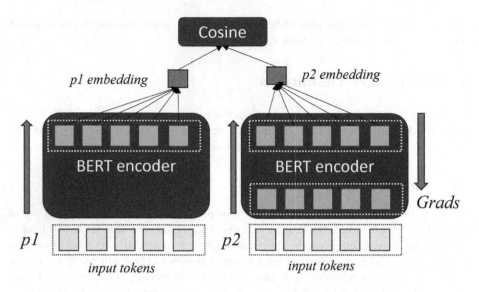

Fig. 1. Method for obtaining the saliency scores according to [6]

$$saliency_i(p_2) = NRM \left(\sum_{k=1}^{|p_2|+1} \Phi(E_i \circ g_i) \right) \ for \ each \ i \in \{0, \ldots, |q_2|\} \qquad (1)$$

Let's explain it in datail. Let's suppose we send as input each sentence p_1 and p_2 to the BERT model. The BERT model can be seen as the function $B : P \longrightarrow \mathbb{R}^{N \times h}$ with N the number of tokens in the input sentence and h the dimension of the hidden layer (768 since we are using BERT models). As a result, we get the tokens embeddings from the first model's layer to generate the sentence embeddings. More specifically, we compute the sentence embedding $F_{p_i} \in \mathbb{R}^h$ as the average of the embeddings of its tokens:

$$F_{p_i} = \frac{1}{|p_i|} \sum_{i=2}^{|p_i|+1} B(p_i)$$

The next step is to compute the cosine between F_{p_1} and F_{p_2} and then to back-propagate this value through the BERT model in which we have send p_2 as input in order to get the gradients of the model, that is to say, we calculate the following expression:

$$g(p1, p2) = \frac{\partial cos(Fp_1, Fp_2)}{\partial E(p_2)}$$

$$cos(Fp_1, Fp_2) = \frac{Fp_1 \cdot Fp_2}{||Fp_1||_2 \cdot ||Fp_2||_2}, \ with \ Fp_1, \ Fp_2 \in \mathbb{R}^h$$

being $E(p_2)$ the embeddings of the last layer of the model. Once we got g_i and E_i we multiply them element-wis.

In the Eq. 1, Φ represents the RELU function, defined as $\Phi(x) = \begin{cases} 0 \ if \ x < 0 \\ x \ if \ x \geq 0 \end{cases}$

Lastly, NRM is the min-max normalization which has the following expression:

$$NRM(a) = \frac{a - a_{min}}{a_{max} - a_{min}}, \ being \ a \in \mathbb{R}^{q_2}$$

Once we have done all these operations, we get a score that highlights the important tokens in p_2. In order to convert the tokens back to words we must join them and choose the maximum saliency of the tokens that form the word. Additionally, to calculate the saliency scores of p_1's words, we must just interchange p_1 and p_2 in the process.

3.3 Integrated Gradients

As an alternative method for measuring the attribution of tokens with respect to a decision, we an apply the method of Integrated Gradients [13].

Let's suppose we have a function $F : R^n \longrightarrow [0,1]$. Specifically, let $x \in \mathbb{R}^n$ be the input of the function F, and $x^0 \in \mathbb{R}^n$ the zero embedding vector. We consider the straight line path (in $\mathbb{R}^n$) from x^0 to x, and compute the gradients at all points along the path. Integrated gradients are obtained by accumulating these gradients. Specifically, integrated gradients are defined as the path integral of the gradients along the straight line path formed. The integrated gradient along the i dimension for an input x and x^0 is defined as follows.

$$IGrads_i(x):: = (x_i - x_i^0) \times \int_0^1 \frac{\partial F(x^0 + \alpha \times (x - x^0))}{\partial x_i} \delta\alpha$$

Being $\frac{\partial F(x)}{\partial x_i}$ the gradient of $F(x)$ along the i dimension.

Notice that this method differs from the previous one in that $IGrads$ operates over all the layers collectively. In the previous method, we only deal with the first layer by keeping its gradients after back-propagation. After testing both methods, we decide to use *saliency* instead of $IGrads$ as it provided us better insights about the relevance of words. Nevertheless, future work should focus on evaluating in an objective way the explainability power of each method.

3.4   Explanations

In this paper, we aim at obtaining explanations that justify the alignment of two concepts through their descriptions. For example, we need to find explanations that justify that one concept is a subgoal of another concept or that an indicator is aimed at a particular (sub)goal. In a more general scenario, we need to explain why and why not two texts express the relation we intend in the knowledge graph.

For this purpose, we envision an explanation as a bipartite graph where the nodes are the words of the two compared sentences, and the edges represent the best alignments between the two sentences. Notice that the same word may appear more than once in the same sentence. Therefore, in these cases, the same word will have multiple nodes as BERT generates different embeddings for them.

Formally, we define a directed bipartite graph $G(V_1, V_S2, E)$, where $V1$ represents the set of nodes corresponding to words occurrences in the description D_1, $V2$ represents the set of nodes corresponding to words occurrences in the description D_2, and E represents the set of edges connecting the aligned word pairs.

$$V1 = \{v_1, v_2, \ldots, v_{n1}\}$$

$$V2 = \{u_1, u_2, \ldots, u_{n2}\}$$

$$E = \{(v_i, u_j) \mid (i, j) \in A\} \cup \{(u_j, v_i) \mid (i, j) \in A\}$$

The edge (v_i, u_j) indicates that the i-th word in D_1 is aligned to the j-th word in D_2, that is the u_j word is the most similar word in D_2 to the word v_i. Similarly, we define the edges from D_2 to D_1. The matrix A accounts for the valid alignments between the nodes of both sides.

Nodes and edges are weighted by the saliency and similarity scores respectively. Word similarity is calculated with the cosine function over average of the output word embeddings (i.e., B_i) returned by the BERT model. As previously mentioned, the saliency of a word is computed as the maximum saliency of its tokens.

Once the explanation graph between two sentences is built, we can look for interesting patterns that justify the decisions made when comparing concepts in the knowledge graph. For example, we can explore if the aligned parts semantically correspond to what is being measured in an indicator and what is being pursued in a goal, and which is the strength of that alignment. Depending on the KG relationship between two concepts, we can extract different types of patterns from the explanation graphs.

Counter-factual explanations (e.g., [12]) are more difficult to be inferred, as they are usually related to missing edges or the weakness of existing ones, which down-score the global similarity between the concepts we intend to link. Figure 2 shows an example of explanation graphs in the case of a false positive association, where we can guess why the similarity score is higher in the wrong link than in the right one. We can notice that the wrong link between the indicator 1.1 and the goal 4 presents a higher level of similarity according to the number of edges.

However, many of these links have a low similarity score. When filtering out low relevant nodes and edges, goal 1 becomes much more similar than goal 4.

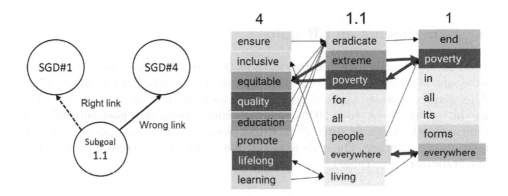

Fig. 2. Example of bipartite graphs as explanation.

From the bipartite graph we can also derive three new textual similarity metrics: the average similarity of the edges from D_1 to D_2 ($sim1$), the average similarity of the edges from D_2 to D_1 ($sim2$) and the whole average similarity of all the edges. These metrics will be usually lower than the similarity of the sentence embeddings, as they only take into account the best word-level alignments. Another interesting property of these metrics is that they account for the direction of the similarity (sim1 vs. sim2) which is important for identifying some KG relationships like specialization (is_a) or containment ($part\_of$).

4   Methodology

We propose the following methodology to explain semantic similarity in a KG where concepts has a text-based description attached.

1. We build a dataset with all the textual descriptions of the concepts in the KG. We attach to each text a code that allows us to know if two descriptions are related to each other.
2. We evaluate the accuracy of the 1-NN approach for linking the KG concepts through their descriptions. Basically, we link each concept with its most similar one in the KG by applying textual semantic similarity. Then we check if both concepts are indeed related in the KG.
3. By using the explanation graphs we measure how well the elements of an upper level relate to the elements of a lower level (top-down) by using STS. We can also identify which words are more prone to produce wrong links.
4. By exploring the explanation graphs we identify the sources that lead to false positives and negatives. False positives (FP) are texts with high similarity but that not related to each other in the KG, and false negatives (FN) are texts with low similarity that are indeed related in the KG.

5. From the analysis of the explanations we aim at improving the 1-NN approach by re-adjusting the similarity metrics provided by BERT in order to better capture the underlying semantics expressed in the KG.

5 Experiments and Results

For the experiments, we use the Agenda 2030 report, which describes goals, subgoals, and indicators related to the Sustainable Development Goals (SDGs). This dataset contains approximately 400 concepts structured in an almost tree-like hierarchy. The Agenda 2030 assigns a unique code to each goal, subgoal, and indicator, which can be directly used to determine if two concepts are directly related to each other. The code of a goal is a prefix of the codes of its subgoals and indicators (see Table 1).

Table 1. Example of Goal-Subgoal-Indicator relation

Goal 2	End hunger, achieve food security and improved nutrition and promote sustainable agriculture
Subgoal 2.2	By 2030, end all forms of malnutrition, including achieving, by 2025, the internationally agreed targets on stunting and wasting in children under 5 years of age, and address the nutritional needs of adolescent girls, pregnant and lactating women and older persons
Indicator 2.2.3	Prevalence of anaemia in women aged 15 to 49 years, by pregnancy status (percentage)

For the sake of simplicity, we stratify this KG into three levels, namely: goals (Level 0), subgoals (Level 1) and indicators (Level 2). Thus, we can evaluate how well the 1-NN approach works for linking each pair of levels, as they account for different semantic relationships. We carried out two experiments to measure the linking capabilities of the STS metrics. In the first one we want to measure the discriminating power of the metrics to link parents to their children, and the second one will measure the precision of linking a node to its parent. Afterwards, we will analyze the corresponding explanations for performing the errors analysis for the concept linking task. Figure 3 shows an example of the stratified KG and the relations between the concepts.

As STS metrics we take the best STS model recommended by the sentence-transformer approach (all-MiniLM-L6-v2), which will be denoted as SBERT. We also evaluate the embeddings provided by OpenAI with the model text-embedding-ada-002, which are being widely used in the academy and the industry for information retrieval with large language models. The corresponding STS metric, denoted ADA, is just obtained with the cosine of the OpenAI embeddings of the compared texts.

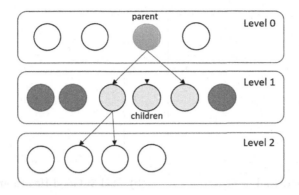

Fig. 3. Stratified Knowledge Graph for Agenda 2030 dataset.

5.1 Experiment 1. Discriminating Power of the STS Metrics

In this experiment we compare the similarities between each parent in a level to its children and other elements of the children's level. The main aim is to measure how well the STS similarity is able to separate the parent's children from the other concepts in the same abstraction level. For this purpose, we use robust statistics to construct box plots representing the [0.25–0.75] quantiles for the STS similarities obtained between levels. One box plot represents the parent-child pairs (in green), and the other one represents the similarities between parents and other nodes (in red). If these box plots overlap, then it is not possible to set a similarity threshold that distinguishes children from other concepts. As an example, in Fig. 4, we show the obtained box plots for two parents and different STS metrics. Apart from the SBERT and ADA, we include the cosine between the CLS token, the similarity between parent to child (sim1) and the similarity between child to parent (sim2). We can see in this figure that Goal 1 is much easier to predict as there are clear limits between the similarities of related concepts and unrelated ones. However, Goal 3 presents a high overlap in all the metrics, which makes it difficult to predict the children of each parent. The SDG goals that present overlap in their box plots are: goal 3 (Good Health and Well-being), 8 (Decent Work and Economic Growth), 16 (Peace, Justice and Strong Institutions) and 17 (Partnership for the Goals).

5.2 Experiment 2. Linking Children to Parents

The usual way to link new elements to a KG is to find a relationship between the new concept and the existing ones. In our scenario, this means that the preferred way to link concepts would be from children to parents. In Table 2 we show the precision of the 1-NN method that assigns each child to its most similar parent. Surprisingly, the SBERT embeddings outperforms the ADA ones in almost all cases, except when linking indicators (Level 2) to top goals (Level 0). Notice that ADA embeddings present higher similarity scores and that they are very

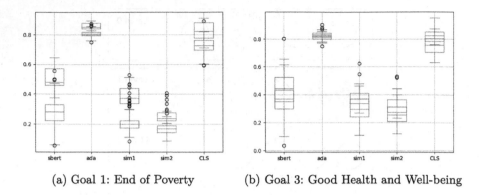

(a) Goal 1: End of Poverty   (b) Goal 3: Good Health and Well-being

Fig. 4. Distributions of similarities between KG levels 0 and 1.

close to each other (see box plots in Fig. 4). We can also see that linking concepts from Level 2 to Level 0 is more difficult than linking concepts between adjacent levels. This is an expected result as the higher the distance in the KG the less the similarity between the concepts.

Table 2. Results of precision of the task children-parent linking.

Compared Levels	Children	Parents	BERT	ADA
Level 1 - Level 0	169	17	61%	57%
Level 2 - Level 1	267	169	62%	59%
Level 2 - Level 0	267	17	43%	55%

Figure 5 shows the confusion matrices of the SBERT and ADA embeddings for the associations of Levels 0 and 1. We can notice that ADA embeddings produce many wrong links in a few goals compared to SBERT.

5.3 Explanations for the Concept Linking Task

Explanations based on bipartite graphs allow us to repair wrong associations by filtering out non-relevant nodes and edges. For this purpose, we set a threshold in the saliency of the words participating in the graph in order to compute a corrected average similarity score based solely on salient words of the indicator we want to link. Taking into account that the saliency is normalized between 0 and 1, we set the threshold in 0.4.

Table 3 shows the results of applying pruning to remove non-relevant information from the explanation graphs. To measure the impact of this method, we consider that a wrong link is repaired when the corrected similarity (after pruning) makes the correct concept more similar to the target concept than the

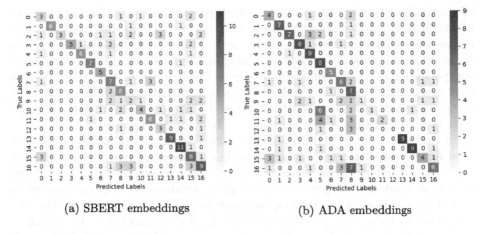

(a) SBERT embeddings   (b) ADA embeddings

Fig. 5. Confusion matrices of the concept linking task between Levels 0 and 1

wrong one. The table includes the resulting precision after applying these corrections. In all cases, the corrections notably improve precision. An interesting finding derived from this table is that ADA embeddings benefit significantly from this pruning, now achieving the best precision in all cases. Moreover, their performance is consistent across all compared levels. This suggests that postprocessing ADA embeddings have great potential for STS, rather than using them directly to compare texts.

Table 3. Repairs from explanations when linking between the KG levels.

		SBERT vectors	ADA vectors
Level 1 - Level 0	Errors	66 (P = 61%)	73 (P = 57%)
	Repairs	39 (P = 84%)	49 (**P = 86%**)
Level 2 - Level 1	Errors	88 (P = 62%)	94 (P = 59%)
	Repairs	19 (P = 71%)	58 (**P = 85%**)
Level 0 - Level 2	Errors	132 (P = 43%)	105 (P = 56%)
	Repairs	59 (P = 69%)	58 (**P = 80%**)

6   Conclusions

In this paper, we have introduced a new XAI method for explaining STS tasks in the context of KGs. We have focused on transformer-based embeddings, which are being massively used for semantic retrieval and other downstream tasks that require comparing texts and documents. We have demonstrated the usefulness of the saliency score proposed in [6], which properly accounts for the relevance of words in the final similarity scores. As a matter of fact, using these scores to adjust the similarity between two texts has been demonstrated as a good way to improve the precision of the concept linking task in a KG. We have also compared the performance of the embeddings of a relatively small model (SBERT) with the embeddings of a LLM (ADA). Surprisingly, SBERT performed better than ADA. However, after applying the proposed XAI approach, ADA embeddings greatly benefited from it and achieved the best precision scores.

As future work, we plan to perform a much in depth evaluation of different transformer encoders. We also need to define a robust method to determine the threshold of saliency scores in order to get the best explanations. Identifying irrelevant words such as punctuation symbols, numbers, etc. also can contribute to reduce the noise when sharing two sentences. We also realize that the way words are tokenised can produce severe issues of information lost. For example, the word "malnutrition" is split into the tokens "mal ##nut ##rit ##ion", which lose completely the original meaning of the word; indeed the word "nutritional" is not split into tokens. We aim in the future at identifying these issues and propose ways to alleviate them. Finally, we also plan to work on the idea of chunks of words, that is to say, there are consecutive words with a high self-attention score that could be aggregated to form multi-word expressions. In this way, the explanations become more meaningful as they show the connection of concepts instead of isolated words.

Acknowledgments. This research has been partially funded by the Spanish Ministry of Science under grants PID2021-123152OB-C22 and PDC2021-121097-I00 both funded by the MCIN/AEI/10.13039/501100011033 and by the European Union and FEDER/ERDF (European Regional Development Funds). Mario Soriano is granted by the Generalitat Valenciana through the project Investigo (INVEST/2022/308).

References

1. Brown, T.B., et al.: Language models are few-shot learners. CoRR, abs/2005.14165 (2020)
2. Conneau, A., Kiela, D.: Senteval: an evaluation toolkit for universal sentence representations. arXiv preprint arXiv:1803.05449 (2018)
3. Devlin, J., Chang, M.-W., Lee, K., Toutanova, K.: Bert: Pre-training of deep bidirectional transformers for language understanding, 2018. cite arxiv:1810.04805Comment, 13 p
4. Khattab, O., Zaharia, M.: Colbert: efficient and effective passage search via contextualized late interaction over bert. In: Proceedings of the 43rd International ACM SIGIR Conference on Research and Development in Information Retrieval, SIGIR 2020, pp. 39–48. Association for Computing Machinery, New York(2020)
5. Lundberg, S.M., Lee, S.-I.: A unified approach to interpreting model predictions. In: Guyon, I., Luxburg, U.V., Bengio, S., Wallach, H., Fergus, R., Vishwanathan, S., Garnett, R. (eds.) Advances in Neural Information Processing Systems 30, pp. 4765–4774. Curran Associates Inc. (2017)

6. Malkiel, I., Ginzburg, D., Barkan, O., Caciularu, A., Weill, J., Koenigstein, N.: Interpreting bert-based text similarity via activation and saliency maps. In: Proceedings of the ACM Web Conference 2022, WWW 2022, pp. 3259–3268. Association for Computing Machinery, New York (2022)
7. Mikolov, T., Chen, K., Corrado, G., Dean, J.: Efficient estimation of word representations in vector space (2013)
8. Miller, T.: Explanation in artificial intelligence: Insights from the social sciences. Artif. Intell. **267**, 1–38 (2019)
9. Pennington, J., Socher, R., Manning, C.D.: Glove: global vectors for word representation. In: Empirical Methods in Natural Language Processing (EMNLP), pp. 1532–1543 (2014)
10. Reimers, N., Gurevych, I.: Sentence-bert: sentence embeddings using siamese bert-networks. In: Proceedings of the 2019 Conference on Empirical Methods in Natural Language Processing. Association for Computational Linguistics, 11 2019
11. Ribeiro, M.T., Singh, S., Guestrin, C.: "why should I trust you?": explaining the predictions of any classifier. In: Proceedings of the 22nd ACM SIGKDD International Conference on Knowledge Discovery and Data Mining, San Francisco, CA, USA, August 13–17, 2016, pp. 1135–1144 (2016)
12. Stepin, I., Alonso, J.M., Catalá, A., Pereira-Fariña, M.: A survey of contrastive and counterfactual explanation generation methods for explainable artificial intelligence. IEEE Access **9**, 11974–12001 (2021)
13. Sundararajan, M., Taly, A., Yan, Q.: Axiomatic attribution for deep networks. In: Precup, D., Whye Teh, Y. (eds.) Proceedings of the 34th International Conference on Machine Learning. Proceedings of Machine Learning Research, vol. 70, pp. 3319–3328. PMLR, 06–11 Aug 2017
14. Thakur, N., Reimers, N., Rücklé, A., Srivastava, A., Gurevych, I.: BEIR: a heterogenous benchmark for zero-shot evaluation of information retrieval models. CoRR, abs/2104.08663 (2021)
15. Tunstall, L., et al.: Efficient few-shot learning without prompts. arXiv (2022)

Active Supervision: Human in the Loop

Ricardo P. M. Cruz[1,2]([✉])[iD], A. S. M. Shihavuddin[3][iD], Md. Hasan Maruf[3], and Jaime S. Cardoso[1,2][iD]

[1] INESC TEC, Porto, Portugal
{ricardo.p.cruz,jaime.cardoso}@inesctec.pt
[2] Faculty of Engineering, University of Porto, Porto, Portugal
[3] Green University of Bangladesh, Dhaka, Bangladesh
{shihav,maruf}@eee.green.edu.bd

Abstract. After the learning process, certain types of images may not be modeled correctly because they were not well represented in the training set. These failures can then be compensated for by collecting more images from the real-world and incorporating them into the learning process – an expensive process known as "active learning". The proposed twist, called **active supervision**, uses the model itself to change the existing images in the direction where the boundary is less defined and requests feedback from the user on how the new image should be labeled. Experiments in the context of class imbalance show the technique is able to increase model performance in rare classes. Active human supervision helps provide crucial information to the model during training that the training set lacks.

Keywords: active learning · deep learning · data augmentation · classification

1 Introduction

In supervised learning, the role of the data practitioner is to collect and label data. After training, the misclassified classes can be identified, and the practitioner can improve performance by (a) coarsely tuning the model by tweaking hyperparameters or by (b) collecting more real-world data to improve the classification of the disadvantaged classes, a costly process known as *active learning* [12].

In the active learning literature, we typically assume that we have a pool of unlabelled data waiting to be labeled if a certain part of the decision boundary is poorly defined. Choosing which unlabelled data is more useful to label can then be done by considering the margin in an SVM model [12] or by considering the samples where the model is most uncertain (e.g., when the output probabilities are lowest) [9], among other sampling techniques. However, in the real world, we may not have a pool of unlabelled data laying around, and the acquisition process may be more costly than the labeling process.

© Springer Nature Switzerland AG 2024
V. Vasconcelos et al. (Eds.): CIARP 2023, LNCS 14469, pp. 540–551, 2024.
https://doi.org/10.1007/978-3-031-49018-7_38

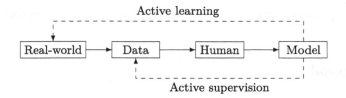

Fig. 1. Active supervision short-circuits the traditional active learning pipeline.

We propose to short-circuit active learning by acquiring data from the model itself – the model can interpolate or extrapolate from the existing data and have the practitioner label the new data. The model itself could be used to identify and manipulate images near the decision boundary in a direction orthogonal to the boundary. We have called this process *active supervision*, as illustrated by Fig. 1.

The proposal proves to be particularly beneficial in the context of class imbalance when there are very few instances of one of the classes. In the experiments, different versions of popular datasets, such as MNIST [7], are produced with very few examples of one class; it is then shown that active supervision significantly improves the recognition of the minority class.

2 Related Work

Previous approaches to introducing humans in the training loop have focused on using interpretability techniques to ascertain if the model is learning useful features [11]. This ignores the fact that the distance of the data relative to the decision boundary has been found to be a significant predictor of overfitting [4].

This problem can be ameliorated in unsupervised ways: by using adversarial training [2,10] or by using manifold interpolations [13]. Both techniques aim to make the boundary smoother and prevent small perturbations from percolating into big decision changes. A typical approach involves minimizing the loss $\mathcal{L}$ for both the data $\mathbf{x}$ and $\mathbf{x} + \boldsymbol{\delta}$, where $\boldsymbol{\delta}$ is Gaussian noise. However, these techniques make the model insensitive to small changes in the input, which is sometimes undesirable – consider breast cancer detection or melanoma classification, which involve minimal changes in tumor morphology. Furthermore, while such techniques produce more resilient models that are harder to fool, more often than not, they come at the expense of performance [2]. It would be highly desirable if the user could actively observe and classify the perturbations during training, but this is obviously not feasible.

Another research line is to use the distance to the decision boundary to ascertain uncertainty and request the data practitioner to focus labeling on the data where uncertainty is highest, a process known as *active learning* [12]. However, unless there is an existing pool of unlabelled data, this process requires collecting more data from the real world, which can be highly expensive. In the next section, we propose combining active learning with data perturbations during training.

3   Proposal

The proposal intends to acquire new images from the model and request the user for a label. The new images are produced by introducing perturbations in the latent space, which are then reversed back to the original image.

Architecture: A traditional classifier can be seen as having two components: an **encoder** E and a **classifier** C component. In the encoding phase, convolutions are applied until a (latent) vector captures the essential semantics of the image. Then, the classifier consists of dense layers that transform this latent vector into K output neurons, at which point a softmax is applied to ensure a probability distribution. Since AlexNet [6], most deep classifiers have been using this formula, including popular architectures such as VGG and ResNet.

Our proposal extends an existing model by adding a **decoder** D, as depicted in blue in Fig. 2 – this decoder is responsible for creating new images from the latent vector. The output of the decoder is an image of the same size as the original image; the goal is that it learns to make the neural network invertible (the encoder-decoder pair forms an autoencoder). Notice that this extra component is an extension that can be trained independently of the rest of the classifier. An alternative to introducing a decoder would have been to back-propagate the latent vector back to the original image; however, in our experience, that typically produces poorer images.

Training: The previous architecture can be trained as follows: The encoder E and the classifier C part are trained for the supervised task. Typically, the classifier produces a probability $\hat{y}_{i,k} = \mathrm{P}(y_{i,k} = 1 \mid \mathbf{x}_i) = C(E(\mathbf{x}_i))$ of each image $\mathbf{x}_i$ being of class k, and the loss to minimize is cross-entropy where each $y_{i,k}$ is a one-hot representation of the real classes,

$$\mathcal{L}_{\mathrm{CE}} = -\sum_{k=1}^{K}\sum_{i=1}^{N} y_{i,k} \log \hat{y}_{i,k}. \tag{1}$$

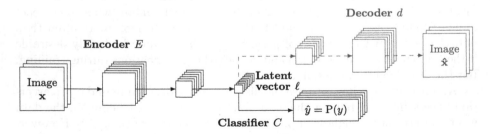

Fig. 2. The classifier (in **solid black**) is extended so that a decoder (in dashed blue) transforms the latent space back into the image space. (Color figure online)

Afterward, the decoder D is trained to produce an invertible representation of the neural network, i.e., the goal is for the decoder to be the inverse function of the encoder, $D = E^{-1}$. Both are optimized so that the output image $\hat{\mathbf{x}}_i$ is the same as the input image $\mathbf{x}_i$, i.e. $\hat{\mathbf{x}}_i = D(E(\mathbf{x}_i)) \approx \mathbf{x}_i$. This correspondence is approximated using L2 as traditionally done when training autoencoders,

$$\mathcal{L}_{\mathrm{D}} = \sum_{i=1}^{N} \|\phi(\mathbf{x}_i) - \phi(\hat{\mathbf{x}}_i))\|_2^2. \tag{2}$$

A function ϕ may be applied to both the original and the produced image in order to compare higher-level features rather than force pixel-wise equivalence. This is known as a perceptual loss [1]. In the experiments, the identity function is used as ϕ.

Active Feedback: After the training phase previously described, user feedback can be used to improve the classifier. If we would like to improve the accuracy of a target class k, then we choose N images of various source classes $k' \neq k$. These images are chosen by proximity to the boundary of the target class k; the distance of an image $\mathbf{x}_i$ relative to the boundary of a class k may be measured by

$$d(\mathbf{x}_i, k) = 1 - \hat{y}_{i,k}. \tag{3}$$

Furthermore, N images of the target class are chosen by random sampling – these two pairs of N images are then interpolated, as per the following process. Both source images $\mathbf{x}_i$ and target images $\mathbf{x}_j$ are converted to the latent space using the encoder, obtaining $\boldsymbol{\ell}_i = E(\mathbf{x}_i)$ and $\boldsymbol{\ell}_j = E(\mathbf{x}_j)$. A linear interpolation of N' cases is produced, $\mathbf{i}_n = (\frac{n}{N'+1})\boldsymbol{\ell}_i + (1 - \frac{n}{N'+1})\boldsymbol{\ell}_j$, for all $n = 1, \ldots, N'$. These vectors are then converted to images using the decoder, $\mathbf{x}_n' = D(\mathbf{i}_n)$. Alternatively, extrapolations could have been produced by computing gradients from the classifier, in which case only source images would have been required, but in our experience, it is harder to obtain good-quality images from extrapolations.

The interpolation procedure is illustrated in Fig. 3, and specific examples of the interpolation are shown in Fig. 4 for several datasets. Furthermore, an illustration of the latent space is depicted in Fig. 5, showing how a random observation moves towards class "8". PCA was used to reduce the feature space to two dimensions.

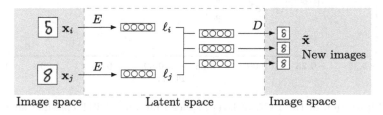

Fig. 3. Illustration of how the encoder and the decoder produce new images by linearly interpolating the latent space.

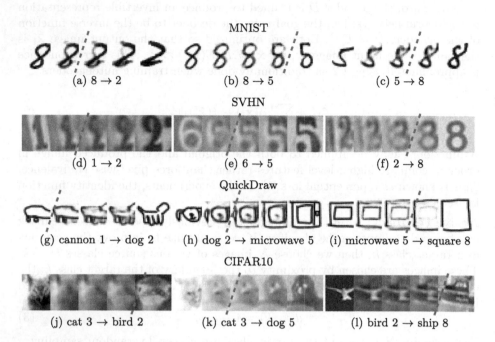

Fig. 4. Examples of interpolations between several source and target classes. The red boundaries represent what the supervisor may consider as the start of the target class. (Color figure online)

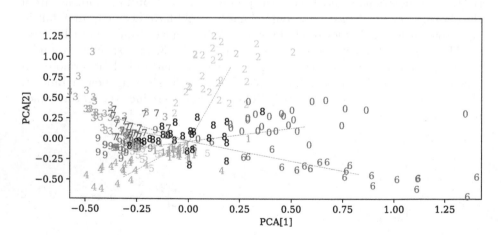

Fig. 5. Illustration of the PCA of latent space from an MNIST classifier showing the transition between each class cluster and class "8".

In the end, $N \times N'$ images are produced that need to be labeled; these labels are then used to re-train the model. The complete procedure is shown in Algorithm 1:

Algorithm 1. Training algorithm using user feedback

1: **Input:** Source images $\mathbf{x}_i$, y_i, Target images $\mathbf{x}_j$, k
   // Sampling of N closest source images to the boundary, see (3)
2: $s \leftarrow [\text{argsort}_i d(\mathbf{x}_i, k)]_{1,\ldots,N}$
   // Random sampling of N target images
3: $t \leftarrow [\mathcal{U}(1, \ldots, |\mathbf{x}_j|)]_{1,\ldots,N}$
   // Produce interpolations between $\mathbf{x}_s$ and $\mathbf{x}_t$
4: **for** $m \leftarrow 1$ **to** N **do**
5:    $\ell_s = E(\mathbf{x}_{s_m})$
6:    $\ell_t = E(\mathbf{x}_{t_m})$
7:    **for** $n \leftarrow 1$ **to** N' **do**
8:       $\mathbf{i}_{m,n} \leftarrow (\frac{n}{N'+1})\ell_s + (1 - \frac{n}{N'+1})\ell_t$
9:       $\tilde{\mathbf{x}}_{m,n} \leftarrow D(\mathbf{i}_{m,n})$
10:   **end for**
11: **end for**
   // Ask the user to accept/reject the new images and respective classes
12: $\bar{\mathbf{x}}, \bar{\mathbf{y}} \leftarrow$ **user_feedback** $(\tilde{\mathbf{x}})$
   // Update weights based on user feedback
13: **for** $it \leftarrow 1$ **to** T **do**
14:    $\mathbf{w}_{E,C} \leftarrow \mathbf{w}_{E,C} - \eta \frac{\partial \mathcal{L}_{CE}(\hat{\mathbf{y}}, \bar{\mathbf{y}})}{\partial \mathbf{w}_{E,C}}$
15: **end for**

Various mechanisms of user feedback (line 12) could have been used, such as asking "yes" or "no", if the image belongs to the target class, ranking images by semblance to the target class, etc. Here, we assume the user specifies the class, which may be the target or any other, or whether he/she is unsure. In the latter case, the new image is not used for training.

4 Experiments

Four datasets were used for the experiments: MNIST [7], SVHN [8], QuickDraw [3] and CIFAR10 [5], as referenced in Table 1. The image resolution of these datasets was either 32 or was resized accordingly. The datasets involve a classification task featuring 10 classes – except for QuickDraw, which has 350 classes, and 10 classes were chosen by evenly-spaced sampling. They all comprise around 100,000 images and are already divided into training and testing folds – again, except QuickDraw for which random sampling was performed with a ratio of 80–20.

It is not uncommon to obtain very high accuracies for these datasets, especially for MNIST, where a 99% accuracy is common. For experimental purposes,

Table 1. Datasets used in the experiments.

Dataset	Ref.	#Images	Examples	Description
MNIST	[7]	70,000	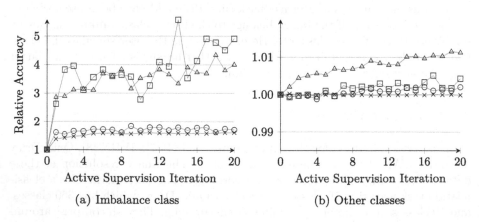	Hand-written digits
SVHN	[8]	99,289		House door numbers
QuickDraw	[3]	137,343		Manually drawn sketches
CIFAR10	[5]	60,000		Photographs of animals and vehicles

undersampling was performed to produce three new datasets, in which a certain class was downsampled to only 10 training images; the classes chosen for undersampling in the experiments were "2", "5" and "8".

The implementation of the architecture, previously illustrated in Fig. 2, consists of an encoder with two blocks of convolution-maxpool layers, with 32 and 64 filters, and a final convolution with 128 filters. Thus, the original images, which start with a resolution of 32×32, are reduced to a latent vector of size $8 \times 8 \times 128$. The classifier applies a convolution-maxpool block with 64 filters followed by dense layers with 32 neurons and, finally, 10 output neurons representing the probability of each class. The decoder was built the same way as the encoder, except that bilinear upsampling was used for upsampling the latent vector. All filters are 3×3, and Adam was used as the optimizer with a learning rate of 10^{-4}. The loss functions were categorical cross-entropy for the classifier, and two loss terms were used for the decoder: mean square error and sigmoid cross-entropy to learn how to reconstruct the image.

(a) Imbalance class

(b) Other classes

Fig. 6. Summary of the results showing accuracy improvement relative to doing nothing (in %), with all experiments averaged for each dataset: MNIST ($\times$), SVHN ($\triangle$), QuickDraw ($\bigcirc$) and CIFAR10 ($\square$). Clearly, active supervision improved the accuracy when evaluating the minority class (left column) without penalizing the accuracy when evaluating the other classes (right column).

Table 2. The impact on accuracy (%) along the active supervision iterations, starting with the base model prior to any iteration (none). The values presented are the average and deviation accuracy (in %) of five repetitions evaluated for the undersampled class.

		Active Supervision Iteration →									
Dataset	none	1	2	3	4	5	6	7	8	9	10
MNIST											
2	39.9±6	52.5±2	53.9±6	58.9±6	55.7±7	60.8±5	59.9±5	60.6±4	60.3±4	60.4±3	**60.9**±5
5	41.4±3	62.7±6	65.5±7	63.3±8	67.4±5	69.5±4	70.2±9	71.3±8	66.8±6	70.0±6	**72.0**±3
8	45.2±3	59.5±3	60.6±5	63.8±3	66.3±3	69.4±3	**70.7**±4	68.6±5	70.6±3	67.6±2	68.4±5
SVHN											
2	4.8±2	12.1±3	11.8±4	13.1±3	10.3±1	12.6±4	14.3±5	14.0±2	**14.5**±5	12.9±3	14.4±4
5	2.2±0	6.7±3	7.2±2	7.3±2	8.4±2	6.8±1	8.3±3	**8.6**±3	8.5±2	6.7±2	6.1±4
8	0.7±1	2.9±1	2.9±2	3.2±2	5.2±4	4.2±3	5.3±2	4.7±3	6.1±3	4.4±2	**6.3**±4
QuickDraw											
2	0.7±1	1.5±1	1.3±1	1.2±1	1.9±0	1.6±1	1.6±1	1.6±1	1.8±1	**2.2**±2	1.2±0
5	15.7±5	30.6±7	29.5±7	30.1±9	31.7±10	33.3±7	33.1±10	29.9±11	29.3±6	**35.9**±10	30.1±12
8	26.2±8	36.9±7	35.1±5	39.3±10	36.9±5	38.7±8	38.8±9	**41.3**±7	37.8±4	40.4±3	40.2±9
CIFAR10											
2	0.1±0	**0.6**±0	0.5±0	0.4±0	0.4±0	0.2±0	0.6±0	0.5±0	0.5±0	0.4±0	0.3±0
5	0.2±0	0.5±1	0.4±0	0.6±1	**0.7**±1	0.6±0	0.5±0	0.7±0	0.5±0	0.4±0	0.4±0
8	0.6±0	1.2±1	**2.6**±2	**2.6**±1	1.6±0	2.4±1	2.4±1	2.1±0	2.3±1	2.4±1	1.8±1

The model is trained until convergence, defined as the loss not lowering for 10 successive epochs. Firstly, the base model is trained; after which, Algorithm 1 is performed: two pairs of 25 images are selected (N), of source and target images, and are interpolated. This interpolation produces 8 new images (N') – at each iteration, the user is asked to choose when the target class starts between the 8 interpolations for the 25 images, similarly to Fig. 4. At most, at each iteration, 200 images (25×8) may be added. This process was repeated for several iterations.

The average and deviation of five repetitions of the experiments are depicted in Table 2, for the four different datasets, under the three different undersample scenarios. As can be seen, it is not always the case where active supervision always increases performance along with the iterations – but that is generally the case. Furthermore, there is no single case where active supervision hurts performance relative to doing nothing. Figure 6 summarizes the results showing only four lines with relative accuracy (relative to doing nothing) for each dataset (averaging over all the experiments for that dataset) – MNIST and QuickDraw improve by more than half, but SVHN and CIFAR10, actually more than quadruple performance.

For the purpose of this work, instead of a human user doing the active supervision, an automated model was used to choose which images to accept or reject – this model was trained in the entire dataset and rejects images for which it thinks the probability is lower than 90%. Smaller experiments were done using human users, but to repeat the experiments several times over and obtain robust results, an automated user was used here.

Table 3. Distribution of the images sampled based on the decision boundary and how many new images were accepted by the user, out of 200 (averaged for five repetitions).

Dataset	Class	Iteration	New images	%0	%1	%2	%3	%4	%5	%6	%7	%8	%9
MNIST	2	1	186 ± 04	0.0	8.0	–	64.0	0.8	2.4	5.6	18.4	0.8	0.0
"	"	2	185 ± 04	2.4	7.2	–	52.8	2.4	0.0	3.2	30.4	0.8	0.8
"	"	3	185 ± 02	0.8	11.2	–	37.6	10.4	0.0	3.2	34.4	1.6	0.8
"	"	4	182 ± 03	0.8	12.8	–	52.8	1.6	0.0	3.2	28.0	0.0	0.8
MNIST	5	1	187 ± 03	0.8	0.0	0.0	58.4	0.0	–	32.8	0.0	8.0	0.0
"	"	2	185 ± 04	0.0	0.0	0.0	51.2	0.0	–	38.4	0.0	10.4	0.0
"	"	3	187 ± 03	0.0	0.0	0.0	63.2	0.0	–	31.2	0.0	4.0	1.6
"	"	4	185 ± 02	0.0	0.0	0.0	45.6	0.0	–	48.0	0.0	6.4	0.0
MNIST	8	1	186 ± 04	0.0	0.0	29.6	50.4	0.0	20.0	0.0	0.0	–	0.0
"	"	2	187 ± 04	0.0	0.0	20.8	63.2	0.0	16.0	0.0	0.0	–	0.0
"	"	3	186 ± 02	0.8	0.0	20.0	60.8	0.8	16.0	0.8	0.0	–	0.8
"	"	4	186 ± 06	0.8	0.0	12.0	58.4	0.0	21.6	0.8	0.0	–	6.4
SVHN	2	1	188 ± 03	0.0	5.6	–	5.6	1.6	0.0	0.0	84.0	0.0	3.2
"	"	2	190 ± 03	0.0	2.4	–	3.2	0.8	0.0	0.0	92.0	0.0	1.6
"	"	3	186 ± 03	0.0	0.8	–	3.2	0.8	0.0	0.0	93.6	0.0	1.6
"	"	4	187 ± 03	0.0	0.8	–	1.6	0.8	0.0	0.0	95.2	0.0	1.6
SVHN	5	1	185 ± 04	0.0	0.8	0.0	57.6	0.8	–	38.4	0.0	1.6	0.8
"	"	2	187 ± 02	0.0	0.8	0.0	55.2	0.0	–	42.4	0.0	0.8	0.8
"	"	3	188 ± 01	0.0	0.8	0.0	54.4	0.0	–	44.8	0.0	0.0	0.0
"	"	4	188 ± 00	0.0	0.8	0.0	52.8	0.0	–	46.4	0.0	0.0	0.0
SVHN	8	1	182 ± 05	11.2	0.0	0.0	12.8	0.0	0.0	76.0	0.0	–	0.0
"	"	2	182 ± 03	2.4	0.0	0.0	16.8	0.0	0.8	80.0	0.0	–	0.0
"	"	3	183 ± 04	2.4	0.0	0.0	12.0	0.0	0.8	84.8	0.0	–	0.0
"	"	4	182 ± 03	1.6	0.0	0.0	11.2	0.0	0.8	86.4	0.0	–	0.0
QuickDraw	2	1	177 ± 05	0.0	74.4	–	7.2	3.2	11.2	2.4	0.0	0.0	1.6
"	"	2	174 ± 07	0.8	80.8	–	5.6	5.6	4.0	1.6	0.0	0.8	0.8
"	"	3	180 ± 05	1.6	92.0	–	2.4	1.6	1.6	0.8	0.0	0.0	0.0
"	"	4	179 ± 03	3.2	88.0	–	3.2	0.8	4.8	0.0	0.0	0.0	0.0
QuickDraw	5	1	180 ± 05	8.0	2.4	0.0	0.0	0.0	–	0.0	0.0	20.8	68.8
"	"	2	181 ± 02	6.4	0.8	0.0	0.0	0.0	–	0.0	0.0	21.6	71.2
"	"	3	181 ± 03	4.8	0.8	0.0	0.0	0.0	–	0.0	0.0	16.8	77.6
"	"	4	182 ± 05	5.6	0.0	0.0	0.0	0.0	–	0.0	0.0	20.8	73.6
QuickDraw	8	1	181 ± 04	33.6	0.0	0.0	0.0	0.0	39.2	0.0	0.0	–	27.2
"	"	2	179 ± 03	32.8	0.0	0.0	0.0	0.0	45.6	0.0	0.0	–	21.6
"	"	3	180 ± 04	32.8	0.0	0.0	0.0	0.0	41.6	0.0	0.0	–	25.6
"	"	4	179 ± 06	40.0	0.0	0.0	0.0	0.0	39.2	0.0	0.0	–	20.8
CIFAR10	2	1	134 ± 16	24.0	1.6	–	15.2	14.4	8.0	36.0	0.8	0.0	0.0
"	"	2	140 ± 20	24.8	0.0	–	16.0	16.0	16.0	26.4	0.0	0.8	0.0
"	"	3	144 ± 21	28.8	0.0	–	16.0	16.8	16.0	22.4	0.0	0.0	0.0
"	"	4	149 ± 11	33.6	0.0	–	12.8	17.6	15.2	20.0	0.0	0.8	0.0
CIFAR10	5	1	155 ± 12	7.2	0.8	10.4	68.8	7.2	–	0.8	2.4	0.8	1.6
"	"	2	159 ± 07	3.2	0.0	5.6	80.8	4.0	–	0.8	4.0	0.8	0.8
"	"	3	164 ± 07	0.0	0.0	5.6	88.8	0.8	–	0.0	2.4	1.6	0.8
"	"	4	160 ± 05	0.0	0.0	6.4	88.8	0.8	–	0.0	2.4	0.8	0.8
CIFAR10	8	1	161 ± 08	4.0	83.2	0.0	0.0	0.0	0.0	0.0	0.0	–	12.8
"	"	2	165 ± 11	9.6	77.6	0.0	0.0	0.0	0.0	0.0	0.0	–	12.8
"	"	3	160 ± 04	14.4	76.0	0.0	0.0	0.0	0.0	0.0	0.0	–	9.6
"	"	4	164 ± 06	19.2	68.0	0.0	0.0	0.8	0.0	0.0	0.0	–	12.0

Table 3 shows how many new images were added at each iteration. Remember that at most 200 images may be added at each iteration (since 8 interpolations are created for 25 images), though not all may be selected by the user for inclusion. In this case, a supervisor model was used and rejected images whenever no class had a probability higher than 90%. Few images were rejected since the interpolation process was able to produce images of high quality (see Fig. 4). Furthermore, the table shows the distribution from which the source images originated – notice that source images were selected based on boundary distance (see (3)). In most cases, at early iterations, the boundary is fuzzy for a certain class (therefore this class has a high percentage), but then the boundary seems to become more well-defined, since the percentage drops.

5 Discussion

The experiments show clear advantages of the proposed method when the data distribution is plentiful for most conditions, but not for some. In these cases, new data of high quality can be generated. Other situations with absolute imbalance could perhaps require transfer learning to train the extrapolation component or changing the way feedback is given and how that feedback is integrated.

In addition, the individual components of the method could be designed differently; some were already alluded to, such as using back-propagation (rather than a decoder) to extrapolate the images. There are also many different conceivable ways of how feedback could be requested and how it could be integrated into the optimization process. Small changes to the feedback process could include asking the user to rank images or even do extrapolations from one image across the boundary rather than interpolations between two images. More interesting would be looking for different ways of how this feedback could be integrated – one possibility is to use only the direction that the user prefers, which would avoid having to create perfect images.

An automated model was used in the production of the results since it was unviable to use a human user for the amount of experiments that were performed. Some people have suggested that this might mean the method is useful for data augmentation. However, this was only possible because we had artificially created an imbalanced dataset, and we still had the complete dataset to train our automated model. Possibly, our method could be used in conjunction with knowledge distillation, so that a larger model (teacher) could refine the decision boundary from a smaller model (student).

6 Conclusion

The main argument behind the proposed approach is that during the training period, any model can face confusion around boundaries in the manifold due to a lack of representative samples. New data is extracted from the model itself by projecting images over the decision boundary so that the user is able to tune the boundary more directly after the model has been trained. Human feedback can then be provided either synchronously or asynchronously. Such specific feedback could help define precise class boundaries and thus improve the performance of the model in terms of classification accuracy.

The proposed method has been shown to be highly beneficial in an experiment where the dataset was augmented with the help of the user in the context of class imbalance. The method seems to work well when enough data is available to train the extrapolation component, while at the same time, a segment of data is poorly represented (e.g., a class is imbalanced) so that active supervision can make a difference.

All in all, active supervision seems to provide a possible mechanism that could help reduce the cost of data collection.

Funding. This work was supported by National Funds through the Portuguese Funding Agency, FCT – Foundation for Science and Technology Portugal, under Project LA/P/0063/2020.

References

1. Chen, Q., Koltun, V.: Photographic image synthesis with cascaded refinement networks. In: Proceedings of the IEEE International Conference on Computer Vision, pp. 1511–1520 (2017)
2. Goodfellow, I., Shlens, J., Szegedy, C.: Explaining and harnessing adversarial examples. In: Proceedings of the International Conference on Machine Learning (ICML), pp. 1–10 (2015)
3. Google: Quick, Draw! https://quickdraw.withgoogle.com/data. Accessed 12 Apr 2021
4. Jiang, Y., Krishnan, D., Mobahi, H., Bengio, S.: Predicting the generalization gap in deep networks with margin distributions. In: International Conference on Learning Representations (2019). https://openreview.net/forum?id=HJlQfnCqKX
5. Krizhevsky, A.: Learning multiple layers of features from tiny images. Tecchnical Report (2009)
6. Krizhevsky, A., Sutskever, I., Hinton, G.E.: ImageNet classification with deep convolutional neural networks. In: Advances in Neural Information Processing Systems (NeurIPS), pp. 1097–1105 (2012)
7. LeCun, Y., Cortes, C., Burges, C.: MNIST handwritten digit database. ATT Labs. http://yann.lecun.com/exdb/mnist 2 (2010)
8. Netzer, Y., Wang, T., Coates, A., Bissacco, A., Wu, B., Ng, A.Y.: Reading digits in natural images with unsupervised feature learning (2011)
9. Settles, B.: Active learning literature survey. University of wisconsin (2010)
10. Shafahi, A., et al.: Adversarial training for free! In: Advances in Neural Information Processing Systems, pp. 3353–3364 (2019)
11. Serrano e Silva, P., Cruz, R., Shihavuddin, A.S.M., Gonçalves, T.: Interpretability-guided human feedback during neural network training. In: Pertusa, A., Gallego, A.J., Sánchez, J.A., Domingues, I. (eds.) Pattern Recognition and Image Analysis. IbPRIA 2023. LNCS, vol. 14062. Springer, Cham (2023). https://doi.org/10.1007/978-3-031-36616-1_22

12. Tong, S., Koller, D.: Support vector machine active learning with applications to text classification. J. Mach. Learn. Res. **2**, 45–66 (2001)
13. Verma, V., et al.: Manifold mixup: better representations by interpolating hidden states. In: Proceedings of the 36th International Conference on Machine Learning, vol. 97, pp. 6438–6447. PMLR, Long Beach, California, USA, 09–15 June 2019. http://proceedings.mlr.press/v97/verma19a.html

Condition Invariance for Autonomous Driving by Adversarial Learning

Diana Teixeira e Silva[2] and Ricardo P. M. Cruz[1,2(✉)]

[1] INESC TEC, Porto, Portugal
ricardo.p.cruz@inesctec.pt
[2] Faculty of Engineering, University of Porto, Porto, Portugal
dianartsilva@gmail.com

Abstract. Object detection is a crucial task in autonomous driving, where domain shift between the training and the test set is one of the main reasons behind the poor performance of a detector when deployed. Some erroneous priors may be learned from the training set, therefore a model must be invariant to conditions that might promote such priors. To tackle this problem, we propose an adversarial learning framework consisting of an encoder, an object-detector, and a condition-classifier. The encoder is trained to deceive the condition-classifier and aid the object-detector as much as possible throughout the learning stage, in order to obtain highly discriminative features. Experiments showed that this framework is not very competitive regarding the trade-off between precision and recall, but it does improve the ability of the model to detect smaller objects and some object classes.

Keywords: Adversarial learning · Autonomous driving · Computer vision · Deep learning · Domain generalization

1 Introduction

Object detection aims to identify and locate relevant objects in an image or video. Using an image/video as an input, this computer vision task outputs the class labels and bounding box coordinates for all the objects that fall into specific categories [27].

Object detection plays an important role in many applications, such as autonomous driving (i.e. environment perception, with the detection and location of road elements [14]). Deep learning techniques regarding object detection have achieved, similarly to other computer vision tasks, a good performance in the benchmark datasets, such as MS COCO [18]. However, in autonomous driving applications, the domain shift between training and test images due to variant conditions (e.g. weather, location, period of the day) can degrade the detector's performance.

For example, the training set may have more people on the street during the day rather than during the night. Or more people are out on the street when it

© Springer Nature Switzerland AG 2024
V. Vasconcelos et al. (Eds.): CIARP 2023, LNCS 14469, pp. 552–563, 2024.
https://doi.org/10.1007/978-3-031-49018-7_39

is sunny versus when it is rainy. If the training set has many more pedestrians when the weather is sunny than when it is rainy, this fact should not condition the model to reduce its sensitivity to pedestrians when the weather is less sunny. The model may pick up on these cues and assume a prior that is unwarranted: the model may then underperform when deployed in a country like the United Kingdom where people are used to the weather being typically rainy and consequently more people are on the streets than when it is rainy in Spain. Thus, an object detection model should be trained in a way that removes a condition's priors so that the model is not modeling $P(object \mid sensors, condition)$ when the desired objective is that it models $P(object \mid sensors)$.

We propose encoding latent variables to not contain information on a condition, such as the weather, so that the decision is not dependent on it. To do that, an adversarial loss is used, penalizing good accuracy in classifying the condition's categories (e.g. in the case of weather invariance, the model will be penalized if the weather is correctly classified).

In addition to the Introduction, this paper is divided as follows: Sect. 2 explores some related work regarding object detection, domain adaptation and feature invariance, Sect. 3 illustrates the proposed method, Sect. 4 details the implementation, Sect. 5 presents the results and its discussion, and Sect. 6 concludes the paper.

2 Related Work

2.1 Object Detection

The existing state-of-the-art detectors are divided into two types: one-stage detectors and two-stage detectors. One-stage detectors' outputs are the classification probabilities and box offsets at each spacial position using a fully convolutional architecture and a pre-defined anchor box. This type of detector prioritizes inference speed and includes models such as YOLO [23], SSD [19], and RetinaNet [17]. Two-stage detectors employ a region proposal network in order to extract the regions more probable to contain an object, feeding them to a region convolutional network. Faster R-CNN [24] and Mask R-CNN [10] are examples of these types of detectors, which are more accurate compared to the one-stage ones.

Studies have been done in order to bridge the gap between these two types of detectors, mainly by trying to overcome the weaknesses of the one-stage detectors [20]. Tian et al. [28] proposed FCOS (Fully Convolutional One-Stage Object Detection), a single-stage, anchor-free and proposal-free object detector. This model minimizes computations associated with anchor boxes, as well as it avoids the hyperparameters related to them, which often worsens performance. This architecture is getting notoriety within object detection tasks regarding autonomous driving [2,15,22,25], thus we use FCOS as our base model.

2.2   Domain Adaptation and Feature Invariance

Domain adaptation has been one potential framework to address the mismatch between the domains and the absence of training data with annotations in not-favorable conditions [12]. This methodology has already been designed for Faster R-CNN by employing an adversarial training strategy to learn robust features indistinguishable in the domains: while the feature extractor must be optimized to maximize the domain classification error, the image- and instance-level domain classifiers (i.e., the classifier used to differentiate between source and target domains using global attributes and particular regional features, respectively) are both developed to minimize error [3].

There is a growing literature on making neural networks invariant to certain conditions. Cruz et al. [4] promote background invariance so that the neural network learns the object independent of the background. Ferreira et al. [6] propose a classification neural network that recognizes sign language used by deaf communities in a way that is invariant to the specific person gesticulating - the goal is to make the model more generalizable by preventing it from associating the person doing the sign with the sign itself (i.e. the model may associate a certain person with a certain sign). The approach for adversarial training proposed in [6] is comparable to that first presented by Ganin et al. [8] for domain adaptation and later by Feutry et al. [7] for learning anonymous representations.

2.3   Domain Adaptation in Autonomous Driving

Domain adaptation in autonomous driving has been studied for the last few years, showing some good results. Hnewa et al. [12] showed that domain adaptation for object detection under rainy conditions performed better when compared to other techniques such as deraining methods to clear the rain residue by restoring an image.

Sindagi et al. [26] did some work regarding weather invariance. They stated that weather conditions, such as rain and haze, can be modeled by a superposition of a clean image and some rain residue or atmospheric light, respectively. Consequently, there will be weather-specific information that will degrade the feature space and result in worse performance. They propose a prior-adversarial loss to train a neural network to predict priors regarding the weather while minimizing weather-specific information in the features at the same time. In order to de-distort the feature space, they additionally incorporate a set of residual feature recovery blocks in the target process.

Li et al. [16] proposed a stepwise adaptation method, where a false target domain is used to bridge the domain gap and the distributions between both domains are aligned at a feature level by an adaptive CenterNet. The process involves two steps: Step 1 involves creating the false target domain by using CycleGAN to train an unpaired image-to-image translator and create synthetic images from the source domain, which are then combined with labels; Step 2 involves feeding the extracted feature maps from the false and real target domains into an encoder network, which will then extract the domain-invariant features using a gradient reversal layer (which changes the sign for the backward pass, but not for the forward pass) and a domain discriminator.

The methodology proposed in this paper is motivated by the works mentioned in the previous paragraphs, but it is mainly inspired by [6].

3   Proposal

Our proposal consists of training a model to learn condition-invariant features that keep suitable information about the object and discard the condition-specific characteristics.

Considering $\mathcal{D} = \{X_i, o_i, c_i\}_{i=1}^N$ as a labeled dataset of N samples, where X_i represents the i-th image, o_i represents the set of object detection labels (e.g. car, truck, pedestrian, etc.) corresponding to image i and c_i represents the image's category of the condition the model should be invariant to (e.g. for weather invariance, c_i could be sunny, rainy or fog). C corresponds to the set of categories of the condition targeted. The proposed model will have 3 sub-networks:

- **Feature Extractor:** learns an encoding function $h(X; \theta_h)$, parameterized by θ_h, that encoded an input image X into a latent representation h;
- **Object-detector:** uses the latent representation h to learn a function $f(h; \theta_f)$, using the parameters θ_f, that yields $p(o|h; \theta_f)$;
- **Condition-classifier:** learns $f(h; \theta_g)$, parameterized by θ_g, that gives $p(c \mid h; \theta_g)$;

The condition-classifier is trained to minimize the negative log-likelihood of the correct condition predictions, as presented in Eq. (1).

$$\min_{\theta_g} \mathcal{L}_{condition}(\theta_h, \theta_g) = -\frac{1}{N} \sum_{i=1}^N \log p(c_i \mid h(X_i; \theta_h); \theta_g) \tag{1}$$

Similarly, the encoder network and the object-detector are trained to minimize the negative log-likelihood of the correct object predictions as presented in Eq. (2).

$$\min_{\theta_h, \theta_f} \mathcal{L}_{object}(\theta_h, \theta_f) = -\frac{1}{N} \sum_{i=1}^N \log p(o_i \mid h(X_i; \theta_h); \theta_f) \tag{2}$$

In order to make features condition-invariant, the weights of the encoder network are adjusted using an adversarial loss, presented in Eq. (3), that tries to force the condition-classifier not to do better than random guessing the condition predictions (i.e. the condition-classifier's probabilities should be close to a uniform distribution).

$$\min_{\theta_h} \mathcal{L}_{adversarial}(\theta_h, \theta_g) = -\frac{1}{N|C|} \sum_{i=1}^N \sum_{c \in C} \log p(c \mid h(X_i; \theta_h); \theta_g) \tag{3}$$

To further promote condition-invariance, a new loss $\mathcal{L}_{transfer}$ was added in order to minimize the distance between the hidden latent representations of different categories of a condition at each layer of the encoder network. Eq. (4) represents the distance between the latent representations of the condition's categories c_α and c_β ($c_\alpha, c_\beta \in C$, $c_\alpha \neq c_\beta$), at the m-th layer, where N_{c_α} and N_{c_β} represent the number of training examples of the conditions c_α and c_β, respectively.

$$D^{(m)}(c_\alpha, c_\beta; \theta_h) = -\left\| \frac{1}{N_{c_\alpha}} \sum_{i:c_i=c_\alpha} h^{(m)}(X_i; \theta_h) - \frac{1}{N_{c_\beta}} \sum_{j:c_j=c_\beta} h^{(m)} X_j; \theta_h) \right\|_2^2 \quad (4)$$

This presupposes that the object labels are evenly distributed across each condition's category. If not, each mini-batch used for training should be created with this criterion in mind.

The transfer loss at the m-th layer is calculated by summing the pairwise distances between all condition's categories and the final $\mathcal{L}_{transfer}$ is given by the sum of the transfer losses for all the layers, weighted by $\beta^{(m)}$, as presented in Eq. (5). In our proposal, all the layers have $\beta^{(m)} = 1$.

$$\mathcal{L}_{transfer}(\theta_h) = \sum_{m=1}^{M} \beta^{(m)} \sum_{c_\alpha \in C} \sum_{c_\beta \in C} D^{(m)}(c_\alpha, c_\beta; \theta_h) \quad (5)$$

An overview of the network architecture and loss functions is shown in Fig. 1.

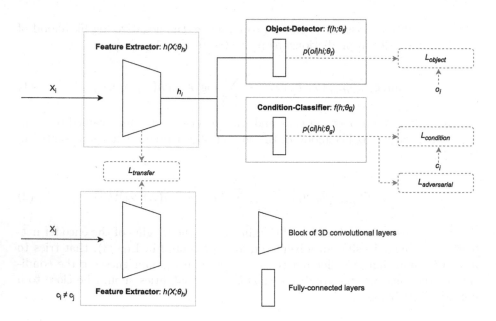

Fig. 1. Architecture for adversarial learning (adapted from [6]).

4   Implementation

The experimental protocol is now elaborated, starting with the datasets used, the model used, how it was optimized, and the metrics used for evaluation.

4.1   Data

(a) Sand   (b) Rain   (c) Snow   (d) Fog

Fig. 2. Examples of the different weather conditions in the DAWN dataset.

(a) City-street, Rainy   (b) Country road, Clear   (c) Highway, Overcast

Fig. 3. Examples of the different conditions in the SODA10M dataset.

To validate our proposal, two datasets designed for object detection in autonomous driving were used: DAWN [13] and SODA10M [9]. These datasets had the particularity to have labels regarding conditions such as weather and scene/location for each image.

The DAWN dataset contains information regarding the weather in each image (sand, rain, snow, fog), as presented in Fig. 2. The SODA10M dataset contains labels regarding the weather (clear, overcast, rainy) and the scene (city-street, country road, highway). Examples of images from the latter dataset are presented in Fig. 3.

Despite the SODA10M dataset being divided into training, validation and test (unlabeled) sets, the distribution of each label is different in both labeled sets (i.e. the training set only contains city-street and clear weather images), hence both sets were merged. In the end, both datasets were randomly split into 70% of the total images being training data and 30% being test data.

4.2   Model

The adversarial training method was applied by starting with a pre-trained FCOS in the MS COCO dataset. The object detection head was altered to suit the dataset used – the number of output classes corresponding to the number of object categories plus the background class – instead of the 90 class outputs from the MS COCO dataset. An additional linear layer was added to the head of the model in order to construct the condition classifier. The experiments were carried out using PyTorch deep learning framework.

4.3   Training and Testing

Ferreira et al. [6] trained the network architecture minimizing the sum of the object detection term (2), adversarial term (3) and transfer term (5) losses, with the last two pondered each by a factor which defines their importance and due to the difference in the order of magnitude. To try to avoid tweaking the term values, our proposal used different optimizers to update the corresponding network weights according to each of the different losses. Studies, such as [11], have shown that this approach results in better performance when there are several terms in the loss. In our method, Adam optimization algorithm was used for all the cases, with a batch size of 8 and a learning rate of 0.00005. DAWN and SODA10M were trained for 50 and 25 epochs, respectively.

The following data augmentation transformations were applied to the training set: horizontal flip; contrast/brightness modification between -0.1 and 0.1; resize to 256×512 and normalization with mean 0 and standard deviation of 1.

The training was performed by providing the model with all the condition categories except one, in which the model would be tested. For example, DAWN has four weather categories (fog/rain/sand/snow), therefore four models were trained: one model with rain/sand/snow and evaluated with fog, another with fog/sand/snow and evaluate with rain, and the same for the remaining two categories. The conditions from the experiments are detailed in Table 1.

Table 1. Experiments details.

Dataset	Experiment	Training Conditions	Test Condition
DAWN	D1	rain, sand, snow	fog
	D2	fog, sand, snow	rain
	D3	fog, rain, snow	sand
	D4	fog, rain, sand	snow
SODA10M Weather	SW1	overcast, raining	clear
	SW2	clear, raining	overcast
	SW3	clear, overcast	raining
SODA10M Scene	SS1	countryroad, highway	citystreet
	SS2	citystreet, highway	countryroad
	SS3	citystreet, countryroad	highway

4.4   Performance Metrics

Mean average precision (mAP), a popular metric used in object detection since its first definition in the PASCAL Challenge 2012 [5], was employed to evaluate the performance of our method. The Precision-Recall (PR) curve is calculated to determine the value of this metric. The prediction is a true positive if its bounding box has an intersection-over-union (IoU) value greater than 0.5 and the same class label as the ground truth. If so, the curve is modified by imposing a monotonic decline in the precision. AP is then the area under the PR-curve and mAP is the mean of the AP among all classes [26].

Another metric used to calculate the assertiveness of an object detection task for a given class is average recall (AR) which, opposite to the AP, does not take into account the confidences of the estimated detections. A range of IoU thresholds [0.5,1] is used to calculate all the recall values, which are then averaged to evaluate the model's performance regarding the detection of true objects [21].

Even though mAP provides a more extensive evaluation of a model by considering both the precision and the recall, mAR might be useful when the interest is situated on maximizing recall and identifying as many objects of interest as possible.

5   Results and Discussion

Table 2 presents the values of mAP and mAR obtained by evaluating the adversarial methodology for different experiments in the DAWN dataset. Bold font indicates the results where a metric's value is higher for the adversarial training in comparison to the baseline.

Table 2. mAP and mAR results for the different experiments in the DAWN dataset.

	D1		D2		D3		D4	
	Baseline	Adversarial	Baseline	Adversarial	Baseline	Adversarial	Baseline	Adversarial
map	0.2550	0.1753	0.3471	0.3361	0.1587	0.1260	0.2434	0.2024
map_50	0.4724	0.3350	0.6184	0.5790	0.2579	0.2186	0.4082	0.3912
map_75	0.2602	0.1583	0.4506	0.3647	0.1640	0.1222	0.2431	0.1967
map_small	0.2181	0.1573	0.3339	0.3338	0.1368	0.1119	0.2624	0.1969
map_medium	0.6071	0.4702	0.4053	0.3283	0.4098	0.3638	0.4051	0.3118
map_large	0.1377	0.0676	1.000	0.9000	0.4000	0.0750	0.3614	0.1776
mar_1	0.1623	0.1180	0.3107	0.2942	0.1289	**0.1363**	0.2004	0.1787
mar_10	0.3548	0.2855	0.4329	**0.4675**	0.2528	**0.2627**	0.3776	0.3559
mar_100	0.3670	0.3175	0.4590	**0.5070**	0.2564	**0.2698**	0.4019	**0.4044**
mar_small	0.3359	0.2786	0.4513	**0.5147**	0.2194	**0.2277**	0.3701	**0.3854**
mar_medium	0.7507	0.6881	0.4958	0.4399	0.5337	**0.5749**	0.5265	0.4909
mar_large	0.2833	0.1167	1.000	0.9000	0.4000	0.2000	0.3800	0.2000

Regarding the mAP values, all of them were lower for the adversarial framework. For the models trained with all the categories except fog (D1), our proposed method showed worse results than the baseline in all metrics. However, for the other models, adversarial training provided higher scores for some mAR metrics.

A high mAR considering a high IoU threshold indicates that the model is successful at detecting a significant fraction of the ground truth objects when subjected to a stricter matching criterion. For the models trained without the rain category (D2), adversarial learning improved this metric by 10.5% and mAR for small objects by 14%. The latter metric reflects the ability of the model to detect small objects, which is higher for 3 out of 4 adversarial models when compared to the baseline.

Analyzing the mAR for each object class, as presented in Table 3, it can be seen that adversarial training allowed better recall for several individual classes in almost all the cases, with the best case being, as expected by the previous paragraph, the model tested on rain data.

The difference in the behavior of these metrics for the same framework but trained with different data categories underlines that some sets of categories allow better feature invariance than others. That is also visible in Tables 4 and 5, where the results for the SODA10M dataset showed an increase in mAP for the adversarial methodology only for the models that did not use overcast or country road images during training.

Table 3. mAR results by class for the different experiments in the DAWN dataset.

mar_class		car	bus	truck	motorcycle	bicycle	person
D1	**Baseline**	0.5052	0.4722	0.3885	0.4167	0.0500	0.3694
	Adversarial	0.5021	0.3167	0.3808	0.3500	0.0000	0.3556
D2	**Baseline**	0.5038	0.3250	0.3829	0.6000	−1.0000	0.4833
	Adversarial	**0.5118**	**0.3750**	**0.4316**	**0.7000**	−1.0000	**0.5167**
D3	**Baseline**	0.3896	0.5308	0.3316	0.0917	0.0000	0.1949
	Adversarial	0.3879	**0.5462**	**0.3754**	**0.1250**	0.0000	0.1846
D4	**Baseline**	0.4165	0.5429	0.4800	−1.0000	−1.0000	0.1682
	Adversarial	**0.4444**	0.4714	0.4700	−1.0000	−1.0000	**0.2318**

Table 4. mAP results for the different experiments in the SODA dataset (weather invariance).

Weather			
mAP	SW1	SW2	SW3
Baseline	0.1680	0.2267	0.2545
Adversarial	0.1642	**0.2419**	0.2126

Table 5. mAP results for the different experiments in the SODA dataset (scene invariance).

Scene			
mAP	SS1	SS2	SS3
Baseline	0.1755	0.2163	0.2038
Adversarial	0.1160	**0.2227**	0.1694

6 Conclusion

In this paper, we propose an adversarial framework to train an object detection model for autonomous driving to be invariant to certain conditions that might result in deceiving priors to the network, such as weather and scene.

The fundamental concept is to train condition-invariant latent features that retain the most object-specific information while removing condition-specific details that may degrade object detection. To do this, an encoder and an object-detector are trained simultaneously over the target object variables while ensuring the latent representations of the encoder are not informative of the condition categories. To further disfavor the priors, another training loss was applied to drive the latent distributions of the distinct condition categories to be as similar as possible.

This framework was applied to two autonomous driving datasets, showing an overall worse mAP than the baseline, but higher mAR for high IoU thresholds and smaller objects, as well as for each class. Even though the trade-off between precision and recall was worse in our method, it is better at detecting relevant objects by providing fewer false negatives.

Some improvements could be done regarding this work. Within this methodology, instead of training the model completely in this framework, the adversarial loss could only be used while the predicting accuracy on the same mini-batch is good, as proposed by [29]. That could be beneficial to the learning process due to the oscillating behavior presented by the evaluating metrics in adversarial learning. In addition, other approaches to condition generalization could be done, such as deep bilevel learning, which has shown promising results regarding activity detection inside vehicles [1].

Acknowledgements. This work was supported by European Structural and Investment Funds in the FEDER component through the Operational Competitiveness and Internationalization Programme (COMPETE 2020) [Project nº 047264; Funding Reference: POCI-01-0247-FEDER-047264].

Funding Information. This work was supported by National Funds through the Portuguese Funding Agency, FCT – Foundation for Science and Technology Portugal, under Project LA/P/0063/2020.

References

1. Capozzi, L., et al.: Toward vehicle occupant-invariant models for activity characterization. IEEE Access **10**, 104215–104225 (2022)
2. Carranza-García, M., Torres-Mateo, J., Lara-Benítez, P., García-Gutiérrez, J.: On the performance of one-stage and two-stage object detectors in autonomous vehicles using camera data. Remote Sens. **13**(1), 89 (2020)
3. Chen, Y., Li, W., Sakaridis, C., Dai, D., Van Gool, L.: Domain adaptive faster R-CNN for object detection in the wild. In: Proceedings of the IEEE Conference on Computer Vision and Pattern Recognition, pp. 3339–3348 (2018)
4. Cruz, R., Prates, R.M., Simas Filho, E.F., Costa, J.F.P., Cardoso, J.S.: Background invariance by adversarial learning. In: 2020 25th International Conference on Pattern Recognition (ICPR), pp. 5883–5888. IEEE (2021)
5. Everingham, M., Van Gool, L., Williams, C.K.I., Winn, J., Zisserman, A.: The PASCAL visual object classes challenge 2012 (VOC2012) results. http://www.pascal-network.org/challenges/VOC/voc2012/workshop/index.html
6. Ferreira, P.M., Pernes, D., Rebelo, A., Cardoso, J.S.: Learning signer-invariant representations with adversarial training. In: Twelfth International Conference on Machine Vision (ICMV 2019), vol. 11433, pp. 918–926. SPIE (2020)
7. Feutry, C., Piantanida, P., Bengio, Y., Duhamel, P.: Learning anonymized representations with adversarial neural networks. arXiv preprint arXiv:1802.09386 (2018)
8. Ganin, Y., Lempitsky, V.: Unsupervised domain adaptation by backpropagation. In: International Conference on Machine Learning, pp. 1180–1189. PMLR (2015)
9. Han, J., et al.: SODA10M: a large-scale 2d self/semi-supervised object detection dataset for autonomous driving (2021)
10. He, K., Gkioxari, G., Dollár, P., Girshick, R.: Mask R-CNN. In: Proceedings of the IEEE International Conference on Computer Vision, pp. 2961–2969 (2017)
11. Hervella, Á.S., Rouco, J., Novo, J., Ortega, M.: End-to-end multi-task learning for simultaneous optic disc and cup segmentation and glaucoma classification in eye fundus images. Appl. Soft Comput. **116**, 108347 (2022)
12. Hnewa, M., Radha, H.: Object detection under rainy conditions for autonomous vehicles: a review of state-of-the-art and emerging techniques. IEEE Sig. Process. Mag. **38**(1), 53–67 (2020)
13. Kenk, M.A., Hassaballah, M.: DAWN: vehicle detection in adverse weather nature dataset. arXiv preprint arXiv:2008.05402 (2020)
14. Khatab, E., Onsy, A., Varley, M., Abouelfarag, A.: Vulnerable objects detection for autonomous driving: a review. Integration **78**, 36–48 (2021)
15. Kim, Y., Hwang, H., Shin, J.: Robust object detection under harsh autonomous-driving environments. IET Image Proc. **16**(4), 958–971 (2022)
16. Li, G., Ji, Z., Qu, X.: Stepwise domain adaptation (SDA) for object detection in autonomous vehicles using an adaptive CenterNet. IEEE Trans. Intell. Transp. Syst. **23**(10), 17729–17743 (2022)
17. Lin, T.Y., Goyal, P., Girshick, R., He, K., Dollár, P.: Focal loss for dense object detection. In: Proceedings of the IEEE International Conference on Computer Vision, pp. 2980–2988 (2017)
18. Lin, T., et al.: Microsoft COCO: common objects in context. CoRR abs/1405.0312, http://arxiv.org/abs/1405.0312 (2014)

19. Liu, W., et al.: SSD: single shot multibox detector. In: Leibe, B., Matas, J., Sebe, N., Welling, M. (eds.) Computer Vision - ECCV 2016. ECCV 2016, LNCS, Part I, vol. 9905, pp. 21–37. Springer, Cham (2016). https://doi.org/10.1007/978-3-319-46448-0_2

20. Lu, X., Li, Q., Li, B., Yan, J.: MimicDet: bridging the gap between one-stage and two-stage object detection. In: Vedaldi, A., Bischof, H., Brox, T., Frahm, J.-M. (eds.) ECCV 2020. LNCS, vol. 12359, pp. 541–557. Springer, Cham (2020). https://doi.org/10.1007/978-3-030-58568-6_32

21. Padilla, R., Passos, W.L., Dias, T.L., Netto, S.L., Da Silva, E.A.: A comparative analysis of object detection metrics with a companion open-source toolkit. Electronics 10(3), 279 (2021)

22. Piao, Z., Wang, J., Tang, L., Zhao, B., Zhou, S.: Anchor-free object detection with scale-aware networks for autonomous driving. Electronics 11(20), 3303 (2022)

23. Redmon, J., Divvala, S., Girshick, R., Farhadi, A.: You only look once: unified, real-time object detection. In: Proceedings of the IEEE Conference on Computer Vision and Pattern Recognition, pp. 779–788 (2016)

24. Ren, S., He, K., Girshick, R., Sun, J.: Faster R-CNN: towards real-time object detection with region proposal networks. In: Advances in Neural Information Processing Systems, vol. 28 (2015)

25. Sha, M., Boukerche, A.: Performance evaluation of CNN-based pedestrian detectors for autonomous vehicles. Ad Hoc Netw. 128, 102784 (2022)

26. Sindagi, V.A., Oza, P., Yasarla, R., Patel, V.M.: Prior-based domain adaptive object detection for hazy and rainy conditions. In: Vedaldi, A., Bischof, H., Brox, T., Frahm, J.-M. (eds.) ECCV 2020. LNCS, vol. 12359, pp. 763–780. Springer, Cham (2020). https://doi.org/10.1007/978-3-030-58568-6_45

27. Sun, T., Chen, J., Ng, F.: Multi-target domain adaptation via unsupervised domain classification for weather invariant object detection. arXiv preprint arXiv:2103.13970 (2021)

28. Tian, Z., Shen, C., Chen, H., He, T.: FCOS: fully convolutional one-stage object detection. In: Proceedings of the IEEE/CVF International Conference on Computer Vision, pp. 9627–9636 (2019)

29. Wu, Z., Suresh, K., Narayanan, P., Xu, H., Kwon, H., Wang, Z.: Delving into robust object detection from unmanned aerial vehicles: A deep nuisance disentanglement approach. In: Proceedings of the IEEE/CVF International Conference on Computer Vision, pp. 1201–1210 (2019)

YOLOMM – You Only Look Once for Multi-modal Multi-tasking

Filipe Campos[1], Francisco Gonçalves Cerqueira[1],
Ricardo P. M. Cruz[1,2](✉), and Jaime S. Cardoso[1,2]

[1] Faculty of Engineering, University of Porto, Porto, Portugal
{up201905609,up201905337}@edu.fe.up.pt, jsc@fe.up.pt
[2] INESC TEC, Porto, Portugal
rpcruz@fe.up.pt

Abstract. Autonomous driving can reduce the number of road accidents due to human error and result in safer roads. One important part of the system is the perception unit, which provides information about the environment surrounding the car. Currently, most manufacturers are using not only RGB cameras, which are passive sensors that capture light already in the environment but also Lidar. This sensor actively emits laser pulses to a surface or object and measures reflection and time-of-flight. Previous work, YOLOP, already proposed a model for object detection and semantic segmentation, but only using RGB. This work extends it for Lidar and evaluates performance on KITTI, a public autonomous driving dataset. The implementation shows improved precision across all objects of different sizes. The implementation is entirely made available: https://github.com/filipepcampos/yolomm.

Keywords: Autonomous driving · RGB camera · Lidar sensor · Object detection · Semantic segmentation

1 Introduction

Autonomous driving has experienced a substantial increase in investment and research efforts [4] as companies strive to achieve the ultimate goal of full driving automation, wherein vehicles can themselves without human intervention. This ambitious endeavor holds the potential to significantly diminish the occurrence of road accidents caused by human error, consequently leading to safer roadways for everyone. A critical component integral to the realization of this vision is the perception unit, which is responsible for gathering and processing essential information pertaining to the environment surrounding the vehicle. By effectively perceiving and understanding its surroundings, an autonomous vehicle can make

This work is supported by European Structural and Investment Funds in the FEDER component, through the Operational Competitiveness and Internationalization Programme (COMPETE 2020) [Project nº 047264; Funding Reference: POCI-01-0247-FEDER-047264].

V. Vasconcelos et al. (Eds.): CIARP 2023, LNCS 14469, pp. 564–574, 2024.
https://doi.org/10.1007/978-3-031-49018-7_40

informed decisions and navigate through complex scenarios with enhanced safety and efficiency.

The two emerging types of sensors are:

- Passive Cameras: RGB cameras are considered passive sensors because they capture light that is already present in the environment. They rely on ambient light or light sources like the sun to illuminate the scene. The camera's lens focuses the light onto a sensor, which converts the light into an electrical signal, creating an image. The camera does not emit any light itself but passively captures the light that reflects off objects.
- Active Lidar: Lidar, on the other hand, is an active sensing technology because it actively emits laser light and measures the reflections. The Lidar sensor emits laser pulses and measures the time it takes for the pulses to return after hitting objects in the environment. By analyzing the properties of the returned light, such as time-of-flight and intensity, Lidar systems can determine the distances to objects and create 3D representations of the scene.

By being independent of ambient light, Lidar is also more resilient to changes in the atmosphere and works noticeably better at night. However, Lidar has several disadvantages; most importantly, the resolution is limited, and data is collected as a sparse collection of 3D points, which may not capture fine details or small objects as effectively as high-resolution cameras. Furthermore, being an active sensor can also be a disadvantage for some objects, such as transparent surfaces or materials with low reflectivity.

Our aim is mainly focused on object detection and semantic segmentation and to evaluate the importance of using each or both of these sensors. The choice for the model was YOLOP, which uses only RGB to perform both of these tasks. A diagram showing our extension with Lidar is shown as Fig. 1.

Section 2 describes existing models and datasets to explain our choice of YOLOP and KITTI, respectively. Section 3 details the changes made on top of YOLOP. Then, Sect. 4 shows the experiments and results performed. The work concludes with Sect. 5.

2   State of the Art

For the choice of the base model and the dataset for the experiments, a review is conducted on existing models (Sub-Sect. 2.1) and datasets (Sub-Sect. 2.2).

2.1   Model Selection

Some models that can perform both object detection and semantic segmentation:

- PANet [16]: proposes a simple model that performs both object detection and semantic segmentation. The most interesting contribution is the Path Aggregation Network substructure, which performs bottom-up path aggregation, similarly to how Feature Pyramid Networks [14] perform top-down aggregation.

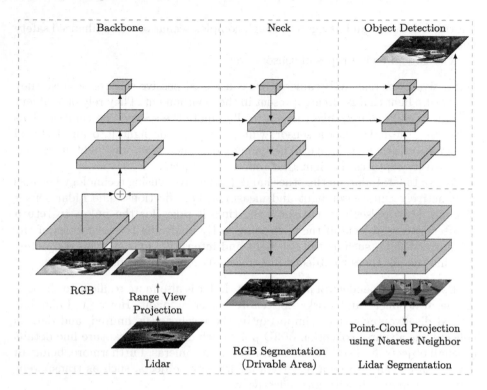

Fig. 1. Extension of YOLOP.

- YOLACT [2]: performs instance-segmentation by learning common representations (prototypes) and merging them to form the segmentation of each object. In this architecture, both tasks are tightly coupled, making it difficult to introduce new tasks or remove existing ones.
- MultiTask-CenterNet [11]: uses a ResNet [10] backbone and performs object detection, semantic segmentation and pose estimation. Its results are inferior to some of the remaining options.
- YOLOP [22]: performs traffic object detection, drivable area segmentation, and lane detection simultaneously. It is composed of one encoder for feature extraction and three decoders to handle specific tasks.
- Joint Semantic Understanding [12]: a multi-task framework for simultaneous traffic object detection, drivable area segmentation, and lane line segmentation. It proposes a network encoder to extract features and three decoders at multilevel branches that handle specific tasks, sharing the feature maps with more similar tasks for joint semantic understanding. It obtained results comparable to those of YOLOP but with a closed-source source code.
- HybridNet [21]: an end-to-end perception network for multi-tasking, focused on traffic object detection, drivable area segmentation, and lane detection. It proposes efficient segmentation head and box/class prediction based on a weighted bidirectional feature network, customized anchor for each level

on the network, and an efficient training loss function and training strategy. This architecture also obtained results similar to YOLOP and is also open-source; the main reason for discarding it was its added complexity, which only reflected in marginal gains in performance.

A summary of the tasks performed by each of the architectures is shown in Table 1.

Table 1. Tasks performed by each model

Paper	Year	Object Detection	Semantic Area Segmentation	Drivable Area Segmentation	Lane Line Segmentation	Pose Estimation
PANet [16]	2018	✓	✓	–	–	–
YOLACT [2]	2019	✓	✓	–	–	–
MCN [11]	2021	✓	✓	–	–	✓
Joint Sem. [12]	2022	✓	–	✓	✓	–
YOLOP [22]	**2022**	✓	–	✓	✓	–
HybridNets [21]	2022	✓	–	✓	✓	–

YOLOP was chosen for three main reasons:

- Performance: YOLOP outperformed the alternatives, delivering excellent results.
- Open-source: YOLOP is open-source, facilitating accessibility, support, and customization.
- Recognition: YOLOP has more citations and has been more scrutinized than the alternatives with similar performance.

Our implementation[1] is developed on top of the existing YOLOP implementation provided by the authors.

2.2 Dataset Selection

The original YOLOP implementation uses BDD100K [22]. This dataset comprises many tasks; however, the only modality collected is RGB images. Since the goal is also to incorporate Lidar, a survey on publicly available datasets is conducted in order to find a dataset suitable for multi-modal multi-task. A comparison between datasets is provided in Table 2. The main criteria were that both RGB and Lidar data had to be available in order to perform object detection and Lidar segmentation.

---

[1] https://github.com/filipepcampos/yolomm.

Table 2. Dataset comparison

Dataset	RGB	Lidar	Object Detection	Semantic Segmentation	Drivable Area Segmentation	Lane Line Segmentation	Lidar Segmentation
Klane [18]	✓	✓	–	–	–	✓	✓
BDD100K [23]	✓	–	✓	✓	✓	✓	–
Cityscapes [6]	✓	–	–	✓	✓	–	–
KITTI [9]	✓	✓	✓	✓	✓	–	✓*
KITTI-360 [13]	✓	✓	✓	✓	✓	–	✓
KITTI-Carla [7]	✓	✓	✓	✓	✓	–	✓
Nuscenes [3]	✓	✓	✓	–	–	–	–
Radiate [19]	✓	✓	✓	–	–	–	–
Waymo [20]	✓	✓	✓	✓	✓	–	✓

* Available through the Semantic KITTI Dataset, which labels the existing Lidar scans from KITTI.

We opted to use the KITTI dataset [9], combined with Semantic KITTI (which extends KITTI with the Lidar semantic ground-truth) [1] because it contained all the required data:

1. Bounding box labels with RGB and Lidar input
2. Drivable area segmentation for RGB input
3. Lidar Segmentation labels for both RGB and Lidar input

One thing to note is that these annotations exist for different image sequences with no overlap. This impacts how we can train our model since it is impossible to train both heads on a single forward pass.

3 Implementation

This section elaborates on how the Lidar was integrated into YOLOP and the subsequent training strategy.

3.1 Multi-modality

Lidar point-clouds can be represented in one of several ways [5]. The three main families of point-cloud discrete representations are:

- **3D voxelization:** a binary 3D-grid, where at each cell (voxel), 1 specifies whether a point is present or 0 otherwise (Fig. 2a). This is typically not used due to the memory and efficiency costs, and it is more sensitive to the resolution that is defined.
- **Bird's eye view:** this 2D view can be seen as a special case of voxelization, whereby the height dimension is ignored (Fig. 2b). It has the advantage that it is easy to merge with map information, and objects are always the same size regardless of range, at the cost of losing information.

- **Range view:** a 2D frustum-view offering a natural representation of how Lidar naturally captures the points (Fig. 2c). Range view provides a compact representation of the scene, which is useful for applications such as object detection and segmentation.

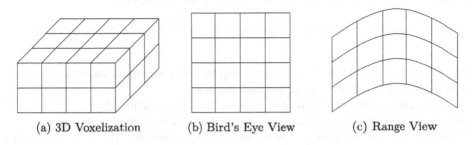

(a) 3D Voxelization   (b) Bird's Eye View   (c) Range View

Fig. 2. Point-cloud discretizations

For this work, the three-dimensional Lidar point-cloud is discretized into a two-dimensional representation using range view (Fig. 2c). We follow the same procedure as RangeNet++ [17], which creates this representation through the use of a spherical projection. A point $p_i = (x, y, z)$ is converted using $\Pi \colon \mathbb{R}^3 \to \mathbb{R}^2$ by using Eq. (1),

$$\begin{pmatrix} u \\ v \end{pmatrix} = \begin{pmatrix} \frac{1}{2} \left[1 - \arctan(y, x)\pi^{-1}\right] w \\ \left[1 - \left(\arcsin(zr^{-1}) + f_{up}\right)f^{-1}\right)\right] h \end{pmatrix} \tag{1}$$

where (u, v) are the image coordinates, (h, w) the height and width of the range image representation, $f = f_{up} + f_{down}$ is the vertical field-of-view of the sensor, and $r = \|p_i\|$ is the range of each point. This outputs a list of (u, v) tuples which contain pairs of image coordinates for each p_i. Using those indexes, for each p_i a range r, a set of x, y, z coordinates, and its remission is extracted and stored in the image which leads to a $[5 \times h \times w]$ tensor. An example projection can be seen in Fig. 3. This backbone change is represented in Fig. 1.

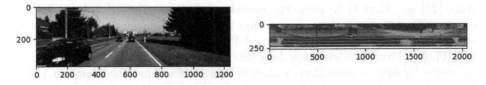

Fig. 3. Spherical projection example. The image on the right has been scaled by a factor of 4 on the y-axis for ease of viewing; the original dimensions are 64×2048.

The Lidar input data to the architecture was fused with the already existing RGB data at a point early in the network's architecture. To perform the fusion,

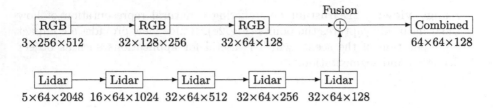

Fig. 4. Diagram of modality-specific branches and their fusion by concatenation.

we experimented with both adding the inputs and concatenating them but later decided to focus on the concatenation operation.

We combined the RGB and Lidar input by concatenating them, as illustrated in Fig. 4. This step was performed in the initial layers before the start of the Feature Pyramid Network [15] since fusing them after that point would lead to a much more convoluted architecture.

3.2 Lidar Segmentation

Using the two-dimensional Lidar representation from the previous sub-section, the neural network was modified to output the segmentation. For this purpose, we adapted one segmentation head from the original YOLOP model in order to have the correct output size ($20 \times 64 \times 2048$). As in RangeNet++, the goal is to predict which of 20 possible classes a Lidar point belongs to, which is reflected in the segmentation output size. This output was then projected back to the point-cloud original space using the nearest neighbor projection algorithm from RangeNet++ [17, Algorithm 1]. This output transformation is represented in the initial diagram in Fig. 1.

3.3 Training Process

The additions mentioned in the previous two sub-sections lead to the model which we will refer to as **YOLOMM** from now on. As mentioned in Sub-Sect. 2.2, each task must be trained separately due to the limitations of the datasets that are publicly available. The training procedure proposed by Hybrid-Nets [21] is: (Step 1) to train the encoder and object detection head for 200 epochs, (Step 2) train the segmentation head for 50 epochs and (Step 3) train the whole network for 50 epochs. We followed the same steps, except that the KITTI object and segmentation data annotations do not overlap (we had bounding boxes for some observations and segmentations for other observations), therefore we had to skip Step 3 (the last step) due to the lack of labels.

4 Experiments and Results

Table 3 shows the object detection comparison between using the vanilla YOLOP as the baseline trained on KITTI against using Lidar and Lidar+RGB. The following metrics have been used, as implemented by the TorchMetrics Python package [8]:

- Mean Average Precision (mAP) is the average Precision for all classes C, $\frac{1}{|C|} \sum_{c \in C} AP_c$, where AP_c is the area under the precision-recall curve for class c. In object detection, for each threshold t, a true positive is whenever an intersection-over-union between a predicted and a true bounding box exists above the threshold, IoU $\geq t$. The recall and precision are thus the number of true positives divided by the number of predicted boxes and ground-truth boxes, respectively.
- Intersection over Union (IoU) between two bounding boxes is the ratio between the intersection and the union of the two bounding boxes.
- mAP@0.5 is when only an IoU threshold t of 0.5 is considered. mAP@0.5:0.95 considers an average of IoU thresholds between 0.5 and 0.95.
- mAP small/medium/large is the mAP computed only for bounding boxes smaller than 32×32 pixels (small), larger than 96×96 pixels (large), or in between (medium). The results show how the size of an object impacted the performance of the models for each modality.

Table 3. Object detection AP comparison according to bounding box sizes

Modality	mAP small (%)	mAP medium (%)	mAP large (%)	mAP@0.5 (%)
RGB	18.98	30.60	45.86	63.05
Lidar	19.34	21.00	37.87	51.95
Lidar+RGB	**24.73**	**37.02**	**55.62**	**68.02**

Overall, RGB outperforms the Lidar-only approach, which is to be expected since the bounding boxes are predicted over the RGB image, so in a Lidar-only approach, the network also has to learn how to map a spherical coordinate onto the camera image space. However, the combined input modalities appear to capture complementary information since, when combined, they outperform the single-modality models.

It should be noted that the KITTI dataset only includes scenes where the atmospheric conditions are near perfect. Perhaps the advantages of Lidar would be highlighted under harsher scenarios such as heavy rain or night-time driving, where the RGB image has much less information.

Table 4 contrasts the approaches on both tasks: object detection and semantic segmentation. Vanilla YOLOP (which features only object detection) is contrasted against YOLOMM with two variations on the fusion operation (see Fusion on Fig. 4). "Add" uses the arithmetic addition between the RGB and Lidar branches, while "Concat" concatenates the two branches. For this purpose, in addition to the Mean Average Precision (mAP), we also report the Precision (P), Recall (R), Accuracy (Acc), and Intersection over Union (IoU).

Table 4. Metric comparison across configurations (%)

	Object Detection (RGB)				Segmentation (Lidar)	
	P	R	mAP@0.5	mAP@0.5:0.95	Acc	IoU
RangeNet++ Paper	–	–	–	–	–	52.2
YOLOP	5.0	**85.9**	66.3	32.6	–	–
YOLOMM Add	35.3	76.6	**67.1**	33.2	**81.3**	**23.7**
YOLOMM Concat	**35.8**	75.7	66.6	32.5	79.9	23.1

The proposed YOLOMM implementation improves results with the exception of recall, albeit not noticeably (perhaps due to the aforementioned observation that KITTI weather is not very adversarial). In addition, the implementation proposed also produces segmentations. It should be noticed that the performance of the YOLOP version trained locally is worse than the one demonstrated in the paper, perhaps due to computing resource limitations.

For comparison, the IoU published by RangeNet++ [17] for Lidar segmentation is also shown in the table. Our implementation is still unable to reach the same IoU, possibly due to their architecture being geared toward LiDAR segmentation. Finally, we can see that for our proposed architecture, the fusion operation does not have a significant impact on performance, so either option is equally viable.

5 Conclusion

In conclusion, we have presented YOLOMM, an addition to YOLOP that integrates multi-modal data. By incorporating Lidar data alongside RGB data, we aimed to enhance the sensory capabilities of autonomous vehicles in complex environments.

Our experimental results demonstrated that YOLOMM improved precision across all objects of different sizes compared to the YOLOP model. Two mid-fusion mechanisms between RGB and Lidar data were considered (addition and concatenation), but neither significantly impacted performance.

Future work could focus on experimenting with different datasets with more diverse atmospheric conditions, such as Radiate [19], or analyzing the impact of the Lidar modality in the uncertainty of the model.

Due to the lack of public datasets that contained both RGB and Lidar for both tasks that we aimed for, we were unable to evaluate more adverse weather conditions and night scenarios. This could be done in the future, if necessary, using synthetic datasets or by introducing these features to existing datasets using generative models.

References

1. Behley, J., et al.: Towards 3D LiDAR-based semantic scene understanding of 3D point cloud sequences: the SemanticKITTI dataset. Int. J. Robot. Res. **40**(8–9), 959–967 (2021). https://doi.org/10.1177/02783649211006735
2. Bolya, D., Zhou, C., Xiao, F., Lee, Y.J.: YOLACT: real-time instance segmentation (2019)
3. Caesar, H., et al.: nuScenes: a multimodal dataset for autonomous driving (2020)
4. Chan, C.Y.: Advancements, prospects, and impacts of automated driving systems. Int. J. Transp. Sci. Technol. **6**(3), 208–216 (2017). https://doi.org/10.1016/j.ijtst.2017.07.008, https://www.sciencedirect.com/science/article/pii/S2046043017300035. safer Road Infrastructure and Operation Management
5. Chen, X., Ma, H., Wan, J., Li, B., Xia, T.: Multi-view 3D object detection network for autonomous driving. In: Proceedings of the IEEE Conference on Computer Vision and Pattern Recognition, pp. 1907–1915 (2017)
6. Cordts, M., et al.: The cityscapes dataset for semantic urban scene understanding (2016)
7. Deschaud, J.E.: KITTI-CARLA: a KITTI-like dataset generated by CARLA Simulator. arXiv e-prints: arXiv:2109.00892 (2021)
8. Detlefsen, N.S., et al.: TorchMetrics - measuring reproducibility in PyTorch. J. Open Sour. Softw. **7**(70), 4101 (2022). https://doi.org/10.21105/joss.04101
9. Geiger, A., Lenz, P., Urtasun, R.: Are we ready for autonomous driving? The KITTI vision benchmark suite. In: Conference on Computer Vision and Pattern Recognition (CVPR) (2012)
10. He, K., Zhang, X., Ren, S., Sun, J.: Deep residual learning for image recognition (2015)
11. Heuer, F., Mantowsky, S., Bukhari, S.S., Schneider, G.: MultiTask-CenterNet (MCN): efficient and diverse multitask learning using an anchor free approach (2021)
12. Lee, D.G., Kim, Y.K.: Joint semantic understanding with a multilevel branch for driving perception. Appl. Sci. **12**(6), 2877 (2022). https://doi.org/10.3390/app12062877
13. Liao, Y., Xie, J., Geiger, A.: KITTI-360: a novel dataset and benchmarks for urban scene understanding in 2D and 3D. Pattern Anal. Mach. Intell. (PAMI) **45**, 3292–310 (2022)
14. Lin, T.Y., Dollár, P., Girshick, R., He, K., Hariharan, B., Belongie, S.: Feature pyramid networks for object detection. In: Proceedings of the IEEE Conference on Computer Vision and Pattern Recognition, pp. 2117–2125 (2017)
15. Lin, T.Y., Dollár, P., Girshick, R., He, K., Hariharan, B., Belongie, S.: Feature pyramid networks for object detection. In: 2017 IEEE Conference on Computer Vision and Pattern Recognition (CVPR), pp. 936–944 (2017). https://doi.org/10.1109/CVPR.2017.106
16. Liu, S., Qi, L., Qin, H., Shi, J., Jia, J.: Path aggregation network for instance segmentation (2018)
17. Milioto, A., Vizzo, I., Behley, J., Stachniss, C.: RangeNet++: fast and accurate LiDAR semantic segmentation. In: IEEE/RSJ International Conference on Intelligent Robots and Systems (IROS) (2019)
18. Paek, D.H., Kong, S.H., Wijaya, K.T.: K-lane: lidar lane dataset and benchmark for urban roads and highways. In: Proceedings of the IEEE/CVF Conference on Computer Vision and Pattern Recognition (CVPR) Workshop on Autonomous Driving (WAD) (2022)

19. Sheeny, M., De Pellegrin, E., Mukherjee, S., Ahrabian, A., Wang, S., Wallace, A.: RADIATE: a radar dataset for automotive perception. arXiv preprint: arXiv:2010.09076 (2020)
20. Sun, P., et al.: Scalability in perception for autonomous driving: Waymo open dataset. In: Proceedings of the IEEE/CVF Conference on Computer Vision and Pattern Recognition (CVPR) (2020)
21. Vu, D., Ngo, B., Phan, H.: HybridNets: end-to-end perception network (2022)
22. Wu, D., et al.: YOLOP: you only look once for panoptic driving perception. Mach. Intell. Res. **19**, 1–13 (2022)
23. Yu, F., et al.: BDD100K: a diverse driving dataset for heterogeneous multitask learning (2020)

Classify NIR Iris Images Under Alcohol/Drugs/Sleepiness Conditions Using a Siamese Network

Juan Tapia[✉] and Christoph Busch

da/sec-Biometrics and Internet Security Research Group, Hochschule Darmstadt,
Darmstadt, Germany
{juan.tapia-farias,christoph.busch}@h-da.de

Abstract. This paper proposes a biometric application for iris capture devices using a Siamese network based on an EfficientNetv2 and a triplet loss function to classify iris NIR images captured under alcohol/drugs/sleepiness conditions. The results show that our model can detect the "Fit/Unfit" alertness condition from iris samples captured after alcohol, drug consumption, and sleepiness conditions robustly with an accuracy of 87.3% and 97.0% for Fit/Unfit, respectively. The sleepiness condition is the most challenging, with an accuracy of 72.4%. The Siamese model uses a smaller number of parameters than the standard Deep learning Network algorithm. This work complements and improves the literature on biometric applications for developing an automatic system to classify "Fitness for Duty" using iris images and prevent accidents due to alcohol/drug consumption and sleepiness.

Keywords: Fitness for Duty · Iris image · alcohol · drugs detection

1 Introduction

Iris recognition systems have been used mainly to recognize and identify cooperative subjects in controlled environments for border control areas, humanitarian areas, buildings, hospitals, and airports using Near-Infra-Red (NIR) capture devices [6]. With the improvements in iris recognition performance and reduction in cost for iris acquisition devices, the technology will witness broader applications and may be confronted with new challenges [6]. One such kind of challenge is identifying if a subject is under alcohol, drug effects, or even under sleep deprivation and sleep restriction conditions. This area is known as "Fitness For Duty" [10,14] and allows us to determine whether the subject is physically able to perform his or her task [14].

Fitness For Duty (FFD) [3,10,14] is a technique used in the context of the occupational test in mining, logistics, health and other markets. These tests are applied to workers to describe a set of tools that help to evaluate a subject's condition considering physical, alertness, and emotional level, which is required for a specific job. Determining if the person is "Fit" or "Unfit" to perform the job is difficult. Stay "Fit" means completing the job's duties in a safe, secure,

© Springer Nature Switzerland AG 2024
V. Vasconcelos et al. (Eds.): CIARP 2023, LNCS 14469, pp. 575–588, 2024.
https://doi.org/10.1007/978-3-031-49018-7_41

productive, and effective manner. In this way, the companies can save resources and keep it a continuous operation.

The iris has been identified as a principal biometric modality compared to fingerprints and faces because removing protection devices used in mining, hospitals, and other industries is unnecessary. Figure 1 shows an example of a worker with safety devices.

In the literature, several approaches have been proposed to estimate FFD based on machine learning and deep learning techniques in the last few years. Previous work has shown the difficulty of estimating Fit and Unfit conditions because, many times, one capture subject presents more than one condition. This means that some subjects are correlated due to alcohol and drugs in labour days, presenting a real condition. It is essential to highlight that today, we have a worldwide dependency rise in alcohol and drug consumption in the workforce, especially among shift workers. Europe is not the exception to this kind of problem[1].

Fig. 1. Example of Iris NIR capture device and worker with safety implements.

Our work proposed a Siamese network based on triplet loss that explores the distance among triplets considering easy examples (one control image with a large margin from alcohol, drugs and sleepiness), semi-hard (average distances among the classes and hard examples (many examples present overlapping or very close distances among the classes.

In summary, the main contributions of this paper are:

– A Siamese network is proposed based on triplet-loss semi-hard distances to classify FFD. Three image inputs are used to take the decision.
– Our proposal can capture multiple correlations among three or four classes based on many subjects under the effect of more than one condition.
– Our results improved the state-of-the-art results in FFD images.

---

[1] https://www.emcdda.europa.eu/publications/data-fact-sheets/european-web-survey-drugs-2021-top-level-findings-eu-21-switzerland_en.

The rest of the manuscript is organised as follows: Sect. 1.1 summarises the related works on FFD. The database description is explained in Sect. 1.4. The metrics are explained in Sect. 1.3. The experiment and results framework is then presented in Sect. 1.5. We conclude the article in Sect. 4.

1.1 Related Work

In the state-of-the-art, several performance tests have been proposed as FFD, including psychomotor tasks, temperature sensor, EEG, finger tapping, pattern comparison, smart band wrist, and in-cab monitoring, among others [12,13,17]. These systems deliver an answer based on the statistical behaviour threshold of "Fit" for the control subject and "Unfit" for alcohol/drug/sleepiness.

In order to address these problems, it is necessary to study which repetitive behaviours or biometric factors manifest through the body to establish the relationship between the cause and effect in a person's behaviour. The Central Nervous System (CNS) controls the iris and pupil movements [1]. In this condition, the subject cannot voluntarily change the pupil's or iris' movement. This action is initiated automatically in response to an external factor, such as light, or alcohol consumption, drug abuse or fatigue. On the other hand, iris recognition allows identifying and distinguishing one worker from another and, thus, guaranteeing that the measurement is from the worker under testing. Therefore, the iris and pupil are highly reliable in measuring the fitness for duty of the subject. Hence, there is a need to develop an automated and reliable FFD model tool based on the iris recognition framework.

It is essential to highlight that FFD does not have any relationship with the manual alcohol test or drug test that measures alcohol/drug percentage in blood. Our proposal is based on the changes in the central nervous system after consuming external factors and their influence on eye behaviour.

Tomeo et al. [22] focused on pupil and iris behaviour induced by the use of alcohol agents. Still, it could also be extended and allied to drug consumption, as it has been shown that drug-induced pupil dilation affects iris recognition performance negatively. However, no solution has been investigated to counteract its effect on iris recognition. Note that an adversary may use alcohol/drugs to mask their identity from an iris recognition system [2]. These agents can be easily obtained online without a medical prescription. Hence, there is a need to understand and counteract the effect of alcohol and drugs on iris recognition.

Also, Tomeo et al. [22] analysed the impact of drugs on pupil dilation and proposed the use of a biomechanical nonlinear iris normalisation scheme along with key point-based feature matching for mitigating the impact of drug-induced pupil dilation on iris recognition. They also investigated the differences between drug-induced and light-induced pupil dilation on iris recognition performance. The authors reported an average ratio p of pupil and iris between $(0.265 < p < 0.515)$.

Tapia et al. [20] proposed a method to detect alcohol consumption from Near-Infra-Red (NIR) periocular eye images based on Capsule-Network. The study focused on determining the effect of external factors such as alcohol on the Central Nervous System (CNS). The aim was to analyse how this impacts iris and pupil movements and if capturing these changes with a standard iris NIR camera is possible. Tapia et al. proposed a novel Fused Capsule Network (F-CapsNet) to classify iris NIR images taken under alcohol consumption subjects. The same author proposed an improved version in a large-scale database based on MobileNetv2.

Causa et al. [4] proposed a behavioural curve model to estimate FFD based on machine learning techniques extracted from 50 handcrafted features. However, this method requires the detection and segmentation of the pupil and iris for each frame and the feature extraction of more than 50 handcrafted measures, determined according to the state-of-the-art. Then, this kind of method demands many resources to be implemented in a mobile device. Segmenting this kind of image is not a trivial task because the method needs to be efficient in the number of parameters implemented in a regular iris capture device. Most of these sensors are self-integrated mobile devices with a limited memory size [21].

Makowski et al. [11] investigated the robustness of oculomotor biometric identification with respect to fatigue and acute alcohol consumption. To this end, they collect the eye-gaze data of users in sleep-deprived and intoxicated states and the baseline state. The Potsdam Binge/Judo dataset of binocular eye movement data (horizontal and vertical gaze coordinates) from 66 subjects aged 18 to 48 years, with a mean age of 24, was used. Eye movements are recorded using a tripod-mounted Eyelink Portable Duo eye tracker at a sampling frequency of 1,000 Hz.

Very recently, Zurita et al. [23] improved one of the previous approaches [4] automating the feature extraction and included the temporal information instead of extracting the 50 handcrafted features as proposed by Causa et al. [4]. The features were extracted by a CNN, which extracts the most relevant spatial characteristics from NIR iris frames, followed by an LSTM that analyses the temporal component of those features. Therefore, a robust system automatically extracts temporal and spatial features from iris sequences. However, the sleepiness condition reached only 53.6%.

1.2   Method

A Siamese network consists of two identical networks which can process different inputs. They are joined at their output layers based on the pre-trained networks by a unique function that calculates a metric between the embedding estimated by each network. Most of these networks have been trained on ImageNet weights with 1,000 classes. In order to extend this traditional approach, we explored using the weights of the 21k version and also reduced the 21k to 1k classes. Imaginet 21k contains about 13M training images with 21,841 classes. This kind of network is ideal for FFD classification because they are primarily designed to find similarities between two inputs.

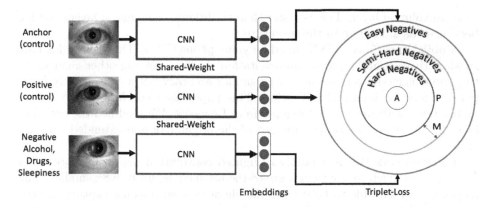

Fig. 2. Siamese Network with Triplet-loss. A: Anchor. P: Positive. M: Margin.

A contrastive loss is typically used to train a siamese network to distinguish between mated (control-Fit) and non-mated (Alcohol, drugs, sleep-Unfit) pairs. Traditionally, the contrastive loss function optimises two identical Convolutional Neural Networks outputs, each operating on a different input image and using a Euclidean distance measure or a Support Vector Machine classifier to make the final decision. At the same time, contrastive representations have achieved state-of-the-art performance on visual recognition tasks and have been theoretically proven effective for binary classification. However, according to the literature [8], the triple-loss function could separate the images more effectively based on the semi-hard-triplet loss function to separate examples that are very close to the control subject (anchor) and Unfit subjects. Then, the contrastive loss is not able to separate these FFD images because we have images from the same subject with several conditions (alcohol, drugs and sleepiness. Conversely, triplet-loss can explore and deal very well with easy, semi-hard and hard examples.

In order to extract the embedding from the iris images, ImageNet (1k and 21k) pre-trained general purpose weight in combination with MobileNetv2 [15], MobileNetv3 [5] and EfficientNetv2 [18] were used. Then, the networks are retrained with an iris FFD database. In the training process, the model is optimised by enforcing the triple loss function to measure the triplets relationships between both classes in a multiclass problem of N classes. Figure 2 shows a representation of the Siamese Network.

Networks. Three state-of-the-art networks have been adapted (modified the loss function) for the task of predicting alcohol, drugs and sleepiness using iris images as a part of the Siamese network.

MobileNetv2 [15] is based on a streamlined architecture to build lightweight deep neural networks for usage in environments with limited resources, such as mobile applications. Depthwise separable convolutions are the basic building blocks of MobileNets, which consist of a depthwise convolution layer and a pointwise convolution layer. This is faster than traditional convolution layers and has been used as a baseline in the literature.

MobileNetv3 [5] is a CNN tuned for smartphone CPUs. This CNN adds hard swish activation and squeeze-and-excitation modules, among other changes, to the previous model version, achieving similar accuracy but a considerably faster performance for image classification. This paper used the MobileNetv3 small architecture with pre-trained weights from Imagenet. We modified the net's last layer to be a two-class output instead of the original 1k and extended it to 21k classes.

EfficientNetv2 [18] is a family of lighted convolution neural networks. The models train much faster than state-of-the-art models, up to 6.8x smaller. This model is very suitable to be used in mobile devices such as iris capture devices.

Triplet Loss. A triplet loss function was used to train a neural network to closely embed features of the same class while maximising the distance between embeddings of different classes. An anchor (control subject) and one negative (potentially under the influence of alcohol, drugs or sleepiness), and one positive sample (control or same class of anchor) are chosen [16].

To formalise this requirement, a loss function is defined over triplets of embeddings:

- An anchor image (a) - For example, control subject.
- A positive image of (p) the same class as the anchor.
- A negative image (n) of a different class - in our example, an alcohol, drugs or sleepiness iris image.

For some distance (d) on the embedding space, the loss of a triplet (a, p, n) is defined as:

$$L_t = \max(d(a, p) - d(a, n) + margin, 0) \qquad (1)$$

We minimise this loss, which pushes $d(a, p)$ to 0 and $d(a, n)$ to be greater than $d(a, p) + margin$. As soon as n becomes an "easy negative", the loss becomes zero. We used a semi-hard triplet, which means that the negative is not closer to the anchor than the positive, but which still has a positive loss:

$$d(a, p) < d(a, n) < d(a, p) + margin \qquad (2)$$

In our case, As a part of the process, a pre-computed template of four random control images is processed in order to compare the embedding distances of the new input iris image, which could potentially be Unfit.

1.3   Metrics

This paper proposed an iris application to predict FFD. Then, we can not consider alcohol, drug and sleepiness as a presentation attack because the goal is to determine if a subject is Fit or Unfit. The subject must be avoided to be detected with unfit conditions.

The False Positive Rate (FPR) and False Negative Rate (FNR) were reported as Error Type I and Error Type II. These metrics effectively measure to what degree the algorithm confuses presentations of Fit and Unfit images with alcohol, Drugs and Sleepiness. The FPR and FNR are dependent on a decision threshold.

A Detection Error Trade-off (DET) curve is also reported for all the experiments. In the DET curve, the Equal Error Rate (EER) value represents the trade-off when the FPR and FNR. Values in this curve are presented as percentages. Additionally, two different operational points are reported. FNR_{10}, which corresponds to the FPR, is fixed at 10% and FNR_{20}, which is when the FPR is fixed at 5%. FNR_{10} and FNR_{20} are independent of decision thresholds.

1.4 Database

In the literature, two different databases have been proposed to classify Fitness for Duty using iris images [19, 20]. Both datasets are available by request. The first one is a novel open-access dataset of 3,000 NIR periocular images from 30 volunteers and was captured with the standard iris-captured device, which contains session control subjects (No-alcohol) and subjects under the influence of alcohol. This database is subject-disjoint, and 3,000 images were used.

A second available database is called the "FFD NIR iris images Stream database" (FFD-NIR-Stream), containing 5-s stream sequences of periocular NIR images organised in four classes with 72,093 images in the testing set: Control, Alcohol, Drugs and Sleepiness. All the image has a size of $640 \times 480 \times 1$. The details of each class and subsets are described as follows:

- Control: healthy subjects that are not under alcohol and/or drug influence and in normal sleeping conditions.
- Alcohol: subjects who have consumed alcohol or are in an inebriation state.
- Drugs: subjects who consumed some drugs (mainly marijuana) or psychotropic drugs (by medical prescription).
- Sleep: subjects with sleep deprivation, resulting in fatigue due to sleep disorders related to occupational factors (shift structures with high turnover).

Table 1. Description of total NIR images recording by condition.

Conditions	Train set	Validation set	Testing set
Control	21,449	3,136	60,222
Alcohol	24,325	3,394	6,998
Drug	8,653	1,253	2,338
Sleep	8,568	1,140	2,535
Total	62,995	8,923	72,093

In order to compare our proposal with the state-of-the-art, we used the FFD-NIR-Stream database organised as a person-disjoint because we did not use a group of frames (5 frames per subject); each image is an unrelated subject. Overall, the same number of images were used. Table 1 describes the total image number. Figure 3 shows an example of iris images for the four conditions.

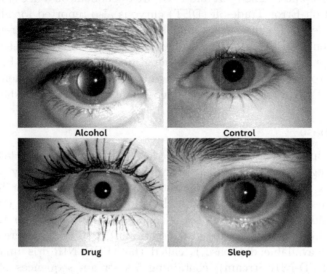

Fig. 3. Eye-image examples of FFD-NIR-Seq database.

1.5   Experiment and Results

Two experiments are conducted based on the Siamese network. A semi-hard triple loss function applied to FFD classification using MobileNetv2, MobileNetv3 and EfficientNetv2 was explored. All the images were resized to $224 \times 224 \times 3$. Aggressive data augmentation was applied based on the imgaug library [7]. Different light changes, hue saturation and blurring operations were applied with a chance of 30%.

The Contrast Limited Adaptive Histogram Equalisation (CLAHE) was also used to highlight the iris textures. All the images were divided into 8×8 sized cells.

Experiment 1 used three classes: control, alcohol, and drugs. It was decided to remove the sleepiness class because it was identified as the most difficult to predict in the literature. This action may help to evaluate its influence on the final predictions. The learning rate was set up based on a grid search from $1e-3$ to $1e-5$. The best results were reached with $1e-5$. An Adam optimiser was used with a batch size of 32.

For Experiment 2, we used four classes: control, alcohol, drugs and sleepiness. The learning rate was set up based on a grid search from $1e-3$ to $1e-5$. The best results were reached with $1e-4$. An Adam optimiser was used with a batch size of 32.

Table 2 shows FNR_{10} and FNR_{10} for three different networks used to extract the embedding on the Siamese network. Overall the EfficientNetv2 with four classes obtained the best results. The three-class models reached higher EER performance (See Fig. 4).

Table 2. Summary results of Siamese four classes network. All the results are in %.

Model	EER	FNR_{10}	FPR_{10}
MobileNetv2 (21k)	20.39	31.64	32.69
MobileNetv3 (21k)	18.21	16.35	16.93
EfficientNetv2 (21k)	**12.74**	**13.04**	**13.54**

Figure 4 shows the DET curve for the three classes model. This model reached a higher EER in comparison with the four classes model. Two black vertical lines show the FNR_{10} and FNR_{20}. The alcohol class is the most difficult to predict, obtaining an EER of 13.94%.

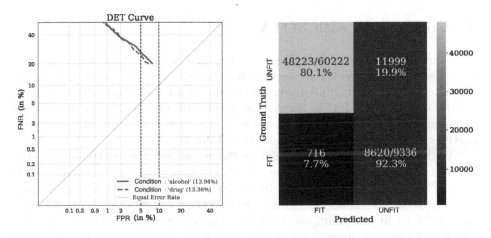

Fig. 4. Experiment 1. Results using three classes. Left DET curves. Right. The confusion Matrix was evaluated in the test set. The Models were trained using control subjects versus alcohol and drug consumption. The EER for the curve is shown in parentheses. The black dashed lines indicate two operational points for FNR_{10} and FNR_{20}. Confusion matrix grouping the alcohol, drugs and sleepiness in Unfit class.

Figure 5 shows the DET curve for the four classes model. Two black vertical lines show the FNR_{10} and FNR_{20}. The sleepiness class is the most difficult to predict, obtaining an EER of 12.74%. The drug and alcohol reached an EER of 6.06% and 7.10%, respectively.

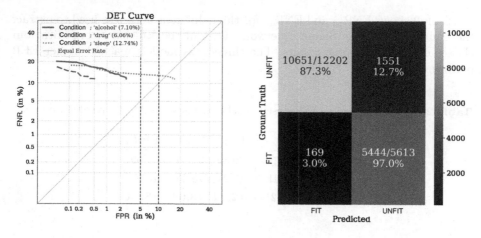

Fig. 5. Experiment 2 results using four classes. The DET and the confusion Matrix were evaluated in the test set. The Models were trained using control subjects versus alcohol, drug consumption and sleepiness. The EER for the curve is shown in parentheses. The black dashed lines indicate two operational points for FNR_{10} and FNR_{20}. Confusion matrix grouping the alcohol, drugs and sleepiness in Unfit class.

2   Comparison with SOTA

We compare our results with those obtained by Causa et al. [4] using machine learning techniques and three classifiers Random Forest (FR), Gradient Boosting Machine (GBM), and Multi-Layer-Perceptron (MLP) and also with Zurita et al. [23] using a CNN-LSTM inspired in a VGG-16 module as both works used iris sequences and the same database with the same train, test and validation partitions. All the results are compared in Table 3.

As shown in Table 3, when analysing the overall accuracy of the models, the CNN-LSTM model obtains a value of 83.6%, which is significantly higher than the methods compared (between 70.8% and 75.5%). Additionally, sensitivity, which provides the model's performance to detect the states separately, is 88.0% and 91.7% for fit and unfit, respectively. This implies that the VGG16-inspired module can extract more relevant features than manual extraction [4]. Additionally, the LSTM uses information from only eight frames, whereas Causa's method uses more than 75 frames to make an inference. Thus, the LSTM is more optimal for determining the dependencies in time, which provides richer information. Overall the Zurita et al. proposal reached The EER achieved 18.59%. The FNR_{10} reached 22.5%, and FNR_{20} reached 25.3%.

Also, as shown in Table 3, we can see that our proposed method based on the Siamese network and triplet-loss outperforms the state-of-the-art by far. This increment is due to the improvement of the sleepiness class. We reduced the EER for the sleepiness class from 18.59% to 12.74%, which can be determined using a semi-hard loss. It is essential to highlight that our method used only one image in a disjoint-subject distribution to predict FFD instead of a sequence. Then, this proposal is more efficient than previous approaches. Our method reached an EER of 7.10% for alcohol, 6.06 drugs% and 12.74% for sleepiness. In summary, two operational points are proposed to obtain an FNR_{10} of 13.04% and FNR_{20} of 13.54%.

Table 3. Comparison with state-of-art. All the models were trained using four classes. Acc: represents Accuracy.

Method	Cond	Sensitivity (%)	Specificity (%)	F1-Score (%)	Acc. (%)
RF [4]	Fit	70.1	94.7	80.5	70.8
	Unfit	75.2	28.5	41.3	
GMB [4]	Fit	73.1	95.8	82.9	73.1
	Unfit	79.8	40.0	45.7	
MLP [4]	Fit	75.3	95.4	84.2	75.3
	Unfit	77.1	33.1	46.3	
CNN-LSTM [23]	Fit	81.4	99.3	89.5	83.6
	Unfit	96.9	46.9	63.3	
Siamese (Ours)	Fit	87.3	97.0	89.5	**87.3**
	Unfit	96.9	90.9	85.3	

3   Visualisation

A t-SNE map projection [9] was used to visualise the projection of the data to a 2D plot. This method shows non-linear connections in the data. The t-SNE algorithm calculates a similarity measure between pairs of instances in high-dimensional and low-dimensional spaces. It then tries to optimise these two similarity measures using a cost function.

Figure 6 shows a projection plot for all the classes. The top image shows the projection of the train and validation set for the four conditions: control (blue), alcohol (green), drugs (red) and sleepiness (light blue). As we can see, the training set is represented very well by the four classes and shows the dataset is properly distributed.

The middle image shows a projection of the test set results with only three classes: control, alcohol and drugs. As we can see, there is an important overlap between the classes; even though the alcohol (green) and drug (red) are projected in separate spaces, the control images are along all projections. Then, the three-class model can not be generalised very well.

The bottom image shows the projection of the test set for four classes (control, alcohol, drugs and sleepiness). As we can see, the sleepiness class helped to improve the results. This allows us to separated better the alcohol (green) and drugs (red) from the control images (blue). A small overlap can be observed between sleep and control subjects.

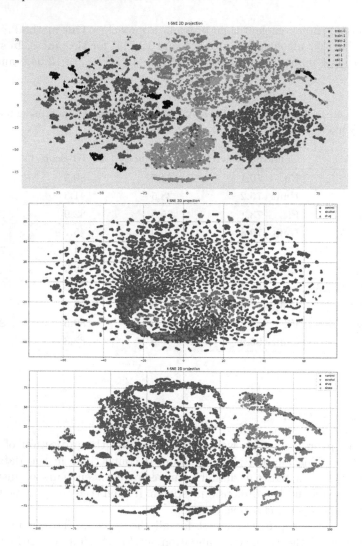

Fig. 6. Top: Data projection of validation set over a training set. Blue represents "Control" (train0/val0), and Green: Alcohol (train1/val1). Red: Drugs (train2/val2) and Light-blue: Sleepiness (train3/val3). Middle: MobileNetV2 test-set and Bottom: test set for the best results with EfficientNetV2. (Color figure online)

4   Conclusion

This paper proposed a Siamese network based on triplet loss and semi-hard-loss approach. The Semi Lard Loss can deal very well with the four classes model to improve state of the art. This is relevant because our problem presents different kinds of easy, semi-hard and hard examples according to the triplet loss scheme. Our proposal can disentangle the overlapped classes well and increase the sleepiness condition's accuracy identified in previous work as harder to detect. The results show that this application may help predict Fitness For Duty using an iris-contactless device with high accuracy. This opens a new insight and application for traditional iris recognition and other biometric modalities.

Acknowledgment. This work is supported by the German Federal Ministry of Education and Research and the Hessen State Ministry for Higher Education, Research and the Arts within their joint support of the National Research Center for Applied Cybersecurity ATHENE.

References

1. Adler, F.H.: Physiology of the eye, vol. 48, 11 ed. Francis Heed Adler, The C. V. Mosby Company, July 1985
2. Arora, S.S., Vatsa, M., Singh, R., Jain, A.: Iris recognition under alcohol influence: a preliminary study. In: 5th IAPR International Conference on Biometrics (ICB), pp. 336–341, March 2012
3. Benderoth, S., Hormann, H.J., Schiebl, C., Elmenhorst, E.M.: Reliability and validity of a 3-min psychomotor vigilance task in assessing sensitivity to sleep loss and alcohol: fitness for duty in aviation and transportation. Sleep **44**(11) (2021)
4. Causa, L., Tapia, J.E., Lopez-Droguett, E., Valenzuela, A., Benalcazar, D., Busch, C.: Behavioural curves analysis using near-infrared-iris image sequences (2022)
5. Howard, A., et al.: Searching for mobilenetv3. In: 2019 IEEE/CVF International Conference on Computer Vision (ICCV), pp. 1314–1324 (2019). https://doi.org/10.1109/ICCV.2019.00140
6. Jain, A.K., Deb, D., Engelsma, J.J.: Biometrics: trust, but verify. IEEE Trans. Biom. Behav. Identity Sci. 1 (2021)
7. Jung, A.B., et al.: Imgaug (2020). https://github.com/aleju/imgaug. Accessed 01 Feb 2020
8. Köhler, M., Eisenbach, M., Gross, H.M.: Few-shot object detection: a comprehensive survey. IEEE Trans. Neural Netw. Learn. Syst. 1–21 (2023). https://doi.org/10.1109/TNNLS.2023.3265051
9. van der Maaten, L., Hinton, G.: Visualizing data using t-SNE. J. Mach. Learn. Res. **9**(86), 2579–2605 (2008)
10. MacQuarrie, A., et al.: Fit for duty: the health status of new south wales paramedics. Ir. J. Paramed. **3** (2018)
11. Makowski, S., Prasse, P., Jäger, L.A., Scheffer, T.: Oculomotoric biometric identification under the influence of alcohol and fatigue. In: 2022 IEEE International Joint Conference on Biometrics (IJCB), pp. 1–9 (2022). https://doi.org/10.1109/IJCB54206.2022.10007970
12. Mardonova, M., Choi, Y.: Review of wearable device technology and its applications to the mining industry. Energies **11**(3) (2018)
13. Miller, J.C.: Fit for duty? Ergon. Des. **4**(2), 11–17 (1996)
14. Murphy, S., Fleming, T.: Fitness for duty in the nuclear power industry: the effects of local characteristics. In: Fifth Conference on Human Factors and Power Plants, pp. 127–132 (1992)
15. Sandler, M., Howard, A., Zhu, M., Zhmoginov, A., Chen, L.C.: MobileNetV 2: inverted residuals and linear bottlenecks. In: Proceedings of the IEEE Conference on Computer Vision and Pattern Recognition (CVPR), June 2018

16. Schroff, F., Kalenichenko, D., Philbin, J.: Facenet: a unified embedding for face recognition and clustering. In: CVPR, pp. 815–823. IEEE Computer Society (2015)
17. Serra, C., Rodriguez, M.C., Delclos, G.L., Plana, M., López, L.I.G., Benavides, F.G.: Criteria and methods used for the assessment of fitness for work: a systematic review. Occup. Environ. Med. **64**(5), 304–312 (2007)
18. Tan, M., Le, Q.: Efficientnetv2: smaller models and faster training. In: Meila, M., Zhang, T. (eds.) Proceedings of the 38th International Conference on Machine Learning. Proceedings of Machine Learning Research, vol. 139, pp. 10096–10106. PMLR, 18–24 July 2021
19. Tapia, J., Benalcazar, D., Valenzuela, A., Causa, L., Droguett, E.L., Busch, C.: Learning to predict fitness for duty using near infrared periocular iris images (2022)
20. Tapia, J., Droguett, E.L., Busch, C.: Alcohol consumption detection from periocular NIR images using capsule network. In: 2022 26th International Conference on Pattern Recognition (ICPR), pp. 959–966 (2022). https://doi.org/10.1109/ICPR56361.2022.9956573
21. Tapia, J.E., Droguett, E.L., Valenzuela, A., Benalcazar, D.P., Causa, L., Busch, C.: Semantic segmentation of periocular near-infra-red eye images under alcohol effects. IEEE Access **9**, 109732–109744 (2021)
22. Tomeo-Reyes, I., Ross, A., Chandran, V.: Investigating the impact of drug induced pupil dilation on automated iris recognition. In: IEEE 8th International Conference on Biometrics Theory, Applications and Systems (BTAS), pp. 1–8, September 2016
23. Zurita, P.C., Benalcazar, D.P., Tapia, J.E.: Fitness-for-duty classification using temporal sequences of iris periocular images (2023)

Bipartite Graph Coarsening for Text Classification Using Graph Neural Networks

Nícolas Roque dos Santos[1]([✉]), Diego Minatel[1], Alan Demétrius Baria Valejo[2], and Alneu de A. Lopes[1]

[1] Institute of Mathematics and Computer Science, University of São Paulo, São Carlos, Brazil
{nrsantos,dminatel}@usp.br, alneu@icmc.usp.br
[2] Department of Computing, Federal University of São Carlos, São Carlos, Brazil
alanvalejo@ufscar.br

Abstract. Text classification is a fundamental task in Text Mining (TM) with applications ranging from spam detection to sentiment analysis. One of the current approaches to this task is Graph Neural Network (GNN), primarily used to deal with complex and unstructured data. However, the scalability of GNNs is a significant challenge when dealing with large-scale graphs. Multilevel optimization is prominent among the methods proposed to tackle the issues that arise in such a scenario. This approach uses a hierarchical coarsening technique to reduce a graph, then applies a target algorithm to the coarsest graph and projects the output back to the original graph. Here, we propose a novel approach for text classification using GNN. We build a bipartite graph from the input corpus and then apply the coarsening technique of the multilevel optimization to generate ten contracted graphs to analyze the GNN's performance, training time, and memory consumption as the graph is gradually reduced. Although we conducted experiments on text classification, we emphasize that the proposed method is not bound to a specific task and, thus, can be generalized to different problems modeled as bipartite graphs. Experiments on datasets from various domains and sizes show that our approach reduces memory consumption and training time without significantly losing performance.

Keywords: Coarsening · Multilevel Optimization · Graph Neural Network · Text Mining

1 Introduction

Text classification is a usual task for many applications, such as spam detection, news filtering, document retrieval, opinion mining, and sentiment analysis [13]. Most approaches for text classification use bag-of-words textual representations, which compute a sparse vector with the count of tokens present in a document,

© Springer Nature Switzerland AG 2024
V. Vasconcelos et al. (Eds.): CIARP 2023, LNCS 14469, pp. 589–604, 2024.
https://doi.org/10.1007/978-3-031-49018-7_42

or n-grams, which are sequences of n words in a sentence [28]. However, these techniques must be improved in different situations, such as when processing large corpus, when semantic aspects are essential, or when word order matters.

Another way to perform text classification is to apply classification algorithms on word representations via embedding techniques, such as the Global Vectors (GloVe) [20]. These methods are pre-trained on thousands of documents and compute vectors to represent words in a multi-dimensional space using machine learning and statistics concepts. Nonetheless, they also have some limitations; for example, they fail to distinguish homonyms [22].

Due to the rapid development of Deep Learning techniques in recent years, new text classification approaches based on Convolutional Neural Networks (CNNs) and Recurrent Neural Networks (RNNs) have also been explored to extract text representations [10,16]. These architectures comprise multiple feature-extracting layers that learn initial representations from the data in the first layer and refine them in the subsequent ones.

Recently, Graph Neural Networks (GNNs), a class of neural networks proposed for graph representation learning, have attracted increasing attention for different graph-based tasks. They have been used in many areas, such as drug discovery, recommendation systems, traffic forecasting, fake news detection, and hardware design [29]. In addition to these applications, various GNN approaches have been proposed for text classification. These architectures model a corpus as a graph, where each node represents a word or a document, and treat this task as a node classification problem. The contributions of these approaches come in diverse forms, including different graph types, learning settings (transductive or inductive), types of learning supervision, and mechanisms to enhance the graph, such as topic enrichment. Some examples encompass heterogeneous graphs [30], homogeneous graphs [33], hypergraphs [5], and multigraphs [17].

Although GNNs have been successful in various domains, they also face some drawbacks that hinder their performance and applicability. One of the primary limitations is their usage in large-scale graphs since training GNNs using gradient descent in environments that require scalability is challenging. Moreover, fitting graphs with thousands of nodes and edges in GPU memory poses a practical issue. Additionally, the computational complexity associated with aggregating neighborhood information for each node is troublesome when dealing with many neighbors. Addressing these challenges has been the subject of extensive research in recent years. Notable approaches include the development of a differentiable graph pooling module [31], utilizing clustering algorithms to obtain subgraphs that are then treated as mini-batches during the GNN's training [3], employing subsampling techniques [32], and exploring graph coarsening methods [2,25].

One way to deal with graphs whose size is computationally prohibitive is through the multilevel optimization approach, which consists of three stages: Firstly, it generates multiple levels of contracted graphs, where each level is a reduced version of the previous one. In this phase, a coarsening technique combines nodes into super-nodes based on a similarity measure and a pre-defined criterion, thus creating a smaller graph. Then, an algorithm of interest is applied to the smallest graph in the last level to obtain a solution. Finally, the result is successively projected back onto the original graph. As a consequence, the cost of the algorithm is lowered, constituting an effective strategy in situations where the original graph's size is a computational challenge.

To the extent of our knowledge, the literature in the representation learning area does not present approaches for text classification via GNNs that use the coarsening step from the multilevel optimization to generate more compact graphs. Therefore, we propose a method to address this gap where we model a corpus as a bipartite graph where its partitions contain, respectively, document and word nodes, and edges are created based on word occurrences within each document. Then, we use the multilevel optimization's coarsening step to reduce the word partition of the input graph based on node similarity. Here, we propose the utilization of the cosine similarity between the representations of the word nodes as the measure employed to contract the graph. Furthermore, we reduce each graph into ten sizes to analyze the results at various contraction levels. Finally, we classify text using the GNN GraphSAGE on each reduced graph. Additionally, we analyze our method's performance, memory consumption, and training time to understand the impact caused by the utilization of graph coarsening on the GNN in text classification.

Thus, our main contributions are as follows: (1) the usage of the multilevel optimization's coarsening step to contract a bipartite graph to perform text classification using GNNs, (2) the utilization of the cosine similarity of the word representations in the coarsening technique, (3) the identification of GraphSAGE's performance, memory consumption, and training time across different coarsened versions of the same graph on the same task, and (4) the determination of the impact of the coarsening we performed on each dataset for text classification.

2 Related Work

Graph Neural Network is a type of neural network designed to operate on graph-structured data that has received increasing attention within the graph representation learning domain. Many proposed GNNs compute representations for nodes, edges, or entire graphs through a mechanism known as *Message Passing* [7]. The main idea of this approach involves exchanging information between neighboring nodes through an aggregation process. Concretely, a message passing GNN computes node representations by stacking multiple layers that perform the following operation:

$$\mathbf{h}_u^{l+1} = \sigma\left(\mathbf{h}_u^l, \bigoplus_{v \in N(u)} \phi(\mathbf{h}_u^l, \mathbf{h}_v^l)\right), \tag{1}$$

where l denotes the layer, u and v are nodes, $\mathbf{h}$ is a node's representation, $N(\cdot)$ is a node's 1-hop neighborhood, $\bigoplus$ is a permutation invariant aggregation function (e.g., $\sum$, max, or average), and σ and ϕ are neural networks, such as a linear layer $\sigma(\mathbf{x}) = ReLU(\mathbf{Wx} + \mathbf{b})$.

One example of the earliest GNN is ConvGNN [1], which adapted the convolution operation from images to graphs using concepts from the graph spectral theory. Since its inception, several modifications have been proposed to it. One well-known example is the Graph Convolutional Network (GCN) [12], which can be interpreted as a mean-pooling aggregation of a node's neighborhood information. Another GNN that was proposed around the same time is GraphSAGE [8]. It uses mean-pooling, max-pooling, or a Long Short-Term Memory (LSTM) as the aggregation function, and it also introduced a neighborhood sampling strategy that is beneficial for learning on large graphs. Later on, [27] introduced the Graph Attention Network (GAT), an architecture that learns weights for each relation in the graph via the attention mechanism.

Among its applications, GNNs have been successfully used in text classification. For example, TextGCN [30] modeled a corpus as a heterogeneous graph and applied a two-layer GCN to perform text classification. Similar work to TextGCN was proposed in [15]. However, the initial node representations are computed using the Bidirectional Encoder Representations from Transformers (BERT) instead of GloVe. In addition, the proposed approach optimizes BERT and GCN together instead of separately.

Rather than building a single graph for the entire corpus, [9] introduced a method that models each document as a graph, processes it with a GNN, and then sums the learned node representations to obtain the document's representation. Similarly, TextING [33] also models one graph per document; however, they used an architecture with gate mechanisms to control information flow during the learning process. According to Ding et al. [5], prior work only considers interactions between pairs of words, which can be problematic when idioms are present in the text. Thus, they proposed the Hypergraph Attention Networks (HyperGAT), a GNN that captures higher-order interactions using hypergraphs.

In [6], a trainable pooling mechanism was introduced to reduce a homogeneous graph after each GNN layer using projection vectors that measures a node's importance. This approach was applied to text classification using one graph per document, where nodes are linked if the distance between them is smaller than a sliding window. A four-layer GCN is employed to compute text representations. Domain-adversarial Graph Neural Networks (DADGNN) [18] is an approach for cross-domain text classification that uses a GCN and a differentiable pooling module to compute document representations.

TensorGCN [17] is a method that performs text classification using multi-graphs. Specifically, three different graphs that capture semantic, syntactic, and sequential information are built. Next, an intra-graph and an inter-graph propagation based on GCN compute node representations, which are then passed through a mean pooling layer to obtain a final representation for each node.

Our work aligns with the papers discussed in this section in that we model a corpus as a graph and apply a GNN to conduct text classification. However, unlike previous works that predominantly utilize homogeneous or heterogeneous graphs, we build a bipartite graph for each corpus. Furthermore, we propose the usage of the multilevel optimization's coarsening step to contract the input into multiple graphs, where each reduced graph is used to obtain the next one. In addition, we also propose the utilization of the cosine similarity of the word representations in order to generate the different smaller graphs. Lastly, we present an extensive analysis regarding the GNN's performance, memory consumption, and training time in text classification as the graph is gradually contracted.

3   Proposed Method

In this section, we introduce our method[1] for text classification. First, we depict how the bipartite graph is built from a corpus. Then, we present how the graph coarsening is performed. Finally, we describe the model we use to predict document classes based on the learned representations. Figure 1 presents an overview of our proposed approach.

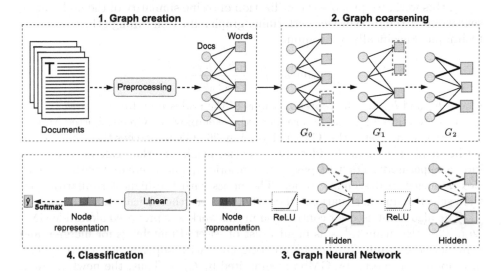

Fig. 1. Overview of our proposed approach. Initially, the corpus is preprocessed as described in Sect. 4.1 and modeled as a bipartite graph, where nodes represent documents and words. Then, a coarsening technique is applied to the word partition of the bipartite graph to hierarchically reduce it into n smaller graphs, where each one is a contracted version of the previous one. In sequence, each coarsened graph is fed to a 2-layer GraphSAGE to obtain a representation for each document node. The target node in this example (colored in purple) receives information from its neighbors via the gray dashed edges. Finally, each document node's label $\hat{y}$ is obtained by passing the node representations through a linear layer followed by a softmax. (Color figure online)

---

[1] Code available at https://github.com/nicolasrsantos/bipartite-coarse-gnn.

3.1   Graph Creation and Coarsening Step

Given a corpus $C = (d_1, d_2, \ldots, d_n)$ with n documents, we build a bipartite graph $G = (U, V, E)$ where $U = (d_1, d_2, \ldots, d_n)$ is the partition of document nodes, $V = (w_1, w_2, \ldots, w_m)$ is the partition of word nodes, and $E = (e_{1,1}, e_{1,2}, \ldots, e_{n,m})$ is the set of edges between a document and a word. Specifically, we create edges between nodes in U and V based on document word occurrence.

After constructing the input bipartite graph G_0, we employ the coarsening step from the multilevel optimization approach to hierarchically reduce G_0 into ten smaller graphs $(G_1, G_2, \ldots, G_{10})$, where each graph is a contracted version of the previous one. Initially, this technique computes a similarity score for each pair of nodes. A matching algorithm then utilizes these scores to determine which nodes should be merged into a super-node based on a criterion. Finally, a contraction algorithm creates super-nodes from the matches obtained in the previous step. The second stage in Fig. 1 illustrates how this technique works.

In this work, we propose the utilization of cosine similarity of the node representations as the measure for performing graph reduction through the coarsening technique. Specifically, we compute:

$$\text{sim}(u, v) = \frac{h_u \cdot h_v}{\|h_u\| \, \|h_v\|}, \tag{2}$$

where h_u and h_v are the representations of the nodes u and v.

To determine which nodes are merged, we employ the Greedy Sorted Matching for Bipartite Networks (GMb) algorithm [26]. GMb searches the 2-hop neighborhood of each node to identify potential neighbors that can be combined into a super-node. For this purpose, a priority queue is created with similarities arranged in descending order. The nodes with the highest similarities are then selected, and in situations where there is a tie, the choice is made randomly.

The algorithm used for contraction in this work was proposed alongside GMb [26]. This algorithm takes as input a list of selected matchings for merging and contracts the graph G_l, thereby creating G_{l+1} with super-nodes. Furthermore, any nodes not chosen by GMb are inherited by G_{l+1}. Thus, the newly created graph consists of super-nodes and unmerged nodes.

After applying the contraction algorithm to reduce the graph, we reconnect each edge that had a merged node as one of its ends to the super-node containing that particular node. Additionally, the representation of each super-node is obtained by taking the element-wise average of the representations of its constituent nodes. The two algorithms used in this stage are available in the Multilevel framework for bipartite networks (MFBN)[2].

To illustrate the method, consider a bipartite graph G_0 consisting of 1500 documents and 1000 words that will be coarsened with a reduction factor of 25%, that is, the desired decrease in size. If we only process G_0's word partition, the first reduced graph (*i.e.*, G_1) ends up with 750 words. Then, G_1 is used as input in the next iteration to produce G_2 with approximately 562 words. This procedure

---

[2] https://github.com/alanvalejo/mfbn.

runs until a stopping criterion is reached, such as the maximum number of coarsened graphs, and the number of documents remains unchanged throughout the entire process. It is worth noting that achieving the exact 25% reduction may not always be possible in smaller versions of the graph due to a potential decrease in node similarity during the coarsening process. Consequently, the matching algorithm may not select enough nodes to reach the 25% mark precisely. Hence, the reduction factor should be considered as a maximum reduction value.

3.2 Graph Representation Learning

In the last stage of our proposed method, a two-layer GraphSAGE[3] is employed to learn node representations. Then, we feed them to a linear layer followed by a softmax function to obtain document labels. It is noteworthy that we perform this step separately for the input and for the coarsened graphs. In this work, we use GraphSAGE with the mean aggregation function in each hidden layer, followed by the ReLU activation function. Formally, this is described as:

$$\mathbf{h}_u^l \leftarrow \sigma\Big(\mathbf{W}_1 \cdot \mathbf{h}_u + \mathbf{W}_2 \cdot \text{mean}_{v \in N(u)} \mathbf{h}_v\Big), \qquad (3)$$

where σ is ReLU, $\mathbf{h}$ denotes a node's representation, $N(\cdot)$ represents a node's neighborhood, and $\mathbf{W}_1$ and $\mathbf{W}_2$ are GraphSAGE's learnable weight matrices. It is worth emphasizing that for a bipartite graph, each hidden layer exclusively enables the flow of information between nodes of different types. This happens because GraphSAGE only performs message-passing between each target node and its 1-hop neighbors. Thus, since we use two hidden layers, information is indirectly exchanged between nodes from the same type only at the second layer because they are two hops away from each other.

In the training step, we minimize the following cross-entropy loss function:

$$-\sum_u y_u \log(\hat{y}_u), \qquad (4)$$

where y_u and $\hat{y}_u$ are, respectively, the ground truth and the network's output for node u.

Since an initial representation for each node is necessary (*i.e.*, $\mathbf{h}_u^0$), we use GloVe's 300-dimensional pre-trained embeddings as word representation [21]. For documents and out-of-vocabulary words, we initialize their $\mathbf{h}_u^0$ using randomly sampled 300-dimensional vectors drawn from a uniform distribution between –0.01 and 0.01.

4   Experiments

This section presents details of the experiments we conducted. First, we introduce the datasets we employed and the number of word nodes per coarsened

---

[3] We use GraphSAGE's implementation provided by the PyTorch Geometric library.

graph. Next, we show the hyperparameters we used in our method. Lastly, we present our experimental results where we evaluate (1) GraphSAGE's performance on text classification with ten coarsened versions of the same bipartite graph, (2) the memory consumption benefits of the coarsening process for Graph-SAGE in our context, and (3) what is the impact caused by the graph reduction on the model's training time.

4.1 Datasets

We conducted the experiments on eight datasets that vary in domain and size. Each dataset is described below, and a summary of their characteristics can be found in Table 1. Additionally, the exact number of word nodes in each contracted graph is presented in Table 2. In our experiments, we follow the preprocessing step performed in TextGCN [30]. Specifically, we clean and tokenize the raw documents. Then, we remove stop words and discard words that appear less than 5 times for all datasets, with the exception of MR.

Table 1. Summary statistics of datasets we used in the experiments.

Dataset	#Docs	#Train	#Test	#Words	#Classes	Avg. Length	#Edges
R8	7674	5485	2189	7688	8	65.72	323670
R52	9100	6532	2568	8892	52	69.82	407084
ohsumed	7400	3357	4043	14157	23	135.82	588958
MR	10662	7108	3554	18764	2	20.39	196826
TREC	5952	5452	500	8783	6	11.05	63253
WebKB	4199	2803	1396	7772	7	133.37	324285
SST-1	11855	9645	2210	17836	5	18.38	199179
SST-2	9613	7792	1821	16188	2	18.52	162660

Dataset Description: R8 and **R52** are two smaller sets derived from the Reuters-21578[4] dataset. The R8 subset comprises 7,674 documents split into eight distinct categories, while the R52 comprises 9,100 documents split into 52 categories. **Ohsumed**[5] is a dataset composed of medical abstracts extracted from the MEDLINE database, which the National Library of Medicine maintains. In this work, we utilize the dataset released by [30]. Specifically, their version comprises 7,400 documents related to cardiovascular diseases split into 23 categories. The **MR** dataset contains 10,662 single-sentence movie reviews equally divided into positive and negative reviews [19]. In this work, we adopt the train/test split provided by [24].

TREC is a dataset composed of 5,952 questions that are categorized into six distinct types [14]. **WebKB** is a collection of 4,199 web pages extracted

---

[4] http://www.daviddlewis.com/resources/testcollections/reuters21578/.
[5] http://disi.unitn.it/moschitti/corpora.htm.

from various computer science departments across different universities [4]. The dataset is divided into seven categories, with an imbalance in their distribution. The **SST-1** and **SST-2** are sentiment analysis datasets with, respectively, 11,855 and 9,613 documents [23]. SST-1 is categorized into five distinct sentiments ranging from very negative to very positive. SST-2 is similar to SST-1, except it excludes all documents labeled as "neutral". Moreover, the remaining data is labeled as either positive or negative.

Table 2. Number of word nodes in the input graph and in its coarsened versions.

Coarse	R8	R52	ohsumed	MR	TREC	WebKB	SST-1	sSST-2
Input	7688	8892	14157	18764	8783	7772	17836	16188
1	5766	6669	10168	14073	6588	5829	13377	12141
2	4325	5002	7964	10555	4941	4372	10033	9106
3	3244	3752	5973	7917	3706	3279	7525	6830
4	2433	2814	4480	5938	2957	2460	5644	5123
5	1825	2111	3360	4454	2601	1845	4263	3843
6	1369	1584	2520	3694	2419	1384	3660	3239
7	1027	1188	1890	3330	2304	1038	3351	2931
8	771	891	1418	3129	2238	779	3177	2756
9	617	669	1064	3010	2195	585	3070	2644
10	543	535	865	2937	2167	439	3000	2576

4.2 Hyperparameter Settings

In our experiments, we randomly select 10% of the training set as the validation set for each dataset. Furthermore, the dimension of both GraphSAGE layers is set to 256, and we train it for 200 epochs using the Adam optimizer [11]. We use a batch size of 64 and set the learning rate as 0.0001 for the MR dataset and 0.001 for the others. If the validation loss does not decrease for 30 consecutive epochs, we terminate the training process. Moreover, we use the best model found during the training step to perform the evaluation on the test set.

In the coarsening step, we generate ten versions of the input graph where we only reduce the number of nodes in the word partition by 25% from the previous version of the coarsened graph. We chose this reduction factor to analyze a fine-grained compression of the graph. Although it is possible to reduce both partitions, we deliberately avoid coarsening the document partition to prevent the mixture of different labeled documents in this work. Consequently, it retains the same number of document nodes as the input graph.

4.3   Experimental Results

Here we present the experimental results of our proposed method. First, we show the performance loss in terms of accuracy and F1-score as the graph is gradually reduced. Next, we present GraphSAGE's memory usage and examine it together with the F1-scores. Lastly, we analyze how the training time reduces from the input graph to the coarsest one.

Performance Analysis: Tables 3 and 4 present, respectively, the average accuracy and F1-score of the classification performed using the test set graph and its ten coarsened versions. As expected, the performance of GraphSAGE gradually declined across all datasets as the graph was coarsened. However, as will be discussed in Sect. 4.3, this decline in performance does not necessarily correspond to GraphSAGE's memory consumption and training time loss.

In our experiments, we observed that the datasets most affected by the graph contraction were those related to news or the medical domain, namely R52 and ohsumed. Additionally, ohsumed, which is known for its challenging nature in text classification, initially exhibited a low performance with a mere 38.37% F1-score, which further dropped to 7.77% after the contractions were performed. This indicates that coarsening does not yield any benefits for this particular dataset.

We also noticed that GraphSAGE's performance declined to a lesser extent for sentiment-related datasets, specifically MR, SST-1, and SST-2. Moreover, the contraction applied to their respective graphs did not reach a 25% decrease from one level to another, starting from the fifth level. Analyzing the overall performance decay of these datasets, we observed that they experienced the least impact on their F1-scores. In addition to these three datasets, TREC exhibited similar behavior and experienced the smallest decrease in size from the input graph to the coarsest one. Particularly, it underwent a reduction of 75.3%.

Memory Consumption: As mentioned earlier, we also analyzed the memory consumption of GraphSAGE for each graph used in the experiments. Surprisingly, as shown in Table 5, not all datasets exhibit a decrease in memory usage from one coarsening level to another. However, it is noteworthy that the consumption decreased between the input graph and the coarsest one.

Analyzing memory consumption in conjunction with the F1-score, it is possible to observe interesting patterns. In certain datasets, we noticed a significant decrease in memory usage, albeit with a slight reduction in performance. For example, Figs. 2a and 2b present, respectively, GraphSAGE's memory consumption and training time curves for the WebKB dataset. Additionally, the F1-score curve is also displayed on both figures to aid our analysis. It is possible to observe that GraphSAGE exhibited a notable 36% reduction in memory usage between the input graph and the third coarsening level, while its F1-score decreased from 77.42% to 76.42%. Moreover, the training time increased by only 5 s between them, as illustrated in Fig. 2b. It is worth highlighting that the applied coarsening technique reduced the number of words by 57.8% for the third coarsest graph, as presented in Table 2.

Another dataset that exhibited a similar behavior was the R8, given that it went through approximately a 36.7% reduction in memory usage between the input graph and the fifth coarsening level. In terms of performance loss, its F1-score decayed from 89.28% to 84.85% between both levels, and the model's training time increased by just three seconds. On the SST-2 dataset, GraphSAGE's memory consumption was lowered by roughly 32.8% between the input and the fifth coarsened graph, while its training time remained relatively stable since it was reduced by one second. The model's performance loss was higher when compared to the results obtained for the WebKB and R8 datasets since the F1-score went from 80.84% to 74.13%. However, it is noteworthy that the coarsening technique shrank SST-2's word partition by 76.26% between both graphs.

Table 3. Document classification accuracy on the test set. We ran the model 10 times and reported the mean accuracy (%) and standard deviation (%).

Coarse	R8	R52	ohsumed	MR	TREC	WebKB	SST-1	SST-2
Input	96.35 (± 0.40)	90.97 (± 0.38)	54.24 (± 0.33)	76.20 (± 0.39)	98.04 (± 0.25)	84.14 (± 0.63)	42.73 (± 2.48)	80.86 (± 0.89)
1	95.77 (± 0.37)	89.10 (± 1.05)	53.29 (± 0.16)	76.16 (± 0.60)	96.68 (± 0.37)	83.38 (± 0.35)	42.67 (± 1.62)	80.05 (± 0.76)
2	95.53 (± 0.26)	87.41 (± 0.63)	51.68 (± 0.23)	73.73 (± 1.06)	96.20 (± 0.72)	83.07 (± 0.63)	40.60 (± 1.80)	78.14 (± 1.04)
3	95.29 (± 0.19)	85.79 (± 0.89)	50.10 (± 0.30)	71.64 (± 2.37)	94.78 (± 0.73)	81.12 (± 0.80)	41.06 (± 1.45)	75.74 (± 2.23)
4	94.53 (+ 0.29)	86.17 (± 0.35)	48.36 (± 0.26)	70.07 (± 1.90)	91.84 (± 0.50)	76.83 (± 0.74)	38.77 (+ 1.26)	75.45 (± 1.28)
5	93.92 (± 0.51)	84.30 (± 0.33)	44.35 (± 0.25)	69.39 (± 1.25)	89.50 (± 0.52)	75.87 (± 0.57)	37.64 (± 1.60)	74.14 (± 0.54)
6	92.47 (± 0.67)	81.91 (± 0.70)	40.70 (± 0.22)	68.54 (± 0.64)	87.60 (± 1.08)	72.11 (± 0.41)	34.44 (± 0.85)	71.03 (± 1.48)
7	91.49 (± 0.34)	79.68 (± 0.37)	35.95 (± 0.24)	66.29 (± 1.37)	83.50 (± 0.30)	69.33 (± 0.52)	34.14 (± 1.20)	69.63 (± 1.31)
8	86.12 (± 0.77)	77.20 (± 0.46)	33.63 (± 0.24)	64.72 (± 1.02)	76.96 (± 0.43)	65.71 (+ 0.41)	33.23 (± 0.79)	68.22 (± 1.66)
9	83.14 (± 0.40)	74.88 (± 0.22)	29.64 (± 0.24)	60.84 (± 1.03)	69.36 (± 0.32)	61.81 (± 0.35)	33.11 (± 0.48)	62.15 (± 1.45)
10	77.93 (± 0.78)	69.59 (± 0.23)	23.49 (± 0.21)	58.38 (± 0.93)	64.80 (± 0.35)	53.22 (± 0.64)	30.65 (± 0.91)	60.93 (± 1.08)

Table 4. Document classification F1-score on the test set. We ran the model 10 times and reported the mean F1-score (%) and standard deviation (%).

Coarse	R8	R52	ohsumed	MR	TREC	WebKB	SST-1	SST-2
Input	89.28 (± 1.12)	56.48 (± 2.28)	38.37 (± 1.17)	76.12 (± 0.38)	98.21 (± 0.24)	77.42 (± 3.58)	39.34 (± 1.39)	80.84 (± 0.90)
1	87.77 (± 1.76)	48.52 (± 3.67)	37.47 (± 0.99)	76.15 (± 0.60)	96.97 (± 0.36)	65.27 (± 1.30)	39.06 (± 0.48)	80.04 (± 0.76)
2	87.70 (± 2.03)	43.85 (± 2.26)	35.00 (± 0.69)	73.64 (± 1.20)	96.50 (± 0.68)	78.04 (± 0.35)	36.54 (± 0.92)	78.08 (± 1.13)
3	87.12 (± 0.90)	41.93 (± 3.84)	32.37 (± 0.64)	71.44 (± 2.84)	95.44 (± 0.58)	76.42 (± 1.13)	35.86 (± 1.09)	75.72 (± 2.24)
4	83.02 (± 3.97)	44.07 (± 1.18)	30.75 (± 0.79)	70.03 (± 1.94)	92.94 (± 0.40)	74.76 (± 0.27)	34.83 (± 0.92)	75.41 (± 1.28)
5	84.85 (± 1.66)	38.79 (± 1.58)	26.44 (± 1.17)	69.37 (± 1.25)	90.28 (± 0.83)	63.37 (± 4.84)	33.52 (± 0.89)	74.13 (± 0.53)
6	81.23 (± 2.68)	31.99 (± 5.02)	21.12 (± 1.31)	68.51 (± 0.65)	87.63 (± 1.31)	56.07 (± 0.68)	30.48 (± 0.62)	71.02 (± 1.49)
7	74.48 (± 2.97)	26.48 (± 2.59)	16.98 (± 0.75)	66.29 (± 1.37)	83.25 (± 0.63)	57.54 (± 1.29)	29.63 (± 0.58)	69.55 (± 1.41)
8	66.92 (± 2.06)	26.23 (± 2.07)	13.72 (± 1.17)	64.65 (± 1.01)	77.56 (± 0.58)	54.93 (± 1.16)	28.16 (± 0.56)	67.98 (± 1.97)
9	58.33 (± 3.13)	20.10 (± 1.17)	10.22 (± 0.58)	60.83 (± 1.04)	69.78 (± 0.57)	52.96 (± 1.67)	27.58 (± 0.73)	60.94 (± 2.29)
10	52.28 (± 0.29)	15.21 (± 1.16)	7.77 (± 0.13)	58.30 (± 0.93)	64.52 (± 0.83)	48.34 (± 0.21)	26.27 (± 0.91)	59.99 (± 1.90)

Table 5. Average memory (in MB) used by GraphSAGE for each input graph and its respective coarsened versions.

Coarse	R8	R52	ohsumed	MR	TREC	WebKB	SST-1	SST-2
Input	1632.0	1672.0	2410.0	640.0	214.0	2400.0	536.0	560.0
1	1410.0	2062.0	1996.0	534.0	210.0	2162.0	554.0	576.0
2	1694.0	2080.0	2044.0	696.0	216.0	2464.0	584.0	484.0
3	1942.0	1732.0	2044.0	730.0	218.0	1540.0	890.0	484.0
4	1096.0	1360.0	1972.0	594.0	172.0	2046.0	590.0	492.0
5	1032.0	1276.0	1882.0	684.0	172.0	1628.0	722.0	376.0
6	490.0	1190.0	1296.0	546.0	214.0	1218.0	432.0	346.0
7	792.0	1012.0	770.0	500.0	156.0	848.0	270.0	322.0
8	362.0	682.0	646.0	470.0	176.0	546.0	472.0	306.0
9	278.0	732.0	518.0	314.0	126.0	458.0	336.0	188.0
10	230.0	462.0	394.0	280.0	138.0	384.0	284.0	214.0

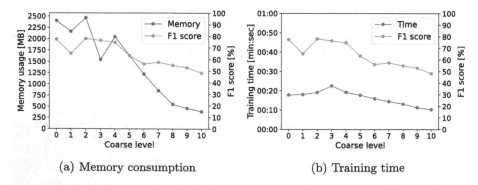

(a) Memory consumption   (b) Training time

Fig. 2. GraphSAGE's memory consumption and training time curves for the input graph and the ten coarsened levels of the WebKB dataset. Here the input graph is represented by the 0 on the horizontal axis.

Training Time: The final analysis performed in this study was regarding the training time of GraphSAGE for each graph. We specifically measured the elapsed time by focusing exclusively on the steps involving the calculation of the network's gradient. Consequently, the time required for validation and the computation of metrics was not taken into account in the values reported in Table 6. It is worth emphasizing that the reported values represent the average training time across the ten runs conducted for each graph.

Table 6. GraphSAGE's average training time for each graph we used. We exclude from this table time information regarding steps that don't involve gradient computation, namely the validation and test steps. These values are shown in the minutes: seconds format.

Coarse	R8	R52	ohsumed	MR	TREC	WebKB	SST-1	SST-2
Input	00:37	01:15	00:34	00:35	00:14	00:17	00:47	00:31
1	00:37	01:02	00:39	00:35	00:14	00:18	00:48	00:32
2	00:39	01:03	00:40	00:35	00:15	00:19	00:49	00:33
3	00:40	01:03	00:38	00:36	00:14	00:22	00:50	00:32
4	00:39	01:06	00:41	00:34	00:13	00:19	00:49	00:30
5	00:40	01:02	00:41	00:33	00:13	00:17	00:49	00:30
6	00:40	00:54	00:38	00:33	00:13	00:15	00:48	00:29
7	00:34	00:50	00:35	00:31	00:12	00:14	00:47	00:28
8	00:30	00:46	00:30	00:30	00:11	00:13	00:45	00:28
9	00:26	00:40	00:25	00:29	00:11	00:11	00:43	00:26
10	00:23	00:36	00:21	00:27	00:11	00:10	00:40	00:24

The data presented in Table 6 reveals that, as anticipated, the training time of GraphSAGE decreases from the input to the coarsest graph for all datasets. However, interestingly, the training time between intermediate levels does not consistently decrease despite the graph reduction. For example, in the case of the WebKB, SST-1, ohsumed, and R8, the training time slightly increases from the input graph to the intermediate coarsened graphs before eventually decreasing. This behavior highlights the non-linear relationship between the graph size and the corresponding training time for these specific datasets.

Finally, the dataset that had the least variation in training time was TREC, and we also observed that it had the fewest merged words. Moreover, Graph-SAGE trained on each of TREC's graphs within a time interval that decreased by only 3 s from the input graph to the coarsest one. Similar behavior is also observed in the SST-1, SST-2, and MR datasets, which, following TREC, underwent the least amount of contraction.

5 Concluding Remarks

In this paper, we introduced a method that employs a GNN to perform text classification on ten coarsened versions of a bipartite graph that were generated through the coarsening step from the multilevel optimization approach. Moreover, we proposed the usage of the cosine similarity of word representations as the measure used in the graph coarsening stage. The method we presented was evaluated on eight datasets with different sizes, class balance, and domains. Our experiments highlight that our approach can effectively reduce GraphSAGE's memory usage and training time without losing a significative F1-score.

However, it is worth noting that our proposed method has some limitations. Firstly, we performed the coarsening step on the word partition of each graph, although it is possible to reduce both. Moreover, we applied a fixed 25% reduction to each graph without further exploring alternative values. Nonetheless, determining the most suitable reduction factor for better solutions is not trivial and requires a thorough investigation. Lastly, we only evaluated our method's capacity on eight textual datasets and did not explore different applications.

In future work, we intend to apply the coarsening technique to the document partition. Moreover, we will conduct experiments with our method using larger graphs from various domains instead of focusing only on text classification. We also plan to employ other GNNs in our pipeline to analyze our approach's capabilities with different message-passing schemes. Furthermore, we will test these models with more than two layers to analyze the impact of adding information from neighbors located more than two hops from each target node. Lastly, we plan to create an end-to-end system to learn graph contraction and perform text classification using a single neural network.

Acknowledgments. This study was supported in part by the Coordenação de Aperfeiçoamento de Pessoal de Nível Superior - Brasil (CAPES) - Finance Code 001; São Paulo Research Foundation (FAPESP) [grants #20/09835-1, #21/06210-3, #22/03090-0, and #22/09091-8]; and Brazilian National Council for Scientific and Technological Development (CNPq) [grant #303588/2022-5].

References

1. Bruna, J., Zaremba, W., Szlam, A., LeCun, Y.: Spectral networks and locally connected networks on graphs. In: ICLR (2014)
2. Cai, C., Wang, D., Wang, Y.: Graph coarsening with neural networks. In: ICLR (2021)
3. Chiang, W.L., Liu, X., Si, S., Li, Y., Bengio, S., Hsieh, C.J.: Cluster-GCN: an efficient algorithm for training deep and large graph convolutional networks. In: SIGKDD, pp. 257–266. Association for Computing Machinery (2019)
4. Craven, M., et al.: Learning to extract symbolic knowledge from the world wide web. In: Conference on Artificial Intelligence, pp. 509–516. AAAI '98, American Association for Artificial Intelligence, USA (1998)
5. Ding, K., Wang, J., Li, J., Li, D., Liu, H.: Be more with less: hypergraph attention networks for inductive text classification. In: EMNLP, pp. 4927–4936 (2020)

6. Gao, H., Chen, Y., Ji, S.: Learning graph pooling and hybrid convolutional operations for text representations. In: The World Wide Web Conference, pp. 2743–2749. WWW '19, Association for Computing Machinery, New York, NY, USA (2019)
7. Gilmer, J., Schoenholz, S.S., Riley, P.F., Vinyals, O., Dahl, G.E.: Neural message passing for quantum chemistry. In: ICML, pp. 1263–1272 (2017)
8. Hamilton, W., Ying, Z., Leskovec, J.: Inductive representation learning on large graphs. Adv. Neural Inf. Process. Syst. **30** (2017)
9. Huang, L., Ma, D., Li, S., Zhang, X., Wang, H.: Text level graph neural network for text classification. In: EMNLP-IJCNLP, pp. 3444–3450. Association for Computational Linguistics, Hong Kong, China, November 2019
10. Kim, Y.: Convolutional neural networks for sentence classification. In: EMNLP, pp. 1746–1751. Association for Computational Linguistics, Doha, Qatar, October 2014
11. Kingma, D.P., Ba, J.: Adam: a method for stochastic optimization. In: ICLR, 2015, San Diego, CA, USA, 7–9 May 2015, Conference Track Proceedings (2015)
12. Kipf, T.N., Welling, M.: Semi-supervised classification with graph convolutional networks. In: ICLR (2017)
13. Kowsari, K., Jafari Meimandi, K., Heidarysafa, M., Mendu, S., Barnes, L., Brown, D.: Text classification algorithms: a survey. Information **10**(4), 150 (2019)
14. Li, X., Roth, D.: Learning question classifiers. In: COLING, pp. 1–7. Association for Computational Linguistics, USA (2002)
15. Lin, Y., et al.: BertGCN: transductive text classification by combining GCN and BERT. Findings of ACL (2021)
16. Liu, P., Qiu, X., Huang, X.: Recurrent neural network for text classification with multi-task learning. In: IJCAI, pp. 2873–2879. IJCAI'16, AAAI Press (2016)
17. Liu, X., You, X., Zhang, X., Wu, J., Lv, P.: Tensor graph convolutional networks for text classification. In: AAAI Conference on Artificial Intelligence, vol. 34, pp. 8409–8416 (2020)
18. Liu, Y., Guan, R., Giunchiglia, F., Liang, Y., Feng, X.: Deep attention diffusion graph neural networks for text classification. In: EMNLP, pp. 8142–8152. Association for Computational Linguistics, November 2021
19. Pang, B., Lee, L.: Seeing stars: exploiting class relationships for sentiment categorization with respect to rating scales. In: ACL, pp. 115–124. Association for Computational Linguistics, Ann Arbor, Michigan, June 2005
20. Pennington, J., Socher, R., Manning, C.: GloVe: global vectors for word representation. In: EMNLP, pp. 1532–1543. Association for Computational Linguistics (2014)
21. Pennington, J., Socher, R., Manning, C.D.: Glove: global vectors for word representation. In: EMNLP, pp. 1532–1543 (2014)
22. Sinoara, R.A., Camacho-Collados, J., Rossi, R.G., Navigli, R., Rezende, S.O.: Knowledge-enhanced document embeddings for text classification. Knowl.-Based Syst. **163**, 955–971 (2019)
23. Socher, R., et al.: Recursive deep models for semantic compositionality over a sentiment treebank. In: EMNLP, pp. 1631–1642. Association for Computational Linguistics, October 2013
24. Tang, J., Qu, M., Mei, Q.: PTE: predictive text embedding through large-scale heterogeneous text networks. In: SIGKDD, pp. 1165–1174. Association for Computing Machinery (2015)
25. Valejo, A., Ferreira, V., Fabbri, R., Oliveira, M.C.F.D., Lopes, A.D.A.: A critical survey of the multilevel method in complex networks. ACM Comput. Surv. **53**(2) (2020)

26. Valejo, A., Ferreira de Oliveira, M.C., Filho, G.P., de Andrade Lopes, A.: Multilevel approach for combinatorial optimization in bipartite network. Knowl.-Based Syst. **151**, 45–61 (2018)
27. Veličković, P., Cucurull, G., Casanova, A., Romero, A., Liò, P., Bengio, Y.: Graph attention networks. In: ICLR (2018)
28. Wang, S.I., Manning, C.D.: Baselines and bigrams: simple, good sentiment and topic classification. In: Annual Meeting of the ACL, pp. 90–94 (2012)
29. Wu, Z., Pan, S., Chen, F., Long, G., Zhang, C., Yu, P.S.: A comprehensive survey on graph neural networks. IEEE Trans. Neural Netw. Learn. Syst. **32**(1), 4–24 (2021)
30. Yao, L., Mao, C., Luo, Y.: Graph convolutional networks for text classification. In: AAAI Conference on Artificial Intelligence, vol. 33, pp. 7370–7377 (2019)
31. Ying, R., You, J., Morris, C., Ren, X., Hamilton, W.L., Leskovec, J.: Hierarchical graph representation learning with differentiable pooling. In: NeurIPS, pp. 4805–4815. NIPS'18, Curran Associates Inc., Red Hook, NY, USA (2018)
32. Zeng, H., Zhou, H., Srivastava, A., Kannan, R., Prasanna, V.: GraphSAINT: graph sampling based inductive learning method. In: ICLR (2020)
33. Zhang, Y., Yu, X., Cui, Z., Wu, S., Wen, Z., Wang, L.: Every document owns its structure: inductive text classification via graph neural networks. In: ACL, pp. 334–339. Association for Computational Linguistics, Online, July 2020

Towards Robust Defect Detection in Casting Using Contrastive Learning

Eneko Intxausti[1]($\boxtimes$)(iD), Ekhi Zugasti[1](iD), Carlos Cernuda[1](iD),
Ane Miren Leibar[2], and Estibaliz Elizondo[3]

[1] Mondragon Unibertsitatea, Loramendi 4, 20500 Mondragon, Spain
{eintxausti,ezugasti,ccernuda}@mondragon.edu
[2] Fagor Ederlan, S. COOP., Torrebaso pasealekua 7, 20500 Mondragon, Spain
a.leibar@fagorederlan.es
[3] Edertek, S. COOP., Isasi Kalea 6, 20500 Mondragon, Spain
e.elizondo@edertek.es

Abstract. Defect detection plays a vital role in ensuring product quality and safety within industrial casting processes. In these dynamic environments, the occasional emergence of new defects in the production line poses a significant challenge for supervised methods. We present a defect detection framework to effectively detect novel defect patterns without prior exposure during training. Our method is based on contrastive learning applied to the Faster R-CNN model, enhanced with a contrastive head to obtain discriminative representations of different defects. By training on an diverse and comprehensive labeled dataset, our method achieves comparable performance to the supervised baseline model, showcasing commendable defect detection capabilities. To evaluate the robustness of our approach, we authentically replicate a real-world use case by deliberately excluding several defect types from the training data. Remarkably, in this new context, our proposed method significantly improves detection performance of the baseline model, particularly in situations with very limited training data, showcasing a remarkable 34.7% enhancement. Our research highlights the potential of the proposed method in real-world environments where the number of available images may be limited or inexistent. By providing valuable insights into defect detection in challenging scenarios, our framework could contribute to ensuring efficient and reliable product quality and safety in industrial manufacturing processes.

Keywords: Defect detection · contrastive learning · casting · optical quality control · deep learning

1 Introduction

In today's global markets, manufacturers are required to prioritize product specialization in order to offer highly reliable goods of superior quality. As manufacturing processes become more complex and quality standards rise, effective quality control becomes essential to ensure product excellence. Given the demand for

© Springer Nature Switzerland AG 2024
V. Vasconcelos et al. (Eds.): CIARP 2023, LNCS 14469, pp. 605–616, 2024.
https://doi.org/10.1007/978-3-031-49018-7_43

rapid production in competitive industries and the limitations associated with human-based image assessment, such as subjectivity and fatigue, the need for automated and reliable image processing methods becomes evident [1].

Defect detection aims to identify patterns in image data that deviate from expected behavior [2], focusing on both classifying the presence of defects and accurately localizing them within the image. In the early stages of defect detection, traditional machine learning methods were widely employed, relying on a well-established pipeline that involved manual feature extraction and subsequent classification. Experts in the field would carefully handcraft features to effectively represent various defect characteristics. However, these methods posed limitations in capturing complex patterns and required domain expertise for feature engineering. Moreover, their performance deteriorates with the emergence of unseen defect types.

In recent years, deep learning-based approaches have revolutionized the field of defect detection, surpassing the limitations of traditional machine learning methods and improving detection performance [3,4]. By automatically extracting intricate features from raw data, deep learning models are able to capture complex patterns in images, leading to improved detection accuracy, especially for subtle defects. However, as traditional methods, these models have certain limitations when it comes to handling new or unseen defects [5]. Supervised deep learning models, trained on labeled data, often face challenges in accurately identifying and classifying novel defects that were not included in their training set.

This paper presents a novel defect detection approach based on contrastive learning, which aims to detect new and infrequent defects in the production line. To assess the robustness of our approach, we conduct an experiment on a real-world dataset by intentionally excluding several defect types from the training data. Remarkably, our proposed method exhibits a substantial improvement in the detection performance of the baseline model, especially in scenarios with limited training data. These findings underscore the potential of our approach in real-world environments with restricted image availability, offering a valuable tool for enhancing defect detection in industrial applications.

2   Related Work

Defect detection in castings is not a new topic in computer vision, as it has been used in many industrial quality control procedures [6,7]. However, the introduction of deep learning techniques has led to increase research in this field. Convolutional Neural Networks (CNNs) have emerged as a powerful tool in defect detection, allowing for more accurate and efficient analysis of casting images. Their ability to automatically learn and extract meaningful features from raw data has revolutionized the detection process.

Numerous studies have explored the use of deep CNNs for image-level classification in defect detection for castings [7,8]. Kuo et al. [9] compared the feature extraction performance of four deep CNN backbones in the detection of sandblasting defects using surface images. Jiang et al. [10] improved residual models accuracy in defect detection presenting a novel activation function to prevent the neuron-death issue. In order to address the challenge of overlooking small defects during the analysis of complete images, Wang et al. [11] proposed a novel strategy for feature extraction in car casting images. Their approach involved using a self-attention guided module in conjunction with a network that extracted general features from X-ray images, incorporating a residual block. By integrating these components, the method successfully captured and represented small defects with high accuracy.

Precise defect localization within the image is crucial for operator support, facilitating efficient identification of the specific areas that require attention. Object detection techniques (Faster R-CNN [12], YOLO [13]) have played a significant role in advancing defect detection capabilities and enabling accurate localization of defects within complex images [14]. Ferguson et al. [15] adapted different object detection architectures to the casting defect detection task and compared them with a sliding-window method based on a classification model. This study showed for the first time the promising performance of such architectures in accurately localizing small objects, highlighting their potential in defect detection applications.

Following this study, subsequent research focused on implementing architectural improvements in object detectors and exploring new strategies to enhance model performance. Du et al. [16] introduced several enhancements to the Faster R-CNN for defect detection, including Feature Pyramid Network [17] (FPN), RoIAlign [18], and data augmentation techniques. Xing et al. [19] proposed improvements to the backbone architecture and introduced new loss functions for IoU (Intersection over Union) and classification. Du et al. [20] proposed an architecture consisting of DetNet as a backbone and PANet to decrease the loss of location information caused by downsampling layers.

However, these supervised approaches require a large and diverse dataset for optimal performance [21]. Additionally, the presence of new and infrequent defects, whether they are underrepresented in the dataset or arise as new instances in the production line, poses a significant challenge for these methods. Moreover, these novel and less frequent defects may be biased by the dominant defect class, leading to difficulties in accurately detecting and classifying them.

Even though some studies have explored domain adaptive strategies to train models with a limited number of samples for object detection [22], it is crucial to demonstrate their discriminative capability among different classes. In the following section, we present our novel approach to tackle the challenge of detecting new or infrequently represented defects in the dataset. Our proposed method aims to effectively reduce the errors in detecting these types of defects, ensuring accurate and reliable defect detection performance.

3   Method

Our defect detection method is based on contrastive learning, an unsupervised approach that enables the model to learn meaningful representations without the need for labels [23]. By leveraging contrastive learning, our method enhances the model discriminative power [5], allowing it to effectively differentiate between various defect categories and accurately distinguish defect from non-defect samples.

Contrastive learning is a technique that leverages the comparison between a real image and its transformed counterpart to learn meaningful representations [24] (see Fig. 1). Upon receiving an image x, it is first encoded using a encoder, resulting in the generation of its representation h. In the subsequent step, the image x undergoes a transformation $(\tilde{x})$, which is then encoded using the same encoder, ultimately leading to the generation of its representation $(\tilde{h})$. The primary goal of the contrastive learning is to maximize the similarity between the features extracted from the both images. In this case, the augmented version of the original sample serves as a positive sample, while the remaining images of the dataset are considered negative samples [23]. By contrasting the augmented positive sample with the negative samples, the model learns to extract discriminative features from samples.

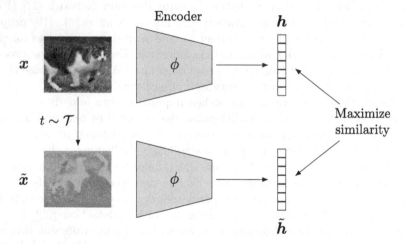

Fig. 1. Enhancing image feature analysis through contrastive learning. Given an image x, a data augmentation operator $t \sim \mathcal{T}$ and an encoder network ϕ, the objective is to maximize similarity between image features h and $\tilde{h}$.

Furthermore, an existing strategy in the literature extends contrastive learning to fully-supervised approach, leveraging label information [25]. This method encourages embeddings from the same class to be pulled closer together, while pushing embeddings from different classes apart. By adopting this strategy, it becomes feasible to obtain additional positive samples, which plays a crucial role in training the model to effectively distinguish between defect and non-defect classes.

Following this approach, we have incorporated improvements into the Faster R-CNN model for more effective defect detection. The baseline Faster R-CNN model serves as the foundation, maintaining its original two-stage architecture. To enhance its performance, we follow the approach proposed by Du et al. [16] and add FPN and RoIAlign to our method. FPN aggregates feature maps from different layers, using top-down and lateral connections to combine high-level semantic information with fine-grained spatial details, enhancing object detection across diverse scales. Furthermore, RoIAlign operation improves RoIPooling by applying bilinear interpolation to obtain the exact location of the object proposals.

Based on the Few-shot Faster R-CNN introduced by Sun et al. [26], our novel defect detection framework incorporates an innovative *contrastive head* into the Faster R-CNN architecture (Fig. 2). The model includes a backbone network for feature extraction, a Region Proposal Network (RPN) for generating region proposals, a shared convolutional network, and along with independent modules for classification, localization, and contrastive learning.

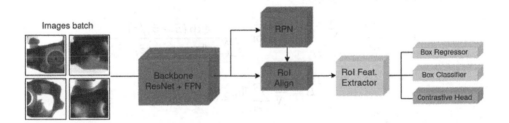

Fig. 2. Architecture of a Faster R-CNN model with a novel contrastive head.

It shares similarities with the supervised contrastive learning approach, but instead of using complete images, it operates on object proposals. In the initial stage, the Region Proposal Network (RPN) identifies several regions of interest where defects might be present. Subsequently, RoIAlign is employed to pool these proposals to a fixed size, while the RoI feature extractor encodes them into h feature vectors, also known as embeddings, where $h \in \mathbb{R}^d$ and $d = 1024$ typically. These embeddings are further processed by the localization head, classification head, and the contrastive head. An illustrative overview is presented in Fig. 3.

The contrastive head focuses on refining the feature embeddings generated by RoIAlign to promote more discriminative representations. For a mini-batch of N, we define $\{z_i, y_i, p_i\}_{i=1}^N$, where z_i represents the features encoded by the contrastive head for the i-th proposal, y_i denotes the ground-truth label and p_i represents the confidence score provided by the RPN. These values are integrated into the Eq. (1) and used to compute the loss function presented in Eq. (2).

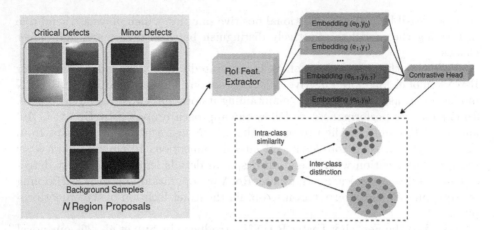

Fig. 3. Leveraging N region proposals processed by the RoI feature extractor, the contrastive head refines the embeddings by minimizing intra-class distances and maximizing inter-class distances, obtaining more compact clusters within classes.

By minimizing this function and applying backpropagation, we optimize the contrastive head module [26]:

$$\ell_i = -\frac{1}{N_{y_i}} \sum_{j=1, j \neq i}^{N} \mathbb{1}_{[y_i = y_j]} \log \frac{\exp \left(\tilde{z}_i \cdot \tilde{z}_j / \tau \right)}{\sum_{k=1}^{N} \mathbb{1}_{[k \neq i]} \exp \left(\tilde{z}_i \cdot \tilde{z}_j / \tau \right)} \tag{1}$$

$$\mathcal{L} = \frac{1}{N} \sum_{i=1}^{N} \mathbb{1}_{[p_i \geq \gamma]} \cdot \ell_i \tag{2}$$

where $\tilde{z}_i \cdot \tilde{z}_j = z_i^{\top} z_j / \|z_i\| \|z_j\|$ denotes the cosine similarity between the normalized features extracted from i-th and j-th proposals and τ is an hyper-parameter known as *temperature*. Additionally, we introduce a confidence regulation function to Eq. (2) that enables the selection of proposals with probabilities of being defects higher than a threshold γ, as determined by the RPN. This function ensures that only the most promising proposals are considered for further processing in the contrastive head.

In this new latent space, the objective is to maximize the separability of embeddings by minimizing the value of the loss function. This step encourages intra-class embeddings to cluster closer together while pushing inter-class embeddings further apart. As a result, proposal vectors of the same class are brought into tighter alignment, while those from different classes are driven farther apart, enabling the model to better discern subtle differences between defect and non-defect regions.

4   Experiments

4.1   Dataset Description

Our dataset comprises 5786 X-ray images collected from a production line in an automotive industry company. These images depict various perspectives of the same product and all possess dimensions of 1024×1024 pixels. The dataset encompasses 19 defect types of diverse sizes, scales, and intensities, exhibiting significant variability. Ground-truth annotations for all defects are provided as bounding-boxes, ensuring accurate reference points for defect detection.

Although the dataset is labeled per defect type and name, we are interested in being able to discriminate between *critical defects* and *minor defects*. This is why we preprocess the dataset and relabel every type of dataset into two big groups, namely *critical* and *minor*, so that we can perform a binary classification and localization of the damages. As their own names suggest, accurate detection of critical defects is essential due to their potential safety implications. While detecting minor defects is also crucial, their severity is comparatively lower, making their identification equally significant but less critical for safety concerns. Additionally, the distribution of defects includes 7 *critical* defect types and 12 *minor* defect types. Thus, proposed scenario poses a challenging defect detection problem with a dual objective. Our aim is twofold: first, to accurately identify defects present in the product, and second, to classify them into two distinct categories based on their severity, differentiating between defects that pose potential safety risks and those of lesser concern.

Regarding the distribution of the dataset, it shows a significant class imbalance with 1784 samples for critical defects and 4002 samples for minor defects. Thus, the dataset partition into training and test sets, while preserving the class proportion, is shown in Table 1.

Table 1. Distribution of images for critical and minor defects in the train and test sets

	Critical defects	Minor defects
Train	1424	3200
Test	360	802

However, as will be showed subsequently, this class imbalance does not pose a major issue for training a defect detector. Furthermore, a comprehensive analysis conducted with the assistance of an expert revealed that the occurrence of defects in the dataset is not homogeneous. Certain defects appear with a high frequency, indicating that they may exert a great influence on the model training. In contrast, some defects are sparsely represented, suggesting that the model might face challenges in detecting those cases.

It is worth mentioning that due to confidentiality concerns, images showing identifiable products and their associated defects are intentionally excluded from this paper.

4.2 Experimental Procedure and Preliminary Results

The objective of this research was to assess whether our method can enhance the performance of an object detector when confronted with novel defects that emerge in the production line, that is, those not previously encountered during model training. To achieve this, we selected the Faster R-CNN with ResNet-101 [27] as the baseline and compared it with our model incorporating the contrastive head. The training of both models followed methodologies employed in previous studies [16,19]. Training process took place on a NVIDIA GeForce GTX 1080 GPU.

As described in the previous section, our model training procedure involved using proposals obtained from RPN. These proposals were then resized to a standardized dimension, and their features were extracted to form an embedding vector. Subsequently, the model generated three outputs through its heads. During backpropagation, both the weights of the heads and the shared RoI feature extractor were updated, allowing the contrastive loss function to effectively influence model training.

The first experiment focused on evaluating the performance of both models using the complete image dataset. The evaluation of the models was performed using two metrics: mean Average Precision (mAP) and recall for the *critical* class. mAP calculates the average precision for both classes, measuring the accuracy of the model in both classification and localization. Furthermore, recall for the *critical* class measures the ability of the model to identify all instances of critical defects correctly. Table 2 shows that both baseline and contrastive models achieved satisfactory and comparable results.

Table 2. Defect detection results obtained using baseline Faster R-CNN and our contrastive approach (*CFaster R-CNN*).

Model	mAP	Recall Critical
Baseline Faster R-CNN	0.848	0.941
CFaster R-CNN	0.868	0.934

Based on the findings presented in Table 2, we could infer that despite the class imbalance within the dataset, it provided sufficient diversity and size for the model to effectively detect the majority of defects in the test set.

4.3 Robust Defect Detection

The primary goal of this study was to evaluate model performance under challenging training conditions, encompassing the emergence of new defects in the production line or the scarcity of images. Consequently, to effectively evaluate model performance in these scenarios, it was necessary to modify the training set accordingly.

After a comprehensive analysis of the defects, we carefully chosen three defect types that encompassed diverse singularities within the image dataset. To assess model robustness in handling unseen defects, we systematically removed all images containing each of these defect types from the training set, one by one. Initially, we selected a critical defect type that appears frequently in the production line. This resulted in a significantly imbalanced training dataset. We chose this defect to evaluate the capability of our model to generalize critical defect characteristics from a limited training set. Subsequently, we selected another critical defect type that appears infrequently in the production line. This defect allowed us to evaluate model performance on rare critical defects. Finally, we deliberately chose a minor defect type with a moderate presence in the training set to thoroughly assess model performance regarding minor defects.

Table 3 presents the comparison results between the Faster R-CNN baseline model and the enhanced model with the contrastive head. The comparison includes the number of defects of each class in the training set, the AP value for the specific defect type, and the Recall for each defect class. For the first defect type, the model with contrastive learning improved the AP by 34.7% compared to the baseline model, while doubling the recall for critical defects. For the second defect type, the results obtained in all three metrics were similar, showing no significant differences between the two models. Lastly, for the third defect type, the contrastive model achieved a 12.8% improvement in defect type AP and a 17.8% enhancement in the recall of minor defects.

Table 3. Detection performance comparison of the baseline (*Faster R-CNN*) and contrastive model (*CFaster R-CNN*) for different defect categories. "*Defect type AP*" represents the Average Precision (AP) metric specifically calculated for each selected defect type.

	Method	Defect class	Training images	Defect type	Recall	
			(Critical, Minor)	AP	Critical	Minor
Type 1	Faster R-CNN	Critical	(258, 3200)	0.613	0.302	0.986
	CFaster R-CNN	Critical	(258, 3200)	**0.826**	**0.604**	0.961
Type 2	Faster R-CNN	Critical	(1274, 3200)	0.947	0.925	0.960
	CFaster R-CNN	Critical	(1274, 3200)	**0.955**	**0.931**	0.972
Type 3	Faster R-CNN	Minor	(1424, 2997)	0.672	0.95	0.783
	CFaster R-CNN	Minor	(1424, 2997)	**0.758**	0.936	**0.922**

4.4 Discussion

Our evaluation revealed that both models exhibited similar and commendable performance when trained on a diverse and extensive dataset that accurately represented the wide range of defects typically encountered in production settings. The integration of a contrastive head into the Faster R-CNN architecture yielded promising results, particularly for specific defect types. However, it is noteworthy that overall, both models consistently performed well in scenarios with abundant and varied data. Therefore, the results showed that comprehensive and diverse datasets are vital for effective defect detection, enabling robust generalization.

In situations with a limited number of images, our model demonstrated remarkable generalization and significant improvement for defects not previously encountered. This capability highlighted the potential of our approach to excel in scenarios with limited and new data. By adopting this approach, the model faced the challenge of encountering previously unseen defect patterns during testing, forcing it to generalize and adapt to new scenarios without prior exposure. This representation of real-world conditions allowed us to thoroughly evaluate model performance and its capacity to effectively detect and classify critical and minor defects.

The contrast between the Faster R-CNN model and our enhanced model became evident when the number of training images for a class was limited (Type 1). However, as the training set size increased (Type 2), the difference between both models converged. This behavior is substantiated by domain expert confirmation, indicating that the excluded defect type in the training set shares similarities with other defect types present during model training, potentially leading to confusion between similar defect patterns.

Finally, it is worth mentioning that the class had no influence on model performance. In other words, for both critical defects (Type 1) and minor defects (Type 3) with limited image for training, the contrastive model showed considerable improvement over the baseline. This preliminary results allow us to embark on a path to create a novel and robust defect detection algorithm tailored for a real industrial use case.

5   Conclusions

Our research focus on developing a defect detection framework for an industrial application, where the emergence of new defects in the production line is an occasional occurrence. In this context, we aim to address the challenge of effectively detecting these novel defect patterns without prior exposure. To achieve this, we introduce a novel defect detection framework based on the integration of a contrastive head into the Faster R-CNN architecture.

Initial evaluation indicates that the model enhanced with contrastive learning achieves a comparable level of performance to the baseline Faster R-CNN model when trained on a diverse and comprehensive dataset, encompassing a wide spectrum of defects commonly encountered in production settings.

Additionally, to assess the robustness of our approach in detecting new defects, we conduct a novel experiment simulating a real-world scenario by intentionally excluding several defect types from the training data. Remarkably, our proposed defect detection framework effectively identifies previously unseen defect patterns, showcasing remarkable generalization capabilities. We observe that the lack of substantial amounts of training data significantly penalizes the performance of baseline model, while the proposed model demonstrates more robustness to the limited data availability.

Our research highlights the potential of the proposed method in real-world environments where the number of available images may be limited. By showing model robustness in detecting defects, even in situations with restricted data, we provide a valuable tool for ensuring product quality and safety in industrial manufacturing processes. Through additional research and validation, this approach could be applied across diverse industrial applications, effectively addressing the dynamic nature of defect patterns and ensuring efficient and reliable defect detection, even under challenging conditions.

Acknowledgment. Eneko intxausti, Carlos Cernuda and Ekhi Zugasti are part of the Intelligent Systems for Industrial Systems research group of Mondragon Unibertsitatea (IT1676-22), supported by the Department of Education, Universities and Research of the Basque Country. They are also supported by the DREMIND project of the Basque Government under Grant KK-2022/00049 from the ELKARTEK program.

References

1. Rafiei, M., Raitoharju, J., Iosifidis, A.: Computer vision on X-ray data in industrial production and security applications: a comprehensive survey. IEEE Access **11**, 2445–2477 (2023)
2. Chandola, V., Banerjee, A., Kumar, V.: Anomaly detection: a survey. ACM Comput. Surv. (CSUR) **41**(3), 1–58 (2009)
3. Krizhevsky, A., Sutskever, I., Hinton, G.E.: ImageNet classification with deep convolutional neural networks. Commun. ACM **60**(6), 84–90 (2017)
4. Simonyan, K., Zisserman, A.: Very Deep Convolutional Networks for Large-Scale Image Recognition (2015)
5. Köhler, M., Eisenbach, M., Gross, H.-M.: Few-Shot Object Detection: A Comprehensive Survey (2022)
6. Da Silva, R.R., Mery, D.: Accuracy estimation of detection of casting defects in X-ray images using some statistical techniques. Insight - Non-Destructive Test. Condition Monit. **49**(10), 603–609 (2007)
7. Mery, D., Arteta, C.: Automatic defect recognition in X-ray testing using computer vision. In: 2017 IEEE Winter Conference on Applications of Computer Vision (WACV), Santa Rosa, CA, USA, pp. 1026–1035. IEEE (2017)
8. Nguyen, T.P., Choi, S., Park, S.-J., Park, S.H., Yoon, J.: Inspecting method for defective casting products with convolutional neural network (CNN). Int. J. Precision Eng. Manuf.-Green Technol. **8**(2), 583–594 (2021)
9. Kuo, J.-K., Wu, J.-J., Huang, P.-H., Cheng, C.-Y.: Inspection of sandblasting defect in investment castings by deep convolutional neural network. Int. J. Adv. Manuf. Technol. **120**(3–4), 2457–2468 (2022)
10. Jiang, X., Wang, X., Chen, D.: Research on defect detection of castings based on deep residual network. In: 2018 11th International Congress on Image and Signal Processing, BioMedical Engineering and Informatics (CISP-BMEI), pp. 1–6. IEEE (2018)

11. Wang, Y., Hu, C., Chen, K., Yin, Z.: Self-attention guided model for defect detection of aluminium alloy casting on X-ray image. Comput. Electr. Eng. **88**, 106821 (2020)
12. Ren, S., He, K., Girshick, R., Sun, J.: Faster R-CNN: towards real-time object detection with region proposal networks. IEEE Trans. Pattern Anal. Mach. Intell. **39**(6), 1137–1149 (2017)
13. Redmon, J., Divvala, S., Girshick, R., Farhadi, A.: You only look once: unified, real-time object detection. In: 2016 IEEE Conference on Computer Vision and Pattern Recognition (CVPR), Las Vegas, NV, USA, pp. 779–788. IEEE (2016)
14. Jiao, L., et al.: A survey of deep learning-based object detection. IEEE Access **7**, 128837–128868 (2019)
15. Ferguson, M., Ak, R., Lee, Y.-T.T., Law, K.H.: Automatic localization of casting defects with convolutional neural networks. In: 2017 IEEE International Conference on Big Data (Big Data), Boston, MA, pp. 1726–1735. IEEE (2017)
16. Du, W., Shen, H., Fu, J., Zhang, G., He, Q.: Approaches for improvement of the X-ray image defect detection of automobile casting aluminum parts based on deep learning. NDT E Int. **107**, 102144 (2019)
17. Lin, T.-Y., Dollar, P., Girshick, R., He, K., Hariharan, B., Belongie, S.: Feature pyramid networks for object detection. In: 2017 IEEE Conference on Computer Vision and Pattern Recognition (CVPR), Honolulu, HI, pp. 936–944. IEEE (2017)
18. He, K., Gkioxari, G., Dollár, P., Girshick, R.: Mask R-CNN. In: Proceedings of the IEEE International Conference on Computer Vision, pp. 2961–2969 (2017)
19. Xing, J., Jia, M.: A convolutional neural network-based method for workpiece surface defect detection. Measurement **176**, 109185 (2021)
20. Du, W., Shen, H., Fu, J., Zhang, G., Shi, X., He, Q.: Automated detection of defects with low semantic information in X-ray images based on deep learning. J. Intell. Manuf. **32**(1), 141–156 (2021)
21. Alzubaidi, L., et al.: A survey on deep learning tools dealing with data scarcity: definitions, challenges, solutions, tips, and applications. J. Big Data **10**(1), 46 (2023)
22. Gong, Y., Luo, J., Shao, H., Li, Z.: A transfer learning object detection model for defects detection in X-ray images of spacecraft composite structures. Compos. Struct. **284**, 115136 (2022)
23. Jaiswal, A., Babu, A.R., Zadeh, M.Z., Banerjee, D., Makedon, F.: A survey on contrastive self-supervised learning. Technologies **9**(1), 2 (2020)
24. Chen, T., Kornblith, S., Norouzi, M., Hinton, G.: A Simple Framework for Contrastive Learning of Visual Representations (2020)
25. Khosla, P., et al.: Supervised contrastive learning. In: Advances in Neural Information Processing Systems, vol. 33, pp. 18661–18673 (2020)
26. Sun, B., Li, B., Cai, S., Yuan, Y., Zhang, C.: FSCE: few-shot object detection via contrastive proposal encoding. In: 2021 IEEE/CVF Conference on Computer Vision and Pattern Recognition (CVPR), Nashville, TN, USA, pp. 7348–7358. IEEE (2021)
27. He, K., Zhang, X., Ren, S., Sun, J.: Deep residual learning for image recognition. In: 2016 IEEE Conference on Computer Vision and Pattern Recognition (CVPR), Las Vegas, NV, USA, pp. 770–778. IEEE (2016)

Development and Testing of an MRI-Compatible Immobilization Device for Head and Neck Imaging

Francisco Zagalo[1,2]([envelope]) [iD], Susete Fetal[3,4] [iD], Paulo Fonte[3,4] [iD], Antero Abrunhosa[2] [iD], Sónia Afonso[2] [iD], Luís Lopes[4] [iD], and Miguel Castelo-Branco[2] [iD]

[1] Department of Physics, University of Coimbra, 3004-516 Coimbra, Portugal
fphzagalo@gmail.com
[2] Institute for Nuclear Sciences Applied to Health (ICNAS), University of Coimbra, 3000-548 Coimbra, Portugal
[3] Polytechnic Institute of Coimbra, Coimbra Institute of Engineering, Rua Pedro Nunes - Quinta da Nora, 3030-199 Coimbra, Portugal
[4] LIP - Laboratory of Instrumentation and Experimental Particle Physics, 3004-516 Coimbra, Portugal

Abstract. MRI imaging with long acquisition times is prone to motion artifacts that can compromise image quality and lead to misinterpretation.

Aiming to address this challenge at the sub-millimeter level, we developed and evaluated a maxilla immobilization approach, which is known to have better performance than other non-invasive techniques, using a personalized mouthpiece connected to an external MRI-compatible frame.

The effectiveness of the device was evaluated by analyzing MRI imagery obtained in different immobilization conditions on a human volunteer. The SURF and Block Matching algorithms were assessed, supplemented by custom software.

Compared with simple cushioning, the immobilizer reduced the amplitudes of involuntary slow-drift movements of the head by more than a factor two in the axial plane, with final values of 0.25 mm and 0.060°. Faster involuntary motions, including those caused by breathing (which were identifiable), were also suppressed, with final standard deviation values below 0.045 mm and 0.025°.

It was also observed a strong restriction of intentional movements, translationally and angularly, by factors from 7.8 to 4.6, with final values of 0.5 mm and 0.2° for moderate forcing.

Keywords: Head Immobilization · Magnetic Resonance Imaging · Image Processing · Movement Detection

1 Introduction

Several systems dedicated to the immobilization of the head and neck are described in the literature, both for imaging purposes or for radiotherapy and radiosurgery.

Promising results for thermoplastic masks were reported in various studies, showing an average intrafraction head motion of around 0.7 mm [1, 2]. Mandija, et al. [3] showed

V. Vasconcelos et al. (Eds.): CIARP 2023, LNCS 14469, pp. 617–629, 2024.
https://doi.org/10.1007/978-3-031-49018-7_44

that while using thermoplastic masks forced movements inside the radiofrequency (RF) coil during MRI acquisitions rarely exceeded 1.5 mm.

Invasive frame-based immobilization devices, such as the Brown-Robert-Wells and Cosman-Robert-Wells frames, assured sub-millimetric immobilization, assuring values of intrafraction motion of 0.4 ± 0.3 mm and 0.3 ± 0.21 mm, respectively, as reported in [1].

For non-invasive frame-based immobilization systems, relying on the immobilization of the patient's maxilla through the utilization of a patient-specific mouthpiece, such as the radiosurgical PinPoint® system by Aktina Medical, the average head-motion magnitude of the patients was 0.3 ± 0.2 mm [4], with other studies showing similar results.

With the immediate exclusion of the invasive techniques, due to impracticability in MRI imaging, and with no guarantee that the thermoplastic masks could assure sub milli-metric immobilization, we chose to explore a non-invasive frame-based immobilization device.

The system had to be adapted to an existing MRI scanner, including its table size, bore diameter and the specific RF coil utilized for head and neck imaging, shown in Fig. 1. The strategy for laying the patient into the coil and inserting the mouthpiece also had to be carefully considered.

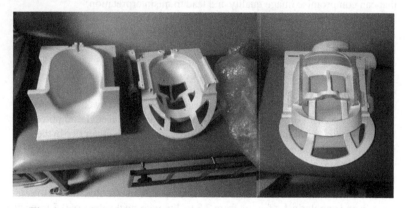

Fig. 1. Head/Neck 64 Coil, open (left) and closed (right), from Siemens.

2   Materials and Methods

2.1   The Immobilization Device

The device is divided into three main detachable parts: the base plate, the external frame, and the mouthpiece. Each of these main blocks can be also decomposed into different components. The external frame was fixed to the base plate, where the patient will be lying, and the mouthpiece was then connected to the external frame above the RF coil, right where the mouth entrance of the coil is. The position of the mouthpiece can be adjusted in an angular range of 180° on the lower joint and 360° on the upper one, and by 85 cm in the vertical translational axis, as seen in Fig. 2, allowing to adjust to each different patient and, possibly, different MRI head and neck coils and scanners. After the patient is positioned, all degrees of freedom can be individually blocked by hand screws, shown in Fig. 3.

The generic design mouthpiece is personalized for each patient by impregnating its grip with a fast-setting silicone moulding paste of the type generally used in dentistry.

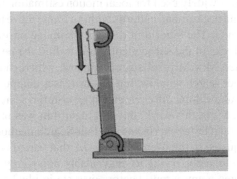

Fig. 2. Schematics of the external frame with its rotational and translational degrees of freedom, indicated by the arrows.

For most MRI applications, PMMA (Poly(methyl methacry-late)) compatible material is used but, with the need of autoclaving the mouthpiece between usages, making it reusable for multiple patients, POM-C (Polyoxymethylene Copolymer) was the next best choice [5]. Besides, POM-C is more suited for mechanical machining, so it was the material chosen to produce the immobilization device.

Fig. 3. Final assembly of the immobilization device with the Head and Neck coil.

2.2   Software Developed to Analyse MRI Images

To assess the existence of movement inside the MRI system while the tests were undergoing, software was developed to analyse the MRI imagery and accurately calculate the translations and rotations present in those sequences. After some preliminary tests, two different algorithms were assessed: the Block Matching (BM) algorithm and the Speeded Up Robust Features (SURF) algorithm. All the algorithms were implemented in MATLAB® [6].

The BM algorithm is widely used for local motion estimation in image sequences. It assumes that objects and background patterns within a frame move to form corresponding objects in the next frame. The algorithm divides the image into blocks and compares them with adjacent blocks in the subsequent frame to find the best match. This search is limited to a specified M x M pixel search window, which can vary in size. Larger motions require a larger search window, but this increases computational complexity.

Various BM-type algorithms differ in matching criteria, search method, and block size determination [7, 8]. In this work, the BM algorithm was utilized with the Mean Squared Error (MSE) matching criteria and the Full-Search method, with varying block and window sizes, adjusted to the overall image size that varied between different MRI acquisitions [8]. Unfortunately, owing to unavoidably noisy grey tone fluctuations, the BM algorithm generates many wrong vectors over static blocks located on the background of the image [9]. We applied a processing algorithm, illustrated in Fig. 4, that we called Histogram Processing (HP), to the data of the BM algorithm. It collects all the motion vectors (MVs) from the MV matrix computed, excluding the low-value MV ones in the background, and visualizes them in a histogram based on vector intensity. The histogram is then analysed and fitted with a Gaussian probability density function. This fit provides valuable information to determine which vectors to accept or remove. Vectors within a tolerance interval of the peak intensity are accepted, using the Gaussian fit as a reliability criterion.

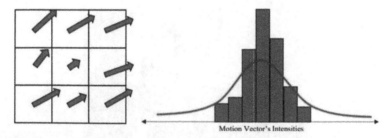

Fig. 4. Histogram Processing principle visualized. The vector intensities from the motion vector matrix on the left are represented in the histogram on the right, where a Gaussian fitting curve was later applied.

The SURF method is an image comparison algorithm that detects and describes scale and rotation-invariant interest points. It involves three main steps: detecting salient features like corners or blobs from a scale-invariant representation, ensuring repeatability across different images and viewing conditions; constructing orientation-invariant descriptors for each interest point, capturing neighbourhood information; and matching the descriptors between images using distance or similarity metrics. Descriptors should possess distinctiveness and robustness to noise, detection errors, and geometric/photometric deformations [10].

The motion estimation results are provided in the form of a transformation matrix, the final parameters consisting of one scale factor, one rotation angle and two translation parameters. However, since the image sequences we are analysing in this work are all collected from the same spatial plan, the scale factor can be neglected and a rigid transformation matrix can be used, providing better results.

We compared two approaches for the motion estimation algorithms. In the first approach, sequential comparison of neighbouring image sequence frames was performed, obtaining the total movement range by their cumulative sum. The second approach, which was the one adopted, compared all image sequence frames to the first frame. Motion estimation was then applied between the selected frames and repeated for each frame until the end of the sequence. The resulting variables represented horizontal (XX) and vertical (YY) translations in pixels, as well as computed rotations in degrees. The data was then converted and presented movement profiles in millimetres or degrees over time for subsequent interpretation and analysis.

The general organization of the analysis is schematized in Fig. 5.

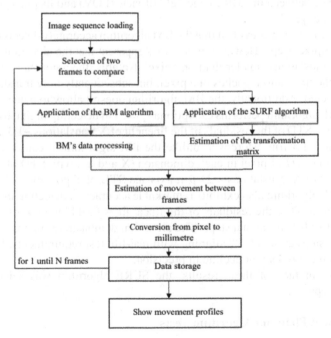

Fig. 5. Flow chart of the software developed to analyze MRI images.

As there are biomechanical limits to the velocity of head movements, it is clear that any large single-frame shifts arising in the data must be of algorithmic origin. Such outliers were discarded by fitting the histogram of the sequential displacement differences with a Gaussian distribution and discarding the data points beyond ±3σ. The fraction of discarded frames was typically lower than 10%.

2.3 Digital Algorithm Tests

The precision and accuracy of the motion detection algorithms was first evaluated by processing mathematical transformations of realistic MRI brain images.

For translational tests, a head and neck MRI image was shifted along the XX axis, simulating translations pixel by pixel up to twenty translations. Subpixel translations were achieved by oversizing the image and applying pixel-by-pixel shifts, followed by downsizing to the original size, resulting in subpixel translation sequences. Rotation tests followed a similar approach, applying known rotations with steps of different angles (2, 1, 0.5, 0.2, 0.1, and 0.05 degrees) to the original MRI image, generating a sequence of nine rotations for each step.

The limits of the algorithms were analyzed through the standard deviation (STD) of the residuals of the linear fit of the computed (detected) translations, in (separately) the XX and YY axis, and the standard deviation of the residuals of the linear fit of the computed rotations. Translation results were presented in millimeters by establishing a correspondence between the MR image field of view (FOV) and its pixel size, resulting in 0.329 mm per pixel.

From these tests, it was evident that the BM algorithm accurately detects digital translations of one-pixel steps. However, it becomes apparent that the algorithm is incapable of detecting translations smaller than one pixel. Notably, the algorithm detects motion even before the translation reaches one pixel, but subsequently only translations of one pixel are detected. Additionally, the BM algorithm doesn't allow to detect rotations.

The SURF algorithm performs exceptionally well in detecting digital translations of one-pixel steps: STD of the residuals of the linear fit of X translations ~1,3 μm and STD of Y positions ~1,4 μm. Moreover, unlike the BM algorithm, it can detect sub-pixel translations up to 1/7 of pixel in each dimension (X and Y), with STD of the residuals of the linear fit of X translations <21 μm axis and STD of Y positions <7 μm.

The SURF algorithm also exhibits remarkable accuracy in detecting induced digital rotations, with STD of the residuals of the linear fit $\leq 0.011°$ with worse results only appearing below 0.1° rotational steps. The algorithm demonstrates impressive precision for rotational steps around 0.1°, aligning with real-life test requirements. However, for smaller steps there is a slight decline in precision.

Therefore, in face of these results, the SURF algorithm was selected for the subsequent steps.

2.4 Movement Phantom Algorithm Tests

With the purpose of generating consistent and predictable displacements, we constructed a phantom composed by water and oil filled balloons enclosed in a container, assembled in a structure where translations and rotations are done interposing spacers as depicted in Fig. 6.

Three image sequences were acquired, each comprising three axial slices of the phantom. One sequence consisted of ten images generated with the phantom in a fixed position, aimed at assessing the intrinsic precision of the algorithms in face of the imaging noise. Another sequence was obtained using 1 mm spacers, resulting in five images with translations of 0, 1, 2, 3 and 4 mm. The last sequence corresponded to rotations (limited to 3 mm of spacing – see Fig. 6) of 0°, 0.39515°, 0.79031° and 1.1855°. All analyses involving translations or rotations were conducted using both the original-sized images and images oversized by a factor of 3 in each direction to facilitate the detection of smaller-scale movements.

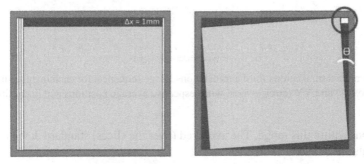

Fig. 6. Translation inducing procedure scheme, on the left, with the introduced spacers to create a 1 mm translation. Rotation inducing procedure scheme, on the right, with the introduced spacer, circled in red, and the rotation angle Θ created.

Analyzing the computed translations returned by the SURF-based algorithm (the best performing algorithm in the previous tests) for the ten frames of the phantom exactly in the same position, we calculated a mean standard deviation of 27.8 μm for the XX axis and 29.7 μm for the YY axis, as well as a rotational standard deviation of 0.0118 degrees, indicating the precision limits of the method.

For the translation sequences, each of the five positions of the phantom were imaged in three different slices of the transversal plane, ten times each to generate statistics, so in the end, the sequence is made up of a hundred and fifty images.

From the translational data, seen in Fig. 7, we can retrieve a linear fit slope for the XX axis of 1.0237 which corresponds to a computed average of 1.0237 mm per translation, establishing the good linearity of the motion analysis. As expected, the YY translations are nearly null.

The averaged (over the slices) standard deviations of the fit residuals are 145.6 μm for the XX axis and 146.7 μm for the YY axis. These values can be considered as indicative of the global accuracy of the algorithm for large movements, when the movement induces a change in the image patterns.

For rotational movements, as for translation, we acquire ten frames in each angle for three different slices of the transversal plane. Figure 8 presents the data retrieved from the computed rotations. From the linear fit of the average of the slices, it can be inferred that although the rotational steps did not precisely match 0.395°, the proximity of this value along with a reasonably good linearity instils confidence in the algorithm

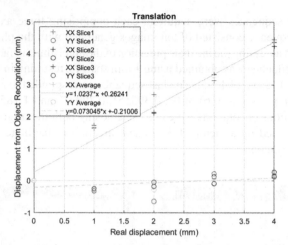

Fig. 7. Computed translations for the translations image sequence, for each imaged slice and for the XX (crosses) and YY (circles) axis, with respective average positions and linear fits.

for rotations within this range. The averaged (over the slices) standard deviations of the fit residuals are $0.0319°$.

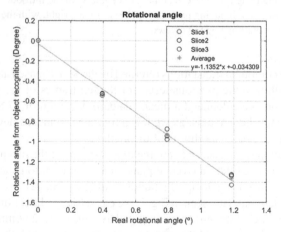

Fig. 8. Computed rotations from the rotations image sequence, with respective average and linear fit.

3   Results

The immobilization tests took place at the National Brain Imaging Network facilities in Coimbra, Portugal, equipped with a MAGNETOM Prisma Fit MRI scanner with a Head/Neck 64 Coil.

A volunteer participated in three main tests: one involving forced movements exerted with maximum force (the "Big" sequence), another with approximately 50% of the maximum force applied (the "Small" sequence), and a final test where the volunteer aimed to remain still (the "Rest" sequence). In total, twelve image sequences were generated to analyze the corresponding "Big", "Small" and "Rest" movements, both with and without the immobilization device, in the sagittal and axial planes. It should be noted that in the non-immobilized situation the head movements were still restricted by lateral cushioning routinely applied between the head and the internal coil wall, so we are comparing the immobilization with this case and not with a fully free head.

All sequences were T1-weighted, and each frame was acquired at 0.35 s intervals. This resulted in a total of 350 frames for each sequence, with duration of 2 min and 3 s per sequence. The movements applied in each plane are aligned with the orientation of the respective plane, as depicted in Fig. 9.

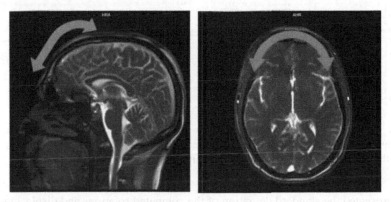

Fig. 9. Sagittal head motion, indicated by the arrows, on the left, and axial head motion, indicated by the arrows, on the right.

The results for the "Rest" image sequences are presented in Fig. 10, for the sagittal and axial planes, with and without the immobilizer. The quantities extracted are the angular rotation in the specified plane and the displacement, computed as:

$$R = \sqrt{x^2 + y^2} \qquad (1)$$

It is apparent that the data shows relatively fast movements, on the level of seconds, superimposed on a slow overall position drift. We characterized separately both types of movement by smoothing the raw data with a moving average of span 20 (8 s) and measuring the amplitude of the drift over the 120 s time span along with the standard deviation of the data relative to the smoothed curve. This last quantity includes the algorithmic noise that was already quantified in Sect. 2.4; the theme will be further developed in the Discussion.

"STD (raw-smoothed)" is the standard deviation of the raw data relative to the smoothed curve.

With the "Small" and "Big" intentional movement sequences we intend to verify the movement restriction limits that are enforced by the use of the immobilizer. In Fig. 11

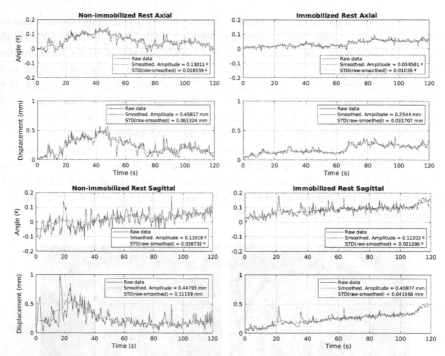

Fig. 10. Computed rotations and displacements for the axial and sagittal "Rest" sequences, with and without the immobilizer. The smoothed line indicates the slow head drift movements during the acquisition. "Amplitude" refers to the full range of the smoothed curve while

it is represented the standard deviation of the computed rotations and displacements for the "Big" and "Small" sequences, in the sagittal and axial planes, with or without the immobilization device.

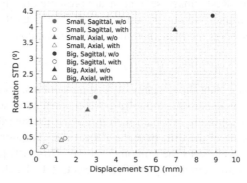

Fig. 11. Standard deviation (STD) of the computed rotations and displacements for the "Big" and "Small" sequences, in the sagittal and axial planes, with or without (w/o) the immobilization device.

4 Discussion

A summary of the quantities extracted from the data displayed in Fig. 10, concerning the "Rest" image sequences, is shown in Fig. 12.

The non-immobilized range of the head drift amplitude is close to 0.45 mm and 0.13° for both the sagittal and axial planes. The introduction of the immobilizer reduces these values to 0.41 mm and 0.12° for the former and to 0.26 mm and 0.059° for the later. There is then an asymmetry on the drift suppression, with the axial movements (head rotations around the neck) being more suppressed than the sagittal ones. This is to be expected from the mechanics of the immobilizer, because the axial rotations are more efficiently blocked by the mouthpiece since the teeth are molded into the silicone filling and can hardly move laterally. On the contrary, sagittal movements can easily arise if the subject loosens her grip on the mouthpiece, creating a small slack that can easily accommodate a rotation.

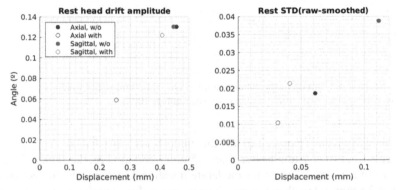

Fig. 12. Summary of the data displayed in Fig. 10. The quantities are defined in the previous section.

On what concerns the standard deviation of the difference between raw and smoothed data - "STD(raw-smoothed)" - corresponding to faster movements, we observe that, on the contrary, the immobilization is more effective in the sagittal plane, where the values go from 0.11 mm, 0.039° to 0.041 mm, 0.021°, while in the axial plane it goes from 0.019 mm, 0.061° to 0.010 mm, 0.032°.

It should be noted that these values are already close to the algorithmic precision limits determined in Sect. 2.4, so one should expect that the values cannot be strongly reduced further by immobilization.

Curiously, the non-immobilized STD value is much larger for the sagittal than for the axial plane. To further understand the phenomenon, we calculated the spectral magnitude of the rotational displacement for both planes, with and without immobilization, as shown in Fig. 13. It can be perceived that for the sagittal plane there is a strong peak around 0.2 Hz, which is absent from the axial plane and which is suppressed by the immobilizer. Such findings are compatible with the movements that might be induced by respiration because, besides the frequency being in the correct range [11], from body mechanics there should be a rotational displacement of the head in the sagittal plane arising from the respiratory movements.

For both planes it can be perceived that the rotation suppression is most effective for the lower frequencies, with the high frequency movements likely corresponding to the algorithmic noise.

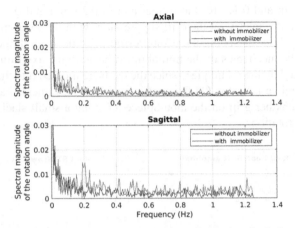

Fig. 13. Spectral magnitude of the rotation angle in the axial and sagittal planes with or without immobilization, calculated from the data displayed in Fig. 10.

From Fig. 11 it can be perceived that the immobilizer restricted the strong intentional movements from over 7 mm to less than 1.5 mm (measured by the standard deviation of the movement series), while for moderate intentional movements the reduction was from over 2.5 mm to less than 0.5 mm. The corresponding angular restriction was from over 3.9° to less than 0.5° and from more than 1.3° to less than 0.2°.

Therefore, there is a strong restriction of the intentional movements, translationally and angularly, by factors from 7.8 to 4.6 compared with simple cushioning. Some part of these residual movements can be imputed to the mechanical elasticity of the frame, which could be reinforced in a future version.

5   Conclusions

We developed and tested a maxilla immobilization approach to head immobilization in MRI imaging. The tests were performed on a human volunteer and compared the device with simple cushioning placed between the head and the RF coil.

For involuntary slow-drift movements the immobilizer reduced the movement amplitudes by more than a factor two in the axial plane, with final values of 0.25 mm and 0.060°. There was little comparative advantage in the sagittal plane, showing values around 0.41 mm and 0.12°.

Faster motions, including those caused by breathing (which were identifiable), were also suppressed, with final values below 0.045 mm and 0.025° (standard deviation).

However, there is indication that these values are close to the limit of precision of the algorithm.

It was observed a strong restriction of intentional movements, translationally and angularly, by factors from 7.8 to 4.6 compared with simple cushioning, with final values of 0.5 mm and 0.2° for moderate forcing. Some part of these residual movements can be imputed to the mechanical elasticity of the frame, which could be reinforced in a future version.

The tests carried out allow us to conclude that the proposed non-invasive immobilization system allows restricting the patient's movements to a sub-millimeter level. This system can be adapted to other head and neck imaging techniques where submillimeter resolution is required, such as brain PET (Positron Emission Tomography).

References

1. Babic, S., et al.: To frame or not to frame? Cone-beam CT-based analysis of head immobilization devices specific to linac-based stereotactic radiosurgery and radiotherapy. J. Appl. Clin. Med. Phys. **19**(2), 111–120 (2018)
2. Ramakrishna, N., et al.: A clinical comparison of patient setup and intra-fraction motion using frame-based radiosurgery versus a frameless image-guided radiosurgery system for intracranial lesions. Radiother. Oncol. **95**(1), 109–115 (2010)
3. Mandija, S., et al.: Brain and head-and-neck MRI in immobilization mask: a practical solution for MR-only radiotherapy. Front. Oncol. **9**(1), 647 (2019)
4. Li, G., et al.: Motion monitoring for cranial frameless stereotactic radiosurgery using video-based three-dimensional optical surface imaging. Med. Phys. **38**(7), 3981–3994 (2011)
5. Waplera, M.C., et al.: Magnetic properties of materials for MR engineering, micro-MR and beyond. J. Magn. Reson. **242**(1), 3981 (2014)
6. The MathWorks, Inc. MATLAB version: 9.11.0 (R2021b) (2021). https://www.mathworks.com. Accessed 01 Sept 2022
7. Tekalp, A.M.: Digital Video Processing. Prentice Hall PTR, Upper Saddle River (1995)
8. Barjatya, A.: Block Matching Algorithms For Motion Estimation (2004)
9. Di Stefano, L., Viarani, E.: Vehicle detection and tracking using the block matching algorithm (1999)
10. Bay, H., et al.: SURF: speeded up robust features. Comput. Vis. Image Underst. **110**(3), 346–359 (2008)
11. Flenady, T., Dwyer, T., Applegarth, J.: Accurate respiratory rates count: so should you! Aust. Emerg. Nurs. J. **20**(1), 45–47 (2017)

DIF-SR: A Differential Item Functioning-Based Sample Reweighting Method

Diego Minatel[⊠], Antonio R. S. Parmezan, Mariana Cúri,
and Alneu de A. Lopes

Institute of Mathematics and Computer Science, University of São Paulo, São Carlos,
Brazil
{dminatel,parmezan}@usp.br, {mcuri,alneu}@icmc.usp.br

Abstract. In recent years, numerous machine learning-based systems have actively propagated discriminatory effects and harmed historically disadvantaged groups through their decision-making. This undesired behavior highlights the importance of research topics such as fairness in machine learning, whose primary goal is to include fairness notions into the training process to build fairer models. In parallel, Differential Item Functioning (DIF) is a mathematical tool often used to identify bias in test preparation for candidate selection; DIF detection assists in identifying test items that disproportionately favor or disadvantage candidates solely because they belong to a specific sociodemographic group. This paper argues that transposing DIF concepts into the machine learning domain can lead to promising approaches for developing fairer solutions. As such, we propose DIF-SR, the first DIF-based Sample Reweighting method for weighting samples so that the assigned values help build fairer classifiers. DIF-SR can be seen as a data preprocessor that imposes more importance on the most auspicious examples in achieving equity ideals. We experimentally evaluated our proposal against two baseline strategies by employing twelve datasets, five classification algorithms, four performance measures, one multicriteria measure, and one statistical significance test. Results indicate that the sample weight computed by DIF-SR can guide supervised machine learning methods to fit fairer models, simultaneously improving group fairness notions such as demographic parity, equal opportunity, and equalized odds.

Keywords: Fairness · Item response theory · Data bias · Preprocessing method · Machine learning

1 Introduction

Over the past few years, we have noticed a significant rise in studies reporting cases where machine learning systems have encouraged unfair decision-making, exacerbating prejudices, stereotypes, and social inequalities in our society. These models have contributed to the propagation of discriminatory effects across

© Springer Nature Switzerland AG 2024
V. Vasconcelos et al. (Eds.): CIARP 2023, LNCS 14469, pp. 630–645, 2024.
https://doi.org/10.1007/978-3-031-49018-7_45

diverse applications, such as web searches, recruitment systems, facial recognition, and court decisions [21].

After the referred cases came to light, authorities and civil society have become more vigilant regarding machine learning-driven decisions and their long-term impacts on historically disadvantaged groups, even when predictive models are not explicitly designed to discriminate [24]. On one hand, a significant challenge for artificial intelligence concerns the development of non-discriminatory machine learning systems [10]. On the other hand, researchers from this field have mobilized to fill the gaps in the research topic of fairness in machine learning, which aims to address and mitigate algorithmic biases and promote equitable outcomes in decision-making processes [20].

Therefore, a way to make predictive models fairer is to embed fairness notions into machine learning algorithms when dealing with tasks that involve people. According to Mehrabi *et al.* [20], there are three potential avenues for inserting fairness notions throughout the machine learning process: (i) preprocessing, (ii) in-processing, and (iii) post-processing. Preprocessing approaches aim to eliminate data bias by modifying the dataset before the training stage [15,19]. In-processing approaches include fairness notions during classifier training, *e.g.*, adding fairness constraints on the optimization function of the classification algorithm [26]. Finally, post-processing techniques modify a model's predictions to make them fairer [23].

Each listed approach provides unique strategies to tackle and alleviate biases in machine learning systems, offering valuable tools to promote fairness and equality in particular applications and domains. This work focuses on preprocessing approaches due to their model agnosticism, especially in sample reweighting methods. That means the same preprocessing algorithm can be applied across various machine learning algorithms and settings, allowing for a more flexible and adaptable approach to combating data biases.

Fairness and unfairness analyses have been studied for a long time in education and employment testing, even before they became a hot topic in machine learning. Indeed, many fairness concepts used in machine learning have already been developed and applied in these areas, particularly after the United States Civil Rights Act was passed in 1964 [14]. One of the mathematical tools employed in educational tests and psychometrics to reduce bias in elaborating tests is Item Response Theory (IRT), which has one of its principles to select the most appropriate questions to analyze a latent trait [9]. Another essential mathematical tool for reducing test bias is Differential Item Functioning (DIF). DIF-based methods detect test items that benefit or harm certain groups based on gender, race, country of origin, and others that should not be relevant in the test evaluation [13].

Therefore, IRT and DIF are valuable instruments to decrease bias in the applicant selection process. Thus, we firmly believe that transposing these concepts into the machine learning domain is crucial. According to Hutchinson and Mitchell [14], using DIF concepts can lead to the development of fairer machine learning solutions, making it a promising direction for fairness in machine learning. This paper handles this gap and proposes a novel sample reweighting method for binary classification called DIF-SR that uses the DIF and IRT concepts to assign weights to sample examples. DIF-SR aims to guide classification algorithms to fit fairer models.

This paper presents four notable contributions. Firstly, it approaches the sample weight by modeling it as a test items problem. Secondly, it applies the IRT and DIF principles to estimate sample examples' weights. Thirdly, it introduces a novel sample reweighting method that guides binary classification algorithms toward developing fairer models. Fourthly, we design an experimental setup based on a multi-criteria evaluation measure never before used in fairness assessment. Herein, this metric combines three fairness measures to help us identify which method generates more impartial results. According to our experimental results, DIF-SR simultaneously improves the central notions of group fairness, making it a powerful preprocessing technique for developing fairer classifiers. With this study, we intend to contribute to fairness in the machine learning community and stimulate discussion on this research topic.

2   Background

This section describes the terminology and fundamental concepts required to understand our proposal in Sect. 3.

2.1   Group Fairness Analysis

Protected attributes refer to features that contain sensitive data, such as gender, race, nationality, religion, and sexual orientation, and require equal treatment regardless of their value. A *group* comprises individuals who share the same protected attributes, such as females and males in the case of gender. When considering race and gender, the groups would be `black men`, `black women`, `white men`, and `white women`. Additionally, a *privileged group* refers to a group or set of groups historically receiving favorable treatment compared with *unprivileged groups*.

Using protected attributes in decision-making, such as hiring and parole tasks, is considered *adverse treatment*. That is often prohibited by law in democratic countries. On the other hand, an *adverse impact* occurs when a particular group is harmed or benefited by outcomes, regardless of whether or not there is adverse treatment [2]. In machine learning, adverse treatment happens when protected attributes are used in model training, while adverse impact refers to disparities in results (*e.g.*, accuracy) among different groups.

Group fairness analysis seeks to identify unfair outcomes among different groups, specifically focusing on adverse impacts. Below are some fundamental group fairness notions commonly applied in binary classification tasks.

Demographic parity: Every group is equally likely to be classified with a positive label (selection rate) [8].

Equal opportunity: Each group exhibits an identical true-positive rate, meaning they all have the same recall score [12].

Equalized odds: All groups share equal true and false-positive rates [12].

We must point out that group fairness notions are determined by comparing group outcomes, enabling the application of group fairness analysis to any performance metric. Disparities in results are usually assessed by calculating the score ratio between privileged and unprivileged groups.

2.2   Item Response Theory

In the field of test assessment, Item Response Theory (IRT) is a collection of mathematical models commonly used in psychometric applications and educational tests like the TOEFL (Test of English as a Foreign Language) [6]. IRT models are designed to enhance evaluation accuracy by describing the relationship between test item responses and test-takers' abilities.

In this work, we are focusing on dichotomous item response models, specifically the two-parameter logistic models. These models help assess tests whose answer to an item (question) is corrected as right or wrong. Table 1 illustrates the data structure for modeling dichotomous items, where $U_{ij} = 1$ denotes a correct response by examinee i on item j, whereas a value of 0 indicates an incorrect answer.

Table 1. Data structure for modeling dichotomous items.

Individual	Item 1	Item 2	$\cdots$	Item k
Examinee 1	1	1	$\cdots$	1
Examinee 2	1	0	$\cdots$	1
$\vdots$	$\vdots$	$\vdots$	$\vdots$	$\vdots$
Examinee n	1	0	$\cdots$	0

Equation 1 shows the two-parameter logistic model (2PL), where the probability of the examinee i correctly responding to item j depends on their ability θ_i. Consequently, the higher the examinee's ability, the more likely the examinee responds correctly to the item [11]. The values of a_j and b_j determine the logistic curve for a particular item j, referred to as the Item Characteristic Curve (ICC).

$$P(U_{ij} = 1 \mid \theta_i) = \frac{1}{1 + e^{-a_j(\theta_i - b_j)}}, \tag{1}$$

with $i = 1, 2, ..., m$ (m = total of examinees), and $j = 1, 2, .., n$ (n = total of items), where:

- U_{ij} represents the observed response that assumes the value 1 when examinee i to respond item j correctly, or 0 otherwise.
- θ_j indicates the ability of individual j.

– a_i is the discrimination parameter of item i.
– b_i is the difficulty parameter of item i.

The ICC's dependence on parameters a and b is depicted in Fig. 1. Parameter b, the difficulty parameter, determines where the ICC is located on the ability scale. An item's difficulty increases as the b's value increases, making it more difficult to answer correctly, as shown in Fig. 1b. Generally, ability values are normally distributed with a zero mean and a unit standard deviation. This implies that θ values usually fall between -4 and 4, while b values fall between -2 and $+2$ [11]. The discrimination parameter a is proportional to the slope of the ICC at point b on the x-axis. As a result, ICCs with high values provide better differentiation between examinees who answer the item correctly and those who do not. In contrast, low values indicate that examinees of varying abilities have similar probabilities of answering the item correctly, as illustrated in Fig. 1a.

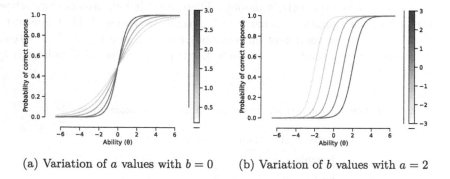

(a) Variation of a values with $b = 0$ (b) Variation of b values with $a = 2$

Fig. 1. Example of Item Characteristic Curves.

2.3   Differential Item Functioning

DIF is a set of methods used to identify which items in a test work differently for distinct groups of examinees [13]. The process, a.k.a. DIF detection, holds significant importance in selection testing. If DIF is detected in a test item, it implies that individuals with similar abilities have varying probabilities of answering correctly solely due to their membership in different demographic groups, such as gender and ethnicity.

As part of DIF detection, an important task involves identifying bias in selection processes via tests, where certain groups may gain an advantage in answering specific items. By detecting DIF, we can create fairer tests that do not unfairly benefit or disadvantage any particular group.

Various methods for detecting DIF in dichotomous items, such as the area method and Lord's chi-square, are available [13]. Figure 2 exhibits examples of DIF detection by the area method. Note that one CCI is generated for each tested group; in this case, a privileged and an unprivileged group. The larger the area delimited by the CCIs, the more partial the item test favors one group over the other.

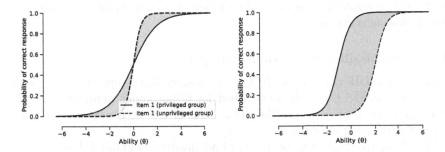

Fig. 2. Example of DIF detection by the area method.

2.4 Related Work

The authors in [18] analyze the use of IRT in classification tasks. They modeled the test set as items and a pool of classifiers as examinees. Their experiments showed a high correlation between the classifiers' ability and accuracy and that the optimal classifier is not necessarily with the greatest ability. Furthermore, they discussed some directions for using IRT in machine learning, such as identifying noisy instances, selecting subsets of more representative instances, and choosing and evaluating classifiers. Some machine learning applications incorporating IRT concepts have emerged in this sense. For instance, the researchers in [5] use IRT concepts to propose novel weighted voting in an ensemble of classifiers. Another case is reported in [4], where the authors use IRT to explain predictive models.

All works reported in this section have one thing in common: They share the exact modeling of items. These studies represent classifiers as examinees and examples as items. The present paper introduces a novel item modeling. We represent classifiers as items and examples as examinees, as described in Sect. 3. This approach is required to allow the use of DIF concepts.

3   Proposed Method: DIF-SR

In Sects. 2.2 and 2.3, we reported how combining IRT and DIF can drive designing less biased tests. Within this context, identifying DIF plays a crucial role in enhancing test fairness and impartiality.

This section introduces our sample reweighting method to binary classification, named DIF-based Sample Reweighting (DIF-SR). The purpose of DIF-SR is to employ the IRT and DIF concepts to assign weight to examples from datasets and consequently guide the classification algorithms in the fit of the models to develop fairer classifiers.

First, we use a set of base classifiers to make predictions from the training set. We then model these predictions as items and the training set as examinees. After modeling the items, we apply the 2PL model to estimate the ICC parameters of each item and θ values. Finally, with θ values, we compute the sample weight. For clarity, we organized our method into four steps and explained each in detail in the following sections.

3.1   Base Classifier Predictions—Step 1

To use IRT and DIF concepts in the binary classification sample reweighting task, we must first model the problem as a test for dichotomous items. A straightforward way to do that is to use the predictions of a base classifier set to model the right and wrong answers.

Therefore, the first step of our method involves training k base classifiers using a training dataset of interest D_n with n examples. Next, we predict the training sample with the k base classifiers. At the end of this step, we have the prediction set $[\hat{Y}_1, \hat{Y}_2, \ldots, \hat{Y}_k]$. We must highlight that it is necessary to satisfy the constraint of $k > 2$ to use the 2PL model, the same used in the IRT calibration [6]. For this reason, the base classifier set needs to consist of at least three classifiers.

3.2   Item Modeling—Step 2

This step is responsible for modeling the items of our problem. It represents the training dataset D_n as examinees and the k base classifiers as dichotomous items. This structure is required because group information is tied to examples. In DIF detection, group information is linked with examinees, as emphasized in Sect. 2.3, so examples must be modeled as examinees. In this way, for each example, we associate whether it belongs to the privileged or unprivileged group. We also use the item modeling U to rearrange the prediction set $[\hat{Y}_1, \hat{Y}_2, \ldots, \hat{Y}_k]$ into items $[I_1, I_2, \ldots, I_k]$. Here, when $U = 1$, classifier j correctly predicted example i; otherwise, $U_{ij} = 0$.

Table 2 schematizes the proposed modeling. We can see from this exemplification that item 2 (I_2) correctly classified examples 3, 4, and 5 and misclassified examples 1 and 2.

Table 2. Modeling the sample weight task as a test. Cell value ij indicates correct (1) or incorrect (0) prediction of example i by model j.

Example	I_1	I_2	...	I_k	Group
1	1	0	...	1	privileged
2	0	0	...	1	unprivileged
3	0	1	...	1	privileged
4	1	1	...	0	privileged
5	1	1	...	1	unprivileged

3.3 IRT Calibration—Step 3

During this step, we apply the 2PL model in the item modeling U to determine the θ value of the examples and the CCIs of the trained classifiers. From now on, these CCIs will be called Classification Characteristic Curve (CCC) for convenience. Our idea is to estimate the CCC parameters a and b so that there is no DIF between the CCCs of the privileged and unprivileged groups. Therefore, we include the following restrictions to IRT calibration given by Eq. 1:

$$a_{privileged} = a_{unprivileged} \tag{2}$$

$$b_{privileged} = b_{unprivileged} \tag{3}$$

The discrimination parameter a and difficulty parameter b must have the same values for both privileged and unprivileged groups in the same item, which is a classifier in our case. This constraint on the calibration of parameters a and b guarantees no DIF between the CCCs of each model belonging to the base classifier set. Therefore, at the end of this step, θ values for the training dataset are estimated without DIF in the analyzed items. To calibrate the CCCs parameters and θ values (both privileged and unprivileged groups), we use the Expectation-Maximization (EM) algorithm [3].

Figure 3 portrays a CCC, with $a = 5$ and $b = 0$, for didactic purposes. We note from this illustration that examples with θ values much smaller than b values (indicated by the red area) are less likely to be predicted correctly. However, examples with θ values much larger than b (indicated by the green area) are highly likely to be classified correctly. Finally, θ values close to b are part of the classifier's uncertainty area (indicated by the white area).

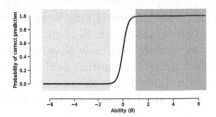

Fig. 3. Exemplification of a CCC: The red area indicates the θ values of examples highly likely to be misclassified. In contrast, the green area indicates a high probability of the classifier correctly predicting the examples. (Color figure online)

3.4 Sample Weighting—Step 4

Finally, after the IRT calibration step, each example is associated with one θ value. Section 2.2 punctuated that θ values range approximately from -4 to 4. Seeking to ease the sample weight assignment formulation, we rescaled the

θ values such that $\theta_{rescaled} \in [1, 8]$. We selected this new scale to eliminate negative values from our sample reweighting formulation and maintain a similar amplitude to the original θ values.

As expressed in Fig. 3, the θ value reflects how difficult it is for the classifiers to label an example accurately. While larger θ values suggest that the models easily classify the example correctly, smaller ones indicate that the example is more difficult to be labeled correctly. Such behavior demonstrates that examples with θ equal to 1 need greater weight in the sample, contrasting those with θ equal to 2 requiring less weight. Following this reasoning, we formulate the sample weight via Eq. 4, where smaller $\theta_{reescaled}$ values have a greater influence on the sample weight and vice versa.

$$DIF - SR = \frac{1}{\theta_{rescaled}} \tag{4}$$

In agreement with Eq. 4, DIF-SR assigns values between $[0.125, 1]$ to each dataset's example.

4   Experimental Setting

We designed an experimental setup to assess the ability of our sample reweighting method to effectively guide classification methods in fitting more impartial models. In this sense, we also applied the classification algorithms without sample reweighting to have a baseline for comparison. Figure 4 summarizes the flowchart of our experimental setting.

According to Fig. 4, the outlined experimental setup provides a dataset describing a binary classification problem for a holdout validation procedure, which splits the original dataset into training (80%) and test (20%) samples. A model selection process employs five-fold cross-validation to estimate the performance of different pairs {sample reweighting method, classification algorithm (hyperparameters)} from the training set. Note that we adopted five-fold cross-validation since the dataset of interest may have few examples. Given a current setting, we apply the sample reweighting method over the training folds so that the resulting sample weights help calibrate the learning of the classification algorithm. The hyperparameterized classification method builds, for each

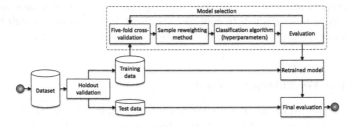

Fig. 4. Summarized flowchart depicting the steps of our experimental setup.

cross-validation iteration, a predictive model using four data partitions, and its performance is evaluated in the remaining partition (validation sample). After building the classifiers, we select the fairest according to a multi-criteria performance measure combining three group fairness metrics (Sect. 2): (i) demographic parity, (ii) equal opportunity, and (iii) equalized odds. Finally, we retrained the identified fairest model using the entire training sample to estimate its generalizability on the test sample.

We conducted experiments using the empirical protocol of Fig. 4 implemented in Python and R programming languages, with the help of the following libraries: `scikit-learn` (classification algorithms), `xgboost`, `aif360` (fairness metrics and reweighing method), and `mirt` (IRT calibration). The following subsections detail the datasets, sample reweighting methods, machine learning algorithms, and evaluation measures used in our experiments. The source code and benchmark datasets are publicly available online[1].

4.1 Datasets

For this study, we selected benchmark binary classification datasets commonly used in the literature on fairness in machine learning. Table 3 lists the datasets, displaying the respective number of examples (#E), number of attributes (#A), the considered protected attributes, the unprivileged group of the task expressed by the data, and its source reference. Figure 5, in turn, shows the ratio of each subset group–class within the datasets.

Table 3. Characteristics of the datasets. Crack and Heroin are the Drug Consumption dataset with a different target class. We split the Recidivism dataset into Recidivism Female and Recidivism Male.

Dataset	#E	#A	Protected attribute	Unprivileged group	Reference
Arrhythmia	452	278	sex	female	[16]
Bank Marketing	45,211	42	age	under 25 years old	[16]
Census Income	48,842	76	race and sex	non-white and female	[16]
Contraceptive	1,473	10	religion	islam	[16]
Crack	1,885	11	race	non-white	[16]
German Credit	1,000	36	sex	female	[16]
Heart	383	13	age	non-middle-aged	[16]
Heroin	1,885	11	race	non-white	[16]
Recidivism Female	1,395	176	race	non-white	[17]
Recidivism Male	5,819	375	race	non-white	[17]
Student	480	46	sex	female	[1]
Titanic	1,309	6	sex	male	[25]

---

[1] https://github.com/diegominatel/dif-based-sample-reweighting-method/.

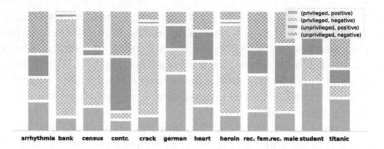

Fig. 5. Ratio of each subset group–class within the datasets.

4.2   Classification Algorithms

We used the following five classification algorithms because they have an argument that makes sample reweighting possible: (i) Adaptive Boosting (ADA), (ii) Classification And Regression Tree (CART), (iii) Random Forest (RF), (iv) Support Vector Machine (SVM), and (v) eXtreme Gradient Boosting (XGB). Table 4 shows each investigated machine learning algorithm and the numerical variation range for its hyperparameters. We tested fifteen hyperparameter settings per classification algorithm.

Table 4. Algorithms and search spaces considered in the hyperparameter estimation step.

Algorithm	Parameters	Fixed value	Variation range (initial:final:step)
ADA	Number of trees	—	100:500:25
CART	Split criterion	Gini	—
	Minimum number of samples to split a node	5	—
	Minimum number of samples to be a leaf node	—	2:30:2
RF	Number of trees	—	100:500:25
	Minimum number of samples to split a node	$\sqrt{\#A}$	—
SVM	Kernel	RBF	—
	Regularization	1	—
	Gamma	—	0.0025:1.075:0.075
XGB	Number of trees	—	100:500:25

4.3   Sample Reweighting Methods

We employed three distinct settings for our method DIF-SR. In all of them, we placed the minimum allowed number of items, three, as described in Sect. 3.1. We opted for a different classification algorithm for each configuration and created three predictive models to compose the base classifier set. Table 5 summarizes these three settings, highlighting which classification algorithm and hyperparameter were varied to generate the classifiers.

Table 5. DIF-SR settings. The acronym not yet defined is K-Nearest Neighbors (KNN).

Instantiated model	Classification algorithm	Hyperparameter	Values
DIF-SR (CART)	CART	Minimum number of samples to be a leaf node	(2, 12, 22)
DIF-SR (KNN)	KNN	Number of nearest neighbours	(3, 11, 23)
DIF-SR (MLP)	MLP	Number of neurons in the hidden layer	(5, 10, 15)

We assessed our method against Reweighing, a well-known preprocessing technique for fairness approaches. Reweighing assigns higher weights to examples of the less frequently occurring group–class based on their pair group–class. We also applied the considered classification algorithms without any sample reweighting to have a baseline for comparison.

4.4   Evaluation Measures

We evaluated the selected classifiers' performance through the macro F1-score due to the class imbalance of some datasets. Additionally, to simplify the categorization of fairness metrics, we use the highest score as the denominator to calculate the ratio of scores between privileged and unprivileged groups for a given group fairness metric, as explained in Sect. 2.1. As a result, the ratio of each fairness metric will always be in the range $[0, 1]$, with the ideal outcome being a score of 1.

We applied a Multi-Criteria Performance Measure (MCPM) [22] to combine three group fairness metrics and thus guide model selection. MCPM reflects the sum of the total area of an irregular polygon whose vertices comprise individual performance measures. Herein, we set demographic parity, equal opportunity, and equalized odd as individual evaluation metrics in MCPM. Higher total area values indicate fairer classification models.

5   Results

This section presents our experimental results obtained in accordance with Sect. 4. We summarize the performances of the sample reweighting methods using the MCPM described in Sect. 4.4.

Table 6 shows the average values of MCPM on the test set, which incorporates three group fairness metrics – (i) demographic parity, (ii) equal opportunity, and (ii) equalized odds – into a single one. Highlighted values correspond to the best result for each dataset, and values in parentheses indicate the standard deviation. The column "Without" refers to results without applying sample reweighting methods.

Table 6. Average MCPM values on the test set reflect the combination of the three following group fairness metrics: (i) demographic parity, (ii) equal opportunity, and (iii) equalized odds.

Dataset	Without	DIF-SR (CART)	DIF-SR (KNN)	DIF-SR (MLP)	Reweighing
Arrhythmia	0.71 (0.07)	0.53 (0.32)	**1.02 (0.24)**	0.62 (0.11)	0.87 (0.12)
Bank Marketing	0.58 (0.06)	0.59 (0.12)	0.63 (0.09)	0.66 (0.15)	**0.75 (0.08)**
Census Income	0.39 (0.03)	0.50 (0.11)	0.58 (0.11)	**0.59 (0.08)**	0.49 (0.06)
Contraceptive	1.11 (0.07)	1.00 (0.20)	0.79 (0.42)	0.79 (0.45)	**1.12 (0.04)**
Crack	0.45 (0.51)	0.46 (0.42)	0.36 (0.42)	**0.81 (0.49)**	0.05 (0.06)
German	0.89 (0.23)	0.94 (0.21)	0.66 (0.37)	0.89 (0.25)	**0.97 (0.20)**
Heart	0.51 (0.22)	0.62 (0.38)	0.67 (0.27)	**0.75 (0.33)**	0.55 (0.04)
Heroin	0.34 (0.23)	0.57 (0.26)	**0.63 (0.40)**	**0.63 (0.43)**	0.24 (0.18)
Recidivism Male	0.77 (0.09)	1.14 (0.08)	**1.17 (0.14)**	1.12 (0.13)	0.83 (0.11)
Recidivism Female	0.89 (0.44)	**1.16 (0.10)**	1.09 (0.13)	1.00 (0.16)	0.88 (0.33)
Student	1.08 (0.16)	1.10 (0.12)	**1.14 (0.10)**	1.03 (0.20)	1.12 (0.11)
Titanic	0.15 (0.02)	0.44 (0.49)	0.39 (0.28)	**0.54 (0.28)**	0.34 (0.22)
Average	0.65 (0.36)	0.75 (0.37)	0.76 (0.37)	**0.78 (0.32)**	0.68 (0.36)

According to Table 6, DIF-SR(MLP) obtained the most promising results in five of the 12 datasets and the best MCPM average. The DIF-SR(KNN) and Reweighing methods achieved the highest performances in four and three datasets, respectively. In contrast, DIF-SR(CART) had the best result in just one dataset. Not using the reweighting method resulted in suboptimal performance across all datasets, with the lowest average outcome. It is important to highlight that the three basic DIF-SR configurations tested resulted in the three best MCPM averages.

We applied Friedman's non-parametric statistical test for paired data and multiple comparisons followed by Nemenyi posthoc test to the MCPM values to verify whether there is a statistically significant difference among the investigated sample reweighting methods. Figure 6 express the results of this statistical validation. The top of the diagram indicates the critical difference (CD), and the horizontal axes indicate the average ranks of the sample reweighting methods, with the best-ranked method to the left; a black line connects the method when it is not detected a significant difference among them [7]. In Figs. 6a and 6b, with a significance level of 5% (p-value < 0.05), the critical difference is 1.76. In contrast, in the diagram of Fig. 6c, with a significance level of 5% (p-value ¡ 0.05), the critical difference is 0.95.

In the validation (Fig. 6a) and the test (Fig. 6b) sets, the three basic settings of DIF-SR obtained the first three positions in the ranking. The non-apply of sample reweighting methods was ranked last, and the Reweighing method was second to last. In the diagram in Fig. 6c, we used the best settings of our method per dataset; thus DIF-SR was ranked first and with a statistically significant difference compared not to applying sample reweighting methods.

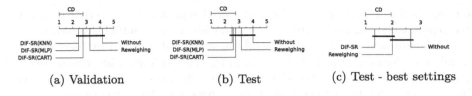

(a) Validation   (b) Test   (c) Test - best settings

Fig. 6. Nemenyi posthoc test applied to the MCPM values.

Table 7 shows the average macro F1-score for the sample reweighting methods. As expected, not using the sample reweighting method had the best macro F1-score. Although DIF-SR led to the worst performances, this fact cannot be seen as a negative point, as we need to decrease the predictive power a little to increase the fairness of the classifier's predictions.

Table 7. Average macro F1-score for the models selected according to the MCPM results.

	Without	DIF-SR (CART)	DIF-SR (KNN)	DIF-SR (MLP)	Reweighing
Macro F1-score	**69.68%**	64.09%	63.42%	63.35%	68.85%

As we can see from the results in Table 6 and Fig. 6, using the DIF-SR can be a great option to improve the main notions of group fairness together. For this experiment, we used the most basic configuration of our method (only three items), and we wanted to show that regardless of the classification algorithm used to generate the base classifiers predictions (first step of our method), DIF-SR is robust enough to improve the fairness group notions. However, using our method had a performance loss compared to not using any sample reweighting method, but this pays off by making the classifiers more impartial in their decisions.

6   Conclusion

This paper introduced DIF-SR, a novel sample reweighting method that uses Item Response Theory and Differential Item Functioning concepts to assign weights to training sets, guiding classification algorithms toward fairer model fitting. Our proposal mitigates bias in tests, minimizing discriminatory effects in classification tasks.

Our experimental results demonstrated that DIF-SR can guide classification algorithms in adjusting models to enhance the central notions of group fairness. It is important to emphasize that these results were achieved with the most basic configuration of DIF-SR, using the minimum number of items regardless of the classification algorithm chosen as a parameter for our method. Such findings underscore the significant potential for improvement inherent in DIF-SR. Finally, our method can be a great option to be used separately or together with other strategies when the goal is to develop fairer binary (or multiclass) classifiers.

In future work, we plan to expand our empirical protocol by increasing the number of items in the DIF-SR modeling, thereby increasing the number of base classifiers. We also intend to introduce classifier heterogeneity by merging predictive models from different algorithms/paradigms to evaluate whether this diversity improves the accuracy of sample weight attribution. With these refinements, we aim to enhance the fairness group metrics and macro F1-score presented in this paper.

Acknowledgments. This study was financed in part by the Coordenação de Aperfeiçoamento de Pessoal de Nível Superior – Brasil (CAPES) – Finance Code 001; the São Paulo Research Foundation [grants #20/09835-1, #22/02176-8, and #22/09091-8]; and the Brazilian National Council for Scientific and Technological Development [grant #303588/2022-5].

References

1. Amrieh, E.A., Hamtini, T., Aljarah, I.: Preprocessing and analyzing educational data set using X-API for improving student's performance. In: IEEE AEECT, pp. 1–5. IEEE (2015)
2. Barocas, S., Selbst, A.D.: Big data's disparate impact. Cal. L. Rev. **104**(3), 671–732 (2016)
3. Bock, R.D., Aitkin, M.: Marginal maximum likelihood estimation of item parameters: application of an EM algorithm. Psychometrika **46**(4), 443–459 (1981)
4. Cardoso, L.F., et al.: Explanation-by-example based on item response theory. In: Xavier-Junior, J.C., Rios, R.A. (eds.) BRACIS 2022. LNAI, vol. 13653, pp. 283–297. Springer, Cham (2022). https://doi.org/10.1007/978-3-031-21686-2_20
5. Chen, Z., Ahn, H.: Item response theory based ensemble in machine learning. Int. J. Autom. Comput. **17**(5), 621–636 (2020)
6. De Ayala, R.J.: The Theory and Practice of Item Response Theory. Guilford Publications, New York (2013)
7. Demšar, J.: Statistical comparisons of classifiers over multiple data sets. J. Mach. Learn. Res. **7**, 1–30 (2006)
8. Dwork, C., Hardt, M., Pitassi, T., Reingold, O., Zemel, R.: Fairness through awareness. In: ITCS, pp. 214–226. ACM (2012)
9. Embretson, S.E., Reise, S.P.: Item Response Theory. Psychology Press (2013)
10. Goodman, B., Flaxman, S.: European union regulations on algorithmic decision-making and a "right to explanation". AI Mag. **38**(3), 50–57 (2017)
11. Hambleton, R.K., Swaminathan, H., Rogers, H.J.: Fundamentals of Item Response Theory, vol. 2. SAGE Publications, Thousand Oaks (1991)
12. Hardt, M., Price, E., Srebro, N.: Equality of opportunity in supervised learning. In: NIPS, pp. 3323–3331. Curran Associates, Inc. (2016)
13. Holland, P.W., Wainer, H.: Differential Item Functioning. Routledge (1993)
14. Hutchinson, B., Mitchell, M.: 50 years of test (un) fairness: lessons for machine learning. In: ACM FAT*, pp. 49–58. ACM (2019)
15. Kamiran, F., Calders, T.: Data preprocessing techniques for classification without discrimination. Knowl. Inf. Syst. **33**(1), 1–33 (2012). https://doi.org/10.1007/s10115-011-0463-8

16. Kelly, M., Longjohn, R., Nottingham, K.: The UCI machine learning repository (2017). http://archive.ics.uci.edu/ml

17. Larson, J., Mattu, S., Kirchner, L., Angwin, J.: How we analyzed the COM-PAS recidivism algorithm (2016). https://www.propublica.org/article/how-we-analyzed-the-compas-recidivism-algorithm

18. Martínez-Plumed, F., Prudêncio, R.B., Martínez-Usó, A., Hernández-Orallo, J.: Item response theory in AI: analysing machine learning classifiers at the instance level. Artif. Intell. **271**, 18–42 (2019)

19. McNamara, D., Ong, C.S., Williamson, R.C.: Costs and benefits of fair representation learning. In: AAAI/ACM AIES, pp. 263–270. ACM (2019)

20. Mehrabi, N., Morstatter, F., Saxena, N., Lerman, K., Galstyan, A.: A survey on bias and fairness in machine learning. ACM Comput. Surv. **54**(6), 1–35 (2021)

21. Minatel, D., dos Santos, N.R., da Silva, A.C.M., Curi, M., Marcacini, R.M., de Andrade Lopes, A.: Unfairness in machine learning for web systems applications. In: Proceedings of the Brazilian Symposium on Multimedia and the Web (2023)

22. Parmezan, A.R.S., Lee, H.D., Wu, F.C.: Metalearning for choosing feature selection algorithms in data mining: proposal of a new framework. Expert Syst. Appl. **75**, 1–24 (2017)

23. Pleiss, G., Raghavan, M., Wu, F., Kleinberg, J., Weinberger, K.Q.: On fairness and calibration. In: NIPS, pp. 5680–5689. Curran Associates, Inc. (2017)

24. Podesta, J.: Big data: seizing opportunities, preserving values. White House, Executive Office of the President, Washington (2014)

25. Vanschoren, J., van Rijn, J.N., Bischl, B., Torgo, L.: OpenML: networked science in machine learning. SIGKDD Explor. **15**(2), 49–60 (2013)

26. Zafar, M.B., Valera, I., Gomez Rodriguez, M., Gummadi, K.P.: Fairness beyond disparate treatment & disparate impact: learning classification without disparate mistreatment. In: WWW, pp. 1171–1180. IW3C2 (2017)

IR-Guided Energy Optimization Framework for Depth Enhancement in Time of Flight Imaging

Amina Achaibou[1,2]([✉]), Filiberto Pla[2], and Javier Calpe[1]

[1] Analog Devices Inc., 46980 Paterna, Spain
{amina.achaibou,javier.calpe}@analog.com
[2] Institute of New Imaging Technologies, University Jaume I, 12071 Castellón, Spain
pla@uji.es

Abstract. This paper introduces an optimization energy framework based on infrared guidance to improve depth consistency in Time of Flight image systems. The primary objective is to formulate the problem as an image energy optimization task, aimed at maximizing the coherence between the depth map and the corresponding infrared image, both captured simultaneously from the same Time of Flight sensor. The concept of depth consistency relies on the underlying hypothesis concerning the correlation between depth maps and their corresponding infrared images. The proposed optimization framework adopts a weighted approach, leveraging an iterative estimator. The image energy is characterized by introducing spatial conditional entropy as a correlation measure and spatial error as image regularization. To address the issue of missing depth values, a preprocessing step is initially applied, by using a depth completion method based on infrared guided belief propagation, which was proposed in a previous work. Subsequently, the proposed framework is employed to regularize and enhance the inpainted depth. The experimental results demonstrate a range of qualitative improvements in depth map reconstruction, with a particular emphasis on the sharpness and continuity of edges.

Keywords: Time of Flight sensor · image fusion · depth enhancement

1 Introduction

Time of Flight (ToF) cameras of 3D sensors are a very competitive 3D sensing choice because of their low cost and relatively high spatial resolution. These sensors consist of an infrared light projector and a depth image sensor. It captures real time depth maps using indirect time of flight technology. However, these devices are unable to correctly estimate depth data in some cases due to larger working distances, occlusions, or low reflective areas. These situations usually lead to get missing depth values regions and unstable boundaries in depth maps as shown in Fig. 1. With the purpose of obtaining fine depth boundaries of

© Springer Nature Switzerland AG 2024
V. Vasconcelos et al. (Eds.): CIARP 2023, LNCS 14469, pp. 646–660, 2024.
https://doi.org/10.1007/978-3-031-49018-7_46

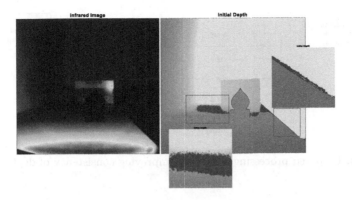

Fig. 1. Corresponding active IR and D map of a ToF camera showing some areas with undesirable/noise effects and missing depth values regions.

objects, by estimating missing depth values and solving the discontinuity problem, some research works propose taking advantage of auxiliary color images captured by an additional RGB sensor, that is, the so-called RGB-D systems. In [2,11,18,19] they propose guided depth enhancement methods based on color images. Other previous works related to this topic used conventional filters. However, depth enhancement techniques that only rely on conventional filtering do not work well when missing depth regions are significant [5].

In order to improve the quality of the estimated depth in ToF cameras, we propose a novel approach which takes advantage from infrared image (IR) captured simultaneously with depth maps (D). The problem is formulated as an energy optimization task with the purpose of maximizing the consistency between the depth map and the corresponding IR image. This formulation is based on the hypothesis that there exists a strong correlation between D and IR. This framework is developed as a weighted estimator based on an iterative conditional modes (ICM) approach. It incorporates conditional entropy information and a spatial error energy term within the image energy model. Additionally, our proposed model introduces a directional weight edges function, which considers all edge directions during the reconstruction and enhancement of object borders. In order to address missing depth values, a pre-processing step is employed, utilizing a depth completion method from a previous work [1], those steps are summarized in Fig. 2.

The main contribution of this work is the use of the active IR image of the same ToF sensor to create a guided depth enhancement, providing more consistency to the depth map. For the best of our knowledge, there is not recent work on IR and depth fusion in ToF systems. The main advantage of IR-D processing rather than RGB-D is to avoid several preprocessing steps, such as calibration, image registration or depth up-scaling.

The diagram of Fig. 2, illustrates the steps of preprocessing and enhancement. The first step involves data denoising, which consists of removing flying pixels and filtering non-confident pixels located around boundaries using morphological

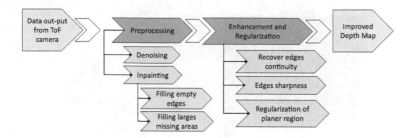

Fig. 2. Proposed processing stages for improving consistency of depth maps.

operators and a confidence map. Subsequently, the inpainting process estimates missing values based on the approach described in [1], aiming to recover missing edges and fill in large missing regions. The final enhancement step enhances sharpness and recover edge continuity through the use of directional edge weights employed in the proposed model. This step also regularizes planar regions by using mutual information extracted from depth and IR images.

The final aim of this work is to use the proposed algorithm in some dedicated hardware build-in the ToF system, and the results achieved so far show that the proposed image fusion-based energy minimization fits better with sensor and hardware resources requirements than using additional RGB sensors. In addition, using a physical model approach avoids learning-based techniques that might need to be fitted for each sensor as a pre-processing or system calibration step. The rest of the paper is organized as follows. First, the problems of depth maps in ToF cameras are introduced, and what are the key issues to be solved. In the Sect. 2, we briefly review some related work. Section 3 explains the reason of using IR as guidance. In Sect. 4 the proposed approach is described in detail. Section 5 presents an overview about experimental data and preprocessing material. Further on, next Sect. 6 shows and discusses the results. Finally, conclusions and further work are summarized.

2   Related Work

The existing approaches for depth enhancement can be broadly categorized into two groups based on the input data: guided depth enhancement and self depth enhancement. On the one hand, the guided depth enhancement category relies on additional information such as color images or depth captured using different sensors like stereo systems. On the other hand, the self depth enhancement category focuses on single depth enhancement techniques. In this work, our primary focus is on techniques that fall under guided depth enhancement. These methods have recently gained significant attention, as evident in several studies such as Diebel et al. [3], Izadi et al. [8], Ferstl et al. [4], Kwon et al. [10], and Lu et al. [13]. Those methods leverage additional depth maps or color images to improve the quality of the final depth map. The most popular solution is to incorporate an additional high resolution color sensor together with a depth sensor for depth image enhancement [3, 4, 13]. Another representative solution is to utilize multiple depth images from the same scene to reconstruct a higher quality depth maps [6]. This method, however, rely heavily on accurate multi-camera or multi-view calibration, and may fail when applied to dynamic environments.

Most methods leverage RGB images to enhance the depth data and assume that there is a strong correlation between the depth map and the corresponding color image. Or et al. [14] fused an intensity image and a depth map to perform a precise shading recovery and albedo estimation for detailed surface reconstruction. Liu et al. [12] proposed an optimization framework that is weighted by color guidance and utilizes a robust penalty function for smoothness modeling. Jiang et al. [9] proposed a method for exploiting correlations in the transformed and spatial domains using a unified model for recovering geometrical structures. Yang et al. [17] extracted scale-independent features from depth maps with the assistance of RGB image and proposed an edge-aware neighbor embedding (NE) framework for facial depth map super-resolution. In addition, several methods have also been proposed for fusion strategies of depth information from multiple frames and other simultaneous sources, in order to produce higher quality depth maps and they have proved its usefulness as a depth enhancement approaches. Deep learning has been introduced recently for depth denoising by using graph networks in [16]. The denoising and enhancement conventional neural network (DE-CNN) proposed in [20] and deep image-guided method [22] have also been adopted for depth enhancement.

In summary, the main guided depth enhancement methods are the filtering-based methods such as (joint bilateral filter), the optimization-based methods (Markov random field, auto regressive model, total variation, graph Laplacian), and the learning-based method (dictionary learning, deep learning). Our focus in this study is related to filtering and optimization-based methods, since one of the final aims is implementing the algorithm on hardware with an easy configuration.

3   IR Image as Guidance

For our experiments, we employed a ToF camera that operates on continuous wave pulse technology and based on indirect time of flight system. It is important to note that this system provides active IR images based on continuous amplitude measurements, while the depth based on phase measurements exhibit discontinuities because of the estimation is based on multiple frequencies agreements (phase unwrapping). It is worth noting that, the amplitude information (active IR images) can offer greater confidence for each corresponding depth value, as we can see from the outcome of a ToF camera when analysing the IR image comparing to the depth map shown in Fig. 3. Note how the depth map shows noise and discontinuous measurements in depth which affect details and important information in some specific areas of the images, such as borders and low reflectance regions.

This can lead to noisy or loss of fine details, whose information could be necessary for accurate interpretation or analysis of the image. However, note that IR measurements preserve the overall structures and details of objects in the scene in a more consistent way, even though pixel values may lose some precision due to noise effect but it usually keeps details of the objects structure. An additional filtering can enhance the IR precision which can not be the case for depth. It is hard to recover the wrong values and lost details in the depth map by using a classical filtering. and that what make active IR used as guidance.

The preprocessing in the pipeline of the camera usually involves filters and contrast stretching that help to improve the overall quality and clarity of the IR image, including the reflectance of low-reflective areas. Looking at the active IR image before and after the preprocessing integrated in the camera pipeline shows that IR images have some interesting properties, such as a better signal-to-noise ratio than depth maps, better defined object boundaries, and they are less sensitive to noise than the corresponding depth maps estimates see Fig. 3.

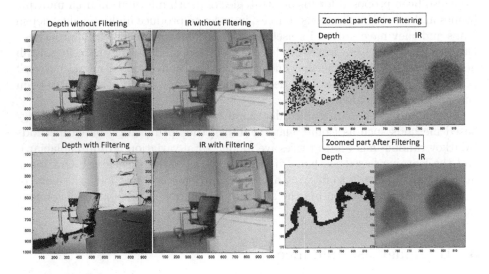

Fig. 3. Discontinues Depth map and its corresponding Continues active IR image captured by a ToF camera before and after filtering in the camera pipeline.

4   Proposed Model

The image energy model is based on a weighted optimization approach to combine the ToF depth maps D and the corresponding active IR images of ToF system. Note that in this case, D and IR images are captured from the same sensor simultaneously, and the active IR is used as a guidance to enhance and improve the consistency of the initial depth.

4.1   Image Energy Function

The proposed energy function is based on two main assumptions: firstly, a strong correlation is assumed between the guiding active IR image and the depth map D; secondly, it assumes a spatial similarity among neighboring depth values, except at depth discontinuities represented by object edge weights. Hence, the image energy Q combines two terms, the spatial error energy Q_S to eliminate local variations in depth, produced by common sensor noise; and the spatial entropy energy term Q_H to enhance depth discontinuities and suppress other non-local noise by maximizing the correlation between the final depth map and the guiding IR image. The resulting enhanced depth map D^* is computed by minimizing the image energy Q expressed as:

$$Q = cQ_S + (1 - c)Q_H. \tag{1}$$

where c is the regularizing parameter used to control the effect of each energy term.

4.2   Spatial Error Energy Term

Spatial error energy term Q_S is introduced to reduce spatial image noise, such as sensor noise. This regularization enforce similarity between neighboring pixels depths. It is expressed as the errors ϵ for each of the 4-neighbours directions as follows,

$$Q_s = \sum_{x,y=1}^{X,Y} \left(\left[WE_{xy}(x-1,y)\epsilon_{xy}^2(x-1,y)\right] + \left[WE_{xy}(x+1,y)\epsilon_{xy}^2(x+1,y)\right] + \right.$$
$$\left. \left[WE_{xy}(x,y-1)\epsilon_{xy}^2(x,y-1)\right] + \left[WE_{xy}(x,y+1)\epsilon_{xy}^2(x,y+1)\right] \right). \tag{2}$$

For each depth $D(x,y)$ at the position (x,y) we define the spatial regularization error ϵ for the 4-neighbour pixels $(x',y') \in N(x,y)$ as:

$$\epsilon_{xy}(x',y') = D(x,y) - D(x',y'). \tag{3}$$

However, enforcing spatial similarity may over-smooth the depth discontinuities at object borders. Therefore, directional edge weights $WE_{xy}(x',y')$ are used to avoid edge blending. The directional weight edge functions are defined in Eq. (4) where $(x',y') \in N(x,y) = \{(x-1,y),(x+1,y),(x,y-1),(x,y+1)\}$.

$$WE_{xy}(x',y') = -\exp[k \cdot EI_{xy}(x',y') \cdot ED(x',y')]. \tag{4}$$

Those weights are based on the depth map gradient module $ED(x',y')$, and directional IR image gradient modules $EI_{xy}(x',y')$. To obtain these gradient modules, IR and depth maps are first smoothed by a Gaussian filter $G(0,\sigma)$ of zero mean and variance σ^2. Then the ED gradient is extracted by directly applying a gradient on depth map D and the directional EI gradients extracted from the active IR image for every 4-neighbour pixels $(x', y') \in N(x, y)$ at a given pixel location (x, y), that is,

$$ED_{xy}(x', y') = |GD(x, y) - GD(x', y')|. \tag{5}$$

$$EI_{xy}(x', y') = |GI(x, y) - GI(x', y')|. \tag{6}$$

where $GD(x', y') = G(0, \sigma) * D(x', y')$ is the depth map D smoothed with a Gaussian filter $G(\mu, \sigma)$ the same for $GI(x', y') = G(0, \sigma) * A(x', y')$, being A the active IR image values. The resulting directional IR gradients EI can include many edges where there are no depth changes. This may lead to unwanted characterization of depth discontinuities in some areas. Hence, to remove edges not corresponding to depth discontinuities, only the IR gradients EI corresponding to depth gradients ED denote depth object edge discontinuities, thus leading to the directional weight edges functions WE_{xy} defined in Eq. (4), With k denoting as a scaling factor.

4.3   Conditional Entropy Energy Term

The rationale behind this energy term is based on the correlation between depth map D and active IR image A. This correlation can be represented by the mutual information between both images $MI(A, D)$, defined as,

$$MI(A, D) = -\sum_{a,d} p(a, d) \log \frac{p(a, d)}{p(a)p(d)}. \tag{7}$$

where $p(a, d)$ is the joint probability of IR image A and depth values D, $p(a)$ and $p(d)$ are the priors for the active IR image and depth map values respectively, and a and d denoting the possible IR and depth values. Therefore, maximizing $MI(A, D)$ would lead to a maximization of the correlation between both images, which can lead to improve the consistency of depth map D with respect to active IR image A. Further on, MI can be defined in terms of conditional entropy as,

$$MI(A, D) = H(D) - H(D/A). \tag{8}$$

Note that, the conditional entropy $H(D/A)$ can be interpreted as how much information remains on D that cannot be explained giving the corresponding active IR values A. Since the derivative of MI is quite complex, and assuming that $H(D)$ is approximately constant, minimizing $H(D/A)$ would be equivalent to maximizing $MI(A, D)$. Thus, the conditional entropy $H(D/A)$ is defined as follows,

$$H(D/A) = -\sum_{a,d} p(a, d) \log p(d/a). \tag{9}$$

Therefore, the corresponding energy term Q_H is defined as:

$$Q_H = H(D/A). \tag{10}$$

where $p(d/a)$ is the conditional probability of depth given the active IR values. Note that in this formulation D must be discretized. In practice, the joint probability $p(a, d)$ is estimated by computing the normalized joint histogram between the initial depth D and the corresponding active IR image A. The conditional in Eq. 9 is computed from the joint probability using the chain rule as $p(d/a) = p(a, d)/p(a)$, and prior $p(a) = \sum_d p(a, d)$ as the marginal of the joint probability $p(a, d)$.

4.4   Image Energy Minimization

The image energy minimization process is based on an Iterative Conditional Modes (ICM) strategy, that is, instead of minimizing Q energy (1) for the entire image, pixel energy $Q(i)$ will be minimized independently for each pixel $i = (x, y)$ in an iterative way, that is,

$$Q(i) = c \cdot Q_S(i) + (1 - c) \cdot Q_H(i) \tag{11}$$

and

$$D(i)^* = argmin \left(c Q_S(i) + (1 - c) Q_H(i) \right). \tag{12}$$

Thus, pixel spatial error term $Q_S(i)$ and pixel conditional entropy $Q_H(i)$ are defined as follows,

$$Q_S(i) = [WE_{xy}(x - 1, y) \cdot \epsilon^2_{xy}(x - 1, y)] + [WE_{xy}(x + 1, y) \cdot \epsilon^2_{xy}(x + 1, y)]$$
$$+ [WE_{xy}(x, y - 1) \cdot \epsilon^2_{xy}(x, y - 1)] + [WE_{xy}(x, y - 1) \cdot \epsilon^2_{xy}(x, y - 1)]; \tag{13}$$

$$Q_H(i) = -WE_{xy}(i) \cdot p(a_i, d_i) \cdot \log p(d_i/a_i). \tag{14}$$

The joint probability $p(a, d)$ can be estimated from the normalized joint histogram between the initial depth map D and the corresponding active IR image A. Where $p(a_i, d_i)$ stands for the joint probability for depth value $d_i = D(i)$ and active IR value a_i at pixel $i = (x, y)$. Edge weights $WE_{xy}(i)$ are also introduced in pixel conditional entropy $Q_H(i)$ to preserve edge pixels regularization. Finally, to minimize the pixel energy $Q(i)$ at each iteration, a gradient descent method is used, with adaptive learning rate λ,

$$\hat{D}(i) = D(i) - \lambda(\partial Q(i)/(\partial D(i)). \tag{15}$$

5   Data and Preprocessing

In this work image data were acquired from a ToF ADI sensor ADTF3175 module based on continuous wave illumination to test the proposed approach, with no depth ground truth available. IR-depth option was used in that sensor to collect the depth aligned with its corresponding active IR image. Both images are geometrically and time aligned, as they are captured by the same sensor with 1024×1024 pixels resolution. Figure 4 illustrates some representative examples where the second column of the first row is the depth map, and the first column of the first row shows the corresponding active IR image. The scenarios were chosen to include both narrow and large missing depth regions as a result from occlusion, dark or low reflective surfaces, or any other factors producing missing depth values.

The initial preprocessing is oriented to estimating missing depth values. Thus, a belief propagation depth completion method was used [1]. Note that this method takes the active IR image as a guidance, particularly for missing edges recovery. The key point of this inpainting approach is that it takes into account the direction of how inpainting is performed around boundaries, where the filling converges from the missing region value boundaries to the objects borders from both depth discontinuity sides, when the missing depth values region contains an object border.

In order to compute the joint probability $p(a, d)$, depth and IR values are discretized. In a further step, the joint histogram of the depth map D and the active IR image A is computed and then normalized to obtain the joint probability and the corresponding conditional $p(d/a)$ using the chain rule.

For the image energy optimization process, the initialization is set to the quantized depth map provided by the ToF sensor and its quantized active IR image. In the gradient descent, the adaptive learning rate is defined as $\lambda_t = \lambda_{(t-1)} - \lambda_c$, being λ_c a constant, and the iterative process is stopped when the change in the MI between the active IR and the discretized version of the estimated depth image is below a given small value.

6   Experiments and Results

The qualitative results of the experiments based on the ADTF3175 data-set are presented in Figs. 4, 5, 6 and 7. These figures provide detailed insights into edges recovering and depth enhancement. Notably, the proposed method significantly improves the consistency of depth maps regarding to infrared image, since missing values around some edges are one of the major problems of ToF cameras the belief propagation depth method (BLFP) used for inpainting as preprocessing. in Fig. 4 the third column of the first row illustrates the result of depth inpainting, while the fourth column showcases the depth enhancement results after integrating the BLFP with the proposed image energy optimization framework. This combination effectively reconstructs missing pixels and enhances depth map, particularly around edges that were initially missing and noisy. This is clearly demonstrated in the second row, first column of Fig. 4.

It is worth noting that even after the depth inpainting process, the depth still exhibits discontinuous edges and some non-homogeneous transitions between objects and the background. For these reasons, the fusion module proposed in this work aims to further refine the inpainted depth. This refinement is illustrated in the analysis of the results presented in Figs. 5, 6 and 7.

Figure 5 shows how the edge profile and continuity were recovered based on optimization framework. Figure 6 illustrate some details of edge continuity when concerning nonlinear borders, the left edge map is extracted from initial depth

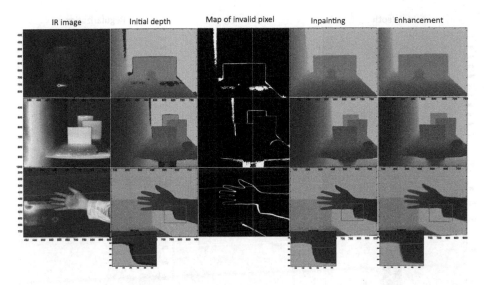

Fig. 4. Depth inpainting and final depth enhancement, large missing depth areas reconstruction, missing edges reconstruction.

and the right one extracted from the enhanced depth after applying the proposed method. The left one shows some discontinuities and double boundaries around the hand, that means the hand in the initial depth was surrounded by a missing area, the right map show the hand border after processing. This significant improvement is attributed to the utilization of the directional weight edge function, which was employed for spatial error regularization in the defined image energy function (4.2). The directional weight edge function plays a crucial role in guiding the optimization process to refine objects boundaries. This edge function provides directional information that helps to preserve edges more accurately during the enhancement process, ensuring smoother and more coherent depth transitions between objects and the background.

Figure 7 represents the edge intensity (EI) [21] of a nonlinear shape, where EI of the inpainted depth is shown in the second column and the EI of the enhanced depth in the third column. The zoomed part from the object boundary is shown in the second row. From these results, it is evident that the inpainted depth still exhibits some artifacts around the object's boundaries. However, the depth enhancement process based on our proposed framework plays a crucial role in regularizing the areas near the edges. This regularization is achieved through the similarity term, which effectively refines non homogeneous depth edges respecting to infrared information.

Figures 8 and 9 shows the depth enhancement comparison results of the proposed approach with respect to other reference guided enhancement methods (joint bilateral filter [15], guided filter [7], and guided anisotropic diffusion [11]). From this figures and from the zoomed patches we can notice that the proposed enhancement model produces regularization on the smooth object surface

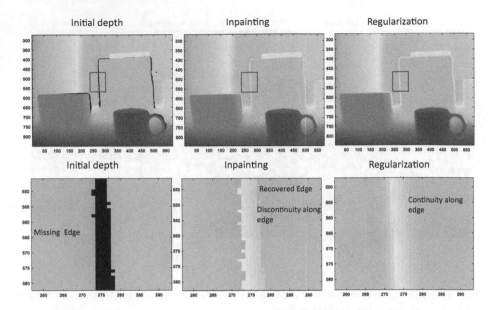

Fig. 5. Edge reconstruction based on belief propagation inpainting method and edges continuity recovering based on the proposed image energy optimization model.

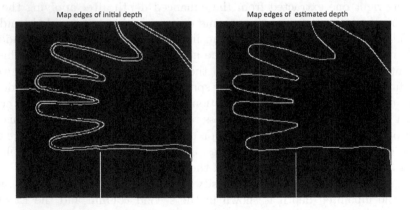

Fig. 6. Edge continuity recovering based on the proposed image energy optimization model.

regions. This is because of the conditional entropy minimization, which also regularizes around edges and improve the sharpness of depth borders due to the directional edge weights introduced in the image energy model.

The most important outcome of the proposed model is the improvement of edges continuity, by maximizing the correlation (mutual information) between the resulting depth and the active IR image. This process tends to change depth values in such a way that they correlate as much as possible with the active IR image values and local spatial distribution.

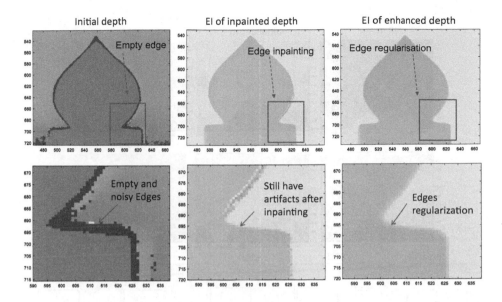

Fig. 7. Edges regularization improvement based on the proposed image energy model. The second and third columns are edge intensity maps.

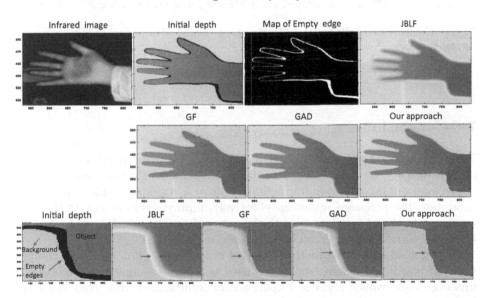

Fig. 8. Comparison of the proposed image energy model with respect to other reference methods.

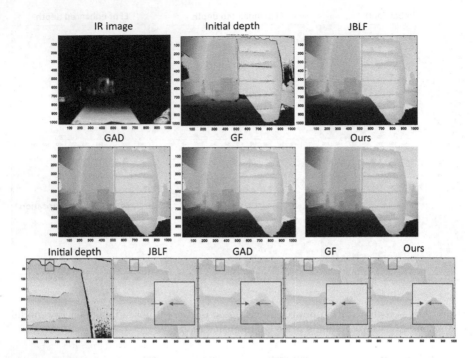

Fig. 9. Comparison of the proposed image energy model with respect to other reference methods.

7   Conclusions

The proposed method is a novel guided approach for dense depth enhancement of ToF imagery, by introducing an image energy model that combines the ToF depth map and its corresponding active IR image as a guidance. The experiments show how the proposed approach effectively improves consistency of the final resulting depth map with respect to the active IR image, which is a proof of the increased correlation and matching between both images.

The guided inpainting method (guided belief propagation for depth completion [1]) based on IR image guide was used as preprocessing step to reconstruct missing depth values regions. Experimental results show that combining the belief propagation depth inpainting as a preprocessing and the proposed image energy model recovers edges and their continuity satisfactorily. The directional edge weights employed in the model provide the enhancement of depth discontinuities at object edges, also outperforming other existing conventional guided filtering methods. As a further work, this method will be extended to an other framework based on multi-sensors fusion, which could also increase the depth accuracy and lead to increase the resolution.

Acknowledgments. This work was partially supported by Analog Devices, Inc. and by the Agencia Valenciana de la Innovacion of the Generalitat Valenciana under program "Plan GEnT. Doctorados Industriales. Innodocto".

References

1. Achaibou, A., Sanmartín-Vich, N., Pla, F., Calpe, J.: Guided depth completion using active infrared images in time of flight systems. In: Pertusa, A., Gallego, A.J., Sánchez, J.A., Domingues, I. (eds.) Pattern Recognition and Image Analysis: 11th Iberian Conference, IbPRIA 2023, Alicante, Spain, 27–30 June 2023, Proceedings, pp. 323–335. Springer, Cham (2023). https://doi.org/10.1007/978-3-031-36616-1_26
2. Ahmed, Z., Shahzad, A., Ali, U.: Enhancement of depth map through weighted combination of guided image filters in shape-from-focus. In: 2022 2nd International Conference on Digital Futures and Transformative Technologies (ICoDT2), pp. 1–7. IEEE (2022)
3. Diebel, J., Thrun, S.: An application of Markov random fields to range sensing. Adv. Neural Inf. Process. Syst. **18** (2005)
4. Ferstl, D., Reinbacher, C., Ranftl, R., Rüther, M., Bischof, H.: Image guided depth upsampling using anisotropic total generalized variation. In: Proceedings of the IEEE International Conference on Computer Vision, pp. 993–1000 (2013)
5. Gong, X., Liu, J., Zhou, W., Liu, J.: Guided depth enhancement via a fast marching method. Image Vis. Comput. **31**(10), 695–703 (2013)
6. Gu, S., Xie, Q., Meng, D., Zuo, W., Feng, X., Zhang, L.: Weighted nuclear norm minimization and its applications to low level vision. Int. J. Comput. Vision **121**, 183–208 (2017)
7. Hui, T.W., Ngan, K.N.: Depth enhancement using RGB-D guided filtering. In: 2014 IEEE International Conference on Image Processing (ICIP), pp. 3832–3836. IEEE (2014)
8. Izadi, S., et al.: Kinectfusion: real-time 3d reconstruction and interaction using a moving depth camera. In: Proceedings of the 24th Annual ACM Symposium on User Interface Software and Technology, pp. 559–568 (2011)
9. Jiang, Z., Hou, Y., Yue, H., Yang, J., Hou, C.: Depth super-resolution from RGB-D pairs with transform and spatial domain regularization. IEEE Trans. Image Process. **27**(5), 2587–2602 (2018)
10. Kwon, H., Tai, Y.W., Lin, S.: Data-driven depth map refinement via multi-scale sparse representation. In: Proceedings of the IEEE Conference on Computer Vision and Pattern Recognition, pp. 159–167 (2015)
11. Liu, J., Gong, X.: Guided depth enhancement via anisotropic diffusion. In: Huet, B., Ngo, C.-W., Tang, J., Zhou, Z.-H., Hauptmann, A.G., Yan, S. (eds.) PCM 2013. LNCS, vol. 8294, pp. 408–417. Springer, Cham (2013). https://doi.org/10.1007/978-3-319-03731-8_38
12. Liu, W., Chen, X., Yang, J., Wu, Q.: Robust color guided depth map restoration. IEEE Trans. Image Process. **26**(1), 315–327 (2016)
13. Lu, S., Ren, X., Liu, F.: Depth enhancement via low-rank matrix completion. In: Proceedings of the IEEE Conference on Computer Vision and Pattern Recognition, pp. 3390–3397 (2014)
14. Or-El, R., Rosman, G., Wetzler, A., Kimmel, R., Bruckstein, A.M.: RGBD-fusion: real-time high precision depth recovery. In: Proceedings of the IEEE Conference on Computer Vision and Pattern Recognition, pp. 5407–5416 (2015)

15. Shen, Y., Li, J., Lü, C.: Depth map enhancement method based on joint bilateral filter. In: 2014 7th International Congress on Image and Signal Processing, pp. 153–158. IEEE (2014)
16. Valsesia, D., Fracastoro, G., Magli, E.: Deep graph-convolutional image denoising. IEEE Trans. Image Process. **29**, 8226–8237 (2020)
17. Yang, S., Liu, J., Fang, Y., Guo, Z.: Joint-feature guided depth map super-resolution with face priors. IEEE Trans. Cybernet. **48**(1), 399–411 (2016)
18. Yi, K., Zhao, Y., Lei, Y., Pan, J.: Depth enhancement with improved inpainting order and smoothing method. J. Phys. Conf. Ser. **1187**, 042065 (2019). IOP Publishing
19. Zhang, L., et al.: Depth enhancement with improved exemplar-based inpainting and joint trilateral guided filtering. In: 2016 IEEE International Conference on Image Processing (ICIP), pp. 4102–4106. IEEE (2016)
20. Zhang, X., Wu, R.: Fast depth image denoising and enhancement using a deep convolutional network. In: 2016 IEEE International Conference on Acoustics, Speech and Signal Processing (ICASSP), pp. 2499–2503. IEEE (2016)
21. Zhang, X., Ye, P., Xiao, G.: Vifb: a visible and infrared image fusion benchmark supplementary document
22. Zhu, J., Zhang, J., Cao, Y., Wang, Z.: Image guided depth enhancement via deep fusion and local linear regularization. In: 2017 IEEE International Conference on Image Processing (ICIP), pp. 4068–4072. IEEE (2017)

Multi-conformation Aproach of ENM-NMA Dynamic-Based Descriptors for HIV Drug Resistance Prediction

Jorge A. Jimenez-Gari[1](✉) , Mario Pupo-Meriño[1] , Héctor R. Gonzalez[1] ,
and Francesc J. Ferri[2]

[1] Universidad de las Ciencias Informáticas (UCI), La Habana, Cuba
{jorgeajg,mpupom,hglez}@uci.cu
[2] Computer Science Department, Universitat de València, Burjassot 46100, Spain
francesc.ferri@uv.es
https://www.uci.cu/

Abstract. Drug resistance is a key factor in the failure of drug therapy, as the antiretroviral therapy against the human immunodeficiency virus (HIV). Due to the high costs of direct phenotypic assays, genotypic assays, based on sequencing of the viral genome or part of it, are commonly used to infer drug resistance via *in silico* predictions. In these approaches, the interpretation of the sequence information constitutes the biggest challenge. The large amount of data linking genotype and phenotype information provides a framework for predicting drug resistance from genotype, based on machine learning methods. Primarily, the sequence based information is used but largely fails to predict resistance in previously unobserved variants. The inclusion of structural and dynamic information is supposed to improve the predictions but has been limited by their computational cost of calculation. This study shows the feasibility of dynamic descriptors derived from normal mode analysis in elastic network models of HIV type 1 (HIV-1) protease in predicting drug resistance. We show that exploring the pre-configuration of dynamic information covering the intrinsic movement spectrum of proteinase in HIV-1 by multiple conformation approach descriptors improve the classification task.

Keywords: Descriptor · Dynamics · ENM-NMA · HIV-1

1 Introduction

Drug resistance (DR) is a growing concern in the field of medicine [15,23]. One area where drug resistance has become particularly problematic is in the treatment of HIV-1. DR occurs when the virus mutates and becomes immune to the drugs used to treat it. This can happen when patients do not take their

Work partially funded by TED2021-809 131003B-C21 and PID2022-137048OB-C41 projects.

medication as prescribed, or when they are exposed to suboptimal drug regimes. HIV-1 DR is a major concern because it leads to treatment failure, disease progression, and the spread of drug-resistant strains of the virus [17,18].

To address the problem of drug resistance, it is important to study the underliying mechanisms of drug-target interactions. This involves understanding the genetic and biochemical processes that allow microorganisms to adapt and survive in the presence of drugs. It also involves developing new drugs and treatment strategies that can overcome DR.

In the case of HIV-1, researchers are working to develop new drugs that target different parts of the virus and are less prone to resistance. They are also exploring new treatment strategies, such as combination therapies, that can reduce the risk of DR. Additionally, efforts are being made to improve patient adherence to medication regimes, which can help preventing the development of DR. In conclusion, DR is a major concern that requires continued attention and investment in research and development.

There are two ways to infer the susceptibility of a mutational variant to an antiretroviral (ARV): phenotype and genotype tests. Genotypic tests are used more frequently than phenotypic tests due to their lower cost, greater availability, simplicity, and shorter turnaround time. These tests are based on the determination of the nucleotide sequence of the HIV-1 genes whose protein products constitute the ARV target, and their interpretation requires a predictive algorithm that describes susceptibility to a variety of ARVs [4]. The computational interpretation of genotypic information leading to a phenotype prediction is an open field in biological sequence analysis and is beyond the problem of HIV-ARV resistance. Computational methods for DR using genotype data are usually divided into two main areas: sequence-based methods and structure-based methods. The difference between them lies in the computational cost-accuracy relationship and feasibility of the data. In one hand, the sequence-based information, wich decreases computational cost but greatly misses predictions due to loss of structural information. In the other hand the structure and dynamic based information, which is highly predictive but te corresponding computational methods are known to be computationally expensive in terms of time and resource consumption. Many computational methodologies have been proposed for the prediction of HIV DR using sequence and structural information (see [22] and references). Several techniques have been tested, among them, the use of (i) decision trees [21], (ii) Support Vector Machine (SVM) [6], (iii) Multilayer Perceptrons(MLP) and (iv) Deep Neural Networks stand out among others [5,25].

Many proteins, such as enzymes, need large conformational movements to perform their function, such as catalytic and regulatory activities. It is well known that these movements depend on collective movements in their domains in contrast to a flexible region. Usually these domains are restrictive (energetic) for movements that involve many aminoacids, so they depend on a region of high flexibility such as loops to achieve their catalytic conformation. Therefore, the dynamics of proteins is not a trivial problem and relies to complex aproaches such as molecular dynamics (MD) simulation. MD simulation is the most established method to extract valuable information of the intrinsic displacement of proteins and many aplications in the HIV DR problem have met the potential of this methodology [13, 30, 32, 34, 35]. However, this method is well known for its computational cost and because of that, many approaches have been proposed to surround a complete atomistic modeling via MD, such as Normal Mode Analysis (NMA).

NMA is a numerical technique used to describe the accessible flexible states of a protein over an equilibrium position. This is based on the physics used to describe small oscillations expressed as eigenvalues and eigenvectors in an orthonormal space [3, 7]. Many methodologies have been built to further simplify this modeling, among them, Elastic Network Models (ENM) stands out as an efficient way for capturing essential dynamics [36]. But the coarse grained characteristics of these methods have a limitation to capture the conformational spectrum involving collective movements in the HIV proteinase (HIVP) and therefore do not exploit the predictive characteristics when a conformation is taken into account. Returning to the HIV problem, it has been shown that protease dynamics is a potentially influential factor in the phenomenon of resistance to its inhibitors [10, 19, 27].

The aim of this work is (1) to analyze the dynamic information coming from ENM-NMA as a potential source of descriptive information in the phenomenon of HIV DR, (2) to address the problem of the loss of resolution coming from ENM-NMA and (3) propose a methodology for the analysis of results and the selection of influential conformations as a baseline model. The structure of this report is as follows. Section 2 shows the general methodology and the specific characteristics of the workflow. Section 3 outlines the results obtained in the study and puts forward some remarks regarding these results. Finally, general conclusions and future directions for new research are given.

2 Materials and Methods

For the computational prediction of DR using HIVP, several factors are taken into account. Starting with the sequence information, since it deals with prediction using descriptors calculated from analising normal modes in proteins, protein-ligand complexes were necessarily constructed. For this, it was necessary to model the structure of the protein via point mutation and to take into account its coupling with the inhibitors and their energetic stabilization. The analysis of the dynamic characterization of HIVP is carried out through the analysis of normal modes in cartesian space with the sequence information. Is well known that NMA techniques rely on a coarse grained explaination of dynamics. Tama et al. [29] shows that ENM-NMA do not represents the full movement spectrum of macromolecules, therefore, a multi-conformation approach was used for the pre-configuration of the natural dynamic description in HIVP. For this task, four conformations of HIVP were evaluated to get statistical independence and significance test results. Taking into account the results, a final multi-conformation descriptor is finally proposed. Figure 1 shows the general methodology. The following sub-sections explain the aforementioned sub-processes in detail.

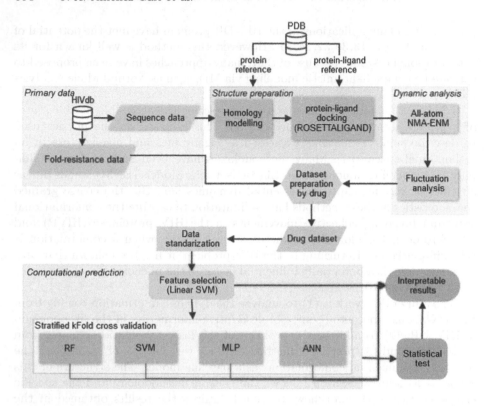

Fig. 1. General workflow corresponding to the study. Different colors are used to highlight different stages of the process.

2.1   Dataset

Sequence Data. The genotype-phenotype pairs were taken from the HIVdb database in relation to the GenoPhenoSense studies for HIVP using the high-quality filtered data as suggested in [11]. Pre-process procedure was made taking into account samples that had missing values, unrecognized data and samples containing mutations that affect the corresponding protein backbone. Seven drugs were used as targets for the classification task with fold-resistance thresholds: 2.0 for TPV, 3.0 for NFV, SQV, IDV, and ATV, 9.0 for LPV, and 10.0 for DRV [26]. Those samples that exceed the threshold are classified as resistant to the drug. From a total of 2,395 samples in the database, 897 were selected for our study as they contained resistence information regargarding some of the drugs considered. The amount of samples with resistence information for each drug is shown in Table 1.

Structure Data. Four conformations were modeled via the ROSETTALI-GAND [8,16] protocol to investigate the influences of HIVP dynamics. These conformations are: the open structure of the HIVP bounded (*open_b*) and unbounded *(open_u)* with ligand and the closed structure bounded *(closed_b)* and unbounded *(closed_u)*. These structures were modeled from the sequence data gathered. For each variant selected from the database, the four aforementioned conformations were modeled. Reference structures were extracted from the RCSB PDB database (accessible at https://www.rcsb.org/). For the open conformation, the structure with pdb 1HHP was used and for each protein-ligand complex the following structures were used as reference: TPV (1D4Y), IDV (1HSG), LPV (1MUI), NFV (1OHR), DRV (1T3R), ATV (2FXE) and SQV (3OXC). Given the extracted reference structure, an energy minimization step was carried out until a stable conformation was found at or near the minimum. For this, the score ref15 [2] function was used.

Dynamics Data. Normal Mode Analysis was computed in each conformation using Bio3d R package version: 2.4–3 [12]. ENM based on all heavy atoms of the input structure (including those in the ligand), which was obtained by fitting to a local energy minimum of a crambin model derived from the AMBER99SB force field. In this way, the vector set of atomic displacements for each alpha carbon was obtained (residue-level) as part of the fluctuation analysis. Finally, the dynamic cross-correlation matrix (DCCM) was calculated, which registers the quantitative coupling of residue-residue atomic displacements.

2.2   Computational Prediction

For the classification task, four classifiers (SVM, RF, MLP, and ANN) were tested with each of the protein inhibitors (PI) to predict antiretroviral resistance, using previously extracted information of dynamic cross-correlation matrices for each conformation for each sample. Since the cross-correlation matrix is symmetric, only unique elements are chosen as variables. This set of variables can be formulated as:

Let $\mathbf{C}$ be a dynamic cross-correlation matrix of atomic displacements at residue level of size 99×99 for HIVP. The set of upper triangular elements of $\mathbf{C}$, excluding the diagonal, can be represented as:

$$\mathbf{X} = \{c_{ij} \mid i < j, 1 \leq i, j \leq n\} \tag{1}$$

where c_{ij} denotes the (i, j)th element of $\mathbf{C}$.

For MLP we define two hidden layers with 60 and 30 neurons consecutively. For the neural network hidden architecture, two blocks of a dense layer followed by a dropout layer with 10 units each and a consecutive dense layer of 5 units.

Stratified k-fold cross-validation with $k = 10$ was used in all experiments. The accuracy metric was used to assess performance in each case. All data was standarized to be a mean = 0 and sd = 1. In addition, a feature selection process was performed with an SVM with an L1-norm loss function and a regularization parameter $C = 0.07$. For comparison of performance, a Friedman test followed by a pairwise Wilcoxon test with Benjamini-Hochberg correction was carried out following the methodology suggested in [9], to check if there are statistical differences in the accuracy of the different algorithms when considering the effect of the conformation-specific descriptor as a factor. Each pair of drug-algorithm was considered for comparison when reporting averaged performance measures.

3   Results and Discussion

Regarding the results for the computational prediction, the open and closed bounded and unbounded conformations were tested for HIV DR. The overall performance was positive, regardless the algorithm and the individual HIVP conformation used. Table 1 shows the results of the computer experiments for the average accuracy for all models in the 10-fold validation procedure.

Table 1. Averaged accuracy (over folds and over models to check the overall convergence) for all drug-conformation pairs and number of samples for each experiment.

drugs	ATV	DRV	IDV	LPV	NFV	TPV	SQV
samples	558	360	849	695	871	393	853
closed-b	0.9553	**0.9639**	**0.9529**	0.9611	0.9529	**0.9363**	0.9520
closed-u	0.9444	**0.9639**	**0.9529**	**0.9612**	**0.9564**	0.9210	**0.9531**
open-b	**0.9569**	0.9611	0.9435	0.9655	0.9323	0.9163	0.9331
open-u	0.9481	0.9583	0.9388	0.9496	0.9403	0.9213	0.9378

Statistics for Conformations. In terms of statistics, Fig. 2 shows the results of the Friedman and Wilcoxon tests. When searching for statistical significance for the observed differences, the Friedman test reported a p-value=5.019e-11, so it can be accepted that there are differences in the average rank in accuracy when considering the effect of DCCM on the studied conformations. However, the results of the Wilcoxon and Friedman rank comparisons showed that the closed-b conformation performed better, contrary to what the visual analysis indicated. Regarding the significant differences, they showed that, regardless of the state of the ligand, the two open and closed conformations show differences as biologically expected, but in both conformations the predictive models did not show differences regarding the presence or absence of the ligand. Said behavior may be related to the molecular coupling process, taking into account that the same references are used, making evident, the need to select more specific starting points, which, for the simplicity of the study, were not taken into account.

Statistics for Predictive Models. Regarding the model test statistics, it can be seen that the best performing model was MLP followed by ANN. When searching for statistical significance for the observed differences, the Friedman test reported a p-value=1.110e-16, so it can be accepted that there are differences

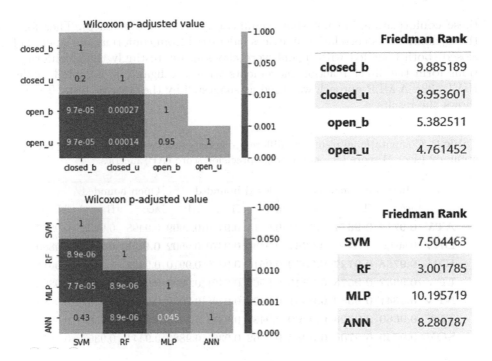

Fig. 2. Statistical results in a row-wise manner for the conformations and the models as instances in the tests. Average rank of the Friedman test and adjusted p-values related to the post-hoc analysis of Wilcoxon signed-rank test with Benjamini/Hochberg correction are presented in both cases.

in the average rank when considering the effect of the models tested. Also, the Wilcoxon signed rank test show that the models SVM and ANN do not show significant differences.

Taking into account the previous results, it can be concluded that: (1) The dynamics of the target protein can be used as a correlative variable in the HIVDR problem, (2) if the dynamics cross-correlation matrix is used as a source of information, favorable results can be noted in terms of what is biologically expected and demonstrated, which presents the bounded conformation as the most related to the problem of HIV DR [24,31], (3) predictive models based on neural networks tend to generalize better in the HIVP problem.

Multi-conformation Aproach. In a natural way to face the problem of prediction using the proposed methodology, a multi-conformation perspective was evaluated in coordination with the statistical results. This assumption comes from the coarse-grained property of the ENM-NMA approximation, which shows a bias regarding the representation of movements between significantly relevant conformations, as exposed by Tama et al. [29]. The statistical test showed that the closed bounded and unbounded are the most correlative, but these conformations do not show significantly relevant differences. Therefore, the dynamic cross-correlation matrices calculated from conformations open and closed, both bounded with ligand were choosen. Interestingly these conformations form the direct reaction movements when the ligand is presented in the HIVP [33]. A MLP approach was used as suggested by the statistic tests. Table 2 shows the results.

Table 2. Computational results for different conformation descriptors showing average accuracy (acc), f1 score (f1) and average-precision score (AP) metrics.

	Multi conformation			Closed bounded			Open bounded		
	acc	f1	AP	acc	f1	AP	acc	f1	AP
ATV	**0.9714**	**0.9671**	**0.9971**	0.9515	0.9419	0.9909	0.9498	0.9408	0.9877
DRV	0.9642	0.9589	**0.9961**	0.9480	0.9370	0.9902	**0.9660**	**0.9600**	0.9889
IDV	**0.9768**	**0.9727**	**0.9970**	0.9516	0.9430	0.9910	0.9462	0.9368	0.9914
LPV	**0.9642**	**0.9578**	**0.9945**	0.9569	0.9491	0.9897	0.9498	0.9410	0.9882
NFV	**0.9641**	**0.9580**	**0.9941**	0.9498	0.9406	0.9920	0.9624	0.9562	0.9888
TPV	**0.9660**	**0.9608**	**0.9969**	0.9481	0.9374	0.9910	0.9641	0.9583	0.9930
SQV	**0.9749**	**0.9706**	**0.9984**	0.9462	0.9361	0.9883	0.9444	0.9357	0.9877

Table 3. Comparison of the multi-conformational approach with state of the art methods with regard to average accuracy and implementation details.

	ATV	DRV	IDV	LPV	NFV	TPV	SQV	Data Expansion	features	Model
Yu et al. [38]	0.955	–	0.960	0.962	0.933	0.961	0.946	Yes	-sequence -structural	sparse dictionary encoder
Pawar et al. [20]	0.983	0.988	0.979	0.984	0.978	0.987	0.969	Yes	-sequence -structural	RBM
Steiner et al. [28]	0.922	0.927	0.932	0.946	0.920	0.884	0.920	No	sequence	CNN
This approach	0.971	0.964	0.977	0.964	0.964	0.966	0.975	No	-dynamic	ANN

In Table 2 it can be seen that the multi-conformation approach was performing better in general in comparison to taking a conformation individually. Note that in most cases this approach showed a circumstantial improvement in terms of the f1 score, which tells us the sensitivity in classifying a sample towards the resistant class. In the case of DRV, the open bounded conformation performed better. Interestingly, it has been proven that in some cases this conformation is relevant for the inhibitors suggesting a non-standard binding [1,39].

Taking into account the results obtained and those found in the state of the art, it can be seen that including only the dynamic information of the protein, the predictive scope improves with respect to those that use sequence and structure information. Only being surpassed by complex methods or with data expansion (See Table 3). An important point to note is the expansion of the primary data of the database. Said expansion is made up of the assumption that if a mixture exists in a position of the sequence, then said sequence can be divided into the combinations of said mixtures. Under this expansion of the sequence, it is also assumed that for the respective position of the mixture, the change of said aminoacids does not constitute a change in its phenotype (resistance ratio) since the different combinations of the mixture are annotated with the same phenotype, that which presented the initial sample. This expansion of the data represents an increase in the amount of raw data used for each drug experiment that is about 15 times the sizes shown in Table 1.

Interpretability of ENM-NMA Based Descriptors. From an interpretability point of view, the use of feature selection methods allows relating the selection process in coordination with the predictive results obtained with biological properties that can elucidate relevant information about the problem to be treated. Under this assumption, the feature selection process used in this research was evaluated. In addition, the use of the dynamic cross-correlation matrix allows, among other things, the construction of protein dynamic correlation networks that exhibit the relative movements between communities and allows the extraction of valuable structural and dynamic metrics to interpret this information. For this, the methodology proposed by Yao et al. [37] can be used to obain the correlation network and the betweenness centrality measure for each residue/node. Figure 3 shows these metrics for the sample *257957* and the results obtained by the feature selection process using the multi-conformation aproach taking into account those variables belonging to each conformation. For simplicity, only the results of IDV was shown.

In the Fig. 3, metrics about the dynamics of the sample 257957 are shown. Firstly, this was chosen because it showed resistance to all the drugs presented. Is shown in the rings of Fig. 3 from outside to inside: (i) the measure of centrality calculated from the betweeness clustering algorithm taking into account a threshold of 0.2 in the DCCM's, which is the count of paths that pass through said node, (ii) the folding patterns calculated by DSSP algorithm [14], (iii) the fluctuations (average of the DCCM) for the four conformations, (iv) the mutation pattern that sample presents and (v) general HIVP regions used in the state of the art. The link plotted are the 20 residue-residue relations that most contribute to the prediction, that was extracted by sorting the coefficients of the support vectors in the LinearSVC feature selection procedure, for the multi-conformation descriptor consisting of open and closed bounded conformations.

It can be seen that in the open conformation, the distal relationships tend to be located between the regions at or neighboring the cantilever and fulcrum but also show relationships between the proximity of the cantilever and the flap tip (49–52). In the closed conformation it is shown that the distal relationships tend to be found between the fireman's grip (including the binding site) and regions neighboring the C-terminal of the protein and the flap.

Taking into account the information from the centrality measure and the relationships chosen by the prediction model when considering features of the closed-b conformation, it seems that those relationships with low centrality are

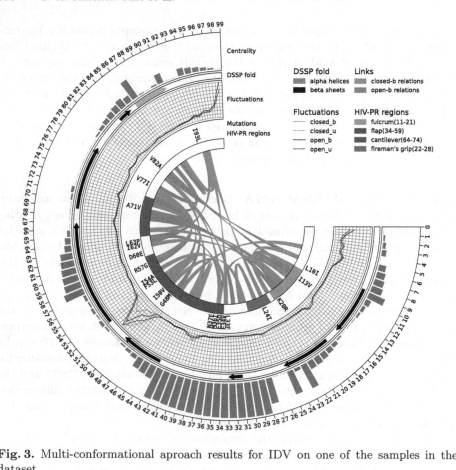

Fig. 3. Multi-conformational aproach results for IDV on one of the samples in the dataset.

potentially those that are related to the HIVDR phenomenon. Such as those of the regions close to the flap tip, that in turn, it is related to movements in the section 73–82 and 33–27 of the HIVP, which present high centrality. These results suggest that relationships distal to the active site are related to the phenomenon of HIVDR. There are collective movements, apart from those belonging to the flaps, that allow the virus to find new ways to evade the stability of the inhibitors. There are relevant relationships for HIVDR not necessarily in the defined mutation sites, but indirectly are the cause of their structural changes.

4 Conclusions and Further Work

For the problem of predicting HIV drug resistance, dynamic information counts as a relevant source of information, mainly for unobserved samples. The inclusion of dynamic information from ENM-NMA of various conformations increases the resolution of the predictions. Potentially, the information in the prediction results can be a source of knowledge base for assumptions and new hypotheses for more complex methodologies and procedures that contribute to finding new inhibitors.

Although the use of multiple conformations when using ENM-NMA is presented in this work, a hyperparameter optimization process has not been carried out both in the prediction process and in the normal modes analysis process. Finding the best set of hyperparameters for ENM-NMA will result in a deeper understanding of the energetic mechanisms surrounding the HIV proteinase drug resistance problem. Although dynamic information is shown as a predictive variable for HIV DR, this is not exclusive of other forms of representation, it is necessary to develop a profile of features that draws on both sequence information, structural information, and dynamic information. This research shows that neural network-based methods tend to work better for HIVP dynamics, building a more complex network architecture that can explore the parameter space deeper and find hidden patterns will improve prediction and provide new insights into subsequent biological interpretability.

References

1. Agniswamy, J., Shen, C.H., Aniana, A., Sayer, J.M., Louis, J.M., Weber, I.T.: Hiv-1 protease with 20 mutations exhibits extreme resistance to clinical inhibitors through coordinated structural rearrangements. Biochemistry **51**(13), 2819–2828 (2012)
2. Alford, R.F., et al.: The Rosetta all-atom energy function for macromolecular modeling and design. J. Chem. Theory Comput. **13**(6), 3031–3048 (2017)
3. Bauer, J.A., Bauerová-Hlinková, V.: Normal mode analysis: a tool for better understanding protein flexibility and dynamics with application to homology models. In: Homology Molecular Modeling-Perspectives and Applications, pp. 1–19. IntechOpen (2020)
4. Bonet, I., Arencibia, J., Pupo, M., Rodriguez, A., Garcia, M.M., Grau, R.: Multiclassifier based on hard instances-new method for prediction of human immunodeficiency virus drug resistance. Curr. Top. Med. Chem. **13**(5), 685–695 (2013). https://doi.org/10.2174/1568026611313050011
5. Bonet, I., García, M.M., Saeys, Y., Van de Peer, Y., Grau, R.: Predicting human immunodeficiency virus (HIV) drug resistance using recurrent neural networks. In: Mira, J., Álvarez, J.R. (eds.) IWINAC 2007. LNCS, vol. 4527, pp. 234–243. Springer, Heidelberg (2007). https://doi.org/10.1007/978-3-540-73053-8_23
6. Cai, Q., Yuan, R., He, J., Li, M., Guo, Y.: Predicting HIV drug resistance using weighted machine learning method at target protein sequence-level. Mol. Diversity **25**, 1541–1551 (2021)
7. Cui, Q., Bahar, I.: Normal Mode Analysis: Theory and Applications to Biological and Chemical Systems. CRC Press (2005)
8. Davis, I.W., Baker, D.: Rosettaligand docking with full ligand and receptor flexibility. J. Mol. Biol. **385**(2), 381–392 (2009)
9. Demšar, J.: Statistical comparisons of classifiers over multiple data sets. J. Mach. Learn. Res. **7**, 1–30 (2006)
10. Ferreiro, D., Khalil, R., Gallego, M.J., Osorio, N.S., Arenas, M.: The evolution of the HIV-1 protease folding stability. Virus Evolution **8**(2), veac115 (2022)

11. Gari, J.A.J.: Assessing ENM-NMA based molecular descriptors of HIV-1 protease for drug resistance prediction by machine learning methods. Rev. Cubana Ciencias Inf. **17**(1) (2023)
12. Grant, B.J., Rodrigues, A.P., ElSawy, K.M., McCammon, J.A., Caves, L.S.: Bio3d: an R package for the comparative analysis of protein structures. Bioinformatics **22**(21), 2695–2696 (2006)
13. Hornak, V., Okur, A., Rizzo, R.C., Simmerling, C.: Hiv-1 protease flaps spontaneously open and reclose in molecular dynamics simulations. Proc. Nat. Acad. Sci. **103**(4), 915–920 (2006)
14. Kabsch, W., Sander, C.: Dictionary of protein secondary structure: pattern recognition of hydrogen-bonded and geometrical features. Biopoly. Orig. Res. Biomolecul. **22**(12), 2577–2637 (1983)
15. Laxminarayan, R., Bhutta, Z., Duse, A.: The International Bank for Reconstruction and Development/The World Bank, Chap. 55. Oxford University Press, Washington, DC; New York (2006)
16. Meiler, J., Baker, D.: Rosettaligand: protein-small molecule docking with full sidechain flexibility. Proteins Struct. Funct. Bioinform. **65**(3), 538–548 (2006)
17. Metzner, K.J.: Technologies for hiv-1 drug resistance testing: inventory and needs. Curr. Opin. HIV AIDS **17**(4), 222–228 (2022)
18. Parikh, U.M., Mellors, J.W.: How could HIV-1 drug resistance impact preexposure prophylaxis for HIV prevention? Curr. Opin. HIV AIDS **17**(4), 213–221 (2022)
19. Paulsen, J.L., Leidner, F., Ragland, D.A., Kurt Yilmaz, N., Schiffer, C.A.: Interdependence of inhibitor recognition in HIV-1 protease. J. Chem. Theory Comput. **13**(5), 2300–2309 (2017)
20. Pawar, S.D., Freas, C., Weber, I.T., Harrison, R.W.: Analysis of drug resistance in HIV protease. BMC Bioinformatics **19**(11), 362 (2018). https://doi.org/10.1186/s12859-018-2331-y
21. Ramon, E., Belanche-Muñoz, L., Pérez-Enciso, M.: Hiv drug resistance prediction with weighted categorical kernel functions. BMC Bioinformatics **20**(1), 1–13 (2019)
22. Riemenschneider, M., Heider, D.: Current approaches in computational drug resistance prediction in HIV. Curr. HIV Res. **14**(4), 307–315 (2016)
23. Saha, M., Sarkar, A.: Review on multiple facets of drug resistance: a rising challenge in the 21st century. J. Xenobiot. **11**(4), 197–214 (2021)
24. Shabanpour, Y., Sajjadi, S., Behmard, E., Abdolmaleki, P., Keihan, A.H.: The structural, dynamic, and thermodynamic basis of darunavir resistance of a heavily mutated HIV-1 protease using molecular dynamics simulation. Front. Molecul. Biosci. **9** (2022)
25. Sheik Amamuddy, O., Bishop, N.T., Tastan Bishop, Ö.: Improving fold resistance prediction of HIV-1 against protease and reverse transcriptase inhibitors using artificial neural networks. BMC Bioinformatics **18**(1), 1–7 (2017)
26. Shen, C., Yu, X., Harrison, R.W., Weber, I.T.: Automated prediction of HIV drug resistance from genotype data. BMC Bioinformatics **17**(8), 563–569 (2016)
27. Simon, V., Ho, D.D.: Hiv-1 dynamics in vivo: implications for therapy. Nat. Rev. Microbiol. **1**(3), 181–190 (2003)
28. Steiner, M.C., Gibson, K.M., Crandall, K.A.: Drug resistance prediction using deep learning techniques on HIV-1 sequence data (2020). https://doi.org/10.3390/v12050560
29. Tama, F., Sanejouand, Y.H.: Conformational change of proteins arising from normal mode calculations. Protein Eng. **14**(1), 1–6 (2001)

30. Tozzini, V., Trylska, J., Chang, C.E., McCammon, J.A.: Flap opening dynamics in HIV-1 protease explored with a coarse-grained model. J. Struct. Biol. **157**(3), 606–615 (2007)

31. de Vera, I.M.S., Smith, A.N., Dancel, M.C.A., Huang, X., Dunn, B.M., Fanucci, G.E.: Elucidating a relationship between conformational sampling and drug resistance in HIV-1 protease. Biochemistry **52**(19), 3278–3288 (2013)

32. Wang, R., Zheng, Q.: Multiple molecular dynamics simulations and energy analysis unravel the dynamic properties and binding mechanism of mutants HIV-1 protease with DRV and CA-P2. Microbiol. Spect. **10**(2), e00748-21 (2022)

33. Weber, I.T., Harrison, R.W.: Tackling the problem of HIV drug resistance. Postępy Biochemii **62**(3), 273–279 (2016)

34. Wittayanarakul, K., et al.: Insights into saquinavir resistance in the g48v HIV-1 protease: quantum calculations and molecular dynamic simulations. Biophys. J . **88**(2), 867–879 (2005)

35. Wittayanarakul, K., et al.: Structure, dynamics and solvation of HIV-1 protease/saquinavir complex in aqueous solution and their contributions to drug resistance: molecular dynamic simulations. J. Chem. Inf. Model. **45**(2), 300–308 (2005)

36. Yang, L., Song, G., Carriquiry, A., Jernigan, R.L.: Close correspondence between the motions from principal component analysis of multiple HIV-1 protease structures and elastic network modes. Structure **16**(2), 321–330 (2008)

37. Yao, X.Q., et al.: Dynamic coupling and allosteric networks in the α subunit of heterotrimeric g proteins. J. Biol. Chem. **291**(9), 4742–4753 (2016)

38. Yu, X., Weber, I.T., Harrison, R.W.: Prediction of HIV drug resistance from genotype with encoded three-dimensional protein structure. BMC Genom. **15**(5), S1 (2014). https://doi.org/10.1186/1471-2164-15-S5-S1

39. Zhang, Y., Chang, Y.C.E., Louis, J.M., Wang, Y.F., Harrison, R.W., Weber, I.T.: Structures of darunavir-resistant HIV-1 protease mutant reveal atypical binding of darunavir to wide open flaps. ACS Chem. Biol. **9**(6), 1351–1358 (2014)

Replay-Based Online Adaptation
for Unsupervised Deep Visual Odometry

Yevhen Kuznietsov[1]([✉]), Marc Proesmans[1], and Luc Van Gool[1,2,3]

[1] KU Leuven, Leuven, Belgium
{yevhen.kuznietsov,marc.proesmans,luc.vangool}@esat.kuleuven.be
[2] ETH Zurich, Zurich, Switzerland
[3] INSAIT Sofia, Sofia, Bulgaria

Abstract. Online adaptation is a promising paradigm that enables dynamic adaptation to new environments. In recent years, there has been a growing interest in exploring online adaptation for various problems, including visual odometry, a crucial task in robotics, autonomous systems, and driver assistance applications. In this work, we leverage experience replay, a potent technique for enhancing online adaptation, to explore the replay-based online adaptation for unsupervised deep visual odometry. Our experiments reveal a remarkable performance boost compared to the non-adapted model. Furthermore, we conduct a comparative analysis against established methods, demonstrating competitive results that showcase the potential of online adaptation in advancing visual odometry.

Keywords: Visual odometry · Online adaptation · Experience replay

1 Introduction

1.1 Unsupervised Deep Visual Odometry

Odometry is one of the essential tasks for such applications as robotics, autonomous driving, advanced driver assistance systems, etc. It refers to the task of estimating the trajectory of a moving agent. The trajectory is determined by the agent's pose at different time points, which constitutes its translation and rotation relative to its initial state. Odometry problems can be categorized according to the type of supervision, sensor setup, and approach to pose inference. In this work, we address unsupervised deep visual odometry.

Supervision Type. While supervised learning relies on labeled data for model training, there are many scenarios where obtaining such annotations can be challenging or impractical. Unsupervised learning has gained popularity due to its ability to learn from raw, unlabeled data by extracting implicit cues without the need for annotations. This trend has become prominent as it allows for the utilization of vast amounts of unannotated data, making it a compelling approach for addressing data-rich domains.

© Springer Nature Switzerland AG 2024
V. Vasconcelos et al. (Eds.): CIARP 2023, LNCS 14469, pp. 674–684, 2024.
https://doi.org/10.1007/978-3-031-49018-7_48

Sensor Setup. Various sensors can be employed for the odometry task, including lidar, inertial measurement units, cameras, and their combinations. Among the different odometry setups, monocular visual odometry stands out as the most challenging one. This approach requires estimating the trajectory solely by analyzing the sequence of images taken by a moving camera. Despite its complexity, this setup also offers the most flexibility and cost efficiency. It is worth noting that this setup with one camera is commonly referred to as visual odometry, while the stereo camera setup is typically called stereo odometry.

Approach to Pose Inference. There are three main directions for pose inference: traditional, deep (deep learning-based), and hybrid. Traditional methods rely on handcrafted features, feature matching, and geometric techniques to estimate camera motion and pose from the matches. Deep approaches assume that the pose is inferred only using a deep neural network, while conventional mechanisms can still be employed for its training. Finally, there are hybrid techniques, which combine the advantages of traditional and deep approaches, typically incorporating learned features. While hybrid algorithms currently yield the best results for visual odometry, deep methods are generally less complicated and allow for smoother integration with other tasks (e.g., semantic segmentation). Furthermore, due to the availability of ever-growing amounts of data for training, deep algorithms have the potential to exceed the performance of hybrid methods.

1.2 Online Adaptation

Deep learning methods often exhibit a performance drop when deployed in new environments. Domain adaptation methods help to mitigate this effect. However, these methods typically operate offline, requiring curated data and extensive amounts of time for training in the new domain. In contrast, both test-time fine-tuning and online adaptation offer ways to improve performance in new environments without the limitations of offline domain adaptation. As a result, these techniques have garnered growing attention in recent years, extending their application to areas such as stereo matching [20,21,31], depth estimation [2,4,11, 30] and completion [3], semantic [12,24] and panoptic [23] segmentation, SLAM [17], and visual odometry [14,15,29].

Both approaches have advantages and limitations. For instance, fine-tuning is generally less challenging, but it tends to be slower compared to adaptation since it requires tens of backpropagation iterations per image. As fine-tuning always starts from a pretrained state and does not leverage the newly available data beyond tiny temporal windows, adaptation has the potential to significantly outperform fine-tuning on long image sequences or streams. However, it also introduces the risk of forgetting the useful information learned while pretraining.

In the context of unsupervised visual odometry, we are aware of three works exploring adaptive algorithms. Li et al. [14] employ meta-learning for online adaptation. Zhang et al. [29] fine-tune the predicted pose without modifying the network parameters, resulting in a much faster runtime compared to whole-model fine-tuning. The hybrid method by Li et al. [15] involves online adaptation of a deep optical flow network. The point correspondences obtained from the predicted flow are further filtered by RANSAC and used to compute the pose.

Experience replay is a promising technique used to enhance online-adaptive scene understanding algorithms [11,12,23] by revisiting previously observed data. However, despite the joint depth and pose optimization performed by these methods, none of the replay-based adaptation works have reported results for pose estimation. As a result, the impact of replay-based online adaptation on visual odometry remains unexplored. In this work, we address this gap by applying this online adaptation approach to the task of visual odometry.

Our main contribution lies in unveiling the profound impact of replay-based online adaptation on visual odometry. We demonstrate that this approach significantly enhances performance compared to the non-adapted model, utilizing the widely adopted KITTI odometry [5] split for evaluation. Additionally, we provide compelling arguments against the use of the five-frame ATE metric [32], which, although misleading, continues to be employed for odometry evaluation [2,6,16,18,25,27].

2   Methodology

For online-adaptive visual odometry, we adopt a modified version of the online adaptation framework proposed by Kuznietsov et al. [11], which leverages self-supervised monocular Structure-from-Motion (SfM) cues and experience replay. Unlike [11], we abstain from incorporating velocity supervision in our loss function. Introducing velocity supervision would lead our method into the realm of visual-inertial odometry, which involves additional sensor inputs and has distinct characteristics from visual odometry.

2.1   Self-supervision

During online adaptation, the parameters of the depth and pose networks are jointly updated using a self-supervision strategy commonly employed in unsupervised monocular depth estimation or deep visual odometry algorithms. This strategy involves two main steps: warping and loss computation. Typically, these steps are performed for several image scales, and the final loss is averaged over all scales. However, for simplicity, we will omit the image scale in further discussion, as the process remains the same for each scale.

Warping. Warping is a technique that allows obtaining a synthesized view $\hat{I}_{2\to1}$ of image I_1 from image I_2 of the same scene using the depth map D_1 predicted for image I_1, the camera motion $M_{1\to2}$ between the two images, and known camera intrinsics K. The process involves several steps. First, the depth map is used to lift the points of image I_1 to a point cloud. This point cloud is then transformed into the coordinate system of image I_2 and subsequently projected onto image

I_2. As a result, it is known where each pixel p_1 of image I_1 is supposed to be in image I_2. Equation 1 establishes the relation between p_1 and its reprojection location $\hat{p}_2$:

$$\hat{p}_2 \sim KZ^{-1}(P)\underbrace{M_{1\to2}D_1(p_1)K^{-1}p_1}_{P}.\tag{1}$$

Here, $Z(P)$ represents the last coordinate of a 3D point P. Finally, the synthesized view $\hat{I}_{2\to1}$ is generated by picking colors at the respective locations of image I_2. After obtaining the synthesized view via warping, the difference between the original image and its synthesized view further serves as supervision for the depth and camera motion estimation networks.

Loss Function. For model optimization, we employ the loss proposed by Godard et al. [6]. This loss relies on two views $\hat{I}_{-1\to0}$ and $\hat{I}_{1\to0}$ of the central frame I_0 being synthesized by warping its neighboring video frames I_{-1} and I_1. In addition to $\hat{I}_{-1\to0}$, $\hat{I}_{1\to0}$, I_{-1}, I_1, and I_0, this loss also takes as input the depth map D_0 predicted for I_0. The total loss is formulated as follows:

$$\mathcal{L} = \mathcal{L}_{IR} + \gamma\mathcal{L}_{sm},\tag{2}$$

where $\mathcal{L}_{IR}$ is the image reconstruction loss, $\mathcal{L}_{sm}$ is the smoothness term, and γ is a hyperparameter.

Image reconstruction loss is computed as the average of the per-pixel minimum of photoconsistency errors $\mathcal{L}_{pe}$:

$$\mathcal{L}_{IR} = \text{avg}_{\text{px}}\min\left[\mathcal{L}_{pe}(I_0, I_1), \mathcal{L}_{pe}(I_0, I_{-1}), \mathcal{L}_{pe}(I_0, \hat{I}_{1\to0}), \mathcal{L}_{pe}(I_0, \hat{I}_{-1\to0})\right].\tag{3}$$

Here, the addition of $\mathcal{L}_{pe}(I_0, I_1)$ and $\mathcal{L}_{pe}(I_0, I_{-1})$ under the minimum corresponds to auto-masking, which allows reducing the ambiguous supervision from the image areas that do not change over time (e.g., if there is a car moving similarly to the ego vehicle). The computation of minimum of $\mathcal{L}_{pe}$ for two synthesized views $\hat{I}_{1\to0}$ and $\hat{I}_{-1\to0}$, in turn, mitigates the effect of occlusions when a point visible in I_0 is occluded in one of its neighbors but visible in the other.

The photoconsistency error $\mathcal{L}_{pe}$ measures the dissimilarity between two images and is computed as follows:

$$\mathcal{L}_{pe}(I, \hat{I}) = \alpha\frac{1 - \text{SSIM}(I, \hat{I})}{2} + (1 - \alpha)|I - \hat{I}|.\tag{4}$$

Here, α is a hyperparameter, and SSIM represents Structural Similarity [26] calculated for each pixel.

The smoothness term $\mathcal{L}_{sm}$ ensures coherence between image edges and the edges of the corresponding depth map:

$$\mathcal{L}_{sm} = \partial_x D_0 \cdot e^{-\partial_x I_0} + \partial_y D_0 \cdot e^{-\partial_y I_0},\tag{5}$$

where ∂_i denotes the gradient magnitudes in the direction i.

2.2   Replay-Based Online Adaptation

In the process of online adaptation, we start with a depth and ego-motion estimation model $\mathcal{M}_0$ that has been pretrained on a specific dataset Ω. This pretrained
model is then adapted on a new video v ($v \cap \Omega = \emptyset$), frame by frame, using the
self-supervision strategy discussed above. Importantly, in order to estimate the
pose between two neighboring frames during online adaptation, we strictly avoid
accessing any future frames or images from other test videos.

One of the most significant challenges of online adaptation is locally limited
data diversity. When observing consecutive images, the data diversity is limited
within a local temporal frame. This lack of diversity can lead to overfitting or
bias towards the narrow range of observations within that window, hindering
the model's ability to generalize and adapt to unseen or diverse situations.

The integration of experience replay into the adaptation pipeline allows to
effectively address the issues arising from locally limited data diversity. At every
adaptation step, the network parameters are optimized not only based on the
triplet of the currently observed frame and its neighbors but also on previously
observed frame triplets (in our case, those used for pretraining). After incorporating experience replay, online adaptation is formalized and illustrated in
Algorithm 1 and Fig. 1. As a note, the pose between the first two frames is estimated as soon as the second frame is available. However, in Algorithm 1, we
moved its estimation inside the loop for simplicity reasons.

Algorithm 1. Replay-Based Online Adaptation

Require: Set of test videos V, model $\mathcal{M}_0$ pretrained on the set of data Ω, number of
    samples to draw from the replay buffer b, intrinsics K, other hyperparameters.
Ensure: Local poses between all pairs of neighboring frames in test videos
1: $\mathcal{E} \leftarrow \Omega$ ▷ Initialize the replay buffer
2: **for** $v \in V$ **do**
3:       $\mathcal{M} \leftarrow \mathcal{M}_0$ ▷ Reset the model parameters for every video
4:       **get** $\mathbf{I}_0^v,\, \mathbf{I}_1^v$
5:       $t \leftarrow 1$
6:       **while** $\neg$**end**(v) **do**   ▷ Loop through video frames
7:           **get** $\mathbf{I}_{t+1}^v$
8:           $[I'_{-1}, I'_0, I'_1] \leftarrow [\mathbf{I}_{t-1}, \mathbf{I}_t, \mathbf{I}_{t+1}]$ ▷ Current triplet
9:           $[I_{-1}, I_0, I_1] \leftarrow [I'_{-1}, I'_0, I'_1] \oplus$ **draw_samples**$(\mathcal{E}, b)$ ▷ Batch of triplets
10:          $M_{-1 \to 0}, M_{0 \to 1}, D_0 \leftarrow \mathcal{M}(I_{-1}, I_0, I_1)$ ▷ Infer depth and ego-motion
11:          **if** $t == 1$ **then**   ▷ Save camera poses for evaluation
12:              **save** $\mathbf{M}_{t-1 \to t}^v$
13:          **save** $\mathbf{M}_{t \to t+1}^v$
14:          $\hat{I}_{-1 \to 0} \leftarrow$ **warp**$(M_{-1 \to 0}^{-1}, D_0, I_{-1}, K)$
15:          $\hat{I}_{0 \to 1} \leftarrow$ **warp**$(M_{0 \to 1}, D_0, I_1, K)$
16:          **compute** $\mathcal{L}(\hat{I}_{-1 \to 0}, \hat{I}_{0 \to 1}, I_{-1}, I_0, I_1, D_0)$ ▷ Eq. 2
17:          $\mathcal{M} \leftarrow$ **optimize**$(\mathcal{M}, \mathcal{L})$
18:          $t \leftarrow t + 1$

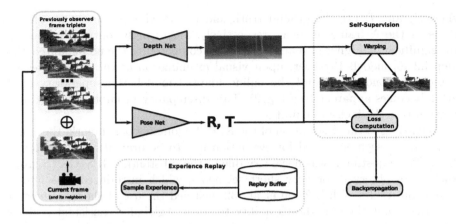

Fig. 1. Illustration of one online adaptation iteration.

2.3 Model Setup

As we perform the adaptation starting from the pretrained Monodepth2 [6] model, we employ the same ResNet-18 [9]-based network architectures for depth and pose estimation. It is important to note that the pose network takes input features from two frames. During online adaptation, we retain the hyperparameters used for Monodepth2 pretraining, setting α to 0.85, γ to 0.001, the learning rate to 0.0001, and utilizing the ADAM optimizer [10]. Similar to [11], we freeze the batch normalization layers and apply no data augmentation during online adaptation. For experience replay, we draw three samples from the replay buffer for every iteration of the adaptation process, using the entire set of pretraining data as the replay buffer. The method runs at 10 FPS with an input resolution of 192×640 pixels on the Nvidia GTX 1080 Ti GPU.

3 Evaluation Protocol

In line with other deep visual odometry works, we selected the KITTI odometry [5] 00-08/09-10 split for evaluation. Specifically, a model pretrained on sequences 00-08 is adapted and evaluated online on sequences 09 and 10. The KITTI dataset consists of ~10 FPS videos capturing driving scenes in and around Karlsruhe, with scene types roughly categorized as urban, suburban, or road.

Initially, we followed the evaluation protocol introduced by Zhou et al. [32], which is still commonly used for evaluating unsupervised deep visual odometry. According to this protocol, the average Absolute Trajectory Error (ATE) is computed over all overlapping five-frame trajectory segments, which are first origin- and scale-aligned. ATE for a trajectory or its part is computed as follows:

$$\sqrt{\frac{1}{N} \sum_{i=1}^{N} \|T_i - \hat{T}_i\|^2}, \tag{6}$$

where T_i and $\hat{T}_i$ are the ground truth and predicted x, y, z poses for the i-th frame in the N-frame segment, respectively. Surprisingly, our five-frame ATE was significantly higher (1.5 and 1.24 times) compared to the non-adapted Monodepth2 [6] model. However, upon visual examination of the trajectories predicted by Monodepth2 and the online-adapted model, the superiority of the latter becomes apparent (see Fig. 2). This discrepancy prompted further investigation into the factors behind it.

According to the description of the KITTI odometry benchmark, the lengths of trajectory segments used for evaluation had to be dramatically increased in 2013. This adjustment was necessary due to the substantial GPS/OXTS ground truth error observed for very short segments, which introduced bias into the evaluation results. While Zhou et al. [32] justified the use of short segments by concerns about their method's ability to learn a globally consistent scale, we argue against evaluating pose estimation performance on five-frame segments (approximately 0.4 s or three-four meters of driving for odometry sequences 09 and 10) with large ground truth errors and per-segment scale alignment. This evaluation approach not only provides limited insight but is also misleading wrt the comparison between different methods.

Instead, we employ evaluation practices more commonly used in hybrid and conventional odometry works. Initially, the trajectories are aligned with the ground truth using the Umeyama algorithm [22]. Subsequently, we utilize three metrics to assess the accuracy of the estimated trajectory. The first metric is ATE computed for the entire video sequence. In addition to ATE, we employ the relative translation T_{rel} and rotation R_{rel} errors. For each 100-, 200-, ..., 800-m segment of the ground truth trajectory and the corresponding prediction, we calculate the relative poses between the ends and starts of the segments. The relative translation error is then computed as the ratio between the difference in predicted and ground truth relative translation and the segment length, averaged over all segments. Similarly, the relative rotation error is computed as the angle between the predicted and ground truth relative rotations per 100 m of the segment length, averaged over all segments. By utilizing these evaluation metrics, we can assess the translation and rotation drift in the estimated trajectory.

4   Results

Table 1 and Fig. 2 demonstrate the results achieved by online adaptation compared to the non-adapted Monodepth2 and other unsupervised deep visual odometry methods.

The quantitative comparison showcases the remarkable superiority of online adaptation over the non-adapted model across all evaluated metrics. The error reduction achieved by our online adaptation method is substantial, ranging from 1.79 to 3.90 times lower errors, depending on the specific metric and sequence. Notably, the gains are more pronounced for the longer sequence 09, where our approach has more time to adapt.

Although our primary objective is to demonstrate the impact of replay-based online adaptation on visual odometry by comparing it to the non-adapted model,

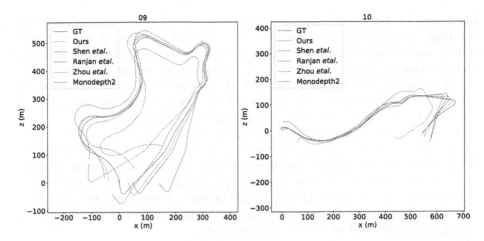

Fig. 2. Ego vehicle trajectories estimated by our algorithm, the non-adapted Monodepth2 [6] model, and other unsupervised deep visual odometry methods for sequences 09 and 10 of KITTI odometry [5] split. "GT" represents the ground truth GPS/OXTS trajectories.

Table 1. Comparison of unsupervised deep visual odometry methods on KITTI odometry [5] sequences 09 and 10. "-" indicates that either there is no data available or our evaluation yielded the results that significantly differ from those reported in other papers. "# frames" denotes the number of frames the method uses for pose inference. "o" and "ft" indicate the use of online adaptation and test-time fine-tuning, respectively.

Method	# frames	09			10		
		T_{rel}	R_{rel}	ATE	T_{rel}	R_{rel}	ATE
Zhou et al. [32]	5	82.97	31.84	79.33	81.63	36.50	84.15
GeoNet [27]	5	28.72	9.77	158.45	23.90	9.04	43.04
Zhan et al. [28]	2	11.93	3.91	–	12.45	3.46	–
Shen et al. [19]	3	9.91	3.77	27.08	12.18	5.90	24.44
Bian et al. [1]	2	11.2	3.35	–	10.1	4.96	–
Casser et al. [2]	3	10.2	2.64	-	29.0	4.28	–
UnDeepVO [13]	2	7.01	3.6	-	10.63	4.6	–
Li et al. [16]	RNN	9.52	3.64	-	6.45	2.41	–
Hariat et al. [8]	–	8.42	2.66	-	7.29	2.14	–
Ranjan et al. [18]	5	6.92	1.77	29.00	7.97	3.11	13.77
Gordon et al. [7]	2	–	–	**19.03**	–	–	14.86
Li et al. [14]^o	RNN	5.89	3.34	–	**4.79**	*0.83*	-
Zhang et al. [29]^{ft}	2	*2.02*	*0.61*	*4.76*	*2.29*	**1.10**	*3.38*
Monodepth2 [6]	2	17.55	3.92	77.72	12.01	5.52	21.30
Ours w/o replay^o	2	9.62	2.74	39.77	12.8	4.02	20.52
Ours^o	2	**5.62**	**1.64**	19.95	6.70	2.02	**8.98**

our approach also yields highly competitive results when compared to other well-established methods. Even though we start the adaptation from a relatively weak model, our online adaptation outperforms all non-adaptive unsupervised deep visual odometry approaches, with the exception of [7] on sequence 09.

When it comes to the adaptive methods, we observe that our algorithm performs better than the meta-learning-based online adaptation approach proposed by Li et al. [14] on sequence 09, while on sequence 10, [14] yields lower errors. The test-time fine-tuning method introduced by Zhang et al. [29] exhibits the best performance on both sequences. Although some of their success can be attributed to the use of higher-resolution input images or stereo pretraining, their algorithm achieves impressive accuracy while only being two times slower than our method. Considering that Zhang et al. [29] do not perform network optimization, their approach and our online adaptation are not mutually exclusive. Combining both could potentially lead to further improvements.

5 Conclusion

In this work, we have presented the replay-based online adaptation method for unsupervised deep visual odometry. Our findings demonstrate the remarkable impact of online adaptation on visual odometry performance. In particular, we achieve a substantial reduction in errors, ranging from 1.79 to 3.90 times, compared to the non-adapted model, across all evaluated metrics for both test sequences of the KITTI odometry split [5]. Moreover, our results showcase the competitiveness of our method when compared to established unsupervised deep visual odometry methods. Lastly, we anticipate that even higher performance levels can be unlocked by exploring the incorporation of recurrent neural networks or combining online adaptation with test-time fine-tuning [29].

References

1. Bian, J., et al.: Unsupervised scale-consistent depth and ego-motion learning from monocular video. In: Advances in Neural Information Processing Systems **32** (2019)
2. Casser, V., Pirk, S., Mahjourian, R., Angelova, A.: Depth prediction without the sensors: leveraging structure for unsupervised learning from monocular videos. In: Proceedings of the AAAI Conference on Artificial Intelligence, vol. 33, pp. 8001–8008 (2019)
3. Chen, Y., Tan, Y.: Self-supervised depth completion with adaptive online adaptation. In: Proceedings of the 2023 7th International Conference on Machine Learning and Soft Computing, pp. 168–174 (2023)
4. Chen, Y., Schmid, C., Sminchisescu, C.: Self-supervised learning with geometric constraints in monocular video: connecting flow, depth, and camera. In: Proceedings of the IEEE/CVF International Conference on Computer Vision, pp. 7063–7072 (2019)
5. Geiger, A., Lenz, P., Urtasun, R.: Are we ready for autonomous driving? the kitti vision benchmark suite. In: Conference on Computer Vision and Pattern Recognition (CVPR) (2012)

6. Godard, C., Mac Aodha, O., Firman, M., Brostow, G.J.: Digging into self-supervised monocular depth estimation. In: Proceedings of the IEEE/CVF International Conference on Computer Vision, pp. 3828–3838 (2019)
7. Gordon, A., Li, H., Jonschkowski, R., Angelova, A.: Depth from videos in the wild: unsupervised monocular depth learning from unknown cameras. In: Proceedings of the IEEE/CVF International Conference on Computer Vision, pp. 8977–8986 (2019)
8. Hariat, M., Manzanera, A., Filliat, D.: Rebalancing gradient to improve self-supervised co-training of depth, odometry and optical flow predictions. In: Proceedings of the IEEE/CVF Winter Conference on Applications of Computer Vision, pp. 1267–1276 (2023)
9. He, K., Zhang, X., Ren, S., Sun, J.: Deep residual learning for image recognition. In: Proceedings of the IEEE Conference on Computer Vision and Pattern Recognition, pp. 770–778 (2016)
10. Kingma, D.P., Ba, J.: Adam: a method for stochastic optimization. arXiv preprint arXiv:1412.6980 (2014)
11. Kuznietsov, Y., Proesmans, M., Van Gool, L.: Comoda: continuous monocular depth adaptation using past experiences. In: Proceedings of the IEEE/CVF Winter Conference on Applications of Computer Vision, pp. 2907–2917 (2021)
12. Kuznietsov, Y., Proesmans, M., Van Gool, L.: Towards unsupervised online domain adaptation for semantic segmentation. In: Proceedings of the IEEE/CVF Winter Conference on Applications of Computer Vision, pp. 261–271 (2022)
13. Li, R., Wang, S., Long, Z., Gu, D.: Undeepvo: monocular visual odometry through unsupervised deep learning. In: 2018 IEEE International Conference on Robotics and Automation (ICRA), pp. 7286–7291. IEEE (2018)
14. Li, S., Wang, X., Cao, Y., Xue, F., Yan, Z., Zha, H.: Self-supervised deep visual odometry with online adaptation. In: Proceedings of the IEEE/CVF Conference on Computer Vision and Pattern Recognition, pp. 6339–6348 (2020)
15. Li, S., Wu, X., Cao, Y., Zha, H.: Generalizing to the open world: Deep visual odometry with online adaptation. In: Proceedings of the IEEE/CVF Conference on Computer Vision and Pattern Recognition, pp. 13184–13193 (2021)
16. Li, S., Xue, F., Wang, X., Yan, Z., Zha, H.: Sequential adversarial learning for self-supervised deep visual odometry. In: Proceedings of the IEEE/CVF International Conference on Computer Vision, pp. 2851–2860 (2019)
17. Loo, S.Y., Shakeri, M., Tang, S.H., Mashohor, S., Zhang, H.: Online mutual adaptation of deep depth prediction and visual slam. arXiv preprint arXiv:2111.04096 (2021)
18. Ranjan, A., et al.: Competitive collaboration: Joint unsupervised learning of depth, camera motion, optical flow and motion segmentation. In: Proceedings of the IEEE/CVF Conference on Computer Vision and Pattern Recognition, pp. 12240–12249 (2019)
19. Shen, T., et al.: Beyond photometric loss for self-supervised ego-motion estimation. In: 2019 International Conference on Robotics and Automation (ICRA), pp. 6359–6365. IEEE (2019)
20. Tonioni, A., Rahnama, O., Joy, T., Stefano, L.D., Ajanthan, T., Torr, P.H.: Learning to adapt for stereo. In: Proceedings of the IEEE/CVF Conference on Computer Vision and Pattern Recognition, pp. 9661–9670 (2019)
21. Tonioni, A., Tosi, F., Poggi, M., Mattoccia, S., Stefano, L.D.: Real-time self-adaptive deep stereo. In: Proceedings of the IEEE/CVF Conference on Computer Vision and Pattern Recognition, pp. 195–204 (2019)

22. Umeyama, S.: Least-squares estimation of transformation parameters between two point patterns. IEEE Trans. Pattern Analy. Mach. Intell. **13**(04), 376–380 (1991)
23. Vödisch, N., Petek, K., Burgard, W., Valada, A.: Codeps: online continual learning for depth estimation and panoptic segmentation. arXiv preprint arXiv:2303.10147 (2023)
24. Wang, D., Shelhamer, E., Liu, S., Olshausen, B., Darrell, T.: Tent: fully test-time adaptation by entropy minimization. In: International Conference on Learning Representations (2020)
25. Wang, F., Cheng, J., Liu, P.: Cbwloss: constrained bidirectional weighted loss for self-supervised learning of depth and pose. IEEE Trans. Intell. Trans. Syst. (2023)
26. Wang, Z., Bovik, A.C., Sheikh, H.R., Simoncelli, E.P.: Image quality assessment: from error visibility to structural similarity. IEEE Trans. Image Process. **13**(4), 600–612 (2004)
27. Yin, Z., Shi, J.: Geonet: unsupervised learning of dense depth, optical flow and camera pose. In: Proceedings of the IEEE Conference on Computer Vision and Pattern Recognition, pp. 1983–1992 (2018)
28. Zhan, H., Garg, R., Weerasekera, C.S., Li, K., Agarwal, H., Reid, I.: Unsupervised learning of monocular depth estimation and visual odometry with deep feature reconstruction. In: Proceedings of the IEEE Conference on Computer Vision and Pattern Recognition, pp. 340–349 (2018)
29. Zhang, J., Sui, W., Wang, X., Meng, W., Zhu, H., Zhang, Q.: Deep online correction for monocular visual odometry. In: 2021 IEEE International Conference on Robotics and Automation (ICRA), pp. 14396–14402. IEEE (2021)
30. Zhang, Z., Lathuiliere, S., Ricci, E., Sebe, N., Yan, Y., Yang, J.: Online depth learning against forgetting in monocular videos. In: Proceedings of the IEEE/CVF Conference on Computer Vision and Pattern Recognition, pp. 4494–4503 (2020)
31. Zhong, Y., Li, H., Dai, Y.: Open-world stereo video matching with deep RNN. In: Ferrari, V., Hebert, M., Sminchisescu, C., Weiss, Y. (eds.) ECCV 2018. LNCS, vol. 11206, pp. 104–119. Springer, Cham (2018). https://doi.org/10.1007/978-3-030-01216-8_7
32. Zhou, T., Brown, M., Snavely, N., Lowe, D.G.: Unsupervised learning of depth and ego-motion from video. In: Proceedings of the IEEE Conference on Computer Vision and Pattern Recognition, pp. 1851–1858 (2017)

Facial Point Graphs for Stroke Identification

Nicolas Barbosa Gomes(ID), Arissa Yoshida(ID),
Guilherme Camargo de Oliveira(ID), Mateus Roder(ID), and João Paulo Papa(✉)(ID)

São Paulo State University (UNESP), Bauru, CEP 17033-360, Brazil
joao.papa@unesp.br

Abstract. Stroke can cause significant damage to neurons, resulting in various sequelae that negatively impact the patient's ability to perform essential daily activities such as chewing, swallowing, and verbal communication. Therefore, it is important for patients with such difficulties to undergo a treatment process and be monitored during its execution to assess the improvement of their health condition. The use of computerized tools and algorithms that can quickly and affordably detect such sequelae proves helpful in aiding the patient's recovery. Due to the death of internal brain cells, a stroke often leads to facial paralysis, resulting in certain asymmetry between the two sides of the face. This paper focuses on analyzing this asymmetry using a deep learning method without relying on handcrafted calculations, introducing the Facial Point Graphs (FPG) model, a novel approach that excels in learning geometric information and effectively handling variations beyond the scope of manual calculations. FPG allows the model to effectively detect orofacial impairment caused by a stroke using video data. The experimental findings on the Toronto Neuroface dataset revealed the proposed approach surpassed state-of-the-art results, promising substantial advancements in this domain.

Keywords: Stroke · Facial paralysis · Deep learning · Facial Point Graph

1 Introduction

Stroke is a neurological disease caused by an abnormality that changes the blood flow in the brain. The incident occurs when vessels that carry blood internally to this organ become clogged or rupture. There are two main types of strokes: ischemic and hemorrhagic. Ischemic strokes lead to insufficient blood supply in the affected area, resulting in a shortage of essential nutrients and oxygen. On the other hand, hemorrhagic strokes involve vessel rupture, causing increased pressure on the surrounding tissues and nerves. The disease is a matter of great concern worldwide and has escalated as the global population undergoes rapid aging [21]. Every year, there are approximately 10.3 to 16.9 million cases worldwide, with 5.9 million of them being fatal and requiring some form of treatment

© Springer Nature Switzerland AG 2024
V. Vasconcelos et al. (Eds.): CIARP 2023, LNCS 14469, pp. 685–699, 2024.
https://doi.org/10.1007/978-3-031-49018-7_49

and therapy for 25.7 to 33 million cases [21]. In the US, the estimated annual cost is 34 billion dollars [2].

As a consequence of the disease, the cells in the affected brain region may incur damage, leading to various sequelae, including motor deficits, muscle atrophy, and facial paralysis [32]. These cells are crucial in coordinating many bodily functions, and their impairment can profoundly affect an individual's overall health and well-being. Patients with facial paralysis experience weakened muscles that affect their ability to produce speech [24], significantly impacting an individual's social communication abilities [30]. Moreover, everyday activities such as eating and drinking become more challenging, highlighting the importance of rehabilitation treatment for patients and creating automated processes to provide accessible treatment monitoring.

To address such a problem, researchers proposed automated image and video analysis techniques for evaluating orofacial impairments resulting from strokes and other diseases. For example, Bandini et al. [2] utilized three-dimensional facial landmarks extracted from video data and applied various feature calculations to assess facial symmetry and movement. Leveraging a Support Vector Machine (SVM) [5], they achieved an impressive 87% precision in classifying orofacial impairments in post-stroke patients. Similarly, Parra-Dominguez et al. [28] proposed a comparable approach involving calculating distances and angles related to facial symmetry. By extracting thirty features from patients' images, they employed a MultiLayer Perceptron (MLP) [15] to detect facial paralysis, achieving an accuracy of over 94% on two different datasets. These advancements may improve treatment outcomes and enhance the overall quality of life for affected individuals.

The results obtained in the previous studies demonstrated impressive performance. However, it is important to consider that all the features employed were manually designed, which presents certain limitations. Handcrafted features heavily rely on human insights and may lack the adaptability required to handle real-world variations. In order to eliminate the need for such calculations, a model called Facial Point Graphs (FPGs) was proposed in [11] to cope with Amyotrophic Lateral Sclerosis (ALS).

The FPG is a model based on graphs formed from the most significant facial landmarks that use Graph Neural Networks (GNNs) [31] for inferring features from them. In this work, the model determines whether an individual has suffered a stroke or not, i.e., if there is orofacial impairment in the patient's video while performing a specific task. To the best of our knowledge, there is currently no existing approach utilizing Facial Point Graph for detecting this impairment resulting from a stroke. Our strong conviction lies in the notion that the landmarks extracted from frames can be more effectively encoded within a non-Euclidean space, facilitating the accurate definition and representation of their unique characteristics. Hence, this paper has two main contributions:

– To introduce Facial Point Graphs for the identification of orofacial impairment arising from a stroke in videos.

– To employ a deep learning approach within this context, eliminating the need for manually crafted features.

The structure of this paper is as follows: Sect. 2 provides a comprehensive review of the relevant literature, and Sect. 3 offers a detailed explanation of the theoretical background. Besides that, Sect. 4 outlines the dataset used, describes the models employed for image cropping and facial feature extraction, presents our proposed approach, and explains the classification method. Section 5 shows the experimental results, while Sect. 6 encompasses discussing the results, drawing conclusions, and outlining avenues for future research.

2   Related Works

In the field of computer vision, the analysis of facial signals in images has aroused great interest and motivated various studies intending to automate the detection of facial paralysis [17,19,28]. Facial paralysis, caused by nerve damage, can arise due to congenital conditions, trauma, or diseases such as Bell's palsy, brain tumor, or stroke. In this context, methods are developed considering facial asymmetry as a key indicator.

Parra-Dominguez et al. [28] proposed an MLP-based model to classify facial paralysis, disregarding the symmetric variations across different facial expressions performed by individuals. Their approach begins by extracting facial landmark features using a publicly available model called MEEshape [14], which accurately identifies 68 intrinsic points representing facial expressions in individual images. To account for potential head tilting in patient images, the authors incorporate a correction for the angle of the extracted facial points. Subsequently, calculations are performed, including measurements such as the distance from the corners of the eyes to the middle of the lower lip. Their MLP-based neural network achieves remarkable accuracy rates: 94.06% on the Massachusetts Eye and Ear Infirmary dataset (MEEI) [12], and 97.22% on the Toronto NeuroFace dataset [3].

Bandini et al. [2] conducted a study aimed at analyzing the impact of stroke on the orofacial musculature. The authors made significant progress in automatically evaluating this musculature. The study involved capturing videos of twelve stroke patients (seven males and five females) with an average age of 62.0 years and a standard deviation of 14.5. As a control group, videos of individuals without the disease were also included (seven males and four females) with an average age of 55.8 years and a standard deviation of 15.7. During the video recording sessions, participants were instructed to perform specific actions, including maintaining a resting position with normal biting and a neutral facial expression for twenty seconds (REST), repeatedly opening their jaw to its maximum five times (OPEN), puckering their lips as if kissing a baby for a total of 4 times (KISS), simulating blowing out a candle five times (BLOW), smiling with closed lips for five times (SPREAD), rapidly repeating the syllable "/pa/" in a single breath (PA), rapidly repeating the word "/pataka/" (PATAKA), and delivering ten repetitions of the sentence "Buy Bobby a puppy" in their usual tone and speaking speed.

Following a standardized procedure, the authors manually segmented each action repetition. For instance, the video of the SPREAD action was divided into five segments since the movement was executed five times. To ensure facial normalization and alignment of facial landmarks in each frame, the authors employed the Supervised Descent Method (SDM) algorithm [34], which aims to minimize the differences between the extracted features from each image. Ten geometric features were derived from the 3D coordinates of the facial landmarks for each repetition. Notably, many of these extracted features are related to facial symmetry, capturing potential signs of facial paralysis. Lastly, a validation method called "Leave One Out" (LOSO-CV) was employed with all 23 patients. This method involved treating one patient as the test set while utilizing the remaining patients as the training set for the machine learning model. The SVM model was employed in this study and achieved an accuracy of 87% in diagnosing the patients.

Pecundo et al. [29] conducted a similar approach. They classified the health status of individuals into three categories: ALS, Orofacial impairment in post-stroke, and healthy patients. The inference dataset used in their study was the Toronto Neuroface dataset [3]. The videos in this dataset showcased individuals performing various tasks, such as maximum jaw opening (OPEN), as already described above in this section. The authors manually segmented the videos since each contained multiple repetitions of the same subtask by an individual. Furthermore, they extracted specific facial landmarks using the Ensemble Regression Tree [18] implemented in the dlib library [20]. This model allowed the extraction of 68 facial landmarks, which were then used to calculate 14 essential features to infer ALS and Stroke and related to Facial Asymmetry, Range of Motion, and Mouth Shape and Geometry. The authors employed machine learning algorithms, including Random Forest [16], SVM, and K-Nearest Neighbors [10], achieving an accuracy of 86%.

As far as we know, the classification of stroke using a graph model formed by facial landmarks is innovative. However, GNNs models have already been used to capture emotional information. According to Ngoc et al. [26], facial landmarks formulated as graphs could result in an appropriate method for extracting valuable information from human expressions. In their model, the authors initially mapped the facial landmarks using Style Aggregated Network (SAN) [8]. It is worth noting that face landmarks formed the graph's vertices; subsequently, the Delaunay triangulation method [6] was employed to establish the edges. Finally, a model called Directed Graph Neural Network was used for facial expression classification on the CK+ [23], MMI [27], and AFEW [7] datasets, achieving accuracies of 96.02%, 69.4%, and 32.64%, respectively. An important point to mention is that damage caused by a stroke to the facial nerve alters the facial expression since that nerve supplies the muscles of facial expression [22]. Given this knowledge, we employed a similar approach based on graphs constructed from facial points to analyze the impairments resulting from a stroke. This decision is based on validating similar methods in distinguishing facial expressions related to emotions described above in this section. Such findings reinforce the GNN model as a promising approach for identifying orofacial impairments.

3   Theoretical Background

3.1   Graph Neural Networks

GNNs revolutionize how patterns are learned from a dataset by leveraging the power of graph representations. In the iterative process of GNNs, each node collects and integrates aggregated messages from its neighboring nodes using specialized aggregation and update functions. Importantly, each node transmits information to its adjacent nodes before updating its features. In the subsequent iteration, this new information (message) is again forwarded to its neighbors. This process of exchanging messages is visually demonstrated in Fig. 1.

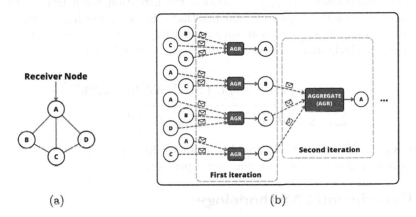

(a)   (b)

Fig. 1. Message aggregation in a bidirectional graph: (a) input graph and (b) working mechanism of a Graph Neural Network. For the sake of clarity, the second iteration focuses exclusively on node 'A'.

In each iteration k of the neural network, a hidden vector $\mathbf{h}_u^{(k)} \in \mathbb{R}^n$ is generated to capture the features of node $u \in \mathcal{V}$, where n represents the number of input features. Notably, the hidden vector $\mathbf{h}_u^{(0)}$ represents the initial encoding of the features before any training takes place. The process involves two steps: first, a node-order invariant function aggregates the features from the neighborhood $\mathcal{N}(u)$ of the node u, allowing the incorporation of relevant information from neighboring nodes. Secondly, the aggregated features are used to update the node's information, leading to the generation of $\mathbf{h}_u^{(k+1)}$. The specific mechanism for this update is described as follows:

$$\mathbf{h}_u^{k+1} = Update^k \left(\mathbf{h}_u^k, \; Agreggate_u^k \left(\{ \mathbf{h}_v^k, \; \forall v \in \mathcal{N}(u) \} \right) \right), \tag{1}$$

Equation 1 reveals the presence of both update and aggregation functions. While various models can be employed for the aggregation function, this paper adopts an attention-based formulation described further.

3.2   Graph Attention Networks

Graph Attention Networks (GATs) offer a powerful technique to enhance aggregation function by assigning different priorities to neighborhood information. This innovative concept was first introduced by Veličković et al. [33], who pioneered the development of a model with the following formulation:

$$
\mathbf{h}_u^{k+1} = \sigma \left(\sum_{v \in \mathcal{N}(u)} \alpha_{v \to u}^k \mathbf{W}^k \mathbf{h}_v^k \right),
\tag{2}
$$

where $\mathbf{W} \in \mathbb{R}^{n' \times n}$ represents the weight matrix, which is a trainable parameter. Here, n' denotes the number of output features, and σ represents the sigmoid function. Additionally, $\alpha_{v \to u} \in \mathbb{R}$ denotes the attention given from node v to node u, indicating the degree of influence that v has on updating the features of u. A higher value of $\alpha_{v \to u}$ indicates a stronger impact of v on the feature update process of u. Mathematically, we can define $\alpha_{v \to u}$ as follows:

$$
\alpha_{v \to u}^k = \frac{exp \left(LeakyReLU \left([\mathbf{a}_u^k]^T \left[\mathbf{W}^k \mathbf{h}_u^k \,\|\, \mathbf{W}^k \mathbf{h}_v^k \right] \right) \right)}{\sum_{v' \in \mathcal{N}(u)} exp \left(LeakyReLU \left([\mathbf{a}_u^k]^T \left[\mathbf{W}^k \mathbf{h}_u^k \,\|\, \mathbf{W}^k \mathbf{h}_{v'}^k \right] \right) \right)},
\tag{3}
$$

where $\mathbf{a}_u \in \mathbb{R}^{2*n'}$ defines a trainable parameter known as the attention vector. The symbol $\|$ denotes the concatenation operator.

4   Experimental Methodology

4.1   Experimental Dataset

Introduced by Bandini et al. [3], Toronto Neuroface is a publicly available dataset designed to assess neurological disorders. The dataset comprises 261 colored videos, each clinically scored by two specialists. It includes oro-facial assessments, making it a valuable resource for tracking and analyzing facial movements in clinical populations affected by amyotrophic lateral sclerosis and post-stroke conditions. To validate the performance of our model, we use a balanced group comprising 11 individuals in the healthy control group (HC) and 11 individuals in the Stroke group. Each participant in the dataset performed multiple repetitions of speech and non-speech tasks. Important to note that not all participants in the dataset complete the tasks[1]. After manually cropping the videos for each participant's task, we obtained 2,509 repetitions. The details of each task and the number of video repetitions are presented in Table 1.

---

[1] For the BLOW subtask, there were 15 participants, consisting of 8 stroke patients and 7 healthy controls (HC).

Table 1. Number of repetitions of each subtask.

Task	Subtask	Description	HC	Stroke
Speech	BBP	Repetitions of the sentence "Buy Bobby a Puppy"	111	104
	PA	Repetitions of the syllables /pa/ as fast as possible in a single breath	884	533
	PATAKA	Repetitions of the syllables /pataka/ as fast as possible in a single breath	275	163
Non-speech	SPREAD	Pretending to smile with tight lips	59	62
	KISS	Pretend to kiss a baby	57	62
	OPEN	Maximum opening of the jaw	55	61
	BLOW	Pretend to blow a candle	39	44

4.2   Pre-processing and Feature Extraction

In the pre-processing stage, we employ the OpenFace 2.0 tool [1] to filter out visual elements outside the subject's face. This technique isolates the region of interest while incorporating head pose estimation to centralize the face. Consequently, the output of the pre-processing phase yields a grayscale cropped image, representing the bounding box region of the centralized main face, with dimensions of 200×200 pixels to each frame in the experimental dataset.

A series of feature extraction steps were essential to create the graph structure used in this study. Firstly, to capture landmark points, we employed Face Alignment Network (FAN) [4], a well-known deep learning approach for face alignment purposes. This method was built upon the hourglass (HG) network, designed initially for human pose estimation [25], using the landmark configuration of the MULTI-PIE 2D 68-point format [13] notation. Since the videos collected from the dataset only featured the frontal faces of individuals performing the tasks, we applied 2D coordinates landmark detection only.

For our study, as stroke facial paralysis can affect one or both sides of the face, we selected points on the face to capture the mirror positions of symmetry, comprising both sides of the forehead, eye, and mouth regions. Consequently, out of the 68 face landmark notations detected by FAN, we removed 20 landmark points irrelevant to our specific area of interest. This process resulted in a final set of 48 points, which we used as nodes in the graph. Each node was represented by its corresponding coordinates [x, y], serving as their features.

The connections among nodes were established using the Delaunay triangulation technique [6]. This method generates a triangular mesh that effectively captures the spatial relationships among data points based on specific criteria. It relies on geometric considerations to determine whether a pair of neighboring triangles represents an optimal choice of connections, resulting in a favorable representation of the spatial graph connection. In addition, a normalization technique was applied to facilitate the communication process among nodes during model training. The final structure of the graph ensures that all nodes were connected to point 31 of the MULTI-PIE notation (nose tip). This normalization point served as a reference to establish a consistent and standardized communication pattern among the nodes in the model. Finally, the Euclidean distance between each pair of nodes was calculated, serving as the weight for each edge in the graph. An overview of the entire process is presented in Fig. 2.

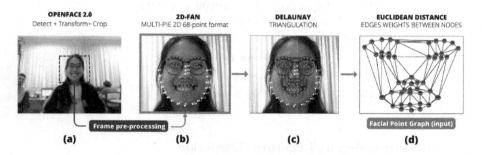

Fig. 2. Pre-processing and feature extraction procedures for constructing the graph structure used in the FPG model: (a) OpenFace 2.0 [1] detection, transformation, and crop of the main face. (b) Landmark extraction using 2D FAN [4] Facial Alignment method; (c) Delaunay triangulation [6] mesh process to connect landmark points of interest; and (d) Euclidean distance between each pair of nodes to calculate the edge weights of the final graph, used as input to the FPG model.

4.3   Classification and Evaluation

As the classification was conducted on a video dataset, we employed the proposed approach by Bandini et al. [2] as our benchmark. Their approach utilized ten geometric features, four representing a range of motion (ROM) and six representing asymmetry, calculated in 3D dimensions. In our experiments, we employed handcrafted models for comparison purposes using the same feature extraction, but in 2D landmark instead due to our dataset limitation: SVM with linear and radial basis function (RBF) and Logistic Regression. For comparison purposes, we also applied the landmark coordinates features to train and evaluate the benchmark models. Moreover, a grid search was employed to find proper values for SVM parameters, i.e., the confidence value C and the RBF kernel scale parameter γ^2.

The classification performance was evaluated using the LOSO-CV approach, ensuring separate splits for the training, validation, and test sets. This methodology effectively addresses concerns such as overfitting or memorizing training data. To maintain a balanced representation of both classes during the selection of the validation set, we randomly chose two subjects, one from the HC category and the other from the Stroke category, during each iteration split.

---

^{2} $C = [2^{-5}, 2^{-3}, ..2^{15}]$, and $\gamma = [2^{-15}, 2^{-13}, ..., 2^3]$.

The evaluation was conducted in two forms, i.e., repetition- and subject-based classification:

1. **Repetition classification:** At each iteration of the LOSO-CV, the repetitions generated by one participant were regarded as individual samples in the test set. In this experiment, we calculated the hit/miss outcomes for each repetition (about each individual) to determine the final classification accuracy, ensuring that HC and Sroke participants and their respective repetitions were assessed in distinct test sets. Under this approach, speech and non-speech repetitions of individuals were classified as either belonging to the HC or Stroke group.
2. **Subject classification:** Following the same method proposed by Bandini et al. [2], we performed the LOSO-CV by treating each subject as a test case and classifying them based on the majority vote among the predicted repetitions, categorizing them as either HC or Stroke. In cases of ties, to ensure a more conservative prediction, the subject was considered HC.

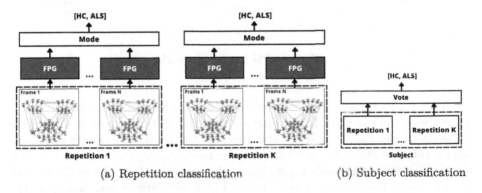

(a) Repetition classification   (b) Subject classification

Fig. 3. Diagram representation of the two experiment modes.

In both repetition- and subject-based classification, we employed the validation set to mitigate bias in the model's hyperparameters and apply the early stopping technique. Additionally, we carefully selected the model's parameters based on empirical evaluation of its performance on the validation set. Therefore, the model is trained for a maximum of 50 epochs, using a batch size of 64 samples and setting the learning rates to 10^{-4} and 10^{-5} for the GAT and linear layers, respectively. The hidden layers have a size of 32, and a dropout rate of 0.2 was applied after some of the LeakyRelu functions following the GAT layers.[3]

---

[3] Note: The experimental evaluation of FPG was conducted using PyTorch Geometric [9] on a GPU. An important point to mention is that, up to the current paper, the implementation of Graph Neural Networks (GNNs) using scatter operation on a GPU introduces non-deterministic behavior.

4.4   Proposed Model

Initially, the proposed model uses fifteen equally spaced frames for each repetition performed by the patients. Therefore, each analyzed frame receives a graph of forty-eight nodes generated by the feature extraction step, as explained in Sect. 4.2. In this process, each of the fifteen graphs from frames of repetition video goes through six GAT layers and two linear layers. As one can notice in Fig. 4, one crucial step before the information enters the linear layers involves pooling, where all feature vectors from the graph are combined into a single vector. This pooling operation is achieved through max pooling, which selects the highest value from each feature node and incorporates them into the pooled node vector.

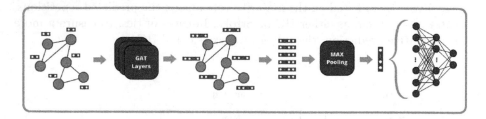

Fig. 4. FPG Model Approach

After pooling, the result moves through two additional linear layers, generating the model's output. It is important to highlight that the error calculation is based on the class of each graph (frame). For repetition classification, the final results are obtained by identifying the mode among the frames used. This means that an individual's repetition is classified based on the majority consensus among the frames, as depicted in Fig. 3. Similarly, when classifying the subject, the majority mode is derived from the binary classifications of each repetition. This approach ensures a robust subject classification by considering the majority consensus across the various repetitions.

5   Results

The experimental results were obtained for each task, as shown in Table 2. It is worth noting that a standard number of frames used per repetition was established for comparing the baseline models with the FPG. Additionally, criteria such as accuracy, specificity, and sensitivity were considered to evaluate the results obtained by the proposed model.

The proposed model achieved an accuracy of approximately 82%, with only three false negatives and one false positive for the 'PA' task, as presented in the confusion matrix in Fig. 5. These findings indicate the development of a conservative model, meaning it is less likely to show the existence of facial paralysis when

Table 2. FPG results for each speech and non-speech tasks.

TASK		Classification	Accuracy	Specificity	Sensitivity
Speech	BBP	Repetition	55.3%	58.6%	51.9%
		Subject	59.1%	63.6%	54.5%
	PA	**Repetition**	**69.3%**	**79.1%**	**53.1%**
		Subject	**81.8%**	**90.9%**	**72.7%**
	PATAKA	Repetition	45.7%	55.3%	29.4%
		Subject	59.1%	63.6%	54.5%
Non-speech	SPREAD	Repetition	52.9%	55.9%	50.0%
		Subject	50.0%	54.5%	45.5%
	KISS	Repetition	64.7%	61.40%	67.7%
		Subject	72.73%	63.64%	81.82%
	OPEN	Repetition	58.6%	38.2%	77.0%
		Subject	63.6%	45.5%	81.8%
	BLOW	Repetition	65.1%	48.7%	79.5%
		Subject	73.3%	57.1%	87.5%

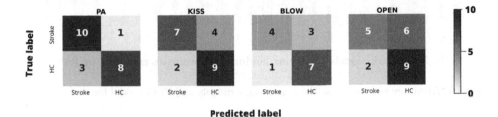

Fig. 5. Confusion matrices of subject-based classification for the most discriminative subtasks in our model.

it does not exist. The binary classification task that stood out was the syllable '/pa/' repetition. It is believed that the superior results in this classification are attributed to a larger quantity of distinct repetitions, as the patient pronounces the same syllable multiple times within a single breath. Deep neural networks, such as GNN, benefit from a larger volume of data, and in this specific case, the PA subtask has the highest number of repetitions compared to the others.

Other noteworthy results come from the BLOW and KISS tasks. The first task was also described by Bandini et al. [2] as one of the best tasks for predicting orofacial impairment caused by a stroke. By observing the confusion matrices presented in Fig. 5, it can be seen that for these two tasks, the model indicated more correct classes for people who suffered a stroke, as also evidenced by the sensitivity of 87.5% for BLOW and 81.82% for KISS in Table 2. We firmly believe that, with the inclusion of more individuals in the healthy control group, the specificity will tend to increase, thus flourishing the accuracy of the model for these tasks.

One justification for the lower results in the other tasks is the classification of videos based on frames where there is not a sufficient extension of the

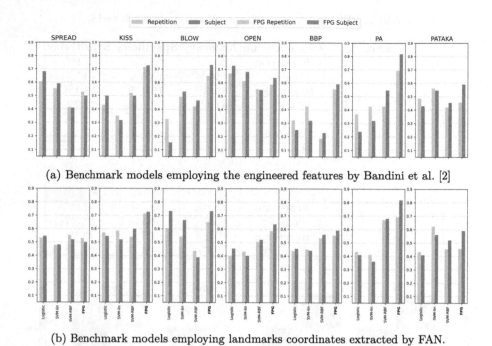

(a) Benchmark models employing the engineered features by Bandini et al. [2]

(b) Benchmark models employing landmarks coordinates extracted by FAN.

Fig. 6. Accuracy scores across the benchmark models and FPG model for each subtask.

orofacial muscles to infer orofacial impairment. Therefore, to validate the FPG, the experiments with SVM and Logistic Regression were conducted under the same conditions as the database used.

The SVM and Logistic Regression results are expected to differ significantly from the findings presented by Bandini et al. [2] due to a slightly different approach. Although Toronto Neuroface shares the same speech and non-speech tasks as Bandini et al.'s study [2], our method exhibits several distinctions. Firstly, the participants in our dataset were different. Additionally, we did not have access to videos containing three-dimensional depth features or samples from the REST subtask, which played a crucial role in normalization for both SVM and Regression models.

FPG achieved superior accuracy in comparison to the handcrafted feature extraction method proposed by Bandini et al. [2] across most classes. However, it's noteworthy that FPG fell short of surpassing the performance of the SVM and Logistic Regression models in the OPEN and SPREAD subtask, as depicted in Fig. 6a. An interesting insight emerges when we compare the outcomes of the benchmark models (Fig. 6b), which rely solely on landmark coordinates as feature inputs. This comparison suggests that, for these specific subtasks, landmark coordinates alone may not offer sufficient discriminative information, clarifying the subtle performance distinctions between the benchmark and FPG modeling.

6   Discussion and Conclusion

This study represents the pioneering effort to evaluate GNNs for identifying Stroke impairments through video classification based on facial expressions. Using GNNs in feature extraction provides a pathway for reducing human bias and offers newfound possibilities for identifying stroke impairments in facial images.

The impact of stroke sequelae on daily activities, such as eating, drinking, and social interactions, can lead to social isolation and requires treatment to restore lost functions. In this pursuit, novel approaches like FPG emerged as promising means for identifying orofacial impairments caused by stroke, which is crucial for effective treatment. Adopting automated identification through FPG carries potential benefits, including improved patient care treatment and cost-effectiveness.

While Deep Neural Networks tend to struggle with generalization in smaller datasets, the graph-based FPG model achieved encouraging results, boasting an accuracy of approximately 82% in classifying orofacial impairments caused by stroke. Moreover, using two-dimensional data for feature extraction allows for more straightforward data collection using simpler cameras, broadening the scope for applications supported by readily accessible devices like smartphones.

In future works, it would be interesting to extend the analysis to other databases, aiming to infer facial impairment caused by paralysis originating from different diseases such as Bell's palsy or brain tumors. This expansion would contribute to enhancing the model's robustness. Another potential approach could involve analyzing only the frames where the most extensive movement occurs, specifically classifying the face image. This would reduce the model's dependence on frames where there is insufficient extension of the muscles affected by the disease, potentially improving the classification accuracy. These explorations could further refine and optimize the FPG model's performance, providing valuable insights for developing more accurate and clinically applicable stroke rehabilitation tools.

References

1. Baltrusaitis, T., Zadeh, A., Lim, Y.C., Morency, L.P.: Openface 2.0: facial behavior analysis toolkit. In: 2018 13th IEEE International Conference on Automatic Face & Gesture Recognition (FG 2018), pp. 59–66. IEEE (2018)
2. Bandini, A., Green, J.R., Richburg, B., Yunusova, Y.: Automatic detection of orofacial impairment in stroke. In: Interspeech, pp. 1711–1715 (2018)
3. Bandini, A., et al.: A new dataset for facial motion analysis in individuals with neurological disorders. IEEE J. Biomed. Health Inform. 25(4), 1111–1119 (2020)
4. Bulat, A., Tzimiropoulos, G.: How far are we from solving the 2d & 3d face alignment problem? (and a dataset of 230,000 3d facial landmarks). In: International Conference on Computer Vision (2017)
5. Cortes, C., Vapnik, V.: Support-vector networks. Mach. Learn. 20, 273–297 (1995)
6. Delaunay, B., et al.: Sur la sphere vide. Izv. Akad. Nauk SSSR, Otdelenie Matematicheskii i Estestvennyka Nauk 7(793-800), 1–2 (1934)

7. Dhall, A., Goecke, R., Lucey, S., Gedeon, T., et al.: Collecting large, richly annotated facial-expression databases from movies. IEEE Multimedia **19**(3), 34 (2012)
8. Dong, X., Yan, Y., Ouyang, W., Yang, Y.: Style aggregated network for facial landmark detection. In: Proceedings of the IEEE Conference on Computer Vision and Pattern Recognition, pp. 379–388 (2018)
9. Fey, M., Lenssen, J.E.: Fast graph representation learning with pytorch geometric. arXiv preprint arXiv:1903.02428 (2019)
10. Fix, E., Hodges, J.L.: Discriminatory analysis. nonparametric discrimination: Consistency properties. Inter. Stat. Review/Revue Internationale de Statistique **57**(3), 238–247 (1989)
11. Gomes, N.B., Yoshida, A., Roder, M., de Oliveira, G.C., Papa, J.P.: Facial point graphs for amyotrophic lateral sclerosis identification. arXiv preprint arXiv:2307.12159 (2023)
12. Greene, J.J., et al.: The spectrum of facial palsy: the meei facial palsy photo and video standard set. Laryngoscope **130**(1), 32–37 (2020)
13. Gross, R., Matthews, I., Cohn, J., Kanade, T., Baker, S.: Multi-pie. Image Vis. Comput. **28**(5), 807–813 (2010)
14. Guarin, D.L., et al.: Toward an automatic system for computer-aided assessment in facial palsy. Facial Plastic Surgery Aesthetic Med. **22**(1), 42–49 (2020)
15. Haykin, S.: Neural networks: a comprehensive foundation. Prentice Hall PTR (1994)
16. Ho, T.K.: Random decision forests. In: Proceedings of 3rd International Conference on Document Analysis and Recognition, vol. 1, pp. 278–282. IEEE (1995)
17. Kaewmahanin, W., et al.: Automatic facial asymmetry analysis for elderly stroke detection by using cosine similarity. In: 2022 19th International Conference on Electrical Engineering/Electronics, Computer, Telecommunications and Information Technology (ECTI-CON), pp. 1–4. IEEE (2022)
18. Kazemi, V., Sullivan, J.: One millisecond face alignment with an ensemble of regression trees. In: Proceedings of the IEEE Conference on Computer Vision and Pattern Recognition, pp. 1867–1874 (2014)
19. Kim, H.S., Kim, S.Y., Kim, Y.H., Park, K.S.: A smartphone-based automatic diagnosis system for facial nerve palsy. Sensors **15**(10), 26756–26768 (2015)
20. King, D.E.: Dlib-ml: a machine learning toolkit. J. Mach. Learn. Res. **10**, 1755–1758 (2009)
21. Lapchak, P.A., Zhang, J.H.: The high cost of stroke and stroke cytoprotection research. Transl. Stroke Res. **8**, 307–317 (2017)
22. Lou, J., Yu, H., Wang, F.Y.: A review on automated facial nerve function assessment from visual face capture. IEEE Trans. Neural Syst. Rehabil. Eng. **28**(2), 488–497 (2019)
23. Lucey, P., Cohn, J.F., Kanade, T., Saragih, J., Ambadar, Z., Matthews, I.: The extended cohn-kanade dataset (ck+): a complete dataset for action unit and emotion-specified expression. In: 2010 IEEE Computer Society Conference on Computer Vision and Pattern Recognition-workshops, pp. 94–101. IEEE (2010)
24. Mullen, M., Loomis, C., et al.: Differentiating facial weakness caused by bell's palsy vs. acute stroke. JEMS (2014)
25. Newell, A., Yang, K., Deng, J.: Stacked hourglass networks for human pose estimation. In: Leibe, B., Matas, J., Sebe, N., Welling, M. (eds.) ECCV 2016. LNCS, vol. 9912, pp. 483–499. Springer, Cham (2016). https://doi.org/10.1007/978-3-319-46484-8_29
26. Ngoc, Q.T., Lee, S., Song, B.C.: Facial landmark-based emotion recognition via directed graph neural network. Electronics **9**(5), 764 (2020)

27. Pantic, M., Valstar, M., Rademaker, R., Maat, L.: Web-based database for facial expression analysis. In: 2005 IEEE International Conference on Multimedia and Expo, pp. 5–pp. IEEE (2005)
28. Parra-Dominguez, G.S., Sanchez-Yanez, R.E., Garcia-Capulin, C.H.: Facial paralysis detection on images using key point analysis. Appl. Sci. **11**(5), 2435 (2021)
29. Pecundo, A.M., Abu, P.A., Alampay, R.: Amyotrophic lateral sclerosis and post-stroke orofacial impairment video-based multi-class classification. In: Proceedings of the 2022 5th Artificial Intelligence and Cloud Computing Conference, pp. 150–157 (2022)
30. Samsudin, W.W., Sundaraj, K.: Image processing on facial paralysis for facial rehabilitation system: A review. In: 2012 IEEE International Conference on Control System, Computing and Engineering, pp. 259–263. IEEE (2012)
31. Scarselli, F., Gori, M., Tsoi, A.C., Hagenbuchner, M., Monfardini, G.: The graph neural network model. IEEE Trans. Neural Netw. **20**(1), 61–80 (2008)
32. Schimmel, M., Ono, T., Lam, O., Müller, F.: Oro-facial impairment in stroke patients. J. Oral Rehabil. **44**(4), 313–326 (2017)
33. Veličković, P., Cucurull, G., Casanova, A., Romero, A., Lio, P., Bengio, Y.: Graph attention networks. arXiv preprint arXiv:1710.10903 (2017)
34. Xiong, X., De la Torre, F.: Supervised descent method and its applications to face alignment. In: Proceedings of the IEEE Conference on Computer Vision and Pattern Recognition, pp. 532–539 (2013)

Fast, Memory-Efficient Spectral Clustering with Cosine Similarity

Ran Li[1]([⊠])[iD] and Guangliang Chen[2][iD]

[1] San José State University, California, USA
ran.li01@sjsu.edu
[2] Hope College, Holland, MI, USA
cheng@hope.edu

Abstract. Spectral clustering is a popular and effective method but known to face two significant challenges: scalability and out-of-sample extension. In this paper, we extend the work of Chen (ICPR 2018) on the speed scalability of spectral clustering in the setting of cosine similarity to deal with massive or online data that are too large to be fully loaded into computer memory. We start by assuming a small batch of data drawn from the full set and develop an efficient procedure that learns both the nonlinear embedding and clustering map from the sample and extends them easily to the rest of the data as they are gradually loaded. We then introduce an automatic approach to selecting the optimal value of the sample size. The combination of the two steps leads to a streamlined memory-efficient algorithm that only uses a small number of batches of data (as they become available), with memory and computational costs that are independent of the size of the data. Experiments are conducted on benchmark data sets to demonstrate the fast speed and excellent accuracy of the proposed algorithm. We conclude the paper by pointing out several future research directions.

Keywords: Spectral clustering · Cosine similarity · Speed scalability · Memory scalability

1 Introduction

Spectral clustering [10,12,16] has emerged as a popular, effective technique in unsupervised learning. Given a set of objects to be clustered according to some kind of similarity, spectral clustering represents the data as a weighted graph and uses the top eigenvectors of the weight matrix (in some normalized form) to embed the objects into a low dimensional space for clustering by simple methods like k-means. It is easy to implement, has rich interpretations, and achieves competitive clustering results in many challenging scenarios where clusters are nonconvex. However, two major limitations of spectral clustering are its scalability to large data sets (due to its high complexity) and out of sample extension (due to the nonlinear embedding).

The authors thank the anonymous reviewers for careful reviews and useful feedback.

V. Vasconcelos et al. (Eds.): CIARP 2023, LNCS 14469, pp. 700–714, 2024.
https://doi.org/10.1007/978-3-031-49018-7_50

Considerable effort has been made in the machine learning and data mining communities to develop fast spectral clustering algorithms that are scalable to large data sets. However, the majority of the current research has focused on the speed scalability of spectral clustering, with the assumption that the full data set can be instantly accessed in computer memory [1–3,5,7,11,13,14,17–20]. In contrast, there is a very scarce amount of work on the memory challenge encountered by spectral tering when data se large to be fully loaded into computer memory [8,9,15]. In [8], the authors introduced a memory-efficient spectral clustering algorithm called Ultra-Scalable Spectral Clustering (U-SPEC). U-SPEC employs a representative selection strategy to build a sparse affinity sub-matrix, represented as a bipartite graph, and derives clusters by partitioning this graph. To enhance performance, the authors introduced Ultra-Scalable Ensemble Clustering (U-SENC), which combines multiple U-SPEC clusters to create a new graph for improved clustering. We point out that memory is a critical consideration at the modern age as there are often millions of observations in a data set. Consequently, it is necessary and helpful to address the memory scalability of spectral clustering as well, in order to apply the method to massive and online data.

In this work we aim to address the memory challenge for spectral clustering in the setting of cosine similarity, which represents our first step along this line of research. There are two important reasons for starting with this particular setting here: (1) As shown in [2], spectral clustering with cosine similarity is very tractable because the similarity matrix has a product form, yet it still has important applications such as documents clustering and customer segmentation. (2) In [3] a landmark-based embedding technique is introduced to transform other similarity functions to cosine similarity. This same technique can be used in the future to address the memory challenge in the general case of spectral clustering, once we have developed a solution for the special case of cosine similarity.

Our strategy is to directly extend the work of Chen [2] to develop a fast, memory-efficient spectral clustering algorithm that can effectively perform clustering on the full data set (which are inaccessible to the computer) based on several small batches of data (which are gradually made available to the computer). We present the algorithm in two steps. First, we assume that a single batch of data have been loaded into computer memory and we derive a base algorithm that uses this fixed batch to find a spectral embedding and perform k-means clustering in the embedding space (both the embedding and clustering steps will be directly generalizable to the rest of the data). Second, we introduce an automatic procedure for choosing the optimal sample size. The technique is to apply the base algorithm repeatedly with more batches of data drawn from the full set and monitor the convergence of estimates of the eigenvectors of a degree-normalized weight matrix. The combination of the two steps leads to a streamlined spectral clustering algorithm with cosine similarity that is scalable, both in speed and memory, to large data sets. We conduct experiments on benchmark data to demonstrate its superior performance while comparing with several other methods in the literature.

The rest of the paper is organized as follows. In Sect. 2 we present our memory-efficient implementation techniques in the setting of massive or online data. Experiments are conducted in Sect. 3 to compare our memory efficient algorithm against the scalable implementation in [2], as well as several other methods. Finally, in Sect. 4, we draw some conclusions and also point out some future directions of research.

2   Methodology

Suppose that we encounter a massive data set, $\mathbf{X} = [\mathbf{x}_1 \, \mathbf{x}_2 \, \dots \, \mathbf{x}_n]^T \in \mathbb{R}^{n \times d}$, that is so large that it cannot be be fully loaded into computer memory, but we have access to small batches of the data, one at a time. Our goal is to divide the data in $\mathbf{X}$ into k clusters by spectral clustering with the cosine similarity. Like Chen [2], we require that $\mathbf{X}$ has some sort of low dimensional structure – either of a moderate dimension ($d \ll n$) or being row-sparse (with average row sparsity $r \ll d$). The former condition is satisfied by collections of small images while the latter by collections of documents under the bag of words model.[1]

In Sect. 2.1, we first introduce a base algorithm which uses a single batch of data to learn the nonlinear embedding and clustering rule associated to spectral clustering with the cosine similarity. We then present in Sect. 2.2 an automatic procedure for determining how large the batch size should be (in order for the base algorithm to work well); the combination of the automatic selection procedure and the base algorithm leads to the complete algorithm that will be tested in the experiment section later. Lastly, in Sect. 2.3, we explain how to extend both the nonlinear embedding and clustering rule learned on the batch data to the rest of the data in $\mathbf{X}$, as they are gradually loaded into computer memory.

2.1   Learning from a Single Batch of Data

Assume a small batch of data of size $s \ll n$, denoted $\mathbf{X}_s \in \mathbb{R}^{s \times d}$, that has become available. In the setting of massive data, this batch can be obtained by sampling from the full data set $\mathbf{X}$; in the setting of online learning, this batch would represent the first s data points that have arrived. In both settings, we assume that $\mathbf{X}_s$ is a random sample from the full set $\mathbf{X}$. Without loss of generality, we assume that $\mathbf{X}_s$ contains the first s rows of $\mathbf{X}$ and that all the rows of $\mathbf{X}$ have already been normalized to be unit vectors in $\mathbb{R}^d$ in the first place.

The scalable spectral clustering algorithm in [2] operates directly on the full data set $\mathbf{X}$ (when it is accessible). It starts by first computing the degrees of the data as follows:

$$\mathbf{D} = \operatorname{diag}(\mathbf{d}), \quad \mathbf{d} = \mathbf{X}\left(\mathbf{X}^T \mathbf{1}_n\right) - \mathbf{1}_n = \mathbf{X} \cdot \sum_{i=1}^{n} \mathbf{x}_i - \mathbf{1}_n, \tag{1}$$

---

[1] When both conditions are violated, one can apply principal component analysis (PCA) to reduce the dimensionality of the data such that the first condition is met.

where $\mathbf{1}_n = (1, \ldots, 1)^T \in \mathbb{R}^n$. Based on these degrees, a small fraction of the data with the lowest degrees are classified as outliers and removed from $\mathbf{X}$. Afterwards, it normalizes the remaining data by using their degrees, $\widetilde{\mathbf{X}} = \mathbf{D}^{-1/2}\mathbf{X}$, and performs the rank-$k$ SVD on the degree-normalized data matrix $\widetilde{\mathbf{X}}$, i.e.,

$$\widetilde{\mathbf{X}}_{n \times d} \approx \widetilde{\mathbf{U}}_{n \times k} \widetilde{\boldsymbol{\Sigma}}_{k \times k} \widetilde{\mathbf{V}}_{d \times k}^T. \tag{2}$$

The rows of $\widetilde{\mathbf{U}}$, the matrix of the top k left singular vectors of $\widetilde{\mathbf{X}}$, represent the nonlinear embedding of $\mathbf{X}$ by spectral clustering with the cosine similarity. Finally, k-means is employed to group the rows of $\widetilde{\mathbf{U}}$ (after being normalized to have unit norm) into k clusters.

When the full data set $\mathbf{X}$ is not accessible (or when it is still accessible but it is too time consuming to run the algorithm), we cannot carry out the procedure outlined above. Our strategy here is to write $\widetilde{\mathbf{U}} = \widetilde{\mathbf{X}}\widetilde{\mathbf{V}}\widetilde{\boldsymbol{\Sigma}}^{-1}$, from which we identify the spectral embedding as $\widetilde{\mathbf{V}}\widetilde{\boldsymbol{\Sigma}}^{-1}$. As a result, we will focus on using the batch data $\mathbf{X}_s$ to estimate the top k singular values $\widetilde{\boldsymbol{\Sigma}}$ and right singular vectors $\widetilde{\mathbf{V}}$ of $\widetilde{\mathbf{X}}$. The above spectral embedding map will be directly applied to both the batch data $\mathbf{X}_s$ and the remaining data in $\mathbf{X}$, after they are degree-normalized.

In order to estimate the right singular vectors of $\widetilde{\mathbf{X}}$, we write

$$\widetilde{\mathbf{X}}^T\widetilde{\mathbf{X}} = \mathbf{X}^T\mathbf{D}^{-1}\mathbf{X} = \begin{bmatrix} \mathbf{x}_1 \cdots \mathbf{x}_n \end{bmatrix} \begin{bmatrix} \frac{1}{d_1} & \\ & \ddots \\ & & \frac{1}{d_n} \end{bmatrix} \begin{bmatrix} \mathbf{x}_1^T \\ \vdots \\ \mathbf{x}_n^T \end{bmatrix} = \sum_{i=1}^{n} \frac{1}{d_i}\mathbf{x}_i\mathbf{x}_i^T. \tag{3}$$

In the above sum, each term $\mathbf{x}_i\mathbf{x}_i^T$ is a rank-1 orthogonal projection matrix (because $\mathbf{x}_i$ is a unit vector), weighted by the inverse degree of the corresponding data point. Since $\mathbf{X}_s$ is a random sample from the full data set $\mathbf{X}$, we obtain the following approximation from (3):

$$\widetilde{\mathbf{X}}^T\widetilde{\mathbf{X}} \approx \frac{n}{s}\sum_{i=1}^{s}\frac{1}{d_i}\mathbf{x}_i\mathbf{x}_i^T = \frac{n}{s}\mathbf{X}_s\mathbf{D}_s^{-1}\mathbf{X}_s = \frac{n}{s}\widetilde{\mathbf{X}}_s^T\widetilde{\mathbf{X}}_s, \tag{4}$$

where

$$\widetilde{\mathbf{X}}_s = \mathbf{D}_s^{-1/2}\mathbf{X}_s, \quad \text{and} \quad \mathbf{D}_s = \text{diag}(\mathbf{d}_s), \ \mathbf{d}_s = (d_1, \ldots, d_s)^T$$

represent the restrictions of $\widetilde{\mathbf{X}}$ and $\mathbf{D}$ to the sample $\mathbf{X}_s$, respectively. Note that the degrees of the sample can be estimated as follows:

$$\mathbf{d}_s = \mathbf{X}_s \cdot \sum_{i=1}^{n}\mathbf{x}_i - \mathbf{1}_s \approx \mathbf{X}_s \cdot \frac{n}{s}\sum_{i=1}^{s}\mathbf{x}_i - \mathbf{1}_s = \frac{n}{s}\mathbf{X}_s(\mathbf{X}_s^T\mathbf{1}_s) - \mathbf{1}_s. \tag{5}$$

In [2], a small fraction of the full data $\mathbf{X}$ that have the lowest degrees are removed in order for the approximation technique introduced there to be valid. We point out that this corresponds to discarding the terms in (3) with the largest weights. Similarly, we remove a small fraction of the sample $\mathbf{X}_s$ that have the lowest degrees estimated in (5), equivalent to discarding the terms in (4) with the largest weights.

Equation (4) shows that $\widetilde{\mathbf{X}}$ and $\widetilde{\mathbf{X}}_s$ (approximately) have the same right singular vectors but the singular values differ by a factor of $\sqrt{\frac{n}{s}}$. That is, letting the rank-k SVD of $\widetilde{\mathbf{X}}_s$ be

$$\widetilde{\mathbf{X}}_s \approx \widetilde{\mathbf{U}}_s \widetilde{\mathbf{\Sigma}}_s \widetilde{\mathbf{V}}_s^T,$$

we have

$$\widetilde{\mathbf{V}}\widetilde{\mathbf{V}}^T \approx \widetilde{\mathbf{V}}_s \widetilde{\mathbf{V}}_s^T \quad \text{and} \quad \widetilde{\mathbf{\Sigma}} \approx \sqrt{\frac{n}{s}}\widetilde{\mathbf{\Sigma}}_s, \tag{6}$$

where the approximations are under the matrix Frobenius norm. Therefore, the nonlinear embedding of the batch $\mathbf{X}_s \in \mathbb{R}^{s \times d}$ by spectral clustering with cosine similarity (on the full set) is

$$\mathbf{Y}_s := \widetilde{\mathbf{X}}_s \widetilde{\mathbf{V}} \widetilde{\mathbf{\Sigma}}^{-1} \approx \widetilde{\mathbf{X}}_s \widetilde{\mathbf{V}}_s \left(\sqrt{\frac{n}{s}}\widetilde{\mathbf{\Sigma}}_s \right)^{-1} = \sqrt{\frac{s}{n}}\widetilde{\mathbf{U}}_s \in \mathbb{R}^{s \times k}. \tag{7}$$

For the clustering step, we normalize the rows of $\mathbf{Y}_s$ to be unit vectors in $\mathbb{R}^k$ and apply k-means to group them into k disjoint clusters, with respective centroids $\boldsymbol{\mu}_s^{(j)}$, $1 \leq j \leq k$.

We display all the steps in Algorithm 0 (which is only a base algorithm). The memory complexity of this algorithm is $\mathcal{O}(s(d+k)+dk)$, and the computational complexity is $\mathcal{O}(sdk)$.

Algorithm 0. The base algorithm (using a single batch of data)

Input: $\mathbf{X}_s \in \mathbb{R}^{s \times d}$ sampled from a massive data set $\mathbf{X}$ (row-sparse or of moderate dimension, with L_2 normalized rows), and number of clusters k

Output: Sample total $\mathbf{X}_s^T \mathbf{1}_s$, embedded sample $\mathbf{Y}_s$, k-means centroids $\{\boldsymbol{\mu}_s^{(j)}\}_{1 \leq j \leq k}$, singular values $\widetilde{\mathbf{\Sigma}}_s$ and right singular vectors $\widetilde{\mathbf{V}}_s$

Steps:

1: Estimate the degrees of the points in $\mathbf{X}_s$ (relative to $\mathbf{X}$) by

$$\mathbf{D}_s = \text{diag}\left(\frac{n}{s}\mathbf{X}_s \left(\mathbf{X}_s^T \mathbf{1}_s \right) - \mathbf{1}_s \right).$$

Remove 10% of the sample that have the smallest degrees, and update $\mathbf{X}_s$ and $\mathbf{D}_s$ accordingly.

2: Use $\mathbf{D}_s$ to normalize $\mathbf{X}_s$ to obtain $\widetilde{\mathbf{X}}_s = \mathbf{D}_s^{-1/2}\mathbf{X}_s$, and then perform rank-$k$ SVD on $\widetilde{\mathbf{X}}_s$ to get $\widetilde{\mathbf{U}}_s, \widetilde{\mathbf{\Sigma}}_s$, and $\widetilde{\mathbf{V}}_s$.

3: Normalize the row vectors of $\mathbf{Y}_s = \sqrt{\frac{s}{n}}\widetilde{\mathbf{U}}_s$ to have unit norm in $\mathbb{R}^k$ and deploy k-means to find k clusters with centroids $\boldsymbol{\mu}_s^{(j)}, j = 1, \ldots, k$.

2.2   How to Choose the Batch Size s

In order for the base algorithm (Alg. 0) to work well, s must be properly set. On one hand, it needs to be large enough so that $\mathbf{X}_s$ can be sufficiently representative of $\mathbf{X}$. On the other hand, for efficiency consideration, we do not want s to be unnecessarily too large. The optimal value of s depends on the complexity of the data (such as the number of clusters, the noise level, and the separation between the clusters), and no specific formula can be used to set its value. Here, we present an automatic approach to choosing the value of s such that it is as small as possible while still working well.

Fix two positive integers s_0, t. Define $s_i = s_0 + i \cdot t$ for each integer $i \geq 1$, and let $\{\mathbf{X}_{s_i}\}_{i \geq 0}$ be a collection of nested batches of increasing sizes that are sampled without replacement from $\mathbf{X}$ in an incremental fashion. That is, at step 0, we sample s_0 points uniformly at random from $\mathbf{X}$ to form an initial batch of data $\mathbf{X}_{s_0}$, and at each additional step i, we sample t new points from the remaining data in $\mathbf{X}$ to append $\mathbf{X}_{s_{i-1}}$ so as to form $\mathbf{X}_{s_i}$.

The proposed technique here is to apply Algorithm 0 with each of these samples $\{\mathbf{X}_{s_i}\}_{i \geq 0}$ separately and focus on the outputs $\widetilde{\mathbf{V}}_{s_i}$, the right singular vectors of $\widetilde{\mathbf{X}}_{s_i}$, because they are expected to converge (in the Grassmannian distance) to $\widetilde{\mathbf{V}}$, the right singular vectors of $\widetilde{\mathbf{X}}$ that we are trying to estimate, i.e.,

$$\lim_{i \to \infty} \left\| \widetilde{\mathbf{V}}_{s_i} \widetilde{\mathbf{V}}_{s_i}^T - \widetilde{\mathbf{V}} \widetilde{\mathbf{V}}^T \right\|_F = 0. \tag{8}$$

However, since we do not know $\widetilde{\mathbf{V}}$ in reality, we instead monitor the Grassmannian distances between consecutive iterations, i.e.,

$$g_i = \left\| \widetilde{\mathbf{V}}_{s_i} \widetilde{\mathbf{V}}_{s_i}^T - \widetilde{\mathbf{V}}_{s_{i-1}} \widetilde{\mathbf{V}}_{s_{i-1}}^T \right\|_F, \quad i = 1, 2, \dots \tag{9}$$

In [4, Appendix], it is shown that

$$g_i = \sqrt{2k - 2 \left\| \widetilde{\mathbf{V}}_{s_i}^T \widetilde{\mathbf{V}}_{s_{i-1}} \right\|_F^2}, \quad i = 1, 2, \dots \tag{10}$$

This will be how we efficiently compute the distances g_i. The smallest working sample size s will be set to s_i, at which the value of g_i has first reached convergence.

The Grassmannian distance between $\widetilde{\mathbf{V}}_{s_i}$ and $\widetilde{\mathbf{V}}_{s_{i-1}}$ is closely related to the principal angles $0 \leq \theta_{i1} \leq \cdots \leq \theta_{ik} \leq \frac{\pi}{2}$ between their column spaces [4, Appendix]:

$$g_i = \sqrt{2 \sum_{j=1}^{k} \sin^2 \theta_{ij}}, \quad i = 1, 2, \dots \tag{11}$$

Empirically, we can set $s = s_i$ such that at this value and onwards, all principal angles between the subspaces spanned by $\widetilde{\mathbf{V}}_{s_i}$ and $\widetilde{\mathbf{V}}_{s_{i-1}}$ have become sufficiently small (less than some small angle θ_0):

$$g_i < \sqrt{2 \cdot k \cdot \sin^2 \theta_0} = \sqrt{2k} \sin \theta_0. \tag{12}$$

We present the combination of the base algorithm (Alg. 0) and the above-introduced automatic procedure for selecting the optimal sample size in Alg. 1 (this is the actual algorithm we will use in practice).

Algorithm 1. Memory-efficient spectral clustering (MESC) with cosine similarity

---

Input: Initial batch size s_0, batch step size t, number of clusters k, principal angle threshold θ_0

Output: Optimal sample size s, singular values $\widetilde{\boldsymbol{\Sigma}}_s$, right singular vectors $\widetilde{\mathbf{V}}_s$, and k-means centroids $\{\boldsymbol{\mu}_s^{(j)}\}_{1 \leq j \leq k}$,

Steps:

1: Sample s_0 points uniformly at random from the massive or online data set ($\mathbf{X}$) to obtain an initial batch $\mathbf{X}_{s_0}$. Apply the base algorithm (Alg. 0) to $\mathbf{X}_{s_0}$ and obtain $\widetilde{\mathbf{V}}_{s_0}$ and other quantities.

2: For each $i \geq 1$, let $s_i = s_0 + i \cdot t$ and do the following in order:

- Sample t new data points and add them to $\mathbf{X}_{s_{i-1}}$ in order to form $\mathbf{X}_{s_i}$.
- Apply Alg. 0 to $\mathbf{X}_{s_i}$ to obtain various quantities including $\widetilde{\mathbf{V}}_{s_i}$.
- Compute the Grassmannian distance between $\widetilde{\mathbf{V}}_{s_{i-1}}$ and $\widetilde{\mathbf{V}}_{s_i}$ using (10).
- Terminate the procedure if $g_i < \sqrt{2k} \sin \theta_0$.

3: Output $s = s_i$, as well as the singular values $\widetilde{\boldsymbol{\Sigma}}_s$ and k-means centroids $\{\boldsymbol{\mu}_s^{(j)}\}_{1 \leq j \leq k}$.

2.3   Out of Sample Extension

To explain how to extend both the nonlinear embedding and clustering to the rest of the data set (as they are gradually loaded into computer memory) or new streaming data, we consider an arbitrary new point, say $\mathbf{x}_{s+1} \in \mathbb{R}^d$. Its degree (relative to the full data set $\mathbf{X}$) is estimated similarly to those in $\mathbf{X}_s$:

$$d_{s+1} = \mathbf{x}_{s+1}^T \sum_{i=1}^{n} \mathbf{x}_i - 1 \approx \frac{n}{s} \mathbf{x}_{s+1}^T (\mathbf{X}_s^T \mathbf{1}_s) - 1. \tag{13}$$

Following the spectral embedding of the initial batch $\mathbf{X}_s$, i.e.,

$$\mathbf{Y}_s = \sqrt{\frac{s}{n}} \widetilde{\mathbf{X}}_s \widetilde{\mathbf{V}}_s \widetilde{\boldsymbol{\Sigma}}_s^{-1},$$

we embed the new point $\mathbf{x}_{s+1}$ as

$$\mathbf{y}_{s+1} = \sqrt{\frac{s}{n}} \left(d_{s+1}^{-1/2} \mathbf{x}_{s+1}^T \right) \widetilde{\mathbf{V}}_s \widetilde{\boldsymbol{\Sigma}}_s^{-1} \in \mathbb{R}^k. \tag{14}$$

For the clustering step, we first normalize $\mathbf{y}_{s+1}$ to have unit norm and then simply assign it to the nearest centroid $\boldsymbol{\mu}_s^{(j)}$, i.e.,

$$\hat{j} = \arg \min_{1 \leq j \leq k} \|\mathbf{y}_{s+1} - \boldsymbol{\mu}_s^{(j)}\|_2. \tag{15}$$

3 Experiments

In this section, we conduct experiments on benchmark data sets to evaluate the performance of the proposed algorithm and compare it with several state-of-the-art scalable spectral clustering algorithms. We record both the clustering accuracy (i.e., percentage of data points that are correctly clustered[2]) and CPU time needed by the relevant methods. All experiments are conducted with MATLAB R2021b on a desktop computer with 32 GB of RAM and a CPU with 4 cores.

3.1 Real-World Benchmark Data Sets

Some image data sets are large in quantity but only have moderate dimensions, and cosine similarity is a reasonable choice for grouping images. Thus, performing clustering on image data serves as an appropriate way to test the effectiveness of our algorithm. To this end, we consider three such benchmark data sets, namely *usps*, *pendigit*, and *mnist*, available on the LIBSVM website[3]. The data sets are initially divided into training and test sets for classification purposes. However, in our unsupervised setting, we combine the two parts together. A summary of their information is presented in Table 1.

Table 1. Summary information of the real-world data sets used in this study.

Data sets	#Instances	#Dimensions	#Classes
usps	9,298	256	10
pendigit	10,992	16	10
mnist	70,000	184	10
20news	18,768	55,570	20
20news100	18,768	100	20
protein	24,387	357	3
covtype	581,012	54	7

Document clustering serves as another suitable test case to evaluate our algorithms. The document-term matrix $\mathbf{X}$, which represents a corpus, is often large in both size and dimension, ranging from thousands to millions. However, $\mathbf{X}$ tends to be highly sparse, aligning with our first assumption regarding the data matrix. Moreover, cosine similarity naturally captures document similarity, making document clustering an ideal application for algorithm testing. We utilized the 20 Newsgroups data[4], comprising 18,774 newsgroup documents nearly

---

[2] To compute this percentage, we need to find the best map between the output labels and the original labels. This is done by using the Kuhn-Munkres algorithm as in [1].

[3] https://www.csie.ntu.edu.tw/~cjlin/libsvmtools/datasets/.

[4] Available at http://qwone.com/~jason/20Newsgroups/; we also used the bydate version.

evenly distributed across 20 categories. The 20 Newsgroups data contain a total of 61,118 unique words (including stop words). We carried out the same statistical operations as in [2] to clean the data without using outside knowledge or text processing software. This resulted in a matrix of 18,768 nonempty documents and 55,570 unique words. The average row sparsity of the matrix is 73.4, indicating that each document contains 73.4 distinct words on average. Additionally, we applied the same SVD projection technique as in [2] to project the data set onto 100 dimensions, and included both versions in our analysis.

In the bioinformatics field, emphasis is often placed on shape similarity rather than magnitude similarity [6]. For instance, in protein and gene sequences, co-expression is defined based on relative shape rather than magnitude. As a result, cosine similarity and correlations are widely used as measures of similarity in bioinformatics. Furthermore, bioinformatics data are typically high-dimensional, posing a challenge for traditional clustering methods but offering a suitable application for our proposed algorithm. We utilized the *protein* and *covtype* data sets, available on the LIBSVM website[5], for our study.

3.2   Experimental Setup

In our experiments, we compare our algorithm with the original spectral clustering algorithm as well as three state-of-the-art scalable spectral clustering algorithms, including Chen's scalable method [2]. The proposed and baseline algorithms considered are as follows:

- MESC: Our fast and memory-efficient spectral clustering algorithm using cosine similarity.
- SC [12]: The original spectral clustering algorithm (with cosine similarity) is the baseline for comparison in this study. It is not scalable in speed or memory.
- SSC [2]: Chen's scalable cosine similarity spectral clustering algorithm, from which our algorithm extends. This method is a speed scalable algorithm and requires the access to the full dataset.
- U-SPEC [8]: A landmark-based memory scalable spectral clustering algorithm.
- U-SENC [8]: An ensemble memory scalable spectral clustering algorithm.

In order to mitigate the potential impact of random chance, we replicate each experiment 20 times and calculate the average scores. Our experimental procedures adhere to the following settings:

- The number of clusters, denoted as k, is set to match the true number of classes in each data set.
- For U-SPEC and U-SENC, we set the parameters based on the guidance from the paper [8]: $p = 1000$, $kNN = 5$, and distance = 'euclidean'.
- In U-SENC, we employ 20 base clusterings as in [8].

---

[5] https://www.csie.ntu.edu.tw/~cjlin/libsvmtools/datasets/.

- In SSC, we follow the approach outlined in [2] to remove outliers. Specifically, we eliminate 1% of the data points as outliers based on the degrees. These outliers are subsequently classified back into the clusters using the nearest centroid classifier in the input data space.
- In the proposed MESC, we set the initial batch size $s_0 = 30k$ and the step size $t = 30k$ because we expect both to be proportional to the number of clusters k in the data set. We oversample each batch by 10% (i.e., $s_i \cdot 1.1$) and then remove the bottom 10% of the batch (those with the lowest degrees) as outliers.

3.3   Choosing the Optimal Sample Size

In Sect. 2.2, we introduced an automatic approach for selecting the optimal batch size s that achieves a balance between accuracy and storage requirements. In this section, we perform experiments on all benchmark data sets using Algorithm 1, incrementally increasing the effective sample size s from $s_0 = 30k$, with a step size of $t = 30k$, where k is the number of clusters in the data set. Our goal is to investigate the impact of varying s and assess the effectiveness of the automatic selection approach. The batch data are regarded as the training set and used for estimating the embedding of $\mathbf{X}$, while the remaining data form the testing set and are clustered by the out-of-sample extension techniques.

The experimental results are presented in Fig. 1. The left column of the figure shows the average accuracy rates obtained from different sample sizes s. For comparison, the blue line represents the results achieved by SSC using the full data set. The middle column displays the Grassmannian distances g_i, calculated using Eq. (10), which measure the distance between $\widetilde{\mathbf{V}}_{s_{i-1}}$ and $\widetilde{\mathbf{V}}_{s_i}$. The red lines represent the thresholds computed using Eq. (12) with $\theta_0 = \pi/60$, indicating the optimal values of s. Finally, the right column presents the CPU time in seconds with varying s.

Based on Figs. 1a, 1b and 1c, increasing the value of s in the *usps*, *pendigits*, and *mnist* data sets leads to higher accuracy rates and CPU times, accompanied by decreasing values of g_i. The linear increase in CPU time is attributed to the outlier removal step. Notably, the proposed MESC algorithm achieves accuracy rates in *usps* and *mnist* that surpass those of the baseline SSC algorithm, even with a smaller sample size. Based on the thresholds represented by the red lines, the optimal values of s for the three data sets are 1,800, 600, and 1,500 respectively. Upon examining the accuracy and CPU time plots, it becomes evident that these optimal values of s yield sufficiently high accuracy rates with minimal computational cost.

In Fig. 1d and 1e, the accuracy rates and g_i values achieved by the proposed MESC algorithm on the original and SVD projected versions of the 20 news-groups data converge when s equals 12,600 and 6,000, respectively, based on the thresholds calculated with $\theta_0 = \pi/60$. These optimal values of s yield accuracy rates close to those of the SSC algorithm with the full data set, while requiring less computational time.

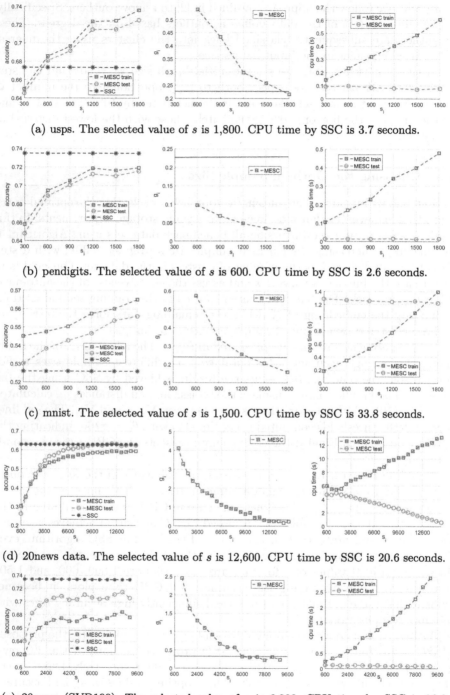

(a) usps. The selected value of s is 1,800. CPU time by SSC is 3.7 seconds.

(b) pendigits. The selected value of s is 600. CPU time by SSC is 2.6 seconds.

(c) mnist. The selected value of s is 1,500. CPU time by SSC is 33.8 seconds.

(d) 20news data. The selected value of s is 12,600. CPU time by SSC is 20.6 seconds.

(e) 20news (SVD100). The selected value of s is 6,000. CPU time by SSC is 16.6 seconds.

Fig. 1. *Left column*: accuracy rates. *Middle column*: Convergence of the Grassmannian distance (g_i). The red line represents the threshold for optimal s. *Right column*: CPU time cost. (Color figure online)

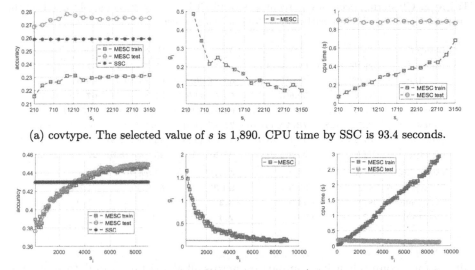

(a) covtype. The selected value of s is 1,890. CPU time by SSC is 93.4 seconds.

(b) protein. The selected value of s is 6,030. CPU time by SSC is 3.4 seconds.

Fig. 2. Figure 1 continued.

In Fig. 2a for the *covtype* data set, the values of g_i decrease rapidly and reach the threshold with fewer than 2,000 data points, which accounts for only about 0.25% of the full data set (consisting of over 500,000 points). The algorithm achieves higher testing accuracy rates compared to the training accuracy rates across different values of s. Moving on to Fig. 2b, the *protein* data set requires a larger number of data points to converge compared to *covtype*. The accuracy rates exhibit an upward trend when increasing s and surpass the accuracy rate achieved by SSC when s exceeds 4,000.

3.4   Method Comparisons

In Table 2, we compare our proposed algorithm with baseline algorithms across various real-world data sets, evaluating their performance based on average clustering accuracy and CPU time. The sample size for each data set in our algorithm is determined as outlined in the last section. Table 2 also presents the average accuracy score (avg. score), normalized average accuracy score (N. avg. score), average CPU time in seconds (avg. time), and average rank (avg. rank) for each algorithm.

The normalized average score is calculated by dividing each entry in a row by the largest entry in that row, followed by averaging the scores for each data set in the column. It provides robustness against outliers. If an algorithm encounters an out-of-memory error for one or more data sets, it will not have an average or normalized average score. However, it will still have an average rank, which is determined by ranking its performance as the lowest for that data set. This

Table 2. Clustering accuracy percentages (and CPU time in seconds) obtained by the proposed and baseline algorithms on the real-world data.

dataset	SC	SSC	MESC	U-SPEC	U-SENC
usps	68.1% (12.3)	67.3% (3.7)	**72.7%** (0.8)	47.0% (1.7)	69.8% (46.7)
pendigits	74.0% (32.9)	73.5% (2.6)	69.4% (0.2)	69.6% (0.9)	**79.3%** (34.5)
mnist	N/A	52.60% (33.8)	55.7% (1.9)	51.0% (6.3)	**67.6%** (137.5)
20news	**66.4%** (120)	62.9% (20.6)	63.0 % (13.4)	5.6% (100.1)	8.2% (1,342.8)
20news100	72.7% (128)	**73.4%** (16.6)	71.0% (1.9)	5.6% (2.9)	10.5 % (28.6)
protein	**45.3** (313.1)	43.1% (3.4)	44.7% (2.8)	43.8% (12.9)	43.8% (226.6)
covtype	N/A	25.8% (93.4)	**27.5%** (1.2)	25.6% (16.6)	22.8 % (254.9)
avg. score	N/A	57.0%	**57.7%**	35.5%	43.1%
N. avg. score	N/A	92.5%	**94.3%**	61.9%	71.7%
avg. time	N/A	24.9	**3.2**	20.2	259.9
avg. rank	2.7	3.0	**2.3**	4.1	2.7

allows for a comprehensive comparison across all algorithms and data sets. Out-of-memory errors are represented by N/A in Table 2.

Due to the requirement of constructing and storing the affinity matrix, the original spectral clustering method is not computationally feasible for large data sets such as mnist and covtype, with 70,000 and 581,012 data points respectively. Our proposed MESC algorithm achieves the highest average and normalized average accuracy rates among all algorithms while maintaining the smallest average CPU time by processing only a fraction of the data set. It also obtains the highest average rank. In comparison, the alternative memory-scalable algorithm U-SPEC demonstrates lower accuracy rates in most benchmark data sets, particularly in the two versions of 20 news data, thereby highlighting MESC's advantage in handling document data. Similarly, the other memory-scalable algorithm, U-SENC, achieves decent accuracy rates but incurs the highest computational cost due to its ensemble of 20 base clusterings. Overall, MESC outperforms the other algorithms in terms of clustering performance in this experiment.

4   Conclusions and Future Work

In this work, we proposed a memory-efficient spectral clustering algorithm that uses small batches of data to effectively learn nonlinear spectral embedding and clustering maps, as well as the out-of-sample extensions. In particular, we introduced an automatic procedure that monitors the Grassmannian distances between consecutive iterations to select the optimal sample size. This combined approach effectively addresses the memory issue associated with spectral clustering with the cosine similarity and drastically reduces the memory needs from the full data set to only a small subset of the data (a few thousand points at most). In the experiments we conducted over benchmark data sets, the proposed method produced similar accuracy rates compared to the original method while using much smaller memory and much less computation time.

There are several questions that need to be studied in our next step. First, the approximations carried out in Sect. 2.1 indicates that there is a statistical perspective underlying the proposed algorithm. For example, in computing the degrees of the sample $\mathbf{X}_s$ (i.e., Eq. (5)), both a population mean ($\frac{1}{n}\sum_{i=1}^{n}\mathbf{x}_i$) and a sample mean ($\frac{1}{s}\sum_{i=1}^{s}\mathbf{x}_i$) are used. It seems that one can assume a mixture distribution for the underlying clusters in the data set $\mathbf{X}$: the population mean would correspond to the distribution mean of the mixture and the sample mean is a way to approximate the distribution mean. Another example is in Eq. (4) where we essentially approximate the global (weighted) mean of the rank-1 orthogonal projection matrices, $\frac{1}{n}\sum_{i=1}^{n}\frac{1}{d_i}\mathbf{x}_i\mathbf{x}_i^T$, by a sample counterpart, i.e., $\frac{1}{s}\sum_{i=1}^{s}\frac{1}{d_i}\mathbf{x}_i\mathbf{x}_i^T$. It will be interesting to determine the distribution of the collection of rank-1 orthogonal projection matrices based on the mixture model assumed for the full data set. Second, there is room to improve the efficiency of the automatic procedure presented in Sect. 2.2 for selecting the optimal value of s, because the base algorithm is applied to nested samples independently without utilizing the information that has been obtained from previous samples. We will develop a more efficient approach that uses the differences between successive samples $\mathbf{X}_{s_i}$ and $\mathbf{X}_{s_{i+1}}$ to directly update the quantities $\widetilde{\mathbf{\Sigma}}_{s_i}$, $\widetilde{\mathbf{V}}_{s_i}$ and $\{\boldsymbol{\mu}_{s_i}^{(j)}\}$ that are associated to the batch $\mathbf{X}_{s_i}$ to their counterparts associated to the batch $\mathbf{X}_{s_{i+1}}$. Lastly, the similarity function in our work is cosine similarity. In the future, we would like to extend the work to other similarities by using the landmark-based technique in [2] to accommodate a wider range of applications.

References

1. Cai, D., Chen, X.: Large scale spectral clustering via landmark-based sparse representation. IEEE Trans. Cybern. **45**(8), 1669–1680 (2015)
2. Chen, G.: Scalable spectral clustering with cosine similarity. In: Proceedings of the 24th International Conference on Pattern Recognition (ICPR), Beijing, China (2018)
3. Chen, G.: A general framework for scalable spectral clustering based on document models. Pattern Recogn. Lett. **125**, 488–493 (2019)
4. Chen, G., Lerman, G.: Foundations of a multi-way spectral clustering framework for hybrid linear modeling. Found. Comput. Math. (2009). https://doi.org/10.1007/s10208-009-9043-7
5. Choromanska, A., Jebara, T., Kim, H., Mohan, M., Monteleoni, C.: Fast spectral clustering via the Nyström method. In: Jain, S., Munos, R., Stephan, F., Zeugmann, T. (eds.) Algorithmic Learning Theory, pp. 367–381. Springer, Berlin, Heidelberg (2013). https://doi.org/10.1007/978-3-642-40935-6_26
6. Everitt, B.S., Landau, S., Leese, M., Stahl, D.: Dissimilarity and distance measures for continuous data, pp. 51–52. Wiley, Boston, MA (2011)
7. Fowlkes, C., Belongie, S., Chung, F., Malik, J.: Spectral grouping using the Nyström method. IEEE Trans. Pattern Anal. Mach. Intell. **26**(2), 214–225 (2004)
8. Huang, D., Wang, C.D., Wu, J.S., Lai, J., Kwoh, C.K.: Ultra-scalable spectral clustering and ensemble clustering. IEEE Trans. Knowl. Data Eng. (TKDE) **32**, 1212–1226 (2020)

9. Li, M., Lian, X.C., Kwok, J.T., Lu, B.L.: Time and space efficient spectral clustering via column sampling. In: CVPR 2011, pp. 2297–2304 (2011). https://doi.org/10.1109/CVPR.2011.5995425

10. Meila, M., Shi, J.: A random walks view of spectral segmentation. In: Proceedings of the Eighth International Workshop on Artificial Intelligence and Statistics (2001)

11. Moazzen, Y., Tasdemir, K.: Sampling based approximate spectral clustering ensemble for partitioning data sets. In: Proceedings of the 23rd International Conference on Pattern Recognition (2016)

12. Ng, A., Jordan, M., Weiss, Y.: On spectral clustering: analysis and an algorithm. In: Advances in Neural Information Processing Systems 14, pp. 849–856 (2001)

13. Pham, K., Chen, G.: Large-scale spectral clustering using diffusion coordinates on landmark-based bipartite graphs. In: Proceedings of the 12th Workshop on Graph-based Natural Language Processing (TextGraphs-12), pp. 28–37. Association for Computational Linguistics (2018)

14. Sakai, T., Imiya, A.: Fast spectral clustering with random projection and sampling. In: Perner, P. (ed.) MLDM 2009. LNCS (LNAI), vol. 5632, pp. 372–384. Springer, Heidelberg (2009). https://doi.org/10.1007/978-3-642-03070-3_28

15. Shaham, U., Stanton, K., Li, H., Basri, R., Nadler, B., Kluger, Y.: Spectralnet: spectral clustering using deep neural networks. In: International Conference on Learning Representations (2018)

16. Shi, J., Malik, J.: Normalized cuts and image segmentation. IEEE Trans. Pattern Anal. Mach. Intell. **22**(8), 888–905 (2000)

17. Tasdemir, K.: Vector quantization based approximate spectral clustering of large datasets. Pattern Recogn. **45**(8), 3034–3044 (2012)

18. Wang, L., Leckie, C., Kotagiri, R., Bezdek, J.: Approximate pairwise clustering for large data sets via sampling plus extension. Pattern Recogn. **44**, 222–235 (2011)

19. Wang, L., Leckie, C., Ramamohanarao, K., Bezdek, J.: Approximate spectral clustering. In: Theeramunkong, T., Kijsirikul, B., Cercone, N., Ho, T.-B. (eds.) PAKDD 2009. LNCS (LNAI), vol. 5476, pp. 134–146. Springer, Heidelberg (2009). https://doi.org/10.1007/978-3-642-01307-2_15

20. Yan, D., Huang, L., Jordan, M.: Fast approximate spectral clustering. In: Proceedings of the 15th ACM SIGKDD International Conference on Knowledge Discovery and Data Mining, pp. 907–916 (2009)

An End-to-End Deep Learning Approach for Video Captioning Through Mobile Devices

Rafael J. Pezzuto Damaceno$^{(\boxtimes)}$ and Roberto M. Cesar Jr.

Institute of Mathematics and Statistics, University of São Paulo, São Paulo, Brazil
rafael.damaceno@ime.usp.br, rmcesar@usp.br

Abstract. Video captioning is a computer vision task that aims at generating a description for video content. This can be achieved using deep learning approaches that leverage image and audio data. In this work, we have developed two strategies to tackle this task in the context of resource-constrained devices: (i) generating one caption per frame combined with audio classification, and (ii) generating one caption for a set of frames combined with audio classification. In these strategies, we have utilized one architecture for the image data and another for the audio data. We have developed an application tailored for resource-constrained devices, where the image sensor captures images at a specific frame rate. The audio data is captured from a microphone for a predefined duration at time. Our application combines the results from both modalities to create a comprehensive description. The main contribution of this work is the introduction of a new end-to-end application that can utilize the developed strategies and be beneficial for environment monitoring. Our method has been implemented on a low-resource computer, which poses a significant challenge.

Keywords: Video captioning · Mobile device · Deep Learning

1 Introduction

Video captioning is a task responsible for converting video content into text. This task can be considered a derivation of image captioning, which involves generating a textual description for an image. When working with videos, a common strategy is to combine a set of frames to generate a caption, often using techniques like 3D convolution. This approach adds more complexity to the computational task as it requires extracting information from the temporal dimension [1].

Another concept related to this task is "Video Summarization", which aims at generating a shortened version of a video, either in its original format or even

The authors would like to thank FAPESP (grants #2015/22308-2, #2022/12204-9, #2022/15304-4), CNPq, CAPES, FINEP and MCTI PPI-SOFTEX (TIC 13 DOU 01245.010222/2022-44) for their financial support for this research.

© Springer Nature Switzerland AG 2024
V. Vasconcelos et al. (Eds.): CIARP 2023, LNCS 14469, pp. 715–729, 2024.
https://doi.org/10.1007/978-3-031-49018-7_51

in text (e.g., paragraph). In this sense, the main idea is to detect important frames and to include them in the resulting content [8]. In our work, our main focus is video captioning, which means we do not depend on detecting significant frames.

Additionally, videos can be accompanied by audio, which brings multimodality to this task [1]. These characteristics, along with the limitations imposed by mobile devices, poses challenges for autonomously conducting video captioning without relying on external resources. In this context, [12] explore the development of an image captioning framework designed for resource-constrained devices. However, when it comes to multimodality and video captioning, little has been explored on these devices.

In the context of mobile devices or of resource-constrained devices, there are many strategies to deal with limiting conditions. Some of these include: distillation, pruning and quantization [13]. There are also numerous deep learning models for tasks as classification and segmentation that can handle them with satisfactory accuracy.

The aim of this work is to develop an end-to-end Deep Learning framework to perform Video Captioning through mobile devices. We employed pre-existing models designed and optimized for mobile devices to accomplish this task. The paper is organized as follows. In Sect. 2, we review recent papers related to Video Captioning, Multimodality, and Mobile Devices. Section 3 presents the strategies we have proposed in our work, while Sect. 4 outlines the experiment setup. In Sect. 5, we report the experimental results we have obtained. Finally, in Sect. 6 we conclude the paper.

2   Related Works

On one hand, many studies have addressed the task of converting video content into text using various strategies, especially in the absence of hardware resource limitations. For example, [7] has developed a model called VPCSum responsible for generating paragraphs from video data. The strategy employed by the authors consists of three parts: 1) image selection; 2) description of the selected images, and 3) summarizing of the descriptions.

The work by [4] has proposed an analytical system for video surveillance capable of identifying objects and their relation to event occurrences. [5] presents a multimodal deep neural approach for generating dense captions for videos. The authors have developed a framework consisting of event detection and two neural networks: a 3D convolutional network for processing sets of frames and a VGG-based network for audio data.

[3] proposed a system to extract textual descriptions from videos based on CNN (encoder) and LSTM (decoder) that focuses on human face features. [9] has developed a strategy based on deep learning and neural networks to generate textual summaries from videos. For each frame of a video, the network computes a score, and the frames with the highest values have their features extracted and utilized in a RNN to generate text.

On the other hand, little has been explored in the context of mobile devices. In this regard, for example, [10] identified and analysed the bottlenecks to implement neural network on low-resource hardware. They have developed a network named MobileOne that is focused on image classification and segmentation.

[6] has developed an application responsible for converting video in audio description, which was implemented in a ARM-based processor hardware. The authors have utilized a set of models, each focused on a category, to perform fine-grained classification of objects, which are then converted to audio through a text-to-speech library.

[11] describes a smart device, based on ARM-based processor hardware, to conduct video surveillance. The system comprises a YOLO network that receives data captured from a video-camera, which is triggered by movements detected through an infrared sensor. The aim was to develop a system capable of detect people and environmental intrusions.

Differently from theses studies, our work focuses on the implementation of an application on a hardware with limited resources and without internet connection. The main objective is to autonomously solve all the processes involved in multimodal video captioning on resource-constrained devices.

3   Video Captioning Frameworks

This work proposes two strategies for performing the task of transforming video into text in the context of mobile devices. Strategy 1 (FW_1) is based on image captioning and audio classification. Strategy 2 (FW_2) replaces the image captioning of FW_1 with a model that leverages the time dimension by generating captions for a set of a frames simultaneously. The remainder of this section provides details of each of these strategies and explains the rationale behind our selection.

3.1 FW_1: Multiple Image Captioning with Audio Classification

In a first experimental architecture, we have chosen not to leverage the time dimension. The system consists of two main networks: one based on image captioning settings, which includes an image features extractor, a sequence decoder, and a sentence generator, while the other utilizes the well-know YAMNet for classifying audio data. Figure 1 illustrates FW_1's general approach to generate a video description by leveraging image and audio data captured from a resource-constrained device.

Regarding the implemented application for the resource-constrained device, a set of images is captured through a camera at a rate of r frames per second. Simultaneously, a microphone records audio for a duration of t seconds. For each frame f out of a total of n frames, the image module generates a caption. This data, combined with the label produced by the audio module, is encapsulated in a text that describes the video's content for the duration of t seconds.

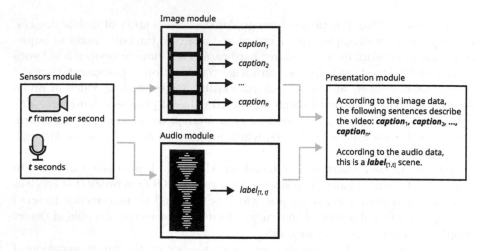

Fig. 1. FW_1 composition: The image module generates one caption per frame, which is later combined with the audio label in the presentation module.

In the image module, we experimented with five image feature extractors that were pre-trained on ImageNet. This module is based on [14] and has been modified to include a 2-layer decoder[1]. We explored different number of the model heads and layers. The intention was to apply it individually at a rate of two frames per second within the mobile application.

3.2 FW_2: Video Captioning with Audio Classification

In the second experimental architecture, we replace the image module with a framework capable of processing with multiple frames simultaneously, utilizing the temporal dimension to generate captions (see Fig. 2). Similar to FW_1, a set of images is collected through a camera in a rate of r frames per second, and a microphone captures audio data in a length of t seconds. However, in this case, we generate a single caption to the entire set of frames.

Since we identified the best backbone to extract features from images in FW_1, we have selected this model for the FW_2 strategy as well. However, instead of extracting features from one image at a time, the model now extracts features from n frames simultaneously and stacks them during both the training and inference phases. Additionally, the audio classifier has remained unchanged.

In this strategy, the image module is responsible for reading a set of frames, extracting their features using a backbone, and stacking all the results into an array. It is important to mention that the videos utilized to train the image module for video captioning are composed of data with various time duration and resolutions. We set the total number of frames of the stacked array to $n = 80$.

---

[1] Part of our developed source code was based on the TensorFlow tutorial available at https://www.tensorflow.org/tutorials/text/image_captioning, accessed on February 21, 2023.

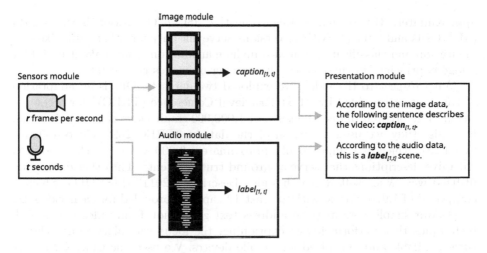

Fig. 2. FW_2 composition.

Regarding the image module part of the end-to-end application implemented on the mobile device, the application captures r frames per second until it accumulates n frames. The features of these frames are extracted and stacked into an array, which is then fed to the model to predict a caption. Another possible approach is to extract a set of features for each frame individually, until reaches 80 frames. Subsequently, an array formed by these features is fed into the model to produce a caption.

4   Experiment Setup

This section presents the datasets and backbones used to train and test the models employed in the two strategies of our work. Additionally, we provide information about the hardware utilized for training the captioning models and the hardware used for deploying the end-to-end application that transforms video content into description.

4.1   Datasets and Backbones

In this work, we explored five datasets related to image captioning and two for video captioning. It is important to mention that none of them leverage or contain descriptions related to audio data. Regarding the image-caption pairs datasets, most of them include five descriptions for each image. An exception is the Conceptual Captions (CC25K5) dataset, which contains only one description per image, totaling up to 3,300,000 image-caption pairs. In our work, we use 30,000 image-caption pairs.

The Flickr8K (FL8K) dataset is composed of eight thousand images, while the Flickr30k (FL30K), an expanded version of the previous dataset, contains approximately 31,000 images. Regarding the MSCOCO dataset, both its 2014 (MSCC14) and 2017 (MSCC17) versions serve not only for other tasks like segmentation and classification but also include annotations that make it useful for image captioning. In these cases, the number of images is 165,000.

With respect to the video, we explored two datasets. One is an adaptation of the content provided by "Text REtrieval Conference (TREC)" for a competition. The original dataset consists of 1,000,000 small video segments in total. For this study, we used a portion of the dataset "TREC 2022 Video-to-Text" (VTT22), which is composed of approximately 2,000 videos, each annotated with five descriptions that serve as ground-truth content[2]. The other dataset we utilized was "Microsoft Research Video Description 2011" (MSVD11), which is composed of 1,970 videos, with at least 12 captions provided for each video [2].

As our simplest strategy to address text generation from video starts with techniques that perform image captioning, the focus was placed on architectures available and optimized for mobile devices. We used the ConvNeXt Tiny (CNTiny), EfficientNet B0 (ENetB0), MobileNet Small (MNSmall), MobileNet Large (MNLarge), and Xception (Xcep) architectures. All these backbones were specific tailored to the context of mobile devices and can be useful in captioning systems that perform this task in resource-constrained environments.

Regarding the audio backbone, as it not the primary focus of this study to explore various audio classifiers, we opted for a pre-trained version of YAMNet, which is based on MobileNet architecture. The selected version is pre-trained on the AudioSet dataset and has the capability to classify audio into 521 distinct classes.

4.2   Hardware, Software and Training Settings

The end-to-end application was implemented on a Raspberry Pi 4 Model B equipped with 8 GB of RAM and a processor based on ARM-v8 architecture. Despite having a large amount RAM, our intention was to use only 1.5 GB of it with a 64-bit Raspberry Pi system loaded Desktop environment. The hardware used during the training phase is composed of dual GPU NVIDIA A5000, each with 24 GB of RAM.

We have used the Python programming language in the deployed application. For training and testing the captioning models, we have used the TensorFlow framework in a Jupyter Notebook environment. Each module of both the strategies is implemented as a Python application.

During the training phase of both strategies, we experimented with various model configurations, including vocabulary size ranging from 1,500 to 5,000, different caption lengths, number of heads, and number of layers. All dataset captions underwent preprocessing to remove punctuation, convert characters to lowercase, and eliminate double spaces. The results present in this paper refer to be best parameters we identified during the experiments.

---

[2] We collected the videos and captions from https://trecvid.nist.gov/trecvid.data. html, accessed on January 10, 2023.

5   Results and Discussion

This section presents the results and discussion regarding the strategies we have employed in our work. In Subsect. 5.1, we elucidate the differences between the captions generated by the image feature extractors. In Subsect. 5.2, we explore the final descriptions generated by the application using the best features extractor and the audio classifier (strategy FW_1), as well as the descriptions presented while using strategy FW_2. Finally, in Subsect. 5.3, we present how much resources the models we developed consume on the mobile device used for deployment.

5.1   Image Feature Extractors Comparison

In this subsection, we present a comparison among the image feature extractors with respect to the different datasets used to train the image modules of FW_1 e FW_2. We utilized the "masked accuracy" metric, which is a suitable measure for Natural Language Processing problems aiming to predict words in text sequences. This metric assesses the model's accuracy in correctly predicting masked words. Specifically, it verifies how many of the predicted words in the masked positions exactly match the words that were originally missing.

Table 1 presents the comparison in terms of accuracy achieved by the architectures composed of each of the five image feature extractors. The purpose was to assess how the backbones influence the generated descriptions and choose the best model to be used in the end-to-end application with the strategies FW_1 and FW_2.

Table 1. Masked accuracy comparison of five image feature extractors used with FW_1.

Backbone	Dataset				
	CC25K5	FL8K	FL30K	MSCC14	MSCC17
MNSmall	0.2416	0.3989	0.3657	0.4227	0.4410
MNLarge	0.2530	0.4048	0.3728	0.4294	0.4654
ENetB0	0.2470	**0.4080**	0.3911	0.4359	0.4720
XCep	0.2174	0.3429	0.3491	0.3747	0.3998
CNTiny	**0.2614**	0.4067	**0.3923**	**0.4405**	**0.4817**

As it can be observed in Table 1, the results show that there are few differences in relation to the performance obtained when using the backbones in general. However, the XCep values were considerably lower than those obtained by the other backbones. For four out of the five tested scenarios, the CNTiny backbone performed better, while for one scenario, the ENetB0 backbone performed the best.

We chose the CNTiny backbone to train the FW_2's image module using the MSVD11 and VTT22 datasets. In this context, we achieved the best masked accuracy of 0.4198 for the MSVD11 dataset, with a maximum caption size of 20 tokens and a vocabulary size of 1,500. When training with the VTT22 dataset, we reached the best masked accuracy of 0.3341, employing a maximum caption size of 50 tokens and a vocabulary size of 5,000.

It is worth mentioning that all the backbones were configured with pretraining enabled on the ImageNet dataset and with an input resolution of 224 by 224 pixels. The chosen versions are the smallest in terms of size, as the objective is to fit them into the application implemented on a mobile device.

5.2   Qualitative Assessment of Video Descriptions

In addition to evaluating the accuracy and quality of the image feature extractors, we also conducted a qualitative assessment of the performance of the two strategies to generate descriptions for videos recorded from a balcony view of a laboratory at the University of São Paulo, Brazil. Another part of the data analysis involved the videos obtained from the dataset VTT22. Importantly, we present results related to the usage of different settings in strategy FW_1 and the optimal configuration for FW_2, which is based on the CNTiny backbone trained with the MSVD11 dataset, restricted to a caption size of 20 tokens, and a vocabulary size of 1,500.

Balcony View Recorded Videos. With the objective of exploring the capabilities of the strategies we developed in this work, we recorded a video using the Raspberry Pi 4B computer and extracted a portion of the captured footage. One possible caption that represents a ten-second segment of one of the events that occurred in the video is "A man and a woman are walking down a path in a park." Fig. 3 shows the ten frames from this particular segment of data.

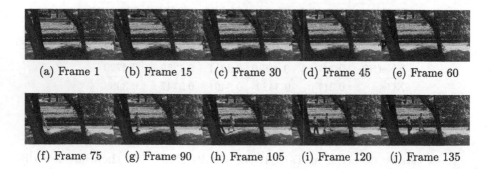

(a) Frame 1   (b) Frame 15   (c) Frame 30   (d) Frame 45   (e) Frame 60

(f) Frame 75   (g) Frame 90   (h) Frame 105   (i) Frame 120   (j) Frame 135

Fig. 3. Ten frames that represents a man and a woman walking down a path in a park.

While the image data provides an idea of people walking along a path, the audio data, represented over the same duration of the ten frames (five seconds of video), contains some background noise (people talking inside the laboratory) and environmental noise (such as cars passing by). When constructing our frameworks, the intention was for these events to be captured to some extent by the models. Table 2 present the captions generated for the frames listed in Fig. 3.

Table 2. Captions generated by the image module for the ten frames shown in Fig. 3. The number in parentheses represents the count of consecutive captions that are identical.

Settings	Captions generated
CNTiny + CC25K5	A view of the garden (10).
ENetB0 + FL8K	A woman is walking down a sidewalk. A man is walking down a sidewalk. A woman is walking down a sidewalk (2). A man is walking down a sidewalk. A woman is walking down a sidewalk (2). A man is walking down a sidewalk (2). A woman is walking down a sidewalk.
CNTiny + FL30K	A man in a green shirt is sitting on a park bench (6). A man in a green shirt and a woman in a park. A man in a park is sitting on a park bench. A man in a blue shirt and a woman are walking down a path. A man in a green shirt and a woman in a park.
CNTiny + MSCC14	A park bench in the middle of a park (2). A park bench in the grass near a tree. A park bench in the middle of a park. A park bench in the grass near a tree. A park bench in the middle of a park. A park bench with a tree in the park. A group of people riding on a path in the park. A group of people standing in a park. A park bench in the middle of a park.
CNTiny + MSCC17	A park with a tree and a tree (6). A park with a large tree in the park. A man flying a kite in a park. A man walking on a path with a dog. A park with a tree and a park bench.

On one hand, regarding the obtained results, the configuration CNTiny + CC25K5 generates ten times the caption "a view of the garden." These captions accurately reflect part of the content of the images; however, they fail to capture the presence of the two people walking (in the last frames). On the other hand, configurations ENetB0 + FL8K and CNTiny + FL30K seem to incorrectly generate captions with people (either man or woman) even for frames that do not contain them. Finally, the settings CNTiny + MSCC14 and CNTiny + MSCC17 correctly identify the presence or absence of people in the frames for the majority of cases.

Consequently, we have chosen the models trained on MSCC14 and MSCC17 to implement in the mobile application with the backbone CNTiny. When fed with the recorded video, whose frames are represented in Fig. 2, the following description is generated: *"According to the image data, the following sentences describe the video: A park bench in the middle of a park, a park bench in the grass near a tree, a park bench in the middle of a park, a park bench in the grass near a tree, a park bench in the middle of a park, a park bench with a tree in the park, a group of people riding on a path in the park, a group of people standing in a park, and a park bench in the middle of a park. According to the audio data, this is a speech scene."*

The generated descriptions can be divided into two groups: one that reflects the initial frames, where the two people are not present in the main area of the images, and another in which two people are more evident. The description model was able to reproduce these occurrences. Some content repetition was generated, which may indicate the need to adopt a lower frame rate. Another solution, still within the scope of the FW_1 strategy, is to use a text summarizer.

As far as the audio module is concerned, in a first usage experiment, we considered a one-second sample, meaning we had one class for every two frames (for $r = 15$). Consequently, the only classification with a reasonable score (above 0.5) was obtained for the data comprising the frames 6–7 shown in Fig. 3, labeled as "music".

In a second experiment, we classified five seconds of audio data at a time, corresponding to the content encompassed by the ten frames presented in Fig. 3. In this case, the classification obtained was "speech". As expected, one second is insufficient to comprehend the recorded data fully. While this type of audio content can be useful to explore pre-processing techniques to enhance label classification, it is not the primary focus of this study.

When applying strategy FW_2, the sentence generated is: *"According to the image data, the following sentence describe the video: a man is running on the ground. According to the audio data, this is a speech scene."* The strategy was able to capture at least one person, but, the action was incorrectly identified (running instead walking). Moreover, the woman present in the video was not perceived by the captioning model. The main characteristics of the scene, which occurred in a park, were not described either.

VTT22 Videos. We decided to explore how our models behave with two videos from the dataset VTT22. The first video, identified by the ID 1943, depicts a woman and a man each accompanying a different cow, with another cow visible in the background - see the main frames of this video in Fig. 4.

One of the five captions provided by the annotators of the dataset describes the scene as follows: "A man and a woman are in the woods pulling two cows and walking next to each other while another man is pulling another cow and coming from the side in the daytime." In this case, we expect the employed strategies to

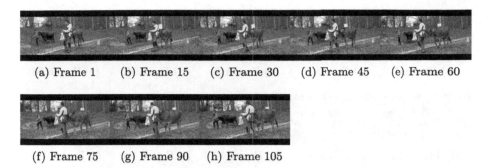

(a) Frame 1   (b) Frame 15   (c) Frame 30   (d) Frame 45   (e) Frame 60

(f) Frame 75   (g) Frame 90   (h) Frame 105

Fig. 4. Key frames depicting the content of video ID 1943.

identify all the people and animals present in the video. Concerning the audio content, a background music is played throughout the entire video, however, it appears that it's not the original sound. Therefore, we expect the audio classifier to accurately categorize it as "music."

The final application, which involves some text processing for presentation purposes, in the case of the strategy FW_1, is formed by the model trained on MSCC17 with CNTiny. In this scenario, the following sentence is generated: *"According to the image data, the following sentences describe the video: A man and a woman standing next to a cow, a man and a woman walking on a horse, a man and a woman standing next to a cow, a man standing on top of a horse, a man and a woman standing next to a cow, a man standing next to a cow on a dirt road, a man and a woman standing next to a cow." According to the audio data, this is a music scene."* We considered the entire audio channel content at once since the total video duration is short (four seconds). And as expected, its classification was "music."

It is important to note that the objective of classifying the audio content is to provide a better understanding of what the video represents. In this case, the video is a presentation of a scene with a background sound that does not correspond to the environmental sound. But in an improved version of our captioning system, this information could be leveraged to generate a more specific description.

When applying strategy FW_2, the sentence generated is: *"According to the image data, the following sentence describe the video: a man is walking on the ground. According to the audio data, this is a music scene."* This result was generated by version FW_2 trained on the MSVD11 dataset, which, in this case, was able to produce grammatically correct descriptions. The model generated by the adapted VTT22 dataset did not produce good results; that is, the generated descriptions did not capture the essential elements of the scenes.

With the intention of assessing the performance of our strategies in different scenarios, we analyzed an additional video. Ten frames from this second video, identified by ID 1953, are presented in Fig. 5. The scene depicts a delightful moment of a dog playfully holding a piece of wood in its mouth.

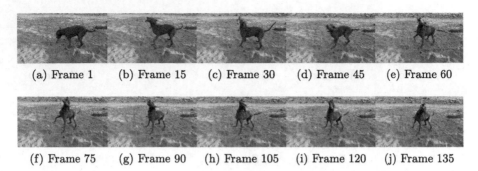

(a) Frame 1   (b) Frame 15   (c) Frame 30   (d) Frame 45   (e) Frame 60

(f) Frame 75   (g) Frame 90   (h) Frame 105   (i) Frame 120   (j) Frame 135

Fig. 5. Key frames depicting the content of video ID 1953.

The ground truth, as reported by the dataset annotators, is "A brown dog in the sand picks up a thick stick with his mouth and tries to maneuver it so it does not fall from his mouth." Our expectation is that the strategies capture the presence of an animal interacting with a piece of wood. However, in the case of this video, this information is present only in the image data because the audio content consists of music. Therefore, we anticipate the classifier to label the audio content as "music".

While fed with this video, the application generates the following sentence: *"According to the image data, the following sentences describe the video: A dog is standing in the sand with a frisbee, a dog is standing on the beach with a frisbee, a dog is standing in the sand next to a beach, and a dog is standing on the beach with a frisbee. According to the audio data, this is a music scene."* In a similar manner to the video 1943, as hypothesized, the audio classifier successfully classified the audio content as "music." We also considered the entire audio channel content, given its duration of only seven seconds.

While applying strategy FW_2, the sentence generated is: *"According to the image data, the following sentence describe the video: a dog is playing with a dog. According to the audio data, this is a music scene."* In this case, the strategy correctly detected the action represented in the scene, but instead of one, two dogs were reported in the description.

5.3   Strategies Resources Consumption

In the analysis of resources consumption, we considered the following elements: the amount of memory occupied during the application's usage, the time required to extract features if necessary, the time taken to generate a description, and vocabulary size. The objective was to verify if the strategies can perform effectively on constrained-resource devices.

One part of this analysis was conducted automatically using the benchmark application provided by the TensorFlow framework[3]. Each TensorFlow

---

[3] The TensorFlow benchmark application utilized in this work is available at https://www.tensorflow.org/lite/performance/measurement, accessed on March 6, 2023.

Lite model was input into this benchmark application. For the CNTiny based model utilized in strategy FW_1, the model size is 25.2 MB, average inference timing was 3.8 ms, and memory utilized at run-time was 25.8 MB.

For the main model (CNTiny) utilized in strategy FW_2, the model size is 25.2 MB, average inference timing was 180 ms, and memory utilized at run-time was 146.1 MB. By comparison, the model with MNSmall is 23.6 MB in size, obtained an average inference timing of 126 ms, and utilized 126.8 MB of memory at run-time. With respect to the model parameters, the best model of strategy FW_1 was implemented with 5,000 tokens, while for FW_2, we used 1,500 tokens.

It is important to note that, on the one hand, the model utilized in the FW_1 strategy takes as input the frame captured by the mobile device's camera, in addition to a predefined segment of audio. The CNTiny backbone is integrated into the model and is responsible for extracting features when using the application. Strategy FW_2, on the other hand, receives both the audio and a file containing previously extracted characteristics from a set of frames, rather than the frames themselves.

Therefore, to fairly assess the time analysis for FW_2 in generating captions, one must sum the time required to extract the features and perform the inference. To describe a video duration of ten-second video, captured at a rate of 30 frames per second, the application will analyze 80 frames out of the total 300 frames in the feature extraction stage.

6   Conclusion and Future Works

The objective of this work was to develop an end-to-end application that utilizes audio and image data captured from sensors on resource-constrained devices to generate video scene descriptions. We employed a heterogeneous solution, incorporating different types of neural networks into two strategies. Specifically, we used a range of architectures, starting from a simple image captioning model combined with an audio classifier, and progressing to a more complex video captioning structure. These approaches allowed us to provide users with descriptions that encompass both audio and image modalities, without consulting external resources (i.e., internet connection), and on low-resource devices.

It is important to mention that in one of the videos, the model trained on the MSCC14 dataset provided more reliable results, while in another video, the best model in qualitative terms was the one trained on the MSCC17 dataset. While dealing with the strategy FW_2, only the model trained on MSVD11 dataset was able of generating acceptable captions. Despite some inconsistencies in the generated captions, the strategies were able to capture the most important occurrences in the scenes, such as the type of setting (for example: park or beach), the presence of people or animals, and elucidating their activities. The models can be further fine-tuned for more specific scenarios.

One limitation of the FW_1 approach is that it does not consider a set of frames when generating captions, resulting in the loss of temporal context. Another limitation is the generation of captions with little difference between them. As for the strategy FW_2, similarly to FW_1, it relies on a classification network to extract information from the audio modality.

These limitations can be be addressed in future research, but the implementation on resource-constrained devices would be challenging, as indicated by the experimental results obtained for FW_1 and FW_2. Moreover, using a video classification network as backbone is one possibility to improve the quality of the captions generated. We intend to modify our network architecture to optimize the execution of the developed strategies, thereby reducing memory usage and processing requirements.

References

1. Abdar, M., et al.: A review of deep learning for video captioning (2023). https://doi.org/10.48550/ARXIV.2304.11431
2. Chen, D.L., Dolan, W.B.: Collecting highly parallel data for paraphrase evaluation. In: Proceedings of the 49th Annual Meeting of the Association for Computational Linguistics (ACL-2011), Portland, OR (2011)
3. Dilawari, A., Khan, M.U.G., Farooq, A., Rehman, Z.U., Rho, S., Mehmood, I.: Natural language description of video streams using task-specific feature encoding. IEEE Access **6**, 16639–16645 (2018)
4. Fonseca, C.M., Paiva, J.G.S.: A system for visual analysis of objects behavior in surveillance videos. In: 2021 34th SIBGRAPI Conference on Graphics, Patterns and Images (SIBGRAPI), pp. 176–183. IEEE (2021)
5. Iashin, V., Rahtu, E.: Multi-modal dense video captioning. In: Proceedings of the IEEE/CVF Conference on Computer Vision and Pattern Recognition (CVPR) Workshops (2020)
6. Karkar, A., Kunhoth, J., Al-Maadeed, S.: A scene-to-speech mobile based application: multiple trained models approach. In: 2020 IEEE International Conference on Informatics, IoT, and Enabling Technologies (ICIoT), pp. 490–497. IEEE (2020)
7. Liu, H., Wan, X.: Video paragraph captioning as a text summarization task. In: Proceedings of the 59th Annual Meeting of the Association for Computational Linguistics and the 11th International Joint Conference on Natural Language Processing (Volume 2: Short Papers), pp. 55–60 (2021)
8. Sabha, A., Selwal, A.: Data-driven enabled approaches for criteria-based video summarization: a comprehensive survey, taxonomy, and future directions. Multimedia Tools Appl., 1–75 (2023)
9. Shah, D., Dedhia, M., Desai, R., Namdev, U., Kanani, P.: Video to text summarisation and timestamp generation to detect important events. In: 2022 2nd Asian Conference on Innovation in Technology (ASIANCON), pp. 1–7. IEEE (2022)
10. Vasu, P.K.A., Gabriel, J., Zhu, J., Tuzel, O., Ranjan, A.: MobileOne: an improved one millisecond mobile backbone. In: Proceedings of the IEEE/CVF Conference on Computer Vision and Pattern Recognition, pp. 7907–7917 (2023)
11. Viswanatha, V., Chandana, R., Ramachandra, A.: IoT based smart mirror using Raspberry Pi 4 and YOLO algorithm: a novel framework for interactive display. Indian J. Sci. Technol. **15**(39), 2011–2020 (2022)
12. Wang, N., et al.: Efficient image captioning for edge devices (2022). https://doi.org/10.48550/ARXIV.2212.08985

13. Wang, Y., et al.: A survey on deploying mobile deep learning applications: a systemic and technical perspective. Digit. Commun. Netw. **8**(1), 1–17 (2022)
14. Xu, K., et al.: Show, attend and tell: neural image caption generation with visual attention. In: International Conference on Machine Learning, pp. 2048–2057. PMLR (2015)

Stingless Bee Classification: A New Dataset and Baseline Results

Matheus H. C. Leme⬥, Vinicius S. Simm⬥, Douglas Rorie Tanno⬥,
Yandre M. G. Costa$^{(\boxtimes)}$⬥, and Marcos Aurélio Domingues⬥

Department of Informatics, State University of Maringá, Maringá, Paraná, Brazil
{pg55308,pg404811,pg404806,ymgcosta,madomingues}@uem.br

Abstract. Bees play an important role as pollinating agents, contributing to the reproduction of many plant species around the world. Brazil is the home for different species of stingless bees, with around 200 registered species out of the more than 500 species classified worldwide. Each species constructs the entrance to its colony in an unique but similar way among colonies of the same species. In this work, we proposed a new dataset created in collaboration with stingless beekeepers from Brazil for the exploration of stingless bee species classification. The dataset consists of 158 samples distributed unequally among the 13 species: *Boca de Sapo, Borá, Bugia, Iraí, Japurá, Jataí, Lambe Olhos, Mandaguari, Mirim Droryana, Mirim Preguiça, Moça Branca, Mandaçaia,* and *Tubuna*. The results presented in this work were obtained using deep learning models (i.e. CNN architectures) such as VGG and DenseNet, which are commonly used for image classification task in different application domains. Pre-trained models from ImageNet were used, along with transfer learning techniques, and due to the small size of the dataset, data augmentation techniques were applied, resulting in an expanded dataset of 1,106 samples. The experimental results demonstrated that the DenseNet model achieved the best results, reaching an accuracy of 95%. The dataset created will be also made available as a contribution of these work. As far as we know, the stingless bee species identification task based on the colony entrance is addressed for the first time in this work.

Keywords: Stingless bees · Colony entrance · Convolutional neural networks · Classification

1 Introduction

All around the world, the reproduction of many plants is carried out by pollinating insects such as beetles, butterflies, and, most importantly, bees. These agents are considered essential for life on the planet as they regulate the agroecosystem [25], in particular, the bees, which are considered the primary pollinators of natural and agricultural ecosystems [16].

From the numerous existing bee species, about 500 of them belong to the genus of stingless bees [15]. These bees play a crucial role in ecology, as already

© Springer Nature Switzerland AG 2024
V. Vasconcelos et al. (Eds.): CIARP 2023, LNCS 14469, pp. 730–744, 2024.
https://doi.org/10.1007/978-3-031-49018-7_52

mentioned, and they also have a significant impact on the economy as a source of income for many families through the commercialization of honey, pollen, and beeswax [5,20]. Additionally, it is worth to mention that the honey produced by stingless bee has remarkable health benefits. Due to its composition and storage location in the colony, the honey is rich in various medicinal properties such as antioxidants, antibacterial, and anti-inflammatory substances. It can prevent oxidative stress, fight infections, and aid in the healing process of burns [1,2].

Despite the undeniable importance of stingless bee, these species have been facing various adversities over the years, such as diseases, parasites, pesticides, climate change, and habitat loss. These factors have significantly compromised the survival of these insects [6]. Additionally, the low popularity and lack of public awareness are also factors that contribute to colony losses. For example, species such as *Tetragonisca angustula* have adapted to urban environments over the years and easily nest in non-natural locations, such as hollow spaces in walls and fences. Due to being unknown to people or confused with other stinging insect species like wasps, they often become targets of pesticides and other harmful measures.

One possible way to contribute to the preservation of stingless bee colonies is by classifying the entry points of colonies in natural or urban environments, tracking which species are successfully adapting to urbanization over the years, and promoting preventive actions for less adaptable species, thereby mitigating the risks of extinction for these species. The entrances of colonies of the same species are generally very similar [21]. This is illustrated in Fig. 1, which presents three examples of four bee species. The noticeable differences are usually related to subsequent stages after swarming, meaning that the complete entrance structure is constructed over time. Another factor that can cause differentiation is the health of the colony. On the other hand, entrances of colonies commonly differ in color, shape, length, composition, ornaments, and the number of sentinel bees positioned at the entrance. These characteristics suggest that by using pattern recognition techniques, we can classify the stingless bee species through their colony entrances. However, there is a lack of researches and datasets in this scenario.

In this work, to the best of our knowledge, we propose the first dataset with stingless bee colony entrances, and the first study to classify stingless bees through the entrance of their colonies. Nowadays, the most popular techniques in the pattern recognition area for image classification and object recognition are convolutional neural networks (CNNs) [10]. Thus, in our study, we use two of the most popular CNN architectures, i.e. VGG [19] and DenseNet [8], to classify stingless bee species from a dataset created with colony entrance images obtained by stingless beekeepers in the southern region of Brazil. To achieve better results and overcome the limitations imposed by the small size of the dataset, we adopted transfer learning in our study, by using pre-trained weights from the extensive ImageNet dataset[1]. The results demonstrate that the DenseNet architecture can reach an accuracy of 95%.

---

[1] https://image-net.org.

Bugia species   Jataí species   Mandaçaia species   Tubuna species

Fig. 1. Examples of *Bugia*, *Jataí*, *Mandaçaia* and *Tubuna* stingless bee entrances organized by columns.

The remaining of this paper is organized as follows. Section 2 presents the related work. In Sect. 3, the methods adopted in this paper are described. Subsection 3.1 details the acquisition of the dataset and shows its division. In Subsect. 3.2, we describe the models used in our work. Subsection 3.3 describes the experimental setup. In Sect. 4, we show and discuss the results obtained in our empirical experiments. Finally, conclusions and future work are presented in Sect. 5.

2   Related Work

In this section, we review some works that use CNNs for image classification. The usage of CNNs combined with transfer learning reduces the computational time and cost, and achieves excellent results in classification tasks, even when a reasonably large dataset is not available [3, 7].

The work presented by Pearline et al. [4] focused on plant classification using two approaches: a traditional method, applying feature extraction of patterns, shapes, textures and colors, and a CNN-based approach. While the traditional methods achieved an accuracy of only 82.38%, the CNN-based approaches surpassed that with 97.17% using VGG16 and 99.41% with VGG19. These results demonstrated that the VGG architectures have the capability to achieve impressive and competitive results compared to traditional classification techniques.

Nakahata [12] and Satake [18] proposed the use of VGG and DenseNet CNN architectures with transfer learning to classify objects with similar classes in the domains of toxic plants and bonsai, respectively. The results achieved in both works demonstrated the capability of the models to perform classification even when the dataset is small, which is not a suitable scenario for deep networks.

In the current literature, there are also some studies aiming to classify certain species of stingless bee using CNNs [14]. Kelley et al. [26] used images of insects with varying brightness, orientation, and cropping to differentiate them. Additionally, regions of interest were delimited for bee localization. With the implemented approach, an average accuracy of 91% was achieved in their study. Rebelo et al. [16] performed classification based on images of insect wings. Several filters and processes were applied to the images to isolate only the skeleton information, where intersections were punctuated. Wing images were also used by Nart et al. [27] for bee species classification. The authors employed cross-validation while maintaining class proportionality, and utilized various well-known CNN models from the literature.

As we can see, although there exist some studies related to the usage of CNNs with stingless bee, to the best of our knowledge, no research has specifically focused on classifying stingless bee species based on the identification of their colony entrances.

3 Methodological Design

The sequence of steps adopted in our work is presented in Fig. 2. Firstly, the acquisition of images and their preprocessing was carried out to create a new dataset containing images from stingless bee colony entrances. Subsequently, data augmentation was performed following standard protocols found in the literature.

The models adopted in our work are derived from three CNN architectures that are highly popular in the literature for feature extraction and classification. In our work, these architectures were tested using cross-validation techniques. Each generated model was evaluated using standard metrics for classification problems, and the average performance was calculated for comparison purposes.

For model implementation, Python[2] and TensorFlow[3] were used. All models were trained on a machine equipped with an Intel Core i7-10750H 2.60GHz processor, 16GB DDR4 3200 MT/s RAM, and an NVIDIA GeForce RTX 2060 graphics card.

3.1 Data Acquisition

As already mentioned, there is no dataset for stingless bee classificaton from colony entrances in the literature. Therefore, we have acquired images of colony

---

[2] https://www.python.org.
[3] https://www.tensorflow.org.

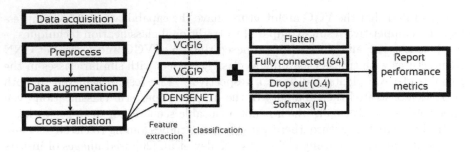

Fig. 2. Methodological flowchart.

entrances from different stingless bee species to compose a new dataset, called StinglessBeeOne. The dataset is available for download upon request to the authors via email.

The acquisition of images was done through requests to stingless beekeepers in the southern region of Brazil, who are part of a group dedicated to the practice of stingless beekeeping. They sent images from their own colonies using mobile devices. As a result, many images had external noise, such as people, multiple colonies in the same image, watermarks, and so on. Therefore, it was necessary to preprocess the images to remove the mentioned noises, and to crop all images to focus on the colony entrances. Additionally, all images were resized to the format of 224 × 224 pixels, which is commonly used by CNNs.

The created dataset consists of 158 images from 13 stingless bee species. These species, named in portuguese, are *Boca-de-Sapo* (*Partamona helleri*), *Borá* (*Tetragona clavipes*), *Bugia* (*Melipona mondury*), *Iraí* (*Nannotrigona testaceicornes*), *Japurá* (*Trigona spinipes*), *Jataí* (*Tetragonisca angustula*), *Lambe-Olhos* (*Leurotrigona muelleri*), *Mandaguari* (*Tetragonisca angustula*), *Mirim-Droryana* (*Plebeia droryana*), *Mirim-Preguiça* (*Friesella schrottkyi*), *Moça-Branca* (*Frieseomelitta doederleini*), *Mandaçaia* (*Melipona quadrifasciata*), and *Tubuna* (*Scaptotrigona bipunctata*).

Table 1 presents the respective quantities of images for each species. It also presents some characteristics for the colony entrances. The characteristics can be used to explain the results of the classifiers. The considered characteristics are the shapes, composition, and the number of sentinel bees. The shapes can be radial, when the entrance structure is radial; or distal, when there is an extension in a particular direction (forward, downward, or sideways). The composition can be pure wax, which means the construction consists exclusively of wax; or composite wax, which includes other organic materials, as soil. Lastly, the category can be rigid, when there are more than five sentinel bees at the entrance; or flexible, when there are fewer than five sentinel bees. Some examples of colony entrance are illustrated in Fig. 3.

In this work, no technique for removing similar images, such as Structural Similarity [24], was adopted. This is due to the nature of the problem addressed,

Table 1. Distribution of the species in the database and their characteristics.

Species	Entrance Characteristic	Quantity
Boca-de-Sapo	Radial, pure wax, and rigid	4 (2.5%)
Borá	Distal, pure wax, and rigid	11 (7%)
Bugia	Radial, composite wax, and flexible	7 (4.4%)
Iraí	Distal, pure wax, and rigid	7 (4.4%)
Japurá	Distal, composite wax, and flexible	6 (3.8%)
Jataí	Distal, pure wax, and rigid	38 (24.1%)
Lambe-Olhos	Distal, pure wax, and flexible	8 (5.1%)
Mandaguari	Distal, pure wax, and rigid	12 (7.6%)
Mirim-Droryana	Distal, pure wax, and flexible	7 (4.4%)
Mirim-Preguiça	Distal, pure wax, and flexible	8 (5.1%)
Moça-Branca	Radial, composite wax, and rigid	6 (3.8%)
Mandaçaia	Radial, composite wax, and flexible	29 (18.4%)
Tubuna	Distal, pure wax, and rigid	15 (9.4%)
Total		158 (100%)

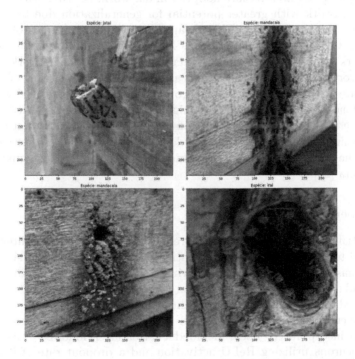

Fig. 3. Examples of colony entrances from the species *Jataí*, two *Mandaçaias*, and *Iraí* used in the study.

as the images are very similar to each other, given that the colony entrances of the stingless bees are highly similar within the same species.

3.2 Classifier Description

Different CNN architectures with densely connected layers can be used to perform feature extraction and image classification tasks. In this work, three architectures were adopted: VGG16, VGG19, and DenseNet.

The VGGNet architecture [19] is commonly used for image classification due to its performance in this task. In its structure, it incorporates small 3×3 convolutional filters with a stride of one pixel and a 2×2 pooling layer after every two or three convolutional layers to reduce the dimensionality, and the number of parameters of the feature maps generated by each convolution. VGGNet also utilizes 1×1 filters as linear transformations of the input. The ReLU activation function is applied to all hidden layers [29]. The models adopted in our work were VGG16 and VGG19.

The VGG16 model consists of 16 layers, including 13 convolutional layers and 3 fully connected layers, with approximately 138 million parameters. The VGG19 model has 19 layers, including 16 convolutional layers and three fully connected layers, with approximately 143 million parameters.

The other CNN architecture adopted in our work, i.e. DenseNet [28], is considered a network with greater potential for generalization due to its progressively hierarchical structure. This means that the input signals of each layer come from the outputs of the previous layers. Additionally, the final output of the network combines all the feature maps produced by the network.

It is worth to mention, that there are three main advantages of using this architecture compared to other ones proposed in the literature: 1) fewer parameters that need to be adjusted, 2) a higher ability to avoid overfitting, and 3) deeper layers. These advantages come from the reuse of feature maps, resulting in fewer convolutional kernels, and the dense connection that creates short paths between the initial and final layers, which makes it suitable for situations with limited data [28].

3.3 Experimental Setup

In the initial phase of our work, a comprehensive review of relevant literature was conducted to identify the commonly utilized hyperparameters in similar problem domains. Through empirical experimentation, multiple configurations were tested, and ultimately, the configuration demonstrating the highest performance was selected for further investigation through our experiments.

At the end of the VGG16, VGG19, and DenseNet architectures, additional layers were incorporated. These included a flatten layer, followed by a dense layer with 64 neurons utilizing ReLU activation and a dropout rate of 0.4. Finally, an output layer consisting of 13 neurons with softmax activation was added, as illustrated in Fig. 2. Pre-trained weights from the ImageNet dataset were utilized for initializing the models. The Adaptive Moment Estimation (Adam) optimizer was employed with a learning rate of 1×10^{-4}. All models underwent training for 50 epochs, with input images of dimensions $224 \times 224 \times 3$, and a batch size of 20 images per iteration.

To mitigate potential overfitting due to the limited number of images in the dataset, synthetic images were created. The data augmentation technique was employed to expand the training set by applying various transformations to the original images. These transformations included horizontal and vertical axis flipping, translation, rotation, and zoom. TensorFlow [29] provided built-in operations for rotation, contrast adjustment, color manipulation, cropping, and resizing. These operations, categorized as geometric transformations, incurred minimal computational cost and solely modified attributes related to the image's dimension and orientation, without affecting color channels and brightness, referred to as photometric transformations [30]. Thus, six additional images were generated for each original image in the dataset, resulting in a total of $1,106$ images used for training and testing the models. The transformation parameters employed for generating these new images are presented in Table 2. Figure 4 illustrates an example of data augmentation applied to our dataset.

A stratified cross-validation was used to assess the performance of each model in the colony entrance classification task. Five folds were utilized, resulting in the creation of five subsets from the StinglessBeeOne dataset. These subsets maintained the same species distribution as the original dataset. In each cross-validation iteration, four subsets were used for training, while one subset was reserved for testing. A 20% split of the training images was allocated for validation purposes.

The cross-validation results were evaluated by calculating the mean and standard deviation of commonly used metrics in the literature for similar problems: Precision, Recall and F1-score. These metrics provide distinct insights: Precision indicates the accuracy of class assignments, Recall measures the correct prediction rate of a class, and F1-score combines Precision and Recall into a harmonic mean [31].

Furthermore, confusion matrices were used in classification models to compare the prediction accuracy for each class, and identify the interactions that lead to higher errors. True classes are represented along one axis, while predicted classes are organized along the other axis. By analyzing the confusion matrix, metrics such as Accuracy, Precision, Recall, and F1-score could be computed to evaluate the model's performance [32].

Table 2. Parameters used for the data augmentation operations.

Operation	Parameter
Zoom	0.3
Vertical translation	0.3
Horizontal translation	0.3
Rotation	$-\frac{\pi}{4}$
Vertical flip	Yes
Horizontal flip	Yes

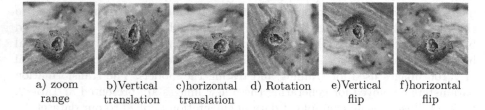

a) zoom range   b)Vertical translation   c)horizontal translation   d) Rotation   e)Vertical flip   f)horizontal flip

Fig. 4. Exemples of data augmentation applied in an image of the *Jataí* species.

4 Results and Discussions

To evaluate the performance of the aforementioned models, the Accuracy metric was used. Confusion matrices were also generated for all folds of each model. These matrices were subsequently aggregated to represent the individual performance of the VGG16, VGG19, and DenseNet models, as shown in Figs. 5, 6 and 7. The training progression is depicted in terms of Accuracy and the number of epochs. The convergence of each model is illustrated through the training of one of the five folds conducted. Figure 8 presents the training and validation processes.

The results obtained for the tested models are presented in Table 3. Table 4 shows Precision, Recall, and F1-score measures for each class from the explored models.

The Accuracy curves displayed over the epochs indicate stability in the training of the models, with minimal variation in the later iterations. This allows the interpretation of the results through confusion matrices. It is noteworthy that the DenseNet model exhibited remarkable stability with low variation after only 20 epochs of training. This can be partly explained by the freezing of pre-trained layers. On the other hand, the VGG16 and VGG19 models yielded inferior results when the number of trainable layers was limited, and thus, the training process encompassed the entire architecture.

Although both VGG models exhibit some confusion between classes, the majority of predictions are accurate. This confusion is reflected in the Precision and Recall of both models as shown in Table 4. On the other hand, the DenseNet

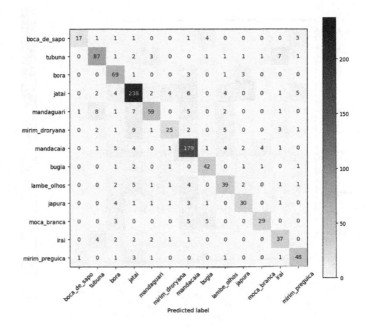

Fig. 5. Confusion matrix for VGG16.

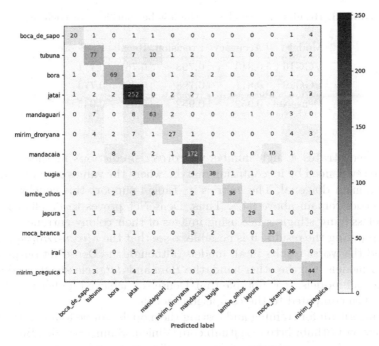

Fig. 6. Confusion matrix for VGG19.

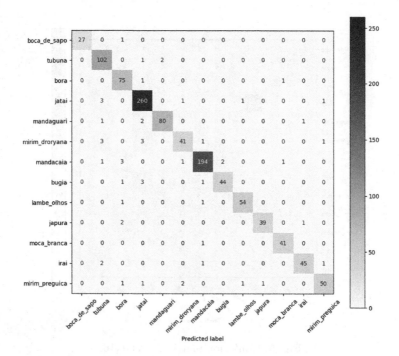

Fig. 7. Confusion matrix for DenseNet.

Table 3. Results of each class (stingless bee species) for each model.

Model	Accuracy	Precision	Recall	F1-Score
VGG16	0.813	0.793	0.761	0.771
VGG19	0.811	0.798	0.758	0.773
DenseNet	0.952	0.952	0.94	0.945

model demonstrates a high number of correct predictions in their respective classes, as evidenced by its confusion matrix, where the values are predominantly along the main diagonal. The metrics in Table 4 support this conclusion, as all three metrics remain above 94%. Thus, DenseNet proves to be efficient in the task of classifying stingless bee using images of their colony entrances.

By analyzing the results, it is possible to see that the *Mirim-Droryana* species presented the worst results in all models, although the DenseNet outperformed the other models. On the other hand, the *Bora* and *Jataí* classes showed positive results for all models. It is notable that DenseNet achieved higher results for all species when compared to the competing models.

The algorithm for training and testing the models can be found in the project repository on Github: https://github.com/ViniciusSimm/StinglessBees.

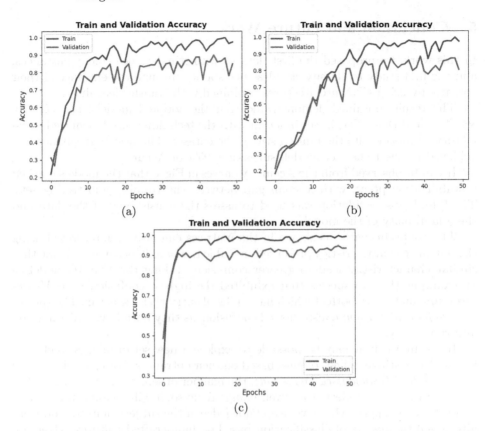

Fig. 8. Training and testing accuracy of VGG16 in (a), VGG19 in (b), and DenseNet in (c) over the course of 50 epochs.

Table 4. Precision, Recall, and F1 score of each class (stingless bee species) obtained for each class.

Species	VGG16			VGG19			DenseNet		
	Precision	Recall	F1-Score	Precision	Recall	F1-Score	Precision	Recall	F1-Score
Boca de sapo	0.89	0.61	0.72	0.83	0.71	0.77	1.00	0.96	0.98
Bora	0.73	0.90	0.80	0.78	0.90	0.83	0.89	0.97	0.93
Bugia	0.78	0.86	0.82	0.83	0.78	0.80	0.96	0.90	0.93
Iraí	0.70	0.76	0.73	0.69	0.73	0.71	0.96	0.92	0.94
Japurá	0.77	0.71	0.74	0.88	0.69	0.77	0.98	0.93	0.95
Jataí	0.87	0.89	0.88	0.84	0.95	0.89	0.96	0.98	0.97
Lambe-olhos	0.68	0.70	0.69	0.95	0.63	0.77	0.96	0.96	0.96
Moça-branca	0.83	0.69	0.75	0.75	0.79	0.77	0.95	0.98	0.96
Mandaguari	0.84	0.70	0.77	0.72	0.75	0.73	0.98	0.95	0.96
Mirim-droryana	0.74	0.51	0.60	0.71	0.55	0.62	0.91	0.84	0.87
Mirim-preguiça	0.80	0.86	0.83	0.77	0.79	0.78	0.94	0.89	0.92
Mandaçaia	0.86	0.89	0.87	0.89	0.85	0.87	0.97	0.96	0.97
Tubuna	0.83	0.83	0.83	0.75	0.73	0.74	0.91	0.97	0.94
Models	0.793	0.761	0.771	0.798	0.758	0.773	0.952	0.94	0.945

5   Conclusion and Future Work

In this work, we proposed the first dataset with stingless bee colony entrances, and the first study to classify stingless bee species through the entrance of their colonies by using CNN architectures combined with transfer learning.

The results obtained demonstrated that the adopted models, i.e. VGG16, VGG19, and DenseNet, in conjunction with the techniques used, achieved satisfactory results despite the limited size of the dataset. The best performance was achieved by the DenseNet model, surpassing 95% of Accuracy.

It can be observed from the Accuracy curves in Fig. 8 that the models reached a stabilization point, with a small gap between the training and testing sets. The 5-fold cross-validation was used to assess the consistency of the data and the adaptability of the models.

The confusion patterns shown in the matrices can also be interpreted using the entrance characteristics presented in Table 1, where it becomes evident that similar characteristics lead to greater confusion. Taking the VGG16 model as an example, the two species that exhibited the highest confusion were *Mirim-Droryana* and *Jataí*, both of which have a distal entry with pure wax. The species *Mandaguari* and *Tubuna* also showed confusion, as they both have a distal, pure, and rigid entry.

In future work, it may be possible to explore a new set of images to classify the health of stingless bee colonies based on their entrances. For example, for the species *Jataí* (*Tetragonisca angustula*), the number of sentinel bees positioned at the colony entrance reflects their well-being. To do so, it will be important to collaborate with experts who have deep knowledge of the subject matter. Additionally, given the results of classification based on handcrafted features extraction techniques such as LBP, LPQ, and LTP [13,33], future work can focus on using these techniques for the classification of stingless bee based on their entrances. The combination of different classifiers can also be explored for further comparison with the results obtained in this work. This would provide a comprehensive analysis of different classification approaches for stingless bee. Finally, we intend to use eXplainable Artificial Intelligence (XAI) techniques aiming at verifying which parts of the image most contribute for the classification.

Acknowledgment. This study was financed in part by the following Brazilian agencies: Coordination for the Improvement of Higher Education Personnel - Brazil (CAPES) - Finance Code 001, Araucária Foundation, and National Council for Scientific and Technological Development (CNPq). We also extend our thanks to each beekeeper who willingly contributed to this work by providing images of their stingless bee colonies. Their contributions were invaluable for this work.

References

1. Abd Jalil, M.A., Kasmuri, A.R., Hadi, H.: Stingless bee honey, the natural wound healer: a review. Skin Pharmacol. Physiol. **30**(2), 66–75 (2017)

2. Ahmed, A.K., Hoekstra, M.J., Hage, J.J., Karim, R.B.: Honey-medicated dressing: transformation of an ancient remedy into modern therapy. Ann. Plast. Surg. **50**(2), 143–148 (2003)
3. Akcay, S., Kundegorski, M., Devereux, M., Breckon, T.P.: Transfer learning using convolutional neural networks for object classification within X-ray baggage security imagery. In: 2016 IEEE International Conference on Image Processing (ICIP), pp. 1057–1061. IEEE (2016)
4. Anubha Pearline, S., Sathiesh Kumar, V., Harini, S.: A study on plant recognition using conventional image processing and deep learning approaches. J. Intell. Fuzzy Syst. **36**(3), 1997–2004 (2019)
5. Ayala, R., Gonzalez, V. H., Engel, M. S.: Mexican stingless bees (Hymenoptera: Apidae): diversity, distribution, and indigenous knowledge. In: Vit, P., Pedro, S., Roubik, D. (eds.) Pot-Honey, pp. 135–152. Springer, New York (2012). https://doi.org/10.1007/978-1-4614-4960-7_9
6. Goulson, D., Nicholls, E., Botiá, C., Rotheray, E.L.: Bee declines driven by combined stress from parasites, pesticides, and lack of flowers. Science **347**(6229), 1255957 (2015)
7. Huang, Z., Pan, Z., Lei, B.: Transfer learning with deep convolutional neural network for SAR target classification with limited labeled data. Remote Sens.-Basel **9**, 907 (2017)
8. Huang, G., Liu, Z., Maaten, L., Weinberger, K. Q.: Densely connected convolutional networks. In: Proceedings of the IEEE Conference on Computer Vision and Pattern Recognition, pp. 4700–4708 (2017)
9. Lavinas, F.C., Macedo, E.H.B.C., Sá, G.B.L., Amaral, A.C.F., et al.: Brazilian stingless bee propolis and geopropolis: promising sources of biologically active compounds. Rev. Bras **29**, 389–399 (2019)
10. Lecun, Y.: Generalization and network design strategies. Connect. Perspect. **19**, 18 (1989)
11. Mohd-Isa, W., Nizam, A., Ali, A.: Image segmentation of meliponine bee using faster R-CNN. In: 2019 Third World Conference on Smart Trends in Systems Security and Sustainablity (WorldS4), pp. 235–238. IEEE (2019)
12. Nakahata, G.H.S., Constantino, A.A., Costa, Y.M.G.: Bonsai style classification: a new database and baseline results. In: 2020 IEEE International Symposium on Multimedia (ISM), pp. 104–110. IEEE (2020)
13. Nanni, L., Ghidoni, S., Brahnam, S.: Handcrafted vs. non-handcrafted features for computer vision classification. Pattern Recogn. **71**, 158–172 (2017)
14. Nizam, A., Mohd-Isa, W., Ali, A.: Identification of the genus of stingless bee via faster R-CNN. In: International Workshop on Advanced Image Technology (IWAIT), pp. 808–813. SPIE (2019)
15. Rasmussen, C., Cameron, S.A.: Global stingless bee phylogeny supports ancient divergence, vicariance, and long distance dispersal. Biol. J. Linn. Soc. **99**, 206–232 (2010)
16. Rabello, A.R., et al. A fully automatic classification of bee species from wing images. Apidologie, 1–15 (2021)
17. Rozman, A.S., Hashim, N., Maringgal, B., Abdan, K.: A comprehensive review of stingless bee products: phytochemical composition and beneficial properties of honey, propolis, and pollen. Appl. Sci. **12**, 6370 (2022)
18. Satake, S.S., Calvo, R., Britto, A.S., Costa, Y.M.G.: Classification of toxic ornamental plants for domestic animals using CNN. In: Rozinaj, G., Vargic, R. (eds.) IWSSIP 2021. CCIS, vol. 1527, pp. 108–120. Springer, Cham (2022). https://doi.org/10.1007/978-3-030-96878-6_10

19. Simonyan, K., Zisserman, A.: Very deep convolutional networks for large-scale image recognition. In: 3rd International Conference on Learning Representations (ICLR 2015), Computational and Biological Learning Society, pp. 1–14 (2015)
20. Slaa, E.J., Chaves, L.A.S., Malagodi-Braga, K.S., Hofstede, F.E.: Stingless bees in applied pollination: practice and perspectives. Apidologie **37**(2), 293–315 (2006)
21. Shanahan, M., Spivak, M.: Resin use by stingless bees: a review. Insects **12**(8), 719 (2021)
22. Theckedath, C., Sedamkar, R.R.: Detecting affect states using VGG16, ResNet50 and SE-ResNet50 networks. SN Comput. Sci. **1**, 1–7 (2020)
23. Xiao, B., Liu, Y., Xiao, B.: Accurate state-of-charge estimation approach for lithium-ion batteries by gated recurrent unit with ensemble optimizer. IEEE Access **7**, 54192–54202 (2019)
24. Wang, Z., Simoncelli, E.P., Bovik, A.C.: Multiscale structural similarity for image quality assessment. In: The Thrity-Seventh Asilomar Conference on Signals, Systems & Computers, pp. 1398–1402. IEEE (2003)
25. Williams, N.M., Isaacs, R., Lonsdorf, E., Winfree, R., Ricketts, T.H.: Building resilience into agricultural pollination using wild pollinators. Agricultural Resilience-perspectives From Ecology and Economics, pp. 109–134 (2019)
26. Kelley, W., Valova, I., Bell, D., Ameh, O., Bader, J.: Honey sources: neural network approach to bee species classification. Procedia Comput. Sci. **192**, 650–657 (2021)
27. De Nart, D., Costa, C., Di Prisco, G., Carpana, E.: Image recognition using convolutional neural networks for classification of honey bee subspecies. Apidologie **53**, 5 (2022)
28. Zhang, J., Lu, C., Li, X., Kim, H., Wang, J.: A full convolutional network based on DenseNet for remote sensing scene classification. Math. Biosci. Eng. **16**(5), 3345–3367 (2019)
29. Abadi, M., et al.: TensorFlow: large-scale machine learning on heterogeneous systems (2015)
30. Taylor, L., Nitschke, G.: Improving deep learning with generic data augmentation. In: 2018 IEEE Symposium Series on Computational Intelligence (SSCI), Bangalore, India, pp. 1542–1547 (2018)
31. Powers, D.M.W.: Evaluation: from precision, recall and F-measure to ROC, informedness, markedness and correlation. Int. J. Mach. Learn. Technol. **2**(1), 37–63 (2020)
32. Deng, X., Liu, Q., Deng, Y., Mahadevan, S.: An improved method to construct basic probability assignment based on the confusion matrix for classification problem. Inf. Sci. **340–341**, 250–261 (2016)
33. Nanni, L., Costa, Y., Brahnam, S.: Set of texture descriptors for music genre classification; Václav Skala-UNION Agency (2014)

Author Index

Printed in the United States
by Baker & Taylor Publisher Services